Mathematical Models in Biology

Mathematical Models in Biology

Leah Edelstein-Keshet
Duke University

The Random House / Birkhäuser Mathematics Series

Random House / New York

First Edition

987654321

 All inquiries should be addressed to Random House, Inc., 201 East 50th Street, New York, N.Y. 10022. Published in the United States by Random House, Inc., and simultaneously in Canada by Random House of Canada, Limited, Toronto.

Library of Congress Cataloging-in-Publication Data

Edelstein-Keshet, Leah.
Mathematical models in biology.

(The Random House / Birkhäuser mathematics series)
Includes bibliographies and index.
1. Biology—Mathematical models. I. Title.
II. Series.
QH323.5.E34 1987 574'.07'24 87–20534
ISBN 0–394–35507–5

Dedicated

To My Parents

and

In fond memory of three teachers

Carrol Cassidy

Samuel Morris

and

James Karr

HGS, 1966–1971

Preface

Mathematical Models in Biology began as a set of lecture notes for a course taught at Brown University. It has since evolved through several years of classroom testing at Brown and Duke Universities. The task of setting down words on paper became a cherished hobby that kept the long process of shaping and reshaping the various manuscripts from becoming an arduous job.

My aim has been to present instances of interaction between two major disciplines, biology and mathematics. The goal has been that of addressing a fairly wide audience. It is my hope that students of biology will find this text useful as a summary of modern mathematical methods currently used in modelling, and furthermore, that students of applied mathematics might benefit from examples of applications of mathematics to real-life problems. As little background as possible (both in mathematics and in biology) has been assumed throughout the book: prerequisites are basic calculus so that undergraduate students, as well as beginning graduate students, will find most of the material accessible.

Other background mathematics such as topics from linear algebra and ordinary differential equations are given in full detail herein as the need arises. Students familiar with this material can advance at a more rapid pace through the book.

There is far more material here than can be taught in a single semester. This leaves some room for personal taste on the part of the instructor as to what to cover. (See table for several suggestions.) While necessitating selectivity in class, the length of the book is intended to encourage independent student reading and exploration of material not formally taught. References to additional sources are included where possible so that the text may be used as a reference source for the more advanced reader.

Features of this book are outlined below.

Organization: Models discussed fall into three broad categories: discrete, continuous, and spatially distributed (forming respectively Parts I, II, and III in the text). The first describes populations that reproduce at fixed intervals; the second pertains to processes that may be viewed as continuous in time; the last treats systems for which distribution over space is an important feature.

Approach: (1) Concepts basic in modelling are introduced in the early chapters and reappear throughout later material. For example *steady states, stability,* and *parameter variations* are first encountered within the context of difference equations and reemerge in models based on ordinary and partial differential equations.

(2) An emphasis is placed on mathematics as a means of unifying related

concepts. For example, we often observe that certain models formulated to describe a given process, whether biological or not, may apply to a different situation. (An illustration of this is the fact that molecular diffusion and migration of a population are describable by the same formal model; see 9.4–9.5, 10.1).

(3) Contrasting modelling approaches or methods are applied to certain biological topics. (For instance a problem on plant-herbivore dynamics is treated in three different ways in Chapters 3, 5, and 10.)

(4) Mathematics is used as a means of obtaining an appreciation of problems that would be hard to understand through verbal reasoning alone. Mathematics is used as a tool rather than as a formalism.

(5) In analyzing models, the emphasis is on qualitative methods and graphical or geometric arguments, not on lengthy calculations.

Scope: The models treated are deterministic and have deliberately been kept simple. In most cases, insight can be acquired by mathematical analysis alone, without the need for extensive numerical simulation. This sometimes restricts realism, but enhances appreciation of broad features or general trends.

Mathematical topics: Material in this book can be used as an introduction to or as a review of topics from linear algebra (matrices, eigenvalues, eigenvectors), properties of ordinary differential equations (classification, qualitative solutions, phase plane methods), difference equations, and some properties of partial differential equations. (This is not, however, a self-contained text on these subjects.)

Biological topics: Biological applications discussed range from the subcellular molecular systems and cellular behavior to physiological problems, population biology, and developmental biology. Previous biological familiarity is not assumed.

Problems: Problems follow each chapter and have different degrees of difficulty. Some are geared towards helping the student practice mathematical techniques. Others guide the student through a modelling topic in which the formulation and analysis of equations are carried out. Certain problems, based on models which have been published elsewhere, are meant to promote an appreciation of the literature and encourage the use of library resources.

Possible usage: The table indicates three possible courses with emphasis on (a) population biology, (b) molecular, cellular and physiological topics, and (c) a general modelling survey, which could be taught using this book. Parentheses ( ) indicate optional material which could be omitted in the interest of saving time. Curly brackets { } denote that some selection of the indicated topics is advisable, at the instructor's discretion. It is possible to omit Chapter 4 and Section 5.10 if Chapter 6 is covered in detail so that methods of Chapter 5 are amply illustrated. While it is advisable to combine material from Parts I through III, there is ample material in Part II alone (Chapters 4–8) for a one-semester course on ordinary differential equation models.

The relationship of various sections in the book is depicted in the following figure. Beginning at the trunk and ascending upwards along various branches, boldface section numbers denote material that is basic and essential for the understanding of topics higher up.

The interrelationships of sections and chapters are shown in this tree. Ascending from the trunk, roman numerals refer to Parts I, II, and III of the book. Boldface section numbers highlight important background material. Branches converge on several topics, as indicated in the diagram.

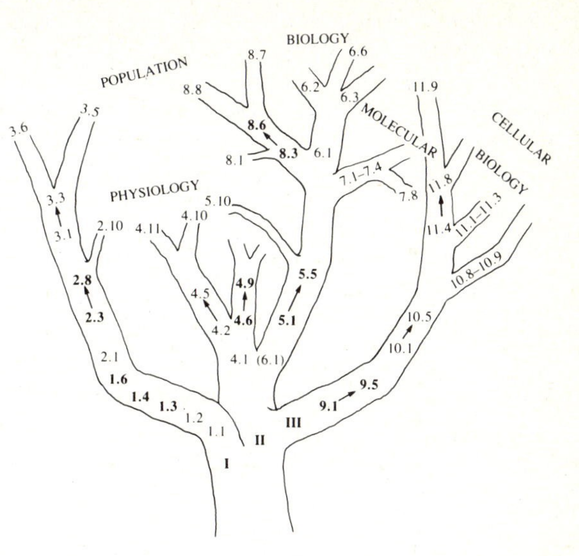

Selected material for three possible courses, with different emphasis.

Part	Chapter	Population Biology	Molecular, Cellular, and Physiological Topics	General Survey
I	1	1.1–1.6 (1.9, 1.10)	1.1, 1.3, 1.4, 1.6–1.9	1.1–1.7, (1.8), 1.9, (1.10)
	2	2.1–2.3, 2.5–2.8	2.1–2.8, 2.10	2.1–2.3, (2.4), 2.5–2.8, (2.9–2.10)
	3	3.1–3.4, (3.5), 3.6	(3.6)	3.1–3.3, (3.6)
II	4	4.1, (4.2–4.10)	all	4.1–4.7, (4.8), 4.9–4.10, (4.11)
	5	all	all	all but (5.3, 5.10–5.11)
	6	all	(6.3, 6.6–6.7)	6.1, {6.2–6.3, 6.6}
	7		7.1–7.4, (7.5–7.9)	7.1–7.3, {7.5–7.8}
	8	8.3, (8.6), 8.7	8.1–8.5, (8.6–8.7), 8.8, (8.9)	8.1–8.5, {8.7–8.9}
III	9	(9.1), 9.2, 9.4–9.5	(9.1), 9.2, (9.3), 9.4–9.8, (9.9)	(9.1), 9.2, (9.3), 9.4–9.5, {9.6–9.9}
	10	10.1, {10.2–10.4} 10.5–10.6, (10.8, 10.9)	10.1–10.2, 10.5–10.6, {10.7–10.10}	10.1–10.2, {10.3–10.4}, 10.5–10.6, {10.7–10.10}
	11	(11.4–11.6, 11.9)	11.1–11.3 or 11.4–11.8 or both	11.1–11.3 or 11.4–11.8, (11.9)

Contents

Acknowledgments

I would like to express my gratitude for the helpful comments of the following reviewers: Carol Newton, University of California, Los Angeles; Robert McKelvey, University of Montana; Herbert W. Hethcote, University of Iowa; Stephen J. Merrill, Marquette University; Stavros Busenberg, Harvey Mudd College; Richard E. Plant, University of California, Davis; and Louis J. Gross, University of Tennessee. I am especially grateful to Charles M. Biles, Humboldt State University, for many detailed suggestions, continual encouragement, and for specific contributions of ideas, improvements, and problems.

The teaching styles, ideas, and specific lectures given by several colleagues and peers have strongly influenced the selection and treatment of many subjects included here: Among these are Peter Kareiva, University of Washington, with whom the original course was designed and taught (Sections 1.10, 3.1–3.4, 3.6, 6.6–6.7, 10.1, 11.8–11.9); Lee A. Segel, Weizmann Institute of Science (4.2–4.5, 4.10, 5.10, 7.1–7.2, 9.3, 10.2, 11.1–11.6); H. Tom Banks, Brown University (9.3); and Michael Reed, Duke University (10.7). I am also greatly indebted to Douglas Lauffenburger and Elizabeth Fisher, University of Pennsylvania, for providing references and prepublication data and results for the material in Section 9.7.

Numerous students have helped at various stages: editing parts of the manuscript—Marjorie Buff, Saleet Jafri, and Susan Paulsen; with research, written reports and other specific contributions which were particularly useful—Marjorie Buff (11.8, Figure 11.16, and the figure for problem 21 of Chapter 11), Laurie Roba (3.1—3.4), David F. Dabbs (3.4 and Figures 3.5–3.8), Reid Harris, Ross Alford, and Susan Paulsen (3.6 and problems 18–20 of Chapter 3), Saleet Jafri (1.9 part 2, 4.11b, and problem 16 of Chapter 1), Richard Fogel (Figure 11.23).

Bertha Livingstone and the Staff of the Biology-Forestry Library (Duke University) were particularly helpful with procurement of research materials. Initial drafts of the manuscript were typed by Dottie Libbuti and Susan Schmidt. I would like to express my sincere appreciation to my greatest helper, Bonnie Farrell, for her speed, elegance, and accuracy in typing many drafts as well as the final manuscript.

While working on final stages of the book, I have been supported by NSF grant no. DMS-86-01644. Two grants from the Duke University Research Council were especially helpful in defraying part of the costs of manuscript preparation.

Finally, I wish to express my appreciation to family members for their help and support from beginning to end.

Any comments from readers on the material, or on errors and misprints would be welcomed.

I Discrete Processes in Biology

1 The Theory of Linear Difference Equations Applied to Population Growth

For we will always have as 5 is to 8 so is 8 to 13, practically, and as 8 is to 13, so is 13 to 21 almost. I think that the seminal faculty is developed in a way analogous to this proportion which perpetuates itself, and so in the flower is displayed a pentagonal standard, so to speak. I let pass all other considerations which might be adduced by the most delightful study to establish this truth.

J. Kepler, (1611). Sterna seu de nive sexangule, *Opera,* ed. Christian Frisch, tome 7, (Frankefurt à Main, Germany: Heyden & Zimmer, 1858–1871), pp. 722–723.

The early Greeks were fascinated by numbers and believed them to hold special magical properties. From the Greeks' special blend of philosophy, mathematics, numerology, and mysticism, there emerged a foundation for the real number system upon which modern mathematics has been built. A preoccupation with aesthetic beauty in the Greek civilization meant, among other things, that architects, artisans, and craftsmen based many of their works of art on geometric principles. So it is that in the stark ruins of the Parthenon many regularly spaced columns and structures capture the essence of the *golden mean,* which derives from the *golden rectangle.*

Considered to have a most visually pleasing proportion, the *golden rectangle* has sides that bear the ratio $\tau = 1{:}1.618033\ldots$ The problem of subdividing a line segment into this so-called *extreme and mean ratio* was a classical problem in Greek geometry, appearing in the *Elements* of Euclid (circa 300 B.C.). It was recognized then and later that this *divine proportion,* as Fra Luca Pacioli (1509) called it, appears in numerous geometric figures, among them the pentagon, and the polyhedral icosahedron (see Figure 1.1).

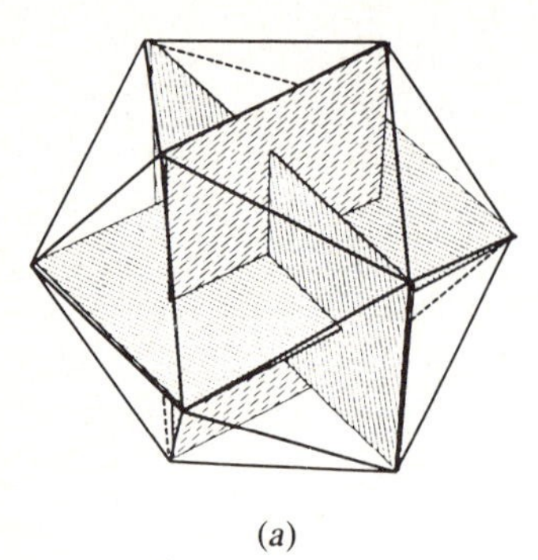

(*a*)

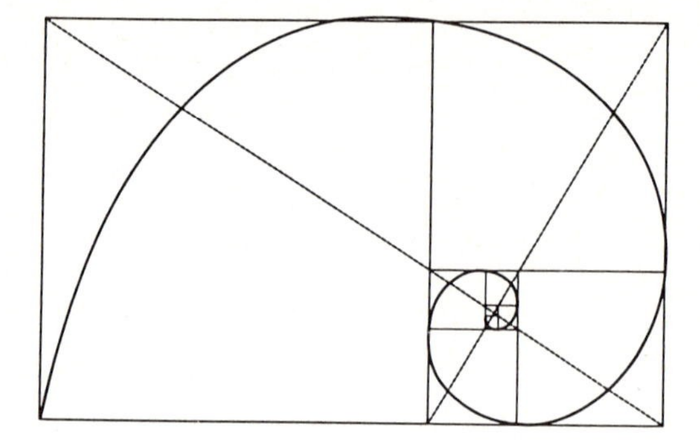

(*b*)

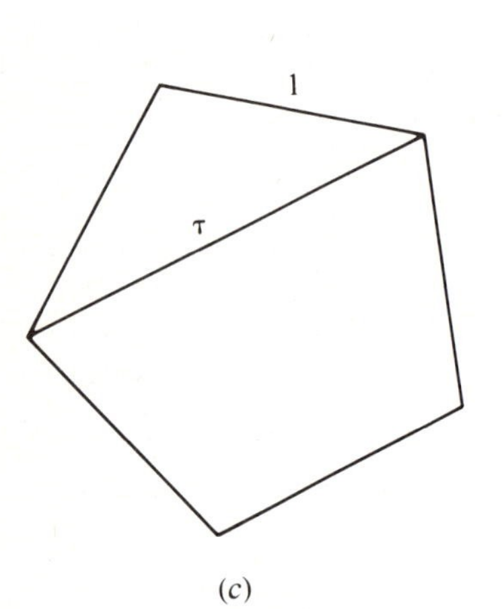

(*c*)

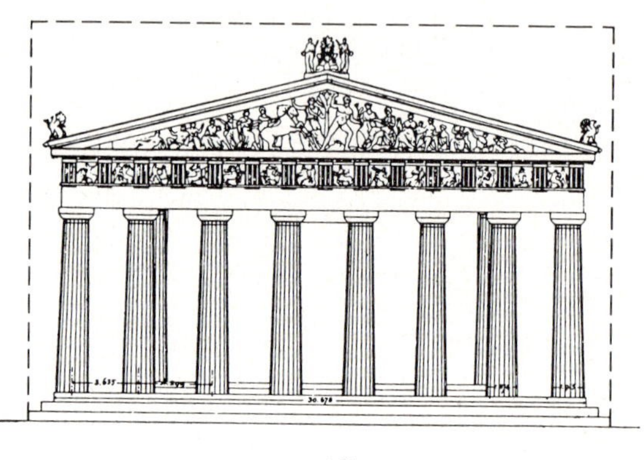

(*d*)

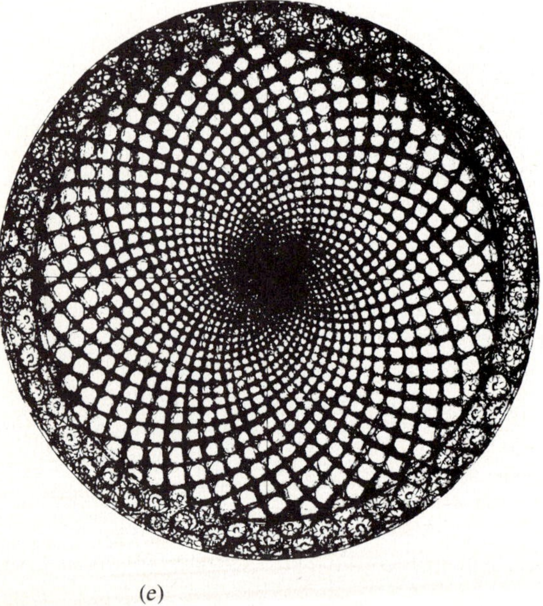

(*e*)

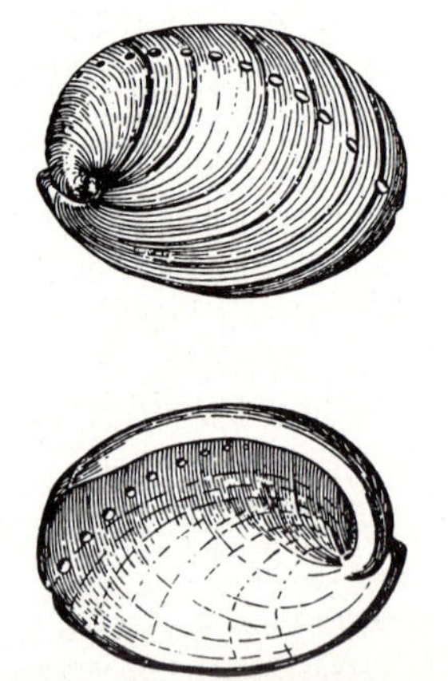

(*f*)

About fifteen hundred years after Euclid, Leonardo of Pisa (1175–1250), an Italian mathematician more affectionately known as Fibonacci ("son of good nature"), proposed a problem whose solution was a series of numbers that eventually led to a reincarnation of τ. It is believed that Kepler (1571–1630) was the first to recognize and state the connection between the *Fibonacci numbers* (0, 1, 1, 2, 3, 5, 8, 13, 21, . . .), the golden mean, and certain aspects of plant growth.

Kepler observed that successive elements of the Fibonacci sequence satisfy the following *recursion relation*

$$n_{k+2} = n_k + n_{k+1}, \tag{1}$$

i.e., each member equals the sum of its two immediate predecessors. He also noted that the ratios 2:1, 3:2, 5:3, 8:5, 13:8, . . . approach the value of τ.[1] Since then, manifestations of the golden mean and the Fibonacci numbers have appeared in art, architecture, and biological form. The *logarithmic spirals* evident in the shells of certain mollusks (e.g., abalone, of the family *Haliotidae*) are figures that result from growth in size without change in proportion and bear a relation to successively inscribed golden rectangles. The regular arrangement of leaves or plant parts along the stem, apex, or flower of a plant, known as *phyllotaxis,* captures the Fibonacci numbers in a succession of helices (called *parastichies*); a striking example is the arrangement of seeds on a ripening sunflower. Biologists have not yet agreed conclu-

1. Certain aspects of the formulation and analysis of the recursion relation (1) governing Fibonacci numbers are credited to the French mathematician Albert Girard, who developed the algebraic notation in 1634, and to Robert Simson (1753) of the University of Glasgow, who recognized ratios of successive members of the sequence as τ and as continued fractions (see problem 12).

Figure 1.1 *The golden mean τ appears in a variety of geometric forms that include: (a) Polyhedra such as the* icosahedron, *a Platonic solid with 20 equilateral triangle faces (τ = ratio of sides of an inscribed golden rectangle; three golden rectangles are shown here). (b) The golden rectangle and every rectangle formed by removing a square from it. Note that corners of successive squares can be connected by a logarithmic spiral). (c) A regular pentagon (τ = the ratio of lengths of the diagonal and a side). (d) The approximate proportions of the Parthenon (dotted line indicates a golden rectangle). (e) Geometric designs such as spirals that result from the arrangement of leaves, scales, or florets on plants (shown here on the head of a sunflower). The number of spirals running in opposite directions quite often bears one of the numerical ratios 2/3, 3/5, 5/8, 8/13, 13/21, 21/34, 34/5, . . . [see R. V. Jean (1984, 86)]; note that these are the ratios of successive Fibonacci numbers. (f) Logarithmic spirals (such as those obtained in (b) are common in shells such as the abalone* Haliotis, *where each increment in size is similar to the preceding one. See D. W. Thompson (1974) for an excellent summary. [(a and b) from M. Gardner (1961),* The Second Scientific American Book of Mathematical Puzzles and Diversions, *pp. 92–93. Copyright 1961 by Martin Gardner. Reprinted by permission of Simon & Schuster, Inc., N.Y., N.Y. (d) from G. Gromort (1947),* Histoire abrégée de l'Architecture en Grèce et à Rome, *Fig 43 on p. 75, Vincent Fréal & Cie, Paris, France. (e) from S. Colman (1971),* Nature's Harmonic Unity, *plate 64, p. 91; Benjamin Blom, N.Y. (reprinted from the 1912 edition). (f) D. Thompson (1961),* On Growth and Form *(abridged ed.) figure 84, p. 186. Reprinted by permission of Cambridge University Press, New York.]*

sively on what causes these geometric designs and patterns in plants, although the subject has been pursued for over three centuries.[2]

Fibonacci stumbled unknowingly onto the esoteric realm of τ through a question related to the growth of rabbits (see problem 14). Equation (1) is arguably the first mathematical idealization of a biological phenomenon phrased in terms of a recursion relation, or in more common terminology, a *difference equation*.

Leaving aside the mystique of golden rectangles, parastichies, and rabbits, we find that in more mundane realms, numerous biological events can be idealized by models in which similar discrete equations are involved. Typically, populations for which difference equations are suitable are those in which adults die and are totally replaced by their progeny at fixed intervals (i.e., generations do not overlap). In such cases, a difference equation might summarize the relationship between population density at a given generation and that of preceding generations. Organisms that undergo abrupt changes or go through a sequence of stages as they mature (i.e., have discrete life-cycle stages) are also commonly described by difference equations.

The goals of this chapter are to demonstrate how equations such as (1) arise in modeling biological phenomena and to develop the mathematical techniques to solve the following problem: given particular starting population levels and a recursion relation, predict the population level after an arbitrary number of generations have elapsed. (It will soon be evident that for a linear equation such as (1), the mathematical sophistication required is minimal.)

To acquire a familiarity with difference equations, we will begin with two rather elementary examples: cell division and insect growth. A somewhat more elaborate problem we then investigate is the propagation of annual plants. This topic will furnish the opportunity to discuss how a slightly more complex model is derived. Sections 1.3 and 1.4 will outline the method of solving certain linear difference equations. As a corollary, the solution of equation (1) and its connection to the golden mean will emerge.

1.1 BIOLOGICAL MODELS USING DIFFERENCE EQUATIONS

Cell Division

Suppose a population of cells divides synchronously, with each member producing a daughter cells.[3] Let us define the number of cells in each generation with a subscript, that is, $M_1, M_2, \ldots, M_n$ are respectively the number of cells in the first, second, $\ldots$, nth generations. A simple equation relating successive generations is

$$M_{n+1} = aM_n. \tag{2}$$

2. An excellent summary of the phenomena of phyllotaxis and the numerous theories that have arisen to explain the observed patterns is given by R. V. Jean (1984). His book contains numerous suggestions for independent research activities and problems related to phyllotaxis. See also Thompson (1942).

3. Note that for real populations only $a > 0$ would make sense; $a < 0$ is unrealistic, and $a = 0$ would be uninteresting.

Let us suppose that initially there are M_0 cells. How big will the population be after n generations? Applying equation (2) recursively results in the following:

$$M_{n+1} = a(aM_{n-1}) = a[a(aM_{n-2})] = \cdots = a^{n+1}M_0. \tag{3}$$

Thus, for the nth generation

$$M_n = a^n M_0. \tag{4}$$

We have arrived at a result worth remembering: The solution of a simple linear difference equation involves an expression of the form (some number)^n, where n is the generation number. (This is true in general for linear difference equations.) Note that the magnitude of a will determine whether the population grows or dwindles with time. That is,

$|a| > 1$ $\quad M_n$ *increases* over successive generations,

$|a| < 1$ $\quad M_n$ *decreases* over successive generations,

$a = 1$ $\quad M_n$ is constant.

An Insect Population

Insects generally have more than one stage in their life cycle from progeny to maturity. The complete cycle may take weeks, months, or even years. However, it is customary to use a single generation as the basic unit of time when attempting to write a model for insect population growth. Several stages in the life cycle can be depicted by writing several difference equations. Often the system of equations condenses to a single equation in which combinations of all the basic parameters appear.

As an example consider the reproduction of the poplar gall aphid. Adult female aphids produce galls on the leaves of poplars. All the progeny of a single aphid are contained in one gall (Whitham, 1980). Some fraction of these will emerge and survive to adulthood. Although generally the capacity for producing offspring (fecundity) and the likelihood of surviving to adulthood (survivorship) depends on their environmental conditions, on the quality of their food, and on the population sizes, let us momentarily ignore these effects and study a naive model in which all parameters are constant.

First we define the following:

a_n = number of adult female aphids in the nth generation,

p_n = number of progeny in the nth generation,

m = fractional mortality of the young aphids,

f = number of progeny per female aphid,

r = ratio of female aphids to total adult aphids.

Then we write equations to represent the successive populations of aphids and use these to obtain an expression for the number of adult females in the nth generation if initially there were a_0 females:

Each female produces f progeny; thus

$$p_{n+1} = fa_n. \tag{5}$$

- p_{n+1}: no. of progeny in $(n + 1)$st generation
- a_n: no. of females in previous generation
- f: no. of offspring per female

Of these, the fraction $1 - m$ survives to adulthood, yielding a final proportion of r females. Thus

$$a_{n+1} = r(1 - m)p_{n+1}. \tag{6}$$

While equations (5) and (6) describe the aphid population, note that these can be combined into the single statement

$$a_{n+1} = fr(1 - m)a_n. \tag{7}$$

For the rather theoretical case where f, r, and m are constant, the solution is

$$a_n = [fr(1 - m)]^n a_0, \tag{8}$$

where a_0 is the initial number of adult females.

Equation (7) is again a first-order linear difference equation, so that solution (8) follows from previous remarks. The expression $fr(1 - m)$ is the per capita number of adult females that each mother aphid produces.

1.2 PROPAGATION OF ANNUAL PLANTS

Annual plants produce seeds at the end of a summer. The flowering plants wilt and die, leaving their progeny in the dormant form of seeds that must survive a winter to give rise to a new generation. The following spring a certain fraction of these seeds germinate. Some seeds might remain dormant for a year or more before reviving. Others might be lost due to predation, disease, or weather. But in order for the plants to survive as a species, a sufficiently large population must be renewed from year to year.

In this section we formulate a model to describe the propagation of annual plants. Complicating the problem somewhat is the fact that annual plants produce seeds that may stay dormant for several years before germinating. The problem thus requires that we systematically keep track of both the plant population and the reserves of seeds of various ages in the seed bank.

Stage 1: Statement of the Problem

Plants produce seeds at the end of their growth season (say August), after which they die. A fraction of these seeds survive the winter, and some of these germinate at the beginning of the season (say May), giving rise to the new generation of plants. The fraction that germinates depends on the age of the seeds.

Stage 2: Definitions and Assumptions

We first collect all the parameters and constants specified in the problem. Next we define the variables. At that stage it will prove useful to consult a rough sketch such as Figure 1.2.

Parameters:

γ = number of seeds produced per plant in August,

α = fraction of one-year-old seeds that germinate in May,

β = fraction of two-year-old seeds that germinate in May,

σ = fraction of seeds that survive a given winter.

In defining the variables, we note that the seed bank changes *several times* during the year as a result of (1) germination of some seeds, (2) production of new seeds, and (3) aging of seeds and partial mortality. To simplify the problem we make the following assumption: Seeds older than two years are no longer viable and can be neglected.

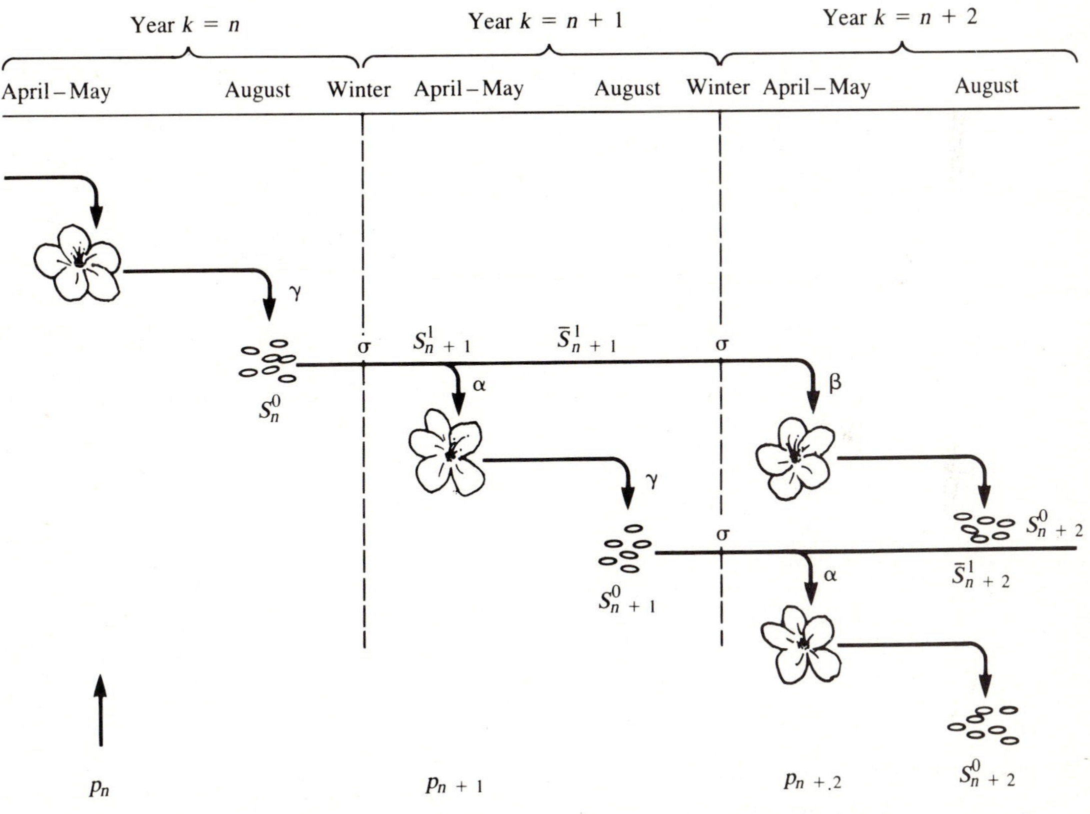

Figure 1.2 *Annual plants produce γ seeds per plant each summer. The seeds can remain in the ground for up to two years before they germinate in the springtime. Fractions α of the one-year-old and β of the two-year-old seeds give rise to a new plant generation. Over the winter seeds age, and a certain proportion of them die. The model for this system is discussed in Section 1.2.*

Consulting Figure 1.2, let us keep track of the various quantities by defining

p_n = number of plants in generation n,
S_n^1 = number of one-year-old seeds in April (before germination),
S_n^2 = number of two-year-old seeds in April (before germination),
$\bar{S}_n^1$ = number of one-year-old seeds left in May (after some have germinated),
$\bar{S}_n^2$ = number of two-year-old seeds left in May (after some have germinated),
S_n^0 = number of new seeds produced in August.

Later we will be able to eliminate some of these variables. In this first attempt at formulating the equations it helps to keep track of all these quantities. Notice that superscripts refer to age of seeds and subscripts to the year number.

Stage 3: The Equations

In May, a fraction α of one-year-old and β of two-year-old seeds produce the plants. Thus

$$p_n = \begin{pmatrix}\text{plants from}\\ \text{one-year-old seeds}\end{pmatrix} + \begin{pmatrix}\text{plants from}\\ \text{two-year-old seeds}\end{pmatrix},$$

$$p_n = \alpha S_n^1 + \beta S_n^2. \tag{9a}$$

The seed bank is reduced as a result of this germination. Indeed, for each age class, we have

$$\text{seeds left} = \begin{pmatrix}\text{fraction not}\\ \text{germinated}\end{pmatrix} \times \begin{pmatrix}\text{original number}\\ \text{of seeds in April}\end{pmatrix}.$$

Thus

$$\bar{S}_n^1 = (1 - \alpha)S_n^1, \tag{9b}$$

$$\bar{S}_n^2 = (1 - \beta)S_n^2. \tag{9c}$$

In August, new (0-year-old) seeds are produced at the rate of γ per plant:

$$S_n^0 = \gamma p_n. \tag{9d}$$

Over the winter the seed bank changes by mortality and aging. Seeds that were new in generation n will be one year old in the next generation, $n + 1$. Thus we have

$$S_{n+1}^1 = \sigma S_n^0, \tag{9e}$$

$$S_{n+1}^2 = \sigma \bar{S}_n^1. \tag{9f}$$

Stage 4: Condensing the Equations

We now use information from equations (9a–f) to recover a set of two equations linking successive plant and seed generations. To do so we observe that by using equation (9d) we can simplify (9e) to the following:

$$S_{n+1}^1 = \sigma(\gamma p_n). \tag{10}$$

Similarly, from equation (9b) equation (9f) becomes

$$S^2_{n+1} = \sigma(1 - \alpha)S^1_n. \tag{11}$$

Now let us rewrite equation (9a) for generation $n + 1$ and make some substitutions:

$$p_{n+1} = \alpha S^1_{n+1} + \beta S^2_{n+1}. \tag{12}$$

Using (10), (11), and (12) we arrive at a *system* of two equations in which plants and one-year-old seeds are coupled:

$$p_{n+1} = \alpha\sigma\gamma p_n + \beta\sigma(1 - \alpha)S^1_n, \tag{13a}$$

$$S^1_{n+1} = \sigma\gamma p_n. \tag{13b}$$

Notice that it is also possible to eliminate the seed variable altogether by first rewriting equation (13b) as

$$S^1_n = \sigma\gamma p_{n-1} \tag{14}$$

and then substituting it into equation (13a) to get

$$p_{n+1} = \alpha\sigma\gamma p_n + \beta\sigma^2(1 - \alpha)\gamma p_{n-1}. \tag{15}$$

We observe that the model can be formulated in a number of alternative ways, as a system of two first-order equations or as one second-order equation (15). Equation (15) is linear since no multiples $p_n p_m$ or terms that are nonlinear in p_n occur; it is second order since two previous generations are implicated in determining the present generation.

Notice that the system of equations (13a and b) could also have been written as a single equation for seeds.

Stage 5: Check

To be on the safe side, we shall further explore equation (15) by interpreting one of the terms on its right hand side. Rewriting it for the nth generation and reading from right to left we see that p_n is given by

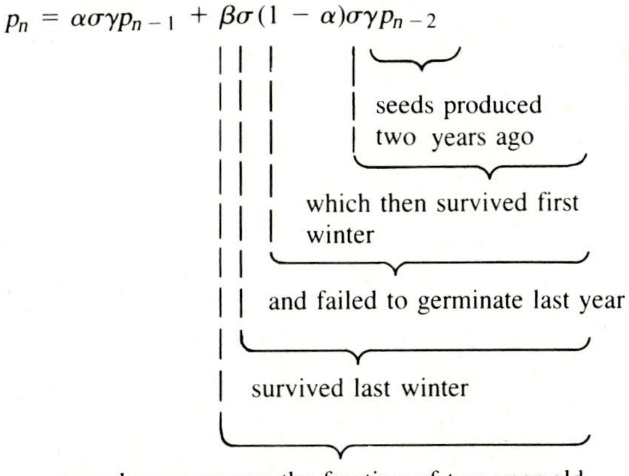

The first term is more elementary and is left as an exercise for the reader to translate.

1.3 SYSTEMS OF LINEAR DIFFERENCE EQUATIONS

The problem of annual plant reproduction leads to a system of two first-order difference equations (10,13), or equivalently a single second-order equation (15). To understand such equations, let us momentarily turn our attention to a general system of the form

$$x_{n+1} = a_{11}x_n + a_{12}y_n, \tag{16a}$$

$$y_{n+1} = a_{21}x_n + a_{22}y_n. \tag{16b}$$

As before, this can be converted to a single higher-order equation. Starting with (16a) and using (16b) to eliminate y_{n+1}, we have

$$\begin{aligned} x_{n+2} &= a_{11}x_{n+1} + a_{12}y_{n+1} \\ &= a_{11}x_{n+1} + a_{12}(a_{21}x_n + a_{22}y_n). \end{aligned}$$

From equation (16a),

$$a_{12}y_n = x_{n+1} - a_{11}x_n.$$

Now eliminating y_n we conclude that

$$x_{n+2} = a_{11}x_{n+1} + a_{12}a_{21}x_n + a_{22}(x_{n+1} - a_{11}x_n),$$

or more simply that

$$x_{n+2} - (a_{11} + a_{22})x_{n+1} + (a_{22}a_{11} - a_{12}a_{21})x_n = 0. \tag{17}$$

In a later chapter, readers may remark on the similarity to situations encountered in reducing a system of *ordinary differential equations* (ODEs) to single ODEs (see Chapter 4). We proceed to discover properties of solutions to equation (17) or equivalently, to (16a, b).

Looking back at the simple first-order linear difference equation (2), recall that solutions to it were of the form

$$x_n = C\lambda^n. \tag{18}$$

While the notation has been changed slightly, the form is still the same: constant depending on initial conditions times some number raised to the power n. Could this type of solution work for higher-order linear equations such as (17)?

We proceed to test this idea by substituting the expression $x_n = C\lambda^n$ (in the form of $x_{n+1} = C\lambda^{n+1}$ and $x_{n+2} = C\lambda^{n+2}$) into equation (17), with the result that

$$C\lambda^{n+2} - (a_{11} + a_{22})C\lambda^{n+1} + (a_{22}a_{11} - a_{12}a_{21})\lambda^n = 0.$$

Now we cancel out a common factor of $C\lambda^n$. (It may be assumed that $C\lambda^n \neq 0$ since $x_n = 0$ is a trivial solution.) We obtain

$$\lambda^2 - (a_{11} + a_{22})\lambda + (a_{22}a_{11} - a_{12}a_{21}) = 0. \tag{19}$$

Thus a solution of the form (18) would in fact work, provided that λ satisfies the quadratic equation (19), which is generally called the *characteristic equation* of (17).

To simplify notation we label the coefficients appearing in equation (19) as follows:

$$\beta = a_{11} + a_{22},$$
$$\gamma = a_{22}a_{11} - a_{12}a_{21}. \tag{20}$$

The solutions to the characteristic equation (there are two of them) are then:

$$\lambda_{1,2} = \frac{\beta \pm \sqrt{\beta^2 - 4\gamma}}{2}. \tag{21}$$

These numbers are called *eigenvalues,* and their properties will uniquely determine the behavior of solutions to equation (17). (*Note:* much of the terminology in this section is common to linear algebra; in the next section we will arrive at identical results using matrix notation.)

Equation (17) is *linear;* like all examples in this chapter it contains only scalar multiples of the variables—no quadratic, exponential, or other nonlinear expressions. For such equations, the *principle of linear superposition* holds: *if several different solutions are known, then any linear combination of these is again a solution.* Since we have just determined that λ_1^n and λ_2^n are two solutions to (17), we can conclude that a general solution is

$$x_n = A_1\lambda_1^n + A_2\lambda_2^n, \tag{22}$$

provided $\lambda_1 \neq \lambda_2$. (See problem 3 for a discussion of the case $\lambda_1 = \lambda_2$.) This expression involves two arbitrary scalars, A_1 and A_2, whose values are not specified by the difference equation (17) itself. They depend on separate constraints, such as particular known values attained by x. Note that specifying any two x values uniquely determines A_1 and A_2. Most commonly, x_0 and x_1, the levels of a population in the first two successive generations, are given (*initial conditions*); A_1 and A_2 are determined by solving the two resulting linear algebraic equations (for an example see Section 1.7). Had we eliminated x instead of y from the system of equations (16), we would have obtained a similar result. In the next section we show that general solutions to the system of first-order linear equations (16) indeed take the form

$$x_n = A_1\lambda_1^n + A_2\lambda_2^n,$$
$$y_n = B_1\lambda_1^n + B_2\lambda_2^n. \tag{23}$$

The connection between the four constants A_1, A_2, B_1, and B_2 will then be made clear.

1.4 A LINEAR ALGEBRA REVIEW[4]

Results of the preceding section can be obtained more directly from equations (16a, b) using linear algebra techniques. Since these are useful in many situations, we will briefly review the basic ideas. Readers not familiar with matrix notation are encour-

4. *To the instructor:* Students unfamiliar with linear algebra and/or complex numbers can omit Sections 1.4 and 1.8 without loss of continuity. An excellent supplement for this chapter is Sherbert (1980).

aged to consult Johnson and Riess (1981), Bradley (1975), or any other elementary linear algebra text.

Recall that a shorthand way of writing the system of algebraic linear equations,

$$\begin{aligned} ax + by &= 0, \\ cx + dy &= 0, \end{aligned} \tag{24}$$

using vector notation is:

$$\mathbf{Mv} = 0,$$

where $\mathbf{M}$ is a matrix of coefficients and $\mathbf{v}$ is the vector of unknowns. Then for system (24),

$$\mathbf{M} = \begin{pmatrix} a & b \\ c & d \end{pmatrix} \quad \text{and} \quad \mathbf{v} = \begin{pmatrix} x \\ y \end{pmatrix}. \tag{25}$$

Note that $\mathbf{Mv}$ then represents matrix multiplication of $\mathbf{M}$ (a 2×2 matrix) with $\mathbf{v}$ (a 2×1 matrix).

Because (24) is a set of linear equations with zero right-hand sides, the vector $\begin{pmatrix} 0 \\ 0 \end{pmatrix}$ is always a solution. It is in fact, a *unique* solution unless the equations are "redundant." A test for this is to see whether the determinant of $\mathbf{M}$ is zero; i.e.,

$$\det \mathbf{M} = ad - bc = 0. \tag{26}$$

When $\det \mathbf{M} = 0$, both the equations contain the same information so that in reality, there is only one constraint governing the unknowns. That means that any combination of values of x and y will solve the problem provided they satisfy any *one* of the equations, e.g.,

$$x = -by/a.$$

Thus there are *many* nonzero solutions when (26) holds.

To apply this notion to systems of difference equations, first note that equations (16) can be written in vector notation as

$$\mathbf{V}_{n+1} = \mathbf{MV}_n \tag{27a}$$

where

$$\mathbf{V}_n = \begin{pmatrix} x_n \\ y_n \end{pmatrix} \tag{27b}$$

and

$$\mathbf{M} = \begin{pmatrix} a_{11} & a_{12} \\ a_{21} & a_{22} \end{pmatrix}. \tag{27c}$$

It has already been remarked that solutions to this system are of the form

$$\mathbf{V}_n = \begin{pmatrix} A\lambda^n \\ B\lambda^n \end{pmatrix}. \tag{28a}$$

Substituting (28a) into (27a) we obtain

$$\begin{pmatrix} A\lambda^{n+1} \\ B\lambda^{n+1} \end{pmatrix} = \begin{pmatrix} a_{11} & a_{12} \\ a_{21} & a_{22} \end{pmatrix} \begin{pmatrix} A\lambda^n \\ B\lambda^n \end{pmatrix}. \tag{28b}$$

We expand the RHS to get

$$A\lambda^{n+1} = a_{11}A\lambda^n + a_{12}B\lambda^n,$$
$$B\lambda^{n+1} = a_{21}A\lambda^n + a_{22}B\lambda^n. \tag{28c}$$

We then cancel a factor of λ^n and rearrange terms to arrive at the following system of equations:

$$0 = A(a_{11} - \lambda) + Ba_{12},$$
$$0 = A(a_{21}) + B(a_{22} - \lambda). \tag{29}$$

This is equivalent to

$$0 = \begin{pmatrix} a_{11} - \lambda & a_{12} \\ a_{21} & a_{22} - \lambda \end{pmatrix} \begin{pmatrix} A \\ B \end{pmatrix}.$$

These are now linear algebraic equations in the quantities A and B. One solution is always $A = B = 0$, but this is clearly a trivial one because it leads to

$$\mathbf{V} = 0,$$

a continually zero level of both x_n and y_n. To have nonzero solutions for A and B we must set the determinant of the matrix of coefficients equal to zero;

$$\det\begin{pmatrix} a_{11} - \lambda & a_{12} \\ a_{21} & a_{22} - \lambda \end{pmatrix} = 0. \tag{30}$$

This leads to

$$(a_{11} - \lambda)(a_{22} - \lambda) - a_{12}a_{21} = 0, \tag{31}$$

which results, as before, in the quadratic characteristic equation for the eigenvalues λ. Rearranging equation (31) we obtain

$$\lambda^2 - \beta\lambda + \gamma = 0$$

where

$$\beta = a_{11} + a_{22},$$
$$\gamma = (a_{11}a_{22} - a_{12}a_{21}).$$

As before, we find that

$$\lambda_{1,2} = \frac{\beta \pm \sqrt{\beta^2 - 4\gamma}}{2}$$

are the two eigenvalues. The quantities β, γ, and $\beta^2 - 4\gamma$ have the following names and symbols:

$\beta = a_{11} + a_{22} = \text{Tr}\,\mathbf{M}$ = the *trace* of the matrix $\mathbf{M}$

$\gamma = a_{11}a_{22} - a_{12}a_{21} = \det \mathbf{M}$ = the *determinant* of $\mathbf{M}$

$\beta^2 - 4\gamma = \text{disc}(\mathbf{M})$ = the *discriminant* of $\mathbf{M}$.

If disc $\mathbf{M} < 0$, we observe that the eigenvalues are complex (see Section 1.8); if disc $\mathbf{M} = 0$, the eigenvalues are equal.

Corresponding to each eigenvalue is a nonzero vector $\mathbf{v}_i = \begin{pmatrix} A_i \\ B_i \end{pmatrix}$, called an *eigenvector,* that satisfies

$$\mathbf{M}\mathbf{v}_i = \lambda_i \mathbf{v}_i$$

This matrix equation is merely a simplified matrix version of (28c) obtained by cancelling a factor of λ^n and then applying the result to a specific eigenvalue λ_i. Alternatively the system of equations (29) in matrix form is

$$\begin{pmatrix} a_{11} - \lambda_i & a_{12} \\ a_{21} & a_{22} - \lambda_i \end{pmatrix} \begin{pmatrix} A_i \\ B_i \end{pmatrix} = 0.$$

It may be shown (see problem 4) that provided $a_{12} \neq 0$,

$$\mathbf{v}_i = \begin{pmatrix} A_i \\ B_i \end{pmatrix} = \begin{pmatrix} 1 \\ \dfrac{\lambda_i - a_{11}}{a_{12}} \end{pmatrix}$$

is an eigenvector corresponding to λ_i. Furthermore, any scalar multiple of an eigenvector is an eigenvector; i.e., if $\mathbf{v}$ is an eigenvector, then so is $\alpha\mathbf{v}$ for any scalar α.

1.5 WILL PLANTS BE SUCCESSFUL?

With the methods of Sections 1.3 and 1.4 at our disposal let us return to the topic of annual plant propagation and pursue the investigation of behavior of solutions to equation (15). The central question that the model should resolve is how many seeds a given plant should produce in order to ensure survival of the species. We shall explore this question in the following series of steps.

To simplify notation, let $a = \alpha\sigma\gamma$ and $b = \beta\sigma^2(1 - \alpha)\gamma$. Then equation (15) becomes

$$p_{n+1} - ap_n - bp_{n-1} = 0, \tag{32}$$

with corresponding characteristic equation

$$\lambda^2 - a\lambda - b = 0. \tag{33}$$

Eigenvalues are

$$\begin{aligned} \lambda_{1,2} &= \frac{1}{2}(a \pm \sqrt{a^2 + 4b}) \\ &= \frac{\sigma\gamma\alpha}{2}(1 \pm \sqrt{1 + \delta}), \end{aligned} \tag{34}$$

where

$$\delta = \frac{4\beta(1 - \alpha)}{\gamma\alpha^2} = \frac{4}{\gamma}\frac{\beta}{\alpha}\left(\frac{1}{\alpha} - 1\right)$$

is a positive quantity since $\alpha < 1$.

We have arrived at a rather cumbersome expression for the eigenvalues. The following rough approximation will give us an estimate of their magnitudes.

Initially we consider a special case. Suppose few two-year-old seeds germinate in comparison with the one-year-old seeds. Then β/α is very small, making δ small relative to 1. This means that at the very least, the positive eigenvalue λ_1 has magnitude

$$\lambda_1 \simeq \frac{\sigma\gamma\alpha}{2}(1 + \sqrt{1}) = 2\frac{\sigma\gamma\alpha}{2} = \sigma\gamma\alpha.$$

Thus, to ensure propagation we need the following conditions:

$$\lambda_1 > 1, \qquad \sigma\gamma\alpha > 1, \qquad \gamma > 1/\sigma\alpha. \tag{35a}$$

By this reasoning we may conclude that the population will grow if the number of seeds per plant is greater than $1/\sigma\alpha$. To give some biological meaning to equation (35a), we observe that the quantity $\sigma\gamma\alpha$ represents the number of seeds produced by a given plant that actually survive and germinate the following year. The approximation $\beta \simeq 0$ means that the parent plant can only be assured of replacing itself if it gives rise to at least *one* such germinated seed. Equation (35a) gives a "strong condition" for plant success where dormancy is not playing a role. If β is not negligibly small, there will be a finite probability of having progeny in the second year, and thus the condition for growth of the population will be less stringent. It can be shown (see problem 17e) that in general $\lambda_1 > 1$ if

$$\gamma > \frac{1}{\alpha\sigma + \beta\sigma^2(1 - \alpha)}. \tag{35b}$$

When $\beta = 0$ this condition reduces to that of (35a). We postpone the discussion of this case to problem 12 of Chapter 2.

As a final step in exploring the plant propagation problem, a simple computer program was written in BASIC and run on an IBM personal computer. The two sample runs derived from this program (see Table 1.1) follow the population for 20 generations starting with 100 plants and no seeds. In the first case $\alpha = 0.5$, $\gamma = 0.2$, $\sigma = 0.8$, $\beta = 0.25$, and the population dwindles. In the second case α and β have been changed to $\alpha = 0.6$, $\beta = 0.3$, and the number of plants is seen to increase from year to year. The general condition (35b) is illustrated by the computer simulations since, upon calculating values of the expressions $1/\alpha\sigma$ and $1/(\alpha\sigma + \beta\sigma^2(1 - \alpha))$ we obtain (a) 2.5 and 2.32 in the first simulation and (b) 2.08 and 1.80 in the second. Since $\gamma = 2.0$ in both cases, we observe that dormancy played an essential role in plant success in simulation b.

To place this linear model in proper context, we should add various qualifying remarks. Clearly we have made many simplifying assumptions. Among them, we have assumed that plants do not interfere with each other's success, that germination and survival rates are constant over many generations, and that all members of the plant population are identical. The problem of seed dispersal and dormancy has been examined by several investigators. For more realistic models in which other factors such as density dependence, environmental variability, and nonuniform distributions of plants are considered, the reader may wish to consult Levin, Cohen, and Hastings

Table 1.1 ***Changes in a Plant Population over 20 Generations: (a) $\alpha = 0.5$, $\beta = 0.25$, $\gamma = 2.0$, $\sigma = 0.8$; (b) $\alpha = 0.6$, $\beta = 0.3$, $\gamma = 2.0$, $\sigma = 0.8$***

Generation	*Plants*	*New seeds*	*One-year-old seeds*	*Two-year old seeds*
0	100.0	0.0	0.0	0.0
1	80.0	200.0	160.0	0.0
2	80.0	160.0	128.0	64.0
3	76.8	160.0	128.0	51.2
4	74.2	153.6	122.8	51.2
5	71.6	148.4	118.7	49.1
6	69.2	143.3	114.6	47.5
7	66.8	138.4	110.7	45.8
8	64.5	133.6	106.9	44.3
9	62.3	129.1	103.2	42.7
10	60.1	124.6	99.7	41.3
11	58.1	120.3	96.3	39.8
12	56.1	116.2	93.0	38.5
13	54.2	112.2	89.8	37.2
14	52.3	108.4	86.7	35.9
15	50.5	104.7	83.7	34.6
16	48.8	101.1	80.8	33.5
17	47.1	97.6	78.1	32.3
18	45.5	94.2	75.4	31.2
19	43.9	91.0	72.8	30.1
20	42.4	87.9	70.3	29.1

Generation	*Plants*	*New seeds*	*One-year-old seeds*	*Two-year-old seeds*
0	100.0	0.0	0.0	0.0
1	96.0	200.0	160.0	0.0
2	107.5	192.0	153.6	51.2
3	117.9	215.0	172.0	49.1
4	129.7	235.9	188.7	55.0
5	142.6	259.5	207.6	60.3
6	156.9	285.3	228.3	66.4
7	172.5	313.8	251.0	73.0
8	189.7	345.1	276.0	80.3
9	208.6	379.5	303.6	88.3
10	229.4	417.3	333.8	97.1
11	252.3	458.9	367.1	106.8
12	277.4	504.6	403.7	117.4
13	305.1	554.9	443.9	129.1
14	335.5	610.3	488.2	142.0
15	369.0	671.1	536.9	156.2
16	405.8	738.0	590.4	171.8
17	446.2	811.6	649.2	188.9
18	490.7	892.5	714.0	207.7
19	539.6	981.4	785.1	228.4
20	593.4	1079.2	863.4	251.2

(1984) and references therein. A related problem involving resistance to herbicides is treated by Segel (1981).

Recent work by Ellner (1986) is relevant to the basic issue of delayed germination in annual plants. Apparently, there is some debate over the underlying biological advantage gained by prolonging the opportunities for germination. Germination is usually controlled exclusively by the *seed coat,* whose properties derive genetically from the mother plant. Mechanical or chemical factors in the seed coat may cause a delay in germination. As a result some of the seeds may not be able to take advantage of conditions that favor seedling survival. In this way the mother plant can maintain some influence on its progeny long after their physical separation. It is held that spreading germination over a prolonged time period may help the mother plant to minimize the risk of losing all its seeds to chance mortality due to environmental conditions. From the point of view of the offspring, however, maternal control may at times be detrimental to individual survival. This *parent-offspring conflict* occurs in a variety of biological settings and is of recent popularity in several theoretical treatments. See Ellner (1986) for a discussion.

1.6 QUALITATIVE BEHAVIOR OF SOLUTIONS TO LINEAR DIFFERENCE EQUATIONS

To recapitulate the results of several examples, linear difference equations are characterized by the following properties:

1. An *m*th-order equation typically takes the form

$$a_0 x_n + a_1 x_{n-1} + \cdots + a_m x_{n-m} = b_n,$$

or equivalently,

$$a_0 x_{n+m} + a_1 x_{n+m-1} + \cdots + a_m x_n = b_n.$$

2. The *order m* of the equation is the number of previous generations that directly influence the value of x in a given generation.

3. When $a_0, a_1, \ldots, a_m$ are constants and $b_n = 0$, the problem is a *constant-coefficient homogeneous linear difference equation*; the method established in this chapter can be used to solve such equations. Solutions are composed of linear combinations of basic expressions of the form

$$x_n = C\lambda^n. \tag{36}$$

4. Values of λ appearing in equation (36) are obtained by finding the roots of the corresponding characteristic equation

$$a_0\lambda^m + a_1\lambda^{m-1} + \cdots + a_m = 0.$$

5. The number of (distinct) basic solutions to a difference equation is determined by its order. For example, a first-order equation has one solution, and a second-order equation has two. In general, an *m*th-order equation, like a system of *m* coupled first-order equations, has *m* basic solutions.

6. The general solution is a linear superposition of the m basic solutions of the equation (provided all values of λ are distinct).

7. For real values of λ the qualitative behavior of a basic solution (24) depends on whether λ falls into one of four possible ranges:

$$\lambda \geq 1, \qquad \lambda \leq -1, \qquad 0 < \lambda < 1, \qquad -1 < \lambda < 0.$$

To observe how the nature of a basic solution is characterized by this broad classification scheme, note that

(a) For $\lambda > 1$, λ^n grows as n increases; thus $x_n = C\lambda^n$ grows without bound.

(b) For $0 < \lambda < 1$, λ^n decreases to zero with increasing n; thus x_n decreases to zero.

(c) For $-1 < \lambda < 0$, λ^n oscillates between positive and negative values while declining in magnitude to zero.

(d) For $\lambda < -1$, λ^n oscillates as in (c) but with increasing magnitude.

The cases where $\lambda = 1$, $\lambda = 0$, or $\lambda = -1$, which are marginal points of demarcation between realms of behavior, correspond respectively to (1) the static (nongrowing) solution where $x = C$, (2) $x = 0$, and (3) an oscillation between the value $x = C$ and $x = -C$. Several representative examples are given in Figure 1.3.

Linear difference equations for which $m > 1$ have general solutions that combine these broad characteristics. However, note that linear combinations of expressions of the form (36) show somewhat more subtle behavior. The *dominant eigenvalue* (the value λ_i of largest magnitude) has the strongest effect on the solution; this means that after many generations the successive values of x are approximately related by

$$x_{n+1} \simeq \lambda_i x_n.$$

Clearly, whether the general solution increases or decreases in the long run depends on whether any one of its eigenvalues satisfies the condition

$$|\lambda_i| > 1.$$

If so, net growth occurs. The growth equation contains an oscillatory component if one of the eigenvalues is negative or complex (to be discussed). However, any model used to describe population growth cannot admit negative values of x_n. Thus, while oscillations typically may occur, they are generally superimposed on a larger-amplitude behavior so that successive x_n levels remain positive.

In difference equations for which $m \geq 2$, the values of λ from which basic solutions are composed are obtained by extracting roots of an mth-order polynomial. For example, a second-order difference equation leads to a quadratic characteristic equation. Such equations in general may have complex roots as well as repeated roots. Thus far we have deliberately ignored these cases for the sake of simplicity. We shall deal with the case of complex (and not real) λ in Section 1.8 and touch on the case of repeated real roots in the problems.

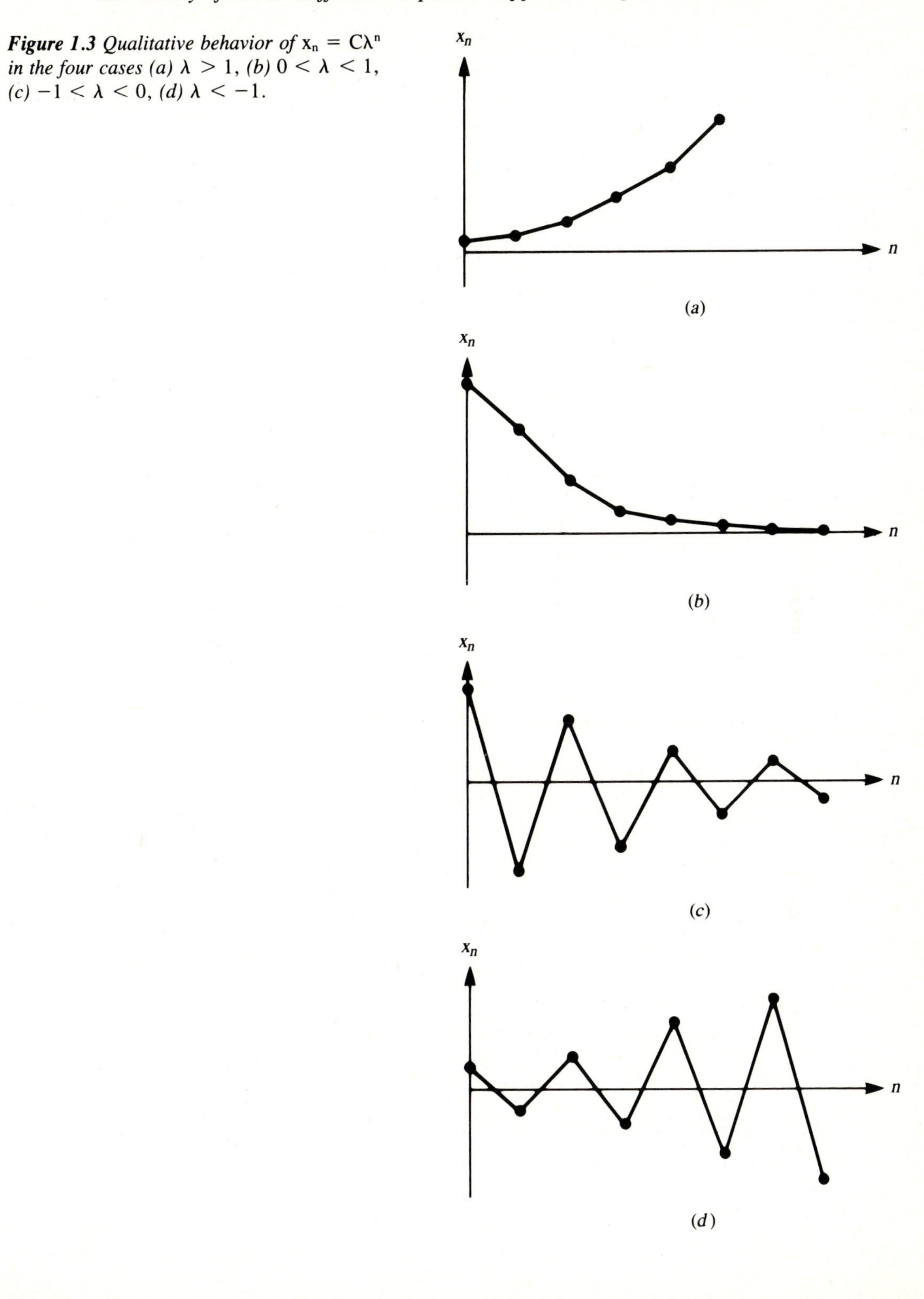

Figure 1.3 *Qualitative behavior of* $x_n = C\lambda^n$ *in the four cases (a)* $\lambda > 1$, *(b)* $0 < \lambda < 1$, *(c)* $-1 < \lambda < 0$, *(d)* $\lambda < -1$.

1.7 THE GOLDEN MEAN REVISITED

We shall apply techniques of this chapter to equation (1), which stems from Fibonacci's work. Assuming solutions of the form (18), we arrive at a characteristic equation corresponding to (1):

$$\lambda^2 = \lambda + 1.$$

Roots are

$$\lambda_1 = (1 - \sqrt{5})/2 \qquad \text{and} \qquad \lambda_2 = (1 + \sqrt{5})/2.$$

Successive members of the Fibonacci sequence are thus given by the formula

$$x_n = A\lambda_1^n + B\lambda_2^n.$$

Suppose we start the sequence with $x_0 = 0$ and $x_1 = 1$. This will uniquely determine the values of the two constants A and B, which must satisfy the following algebraic equations:

$$0 = A\lambda_1^0 + B\lambda_2^0 = A + B,$$
$$1 = A\lambda_1 + B\lambda_2 = \tfrac{1}{2}[A(1 - \sqrt{5}) + B(1 + \sqrt{5})].$$

It may be shown that A and B are given by

$$A = -1/\sqrt{5} \qquad \text{and} \qquad B = +1/\sqrt{5}.$$

Thus the solution is

$$x_n = -\frac{1}{\sqrt{5}}\left(\frac{1 - \sqrt{5}}{2}\right)^n + \frac{1}{\sqrt{5}}\left(\frac{1 + \sqrt{5}}{2}\right)^n.$$

Observe that $\lambda_2 > 1$ and $-1 < \lambda_1 < 0$. Thus the dominant eigenvalue is $\lambda_2 = (1 + \sqrt{5})/2$, and its magnitude guarantees that the Fibonacci numbers form an increasing sequence. Since the second eigenvalue is negative but of magnitude smaller than 1, its only effect is to superimpose a slight oscillation that dies out as n increases. It can be concluded that for large values of n the effect of λ_1 is negligible, so that

$$x_n \simeq (1/\sqrt{5})\lambda_2^n.$$

The ratios of successive Fibonacci numbers converge to

$$\frac{x_{n+1}}{x_n} = \lambda_2 = \frac{1 + \sqrt{5}}{2}.$$

Thus the value of the golden mean is $(1 + \sqrt{5})/2 = 1.618033\ldots.$

1.8 COMPLEX EIGENVALUES IN SOLUTIONS TO DIFFERENCE EQUATIONS

The quadratic characteristic equation (19) can have complex eigenvalues (21) with nonzero imaginary parts when $\beta^2 < 4\gamma$. These occur in conjugate pairs,

$$\lambda_1 = a + bi \qquad \text{and} \qquad \lambda_2 = a - bi,$$

where $a = \beta/2$ and $b = \frac{1}{2}|\beta^2 - 4\gamma|^{1/2}$. A similar situation can occur in linear difference equations of any order greater than 1, since these are associated with polynomial characteristic equations.

When complex values of λ are obtained, it is necessary to make sense of general solutions that involve powers of complex numbers. For example,

$$x_n = A_1(a + bi)^n + A_2(a - bi)^n . \tag{37}$$

To do so, we must first review several fundamental properties of complex numbers.

Review of Complex Numbers

A complex number can be represented in two equivalent ways. We may take $a + bi$ to be a point in the complex plane with coordinates (a, b). Equivalently, by specifying an angle ϕ in *standard position* (clockwise from positive real axis to $a + bi$) and a distance, r from (a, b) to the origin, we can represent the complex number by a pair (r, ϕ). These coordinates can be related by

$$a = r \cos \phi, \tag{38a}$$

$$b = r \sin \phi. \tag{38b}$$

Equivalently

$$r = (a^2 + b^2)^{1/2}, \tag{39a}$$

$$\phi = \arctan (b/a). \tag{39b}$$

The following identities, together known as *Euler's theorem*, summarize these relations; they can also be considered to define $e^{i\phi}$:

$$a + bi = r(\cos \phi + i \sin \phi) = re^{i\phi}, \tag{40a}$$

$$a - bi = r(\cos \phi - i \sin \phi) = re^{-i\phi}. \tag{40b}$$

This leads to the conclusion that raising a complex number to some power can be understood in the following way:

$$(a + bi)^n = (re^{i\phi})^n = r^n e^{in\phi} = c + di,$$

where

$$c = r^n \cos n\phi, \tag{41a}$$

$$d = r^n \sin n\phi. \tag{41b}$$

Graphically the relationship between the complex numbers $a + bi$ and $c + di$ is as follows: the latter has been obtained by rotating the vector (a, b) by a multiple n of the angle ϕ and then extending its length to a power n of its former length. (See Figure 1.4.) This rotating vector will be lead to an oscillating solution, as will be clarified shortly.

Proceeding formally, we rewrite (37) using equations (40a,b):

$$\begin{aligned} x_n &= A_1(a + bi)^n + A_2(a - bi)^n \\ &= A_1 r^n(\cos n\phi + i \sin n\phi) + A_2 r^n(\cos n\phi - i \sin n\phi) \\ &= B_1 r^n \cos n\phi + iB_2 r^n \sin n\phi, \end{aligned}$$

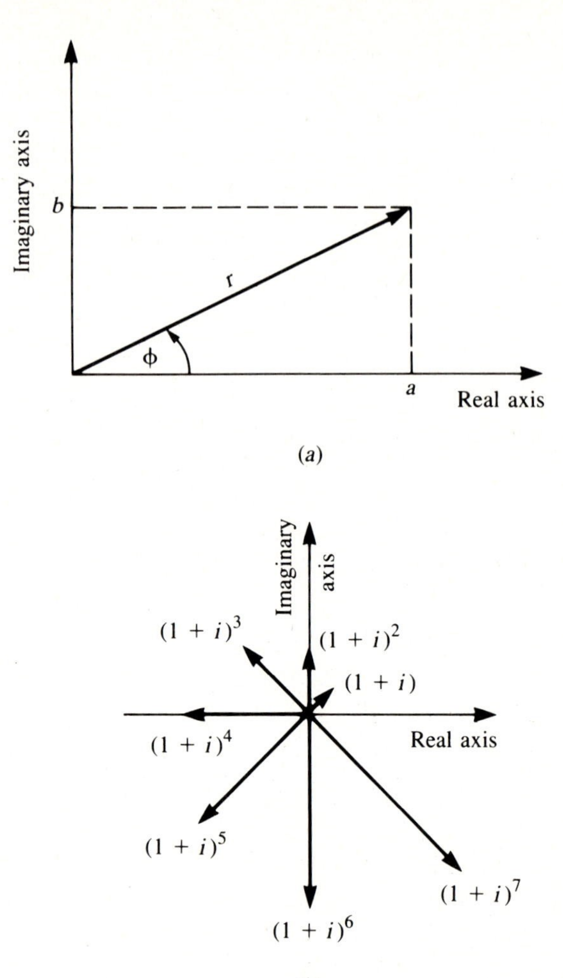

Figure 1.4 *(a) Representation of a complex number as a point in the complex plane in both cartesian* (a, b) *and polar* (r, ϕ) *coordinates. (b) A succession of values of the complex numbers* $(1 + i)^n$. *The radius vector rotates and stretches as higher powers are taken.*

where $B_1 = A_1 + A_2$ and $B_2 = (A_1 - A_2)$. Thus x_n has a real part and an imaginary part. For

$$u_n = r^n \cos n\phi, \tag{42a}$$

$$v_n = r^n \sin n\phi, \tag{42b}$$

we have

$$x_n = B_1 u_n + iB_2 v_n. \tag{43}$$

Because the equation leading to (43) is linear, it can be proved that the real and imaginary parts of this complex solution are themselves solutions. It is then customary to define a *real-valued solution* by linear superposition of the real quantities u_n and v_n:

$$\begin{aligned} x_n &= C_1 u_n + C_2 v_n \\ &= r^n(C_1 \cos n\phi + C_2 \sin n\phi) \end{aligned} \tag{44}$$

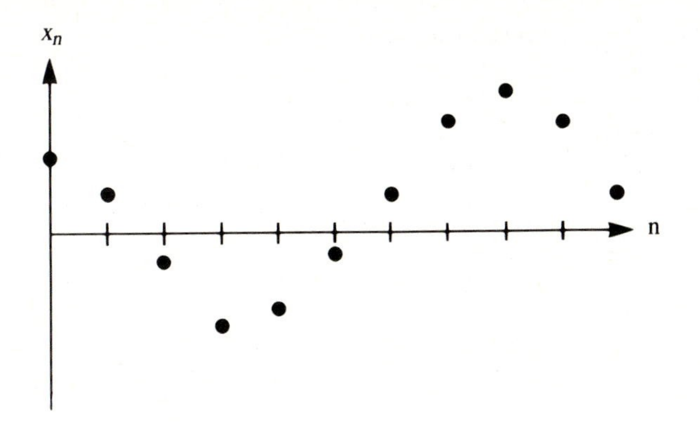

Figure 1.5 A "time sequence" of the real-valued solution given by equation (44) would display oscillations as above. Shown are values of x_n for $n = 0, 1, \ldots, 10$. The amplitude of oscillation is r^n, and the frequency is $1/\phi$ where r and ϕ are given in equation (39).

where r and ϕ are related to a and b by equations (38a,b) or (39a,b). (See Figure 1.5.)

Example

The difference equation

$$x_{n+2} - 2x_{n+1} + 2x_n = 0 \tag{45}$$

has a characteristic equation

$$\lambda^2 - 2\lambda + 2 = 0,$$

with the complex conjugate roots $\lambda = 1 \pm i$. Thus $a = 1$ and $b = 1$, so that

$$r = (a^2 + b^2)^{1/2} = \sqrt{2},$$

$$\phi = \arctan (b/a) = \pi/4.$$

Thus the real-valued general solution to equation (45) is

$$x_n = \sqrt{2}^n[C_1 \cos (n\pi/4) + C_2 \sin (n\pi/4)]. \tag{46}$$

We conclude that complex eigenvalues $\lambda = a \pm bi$ are associated with oscillatory solutions. These solutions have growing or decreasing amplitudes if $r = \sqrt{(a^2 + b^2)} > 1$ and $r = \sqrt{(a^2 + b^2)} < 1$ respectively and constant amplitudes if $r = 1$. The frequency of oscillation depends on the ratio b/a. We note also that when (and only when) arctan (b/a) is a rational multiple of π and $r = 1$, the solution will be truly periodic in that it swings through a finite number of values and returns to these exact values at every cycle.

1.9 RELATED APPLICATIONS TO SIMILAR PROBLEMS

In this section we mention several problems that can be treated similarly but leave detailed calculations for independent work in the problems.

Problem 1: Growth of Segmental Organisms

The following hypothetical situation arises in organisms such as certain filamentous algae and fungi that propagate by addition of segments. The rates of growth and branching may be complicated functions of densities, nutrient availability, and internal reserves. However, we present here a simplified version of this phenomenon to illustrate the versatility of difference equation models.

A segmental organism grows by adding new segments at intervals of 24 h in several possible ways (see Figure 1.6):

1. A terminal segment can produce a single daughter with frequency p, thereby elongating its branch.
2. A terminal segment can produce a pair of daughters (dichotomous branching) with frequency q.
3. A next-to-terminal segment can produce a single daughter (lateral branching) with frequency r.

The question to be addressed is how the numbers of segments change as this organism grows.

In approaching the problem, it is best to define the variables depicting the number of segments of each type (terminal and next-to-terminal), to make several

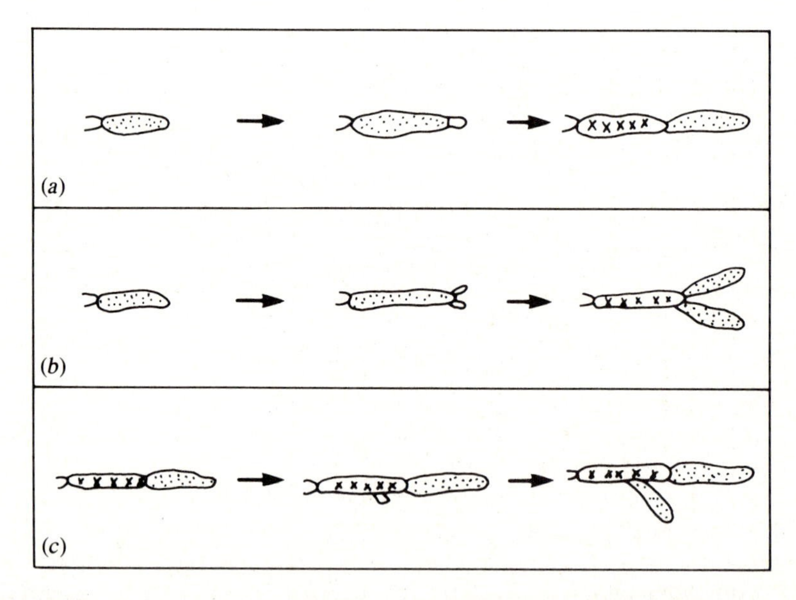

***Figure 1.6** A hypothetical segmental organism can grow in one of three ways: (a) by addition of a new segment at its terminal end; by (b) dichotomous branching, in which two new segments are added at the apex; and by (c) lateral branching, which occurs at a next-to-terminal segment. Terminal segments are dotted, and next-to-terminal segments are marked with x's. Problem 1 proposes a modeling problem based on branching growth.*

assumptions, and to account for each variable in a separate equation. The equations can then be combined into a single higher-order equation by means analogous to those given in Section 1.2. Problem 15 gives a guided approach.

For a more advanced model of branching that deals with transport of vesicles within the filaments, see Prosser and Trinci (1979).

Problem 2: A Schematic Model of Red Blood Cell Production

The following problem[5] deals with the number of red blood cells (RBCs) circulating in the blood. Here we will present it as a discrete problem to be modeled by difference equations, though a different approach is clearly possible.

In the circulatory system, the red blood cells (RBCs) are constantly being destroyed and replaced. Since these cells carry oxygen throughout the body, their number must be maintained at some fixed level. Assume that the spleen filters out and destroys a certain fraction of the cells daily and that the bone marrow produces a number proportional to the number lost on the previous day. What would be the cell count on the nth day?

To approach this problem, consider defining the following quantities:

R_n = number of RBCs in circulation on day n,

M_n = number of RBCs produced by marrow on day n,

f = fraction RBCs removed by spleen,

γ = production constant (number produced per number lost).

It follows that equations for R_n and M_n are

$$\begin{aligned} R_{n+1} &= (1 - f)R_n + M_n, \\ M_{n+1} &= \gamma f R_n. \end{aligned} \qquad (47)$$

Problem 16 discusses this model. By solving the equations it can be shown that the only way to maintain a nearly constant cell count is to assume that $\gamma = 1$. Moreover, it transpires that the delayed response of the marrow leads to some fluctuations in the red cell population.

As in problem 2, we apply a discrete approach to a physiological situation in which a somewhat more accurate description might be that of an underlying continuous process with a time delay. (Aside from practice at formulating the equations of a discrete model, this will provide a further example of difference equations analysis.)

Problem 3: Ventilation Volume and Blood CO_2 Levels

In the blood there is a steady production of CO_2 that results from the basal metabolic rate. CO_2 is lost by way of the lungs at a ventilation rate governed by CO_2-sensitive chemoreceptors located in the brainstem. In reality, both the rate of breathing and

5. I would like to acknowledge Saleet Jafri for proposing this problem.

the depth of breathing (volume of a breath) are controlled physiologically. We will simplify the problem by assuming that breathing takes place at constant intervals $t, t + \tau, t + 2\tau, \ldots$ and that the volume $V_n = V(t + n\tau)$ is controlled by the CO_2 concentration in the blood in the previous time interval, $C_{n-1} = C[t + (n - 1)\tau]$.

Keeping track of the two variables C_n and V_n might lead to the following equations:

$$C_{n+1} = \begin{pmatrix} \text{amount of } CO_2 \\ \text{previously in} \\ \text{blood} \end{pmatrix} - (\text{amount lost}) + \begin{pmatrix} \text{constant production} \\ \text{due to metabolism} \end{pmatrix},$$

$$V_{n+1} = \text{volume determined by } CO_2 \text{ at time } n. \tag{48}$$

We shall fill in these terms, thereby defining the quantities $\mathcal{L}$ (amount CO_2 lost), $\mathcal{S}$ (sensitivity to blood CO_2), and m (constant production rate of CO_2 in blood), as follows:

$$\begin{aligned} C_{n+1} &= C_n - \mathcal{L}(V_n, C_n) + m, \\ V_{n+1} &= \mathcal{S}(C_n). \end{aligned} \tag{49}$$

Before turning to the approach suggested in problem 18, it is a challenge to think about possible forms for the unspecified terms and how the resulting equations might be solved. A question to be addressed is whether the model predicts a *steady* ventilation rate (given a constant rate of metabolism) or whether certain parameter regimes might lead to fluctuations in the basal ventilation rate. This problem is closely related to one studied by Mackey and Glass (1977) and Glass and Mackey (1978), who used a differential-delay equation instead of a difference equation. The topic is thus suitable for a longer project that might include in-class presentation of their results and a comparison with the simpler model presented here. (See also review by May (1978) and Section 2.10.)

1.10 FOR FURTHER STUDY: LINEAR DIFFERENCE EQUATIONS IN DEMOGRAPHY

Demography is the study of age-structured populations. Factors such as death and birth rates correlate closely with age; such parameters also govern the way a population evolves over generations. Thus, if one is able to accurately estimate the age-dependent fecundity and mortality in a population, it is in principle possible to deduce the age structure of all successive generations. Such estimates have been available since the mid 1600s, when John Graunt used the London Bills of Mortality, an account of yearly deaths and their causes compiled in the 1600s, to infer mortality rates [see reprint of Graunt (1662) in Smith and Keyfitz (1977), and Newman (1956)]. Later, in 1693, Edmund Halley produced somewhat better estimates for the population of Breslaw, for which more complete data were available.

The mathematics of demography developed gradually over several centuries with contributions by Léonard Euler, 1760; Joshua Milne, 1815; Benjamin Gompertz, 1825; Alfred Lotka, 1907; H. Bernardelli, 1941; E. G. Lewis, 1942 and P. H.

Leslie (1945). See Smith and Keyfitz (1977) for a historical anthology. Bernardelli and Lewis were among the first to apply matrix algebra to the problem. They dealt with age structure in a population with nonoverlapping generations that can thus be modeled by a set of difference equations, one for each age class. Later Leslie formalized the theory; it is after him that Leslie matrices, which govern the succession of generations, are named.

This topic is an excellent extension of methods discussed in this chapter and is thus suitable for independent study (especially by students who have a good background in linear algebra and polynomials). See Rorres and Anton (1977) for a good introduction. Other students less conversant with the mathematics might wish to focus on the topic from a historical perspective. Smith and Keyfitz (1977) gives an excellent historical account. Several examples are given as problems at the end of this chapter and a bibliography is suggested.

PROBLEMS*

1. Consider the difference equation $x_{n+2} - 3x_{n+1} - 2x_n = 0$.

(a) Show that the general solution to this equation is

$$x_n = A_1 + 2^n A_2.$$

Now suppose that $x_0 = 10$ and $x_1 = 20$. Then A_1 and A_2 must satisfy the system of equations

$$A_1 + 2^0 A_2 = x_0 = 10,$$
$$A_1 + 2^1 A_2 = x_1 = 20.$$

(b) Solve for A_1 and A_2 and find the solution to the above *initial value problem*.

2. Solve the following difference equations subject to the specified x values and sketch the solutions:

(a) $x_n - 5x_{n-1} + 6x_{n-2} = 0$; $x_0 = 2$, $x_1 = 5$.
(b) $x_{n+1} - 5x_n + 4x_{n-1} = 0$; $x_1 = 9$, $x_2 = 33$.
(c) $x_n - x_{n-2} = 0$; $x_1 = 3$, $x_2 = 5$.
(d) $x_{n+2} - 2x_{n+1} = 0$; $x_0 = 10$.
(e) $x_{n+2} + x_{n+1} - 2x_n = 0$; $x_0 = 6$, $x_1 = 3$.
(f) $x_{n+1} = 3x_n$; $x_1 = 12$.

3. ***(a)** In Section 1.3 it was shown that the general solution to equations (16a,b) is (22) provided $\lambda_1 \neq \lambda_2$. Show that if $\lambda_1 = \lambda_2 = \lambda$ then the general solution is

$$A_1\lambda^n + A_2 n\lambda^n.$$

(b) Solve and graph the solutions to each of the following equations or systems

(i) $x_{n+2} - x_n = 0$,

*Problems preceded by asterisks (*) are especially challenging.

(ii) $x_{n+2} - 2x_{n+1} + x_n = 0,$

(iii) $x_{n+1} = -3x_n - 2y_n,$
$y_{n+1} = 2x_n + y_n.$

(iv) $x_{n+1} = 5x_n - 4y_n,$
$y_{n+1} = x_n + y_n.$

4. In Section 1.4 we determined that there are two values λ_1 and λ_2 and two vectors $\begin{pmatrix} A_1 \\ B_1 \end{pmatrix}$ and $\begin{pmatrix} A_2 \\ B_2 \end{pmatrix}$ called *eigenvectors* that satisfy equation (29).

(a) Show that this equation can be written in matrix form as

$$\lambda\begin{pmatrix} A \\ B \end{pmatrix} = \mathbf{M}\begin{pmatrix} A \\ B \end{pmatrix}$$

where **M** is given by equation (27c).

(b) Show that one way of expressing the eigenvectors in terms of a_{ij} and λ is:

$$\begin{pmatrix} A_i \\ B_i \end{pmatrix} = \begin{pmatrix} 1 \\ \dfrac{\lambda_i - a_{11}}{a_{12}} \end{pmatrix}$$

for $a_{12} \neq 0$.

(c) Show that eigenvectors are defined only up to a multiplicative constant; i.e., if **v** is an eigenvector corresponding to the eigenvalue λ, then $\alpha\mathbf{v}$ is also an eigenvector corresponding to λ for all real numbers α.

5. The *Taylor series* expansions of the functions $\sin x$, $\cos x$, and e^x are

$$\sin x = x - \frac{x^2}{2!} + \frac{x^4}{4!} - \frac{x^6}{6!} + \frac{x^8}{8!} + \cdots$$

$$\cos x = 1 - x + \frac{x^3}{3!} - \frac{x^5}{5!} + \frac{x^7}{7!} - \cdots$$

$$e^x = 1 + x + \frac{x^2}{2!} + \frac{x^3}{3!} + \frac{x^4}{4!} + \cdots .$$

Use these identities to prove Euler's theorem (equations 40). (*Note:* More details about Taylor series are given in the appendix to Chapter 2.)

6. Convert the following systems of difference equations to single higher-order equations and find their solutions. Sketch the behavior of each solution, indicating whether it increases or decreases and whether oscillations occur.

(a) $x_{n+1} = 3x_n + 2y_n$
$y_{n+1} = x_n + 4y_n$

(b) $x_{n+1} = \dfrac{x_n}{4} + y_n$
$y_{n+1} = \dfrac{3x_n}{16} - \dfrac{y_n}{4}$

(c) $x_{n+1} = \sigma_1 x_n + \sigma_2 y_n$
$y_{n+1} = \beta x_n$

(d) $x_{n+1} = x_n + y_n$
$y_{n+1} = 2x_n$

(e) $x_{n+1} = -x_n + 3y_n$
$y_{n+1} = \dfrac{y_n}{3}$

(f) $x_{n+1} = \dfrac{x_n}{4} + 3y_n$
$y_{n+1} = -\dfrac{x_n}{8} + y_n$

7. For each system in problem 6 determine the eigenvalues and eigenvectors and express the solutions in vector form (see problem 4).

8. The following complex numbers are expressed as $\lambda = a + bi$, where a is the real part and b the imaginary part. Express the number in polar form $\lambda = re^{i\theta}$, and use your result to compute the indicated power λ^n of this complex number. Sketch λ^n, for $n = 0, 1, 2, 3, 4$, as a function of n.

(a) $1 + i$
(b) $1 - i$
(c) $10i$
(d) $-1 + \sqrt{3}i$
(e) $-\frac{1}{2} - \frac{i}{2}$

9. *Complex eigenvalues.* Solve and graph the solutions to the following difference equations

(a) $x_{n+2} + x_n = 0,$
(b) $x_{n+2} - x_{n+1} + x_n = 0,$
(c) $x_{n+2} - 2x_{n+1} + 2x_n = 0,$
(d) $x_{n+2} + 2x_{n+1} + 3x_n = 0.$

10. **(a)** Consider the growth of an aphid population described in Section 1.1. If the fractional mortality of aphids is 80% and the sex ratio (ratio of females to the total number of aphids) is 50%, what minimum fecundity f is required to prevent extinction?
(b) Establish a general condition on the fecundity of aphids to guarantee population growth given a fixed survivorship and a known sex ratio.

11. **(a)** Solve the first order equation $x_{n+1} = ax_n + b$ (a, b constants).
(b) Consider the difference equation

$$aX_{n+2} + bX_{n+1} + cX_n = d.$$

If $d \neq 0$, the equation is called *nonhomogeneous*. Show that $X_n = K$, where K is a particular constant, will solve this second order nonhomogeneous equation (provided $\lambda = 1$ is not an eigenvalue) and find the value of K. (This is called a *particular solution.*)
(c) Suppose the solution to the corresponding homogeneous equation $aX_{n+2} + bX_{n+1} + cX_n = 0$ is

$$X_n = c_1\lambda_1^n + c_2\lambda_2^n,$$

where c_1 and c_2 are arbitrary constants. (This is called the *complementary* or *homogeneous solution.*) Show that

$$X_n = K + c_1\lambda_1^n + c_2\lambda_2^n$$

will be a solution to the nonhomogeneous problem (provided $\lambda_1 \neq \lambda_2$ and $\lambda_i \neq 1$). This solution is called the *general solution. Note:* As in linear nonhomogeneous differential equations, one obtains the result that for nonhomogeneous difference equations:

general solution = complementary solutions + particular solution.

12. **(a)** Show that equation (1) implies that τ, the golden mean, satisfies the relation

$$1 + \frac{1}{\tau} = \tau.$$

(b) Use this result to demonstrate that τ can be expressed as a nonterminating continued fraction, as follows:

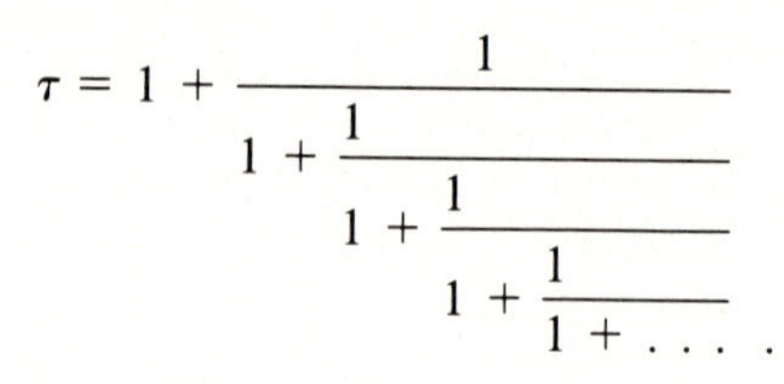

$$\tau = 1 + \cfrac{1}{1 + \cfrac{1}{1 + \cfrac{1}{1 + \cfrac{1}{1 + \cdots}}}}.$$

***13. (a)** Solve equation (1) subject to the initial values $n_0 = g_0 = 2$ and $n_1 = g_1 = 1$. The sequence of numbers you will obtain was first considered by E. Lucas (1842 – 1891). Determine the values of g_2, g_3, g_4, . . . , g_{10}.

(b) Show that in general

$$g_n = f_{n-1} + f_{n+1},$$

where f_n is the nth Fibonacci number and g_n is the nth Lucas number.

(c) Similarly show that

$$f_{2n} = f_n g_n.$$

14. *The Rabbit problem*. In 1202 Fibonacci posed and solved the following problem. Suppose that every pair of rabbits can reproduce only twice, when they are one and two months old, and that each time they produce exactly one new pair of rabbits. Assume that all rabbits survive. How many pairs will there be after n generations? (See figure.)

(a) To solve this problem, define

R_n^0 = number of newborn pairs in generation n,

R_n^1 = number of one-month-old pairs in generation n,

R_n^2 = number of two-month-old pairs in generation n.

Show that R_n^0 satisfies the equation

$$R_{n+1}^0 = R_n^0 + R_{n-1}^0.$$

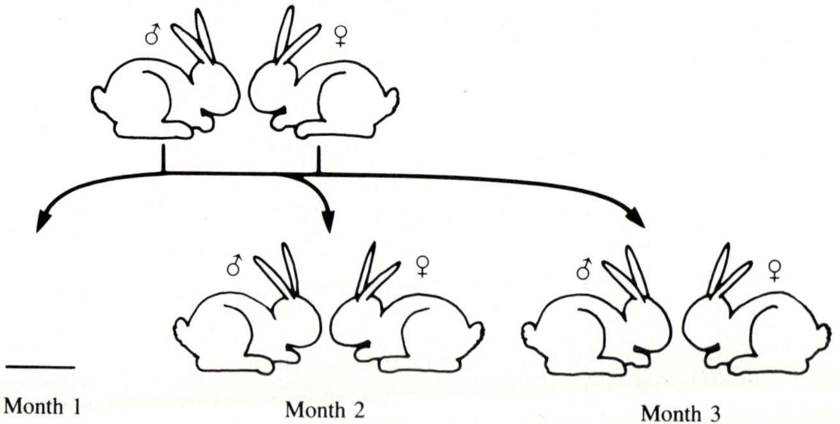

Figure for problem 14.

(b) Suppose that Fibonacci initially had one pair of newborn rabbits, i.e., that $R_0^0 = 1$ and $R_1^0 = 1$. Find the numbers of newborn, one-month-old, and two-month-old rabbits he had after n generations.

(*Historical note to problem 14:* F. C. Gies quotes the following wording of Fibonacci's question, reprinted from the *Liber abaci* in the 1981 edition of the *Encyclopaedia Britannica:* "A certain man put a pair of rabbits in a place surrounded on all sides by a wall. How many pairs of rabbits can be produced from that pair in a year if it is supposed that every month each pair begets a new pair which from the second month on becomes productive?")

15. In this problem we pursue the topic of segmental growth suggested in Section 1.9.

(a) Let a_n = number of terminal segments,
b_n = number of next-to-terminal segments,
s_n = total number of segments.

Assume that all daughters are terminal segments; that all terminal segments participate in growth ($p + q = 1$) and thereby become next-to-terminal segments in a single generation; and that all next-to-terminal segments are thereby displaced and can no longer branch after each generation. Write equations for these three variables.

(b) Show that equations for a_n and b_n can be combined to give

$$a_{n+1} - (1 + q)a_n - ra_{n-1} = 0,$$

and explain the equation.

(c) If initially there is just one terminal segment, how many terminal segments will there be after 10 days? What will be the total number of segments?

16. *Red blood cell production*

(a) Explain the equations given in Section 1.9, and show that they can be reduced to a single, higher-order equation.

(b) Show that eigenvalues are given by

$$\lambda_{1,2} = \frac{(1 - f) \pm \sqrt{(1 - f)^2 + 4\gamma f}}{2},$$

and determine their signs and magnitudes.

(c) For homeostasis in the red cell count, the total number of red blood cells, R_n, should remain roughly constant. Show that one way of achieving this is by letting $\lambda_1 = 1$. What does this imply about γ?

(d) Using the results of part (c) show that the second eigenvalue is then given by $\lambda_2 = -f$. What then is the behavior of the solution

$$R_n = A\lambda_1^n + B\lambda_2^n?$$

17. *Annual plant propagation*

(a) The model for annual plants was condensed into a single equation (15) for p_n, the number of plants. Show that it can also be written as a single equation in S_n^1.

(b) Explain the term $\alpha\sigma\gamma p_n$ in equation (15).

(c) Let $\alpha = \beta = 0.001$ and $\sigma = 1$. How big should γ be to ensure that the plant population increases in size?
(d) Explain why the plant population increases or decreases in the two simulations shown in Table 1.1.
(e) Show that equation (35b) gives a more general condition for plant success. (*Hint:* Consider $\lambda_1 = \frac{1}{2}(a \pm \sqrt{a^2 + 4b})$ and show that $\lambda_1 > 1$ implies $a + b < 1$.)

18. *Blood CO_2 and ventilation.* We now consider the problem of blood CO_2 and the physiological control of ventilation.
(a) As a first model, assume that the amount of CO_2 lost, $\mathscr{L}(V_n, C_n)$, is simply proportional to the ventilation volume V_n with constant factor β (and does not depend on C_n). Further assume that the ventilation at time $n + 1$ is directly proportional to C_n (with factor α), i.e., that $\mathscr{S}(C_n) = \alpha C_n$. (This may be physiologically unrealistic but for the moment it makes the model linear.) Write down the system of equations (49) and show that it corresponds to a single equation

$$C_{n+1} - C_n + \alpha\beta C_{n-1} = m.$$

(b) For $m \neq 0$ the equation in part (a) is a nonhomogeneous problem. Use the steps outlined in problem (11) to solve it.
(1) Show that $C_n = m/\alpha\beta$ is a particular solution.
(2) Find the general solution.
(c) Now consider the nature of this solution in two stages.
(1) First assume that $4\alpha\beta < 1$. Interpret this inequality in terms of the biological process. Give evidence for the assertion that under this condition, a steady blood CO_2 level C equal to $m/\alpha\beta$ will eventually be established, regardless of the initial conditions. What will the steady ventilation rate then be? (*Hint:* Show that $|\lambda_i| < 1$.)
(2) Now suppose $4\alpha\beta > 1$. Show that the CO_2 level will undergo oscillations. If $\alpha\beta$ is large enough, show that the oscillations may *increase* in magnitude. Find the frequency of the oscillations. Comment on the biological relevance of this solution. May (1978) refers to such situations as *dynamical diseases.*
(d) Suggest how the model might be made more realistic. The equations you obtain may be nonlinear. Determine whether they admit *steady-state solutions* in which $C_{n+1} = C_n$ and $V_{n+1} = V_n$.

19. This problem demonstrates an alternate formulation for the annual plant model in which we define the beginning of a generation at the time when seeds are produced. The figure shows the new census method.
(a) Define the variables and write new equations linking them.
(b) Show that the system can be condensed to the following two equations,

$$S^0_{n+1} = \gamma(\beta S^1_n + \alpha S^0_n),$$
$$S^1_{n+1} = \sigma(1 - \alpha)S^0_n,$$

and that in matrix form the system can be represented by

$$\begin{pmatrix} S^0 \\ S^1 \end{pmatrix}_{n+1} = \begin{pmatrix} \gamma\alpha & \gamma\beta \\ \sigma(1-\alpha) & 0 \end{pmatrix} \begin{pmatrix} S^0 \\ S^1 \end{pmatrix}_n.$$

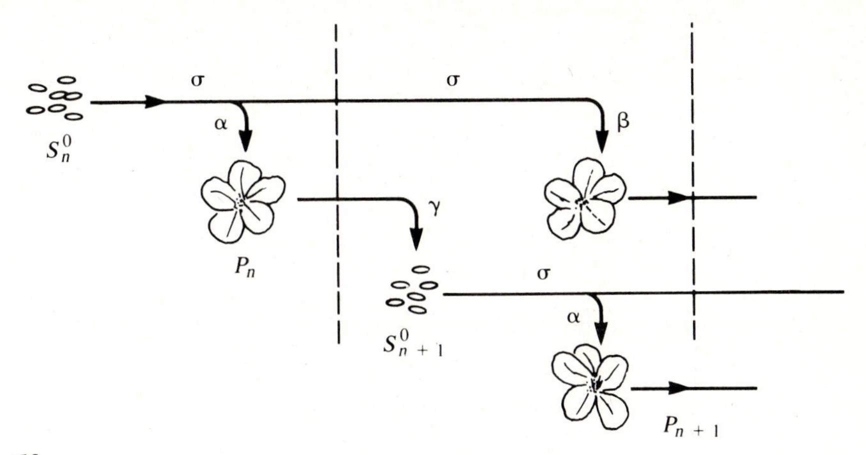

Figure for problem 19.

(c) (i) Solve the system given in (b).
(ii) Show that $p_{n+2} - \alpha\sigma\gamma p_{n+1} - \beta\sigma\gamma(1 - \alpha)p_n = 0$.
(iii) Show that the adult plant population is preserved (i.e., $\lambda_1 \geq 1$) if and only if

$$\gamma \geq \frac{1}{\alpha\sigma + \beta\sigma^2(1 - \alpha)}.$$

(d) Similarly consider this problem in which now seeds S_n^0 can survive and germinate in generations n, $n + 1$, $n + 2$, and $n + 3$, with germination fractions α, β, δ, and ε. Show that the corresponding system of equations in matrix form would be

$$\begin{pmatrix} S^0 \\ S^1 \\ S^2 \\ S^3 \end{pmatrix}_{n+1} = \begin{pmatrix} \gamma\alpha & \gamma\beta & \gamma\delta & \gamma\varepsilon \\ \sigma(1 - \alpha) & 0 & 0 & 0 \\ 0 & \sigma^2(1 - \alpha - \beta) & 0 & 0 \\ 0 & 0 & \sigma^3(1 - \alpha - \beta - \delta) & 0 \end{pmatrix} \begin{pmatrix} S^0 \\ S^1 \\ S^2 \\ S^3 \end{pmatrix}_n .$$

Interpret the terms in this matrix. You may wish to write out explicitly the system of equations.

20. **(a)** Consider a population with m age classes, and let $p_n^1, p_n^2, \ldots, p_n^m$ be the numbers of individuals within each class such that p_n^0 is the number of newborns and p_n^m is the number of oldest individuals. Define

$\alpha_1, \ldots, \alpha_k, \ldots, \alpha_m$ = number of births from individuals of a given age class,

$\sigma_1, \ldots, \sigma_k, \ldots, \sigma_{m-1}$ = fraction of k year olds that survive to be $k + 1$ year olds.

Show that the system can be described by the following matrix equation:

$$\mathbf{P}_{n+1} = \mathbf{A}\,\mathbf{P}_n,$$

where

$$\mathbf{P} = \begin{pmatrix} p^1 \\ p^2 \\ \vdots \\ p^m \end{pmatrix}, \qquad \mathbf{A} = \begin{pmatrix} \alpha_1 & \alpha_2 & \cdots & \alpha_m \\ \sigma_1 & 0 & \cdots & 0 \\ 0 & \sigma_2 & \ddots & \\ 0 & 0 & \cdots & \sigma_{m-1} 0 \end{pmatrix}.$$

A is called a *Leslie matrix*. (See references on demography for a summary of special properties of such matrices.) *Note:* For biological realism, we assume at least one $\alpha_i > 0$ and all $\sigma_i > 0$.)

***(b)** The characteristic equation of a Leslie matrix is

$$p_n(\lambda) = \det(\lambda\mathbf{I} - \mathbf{A}) = 0.$$

Show that this leads to the equation

$$\lambda^n - \alpha_1\lambda^{n-1} - \alpha_2\sigma_1\lambda^{n-2} - \alpha_3\sigma_1\sigma_2\lambda^{n-3} - \cdots - \alpha_n\sigma_1\sigma_2 \cdots \sigma_{n-1} = 0.$$

***(c)** A Leslie matrix has a unique positive eigenvalue λ^*. To prove this assertion, define the function

$$f(\lambda) = 1 - \frac{p_n(\lambda)}{\lambda^n}.$$

Show that $f(\lambda)$ is a monotone decreasing function with $f(\lambda) \to +\infty$ for $\lambda \to 0$, $f(\lambda) \to 0$ for $\lambda \to \infty$. Conclude that there is a unique value of λ, λ^* such that $f(\lambda^*) = 1$, and use this observation in proving the assertion. (*Note:* This can also be easily proved by Descartes' Rule of Signs.)

***(d)** Suppose that $\mathbf{v}^*$ is the eigenvector corresponding to λ^*. Further suppose that $|\lambda^*|$ is strictly greater than $|\lambda|$ for any other eigenvalue λ of the Leslie matrix. Reason that successive generations will eventually produce an age distribution in which the ages are proportional to elements of $\mathbf{v}^*$. This is called a *stable age distribution*. (*Note:* Some confusion in the literature stems from the fact that Leslie matrices are formulated for systems in which a census of the population is taken *after* births occur. Compare with the plant-seed example, which was done in this way in problem 19 but in another way in Section 1.2. A good reference on this point is M. R. Cullen (1985), *Linear Models in Biology*, Halstead Press, New York.

REFERENCES

Several topics mentioned briefly in this chapter are suggested as excellent subjects for further independent study.

The Golden Mean and Fibonacci Numbers

Archibald, R. C. (1918). Golden Section, *American Mathematical Monthly*, *25*, 232–238.
Bell, E. T. (1940). *The Development of Mathematics*. McGraw-Hill, New York.

Coxeter, H. S. M. (1961). *Introduction to Geometry*. Wiley, New York.
Gardner, M. (1961). *The Second Scientific American Book of Mathematical Puzzles and Diversions*. Simon & Schuster, New York.
Huntley, H. E. (1970). *The Divine Proportion*. Dover, New York.

Patterns in Nature and Phyllotaxis

Jean, R. V. (1984). *Mathematical Approach to Pattern and Form in Plant Growth*. Wiley, New York.
Jean, R. V. (1986). A basic theorem on and a fundamental approach to pattern formation on plants. *Math. Biosci.*, 79, 127–154.
Thompson, D'Arcy (1942). *On Growth and Form* (unabridged ed.). The University Press, Cambridge, U.K.

Linear Algebra

Bradley, G. L. (1975). *A Primer of Linear Algebra*. Prentice-Hall, Englewood Cliffs, N. J.
Johnson, L. W., and Riess, R. D. (1981). *Introduction to Linear Algebra*. Addison-Wesley, Reading, Mass.

Difference Equations

Brand, L. (1966). *Differential and Difference Equations*. Wiley, New York.
Grossman, S. I., and Turner, J. E. (1974). *Mathematics for the Biological Sciences*. Macmillan, New York, chap. 6.
Sherbert, D. R. (1980). *Difference Equations with Applications*. UMAP Module 322, COMAP, Lexington, Mass. Published also in *UMAP Modules, Tools for Teaching*. Birkhauser, Boston (1982).
Spiegel, M. R. (1971). *Calculus of Finite Differences and Difference Equations* (Schaum's Outline Series). McGraw-Hill, New York.

Miscellaneous

Ellner, S. (1986). Germination Dimorphisms and Parent-Offspring Conflict in Seed Germination, *J. Theor. Biol., 123*, 173–185.
Glass, L., and Mackey, M. C. (1978). Pathological conditions resulting from instabilities in physiological control systems. *Ann. N. Y. Acad. Sci., 316*, 214–235 (published 1979).
Hoppensteadt, F. C. (1982). *Mathematical Methods of Population Biology*. Cambridge University Press, New York.
Levin, S. A.; Cohen, D.; and Hastings, A. (1984). Dispersal strategies in patchy environments. *Theor. Pop. Biol., 26*, 165–191.
Mackey, M. C., and Glass, L. (1977). Oscillation and chaos in physiological control systems. *Science, 197*, 287–289.
May, R. M. (1978). Dynamical Diseases. *Nature, 272*, 673–674.

Prosser, J. I., and Trinci, A. P. J. (1979). A model for hyphal growth and branching. *J. Gen. Microbiol., 111,* 153–164.

Segel, L. A. (1981). A mathematical model relating to herbicide resistance. In W. E. Boyce, ed., *Case Studies in Mathematical Modeling*. Pitman Advanced Publishing, Boston.

Whitham, T. G. (1980). The theory of habitat selection: examined and extended using *Pemphigus* aphids. *Am. Nat., 115,* 449–466.

Demography (For Further Study)[6]

Historical accounts

Burton, D. M. (1985). *The History of Mathematics: An Introduction*. Allyn and Bacon, Inc., Boston.

Newman, J. R. (1956). *The World of Mathematics,* vol. 3. Simon & Schuster, New York, chaps. 1 and 2.

Smith, D., and Keyfitz, N. (1977). *Mathematical Demography: selected papers*. Springer-Verlag, New York.

Leslie matrices and related models

Fennel, R. E. (1976). Population growth: An age structured model. Chap. 2 in H. Marcus-Roberts and M. Thompson, eds., *Life Science Models*. Springer-Verlag, New York.

Keller, E. (1978). *Population Projection*. UMAP Module 345, COMAP, Lexington, Mass.

Leslie, P. H. (1945). On the use of matrices in certain population mathematics. *Biometrika, 33;* 183–212.

Rorres, C., and Anton, H. (1977). *Applications of Linear Algebra*. Wiley, New York, chap. 13.

Vandermeer, J. (1981). *Elementary Mathematical Ecology*. Wiley, New York.

Other references

Charlesworth, B. (1980). *Evolution in Age-Structured Populations*. Cambridge University Press, New York.

Impagliazzo, J. (1985). *Deterministic Aspects of Mathematical Demography*. Springer-Verlag, New York.

6. Compiled in part by Mike Halverson, Dale Irons, and Bill Shew.

2 Nonlinear Difference Equations

. . . It could be argued that a study of very simple nonlinear difference equations . . . should be part of high school or elementary college mathematics courses. They would enrich the intuition of students who are currently nurtured on a diet of almost exclusively linear problems.

R. M. May and G. F. Oster (1976)

Before reading through this chapter, you are invited to do some exploration using a calculator and some graph paper. The problem is to understand the behavior of the rather innocent-looking difference equation shown below:

$$x_{n+1} = rx_n(1 - x_n). \tag{1}$$

Set $r = 2.5$, let $x_0 = 0.1$, and find $x_1, x_2, \ldots, x_{20}$ using equation (1). Now repeat the process for $r = 3.3$, 3.55, and 3.9. As r is increased from 3 to 4, you should notice some changes in the sort of solutions you get.

In this chapter we will devote some time to understanding this equation while developing some concepts and techniques of more general applicability.

The first thing to notice about equation (1) is that it is *nonlinear,* since it involves a term x_n^2. Attempting to "solve" (1) by setting $x_n = \lambda^n$ as for linear problems leads nowhere. Clearly this problem cannot be understood directly by methods used in Chapter 1. Indeed, nonlinear difference equations must be handled with special methods, and many of them, despite their apparent simplicity, to this day puzzle mathematicians.

Why then should we study nonlinear difference equations? Mainly because almost all biological processes are truly nonlinear. In Chapter 3 many examples of difference equation models drawn from population biology illustrate the fact that self-regulation of a population or interactions of competing species lead to nonlinearity. For example, the per capita growth rate of a population often depends on its size, so that as density increases, the birth rate or survivorship declines. As a second example, the proportion of a prey population killed by predators varies depending on the predator population. Even the problem of annual plants is much more complicated than implied in Chapter 1 since seed germination and plant survival may be regulated by the competition for available resources.

On the other hand, transcending the immediate application of difference equation models are some rather deep philosophical issues. For example, an important discovery in the last decade is that what may appear to be totally random fluctuations in a population with discrete (nonoverlapping) generations could in fact arise from a purely deterministic rule such as equation (1) (May, 1976). At least one example of this effect is recognized in real populations (see Section 3.1), though broader application is regarded with some doubt. Before delving more deeply into these exotic results, some of the methods of tackling equations such as (1) analytically deserve consideration.

2.1 RECOGNIZING A NONLINEAR DIFFERENCE EQUATION

A nonlinear difference equation is any equation of the form

$$x_{n+1} = f(x_n, x_{n-1}, \ldots), \tag{2}$$

where x_n is the value of x in generation n and where the recursion function f depends on nonlinear combinations of its arguments (f may involve quadratics, exponentials, reciprocals, or powers of the x_n's, and so forth). A solution is again a general formula relating x_n to the generation n and to some initially specified values, e.g., x_0, x_1, and so on. In relatively few cases can an analytic solution be obtained directly when equation (2) is nonlinear. Thus we must generally be satisfied with determining something about the nature of solutions or with exploring solutions with the help of the computer.

While the methods of Chapter 1 cannot be applied directly to solving nonlinear difference equations, we shall soon see their usefulness in understanding the characteristics of special classes of solutions. Before proceeding to demonstrate these *linear techniques,* certain key concepts must be established to prepare the way. In the following section we discuss specifically the case of *first-order difference equations,* which take the form

$$x_{n+1} = f(x_n). \tag{3}$$

Properties of solutions of equation (3) will be encountered again in many related situations.

2.2 STEADY STATES, STABILITY, AND CRITICAL PARAMETERS

The concepts of *homeostasis, equilibrium,* and *steady state* relate to the absence of changes in a system. An important question stemming from many problems in the natural sciences is whether constant solutions representing these static situations exist.

In some cases steady-state solutions are of intrinsic interest: for example, most living organisms function well in rather narrow ranges of temperature, acidity, or salinity. (More highly evolved organisms have developed intricate internal mechanisms for maintaining body temperatures and other factors at their appropriate constant levels.) On the other hand, steady-state solutions may seem of marginal interest in problems involving dynamic events such as growth, propagation, or reproduction of a population. Nevertheless, it is often true that by examining carefully what happens in a steady state, we can better understand the behavior of a system, as will be demonstrated shortly.

In the context of difference equations, a *steady-state solution* $\bar{x}$ is defined to be the value that satisfies the relations

$$x_{n+1} = x_n = \bar{x}, \tag{4}$$

so that no change occurs from generation n to generation $n + 1$. From equation (3) it follows that $\bar{x}$ also satisfies the relation

$$\bar{x} = f(\bar{x}) \tag{5}$$

and is thus frequently referred to as a *fixed point* of the function f (a value that f leaves unchanged). While not always the case, it is often true that solving an equation such as (5) for the steady-state value is simpler than finding a general solution to a full nonlinear difference equation problem such as equation (3).

We now distinguish between two types of steady-state solutions. Here the concept of *stability* must be introduced. Since this is best described by analogy, see Figure 2.1, which exemplifies three situations, two of which are steady states; one steady state is stable, the other unstable.

A steady state is termed stable if neighboring states are attracted to it and unstable if the converse is true. As shown in Figure 2.1, while an object balanced precariously on a hill may be in steady state, it will not return to this position if disturbed slightly. Rather, it may proceed on some lengthy excursion leading possibly to a second, more stable situation.

Such distinctions are of interest in biology. When steady states are unstable, great changes may be about to happen: a population may crash, homeostasis may be disrupted, or else the balance in a number of competing groups may shift in favor of

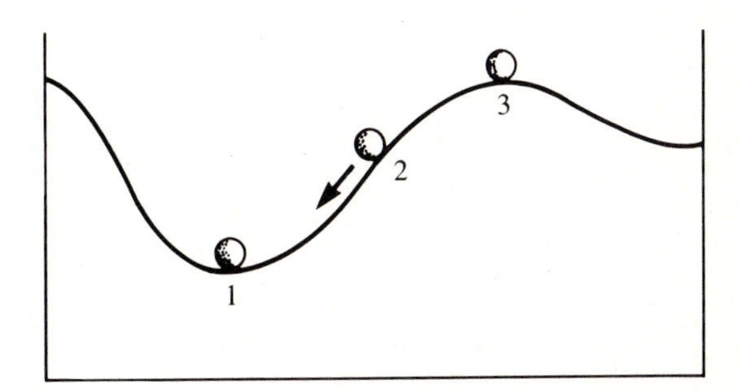

Figure 2.1 *In this landscape balls 1 and 3 are at rest and represent steady-state situations. Ball 1 is stable; if moved slightly it will return to its former position. Ball 3 is unstable. The slightest disturbance will cause it to fall into one of the adjoining valleys. Ball 2 is not in a steady state, since its position and speed are continually changing.*

the few. Thus, even if an exact mathematical solution is not easy to come by, qualitative information about whether change is imminent is of potential importance. With this motivation behind us, we turn to the analysis that permits us to make such predictions.

Let us assume that, given equation (3), we have already determined $\bar{x}$, a steady-state solution according to equation (5). We now proceed to explore its stability by asking the following key question: Given some value x_n *close* to $\bar{x}$, will x_n tend toward or away from this steady state? To address this question we start with a solution

$$x_n = \bar{x} + x_n', \tag{6}$$

where x_n' is a small quantity termed a *perturbation* of the steady state $\bar{x}$. We then determine whether x_n' gets smaller or bigger. As we will show presently, these steps reduce the problem to a linear difference equation, so that we can apply methods developed in the previous section.

From equations (5) and (6) it follows that the perturbation x_n' satisfies

$$x_{n+1}' = x_{n+1} - \bar{x} = f(x_n) - \bar{x} = f(\bar{x} + x_n') - \bar{x}. \tag{7}$$

Equation (7) is still not in a form from which direct information can be gleaned because the RHS involves the function f evaluated at $\bar{x} + x_n'$, a value that often is not known. Now we resort to a classic step that will be used again in many nonlinear problems; the value of f will be *approximated* by exploiting the fact that x_n' is a small quantity. That is, in writing a Taylor series expansion, we note that for a suitable function f,

$$f(\bar{x} + x_n') = f(\bar{x}) + \left(\frac{df}{dx}\bigg|_{\bar{x}}\right)x_n' + \underbrace{O(x_n'^2)}_{\text{very small terms}}.$$

The "very small terms" can be neglected, at least close to the steady state. This approximation results in some cancellation of terms in (7) because $f(\bar{x}) = \bar{x}$ according to equation (5). Thus the approximation

$$x_{n+1}' \simeq f(\bar{x}) - \bar{x} + \left(\frac{df}{dx}\bigg|_{\bar{x}}\right)x_n'$$

can be written as

$$x_{n+1}' = ax_n', \tag{8}$$

where

$$a = \left(\frac{df}{dx}\bigg|_{\bar{x}}\right).$$

The nonlinear problems in equations (3) or (7) have led to a linear equation (8) that describes what happens close to some steady state. Note that the constant a is a known quantity, obtained by computing the derivative of f and evaluating it at $\bar{x}$. Thus to understand whether small deviations from steady state increase or decrease, we can now apply the methods of linear difference equations. From Chapter 1 we know that the solution of equation (8) will be decreasing whenever $|a| < 1$. We conclude that

Condition for Stability

$$\bar{x} \text{ is a stable steady state of (3)} \Leftrightarrow \left| \frac{df}{dx} \bigg|_{\bar{x}} \right| < 1. \tag{9}$$

Example 1

Consider the following nonlinear difference equation for population growth:

$$x_{n+1} = \frac{kx_n}{b + x_n}, \qquad b, k > 0. \tag{10}$$

(1) Does equation (10) have a steady state? (2) If so, is that steady state stable?

Solutions: (1) To compute a steady-state value, let

$$\bar{x} = x_{n+1} = x_n.$$

Then

$$\bar{x} = \frac{k\bar{x}}{b + \bar{x}}, \qquad \bar{x}(b + \bar{x}) = k\bar{x}, \qquad \bar{x}(\bar{x} + b - k) = 0.$$

So

$$\bar{x} = k - b \qquad \text{or} \qquad \bar{x} = 0.$$

The steady state makes sense only if $k > b$, since a negative population $\bar{x}$ would be biologically meaningless.

(2) As previously mentioned, any small deviation must satisfy

$$x'_{n+1} = ax'_n, \tag{8}$$

where now

$$a = \frac{df}{dx}\bigg|_{\bar{x}} = \frac{d}{dx}\left(\frac{kx}{b + x}\right)\bigg|_{\bar{x}} = \frac{kb}{(b + \bar{x})^2} = \frac{b}{k}.$$

Thus, by the stability condition, the steady state is stable if and only if $|b/k| < 1$. Since both b and k are positive, this implies that

$$k > b.$$

Thus, the nontrivial steady state is stable whenever it exists. Stability of $\bar{x} = 0$ is left as an exercise.

In example 2, stability of one of the steady states is conditional on a *parameter* r. If r is greater or smaller than certain critical values (here 1 or 3), the steady state $\bar{x}_2$ is not stable. Such critical parameter values, often called *bifurcation values,* are points of demarcation for abrupt changes in qualitative behavior of the equation or of the system that it models. There may be a multitude of such transitions, so that as increasing values of the parameter are used, one encounters different behaviors.

Example 2

Now consider the following equation:

$$x_{n+1} = rx_n(1 - x_n), \tag{11}$$

and determine stability properties of its steady state(s).

Solution:

Again, steady states are computed by setting

$$\bar{x} = r\bar{x}(1 - \bar{x}),$$

so that

$$r\bar{x}^2 - \bar{x}(r - 1) = 0.$$

This time two steady states are possible:

$$\bar{x}_1 = 0 \qquad \text{and} \qquad \bar{x}_2 = 1 - 1/r.$$

Perturbations about $\bar{x}_2$ satisfy

$$x'_{n+1} = ax'_n,$$

where here

$$a = \left.\frac{df}{dx}\right|_{\bar{x}_2} = r(1 - 2x)\Big|_{\bar{x}_2} = (2 - r).$$

Thus x_2 will be stable whenever $|a| < 1$ according to our linear theory. For stability of this steady state we conclude that the parameter r must satisfy the condition that $1 < r < 3$.

Equation (11), which will be the subject of some scrutiny in Section 2.3, provides a striking example of bifurcations.

One of the particularly important underlying ideas here is that we can think of a particular difference equation as a rule that governs the behavior of many classes of systems (e.g., different populations, distinct species, or one species at different stages of evolution or at different developmental stages). The rules can have shades of meaning depending on critical parameters that appear in the equation. This point, which we will amply illustrate and exploit in a variety of settings, will reappear in discussions of almost all models.

2.3 THE LOGISTIC DIFFERENCE EQUATION

Equation (11) encountered in example 2 has been known for some time to possess interesting behavior, but it first received public attention as an outcome of a classic paper by May (1976) which provided an exposition to some of the perhaps unexpected properties of simple difference equations. Sometimes known as the *discrete logistic equation,* (11) is an equivalent version of

$$y_{n+1} = y_n(r - dy_n), \tag{12}$$

where r and d are constants. To see this, we redefine variables as follows. Let

$$x_n = \left(\frac{d}{r}\right) y_n.$$

This just means that quantities are measured in units of d/r. This results in a reduction of parameters based on consideration of scale, which we will discuss at great length in later chapters. This type of first step proves an aid in both formulating and analyzing complicated models.

The resulting equation,

$$x_{n+1} = rx_n(1 - x_n), \tag{11}$$

is one of the simplest nonlinear difference equations, containing just one parameter, r, and a single quadratic nonlinearity. While (11) could be a description of a population whose reproductive rate is density-regulated, there are practical problems with this interpretation. (In particular, it is necessary to restrict x and r to the intervals $0 < x < 1$, $1 < r < 4$ since otherwise the population becomes extinct; see May, 1976.)

Comparison with a Continuous Equation

Consider the following ordinary differential equation:

$$\frac{dx}{dt} = rx(1 - x), \qquad x \geq 0.$$

Sometimes called the *Pearl-Verhulst* or logistic equation, the above is often used to describe continuous density-dependent growth rates of populations. Its solutions are shown in Figure 2.2.

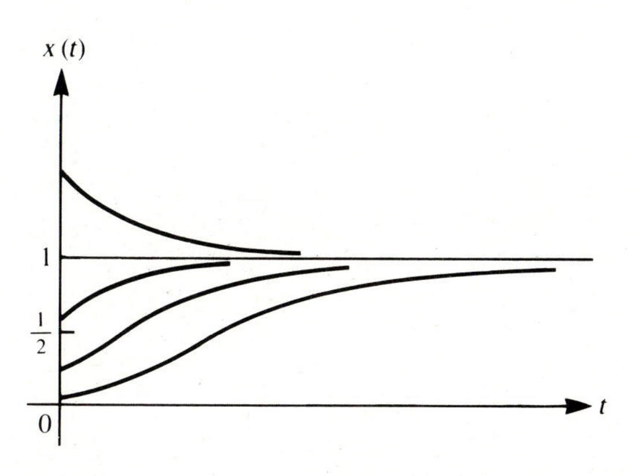

Figure 2.2 *Solutions to the* logistic differential *equation are characterized by a single nontrivial steady state* $\bar{x} = 1$, *which is stable regardless of the parameter value chosen for* r. *Thus* x(t) *tends toward a limiting value of* $x = 1$ *for all positive starting values.*

In example 2 of Section 2.2 we showed that equation (11) has two possible steady states, only one of which is *nontrivial* (nonzero): $\bar{x}_2 = 1 - 1/r$. The steady state is stable only when the parameter r satisfies $1 < r < 3$. (Notice that the parameter d in equation (12) only influences scaling of the qualitative behavior.)

What happens beyond the value of $r = 3$ and up to the permitted maximal value of r? Let us cautiously turn the knob on our metaphorical dial (Figure 2.3) and find out.

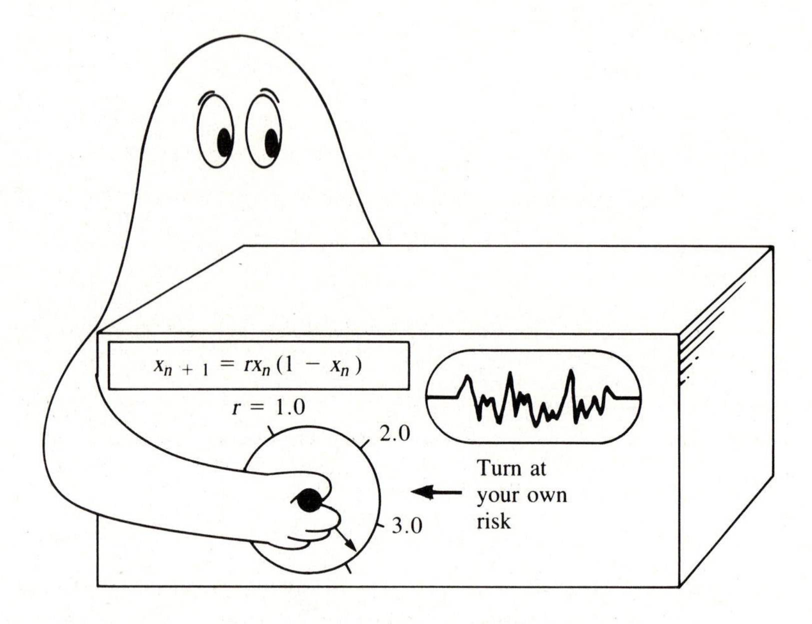

Figure 2.3

2.4 BEYOND $r = 3$

We shall resort to a clever trick (May, 1976) to prove that as r increases slightly beyond 3 in equation (11) *stable oscillations* of period 2 appear. A stable oscillation is a periodic behavior that is maintained despite small disturbances. Period 2 implies that successive generations alternate between two fixed values of x, which we will call $\bar{\bar{x}}_1$ and $\bar{\bar{x}}_2$. Thus period 2 oscillations (sometimes called *two-point cycles*) simultaneously satisfy two equations:

$$x_{n+1} = f(x_n), \tag{13a}$$

$$x_{n+2} = x_n. \tag{13b}$$

Now observe that these can be combined,

$$f(x_{n+1}) = f(f(x_n)),$$

so that

$$x_{n+2} = f(f(x_n)). \tag{14}$$

Let us call the composite function by the new name g,

$$g(x) = f(f(x)),$$

and let k be a new index that skips every other generation:

$$k = n/2, \qquad n \text{ even.}$$

Then equation (14) becomes

$$x_{k+1} = g(x_k), \tag{15}$$

and a steady state of equation (15), $\bar{\bar{x}}$ (or a fixed point of g), is really a period 2 solution of (13a). Note that there must be two such values, $\bar{\bar{x}}_1$ and $\bar{\bar{x}}_2$ since by assumption $\bar{\bar{x}}$ oscillates between two fixed values.

By this trick we have reduced the new problem to one with which we are familiar. That is, stability of a period 2 oscillation can be determined by using the methods of Section 2.2 on equation (15). Briefly, suppose an initial situation is created whereby $x_0 = \bar{\bar{x}}_1 + \epsilon_0$, where ϵ_0 is a small quantity. Stability of $\bar{\bar{x}}_1$ implies that periodic behavior will be reestablished, i.e., that the deviation ϵ_0 from this behavior will grow small. This happens whenever

$$\left| \left(\frac{dg}{dx}\right)\bigg|_{\bar{\bar{x}}_i} \right| < 1. \tag{16}$$

It is a straightforward calculation (see problem 5) to prove that this condition is equivalent to stating the following:

$$x_i \text{ is a stable 2-point cycle} \Leftrightarrow \left| \left(\frac{df}{dx}\bigg|_{\bar{\bar{x}}_1}\right)\left(\frac{df}{dx}\bigg|_{\bar{\bar{x}}_2}\right) \right| < 1 \tag{17}$$

From equation (17) we conclude that the stability of period 2 oscillations depends on the size of df/dx at $\bar{\bar{x}}_i$. The results will now be applied to further exploration of equation (11). Steps will include (1) determining $\bar{\bar{x}}_1$ and $\bar{\bar{x}}_2$, the steady two-period oscillation values, and (2) exploring their stability.[1]

1. *To the instructor:* This section may be skipped without loss of continuity in the discussion.

Example 3
Find $\bar{\bar{x}}_1$ and $\bar{\bar{x}}_2$ for the two-point cycles of equation (11).

Solution
To do so, first determine the composite map $g(x) = f(f(x))$:

$$\begin{aligned} g(x) &= r[rx(1-x)](1-[rx(1-x)]) \\ &= r^2x(1-x)[1-rx(1-x)]. \end{aligned} \tag{18}$$

Next, in equation (18) set $\bar{\bar{x}}$ equal to $g(\bar{\bar{x}})$ to obtain

$$1 = r^2(1-\bar{\bar{x}})[1-r\bar{\bar{x}}(1-\bar{\bar{x}})]. \tag{19}$$

Here it is necessary to be slightly resourceful, for the expression obtained is a third-order polynomial. We look back at the information at hand and use an important fact in solving this problem.

We notice that any steady-state values of the equation $x_{n+1} = f(x_n)$ are automatically steady states also of $x_{n+1} = f(f(x_n))$ or of any higher composition of f with itself. (In other words, $\bar{x}$ is also a periodic solution in the trivial sense.) This means that $\bar{x}$ satisfies the equation $\bar{x} = g(\bar{x})$. To see this, note that

$$\bar{x} = x_n = x_{n+1} = x_{n+2},$$

so

$$\bar{x} = f(\bar{x}) = f(f(\bar{x})) = g(\bar{x}).$$

Continuing the analysis of example 3, we now exploit the fact that $x = 1 - 1/r$ must be one solution to equation (19). This enables us to factor the polynomial so that the problem is reduced to solving a quadratic equation. To do this, we expand equation (19):

$$p(x) = x^3 - 2x^2 + \left(1 + \frac{1}{r}\right)x + \left(\frac{1}{r^3} - \frac{1}{r}\right) = 0. \tag{20}$$

Now divide by the factor $\{x - [1 - (1/r)]\}$, to get

$$\left[x - \left(1 - \frac{1}{r}\right)\right]\left[x^2 - \left(1 + \frac{1}{r}\right)x + \left(\frac{1}{r} + \frac{1}{r^2}\right)\right] = p(x) = 0. \tag{21}$$

This can be done by standard long division of polynomials.

The second factor is a quadratic expression whose roots are solutions to the equation

$$x^2 - \left(\frac{r+1}{r}\right)x + \left(\frac{r+1}{r^2}\right) = 0.$$

Hence

$$\bar{\bar{x}} = \frac{1}{2}\left[\left(\frac{r+1}{r}\right) \pm \sqrt{\left(\frac{r+1}{r}\right)^2 - \frac{4(r+1)}{r^2}}\right],$$

$$\bar{\bar{x}}_1, \bar{\bar{x}}_2 = \frac{r + 1 \pm \sqrt{(r-3)(r+1)}}{2r}. \tag{22}$$

The two possible roots, denoted $\bar{\bar{x}}_1$ and $\bar{\bar{x}}_2$, are real if $r < -1$ or $r > 3$. Thus for positive r, steady states of the two-generation map $f(f(x_n))$ exist only when $r > 3$. Note that this occurs when $\bar{x} = 1 - 1/r$ ceases to be stable.

With $\bar{\bar{x}}_1$ and $\bar{\bar{x}}_2$ computed, it is a straightforward (albeit algebraically messy) task to test their stability. To do so, it is necessary to compute (df/dx) and evaluate at the values $\bar{\bar{x}}_1$ and $\bar{\bar{x}}_2$. When this is done, we obtain a second range of behavior: stability of the two-point cycles for $3 < r < r_2$ with $r_2 \simeq 3.3$. Again we could pose the question, What happens beyond $r = r_2$?

It should be emphasized that the trick used in exploring period 2 oscillations could be used for any higher period n: $n = 3, 4, \ldots$. Because the analysis becomes increasingly cumbersome, this method will not be further applied. Instead, we will explore some underlying geometric ideas that make the process of "tuning" a parameter more immediately significant.

2.5 GRAPHICAL METHODS FOR FIRST-ORDER EQUATIONS

In this section we examine a simple technique for visualizing the solutions of first-order difference equations that can be used for gaining insight into the stability of steady states and the effects of parameter variations. As an example, consider equation (11). Let us draw a graph of $f(x)$, the next-generation function (Figure 2.4). In this case $f(x) = rx(1 - x)$, so that f describes a parabola passing through zero at $x = 1$ and $x = 0$ and with a maximum at $x = \frac{1}{2}$.

Choosing an initial value x_0, we can read off $x_1 = f(x_0)$ directly from the parabolic curve. To continue finding $x_2 = f(x_1)$, $x_3 = f(x_2)$ and so on, we need to similarly evaluate f at each succeeding value of x_n. One way of achieving this is to use the line $y = x$ to reflect each value of x_{n+1} back to the x_n axis (Figure 2.4). This process, which is equivalent to bouncing between the curves $y = x$ and $y = f(x)$ (Figure 2.5) is a recursive graphical method for determining the population level. In Figure 2.5, a time sequence of x_n values is also shown. (This method should be compared to the one outlined in problem 13.)

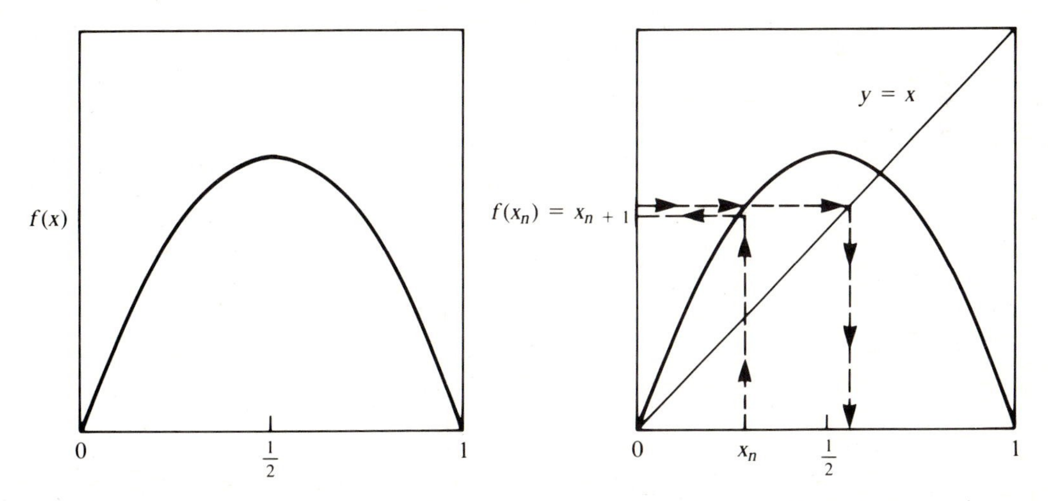

Figure 2.4 *The parabola* y = f(x) *and the line* y = x *can be used to graph the successive values of* x_n, n = 0, 1, 2, ... *This is known as the* cobwebbing method *(see Figure 2.5).*

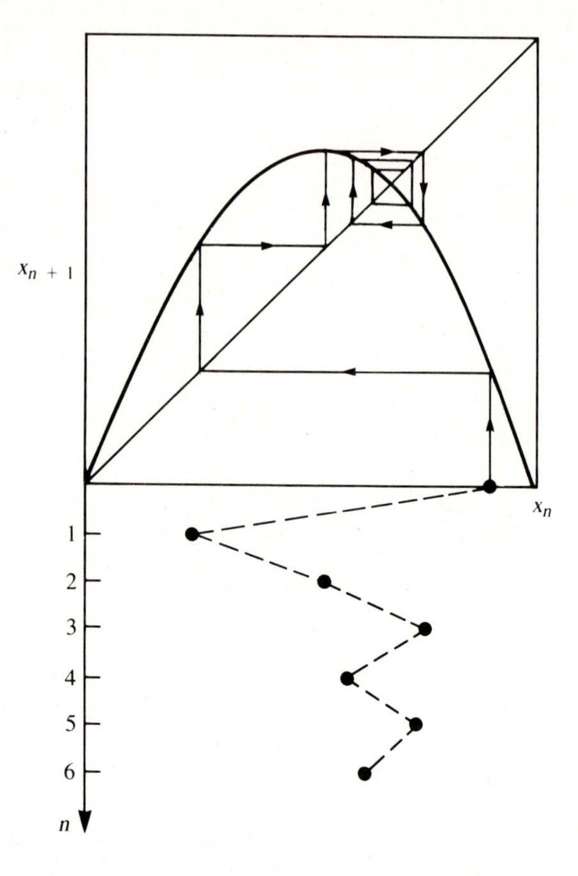

Figure 2.5 *A number of values of* x_n, $n = 0, 1, \ldots, 7$ *are shown. Below is the corresponding time sequence, where the vertical axis shows time* n.

In Figure 2.5 the sequence of points converges to a single point at the intersection of the curve with the line $y = x$. This point of intersection satisfies

$$x_{n+1} = x_n \qquad \text{and} \qquad x_{n+1} = f(x_n).$$

It is by definition a steady state of the equation. The present example illustrates a particular regime for which this steady state is stable. Recall that the condition for stability is that

$$|a| = \left| \frac{df}{dx}\bigg|_{\bar{x}} \right| < 1.$$

Interpreted graphically, this condition means that the tangent line $\mathscr{L}$ to $f(x)$ at the steady state value $\bar{x}$ has a slope not steeper than 1 (see Figure 2.6). Notice what happens when the parameter r is increased. This effectively increases the steepness of

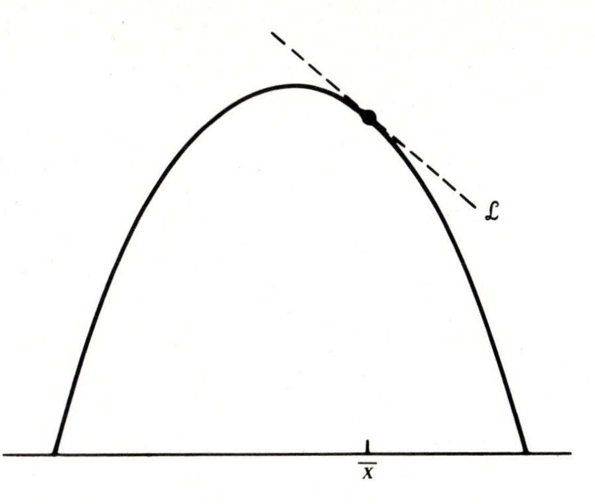

Figure 2.6 *The steady state* $\bar{x}$ *is always located at the intersection of* $y = f(x)$ *with* $y = x$. $\bar{x}$ *is stable if the tangent to the curve* $y = f(x)$ *is not too steep (a slope of magnitude less than* 1*).*

the parabola, which makes its tangent line $\mathcal{L}$ at $\bar{x}$ steeper, so that eventually the stability condition is violated (see Figure 2.7).

A similar procedure can be applied to the case of period 2 solutions discussed analytically in example 3, or indeed solutions of higher period. Here one is required to construct a graph of $f(f(x))$ [or $f^n(x)$]. This task can be accomplished once the maxima and minima of these composed functions are identified, for instance by setting their derivatives equal to zero. (It can also be done computationally.) One can verify that the function $f(f(x))$ given in example 3 undergoes the following sequence of shape changes as the parameter r is increased. Initially, $g(x) = f(f(x))$ has a flat graph, but as r increases, two humps appear and grow in size. Eventually these are big enough to intersect the line $y = x$ at three locations: $\bar{x}$ (as before) and $\bar{\bar{x}}_1$, $\bar{\bar{x}}_2$. At this point $\bar{x}$ loses its stability to the two-point cycle consisting of $\bar{\bar{x}}_1$, $\bar{\bar{x}}_2$.

As r is increased even further, $\bar{\bar{x}}_1$ and $\bar{\bar{x}}_2$ in turn lose their property of stability to other periodic states (with periods 4, 8, etc.). Each time a transition point of bifurcation value is reached, some new qualitative behavior is established. One can see these effects graphically or by using the simple pocket calculator computations, as suggested in the introduction to this chapter.

One way of summarizing the range of behaviors is shown in Figure 2.8, a *bifurcation diagram,* which depicts the locations and stability properties of periodic states. In the case of the logistic difference equation (11), since all of the relevant cases must fall within the interval $1 < r < 4$ (see May, 1976), the intervals of stability of any of the 2^n periodic solutions become increasingly smaller.

The first significantly new thing that happens (when $r \simeq 3.83$) is that a solution of period 3 suddenly appears. Li and Yorke (1975) proved that period 3 orbits were harbingers of a somewhat perplexing phenomenon they called *chaos*. This is a solution that appears to undergo large random fluctuations with no inherent periodicity or order whatsoever. An example of this type of behavior is given in Figure 2.9.

Figure 2.7 *(a)* f(f(x)) *initially has a flat graph with a single intersection of* $x_{n+2} = x_n$ *at* $\bar{x}$, *the steady state of* $\bar{x} = f(\bar{x})$. *(b) As* r *increases, two humps appear. As yet a single intersection is maintained.* $\bar{x}$ *remains stable as the slope of the tangent line is still less than* 1. *(c) As the two humps grow and as* r *increases beyond* 3, *two new intersections appear. Simultaneously,* $\bar{x}$ *becomes unstable.* $\bar{\bar{x}}_1$ *and* $\bar{\bar{x}}_2$ *are now stable as fixed points of* f(f(x)). *They thus form the period* 2 *orbit of this system.*

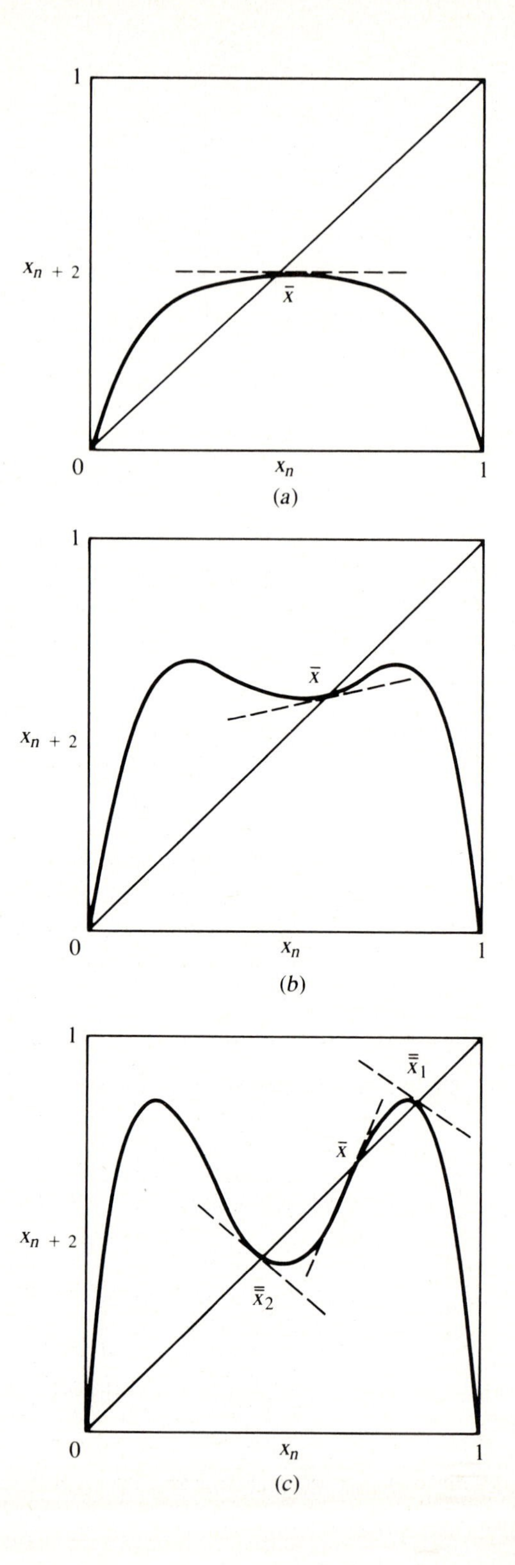

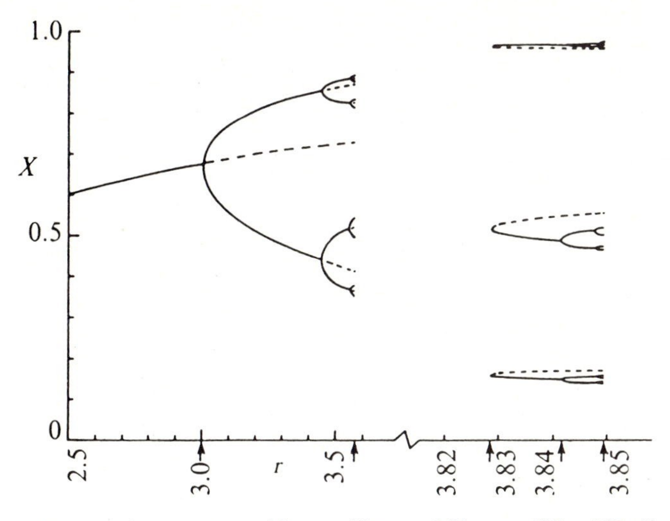

Figure 2.8 *Bifurcation diagram for equation (11). [From May (1976). Reprinted by permission from* Nature, 261, *pp. 459–467. Copyright © 1976 Macmillan Journals Limited.]*

A disturbing aspect of this type of solution is that two very close initial values x_0 and y_0 will in general grow very dissimilar in a few iterations. May (1976) remarked that it may be visually impossible to distinguish between a totally random sequence of events and the chaotic solution of a deterministic equation such as (11).

Does the chaotic regime of a difference equation have any biological relevance? In Chapter 3 one example of a chaotically fluctuating population, the blowflies, will be discussed. The majority of biologists feel, however, that most biological systems operate well within the range of stability of the low order periodic solutions. Thus the rather bizarre chaos remains largely a mathematical curiosity.

Bifurcation Diagrams

The effect of a parameter variation on existence and stability properties of steady states in equations such as (11) are often represented on a bifurcation diagram, such as the one in Figure 2.8. The horizontal axis gives the parameter value (e.g., r in equation (11)). The vertical axis represents the magnitudes of steady state(s) of the equation. A branch of the diagram (shown in solid lines) represents the dependence of the steady state level on the parameter. At points of bifurcation (e.g., at $r = 3.0$, $r = 3.5, \ldots$) new steady states come into existence and old ones lose their stability (henceforth shown dotted). Two stable branches imply that a stable period 2 orbit exists; four stable branches imply that a period 4 orbit exists; and so on. Note that the succession of bifurcations may occur at ever closer intervals. The values of r at which such transitions occur are called bifurcation values.

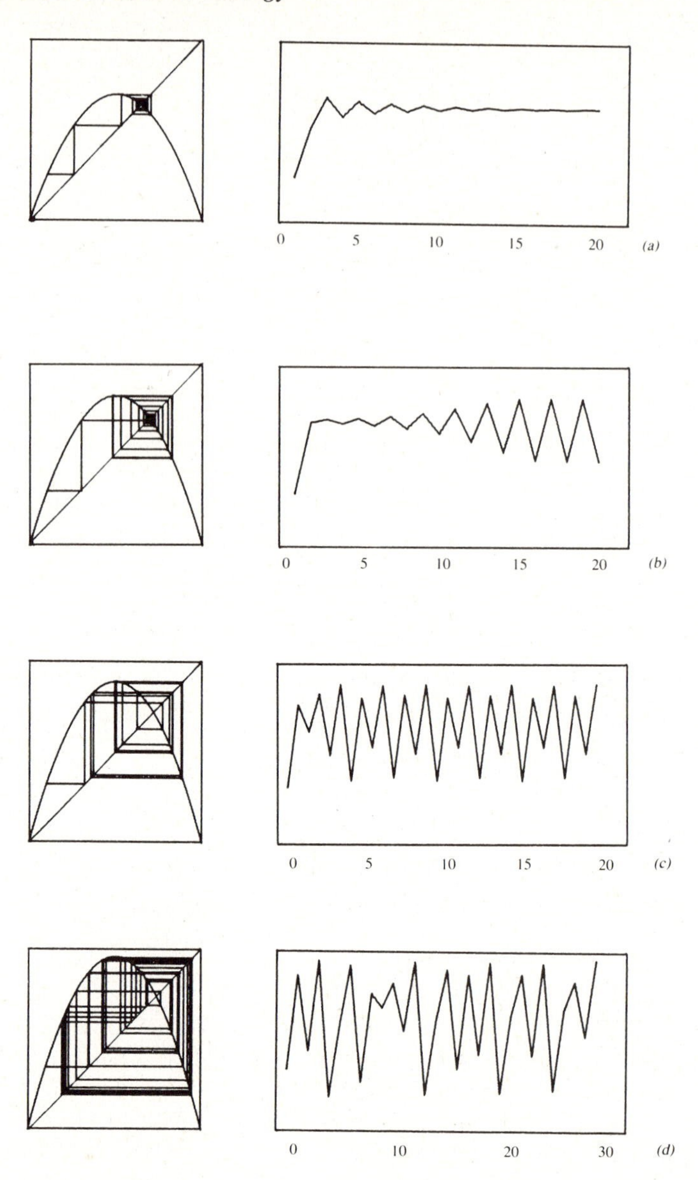

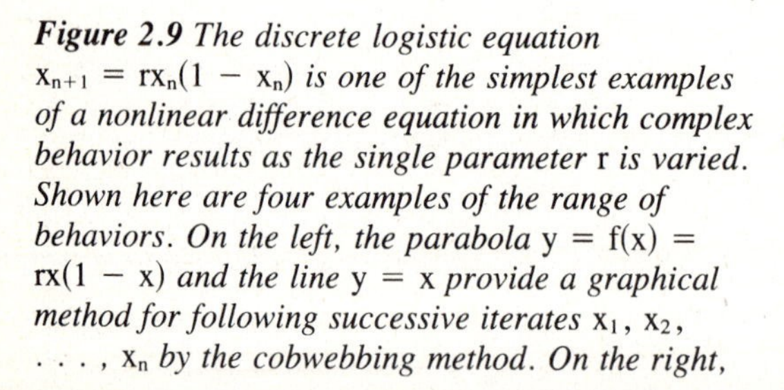

Figure 2.9 *The discrete logistic equation $x_{n+1} = rx_n(1 - x_n)$ is one of the simplest examples of a nonlinear difference equation in which complex behavior results as the single parameter r is varied. Shown here are four examples of the range of behaviors. On the left, the parabola $y = f(x) = rx(1 - x)$ and the line $y = x$ provide a graphical method for following successive iterates $x_1, x_2, \ldots, x_n$ by the cobwebbing method. On the right, successive values of x_n are displayed. r values are (a) 2.8, (b) 3.3, (c) 3.55, and (d) 3.829. Note that as r increases, the parabola steepens. There are transitions from a stable steady state (a), to a two-point (b), four-point (c) cycle, and eventually to chaos (d). See text for explanation. These figures were produced by a FORTRAN program on an IBM 360 digital computer.*

Summary and Applications of the Logistic Difference Equation (11)

Equation (11) has been used as a convenient example for illustrating a number of key ideas. First, we saw that the number of parameters affecting the qualitative features of a model may be smaller than the number that initially appear. Further, we observed that existence and stability of steady states and periodic solutions changed as the critical parameter was varied (or "tuned"). Finally, we had a brief exposure to the fact that difference equations can produce somewhat unusual solutions quite unlike their "tame" continuous counterparts.

Equation (11) is seldom used as an honest-to-goodness biological model. However, it serves as a useful pedagogical example of calculations and results that also hold for other, more realistic models, some of which will be described in Chapter 3. For a more detailed and thorough analysis of this equation, turn to the lucid review by May (1976).

2.6 A WORD ABOUT THE COMPUTER

Perhaps one of the most pleasing properties of difference equations is that they readily yield to numerical exploration, whether by calculator or with a digital computer. This property is not shared with the continuous differential equations. Solutions to difference equations are obtainable by sequential arithmetic operations, a task for which the computer is precisely suited. Indeed, the key strategy in tackling the more problematic differential equations by numerical computations is to find a reliable approximating difference equation to solve instead. This makes it particularly important to appreciate the properties and eccentricities of these equations.

2.7 SYSTEMS OF NONLINEAR DIFFERENCE EQUATIONS

To conclude this chapter, we will extend the methods developed for single equations to systems of n difference equations for arbitrary n. For simplicity of notation we will discuss here the case where $n = 2$. Assume therefore that two independent variables x and y are related by the system of equations

$$\begin{aligned} x_{n+1} &= f(x_n, y_n), \\ y_{n+1} &= g(x_n, y_n), \end{aligned} \tag{23}$$

where f and g are nonlinear functions. Steady-state values $\bar{x}$ and $\bar{y}$ satisfy

$$\begin{aligned} \bar{x} &= f(\bar{x}, \bar{y}), \\ \bar{y} &= g(\bar{x}, \bar{y}). \end{aligned} \tag{24}$$

We now explore the stability of these steady states by analyzing the fate of small deviations. As before, this will result in a linearized system of equations for the small perturbations x' and y'. To achieve this we must now use Taylor series expansions of the functions of two variables, f and g (see Appendix at end of chapter). That is, we approximate $f(\bar{x} + x', \bar{y} + y')$ by the expression

$$f(\bar{x} + x', \bar{y} + y') = f(\bar{x}, \bar{y}) + \left.\frac{\partial f}{\partial x}\right|_{\bar{x},\bar{y}} x' + \left.\frac{\partial f}{\partial y}\right|_{\bar{x},\bar{y}} y' + \dots, \tag{25}$$

and do the same for g. You should verify that when other steps identical to those made in the one-variable case are carried out, the result is

$$\begin{aligned} x'_{n+1} &= a_{11}x'_n + a_{12}y'_n, \\ y'_{n+1} &= a_{21}x'_n + a_{22}y'_n, \end{aligned} \tag{26}$$

where now

$$\begin{aligned} a_{11} &= \left.\frac{\partial f}{\partial x}\right|_{\bar{x},\bar{y}}, & a_{12} &= \left.\frac{\partial f}{\partial y}\right|_{\bar{x},\bar{y}}, \\ a_{21} &= \left.\frac{\partial g}{\partial x}\right|_{\bar{x},\bar{y}}, & a_{22} &= \left.\frac{\partial g}{\partial y}\right|_{\bar{x},\bar{y}}. \end{aligned} \tag{27}$$

The matrix consisting of these four coefficients, i.e.,

$$\mathbf{A} = \begin{pmatrix} a_{11} & a_{12} \\ a_{21} & a_{22} \end{pmatrix}, \tag{28}$$

is called the *Jacobian* of the system of equations (23). One often encounters the matrix notation

$$\mathbf{x}'_{n+1} = \mathbf{A}\mathbf{x}'_n, \tag{29}$$

as a shorthand representation of equations (26), where

$$\mathbf{x}'_n = \begin{pmatrix} x'_n \\ y'_n \end{pmatrix}. \tag{30}$$

The problem has again been reduced to a linear system of equations for states that are in proximity to the steady state $(\bar{x}, \bar{y})$. Thus we can determine the stability of $(\bar{x}, \bar{y})$ by methods given in Chapter 1. To briefly review, this would entail the following:

1. Finding the characteristic equation of (26) by setting

$$\det(\mathbf{A} - \lambda\mathbf{I}) = 0.$$

The result is always the quadratic equation

$$\lambda^2 - \beta\lambda + \gamma = 0,$$

where

$$\begin{aligned} \beta &= a_{11} + a_{22}, \\ \gamma &= a_{11}a_{22} - a_{12}a_{21}. \end{aligned}$$

2. Determining whether the roots of this equation (the eigenvalues) are of magnitude smaller than 1.

If the answer to (2) is affirmative, we can conclude that small deviations from the steady state will decay, i.e., that the equilibrium is stable. In the next section we show that it is not always necessary to compute the eigenvalues explicitly in order to determine their magnitudes. Rather, as we will demonstrate, it is sufficient to test whether the following condition is satisfied:

$$2 > 1 + \gamma > |\beta| \rightarrow \begin{cases} \text{both eigenvalues } |\lambda_i| < 1 \\ \text{steady state } (\bar{x}, \bar{y}) \text{ is stable} \end{cases}$$

2.8 STABILITY CRITERIA FOR SECOND-ORDER EQUATIONS

It is possible to formulate stability criteria for systems of difference equations in terms of the coefficients in the corresponding characteristic equation. May, et al. (1974) were among the first to derive explicitly a necessary and sufficient condition for second-order systems:

Given the characteristic equation

$$\lambda^2 - \beta\lambda + \gamma = 0, \tag{31}$$

both roots will have magnitude less than 1 if

$$2 > 1 + \gamma > |\beta|. \tag{32}$$

Now we examine how this condition is derived.

Derivation of Condition (32)

Recall that roots of equation (31) are

$$\lambda_{1,2} = \frac{\beta \pm \sqrt{\beta^2 - 4\gamma}}{2}. \tag{33}$$

For stability it is necessary that

$$|\lambda_1| < 1 \quad \text{and} \quad |\lambda_2| < 1. \tag{34}$$

Figure 2.10 shows the desired result geometrically when equation (31) has real roots. Notice that roots of equation (31) are equidistant from the value $\beta/2$. Thus it is first necessary that this midpoint should be within the interval $(-1, 1)$:

$$-1 < \beta/2 < 1, \quad \text{or} \quad \left|\frac{\beta}{2}\right| < 1. \tag{35}$$

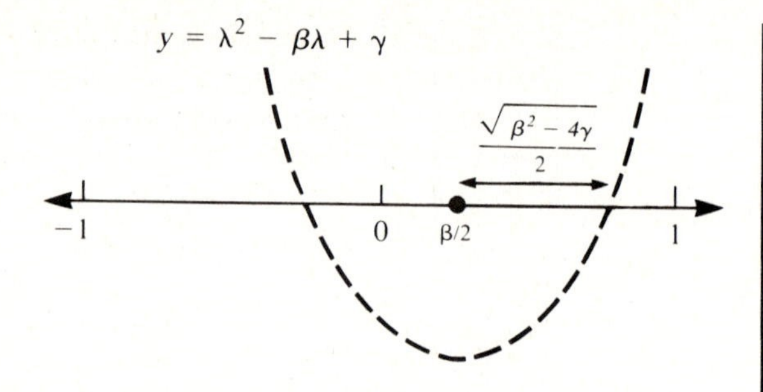

Figure 2.10 *For stability of a second-order system such as (23), both eigenvalues corresponding to roots of the characteristic equation* $y = \lambda^2 - \beta\lambda + \gamma$ *must lie within the interval* $(-1, 1)$.

Furthermore, it is necessary that the distance from $\beta/2$ to either root be *smaller* than the distance to an endpoint of the interval. In the situation graphed shown in Figure 2.10 this implies that

$$1 - \left|\frac{\beta}{2}\right| > \frac{\sqrt{\beta^2 - 4\gamma}}{2}.$$

Squaring both sides does not change the inequality since each side is positive.

$$\left(1 - \left|\frac{\beta}{2}\right|\right)^2 > \frac{\beta^2 - 4\gamma}{4},$$

$$1 - |\beta| + \frac{\beta^2}{4} > \frac{\beta^2}{4} - \gamma.$$

Cancelling $\beta^2/4$ and rearranging terms gives

$$1 + \gamma > |\beta|. \qquad (36)$$

The final step of combining the inequalities in equations (35) and (36) is left as a problem.

2.9 STABILITY CRITERIA FOR HIGHER-ORDER SYSTEMS

Thus far we have examined only the relatively straightforward case of two coupled difference equations. In theory the techniques of analyzing stability of steady-state solutions are similar for bigger systems, e.g., a set of k equations. However, computationally one can encounter problems in estimating magnitudes of eigenvalues because the characteristic equation associated with this system is a polynomial of degree k. In general it is then impossible to actually determine the eigenvalues, and it is necessary to use certain criteria to obtain information about their magnitudes.

One such criterion, called *the Jury test* as developed by Jury (1971), is described in this section. See Lewis (1977), for a broader discussion.

Consider the polynomial

$$P(\lambda) = \lambda^n + a_1\lambda^{n-1} + a_2\lambda^{n-2} + \cdots + a_{n-1}\lambda + a_n.$$

Define the following combinations of parameters:

$$
\begin{array}{lll}
b_n = 1 - a_n^2, & c_n = b_n^2 - b_1^2, & d_n = c_n^2 - c_2^2, \\
b_{n-1} = a_1 - a_n a_{n-1}, & c_{n-1} = b_n b_{n-1} - b_1 b_2, & d_{n-1} = c_n c_{n-1} - c_2 c_3, \\
\vdots & \vdots & \vdots \\
b_{n-k} = a_k - a_n a_{n-k}, & c_{n-k} = b_n b_{n-k} - b_1 b_{k+1}, & d_{n-k} = c_n c_{n-k} - c_2 c_{k+2}, \\
\vdots & \vdots & \vdots \\
b_1 = a_{n-1} - a_n a_1, & c_2 = b_n b_2 - b_1 b_{n-1}, & d_3 = c_n c_3 - c_2 c_{n-1}.
\end{array}
$$

The list grows shorter at each stage until there are just three quantities that relate to their predecessors by the rule

$$q_n = p_n^2 - p_{n-3}^2,$$

$$q_{n-1} = p_n p_{n-1} - p_{n-3} p_{n-2},$$

$$q_{n-2} = p_n p_{n-2} - p_{n-3} p_{n-1}.$$

Now we formulate the criterion as follows:

Jury Test

Necessary and sufficient conditions for all roots of $P(\lambda)$ to satisfy the condition that $|\lambda| < 1$ are the following:

1. $P(1) = 1 + a_1 + \cdots + a_{n-1} + a_n > 0.$
2. $(-1)^n P(-1) = (-1)^n[(-1)^n + a_1(-1)^{n-1} + \cdots + a_{n-1}(-1) + a_n] > 0,$
3. (a) $|a_n| < 1,$
   $|b_n| > |b_1|,$
   $|c_n| > |c_2|,$
   $|d_n| > |d_3|,$
   $\vdots$
3. (q) $|q_n| > |q_{n-2}|.$

Example
Apply the Jury test to

$$\lambda^4 + \lambda^3 + \lambda^2 + \lambda + 1 = 0.$$

Solution
We note that $n = 4$ and examine conditions 1 and 2:

1. $P(1) = 1 + 1 + 1 + 1 + 1 > 0$
2. $(-1)^4 P(1) = 1(1 - 1 + 1 - 1 + 1) > 0$

Since these are both satisfied, we go on to the first part of condition 3, namely that $|a_n| < 1$. We see that this condition is not satisfied since $a_4 = 1$. Thus the polynomial will have at least one root whose magnitude is not smaller than 1.

2.10 FOR FURTHER STUDY: PHYSIOLOGICAL APPLICATIONS

In recent years several problems in physiology have been modeled by the use of difference equations or by a combination of difference-differential equations (called *delay equations*) that share certain properties. One of the earliest review articles on this subject, entitled "Oscillation and Chaos in Physiological Control Systems," dates back to 1977 (see Mackey and Glass, 1977). Students who are interested in such applications should consult the references provided. Two of these are briefly highlighted here.

1. In their 1978 article, Glass and Mackey describe several respiratory disorders in which the pattern of breathing is irregular. In normal resting humans, ventilation volume is approximately constant from one breath to another. *Cheyne-Stokes breathing* consists of a repeated waxing and waning of the depth of breathing, with an amplitude of ventilation volume that oscillates over intervals of 0.5 to 1 min. Other disorders (such as *biot breathing* and *infant apnea*) also seem to indicate a problem in the control system governing ventilation.

A simple linear model for ventilation volumes and chemoreceptor sensitivity to CO_2 in the blood was described in Section 1.9 and outlined in problem 18 of Chapter

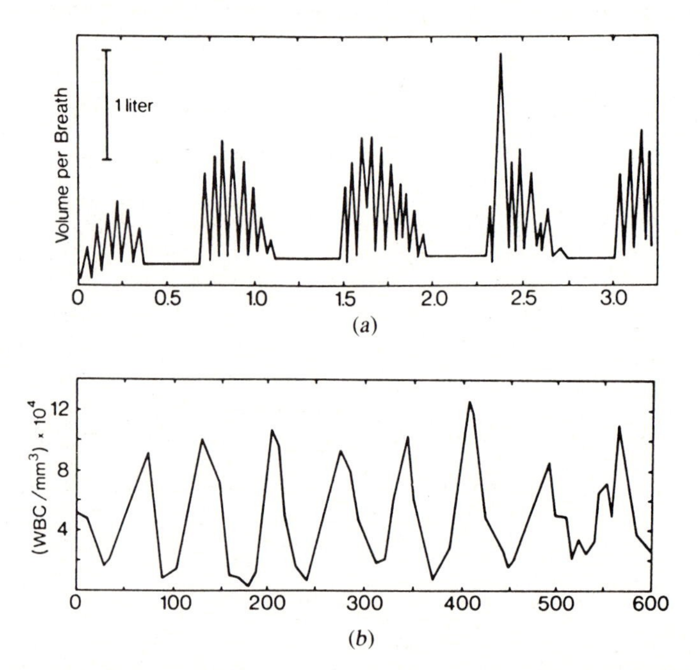

Figure 2.11 *Two dynamical diseases: (a) In Cheyne-Stokes respiration, the volume per breath undergoes periodic cycles of deep breathing interspersed by intervals of apnea (no breathing). (b) Chronic granulocyclic leukemia is associated with characteristic oscillations in the level of circulating white blood cells (WBCs) over periods of several months. [From Mackey and Glass (1977), "Oscillations and Chaos in Physiological Control Systems,"* Science, *197, 287–289. Reprinted by permission. Copyright 1977 by the AAAS.]*

1. We continue to develop the model in problem 17 of this chapter. The articles by Glass and Mackey are recommended for good summaries of the phenomena and for a more sophisticated approach using delay equations. The authors apply similar models to *hematopoiesis* (the control and formation of circulating blood components). This makes for a more detailed model of a process described rather crudely in Section 1.9.

2. Cardiac and neurological disorders have also been modeled with discrete equations. These processes too are characterized by events that recur at some regular time intervals. The beating pattern of the heart is regular in normal resting humans. Specialized tissues at sites called the sinoatrial (SA) node and the atrioventricular (AV) node act as pacemakers to set the rhythm of contraction of the atria and ventricles. When one of these nodes is not functioning properly, *arrhythmia* (irregular rhythms) may result. The papers by Keener (1981) and Ikeda et al. (1983) outline models for arrhythmia based on the interaction of the SA node with some secondary pacemaker in the ventricle.

Collectively, physiological disorders in which a generally adequate control system becomes unstable have been called *dynamical diseases* (Figure 2.11). Some of the very simple abstract models given in this chapter have revealed that such phenomena can arise spontaneously when one or several parameters of a system have values slightly beyond certain threshold bifurcation points.

PROBLEMS*

1. Indicate whether each of the following equations is linear or nonlinear. If linear, determine the solution; if nonlinear, find any steady states of the equation.

(a) $x_n = (1 - \alpha)x_{n-1} + \beta x_n$, α and β are constants

(b) $x_{n+1} = \dfrac{x_n}{1 + x_n}$

(c) $x_{n+1} = x_n e^{-ax_n}$, a is a constant

(d) $(x_{n+1} - \alpha)^2 = \alpha^2(x_n^2 - 2x_n + 1)$, α is a constant

(e) $x_{n+1} = \dfrac{K}{k_1 + k_2/x_n}$, k_1, k_2 and K are constants

2. Determine when the following steady states are stable:

(a) $x_{n+1} = rx_n(1 - x_n)$, $\bar{x} = 0$

(b) $x_{n+1} = -x_n^2(1 - x_n)$, $\bar{x} = (1 + \sqrt{5})/2$

(c) $x_{n+1} = 1/(2 + x_n)$, $\bar{x} = \sqrt{2} - 1$

(d) $x_{n+1} = x_n \ln x_n^2$, $\bar{x} = e^{1/2}$

Sketch the functions $f(x)$ given in this problem. Use the cobwebbing method to sketch the approximate behavior of solutions to the equations from some initial starting value of x_0.

*Problems preceded by an asterisk (*) are especially challenging.

3. In population dynamics a frequently encountered model for fish populations is based on an empirical equation called the *Ricker equation* (see Greenwell, 1984):

$$N_{n+1} = \alpha N_n e^{-\beta N_n}.$$

In this equation, α represents the maximal growth rate of the organism and β is the inhibition of growth caused by overpopulation.

(a) Show that this equation has a steady state

$$\overline{N} = \frac{\ln \alpha}{\beta}.$$

(b) Show that the steady state in (a) is stable provided that

$$|1 - \ln \alpha| < 1.$$

4. Consider the equation

$$N_{t+1} = N_t \exp[r(1 - N_t/K)].$$

This equation is sometimes called an analog of the logistic differential equation (May, 1975). The equation models a single-species population growing in an environment that has a *carrying capacity* K. By this we mean that the environment can only sustain a maximal population level $N = K$. The expression

$$\lambda = \exp[r(1 - N_t/K)]$$

reflects a density dependence in the reproductive rate. To verify this observation, consider the following steps:

(a) Sketch λ as a function of N. Show that the population continues to grow and reproduce only if $N < K$.

(b) Show that $\overline{N} = K$ is a steady state of the equation.

(c) Show that the steady state is stable. (Are there restrictions on parameters r and K?)

(d) Using a hand calculator or simple computer program, plot successive population values N_t for some choice of parameters r and K.

***5.** Show that a first-order difference equation

$$x_{n+1} = f(x_n)$$

has stable two-point cycles if condition (17) is satisfied.

6. Show that by using a Taylor series expansion for the functions f and g in equations (23) one obtains the linearized equations (26) for perturbations (x', y') about the steady state $(\overline{x}, \overline{y})$.

7. In Section 2.8 we demonstrated that conditions (35) and (36) are necessary for both roots of $\lambda^2 - \beta\lambda + \gamma = 0$ to be negative.

(a) Derive an additional constraint that $\gamma < 1$ and hence show that condition (32) must be satisfied.

(b) Show that the same result is obtained when $\beta/2$ is negative.

(c) Equation (31) admits complex conjugate roots $\lambda_{1,2} = a \pm bi$ when $\beta^2 < 4\gamma$. In this situation it is necessary that the modulus $(a^2 + b^2)^{1/2}$ be smaller than 1 for the quantities λ_i^n to decay with increasing n (see

Figure 2.2). Show that this will be true whenever condition (32) is satisfied.

8. The equation

$$N_{t+1} = \lambda N_t(1 + aN_t)^{-b},$$

where $\lambda, a, b > 0$ is often encountered in the biological literature as an empirical description of density-limited population growth. For example, see M. P. Hassell (1975). λ is a growth rate, and a and b are parameters related to the density feedback rate.

(a) Show that by rescaling the equation one can reduce the number of parameters.

(b) Find the steady states of the equation $N_{t+1} = \lambda N_t(1 + aN_t)^{-b}$ and determine the conditions for stability of each steady state.

(c) Draw the function $f(x) = \lambda x(1 + ax)^{-b}$ and use it to graph $N_1, \ldots, N_{10}$ for some starting value N_0.

9. The difference equation

$$N_{t+1} = \left[\lambda_1 + \frac{\lambda_2}{1 + \exp A(N_t - B)}\right] N_t$$

is discussed by C. Pennycuik, R. Compton, and L. Beckingham (1968). A computer model for simulating the growth of two interacting populations. *J. Theor. Biol. 18,* 316 – 329; and by M. B. Usher (1972). Developments in the Leslie Matrix Model, in J. N. R. Jeffers, ed., *Mathematical Models in Ecology,* Blackwells, Oxford. Determine the behavior of solutions to this equation. (Assume $\lambda_1, \lambda_2, A, B > 0$.)

10. Graph the function

$$f(x) = \frac{\lambda x}{1 + (ax)^b},$$

for $\lambda, a, b > 0$ and use this to deduce the properties of the equation

$$N_{t+1} = \frac{\lambda N_t}{1 + (aN_t)^b}.$$

This is one example of a class of equations discussed by Maynard Smith (1974). What happens when $b = 1$?

11. In May (1978), Host-parasitoid systems in patchy environments: a phenomenological model. *J. Anim. Ecol.*, *47*, 833 – 843, the following system of equations appear:

$$H_{t+1} = FH_t(1 + aP_t/k)^{-k},$$
$$P_{t+1} = H_t - H_{t+1}/F.$$

Determine the steady states and their stability for this system. (The parameters F, a, k are positive.)

12. In Chapter 1 a problem for annual plant propagation led to a set of linear difference equations whose eigenvalues were given by equation (34). Using the criteria in Section 2.8, determine a general condition that ensures growth of the plant population.

13. In Levine (1975), a single species population is assumed to be governed by the difference equation

$$n_{k+1} = F(n_k),$$

where the function F has the shape shown in part (a) of the figure.

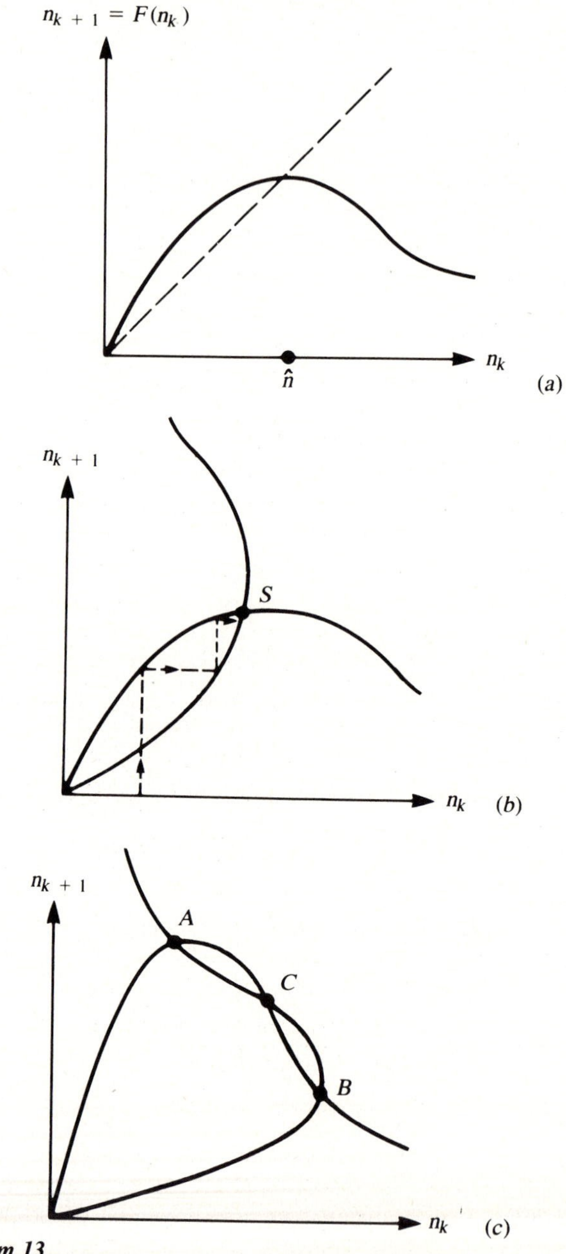

Figure for problem 13.

(a) Levine shows a variant of the cobwebbing method discussed in this chapter, in which a graph of $F(n)$ and its reflection $\overline{F}(n)$ are used to produce successive iterates n_k (see part (*b*) of the figure). What is the point marked S in this figure? A somewhat different situation is shown in part (*c*) of the figure. Interpret the points A, B, C, and discuss what happens to the population n_k.

(b) Levine describes the problem of stability in enriched ecosystems by suggesting that "enrichment should tend to elevate the curve $F(n_k)$ though not necessarily in a simple fashion." Explain this statement. Illustrate how a sequence of ecosystems enriched to different degrees might correspond to functions $F(n_k)$, which yield changes in stability of the steady state, $\hat{n}$.

14. Show that the steady state $\bar{x} = 0$ of equation (10) is unstable whenever the nontrivial steady state is stable.

15. *Advanced exercise or project.* Analyze the model

$$x_{n+1} = ax_n^2(1 - x_n)$$

for $a > 0$. See Marotta (1982).

16. *Waves of disease.* In a popular article that appeared in the *New Scientist,* Anderson and May (1982)[2] suggest a simple discrete model for the spread of disease that demonstrates how regular cycles of infection may arise in a population. Taking the average period of infection as the unit of time, they write equations for the number of disease cases C_t and the number of susceptible individuals S_t in the tth time interval. They make the following assumptions: (i) The number of new cases at time $t + 1$ is some fraction f of the product of current cases C_t and current susceptibles S_t; (ii) a case lasts only for a single time period; (iii) the current number of susceptibles is increased at each time period by a fixed number of births B and decreased by the number of new cases ($B \neq 0$). (iv) Individuals who have recovered from the disease are immune.

(a) Explain assumption (i).

(b) Write the equations for C_{t+1} and S_{t+1} based on the above information.

(c) Show that $\hat{S} = 1/f$, $\hat{C} = B$ is a steady state of the equations.

(d) Use stability analysis to show that a small deviation away from steady state may result in oscillatory behavior.

(e) What happens when $f = 2/B$?

(f) Using a hand calculator or a simple computer program, show how solutions to the equations depict waves of incidence of the disease. Typical parameter values given by Anderson and May (1982) are

B = 12 births per 1000 people for the U.K.,
or 36 births per 1000 in a third-world country.
$f = 0.3 \times 10^{-4}$.

2. *Note to the instructor:* The article by Anderson and May (1982), which deals with the control of disease by vaccination, is accessible to students at this stage and may be considered as a topic for class presentation. More advanced models for epidemics will be discussed in Chapter 6.

Typical population data are

$$S_0 = 2000, \qquad C_0 = 20.$$

17. This problem pursues further the topic of blood CO_2 and ventilation volume first described in Section 1.9 and problem 16 of Chapter 1. There we studied a *linear* model for the process; we now examine a nonlinear extension.

(a) Whereas in problem 1.16 we assumed that the amount of CO_2 lost, $\mathcal{L}(V_n, C_n)$, was simply proportional to V_n and independent of C_n, let us now consider the case in which

$$\mathcal{L}(V_n, C_n) = \beta V_n C_n.$$

Explain the biological difference between these distinct hypotheses.

(b) Further assume that the ventilation volume V_{n+1} is simply proportional to C_n, so that

$$V_{n+1} = \alpha C_n$$

as before. Write down the system of (nonlinear) equations for C_n and V_n.

(c) Show that the steady state of this system is

$$C_n = V_n/\alpha = (m/\beta\alpha)^{1/2}$$

and determine its stability.

(d) Are oscillations in V_n and C_n possible for certain ranges of the parameters?

(e) Now consider a more realistic model (based on Mackey and Glass, 1977, 1978). Assume that the sensitivity of chemoreceptors to CO_2 is not linear but rather *sigmoidal*, i.e.,

$$\mathcal{S}(C) = \frac{C^\ell}{K^\ell + C^\ell}$$

where

ℓ = an integer,
K = some real number.

Further suppose that

$$V_{n+1} = V_{\max}\,\mathcal{S}(C_n),$$

with $\mathcal{L}(V_n, C_n)$ as before. Write down the system of equations, and show that it can be reduced to the following single equation for C_n:

$$C_{n+1} = C_n - \beta V_{\max} C_{n-1}^\ell\, C_n/(K^\ell + C_{n-1}^\ell) + m.$$

(f) Sketch $\mathcal{S}(C)$ as a function of C for $\ell = 0, 1, 2$. The integer ℓ can be described as a *cooperativity parameter*. (For $\ell > 1$ the binding of a single CO_2 molecule to its chemoreceptor *enhances* additional binding; this type of kinetic assumption is described in Chapter 7.)

(g) Determine what equation is satisfied by the steady states C_n and V_n.

(h) For $\ell = 1$ find the value of the physiological steady state(s). Give a condition for their stability.

(i) Investigate the model by writing a computer simulation or by further analysis. Are oscillations possible for $\ell = 1$ or for higher ℓ values?

Comment: After thinking about this problem, you may wish to refer to the articles by Mackey and Glass. Their model combines differential and difference equations so that the details of the analysis are different. See references in Chapter 1.

REFERENCES

Stability, Oscillations, and Chaos in Difference Equations

Frauenthal, J. C. (1979). *Introduction to Population Modelling* (UMAP Monograph Series). Birkhauser, Boston, 59–73.

Frauenthal, J. C. (1983). Difference and differential equation population growth models. Chap. 3 in (M. Braun, C. S. Coleman, and D. A. Drew, eds.), *Differential Equation Models*. Springer-Verlag, New York.

Guckenheimer, J.; Oster, G.; and Ipaktchi, A. (1977). The dynamics of density-dependent population models. *J. Math. Biol., 4,* 101–147.

Hofstadter, D. (1981). Strange attractors: Mathematical patterns delicately poised between order and chaos. *Sci. Am.*, November 1981 vol. 245, pp. 22–43.

Kloeden, P. E., and Mees, A. I. (1985). Chaotic phenomena. *Bull. Math. Biol., 47,* 697–738.

Li, T. Y., and Yorke, J. A. (1975). Period three implies chaos. *Amer. Math. Monthly, 82,* 985–992.

May, R. M. (1975). Biological populations obeying difference equations: Stable points, stable cycles, and chaos. *J. Theor. Biol., 51,* 511–524.

May, R. M. (1976). Simple mathematical models with very complicated dynamics. *Nature, 261,* 459–467.

May, R. M., and Oster, G. (1976). Bifurcations and dynamic complexity in simple ecological models. *Am. Nat., 110,* 573–599.

Perelson, A. (1980). Chaos. In L. A. Segel, ed., *Mathematical Models in Molecular and Cellular Biology*. Cambridge University Press, Cambridge.

Rogers, T. D. (1981). Chaos in Systems in Population Biology, in R. Rosen ed., *Progress in Theoretical Biology, 6,* 91–146. Academic Press, New York.

Physiological Applications

Glass, L., and Mackey, M. C. (1978). Pathological conditions resulting from instabilities in physiological control systems. *Ann. N. Y. Acad. Sci., 316,* 214–235, published 1979.

Ikeda, N., Yoshizawa, S., and Sato, T. (1983). Difference equation model of ventricular parasystole as an interaction between cardiac pacemakers based on the phase response curve. *J. Theor. Biol., 103,* 439–465.

Keener, J. P. (1981). Chaotic cardiac dynamics. Pp. 299–325 in F. C. Hoppenstaedt, ed. in *Mathematical Aspects of Physiology*. American Mathematical Society, Providence, R.I.

Mackey, M. C., and Glass, L. (1977). Oscillation and chaos in physiological control systems. *Science, 197,* 287–289.

Mackey, M. C., and Milton, J. (1986). Dynamical Diseases, *Ann, N.Y. Acad. Sci.*, in press.

May, R. M. (1978). Dynamical Diseases. *Nature, 272,* 673–674.

(See also miscellaneous papers in New York Academy of Sciences (1986). *Perspectives on Biological Dynamics and Theoretical Medicine;* and in *Ann. N.Y. Acad. Sci.*, vol. 316.)

Miscellaneous

Anderson, R., and May, R. (1982). The logic of vaccination. *New Scientist,* November 1982.

Greenwell, R. (1984). The Ricker salmon model. UMAP Unit 653. Reprinted in *UMAP Journal, 5*(3), 337–359.

Hassell, M. P. (1975). Density dependence in single-species populations. *J. Anim. Ecol., 44,* 283–295.

Jury, E. I. (1971). The inners approach to some problems of system theory. *IEEE Trans. Automatic Contr., AC–16,* 233–240.

Levine, S. H. (1975). Discrete time modeling of ecosystems with applications in environmental enrichment. *Math. Biosci., 24,* 307–317.

Lewis, E. R. (1977). *Network Models in Population Biology* (*Biomath.,* vol. 7). Springer-Verlag, New York.

Marotta, F. R. (1982). The dynamics of a discrete population model with threshold. *Math. Biosci., 58,* 123–128.

May, R. M., Conway, G. R., Hassell, M. P., and Southwood, T. R. E. (1974). Time delays, density dependence and single species oscillations. *J. Anim. Ecol., 43,* 747–770.

Maynard Smith, J. (1974). *Models in Ecology.* Cambridge University Press, Cambridge, Eng.

APPENDIX TO CHAPTER 2: TAYLOR SERIES

PART 1: FUNCTIONS OF ONE VARIABLE

Technical Matters

In order for a function $F(x)$ to have a Taylor series at some point x_0, it must have derivatives of all orders at that point. (The function $F(x) = |x|$ at $x = 0$ does not qualify because of the sharp corner it makes at $x = 0$.)

A Taylor series about x_0 is an expression involving powers of $(x - x_0)$, where x_0 is a point at which F and all its derivatives are known. Written in the form

$$T(x) = \sum_{n=0}^{\infty} a_n(x - x_0)^n, \tag{37}$$

this expression is called a *power series* and may be thought of as a polynomial with infinitely many terms. Equation (37) makes sense only for values of x for which the infinite sum is some finite number (i.e., when the series *converges*). There are precise tests (e.g., the *ratio test*) that determine whether a given power series converges.

Suppose $T(x)$ is a power series that converges whenever $|x - x_0| < r$ (r is then called the *radius of convergence*). It can be shown that

$$F(x) = T(x) = \sum_{n=0}^{\infty} a_n(x - x_0)^n$$

provided the coefficients a_n are of the following form:

$$a_0 = F(x_0),$$

$$a_1 = \left.\frac{dF}{dx}\right|_{x_0},$$

$$a_2 = \frac{1}{2}\frac{d^2F}{dx^2}\bigg|_{x_0},$$

$$\vdots$$

$$a_n = \frac{1}{n!}\frac{d^nF}{dx^n}\bigg|_{x_0}, \tag{38}$$

$$\vdots$$

where these expressions are derivatives of F evaluated at the point x_0. A technical question is whether the sum (37) with the coefficients in (38) actually equals the value of F at points x other than x_0. Below we assume this to be the case.

Practical Matters

The Taylor series of a function of one variable can be written as follows:

$$F(x) = F(x_0) + \frac{dF}{dx}\bigg|_{x_0}(x - x_0) + \frac{1}{2}\frac{d^2F}{dx^2}\bigg|_{x_0}(x - x_0)^2 + \frac{1}{3!}\frac{d^3F}{dx^3}\bigg|_{x_0}(x - x_0)^3$$
$$+ \cdots + \frac{1}{n!}\frac{d^nF}{dx^n}\bigg|_{x_0}(x - x_0)^n + \cdots.$$

When $x - x_0$ is a very small quantity, the terms $(x - x_0)^k$ for $k > 1$ can usually be neglected (provided the derivative coefficients are not very large; i.e., the function does not make abrupt changes). In this case

$$F(x) \simeq F(x_0) + \frac{dF}{dx}\bigg|_{x_0}(x - x_0) + \cdots + \text{(neglected terms)}.$$

Figure 2.12 demonstrates how this expression approximates the actual function.

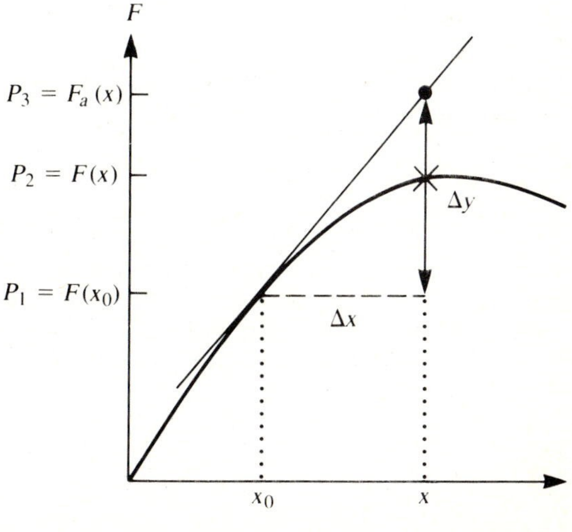

Figure 2.12 By retaining the first two terms in a Taylor series expansion for a function, the value of the function at a point x, $P_2 = F(x)$, *is approximated by the expression*

$$P_3 = F_a(x) = F(x_0) + \Delta y = F(x_0) + (\text{slope})\,\Delta x.$$

The quantity $\frac{dF}{dx}\Big|_{x_0}$ *is the slope of the tangent line, and* $\Delta x = x - x_0$. *Using these terms one arrives at an approximation that is accurate only when* x_0 *and* x *are close.*

PART 2: FUNCTIONS OF TWO VARIABLES

Technical Matters

Consider the function of two variables $F(x, y)$ that has partial derivatives of all orders with respect to x and y at some point (x_0, y_0) such that $P_1 = F(x_0, y_0)$. Then the value of F at a neighboring point $P_2 = F(x, y)$ can be calculated using a Taylor series expansion for F, provided this expansion converges to the value of the function. Since variation in both x and y must be taken into account, the series involves partial derivatives. It proves convenient to define the following quantity:

$$\hat{\Delta}F = \left(h\frac{\partial}{\partial x} + k\frac{\partial}{\partial y}\right)F(x, y) \stackrel{\text{def}}{\equiv} (x - x_0)\frac{\partial F}{\partial x} + (y - y_0)\frac{\partial F}{\partial y}.$$

By nth power of the above expression we mean the following:

$$\hat{\Delta}^n F = \left(h\frac{\partial}{\partial x} + k\frac{\partial}{\partial y}\right)^n F(x, y) \equiv (x - x_0)^n\frac{\partial^n F}{\partial x^n} + n(x - x_0)^{n-1}(y - y_0)\frac{\partial^n F}{\partial x^{n-1}\partial y}$$

$$+ \cdots + n(x - x_0)(y - y_0)^{n-1}\frac{\partial^n F}{\partial x\,\partial y^{n-1}}$$

$$+ (y - y_0)^n\frac{\partial^n F}{\partial y^n}.$$

Using this notation, we can now express the Taylor series of F as follows:

$$F(x, y) = F(x_0, y_0) + \hat{\Delta}F(x_0, y_0) + \frac{1}{2!}\hat{\Delta}^2 F(x_0, y_0) + \cdots + \frac{1}{n!}\hat{\Delta}^n F(x_0, y_0) + \cdots$$

In this shorthand, the similarity to functions of one variable is apparent. Note that the expressions $\hat{\Delta}^n F$ are binomial expansions involving mixed partial derivatives of F that are evaluated at (x_0, y_0).

Practical Matters

When (x, y) is close to (x_0, y_0), the function F can be approximated by retaining the first two terms of a Taylor series expansion, as follows:

$$F(x, y) \simeq F(x_0, y_0) + \hat{\Delta}F(x_0, y_0)$$

$$= F(x_0, y_0) + (x - x_0)\left.\frac{\partial F}{\partial x}\right|_{x_0, y_0} + (y - y_0)\left.\frac{\partial F}{\partial y}\right|_{x_0, y_0}.$$

Figure 2.13 illustrates how this expression approximates the value of the function. The geometric ideas underlining this expression are best understood by readers who have had some exposure to vector calculus. For others it suffices to remark that this is a natural extension of the one-variable case to a higher dimension.

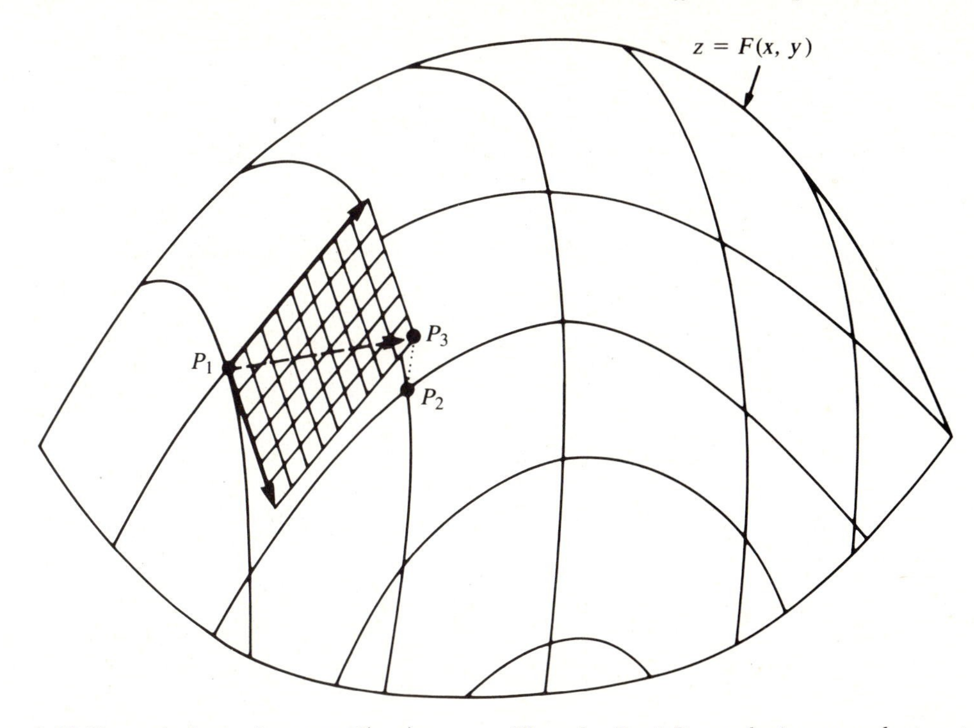

Figure 2.13 *Shown is the surface* $z = F(x, y)$, *where* F *is a function of two variables. Assuming that the value of* F *at* (x_0, y_0), $P_1 = F(x_0, y_0)$, *is known, we approximate the value of* F *at some neighboring point* (x, y), $P_2 = F(x, y)$ *by a Taylor series expansion. If only terms up to* order 1 *are taken, the expression yields*

$$P_3 = F(x_0, y_0) + \left.\frac{\partial F}{\partial x}\right|_{x_0, y_0} (x - x_0) + \left.\frac{\partial F}{\partial y}\right|_{x_0, y_0} (y - y_0).$$

The value $P_3 \neq P_2$ *can be interpreted geometrically.* P_3 *is actually the height* [*at* (x_0, y_0)] *of a tangent plane (shaded) that approximates the surface* $z = F(x, y)$. *The partial derivatives of* F *are slopes of the vectors that are shown as bold arrows.*

3 Applications of Nonlinear Difference Equations to Population Biology

Great fleas have little fleas upon their backs to bite 'em.
And little fleas have lesser fleas, and so ad infinitum.
The great fleas themselves in turn have greater fleas to go on,
While these again have greater still, and greater still, and so on.

Anonymous (1981). *The brand x anthology of poetry,* Burnt Norton edition, William Zaranka, ed. Apple-wood Books, Cambridge, Mass.

They hop in tens and in hundreds they fly
Thousands lie still while millions go by
That makes billions of bugs who jump and who crawl
So try as you will, you can't count them all.

Haris Petie (1975). *Billions of bugs,* Prentice-Hall, Englewood Cliffs, N.J.

Methods developed in Chapter 2 prove useful in addressing the population dynamics of organisms that have distinct breeding periods and life-cycle stages, notably insects and other *arthropods* (the phylum consisting of segmented-leg organisms). Since insects often compete with humans for crops or natural resources, efforts at pest management have been of fundamental economic interest.

Recognizing that chemical sprays and toxins are as unhealthy in the long run for people as for pests, recent efforts have been directed at biological control of pest species using other potential competitors, predators, or parasites. Folk tradition has held that this would lead to eradication of the undesirable pest. In fact, however, the outcomes of biological intervention are not always as clear-cut as we might naively expect. We discover time and again that interactions of a species with the environment, with other members of its own species, or with another population can be potentially complex and bizarre.

To understand some of the outcomes of population interactions, an examination of fairly elementary mathematical models proves quite illuminating. Here again, we make no claim to describe the full intricacies of a given situation. Rather, the approach is to examine the consequences of several simplifying assumptions. Often these assumptions lead us to predictions that are biologically unrealistic (see the Nicholson-Bailey model, for example, as described in Section 3.3). It is then instructive to consider how modifying the underlying assumptions tends to change the predictions.

To introduce discrete models into population dynamics we first consider single-species populations in Section 3.1. As indicated in Chapter 2, nonlinear difference equations arise rather naturally in models for populations that have nonoverlapping generations. Many of these models are based on the observation that the net growth rate of the population depends in some way on its density. These effects may stem from competition of individuals for limited resources or from numerous other environmental considerations including predation, disease, and so forth. Single-species models are often based on empirical formulae rather than on detailed interactions in the population. We examine several of these in Section 3.1.

In many ecological settings one finds that two or more species are intimately related in their influence on each other. A classic example, that of the host-parasitoid system, is outlined in Sections 3.2 through 3.4. Here one species (the parasitoid) can reproduce only in the presence and at the expense of the other (its host). We observe that under the simplest reasonable assumptions, a model for such systems (the Nicholson-Bailey model) predicts growing population oscillations in the two species. Section 3.4 summarizes a number of potentially stabilizing influences.

Models described in Sections 3.1 through 3.4 proceed largely from detailed choices for functions that represent growth rates, fraction of hosts parasitized, or survivorships. In a model for plant-herbivore interactions (Section 3.5) we depart somewhat from this traditional approach. We observe that certain logical deductions about population behavior can be made even when only broad features of the system are known. For would-be modelers this approach is an instructive one and reappears in later material. Sections 3.5 and 3.3 as well as 3.1 contain several explicit examples of how to apply the stability criteria derived in Chapter 2. As such, these examples may enhance your appreciation of the techniques previously developed.

The concluding section of this chapter contains some introductory material on population genetics. This topic provides an excellent example of yet another realm in which discrete difference equations are important.

Note to the instructor: For rapid coverage of this chapter, include only Sections 3.1 through 3.3 and 3.6, leaving Sections 3.4 and 3.5 for further independent study by more advanced students.

3.1 DENSITY DEPENDENCE IN SINGLE-SPECIES POPULATIONS[1]

An assumption that growth rate, reproductive rate, or survivorship depends on the density of the population leads us to consider models of the following form:

$$N_{t+1} = f(N_t) \tag{1}$$

where $f(N_t)$ is some (nonlinear) function of the population density.

Quite often single-species populations (of insects, for example) are described by such equations, where f is a function that is fit to data obtained by following successive generations of the population. Here we consider several models of this type and demonstrate their properties.

1. A model by Varley, Gradwell, and Hassell (1973) consists of the single equation

$$N_{t+1} = \frac{\lambda}{\alpha} N_t^{1-b}. \tag{2}$$

Here λ is the reproductive rate, assumed to be greater than 1, and $1/\alpha\, N_t^{-b}$ is the fraction of the population that survives from infancy to reproductive adulthood. The equation is thus best understood in the form

$$N_{t+1} = \left(\frac{1}{\alpha} N_t^{-b}\right)(\lambda N_t). \tag{3}$$

↑ no. of progeny at generation t

↑ fraction that survives to generation $t + 1$

where $\alpha, b, \lambda > 0$. Since the fraction of survivors can at most equal but not exceed 1, we find that the population must exceed a certain size, $N_t > N_c$ for this model to be biologically reasonable (see problem 1).

Populations satisfying equation (3) can be maintained at steady density levels. To observe this we look for the steady-state solutions to (3) by setting

$$\overline{N} = N_{t+1} = N_t.$$

Substituting into (3) we find that

$$\overline{N} = \frac{\lambda}{\alpha} \overline{N}^{1-b}. \tag{4}$$

Cancelling the common factor $\overline{N}$ and rearranging terms gives us

$$\overline{N} = \left(\frac{\lambda}{\alpha}\right)^{1/b} \tag{5}$$

Next, we let

$$f(N) = \frac{\lambda}{\alpha} N^{1-b}$$

1. This section contains material compiled by Laurie Roba.

and proceed to test for the stability of $\bar{N}$. We find that perturbations δ_t from this steady state must satisfy

$$\bar{N} + \delta_{t+1} = f(\bar{N} + \delta_t)$$
$$\simeq f(\bar{N}) + \left.\frac{df}{dN}\right|_{\bar{N}} \delta_t + \ldots .$$

Since $\bar{N} = f(\bar{N})$, recall that this simplifies to

$$\delta_{t+1} \simeq \left.\frac{df}{dN}\right|_{\bar{N}} \delta_t.$$

But

$$\left.\frac{df}{dN}\right|_{\bar{N}} = \left.\frac{\lambda}{\alpha}(1 - b)N^{-b}\right|_{\bar{N}} = 1 - b. \tag{6}$$

Thus stability of $\bar{N}$ hinges on whether the quantity $1 - b$ is of magnitude smaller than 1; that is, $\bar{N}$ will be stable provided that

$$-1 < 1 - b < 1,$$

or

$$0 < b < 2. \tag{7}$$

It is clear that $b = 0$ is a situation in which survivorship is not density-dependent; that is, the population grows at the rate λ/α. Thus the lower bound for the stabilizing values of b makes sense. It is at first less clear from an intuitive point of view why values of b greater than 2 are not consistent with stability; it appears that density dependence that is *too* strong is destabilizing due to the potential for boom-and-bust cycles.

2. A second model cited in the literature (for example, May, 1975) consists of the equation

$$N_{t+1} = N_t \exp r(1 - N_t/K), \tag{8}$$

where r, K are positive constants. The quantity $\lambda = \exp r(1 - N_t/K)$ could be considered the density-dependent reproductive rate of the population. Again, by carrying out stability analysis we observe that

$$N_t = N_{t+1} = K$$

is the nontrivial steady state. To analyze its stability properties we remark that for

$$f(N) = N \exp r(1 - N/K) \tag{9}$$

we have

$$f'(N) = [\exp r(1 - N/K)](1 - Nr/K). \tag{10}$$

Evaluated at $\bar{N} = K$, (10) leads to

$$f'(K) = 1 - K(r/K) = 1 - r. \tag{11}$$

Thus stability is obtained when

$$|1 - r| < 1, \qquad \text{or} \qquad 0 < r < 2. \tag{12}$$

We observe that when $N < K$ the reproductive rate $\lambda > 1$, whereas when $N > K$, $\lambda < 1$ (see problem 3). This property is shared with equation (11) of Chapter 2 where $K = 1$. K is said to be the *carrying capacity* of the environment for the population. In the next chapter we shall see examples of similar density-dependent relationships within the framework of continuous populations.

3. Yet a third model, proposed by Hassell (1975), is given by the equation

$$N_{t+1} = \lambda N_t(1 + aN_t)^{-b}, \tag{13}$$

for λ, a, b positive constants. Analysis of this equation is left as a problem for the reader.

One generally observes with models such as 1, 2, and 3 (and with other discrete equations such as the prototype given in Chapter 2) that the dynamical behavior depends in a sensitive way on parameter settings. Typically such equations have stable cycles of arbitrary periods as well as chaotic behavior. Each model thus describes a highly complex range of dynamic behavior if parameter values are pushed to high values. For example, equation (13) has the behavioral regimes mapped out on the *λb parameter plane* shown in Figure 3.1. The values $\lambda = 100$ and $b = 6$ fall

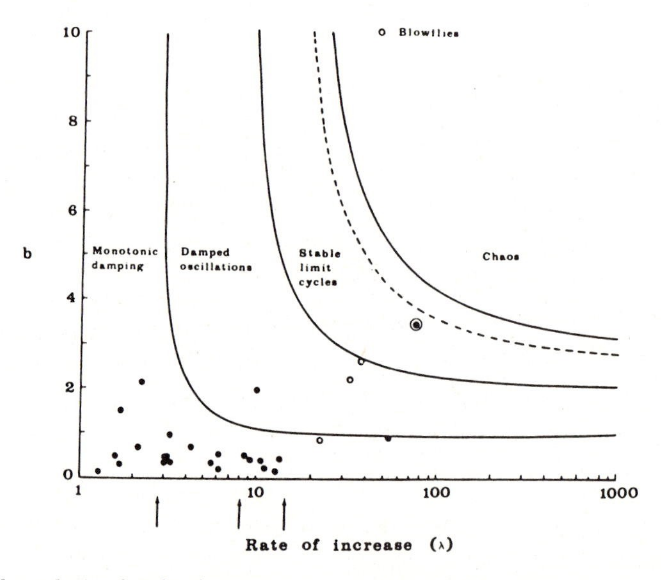

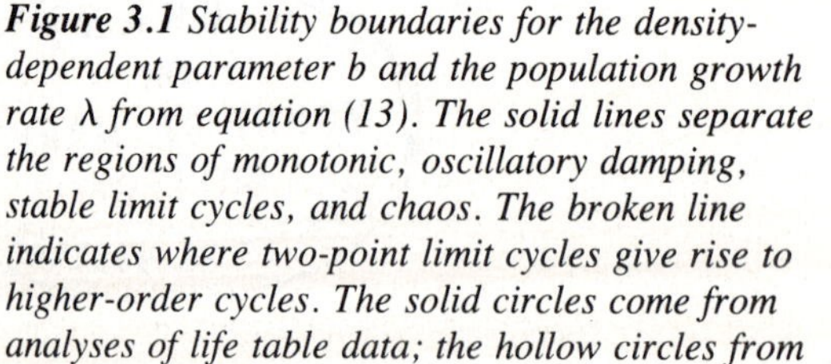

***Figure 3.1** Stability boundaries for the density-dependent parameter b and the population growth rate λ from equation (13). The solid lines separate the regions of monotonic, oscillatory damping, stable limit cycles, and chaos. The broken line indicates where two-point limit cycles give rise to higher-order cycles. The solid circles come from analyses of life table data; the hollow circles from analyses of laboratory experiments. [Reproduced from Michael P. Hassell,* The Dynamics of Anthropod Predator-Prey Systems. *Monographs in Population Biology 13. Copyright © 1978 by Princeton University Press. Fig. 2.5 (after Hassell, Lawton, May 1976) reprinted by permission of Princeton University Press.]*

in the chaotic domain, so that populations fluctuate wildly. The values $\lambda = 100$ and $b = 0.5$, correspond to a stable steady state, so that a perturbed population undergoes monotonic damping back to its steady-state level.

For a given single-species population, density fluctuations may or may not be described well by a model such as equation (13). If so, parameters such as b and λ can be estimated by following the observed levels of the population over successive generations. Such observations are called *life table data*. Studies of this sort have been carried out under a variety of conditions, both in the field and in laboratory settings (see Hassell et al., 1976). Typical species observed in the field have included insects such as the moth *Zeiraphera diniana* and the parasitoid fly *Cyzenis albicans*. Laboratory data on beetles and on the blowfly *Lucilia cuprina* (Nicholson, 1954) have also been collected.

Pooling results of many observations in the literature and in their own experiments, Hassell et al. (1976) plotted the parameter values b and λ of some two dozen species on the $b\lambda$ parameter plane. In all but two of these cases, the values of b and λ obtained were well within the region of stability; that is, they reflected either monotonic or oscillatory return to the steady states.

Hassell et al. (1976) found two examples of unstable populations. The only one occurring in a natural system was that of the colorado potato beetle (shown as a circled dot in Figure 3.1), which is known to fluctuate periodically in certain situations. A single laboratory population, that of the blowfly (Nicholson, 1954), was found to have (λ, b) values corresponding to the chaotic regime in Figure 3.1. Some controversy surrounds the acceptance of this single example as a true case of chaotic population dynamics.

From their particular set of examples, Hassel et al. (1976) concluded that complex behavioral regimes typical of discrete difference equations are not frequently observed in reality. Of course, to place this deduction in its proper context, we should remember that only a relatively small sample of species has been sufficiently well studied to be represented, and that Figure 3.1 describes the fit to one particular model, chosen somewhat arbitrarily from many equally plausible ones.

One of the contributions of mathematical modeling and analysis to the study of population behavior has been in bringing forward questions that might otherwise have been of lesser interest. Comparison between observations and model predictions indicate that many dynamical behavior patterns, which are theoretically possible, are not observed in nature. We are thereby led to inquire which effects in natural systems have stabilizing influences on populations that might otherwise behave chaotically.

Hassell et al. (1976) comment on some of the key elements of studies based on data collected in the field versus those collected under controlled laboratory conditions. In the former, the survival of a population may depend on multiple factors including predation, parasitism, competition, and environmental conditions (see Sections 3.2–3.4). Thus a description of the population by a single-species model is, at best, a crude approximation.

Laboratory experiments on the other hand, can provide conditions in which a population is truly isolated from other species. In this sense, such data is more suitable for interpretation by single-species models. However, the influence of a somewhat artificial setting may result in effects (such as competition in close confinement) that are not significant in the natural setting. Thus, data for laboratory studies such as those of Nicholson's blowflies, in which erratic chaotic behavior is observed, may reflect not a realistic trend but rather an artifact observed only in the laboratory.

3.2 TWO-SPECIES INTERACTIONS: HOST-PARASITOID SYSTEMS

Discrete difference-equation models apply most readily to groups such as insect populations where there is a rather natural division of time into discrete generations. In this section we examine a particular two-species model that has received considerable attention from experimental and theoretical population biologists, that of the *host-parasitoid system*.

Found almost entirely in the world of insects, such two-species systems have several distinguishing features. Typical of insect species, both species have a number of life-cycle stages that include *eggs, larvae, pupae* and *adults*. One of the species, called the *parasitoid,* exploits the second in the following way: An adult female parasitoid searches for a *host* on which to *oviposit* (deposit its eggs). In some cases eggs are attached to the outer surface of the host during its larval or pupal stage. In other cases the eggs are injected into the host's flesh. The larval parasitoids develop and grow at the expense of their host, consuming it and eventually killing it before they pupate. The life cycles of the two species, shown in Figure 3.2, are thus closely intertwined.

A simple model for this system has the following common set of assumptions:

1. Hosts that have been parasitized will give rise to the next generation of parasitoids.
2. Hosts that have not been parasitized will give rise to their own progeny.
3. The fraction of hosts that are parasitized depends on the rate of *encounter* of the two species; in general, this fraction may depend on the densities of one or both species.

While other effects causing mortality abound in any natural system, it is instructive to consider only this minimal set of interactions first and examine their consequences. We therefore define the following:

N_t = density of host species in generation t,

P_t = density of parasitoid in generation t,

$f = f(N_t, P_t)$ = fraction of hosts not parasitized,

λ = host reproductive rate,

c = average number of viable eggs laid by a parasitoid on a single host.

Then our three assumptions lead to:

N_{t+1} = number of hosts in previous generation $\times$ fraction not parasitized $\times$ reproductive rate (λ),

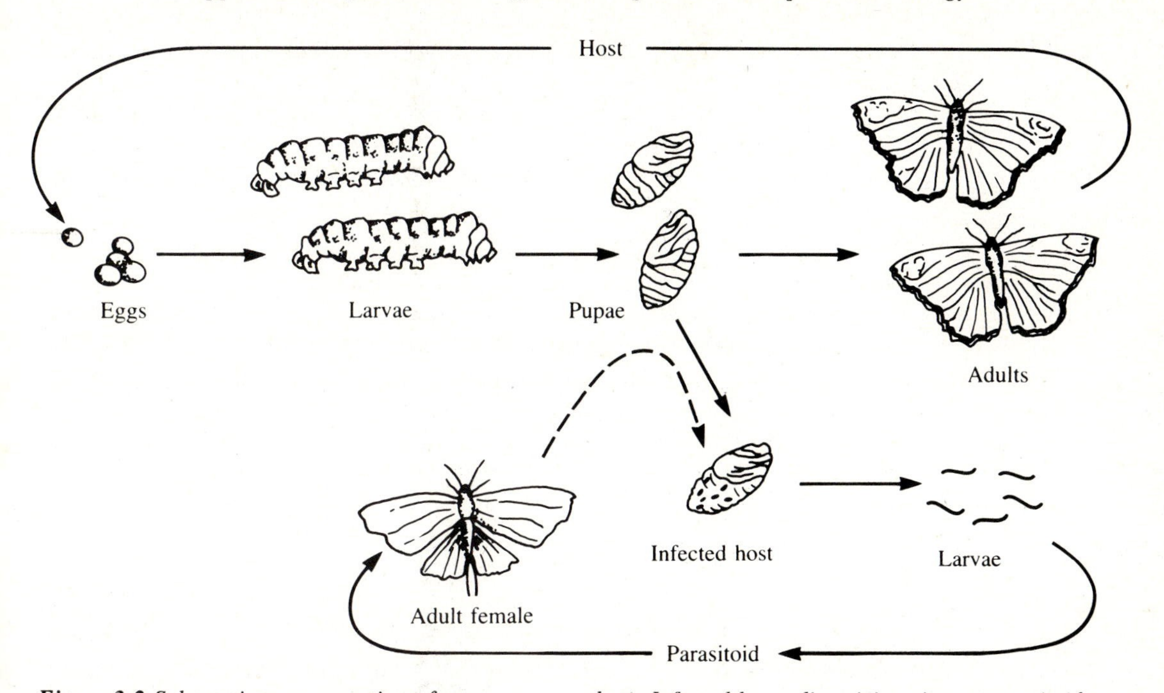

Figure 3.2 *Schematic representation of a host-parasitoid system. The adult female parasitoid deposits eggs on or in either larvae or pupae of the host. Infected hosts die, giving rise to parasitoid progeny. Uninfected hosts may develop into adults and give rise to the next generation of hosts.*

P_{t+1} = number of hosts parasitized in previous generation × fecundity of parasitoids (c).

Noting that $1 - f$ is the fraction of hosts that are parasitized, we obtain

$$N_{t+1} = \lambda N_t f(N_t, P_t), \tag{14a}$$

$$P_{t+1} = cN_t[1 - f(N_t, P_t)]. \tag{14b}$$

These equations outline a general framework for host-parasitoid models. To proceed further it is necessary to specify the term $f(N_t, P_t)$ and how it depends on the two populations. In the next section we examine one particular form suggested by Nicholson and Bailey (1935).

3.3 THE NICHOLSON-BAILEY MODEL

A. J. Nicholson was one of the first biologists to suggest that host-parasitoid systems could be understood using a theoretical model, although only with the help of the physicist V. A. Bailey were his arguments given mathematical rigor. (See Kingsland, 1985 for a historical account.)

Nicholson and Bailey made two assumptions about the number of encounters and the rate of parasitism of a host:

4. Encounters occur randomly. The number of encounters N_e of hosts by parasitoids is therefore proportional to the product of their densities.

$$N_e = aN_tP_t, \tag{15}$$

where a is a constant, which represents the searching efficiency of the parasitoid. (This kind of assumption presupposes random encounters and is known as the *law of mass action*. It is a common approximation which will reappear in many mathematical models; see Chapters 4, 6, and 7.)

5. Only the first encounter between a host and a parasitoid is significant. (Once a host has been parasitized it gives rise to exactly c parasitoid progeny; a second encounter with an egg-laying parasitoid will not increase or decrease this number.)

The Poisson Distribution and Escape from Parasitism

The *Poisson distribution* is a probability distribution that describes the occurrence of discrete, random events (such as encounters between a predator and its prey). The probability that a certain number of events will occur in some time interval (such as the lifetime of the host) is given by successive terms in this distribution. For example, the probability of r events is

$$p(r) = \frac{e^{-\mu}\mu^r}{r!} \tag{16}$$

where μ is the average number of events in the given time interval. (For more details on the Poisson distribution consult any elementary text in statistics, for example, Hogg and Craig, 1978.) In the case of host-parasitoid encounters, the *average* number of encounters per host per unit time is

$$\mu = \frac{N_e}{N_t}. \tag{17}$$

Note that by equation (15) this is the same as

$$\mu = aP_t. \tag{18}$$

Thus, for example, the probability of exactly two encounters would be given by

$$p(2) = \frac{e^{-aP_t}}{2!}(aP_t)^2.$$

The likelihood of escaping parasitism is the same as the probability of zero encounters during the host lifetime, or $p(0)$. Thus

$$f(N_t, P_t) = p(0) = \frac{e^{-aP_t}}{0!}(aP_t)^0 = e^{-aP_t}. \tag{19}$$

Based on the latter assumption it proves necessary to distinguish only between those hosts that have had no encounters and those that have had n encounters, where $n \geq 1$.

Because the encounters are assumed to be random, one can represent the probability of r encounters by some distribution based on the average number of encounters that take place per unit time. It transpires that an appropriate probability distribution for describing this situation is that of Poisson, highlighted briefly in the box on page 80.

Combining assumptions 4 and 5 with the comments about the Poisson distribution leads us to the expression for the fraction that *escapes* parasitism,

$$f(N_t, P_t) = p(0) = e^{-aP_t}, \tag{20}$$

given by the zero term of the Poisson distribution.

Thus the assumption that parasitoids search independently and randomly and that their searching efficiently is constant (depicted by the parameter a) leads to the Nicholson-Bailey equations:

$$N_{t+1} = \lambda N_t\, e^{-aP_t}, \tag{21a}$$

$$P_{t+1} = cN_t(1 - e^{-aP_t}). \tag{21b}$$

We now analyze this model using the methods developed in Chapter 2. The steps include:

1. Solving for steady states.
2. Finding the coefficients of the Jacobian matrix (for the system linearized about the steady state).
3. Checking the stability condition derived in Section 2.8.

Nicholson-Bailey Model: Equilibrium and Stability

Let

$$F(N, P) = \lambda N e^{-aP}, \tag{22a}$$

$$G(N, P) = cN(1 - e^{-aP}). \tag{22b}$$

Solving for steady states, we obtain the trivial solution $\bar{N} = 0$, or

$$\bar{N} = F(\bar{N}, \bar{P}) = \lambda \bar{N}\, e^{-a\bar{P}}, \tag{23a}$$

$$\bar{P} = G(\bar{N}, \bar{P}) = c\bar{N}(1 - e^{-a\bar{P}}). \tag{23b}$$

These imply that

$$\bar{P} = \frac{\ln \lambda}{a}, \tag{24a}$$

$$e^{-a\bar{P}} = 1/\lambda, \tag{24b}$$

$$\bar{N} = \frac{\lambda \ln \lambda}{(\lambda - 1)ac}. \tag{24c}$$

From these equations we observe that $\lambda > 1$ is required, since otherwise $\overline{N}$ would be a negative quantity. Computing the coefficients a_{ij} of the Jacobian, we obtain

$$a_{11} = F_N(\overline{N}, \overline{P}) = \lambda e^{-a\overline{P}} = 1, \quad (25a)$$

$$a_{12} = F_P(\overline{N}, \overline{P}) = -a\lambda\overline{N}e^{-a\overline{P}} = -a\overline{N}, \quad (25b)$$

$$a_{21} = G_N(\overline{N}, \overline{P}) = c(1 - e^{-a\overline{P}}) = c(1 - 1/\lambda), \quad (25c)$$

$$a_{22} = G_P(\overline{N}, \overline{P}) = ca\overline{N}e^{-a\overline{P}} = ca\overline{N}/\lambda. \quad (25d)$$

(*Comment:* The notation $F_N(\overline{N}, \overline{P})$ is shorthand for $\partial F/\partial N|_{(\overline{N},\overline{P})}$.) To check the stability of $(\overline{N}, \overline{P})$ the quantities we need to examine are thus

$$\beta = a_{11} + a_{22} = 1 + ca\overline{N}/\lambda = 1 + \frac{\ln \lambda}{\lambda - 1}, \quad (26a)$$

$$\gamma = a_{11}a_{22} - a_{12}a_{21} = ca\overline{N}/\lambda + ca\overline{N}(1 - 1/\lambda) = ca\overline{N} = \frac{\lambda \ln \lambda}{\lambda - 1}. \quad (26b)$$

We now show that $\gamma > 1$. To do so we need to verify that $\lambda(\ln \lambda)/(\lambda - 1) > 1$ or $S(\lambda) \equiv \lambda - 1 - \lambda \ln \lambda < 0$. Observe that $S(1) = 0$, $S'(\lambda) = 1 - \ln \lambda - \lambda(1/\lambda) = -\ln \lambda$. So $S'(\lambda) < 0$ for $\lambda \geq 1$. Thus $S(\lambda)$ is a decreasing function of λ and consequently $S(\lambda) < 0$ for $\lambda \geq 1$.

We have verified that $\gamma > 1$ and so the stability condition given in Chapter 2, equation (32), is violated. We conclude that the equilibrium $(\overline{N}, \overline{P})$ can never be stable.

From the analysis we observe that the Nicholson-Bailey model has a single equilibrium

$$\overline{N} = \frac{\lambda \ln \lambda}{(\lambda - 1)ac}, \quad (27a)$$

$$\overline{P} = \frac{\ln \lambda}{a}, \quad (27b)$$

and that the equilibrium is never stable; small deviations of either species from the steady-state level leads to diverging oscillations. Curiously enough, a host-parasitoid system consisting of a greenhouse whitefly and its parasitoid was found to have such dynamics when grown under particular, albeit somewhat contrived, laboratory conditions (see Hassell, 1978, for details). Figure 3.3 demonstrates the fluctuations observed in this laboratory system and a comparison with the predictions of the Nicholson-Bailey model.

Most natural host-parasitoid systems are certainly more stable than the Nicholson-Bailey model seems to indicate. It would therefore seem that the model is not a satisfactory representation of real host-parasitoid interactions. However, before dismissing it as an ineffective model we shall exploit this theoretical tool to experiment with a number of conjectures on the effects (in natural systems) that might act as stabilizing influences. In the following section we therefore focus on more realistic assumptions about the searching behavior of the parasitoids and the host survival rate.

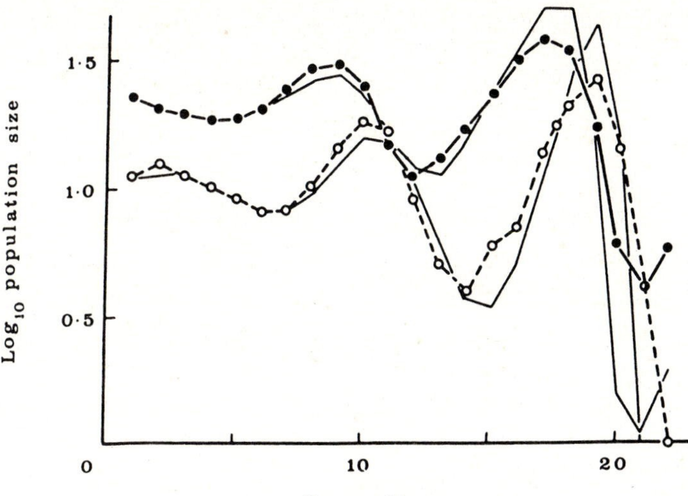

Figure 3.3 The Nicholson-Bailey model, given by equations (21a,b), predicts unstable oscillations in the dynamics of a host-parasitoid system. The fluctuations of a greenhouse whitefly Trialeurodes vaporariorum *(●) and its chalcid parasitoid* Encarsia formosa *(○) give evidence for such behavior. The solid lines are predictions of equations (21) where* a = 0.068, c = 1, *and* λ = 2. *[From Michael P. Hassell,* The Dynamics of Arthropod Predator-Prey Systems. *Monographs in Population Biology 13. Copyright © 1978 by Princeton University Press. Fig. 2.3 (after Burnett, 1958) reprinted by permission of Princeton University Press.]*

3.4 MODIFICATIONS OF THE NICHOLSON-BAILEY MODEL[2]

Density Dependence in the Host Population

Since the Nicholson-Bailey model is unstable for *all* parameter values, we consider first a modification of the assumptions underlying the host population dynamics and investigate whether these are potentially stabilizing factors. Thus, consider the following assumption:

6. In the absence of parasitoids, the host population grows to some limited density (determined by the carrying capacity K of its environment).

Thus the equations would be amended as follows:

$$N_{t+1} = N_t \lambda(N_t) e^{-aP_t},$$
$$P_{t+1} = N_t(1 - e^{-aP_t}).$$

For the growth rate $\lambda(N_t)$ we might adopt

$$\lambda(N_t) = \exp r(1 - N_t/K),$$

2. This section is based on a review by David F. Dabbs.

as in equation (8). Thus if $P = 0$, the host population grows up to density $N_t = K$ and declines if $N_t > K$. The revised model is

$$N_{t+1} = N_t \exp\left[r(1 - N_t/K) - aP_t\right], \tag{28a}$$

$$P_{t+1} = N_t(1 - e^{-aP_t}). \tag{28b}$$

This model was studied in some detail by Beddington et al. (1975). They found it convenient to discuss its behavior in terms of the quantity q where

$q = \overline{N}/K =$ the ratio of steady-state host density with and without parasitoids present.

The value of q indicates to what extent the steady-state population is depressed by the presence of parasitoids.

Equations (28a,b) are sufficiently complicated that it is impossible to derive explicit expressions for the states $\overline{N}$ and $\overline{P}$. However, these can be expressed in terms of q and $\overline{P}$ as follows:

$$\overline{P} = \frac{r}{a}(1 - \overline{N}/K) = \frac{r}{a}(1 - q), \tag{29a}$$

$$\overline{N} = \overline{P}/(1 - e^{-a\overline{P}}). \tag{29b}$$

It transpires that the resulting model is stable for a fairly wide range of realistic parameter values, as desired. Even so, the return to equilibrium in these ranges is typically rather complex. As parameters are changed, the equilibrium does lose its stability property, so that cycles and other more complicated dynamics ensue. Beddington et al. (1975) demonstrate that stability depends on r and the quantity $q = \overline{N}/K$ with the system stable within the shaded range in Figure 3.4. We see that for each value of r, there exists a range of q values for which the model is stable; the larger the value of r, the narrower the range.

When the equations of a model are difficult to analyze explicitly, computer simulations can prove particularly revealing. In Figures 3.5 through 3.8 the behavior

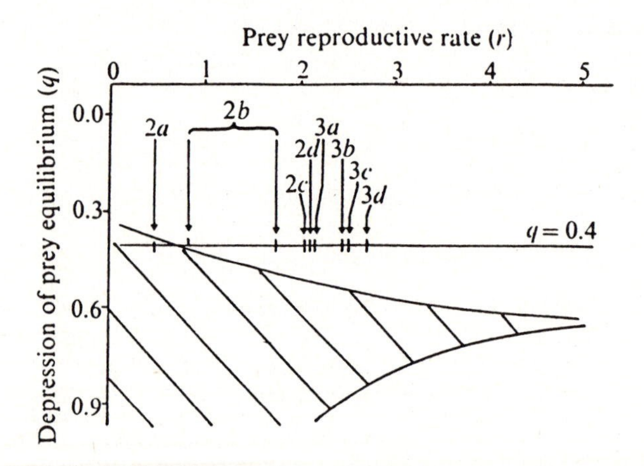

Figure 3.4 *The density-dependent Nicholson-Bailey model (equations 28) is stable within the hatched area. Note how the area of stability narrows for high values of* r. *[Reprinted by permission from* Nature, 255, *58–60. Copyright © 1975 Macmillan Journals Limited.]*

of solutions to equations (28a,b) obtained with a simple computer simulation are displayed for a variety of parameter choices. A TURBO – PASCAL program, written by David F. Dabbs and run on a personal computer, was used to generate successive values of N_t and P_t and to plot these simultaneously. What is somewhat novel about these plots is that in this two-variable system, time is suppressed and (N_t, P_t) values are plotted in the plane sometimes referred to as the *NP phase plane*. (In a later chapter a similar technique will be applied to systems of two differential equations in two variables.)

To interpret these figures, note that a central cross indicates the position of the steady state of the equations. The initial values (N_0, P_0) are specified at the top right-hand corner of the graph. In Figure 3.5 successive values proceed in a counterclockwise manner, visiting each of the arms of the "spiral galaxy" in succession. In Figures 3.6 through 8, $q = 0.40$ is kept fixed, while r is given the values 0.50, 2.00, 2.20, and 2.65.

For small values of r, the equilibrium point $(\overline{N}, \overline{P})$ is stable; any initial value spirals in toward it (Figure 3.5) and will eventually reach it. As r grows past a certain value, the equilibrium becomes unstable and new patterns emerge.

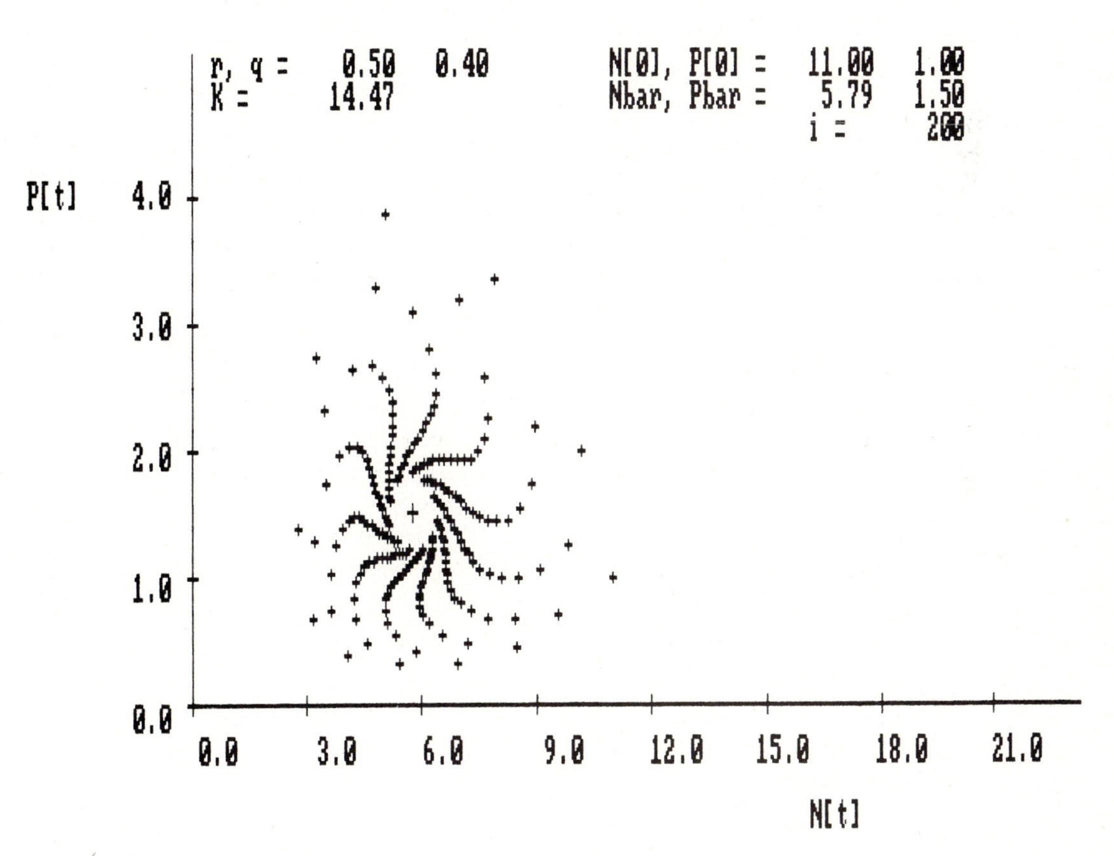

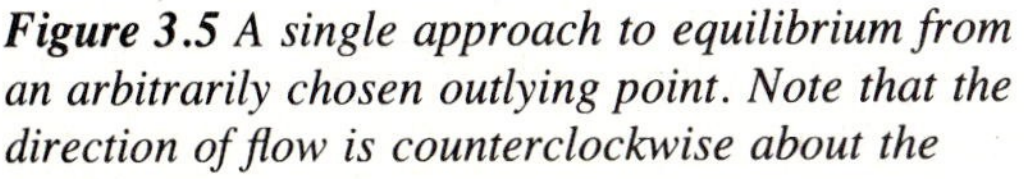
Figure 3.5 *A single approach to equilibrium from an arbitrarily chosen outlying point. Note that the direction of flow is counterclockwise about the steady-state point, not inward along the spiral arms. [Computer-generated plot made by David F. Dabbs.]*

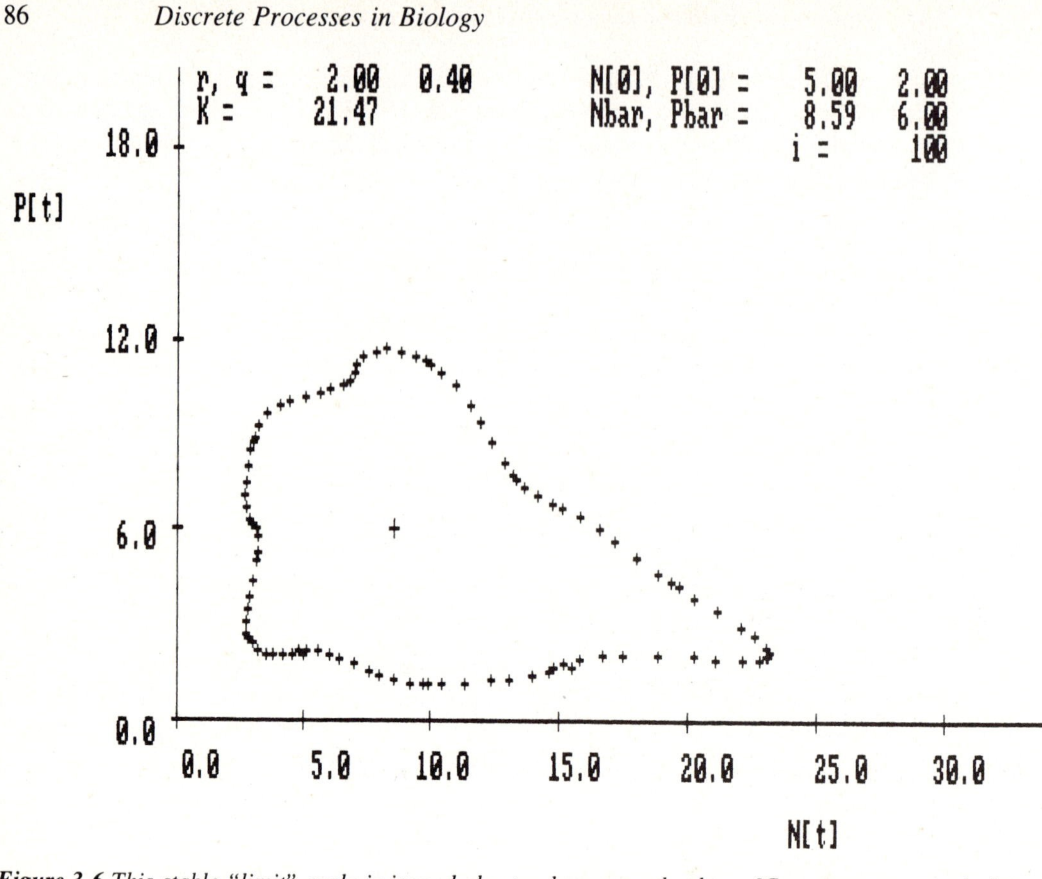

Figure 3.6 *This stable "limit" cycle is jagged about the edges. Similar cycles for smaller values of* r *have smooth edges. [Computer-generated plot made by David F. Dabbs.]*

Away from the single equilibrium point, the model will settle into a stable limit cycle around the equilibrium point, as shown in Figure 3.6. Larger values of r result in larger and larger cycles. Beyond a certain point there appear cycles whose periods are multiples of 5 (Figure 3.7). Still larger values of r yield either chaos or cycles of extremely high integral period. For large enough values of r, this chaotic behavior will seem to fill in a sharply bounded area (Figure 3.8).

As Figure 3.4 indicates, q and r are both involved in determining the dynamic population behavior. This figure can be interpreted to mean that the greater the depressing influence of parasitoids on their hosts, the lower the growth rate r that suffices to induce chaotic dynamics.

Other Stabilizing Factors

As we have just shown, the Nicholson-Bailey model is rendered more stable, and hence more realistic for most natural host-parasitoid systems, by taking into account the limitations of the environment and the fact that populations are not capable of infinite growth. Several other effects have been studied (notably by Beddington et

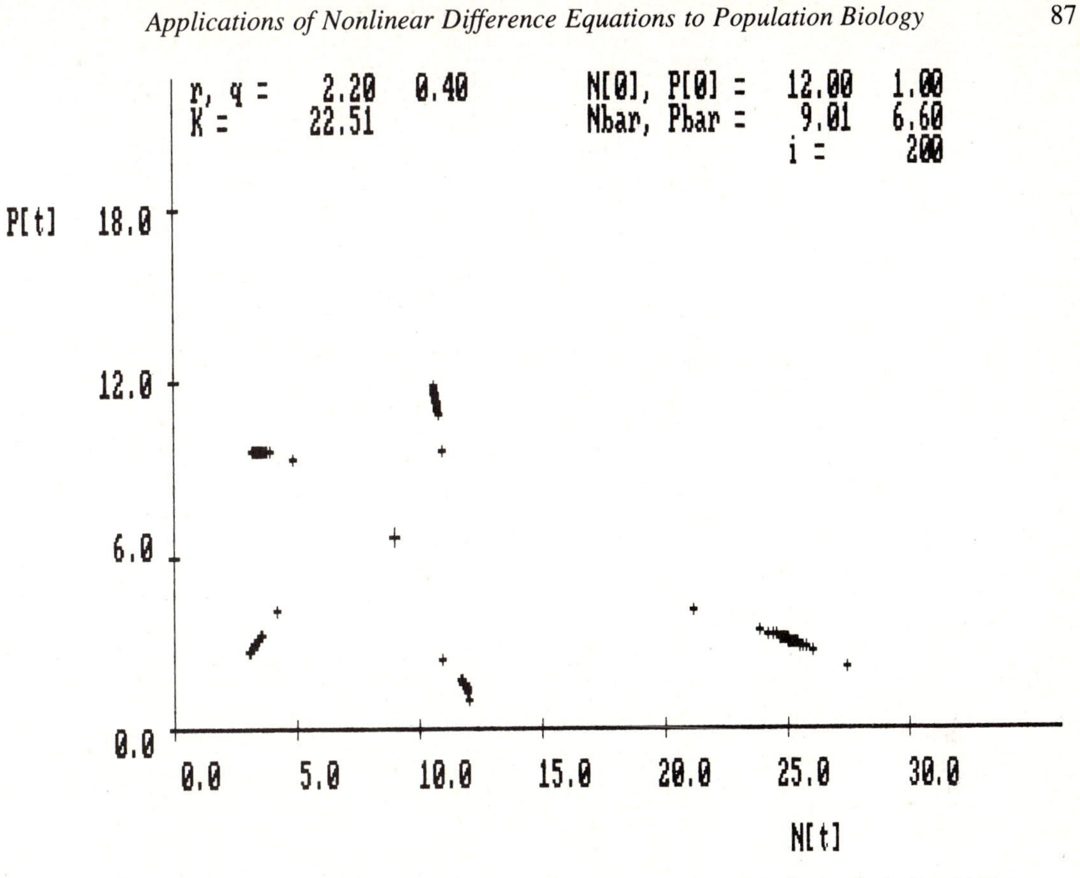

Figure 3.7 *This cycle shows a cycle whose period is 5. Further increasing* r *slowly would produce cycles of periods 10, 20, 40, and so on.* *[Computer-generated plot made by David F. Dabbs.]*

al., 1978) in further exploring the interactions that stabilize the host-parasitoid populations. Two of these are as follows.

1. *Efficiency of the parasitoids*
   The density of the attacking parasitoids may have some effect on their efficiency in searching for hosts. It is observed that efficiency generally decreases somewhat when the parasitoid population is too large. This effect is modeled by changing the assumed form of $f(N_t, P_t)$. One version studied by Beddington et al. (1978) is

$$f(N_t, P_t) = \exp -(aP_t)^{1-m}, \tag{30}$$

   where $m < 1$. (See Beddington et al., 1978, for a discussion of this assumption and predictions of the model.)

2. *Heterogeneity of the environment (refuges)*
   A second factor that has been brought into closer scrutiny in recent years is the supposed homogeneity of the environment. Researchers recognize that the physical set-

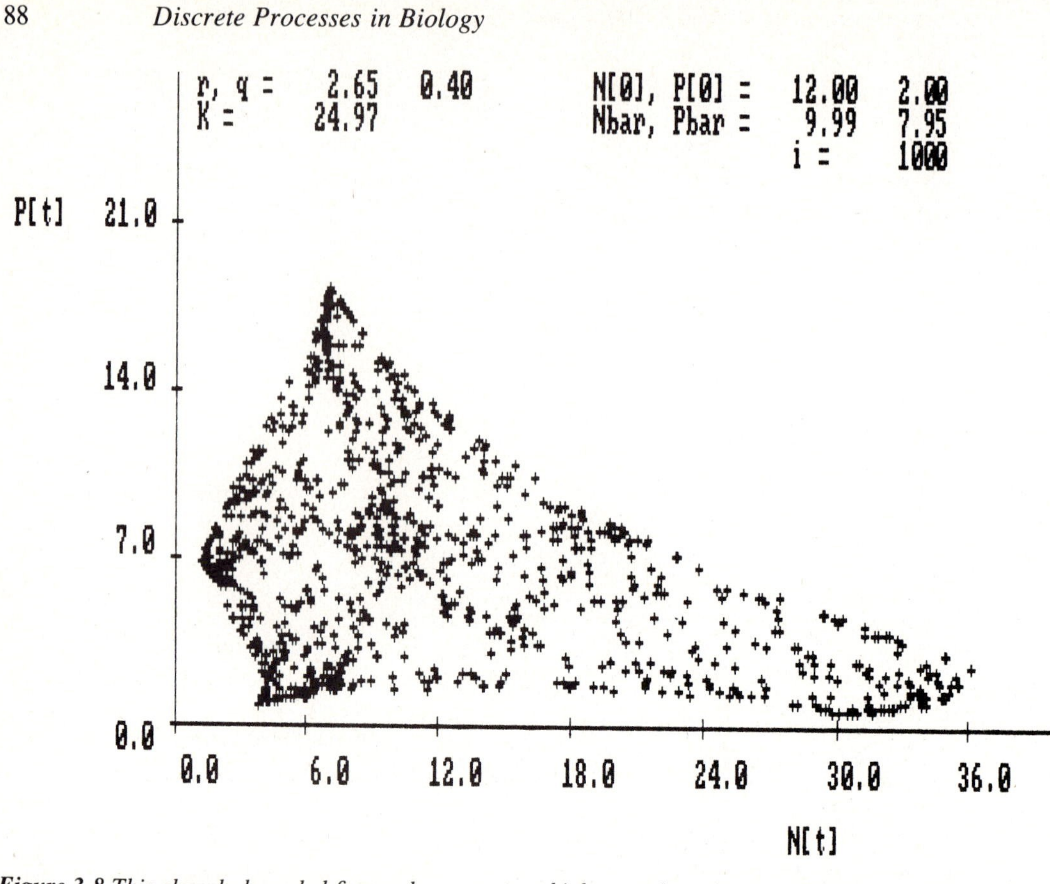

Figure 3.8 This sharply bounded figure shows definable areas without any points. For slightly lower r *values these areas are better defined; for higher* r *values they tend to fill in. [Computer-generated plot made by David F. Dabbs.]*

ting is never perfectly uniform, so part of the host population may be less exposed and thus less vulnerable to attack. It has become popular to refer to *patchy* environments, which are spatially as well as temporally heterogeneous.

While a full treatment of spatial variation would lead us to models that involve several independent variables (for example, time as well as physical position), it is possible to consider a simple example that gives some broad indication of the effects. This is generally done by assuming that there exists a small refuge, representing some physical location to which some fraction of the attacked population can retreat for safety from the attackers. For example, let us assume that E is the fraction of the carrying capacity population K that can be accommodated in safe refuges. Then at any given time

$$EK/N_t = \text{the fraction of the population that can retreat to a refuge,}$$

$$1 - EK/N_t = \text{the fraction vulnerable to attack.}$$

The equations would then be modified accordingly (see problem 11 and the original articles cited by the sources listed in the References). It has been a recurrent theme of articles by Hassell, May, and others that the patchiness of ecosystems leads to stabilization. Part of the argument is that refuges serve as sites for maintaining vulnerable species that might otherwise become extinct. Such sites also indirectly benefit the exploiting species since a constant spillover of victims into the unprotected areas guarantees a constant food source. You are encouraged to pursue these topics by reading the excellent summaries and reviews and through further independent research.

3.5 A MODEL FOR PLANT-HERBIVORE INTERACTIONS

Outlining the Problem

In Sections 3.1 to 3.4 we saw numerous models that describe particular responses of a population to its environment, to another species, or to intraspecific competition. A notable feature of many such models is that they contain functions that are chosen to fit empirical data and that may or may not reveal any basic insight into underlying population behavior. What does one do when a plethora of empirical data is unavailable and one knows only vague, general properties of the processes? Is it always necessary to restrict attention to well-defined functional relationships when proceeding with a model?

As the model in this section will demonstrate, often even when data are available, it may be an advantage to study the problem in a rather general framework before fitting exact functional forms to the empirical observations. In this section we introduce a problem stemming from plant-herbivore systems and then use this general approach to study its properties. The problem to be considered here is hypothetical but sufficiently general to apply to a variety of cases. We use it to illustrate a technique and later comment on its applicability.

Consider herbivores that feed on a vegetation and consume part of its *biomass*.[3] Unlike predation it need not be true that the damage or consumption inflicted by the herbivore, commonly called *herbivory*, necessarily leads to death of the victim, which in this case is the host plant. Rather, herbivores might reduce the biomass of vegetation they consume, possibly also causing other qualitative changes in the plant. In this first attempt at modeling plant-herbivore interactions we will focus only on quantitative changes, i.e., changes in the biomass of the populations. Some comments about plant quality will be made at the end of this section.

To give structure to the problem, we make the following broad assumptions:

1. Herbivores have discrete generations that correspond to the seasonality of the vegetation.

   (*Comment:* We can thus treat the problem using a set of difference equations; the generation span will be identical for the two participants. This assumption is fairly realistic. Many herbivores have *coevolved* with their host

3. The term *biomass* is often used as a measure of population size in units of mass rather than, say, density or numbers of individuals.

plant species and have life cycles that are closely linked to seasonality or to stages of growth of the vegetation.)

2. The availability of vegetation and the current population density of herbivores are the main factors that determine fecundity and survivorship of the herbivores.

   (*Comment:* While this statement seems plausible, we are surreptitiously assuming that other factors *do not* play major roles. This is an arguable point, to which we shall return.)

3. The abundance of the vegetation depends on the extent of herbivory to which the plant was subjected in the previous season as well as on the previous biomass of the vegetation.

   (*Comment:* This too seems to be a reasonable basic hypothesis. To take one example, leaf biomass in deciduous trees contributes to production and storage of substances that will eventually be used to produce the next season's crop of leaves. Thus the plant biomass contributes in a positive sense to its future abundance. On the other hand, defoliation or herbivory might reduce the potential of a plant to grow and so would contribute negatively to the abundance of the vegetation in the next season.)

The model will be written in terms of the following two variables:

v_n = vegetation biomass in generation n,

h_n = number of herbivores in generation n.

From our three assumptions we infer that the most general framework for a model of this plant-herbivore system would consist of the two equations

$$v_{n+1} = F(v_n, h_n), \tag{31a}$$

$$h_{n+1} = G(v_n, h_n). \tag{31b}$$

The functions F and G, which govern the population levels of the vegetation and herbivores respectively, will not be assigned particular mathematical expressions. Rather, we will use certain qualitative features of these functions to reason further. Our purpose below is to shed light on the following question: *Under what assumptions will it be true that the herbivores and plant populations are mutually regulating?* (The question is deliberately phrased in a vague way and bears more careful discussion. Before proceeding, you are encouraged to attempt to decipher this question independently.)

For a "mutually" regulated situation it must be true, first of all, that the model consisting of equations (31a,b) admits a nonzero steady-state solution $(\bar{v}, \bar{h})$. That is, the populations can coexist at some constant levels at which neither increase nor decrease occurs. We recall that these values must, by definition, satisfy the relations

$$\bar{v} = F(\bar{v}, \bar{h}), \qquad \bar{h} = G(\bar{v}, \bar{h}). \tag{32}$$

In dealing with even the simplest model, it often proves enormously useful to scale the equations in terms of quantities that are inherent to the process. In this way we will reduce the number of parameters to consider. For those of you who do not wish to scrutinize the details of this rescaling procedure (in the following subsection), the idea is equivalent to *assuming that the values* $\bar{v}$ *and* $\bar{h}$ *are both equal to 1.*

Rescaling the Equations

We define new variables that are ratios of the old variables and their steady-state values as follows:

$$v_n^* = v_n/\bar{v}, \tag{33a}$$

$$h_n^* = h_n/\bar{h}. \tag{33b}$$

The equations rewritten in terms of the new, scaled variables will have steady-state values $\bar{v}_n^* = \bar{h}_n^* = 1$.

Example

The equations

$$x_{n+1} = \alpha x_n y_n^\beta, \tag{34a}$$

$$y_{n+1} = \gamma \frac{y_n}{\delta + x_n}, \tag{34b}$$

have the corresponding steady states

$$\bar{y} = \alpha^{-1/\beta}, \qquad \bar{x} = \gamma - \delta, \quad \gamma > \delta. \tag{35}$$

Defining

$$x_n^* = x_n/(\gamma - \delta), \qquad y_n^* = \alpha^{1/\beta} y_n, \tag{36}$$

we obtain the rescaled form

$$x_{n+1}^* = x_n^* y_n^{*\beta}, \qquad y_{n+1}^* = \frac{y_n^*}{\epsilon + x_n^*(1 - \epsilon)}, \tag{37}$$

where $\epsilon = \delta/\gamma$. It can be verified (see problem 14) that the steady-state solutions to (37) are

$$\bar{x}_n^* = \bar{x}_{n+1}^* = 1,$$

$$\bar{y}_n^* = \bar{y}_{n+1}^* = 1.$$

Notice that two parameters rather than the previous four appear in equations (37).

With the justification given in this example we may at this point assume that equations (31a,b) have been rescaled and are written in terms of new variables, as follows:

$$v_{n+1}^* = F^*(v_n^*, h_n^*),$$
$$h_{n+1}^* = G^*(v_n^*, h_n^*), \tag{38}$$

with steady-state solutions

$$\bar{v}_{n+1}^* = \bar{h}_{n+1}^* = 1. \tag{39}$$

For convenience of notation, we shall drop the asterisks, return to the form shown in (31), and simply assume that both steady-state levels are unity. The reasons for making this simplification will emerge.

Further Assumptions and Stability Calculations

Assuming that equations (31a,b) have steady states $\bar{v} = 1$ and $\bar{h} = 1$ will simplify analysis of the model. To continue defining the problem we use several assumptions to deduce features of the functions F and G. From the comment on assumption 3 we may infer that the following two statements are true.

3a. The greater the herbivory, the lower will be the abundance of vegetation in the next season.

3b. The greater the current vegetation biomass, the greater will be next season's vegetation biomass.

Since the vegetation level is governed by equation (31a), these statements lead us to deduce that F is a decreasing function of h and an increasing function of v. Mathematically this means that

$$\frac{\partial F}{\partial h} < 0 \qquad \text{(from 3a)},$$

$$\frac{\partial F}{\partial v} > 0 \qquad \text{(from 3b)}.$$

In other words, assumptions (3a) and (3b) determine the signs of the partial derivatives of the function F (see Figure 3.9).

We shall treat the second function in a slightly different way, first noting that it represents herbivore recruitment. Assuming the herbivore population changes by reproduction or mortality (and not by migration), we shall rewrite G as a product of the number of herbivores and the net recruitment per individual; that is, assume G has the form

$$G(v_n, h_n) = h_n R(v_n, h_n). \tag{40}$$

The function R is the net number of adult herbivores that are progeny of a single individual herbivore when both fecundity and survivorship are taken into account. It

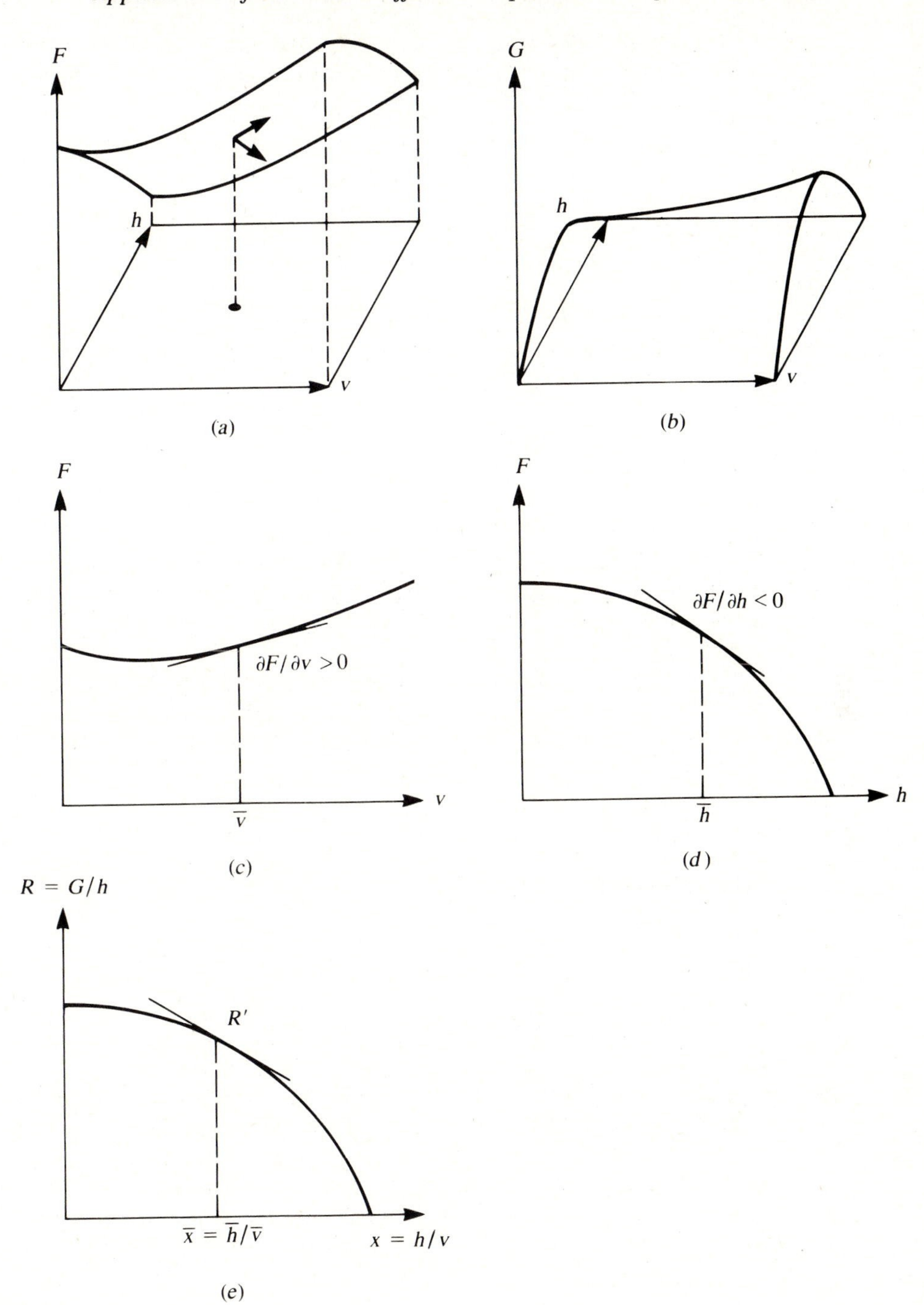

Figure 3.9 *The functions* F *and* G, *which depict plant and herbivore responses, may depend on* v *and* h *in some complicated way as shown in (a) and (b). (c–e) In analyzing the plant-herbivore model we have only used information about the slopes (partial derivatives) of* F *and* $R = G/h$ *at the steady-state point* $(\bar{v}, \bar{h})$.

follows from equation (40) that

$$\frac{\partial G}{\partial v} = h\frac{\partial R}{\partial v}, \tag{41a}$$

$$\frac{\partial G}{\partial h} = h\frac{\partial R}{\partial h} + R. \tag{41b}$$

We now complete the formulation of this model by making assumptions about the function R.

2a. Recruitment of herbivores depends on the extent of competition for the vegetation, i.e., on the average number of herbivores that must share a unit of available vegetation.

2b. The greater the competition in generation n, the lower the recruitment for the next generation.

(*Comment:* This will be true of some but not all plant-herbivore systems. A number of leading biologists maintain that competition for resources plays a minimal role, if any, in herbivory. See, for example, Strong et al., 1984. The question is a controversial one.)

By assumption (2a), R is a function of the ratio $x = h_n/v_n$ of the two variables; by assumption (2b), it is a decreasing function of its argument. To summarize:

$$R = R(x), \qquad \text{where } x = \frac{h_n}{v_n} \qquad \text{(from 2a)}$$

$$\frac{dR}{dx} < 0. \qquad \text{(from 2b)}$$

Using the *chain rule* of elementary calculus we notice that

$$\frac{\partial R}{\partial h} = \frac{dR}{dx}\frac{\partial x}{\partial h} = \frac{R'}{v}, \tag{42a}$$

$$\frac{\partial R}{\partial v} = \frac{dR}{dx}\frac{\partial x}{\partial v} = \frac{-R'h}{v^2}. \tag{42b}$$

where $R' = dR/dx$.

Collecting all equations and assumptions of the model so far, we have the following two equations:

$$V_{n+1} = F(V_n, h_n), \tag{43}$$

$$h_{n+1} = G(V_n, h_n) = h_n R(x_n), \tag{44}$$

where $x_n = h_n/V_n$, $R'(x) < 0$, and the partial derivatives of F satisfy $F_h < 0$, $F_v > 0$.

We now explore the outcome and predictions of the model. We return to a question posed earlier in this section to determine when the plant-herbivore system is self-regulating in the sense that the nonzero steady state is stable. To determine when this holds, we resort to the stability criteria outlined previously. To formulate these in a convenient way, we use the following symbols to represent quantities that enter into a stability calculation:

Let

$$\nu = \left.\frac{\partial F}{\partial v}\right|_{ss}, \tag{45a}$$

$$-\mu = \left.\frac{\partial F}{\partial h}\right|_{ss}, \tag{45b}$$

$$-\epsilon = \left.\frac{dR}{dx}\right|_{ss}, \qquad x = h/v, \tag{45c}$$

where ss means the steady-state value, $\bar{v} = \bar{h} = 1$. The constants ν, μ, and ϵ are positive, and signs preceding them depict our conclusions about the derivatives of the functions F and R. Note that partial derivatives are all evaluated at the steady state $\bar{v} = \bar{h} = 1$. Using the last definition one can show that

$$\left.\frac{\partial G}{\partial h}\right|_{ss} = 1 - \epsilon, \tag{46}$$

$$\left.\frac{\partial G}{\partial v}\right|_{ss} = \epsilon.$$

[See problem 12(b).] Thus the Jacobian of (43) and (44) at the steady state $\bar{v} = \bar{h} = 1$ is

$$\mathbf{J} = \begin{pmatrix} \dfrac{\partial F}{\partial v} & \dfrac{\partial F}{\partial h} \\ \dfrac{\partial G}{\partial v} & \dfrac{\partial G}{\partial h} \end{pmatrix}_{ss} = \begin{pmatrix} \nu & -\mu \\ \epsilon & 1 - \epsilon \end{pmatrix}. \tag{47}$$

Comment: The fact that equations were scaled in terms of the steady-state values and that herbivore recruitment depends only on the ratio $\bar{h}/\bar{v}$ results in a total of three independent parameters in equation (47).

The characteristic equation for (47) is

$$\lambda^2 - \beta\lambda + \gamma = 0, \qquad \text{where} \qquad \begin{aligned} \beta &= \nu + (1 - \epsilon), \\ \gamma &= \nu(1 - \epsilon) + \mu\epsilon. \end{aligned} \tag{48}$$

For stability of the steady state, both roots of the characteristic equation must satisfy $|\lambda| < 1$. Using the stability criteria (see Chapter 2, equations 32), we conclude that for this to be true

$$|\beta| < 1 + \gamma < 2,$$

that is,

$$|\nu + (1 - \epsilon)| < 1 + \nu + \epsilon(\mu - \nu) < 2. \tag{49}$$

Deciphering the Conditions for Stability

Although the mathematical steps leading to a stability condition are now complete, there is still work to be done; we must now extract some meaningful information from the murky inequality in (49). To do so it is necessary to "unravel" the tangle of parameters to reach a clearer statement and then to interpret the results in their biological context by using the original definitions. The process of manipulating the appropriate inequalities is illustrated in the following:

$$|\beta| < 1 + \gamma < 2:$$

1. $|\beta| < 2$ means that $-2 < \nu + 1 - \epsilon < 2$, so

$$-3 < \nu - \epsilon < 1. \tag{50a}$$

2. $\beta < 1 + \gamma$ means that $\nu + 1 - \epsilon < 1 + \nu + \epsilon(\mu - \nu)$. Therefore,

$$-\epsilon < \epsilon(\mu - \nu),$$

$$-1 < \mu - \nu,$$

$$\nu - \mu < 1. \tag{50b}$$

3. Subtracting (50b) from (50a) produces $\mu - \epsilon < 0$, or

$$\mu < \epsilon. \tag{50c}$$

4. $1 + \gamma < 2$ means that

$$\nu - \mu < 1. \tag{50d}$$

(See problem 13.) *Note:* This is covered by the results of (50b), so no new information is obtained.

5. $|\beta| < 1 + \gamma$ means that

$$-1 - \gamma < \beta < 1 + \gamma$$

so by problem 13,

$$\epsilon < \frac{2(1 + \nu)}{1 - \mu + \nu}. \tag{50e}$$

The above procedure yields a set of three relationships (50a–c) linking pairs of parameters and one inequality of a more complicated form (50e). To interpret these conditions for stability we now summarize what the parameters represent.

$$\nu = F_v\big|_{ss} \simeq \frac{\Delta v_{n+1}}{\Delta v_n} \quad \text{at} \quad (\bar{v}, \bar{h})$$ = the change in next year's vegetation biomass caused by a change in the current vegetation biomass.

$$-\mu = F_h\big|_{ss} \simeq \frac{\Delta v_{n+1}}{\Delta h_n} \quad \text{at} \quad (\bar{v}, \bar{h})$$ = change in next year's vegetation biomass due to an increment in current herbivore level.

$$\epsilon = G_v\big|_{ss} \simeq \frac{\Delta h_{n+1}}{\Delta v_n} \quad \text{at} \quad (\bar{v}, \bar{h})$$ = change in next year's herbivore population h_{n+1} due to increment in current plant biomass.

$$1 - \epsilon = G_h\big|_{ss} \simeq \frac{\Delta h_{n+1}}{\Delta h_n} \quad \text{at} \quad (\bar{v}, \bar{h})$$ = change in h_{n+1} due to increment in current herbivore population.

Note: all these quantities are computed when the populations are close to their steady-state levels.

We see that from equations (50a–e) we can reach the following conclusions:

1. $\mu < \epsilon$ (50c) means that, close to the steady state,

$$-\frac{\partial F}{\partial h} < \frac{\partial G}{\partial v}. \tag{51a}$$

When paraphrased, equation (51a) implies that

$$\left.\begin{matrix}\text{the decline in vegetation}\\ \text{biomass due to an}\\ \text{increment in herbivory}\end{matrix}\right\} \quad \text{is less than} \quad \left\{\begin{matrix}\text{the increase in the herbi-}\\ \text{vore population due to an}\\ \text{increase in vegetation mass}\end{matrix}\right. .$$

2. $\nu - \mu < 1$ (50b) means that, close to the steady state,

$$\frac{\partial F}{\partial v} + \frac{\partial F}{\partial h} < 1. \tag{51b}$$

To interpret, this would mean that the combined changes in plant biomass due to slight increases in both v and h are not too drastic. It is interesting to note that if F is independent of h, the condition reduces to the stability condition for the plant in isolation (i.e., $\partial F/\partial v\big|_{ss} < 1$).

3. $-3 < \nu - \epsilon < 1$ (50a) means that, close to the steady state,

$$-3 < \frac{\partial F}{\partial v} - \frac{\partial G}{\partial v} < 1. \tag{51c}$$

Thus changes in v and h caused by an increment in the vegetation biomass should be roughly balanced within the indicated bounds.

Other conclusions are left as an exercise in the problems.

To understand why some of these conditions are a prerequisite for stability, consider a hypothetical situation in which v_n and h_n are close to (but not at) the steady state; for example, $v_n = 1 + \Delta v$, $h_n = 1$.

If condition (51a) is not satisfied, the following chain of events might occur: the biomass increment causes herbivores to proliferate ($h_{n+1} > 1$). This causes a *large* decline in plant biomass ($v_{n+2} < 1$), which leads to a drop in herbivores ($h_{n+3} < 1$). The plant biomass increases again ($v_{n+4} > 1$), and the cycle repeats. By this means a periodic behavior could be established with both populations cycling about their steady-state values. By considering several other scenarios, the student should be able to give similar justification for these stability conditions.

Comments and Extensions

Perhaps the most important conclusion to be drawn from the example discussed above in the previous subsection is that it often makes good sense to treat a problem in an "impressionistic" way. Rather than adopting particular functional forms for describing the population growth, fecundity, and interactions, one might consider first trying rather broad assumptions about their dependence on population levels.

What makes this approach attractive is that it can ease the burden of manipulating complicated mathematical expressions. After the appropriate inequalities are derived, it is generally straightforward to determine when particular functions are likely to satisfy these conditions and so lead to stability in the population. Moreover, given a whole *class* of growth functions or fecundity functions, one can identify the particular *feature* that contributes to stabilizing the population. [For example, inequality (51b) tells us that F cannot be a very steeply varying function of its arguments: small changes in the population levels should not engender large changes in the predicted vegetation biomass.]

Yet a third positive feature of this general analysis is that it leads to much greater ease of experimentation with the model, as suggested by several problems at the end of the chapter. For example, we might like to determine how changing one assumption alters the conclusions. This is rather easy to do in the abstract and usually does not require a repetition of all the calculations.

As a conclusion to this model, it may be wise to shed a somewhat broader perspective on the topic of plant-herbivore interactions. Recent biological research on this problem has revealed that interactions between herbivores and their vegetation may be extremely diverse, subtle, and full of surprises. This is especially true of insect herbivores, whose evolution may be closely linked to those of their plants. In many cases it has been discovered that plants have active defense strategies (using many forms of chemical weapons) or are able to undergo qualitative changes that may adversely affect their attackers.

Models that treat only the quantitative aspects of herbivory such as reduction in plant biomass are relatively naive and rarely accurate in portraying the situation. For this reason, we would want to develop a different type of model that treats more fully the changes in character of the plant and its host. (For a hint on how this might be done, see problem 17.)

Several recent books and articles provide excellent summaries of current thoughts on plant-herbivore interactions. Among these, Crawley (1983) and Rhoades (1983) are strongly recommended sources for further reading and research.

3.6 FOR FURTHER STUDY: POPULATION GENETICS

The genetics of populations with discrete generations is yet another topic well suited for difference equation techniques. Many excellent books and articles on this subject can be recommended for independent research. (A partial list is given in the References.) This section is an elementary introduction to Hardy-Weinberg genetics, which is explored further in the problems following the chapter.

The genetic material in *eukaryotes*[4] is made up of units called *chromosomes*. Organisms (such as humans) that are *diploid* have two sets of chromosomes, one obtained from each parent. A *locus* (a given location on a chromosome) may contain the blueprint instructions for some physical trait (such as eye color), which is determined by the combination of genes derived from each of the parents. A given gene may have one of several forms, called *alleles*. [It is not always known how many alleles of a given gene are present in the *genetic pool* (i.e., total genetic material) of a population.]

Suppose that there are two alleles, denoted by a and A, and that these are passed down in the population from one generation to the next. A given individual could then have one of three combinations: AA, aa, or aA. (The first two combinations are called *homozygous,* the last one *heterozygous*.) It is of interest to follow the distribution of genes in a population over the course of many generations.

A question we might explore is whether the relative frequencies of genes will change, and, if so, whether some new stable distribution will emerge. Until 1914 it was believed that any rare allele would gradually disappear from a population. After a more rigorous treatment of the problem it was shown that if *mating is random* and all *genotypes* (combinations of alleles, which in this case are aa, AA, and aA) are *equally fit* (have an equal likelihood of surviving to produce offspring), then gene frequencies do not change. This fact is now known as the *Hardy-Weinberg law*.

In the problems we investigate how the Hardy-Weinberg law can be demonstrated, and then explore other areas in which the theory can be extended. For further independent reading, Li (1976), Roughgarden (1979), Ewens (1979), and Crow

4. *Eukaryotes* are organisms whose cells have well-defined nuclei in which all the genetic material is contained.

and Kimura (1970) are recommended. A mathematical treatment of the case of unequal genotype fitness is given by Maynard Smith (1968) and in a more expanded form by Segel (1984).

To outline the problem, several definitions and assumptions are needed. By convention we shall define the frequencies of the alleles A and a in the nth generation as follows:

$$p = \text{frequency of allele } A = \frac{\text{total number of } A \text{ alleles}}{2N},$$

$$q = \text{frequency of allele } a = \frac{\text{total number of } a \text{ alleles}}{2N},$$

where $p + q = 1$, and $N =$ the population size.

We now incorporate the following assumptions:

1. Mating is random.
2. There is no variation in the number of progeny from parents of different genotypes.
3. Progeny have equal fitness (that is, are equally likely to survive).
4. There are no mutations at any step.

We define the genotype frequencies of AA, aA, and aa in a given population to be:

$u =$ frequency of AA genotype,

$v =$ frequency of aA genotype,

$w =$ frequency of aa genotype.

Then $u + v + w = 1$. Since aA is equivalent to Aa, it is clear that

$$p = u + \tfrac{1}{2}v, \tag{52a}$$

$$q = \tfrac{1}{2}v + w. \tag{52b}$$

The next step is to calculate the probability that parents of particular genotypes will mate. If mating is random, the mating likelihood depends only on the likelihood of encounter. This in turn depends on the product of the frequencies of the two parents. The *mating table* (Table 3.1) summarizes these probabilities. (Six missing entries are left as an exercise for the reader.)

Table 3.1 ***Mating Table***

			Fathers		
	Genotype		*AA*	*Aa*	*aa*
		Frequency %	u	v	w
Mothers	*AA*	u	u^2	uv	uw
	Aa	v	—	v^2	—
	aa	w	—	—	—

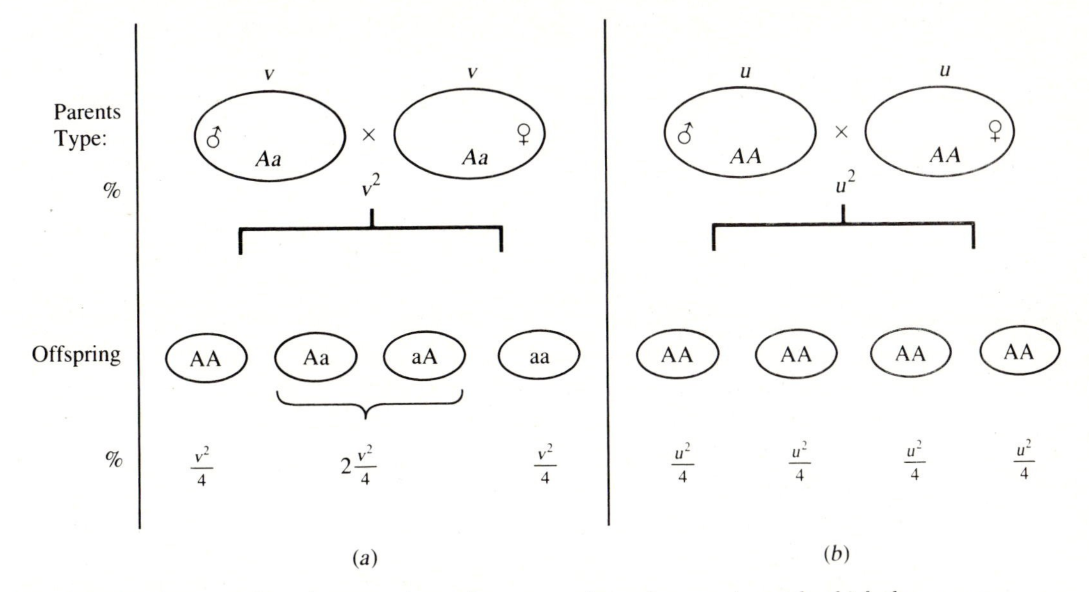

Figure 3.10 *Offspring of (a) the cross* Aa × Aa *and (b) the cross* AA × AA *are shown with the relative frequencies with which they occur.*

Based on the likelihood of mating, one can determine the probability of a given match resulting in offspring of a given genotype. To do this, it is necessary to take into account the four possible combinations of alleles derived from a given pair of parents. Figure 3.10 demonstrates what happens in the case of two heterozygous or two homozygous parents. Other cases are left for the reader. The *offspring table* (Table 3.2) is a convenient way to summarize the information. (Some entries have been left blank for you to fill in—see problem 18.)

Table 3.2 ***Offspring Table***

		Offspring Genotype Frequencies		
Type of Parents	*Frequency*	*AA*	*aA*	*aa*
AA × *AA*	u^2	u^2	—	—
AA × *Aa*	$2uv$	uv	uv	—
AA × *aa*	—	0	—	—
Aa × *Aa*	v^2	$v^2/4$	$v^2/2$	$v^2/4$
Aa × *aa*	—	0	—	—
aa × *aa*	—	0	—	—
	Total	$(u^2 + uv + v^2/4)$	—	—

Note: All genotypes are assumed to have equal likelihood of surviving to reproduce. If *fitness* depended on genotype, these entries would be weighted by the probability that a given parent survived to mate.

By counting up the total frequency of offspring of each type, we get values for u, v, and w for the next generation, which we shall call u_{n+1}, v_{n+1}, and w_{n+1}. It can be shown (see problem 18e) that these are related to the nth-generation frequencies by the equations

$$u_{n+1} = u_n^2 + u_n v_n + \tfrac{1}{4} v_n^2, \tag{53a}$$

$$v_{n+1} = u_n v_n + 2u_n w_n + \tfrac{1}{2} v_n^2 + v_n w_n, \tag{53b}$$

$$w_{n+1} = \tfrac{1}{4} v_n^2 + v_n w_n + w_n^2. \tag{53c}$$

As an example, consider equation (53a) and note that the cumulative number of offspring of genotype *AA* (in the vertical column of Table 3.2) leads to the given relationship.

In the problems, you are asked to show that these equations lead to the conclusion that gene frequencies p and q, as well as genotype frequencies u, v, and w, do not change under random mating with equal fitness. In addition, Hardy-Weinberg equilibrium is attained after a single generation. This conclusion depends on an assumption that the genes are not located on the chromosomes that determine the sex of an individual.

PROBLEMS*

1. The model for density-dependent growth due to Varley, Gradwell, and Hassell (1973) given by equation (3), has the undesirable feature that, for low population levels, the fraction of survivors $f = N_s/N_t = \frac{1}{\alpha}(N_t)^{-b}$ is greater than 1.

(a) Explain why this is faulty.

(b) Find the critical population level, N_c, below which this happens. (N_c should be expressed in terms of α and b.)

(c) Suggest how the model might be modified to alleviate this problem.

2. Sketch the fraction of survivors as a function of population size for the following models:

(a) Equation (3).

(b) Equation (13).

3. **(a)** In the model for single-species populations given by equation (8), show that $N = 0$ is one steady state and $N = K$ is another (the only *nontrivial* steady state).

(b) Verify equation (10).

(c) Show that the population increases if $N < K$ and decreases if $N > K$ by considering the magnitude of the reproductive rate

$$\lambda = \exp r(1 - N_t/K).$$

4. In this problem we investigate the model for density dependence due to Hassell (1975) given by equation (13).

*Problems preceded by an asterisk (*) are especially challenging.

(a) What are the effects of the three parameters, λ, a, and b? Do increased values of these parameters strengthen or weaken the growth of the population?

(b) For certain parameter ranges, this equation has a biologically meaningful nonzero steady state. Show that this steady state is given by

$$\bar{N} = \frac{\lambda^{1/b}}{a},$$

and discuss the appropriate parameter constraints. (*Hint:* Recall that $\bar{N}$ has to be positive.)

(c) Determine conditions for stability of the above steady state.

5. Hassell et al. (1976) give the following estimates for the parameters λ and b of equation (13) for several insect populations.

	b	λ
Moth: *Zeiraphera diniana*	0.1	1.3
Bug: *Leptoterna dolobrata*	2.1	2.2
Mosquito: *Aedes aegypti*	1.9	10.6
Potato Beetle: *Leptinotarsa decemlineata*	3.4	75.0
Parasitoid wasp: *Bracon hebetor*	0.9	54.0

(a) Plot these values on a λb-parameter plane. (Recommendation: use a log scale for the λ axis.)

(b) Use your results from problem 4 to determine which of these species will have a stable steady-state population level.

6. **(a)** Give two examples of physical processes that can be described by a Poisson distribution. (You may wish to consult a text on probability and statistics.)

(b) Sketch $p(r)$ as a function of r.

(c) Support the claim that in a host-parasitoid system the average number of encounters per host per unit time is

$$\mu = \frac{N_e}{N_t} = aP_t.$$

(d) Suppose we relax the assumption 5 of Section 3.3 and instead assume that if a host is encountered *once,* it gives rise to c parasitoid progeny; if a host is encountered *twice* or more, it gives rise to $2c$ parasitoid progeny. How would the Nicholson-Bailey model change?

7. **(a)** From the stability calculations for the Nicholson-Bailey model we observe that the predictions are independent of the parameters a and c and depend only on λ. Explain the following observation: Consider a somewhat different formulation of the model. Define

$$n_t = acN_t,$$

$$p_t = aP_t$$

where a, c are the constants defined in Section 3.3. Substitute the new variables into equations (21a, b) to obtain equations in terms of n_t and p_t. How many (new) parameters do these equations contain?

(b) Interpret n_t and p_t.

***(c)** Determine whether the instability of the Nicholson-Bailey model is accompanied by oscillations by determining whether $\beta^2 - 4\gamma < 0$ for β, γ given by equations (26).

8. *Project:* One problem that leads to the oscillations in the Nicholson-Bailey model (equations 21a, b) is the fact that at low parasitoid densities the host population behaves approximately as follows:

$$N_{t+1} = \lambda N_t;$$

that is, it grows at an unchecked rate. Notice that this eventually causes the parasitoids to increase in number until their hosts are overwhelmed by the attack. This is what sets up the increasing oscillations in the system.

Consider the effect of introducing other types of density dependence into the host population. Two examples include

$$N_{t+1} = \lambda N_t \frac{(K - N_t)}{K} e^{-aP_t}, \tag{54}$$

and,

$$N_{t+1} = \lambda N_t^{1-b} e^{-aP_t}, \tag{55}$$

(see Varley and Gradwell (1963) and Hassell (1978)). Use the literature, computer simulations, and your own analysis to compare the predictions of these models and to comment on their applicability and/or special features. (*Note:* Stability analysis may be algebraically messy in such models.)

9. (a) Show that equations (28a, b) have the steady state given by (29), where q is defined as $q = \bar{N}/K$.

(b) Find the Jacobian matrix that corresponds to the linearized version of (28).

***(c)** Find the stability conditions in terms of the parameter q and other parameters appearing in equations (28a, b).

***(d)** Reason that Figure 3.4 is the expected stability domain in the qr parameter plane.

10. In this problem we consider the effect of parasitoid searching efficiency in the Nicholson-Bailey model.

(a) Explain the restriction $m < 1$ in equation (30).

(b) Write a set of host-parasitoid equations that incorporate equation (30).

(c) Investigate the effect of the new assumption about $f(N_t, P_t)$. Determine whether steady states are elevated or depressed and whether stability is affected.

(d) Suggest other forms of $f(N_t, P_t)$ that might be reasonable in view of the biological assumption that parasitoids interfere with each other less and are hence more efficient at low densities.

11. In this problem we consider a host population that has refuges from parasitoid attack.

(a) Explain the assertion that EK/N_t is the fraction of the population that can retreat to a refuge.

(b) Explain why the fraction of hosts *not* parasitized is

$$f(N_t, P_t) = \frac{EK}{N_t} + \left(1 - \frac{EK}{N_t}\right)e^{-aP_t}.$$

(c) Write an equation for N_{t+1}.

(d) Explain why at time $t + 1$ the parasitoid density is given by

$$P_{t+1} = (N_t - EK)(1 - e^{-aP_t}).$$

***(e)** *Longer Project:* Consult the literature and use computer simulations and/or other analysis to explore the results and predictions of this model.

Questions 12 through 17 are based on the model for plant-herbivore interactions described in Section 3.5.

12. **(a)** Show that the function R takes on the value

$$R(\bar{h}/\bar{v}) = 1$$

for $\bar{h} = 1$ and $\bar{v} = 1$, the steady state of equations (31a, b).

(b) Use part (a) and the results of Section 3.5 (subsection "Further Assumptions and Stability Calculations") to show that for $x = h/v$ and

$$-\epsilon = \left.\frac{dR}{dx}\right|_{ss}$$

it follows that

$$\left.\frac{\partial G}{\partial h}\right|_{ss} = 1 - \epsilon,$$

$$\left.\frac{\partial G}{\partial v}\right|_{ss} = \epsilon.$$

13. **(a)** Demonstrate that the inequalities (50d) and (50e) are correct.

(b) Interpret the biological meaning of (50e).

(c) Discuss why conditions (50d) and (50e) might be necessary to avoid instability of the plant-herbivore system.

14. Verify that steady-state solutions of equations (37) have the values $\bar{x} = 1$ and $\bar{y} = 1$.

15. Consider the following model for leaf-eating herbivores whose population size (number of individuals) is h_n on a tree whose leaf mass is v_n:

$$v_{n+1} = fv_n(e^{-ah_n}),$$

$$h_{n+1} = rh_n\left(\delta - \frac{h_n}{v_n}\right), \qquad \text{where } v_n \neq 0$$

and where f, a, r, δ are positive constants.

(a) Find the steady state(s) $\overline{v}$ and $\overline{h}$ of this system. What happens if $f = 1$? What restrictions on the parameters should be met for a biologically reasonable steady state?

***(b)** Show that by rescaling the equations, it is possible to reduce the number of parameters. To do this, define

$$V_n = v_n/\overline{v}, \qquad H_n = h_n/\overline{h}.$$

Show that the system of equations can then be converted to the following form:

$$V_{n+1} = V_n \exp k(1 - H_n),$$

$$H_{n+1} = bH_n\left(1 + \frac{1}{b} - \frac{H_n}{V_n}\right).$$

What is the connection between b, k, and the previous four parameters f, a, r, and δ?

(c) Show that the equations in part (b) now have steady-state solutions

$$\overline{Q} = \overline{H} = 1.$$

(d) Determine whether the functions in the equations of the model fall into the general category described in Section 3.5.

(e) Determine when the steady state will be stable.

16. In this question we explore several variants of the model outlined in Section 3.5.

(a) Suppose that for low levels of herbivory the function F, which represents the vegetation response, does not depend on the herbivore population and that $\partial F/\partial h < 0$ only beyond some value $h = \hat{H}$. How would this affect the conclusions of the model?

(b) In some plants it is observed that moderate to low herbivory actually promotes enhanced growth. How would this be incorporated into the model and what would be the results?

(c) Suggest other ways of incorporating the availability of the vegetation into the response of the herbivore population and explore the outcomes of these assumptions.

(d) Suppose that migration into the patch of vegetation takes place at some rate that is enhanced by greater plant abundance. How would this be modeled, and how would it change the problem?

17. (*Note to the instructor:* Problem 17 gives the student good practice at formulating a model and gradually increasing its complexity. Do not expect the later stages of this problem to be as amenable to further direct analysis as the more elementary model. More advanced students may wish to implement their models in computer simulations.) As indicated near the end of Section 3.5, an important influence on herbivores is the quality of the host vegetation, not just its abundance. If the plants have induced chemical defenses or other protective responses, the population of attackers may suffer from increased mortality, decreased fecundity, and lower growth rates.

(a) Define a new variable $\hat{q}_n$ as the *average quality* of the vegetation in generation n. Suggest what general equations might then be used to model the v_n, $\hat{q}_n$, h_n system.

(b) Instead of treating the vegetation as a uniform collection of identical plants, consider making a distinction between plants that are nutritious (or chemically undefended) and those that are not. For example, with h_n as before, let

u_n = number of undefended plants,

v_n = number of defended plants.

Now suppose that plants make the transition from u to v after attack by herbivores and back from v to u at some constant rate that represents the loss of chemical defenses. Formulate a model for v_n, u_n, and h_n.

(c) As a final step, consider vegetation in which plants can range in quality q from 0 (very hostile to herbivores) to 1 (very nutritious to herbivores). Assume that the change in quality of a given plant takes place in very small steps every $\Delta\tau$ days at a rate that depends on herbivory in time interval n and on previous vegetation quality. That is,

$$q_{n+1} = q_n + F(q_n, h_n)\Delta\tau,$$

where F can be positive or negative. Let $v_n(q)$ be the percentage of the vegetation whose quality is q at the nth step of the process, and assume that the total biomass

$$V_n = \int_0^1 v_n(q)\,dq,$$

does not change over the time scale of the problem. Can you formulate an equation that describes how the distribution of vegetation quality $v_n(q)$ changes as each plant undergoes the above defense response?

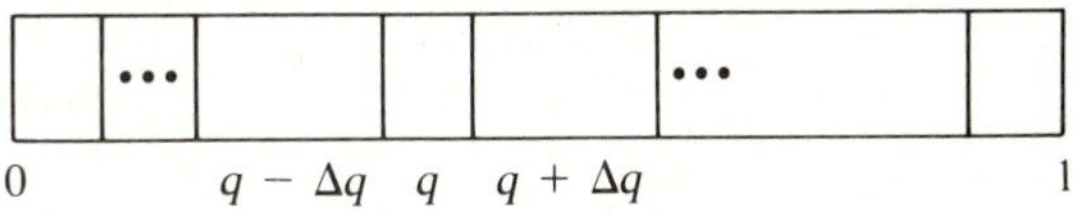

...   ...

0   $q - \Delta q$ q $q + \Delta q$ 1

Figure for problem 17(c).

Hint: Consider subdividing the interval (0, 1) into n quality classes, each of range Δq. How many plants leave or enter a given quality class during time $\Delta\tau$?

Questions 18 through 20 deal with the topic of population genetics suggested in Section 3.6.

18. (a) From the definitions of u, v, and w it is clear that

$$u + v + w = 1.$$

Show that $p + q = 1$.

(b) Fill in the remaining six entries in Table 3.1.

(c) Draw a diagram similar to that of Figure 3.10 for the mating of parents of type $AA \times aa$ and $Aa \times AA$.

(d) Fill in the remaining entries in Table 3.2.

(e) Verify that the cumulative frequencies of genotypes AA, Aa, and aa in the offspring are governed by equations (53a, b).

(f) Show that

$$u_{n+1} + v_{n+1} + w_{n+1} = 1,$$

i.e., that u, v, and w will always sum up to 1.

(g) Suppose that $(\overline{u}, \overline{v}, \overline{w})$ is a steady state of equations (53a, b). Use the information in part (f) to show that

$$\overline{u} = \overline{v}^2/4\overline{w}.$$

(h) Since w_{n+1} is related to u_{n+1} and v_{n+1} (similarly for w_n, u_n, and v_n) by the identity in part (f), we can eliminate one variable from equations (53a, b). Rewrite these equations in terms of u and v.

(i) Using part (h) show that

$$u_{n+1} = (u_n + \tfrac{1}{2}v_n)^2,$$
$$v_{n+1} = (u_n + \tfrac{1}{2}v_n)[2 - 2(u_n + \tfrac{1}{2}v_n)].$$

(j) Now show that

$$(u_{n+1} + \tfrac{1}{2}v_{n+1}) = (u_n + \tfrac{1}{2}v_n).$$

***(k)** Show that this implies that the frequencies p and q do not change from one generation to the next; i.e., that

$$q_{n+1} = q_n, \qquad p_{n+1} = p_n.$$

Hint: you must use the fact that $p^2 + 2pq + q^2 = 1$. When you succeed at this problem, you have proved the Hardy-Weinberg law.

19. Now consider the following modification of the random mating assumption: Suppose that individuals mate only with those of like genotype (e.g., Aa with Aa, AA with AA, and so forth). This is called *positive assortative mating*. How would you set up this problem, and what conclusions do you reach?

20. In *negative assortative matings,* like individuals do not mate with each other. Different types of models may be obtained, depending on assumptions made about the permissible matings. In the questions that follow it is assumed that homozygous females mate only with homozygous males of opposite type and that heterozygous females (Aa) mate with AA and aa males depending on their relative prevalence. The permitted matings are then as follows:

Females		*Males*
AA	×	aa
aa	×	AA
Aa	×	AA
Aa	×	aa

Assuming that a single male can fertilize any number of females results in a mating table that depends largely on the female frequencies, as shown in the mating table.

			Males		
	Genotype		*AA*	*Aa*	*aa*
		Frequency	u	v	w
Females	*AA*	u	0	0	u
	Aa	v	$\frac{vu}{u+w}$	0	$\frac{vw}{u+w}$
	aa	w	w	0	0

It is assumed that $u + v + w = 1$.

(a) Explain the entries in the table.

(b) Derive an offspring table by accounting for all possible products of the matings shown in the above mating table.

(c) Show that the fractions of *AA*, *Aa*, and *aa* offspring denoted by u_{n+1}, v_{n+1}, and w_{n+1} satisfy the equations

$$u_{n+1} = \frac{1}{2}v_n\frac{u_n}{u_n + w_n},$$

$$v_{n+1} = u_n + w_n + \frac{1}{2}v_n,$$

$$w_{n+1} = \frac{1}{2}v_n\frac{w_n}{u_n + w_n}.$$

(d) Show that $u + v + w = 1$ in the $(n + 1)$st generation. Use this fact to eliminate w from the equations.

(e) Show that the equations you obtain have a steady state with $\overline{v} = \frac{2}{3}$. Is there a unique value of $(\overline{u}, \overline{w})$ for this steady state? Is this steady state stable?

(f) Show that the ratio u/w does not change from one generation to the next.

PROJECTS

1. Write a short computer program to simulate the Nicholson-Bailey model and display the oscillatory behavior of its solutions.

2. *Journal Article Report Difference Equations*
The references contain a list of journal articles, grouped into several general topics. Select one topic and write a short, concise review of the material presented in these articles. The objective of this project is to get you to read and think about some of the original work in the field. Thus your summary should deal critically with the ideas, methods, and presentation in these articles, rather than merely restating the contents. Following are some questions you may wish to address:

1. What is the main focus of the article(s); is a particular question being addressed?
2. Do the mathematical models help in illuminating the topic; if so, in what ways?
3. Are there alternative methods or approaches that might have been suitable for answering the questions the authors addressed?

In certain cases you may need to fill in background details by consulting the sources cited by the authors of these articles.

Note: The reference list at the end of this chapter is by no means a complete survey. You may wish to propose your own topic and search through the literature for relevant papers. Possible sources include the *Journal of Theoretical Biology, American Naturalist,* the *Journal of Animal Ecology, Ecology, Theoretical Population Biology,* and the *Journal of Mathematical Biology.*

REFERENCES

Theory of Difference Equations

May, R. M. (1975). Biological populations obeying difference equations: Stable points, stable cycles, and chaos. *J. Theor. Biol., 51,* 511–524.

May, R. M. (1975). Deterministic models with chaotic dynamics. *Nature, 256,* 165–166.

May, R. M. (1976): Simple mathematical models with very complicated dynamics. *Nature, 261,* 459–467.

May, R. M. and Oster, G. F. (1976). Bifurcations and dynamic complexity in simple ecological models. *Am. Nat., 110,* 573–599.

Difference Equations Applied to Single-Species Populations

Hassell, M. P. (1975). Density dependence in single-species populations. *J. Anim. Ecol., 44,* 283–295.

Hassell, M. P.; Lawton, J. H.; and May, R. M. (1976). Patterns of dynamical behaviour in single-species populations. *J. Anim. Ecol., 45.* 471–486.

Varley, G. C., and Gradwell, G. R. (1960). Key [illegible]rs in population studies. *J. Anim. Ecol., 29,* 399–401.

Varley, G. C.; Gradwell, G. R.; and Hassell, M. P. (1973). *Insect Population Ecology.* Blackwell Scientific Publications, Oxford.

Predator-Prey Models

Beddington, J. R.; Free, C. A.; and Lawton, J. H. (1975). Dynamic complexity in predator-prey models framed in difference equations. *Nature, 255,* 58–60.

Beddington, J. R.; Free, C. A.; and Lawton, J. H. (1978). Characteristics of successful natural enemies in models of biological control of insect pests. *Nature, 273,* 513–519.

Host-Parasite Models

Burnett, T. (1958). A model of host-parasite interaction. *Proc. 10th Int. Congr. Ent., 2,* 679–686.

Hassell, M. P., and May, R. M. (1973). Stability in insect host-parasite models. *J. Anim. Ecol., 42,* 693–726.

Hassell, M. P., and May, R. M. (1974). Aggregation of predators and insect parasites and its effect on stability. *J. Anim. Ecol., 43,* 567–594.

Nicholson, A. J., and Bailey, V. A. (1935). The balance of animal populations. Part I. *Proc. Zool. Soc. Lond., 3,* 551–598.

Varley, G. C., and Gradwell, G. R. (1963). The interpretation of insect population changes. *Proc. Ceylon Ass. Advmt. Sci., 18,* 142–156.

General Sources

Hassell, M. P. (1978). *The Dynamics of Arthropod Predator-Prey Systems,* Monographs in Population Biology 13, Princeton University Press, Princeton, N. J.

Hassell, M. P. (1980). Foraging strategies, population models, and biological control: a case study. *J. Anim. Ecol., 49,* 603–628.

Hogg, R. V., and Craig, A. T. (1978). *Introduction to Mathematical Statistics.* 4th ed. Macmillan, New York.

Nicholson, A. J. (1954). An outline of the dynamics of animal populations. *Aust. J. Zool., 2,* 9–65.

Plant-Herbivore Systems

Crawley, M. J. (1983). *Herbivory: The Dynamics of Animal-Plant Interactions.* University of California Press, Berkeley, 1983.

Rhoades, D. (1983). Herbivore population dynamics and plant chemistry. Pp. 155–220 in R. F. Denno and M. S. McClure, eds., *Variable Plants and Herbivores in Natural and Managed Systems.* Academic Press, New York.

Strong, B. R.; Lawton, J. H.; and Southwood, R. (1984). *Insects on plants: Community Patterns and Mechanisms.* Harvard University Press, Cambridge, Mass.

Population Genetics

Crow, J. F., and Kimura, M. (1970). *An Introduction to Population Genetics Theory.* Harper & Row, New York.

Ewens, W. J. (1979). *Mathematical Population Genetics.* Springer-Verlag, New York.

Li, C. C. (1976). *A First Course in Population Genetics.* Boxwood Press, Pacific Grove, Calif.

Roughgarden, J. (1979). *Theory of Population Genetics and Evolution Ecology: An Introduction.* Macmillan, New York.

Segel, L. A. (1984). *Modeling Dynamic Phenomena in Molecular and Cellular Biology.* Cambridge University Press, Cambridge, chap. 3.

Smith, J. M. (1968). *Mathematical Ideas in Biology.* Cambridge University Press, Cambridge.

Historical Account

Kingsland, S. (1985). *Modeling Nature: Episodes in the History of Population Ecology.* University of Chicago Press, Chicago.

II Continuous Processes and Ordinary Differential Equations

4 An Introduction to Continuous Models

The mathematics of uncontrolled growth are frightening. A single cell of the bacterium E. coli would, under ideal circumstances, divide every twenty minutes. That is not particularly disturbing until you think about it, but the fact is that bacteria multiply geometrically: one becomes two, two become four, four become eight, and so on. In this way, it can be shown that in a single day, one cell of E. coli could produce a super-colony equal in size and weight to the entire planet earth.

M. Crichton (1969), *The Andromeda Strain* (Dell, New York, p. 247).

This chapter introduces the topic of ordinary differential equation models, their formulation, analysis, and interpretation. A main emphasis at this stage is on how appropriate assumptions simplify the problem, how important variables are identified, and how differential equations are tailored to describing the essential features of a continuous process.

Because one of the most challenging parts of modeling is writing the equations, we dwell on this aspect purposely. The equations are written in stages, with appropriate assumptions introduced as they are needed. We begin with a rather simple ordinary differential equation as a model for bacterial growth. Gradually, more realistic aspects of the situation are considered, eventually leading to a system of two ordinary differential equations that describe the way that the microorganisms reproduce at the expense of nutrient consumption in a device called the *chemostat*.

Some of the simplest models are analytically solvable. However, as complexity increases, even formulating the equations correctly can be tricky. One technique that proves useful in detecting potential errors in the equations is *dimensional analysis*. This method forms an underlying secondary theme in the chapter.

After defining the problem and formulating a consistent set of equations, we turn to analysis of solutions. In the chemostat model we find that, due to complexity of the system, the only solutions that can be found analytically are the steady states. Their stability properties are of particular importance, and are explored in Section 4.10.

In a brief digression (Sections 4.7–4.9), we review some aspects of the mathematical background. Those of you familiar with differential equations may skip or skim over Section 4.8. Others who have had no previous exposure may find it helpful to supplement this terse review with readings from any standard text on ordinary differential equations (for example, Boyce and DiPrima, 1977 or Braun, 1979).

Culminating the analysis of the model in Section 4.10 is an interpretation of the various mathematical results in terms of the biological problem. We shall see that in this example predictions can be made about how the chemostat is to be operated for successful harvesting. Treatments of this problem with slightly different flavors are also to be found in Segel (1984), Rubinow (1975), and Biles (1982).

Methods applied to one situation often prove useful in a host of related or unrelated problems. Three such examples are described in a concluding section for further independent study.

4.1 WARMUP EXAMPLES: GROWTH OF MICROORGANISMS

One of the simplest experiments in microbiology consists of growing unicellular microorganisms such as bacteria and following changes in their population over several days. Typically a droplet of bacterial suspension is introduced into a flask or test tube containing *nutrient medium* (a broth that supplies all the essentials for bacterial viability). After this process of *inoculation,* the *culture* is maintained at conditions that are compatible with growth (e.g., at suitable temperatures) and often kept in an agitated state. The bacteria are then found to reproduce by undergoing successive cell divisions so that their numbers (and thus density) greatly increase.[1]

In such situations, one typically observes that the graph of log bacterial density versus time of observation falls along a straight line at least for certain phases of growth: after the initial adjustment of the organism and before its nutrient substrate has been depleted. Here we investigate more closely why this is true and what limitations to this general observation should be pointed out.

Let

$$N(t) = \text{bacterial density observed at time } t.$$

Suppose we are able to observe that over a period of one unit time, a single bacterial cell divides, its daughters divide, and so forth, leading to a total of K new bacterial cells. We define the reproductive rate of the bacteria by the constant K, $(K > 0)$ that is,

$$K = \text{rate of reproduction per unit time.}$$

1. There are several ways of ascertaining bacterial densities in a culture. One is by successive cell counts in small volumes withdrawn from the flask. Even more convenient is a determination of the *optical density* of the culture medium, which correlates with cell density.

Now suppose densities are observed at two closely spaced times t and $t + \Delta t$. Neglecting death, we then expect to find the following relationship:

$$N(t + \Delta t) \simeq N(t) + KN(t)\,\Delta t,$$

$$\begin{matrix} \text{total density at} & & \text{density at} & & \text{increase in density} \\ \text{time} & & \text{time} & & \text{due to reproduction} \\ & & & & \text{during time interval} \\ t + \Delta t & \simeq & t & + & \Delta t \end{matrix}.$$

This implies that

$$\frac{N(t + \Delta t) - N(t)}{\Delta t} = KN(t). \tag{1a}$$

We now approximate N (strictly speaking a large integer, e.g., $N = 10^6$ bacterial cells/ml) by a continuous dependent variable $N(t)$. Such an approximation is reasonable provided that (1) N is sufficiently large that the addition of one or several individuals to the population is of little consequence, and (2) the growth or reproduction of individuals is not correlated (i.e., there are no distinct population changes that occur at timed intervals).

Then in the limit $\Delta t \to 0$ equation (1) can be approximated by the following ordinary differential equation:

$$\frac{dN}{dt} = KN. \tag{1b}$$

This simple equation is sometimes known as the *Malthus law*. We can easily solve it as follows: multiplying both sides by dt/N we find that

$$\frac{dN}{N} = K\,dt.$$

Integrating both sides we obtain

$$\int_0^t \frac{dN}{N} = \int_0^t K\,ds,$$

$$\ln N \Big|_0^t = Kt,$$

$$\ln N(t) - \ln N(0) = Kt,$$

$$\ln N(t) = Kt + a, \tag{2a}$$

where $a = \ln N(0)$. This explains the assertion that a log plot of $N(t)$ is linear in time, at least for that phase of growth for which K may be assumed to be a constant.

We also conclude from equation (2a) that

$$N(t) = N_0 e^{Kt}, \tag{2b}$$

where $N_0 = N(0) =$ the initial population. For this reason, populations that obey equations such as (1b) are said to be undergoing *exponential growth*.

This constitutes the simplest minimal model of bacterial growth, or indeed, growth of any reproducing population. It was first applied by Malthus in 1798 to human populations in a treatise that caused sensation in the scientific community of his day. (He claimed that barring natural disasters, the world's population would grow exponentially and thereby eventually outgrow its resources; he concluded that mass starvation would befall humanity.) These deductions are discussed at greater length in problem 1.

Equation (1b), while disarmingly simple, turns up in a number of natural processes. By reversing the sign of K one obtains a model of a population in which a fraction K of the individuals is continually removed per unit time, such as by death or migration. The solution

$$N(t) = N_0 e^{-Kt}, \qquad (K > 0) \tag{2c}$$

thus describes a *decaying* population. This equation is commonly used to describe radioactive decay.

One defines a population *doubling time* τ_2 (for K), or *half-life* $\tau_{1/2}$ (for $-K$) in the following way. For growing populations, we seek a time τ_2 such that

$$\frac{N(\tau_2)}{N_0} = 2.$$

Substituting into equation (2b) we obtain

$$\frac{N(\tau)}{N_0} = 2 = e^{K\tau},$$

$$\ln 2 = K\tau,$$

$$\tau = \frac{\ln 2}{K}. \tag{3}$$

The doubling time τ is thus inversely proportional to the reproductive constant K. In problem 3 a similar conclusion is obtained for the half-life of a decaying population.

Returning to the biological problem, several comments are necessary:

1. We must avoid the trap of assuming that the model consisting of equation (1b) is accurate for all time since, realistically, the growth of bacterial populations in the presence of a limited nutrient supply always *decelerates* and eventually stops. This would tend to imply that K is not a constant but changes with time.

2. Suppose we knew the bacterial growth rate $K(t)$ as a function of time. Then a simple extension of our previous calculations leads to

$$N(t) = N(0) \exp\left(\int_0^t K\, ds\right). \tag{4}$$

(See problem 4.) For example, if K itself decreases at an exponential rate, the population eventually ceases to grow. (This assumption, known as the *Gompertz law,* will be discussed further in Section 6.1.)

3. Generally we have no knowledge of the exact time dependence of the reproductive rate. However, we may know that it depends directly or indirectly on the density of the population, as in previous density-dependent models explored in connection with discrete difference equations. This is particularly true in populations that are known to regulate their reproduction in response to population pressure. This phenomenon will be discussed in more detail in Chapter 6.

4. Another possibility is that the growth rate depends directly on the resources available to the population (e.g., on the level of nutrient remaining in the flask). Suppose we assume that the reproductive rate K is simply proportional to the nutrient concentration, C:

$$K(C) = \kappa C. \tag{5}$$

Further assume that α units of nutrient are consumed in producing one unit of population increment ($Y = 1/\alpha$ is then called the *yield*). This then implies that bacterial growth and nutrient consumption can be described by the following pair of equations:

$$\frac{dN}{dt} = K(C)N = \kappa CN, \tag{6a}$$

$$\frac{dC}{dt} = -\alpha \frac{dN}{dt} = -\alpha\kappa CN. \tag{6b}$$

This *system* of ordinary differential equations is solvable as follows:

$$\frac{dC}{dt} = -\alpha \frac{dN}{dt},$$

so

$$\int dC = -\alpha \int dN,$$

$$C(t) = -\alpha N(t) + C_0, \tag{7}$$

where $C_0 = C(0) + \alpha N(0)$ is a constant. If the population is initially very small, C_0 is approximately equal to the initial amount of nutrient in the flask. By substituting (7) into equation (6a) we obtain

$$\frac{dN}{dt} = \kappa(C_0 - \alpha N)N. \tag{8}$$

[*Comment:* Observe that the assumptions set forth here are thus mathematically equivalent to assuming that reproduction is density-dependent with

$$K(N) = \kappa(C_0 - \alpha N). \tag{9}$$

This type of growth law, on which we shall comment further, is known as *logistic growth;* it appears commonly in population dynamics models in the form $dN/dt = r(1 - N/B)N$.] The solution to equation (8), obtained in a straightforward way, is

$$N(t) = \frac{N_0 B}{N_0 + (B - N_0)e^{-rt}}, \tag{10}$$

where $N_0 = N(0)$ = initial population,

$r = (\kappa C_0)$ = intrinsic growth parameter,

$B = (C_0/\alpha)$ = carrying capacity.

(See problem 5 for a discussion of this equation and problem 13 for another approach to the problem.) A noteworthy feature of equation (10) is that for large values of t the population approaches a level $N(\infty) = B = C_0/\alpha$, whereas at very low levels the population grows roughly exponentially at a rate $r = \kappa C_0$. It is interesting to compare this prediction with the experimental data given by Gause (1969) for the cultivation of the yeast *Schizosaccharomyces kephir*. (See Figure 4.1.)

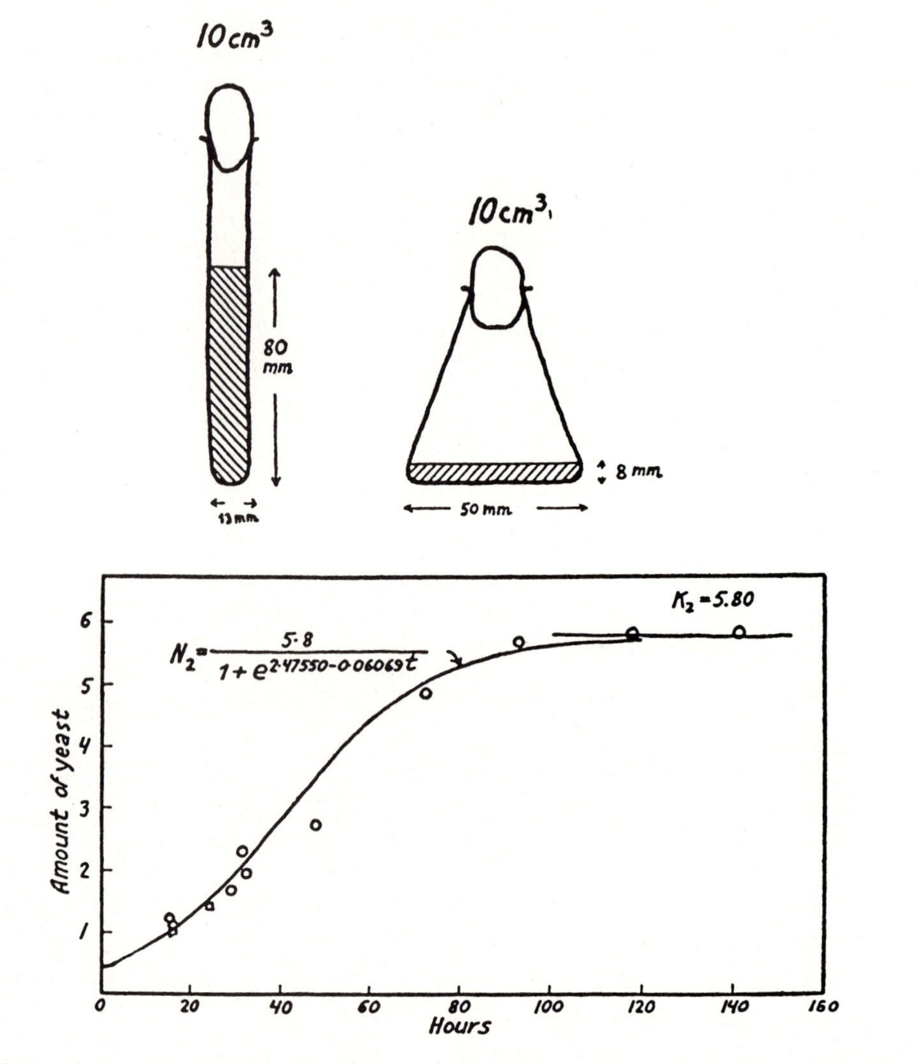

Figure 4.1 Vessels for cultivating yeast or other microorganisms: (a) test tube, (b) Erlenmeyer flask. (c) Growth of the yeast Schizosaccharomyces kephir *over a period of 160 h. The circles are experimental observations. The solid line is the curve*

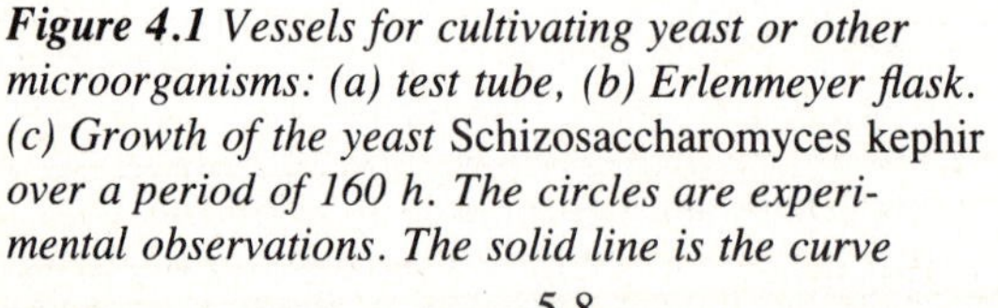

$$N(t) = \frac{5.8}{1 + e^{2.47-0.0607t}}$$

[From Gause, G. F. (1969), Figs. 8 and 15. The Struggle for Existence, *Hafner, New York.* K_2 *in the figure is equivalent to* $B = C_0/\alpha$, *the carrying capacity in equations (8)–(10).*

In the following sections we consider a somewhat more advanced model for bacterial growth in a chemostat.

4.2 BACTERIAL GROWTH IN A CHEMOSTAT[2]

In experiments on the growth of microorganisms under various laboratory conditions, it is usually necessary to keep a stock supply of the strain being studied. Rather than use some dormant form, such as spores or cysts, which would require time to produce active cultures, a convenient alternative is to maintain a continuous culture from which actively growing cells can be harvested at any time.

To set up this sort of culture, it is necessary to devise a means of replenishing the supply of nutrients as they are being consumed and at the same time maintain some convenient population levels of the bacteria or other organism in the culture. This is usually done in a device called a *chemostat,* shown in Figure 4.2.

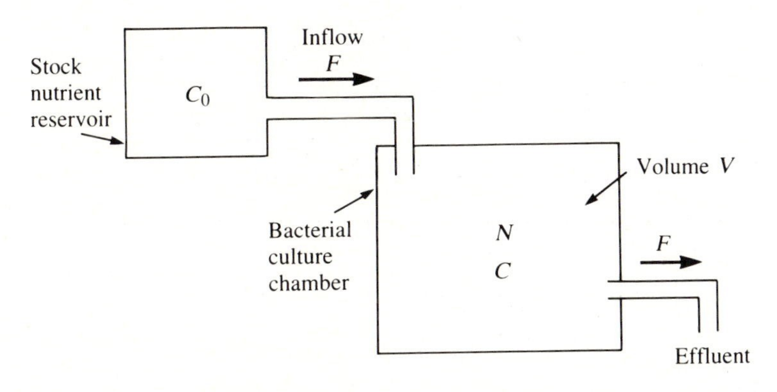

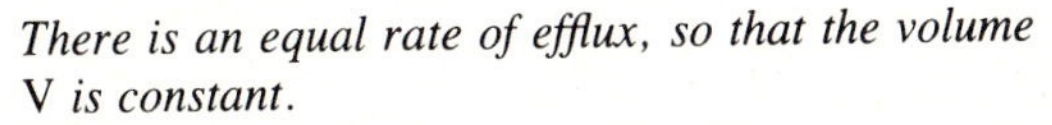

Figure 4.2 *The chemostat is a device for harvesting bacteria. Stock nutrient of concentration* C_0 *enters the bacterial culture chamber with inflow rate* F. *There is an equal rate of efflux, so that the volume* V *is constant.*

A stock solution of nutrient is pumped at some fixed rate into a growth chamber where the bacteria are being cultivated. An outflow valve allows the growth medium to leave at the same rate, so that the volume of the culture remains constant.

Our task is to design the system so that

1. The flow rate will not be so great that it causes the whole culture to be washed out and eliminated.

2. Portions of this material were adapted from the author's recollection of lectures given by L. A. Segel to students at the Weizmann Institute. It has also appeared recently in Segel (1984).

2. The nutrient replenishment is sufficiently rapid so that the culture continues to grow normally.

We are able to choose the appropriate stock nutrient concentration, the flow rate, and the size of the growth chamber.

In this example the purpose of the model will be twofold. First, the progression of steps culminating in precise mathematical statements will enhance our understanding of the chemostat. Second, the model itself will guide us in making appropriate choices for such parameters as flow rates, nutrient stock concentration, and so on.

4.3 FORMULATING A MODEL

A First Attempt

Since a number of factors must be considered in keeping track of the bacterial population and its food supply, we must take great care in assembling the equations. Our first step is to identify quantities that govern the chemostat operation. Such a list appears in Table 4.1, along with assigned symbols and dimensions.

Table 4.1 ***Chemostat Parameters***

Quantity	*Symbol*	*Dimensions*
Nutrient concentration in growth chamber	C	Mass/volume
Nutrient concentration in reservoir	C_0	Mass/volume
Bacterial population density	N	Number/volume
Yield constant	$Y = 1/\alpha$	(See problem 6)
Volume of growth chamber	V	Volume
Intake/output flow rate	F	Volume/time

We also keep track of assumptions made in the model; here are a few to begin with:

1. The culture chamber is kept well stirred, and there are no spatial variations in concentrations of nutrient or bacteria. (We can describe the events using ordinary differential equations with time as the only independent variable.)

At this point we write a preliminary equation for the bacterial population density N. From Fig. 4.2 it can be seen that the way N changes inside the culture chamber depends on the balance between the number of bacteria formed as the culture reproduces and the number that flow out of the tank. A first attempt at writing this in an equation might be,

$$\frac{dN}{dt} = KN - FN \tag{11}$$

rate of change of bacteria — reproduction — outflow

where K is the reproduction rate of the bacteria, as before.

To go further, more assumptions must be made; typically we could simplify the problem by supposing that

2. Although the nutrient medium may contain a number of components, we can focus attention on a single growth-limiting nutrient whose concentration will determine the rate of growth of the culture.

3. The growth rate of the population depends on nutrient availability, so that $K = K(C)$. This assumption will be made more specific later, when we choose a more realistic version of this concentration dependence than that of simple proportionality.

Next we write an equation for changes in C, the nutrient level in the growth chamber. Here again there are several influences tending to increase or decrease concentration: inflow of stock supply and depletion by bacteria, as well as outflow of nutrients in the effluent. Let us assume that

4. Nutrient depletion occurs continuously as a result of reproduction, so that the rule we specified for culture growth and that for nutrient depletion are essentially going to be the same as before. Here α has the same meaning as in equation (6b).

Our attempt to write the equation for rate of change of nutrient might result in the following:

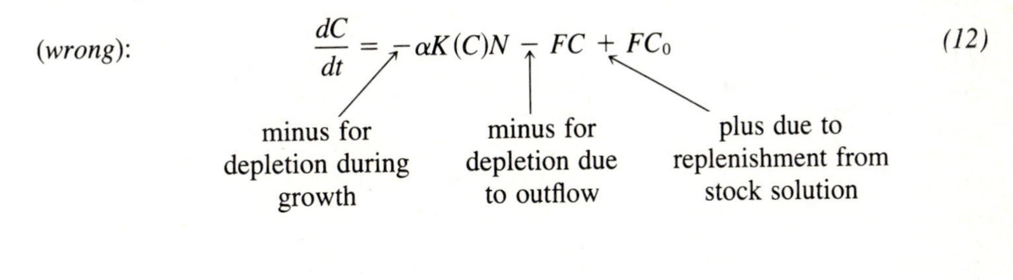

Corrected Version

Equations (11) and (12) are not quite correct, so we now have to uncover mistakes made in writing them. A convenient way of achieving this is by comparing the *dimensions* of terms appearing in an equation. These have to match, clearly, since it would be meaningless to equate quantities not measured in similar units. (For example 10 msec^{-1} can never equal 10 lb.)

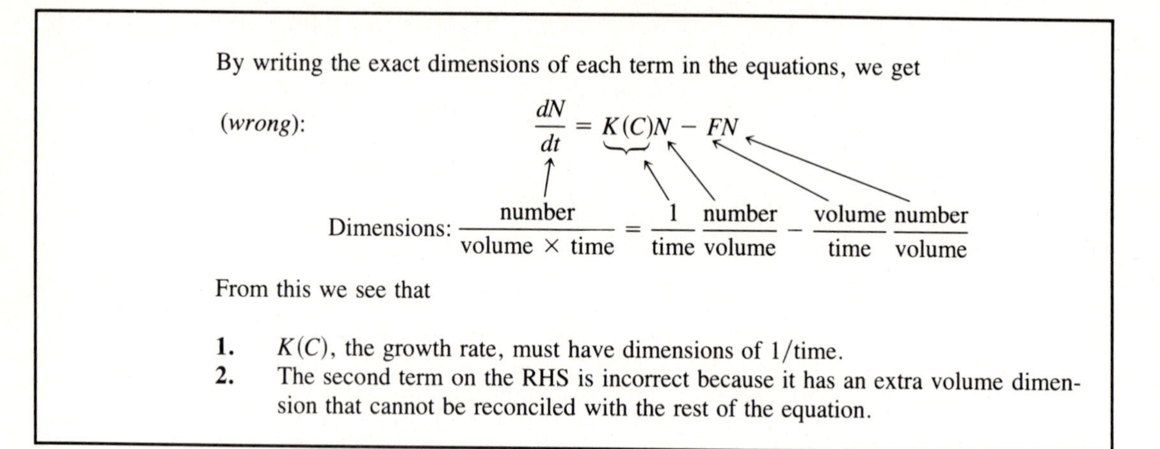

By considering dimensions, we have uncovered an inconsistency in the term FN of equation (11). A way of correcting this problem would be to divide FN by a quantity bearing dimensions of volume. Since the only such parameter available is V, we are led to consider FN/V as the appropriate correction. Notice that FN is the *number* of bacteria that leave per minute, and FN/V is thus the effective *density* of bacteria that leave per minute.

A similar analysis applied to equation (12) reveals that the terms FC and FC_0 should be divided by V (see problem 6). After correcting by the same procedure, we arrive at the following two corrected versions of equations (11) and (12):

$$\frac{dN}{dt} = K(C)N - \frac{FN}{V}, \tag{13a}$$

$$\frac{dC}{dt} = -\alpha K(C)N - \frac{FC}{V} + \frac{FC_0}{V}, \tag{13b}$$

As we have now seen, the analysis of dimensions is often helpful in detecting errors in this stage of modeling. However, the fact that an equation is dimensionally consistent does not always imply that it is correct from physical principles. In problems such as the chemostat, where substances are being transported from one compartment to another, a good starting point for writing an equation is the physical principle that *mass is conserved*. An equivalent conservation statement is that *the number of particles is conserved*. Thus, noting that

NV = number of bacteria in the chamber,

CV = mass nutrient in the chamber,

we obtain a mass balance of the two species by writing

$$\frac{d(NV)}{dt} = K(C)NV - FN, \tag{14a}$$

$$\frac{d(CV)}{dt} = -\alpha K(C)NV - FC + FC_0, \tag{14b}$$

(problem 9). Division by the constant V then leads to the correct set of equations (13a, b).

For further practice at formulating differential-equation models from word problems an excellent source is Henderson West (1983) and other references in the same volume.

4.4 A SATURATING NUTRIENT CONSUMPTION RATE

To add a degree of realism to the model we could at this point incorporate the fact that bacterial growth rates may depend on nutrient availability. For low nutrient abundance, growth rate typically increases with increasing nutrient concentrations. Eventually, when an excess of nutrient is available, its uptake rate and the resultant reproductive rate of the organisms does not continue to increase indefinitely. An appropriate assumption would thus be one that incorporates the effect of a *saturating* dependence. That is, we will assume that

5. The rate of growth increases with nutrient availability only up to some limiting value. (The individual bacterium can only consume nutrient and reproduce at some limited rate.)

One type of mechanism that incorporates this effect is Michaelis-Menten kinetics,

$$K(C) = \frac{K_{\max} C}{K_n + C}, \tag{15}$$

shown in Figure 4.3. Chapter 7 will give a detailed discussion of the molecular events underlying saturating kinetics. For now, it will suffice to note that $K_{\max}$ represents an upper bound for $K(C)$ and that for $C = K_n$, $K(C) = \frac{1}{2}K_{\max}$.

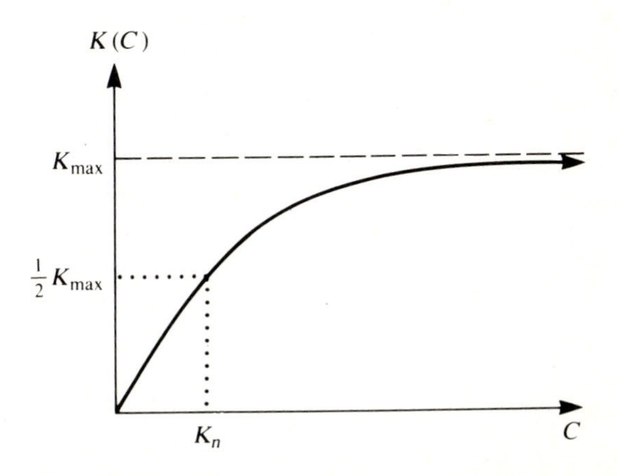

Figure 4.3 *Michaelis-Menten kinetics: Bacterial growth rate and nutrient consumption* K(C) *is assumed to be a saturating function of nutrient concentration. See equation (15).*

Our model equations can now be summarized as follows:

$$\frac{dN}{dt} = \left(\frac{K_{max}C}{K_n + C}\right)N - \frac{FN}{V} \tag{16a}$$

$$\frac{dC}{dt} = -\alpha\left(\frac{K_{max}C}{K_n + C}\right)N - \frac{FC}{V} + \frac{FC_0}{V}. \tag{16b}$$

In understanding these statements we draw a distinction between quantities that are variables, such as N and C and those that are parameters. There is little we can do to control the former *directly*, as they undergo changes in response to their inherent dynamics. However, we may be able to select values of certain parameters (such as F, C_0, and V) that will influence the process. (Other parameters such as K_{max} and K_n depend on the types of bacteria and nutrient medium selected in the experiment.)

It is of interest to determine what happens as certain combinations of parameters are varied over a range of values. Conceivably, an increase in some quantities could just compensate for a decrease in others so that, qualitatively, the system as a whole remains the same. Thus, while a total of six parameters appear in equations (16a,b) the chemostat may indeed have fewer than six *degrees of freedom*. This idea can be made more precise through further *dimensional analysis* of the equations in order to rewrite the model in terms of dimensionless quantities.

4.5 DIMENSIONAL ANALYSIS OF THE EQUATIONS

As shown in Table 4.1, quantities measured in an experiment such as that of the chemostat are specified in terms of certain conventional units. These are, to a great extent, arbitrary. For example a bacterial density of 10^5 cells per liter can be written in any one of the following equivalent ways:

$$\begin{aligned} N &= 10^5 \text{ cells/liter}, \\ &= 1 \text{ (unit of } 10^5 \text{ cells)/liter}, \\ &= 100 \text{ cells/milliliter}, \\ &= N^*\hat{N}. \end{aligned}$$

Here we have distinctly separated the measured quantity into two parts: a number N^*, which has no dimensions, and a quantity $\hat{N}$, which represents the units of measurement and carries the physical dimensions. The values 10^5, 1, 100, and N^* all refer to the same observation but in terms of different scales. As time evolves, N and N^* might change, but $\hat{N}$ is a constant, reflecting the fact that the scale of measurement does not change.

All of the original variables can be expressed similarly, as follows:

measured quantity	=	scalar multiple	×	unit carrying dimensions,
N	=	N^*	×	$\hat{N}$,
C	=	C^*	×	$\hat{C}$,
t	=	t^*	×	τ.

We shall see presently that advantage is gained by expressing the equations in terms of such *dimensionless quantities* as N^*, C^*, and t^*. To do so, we first substitute the expressions $N^*\hat{N}$, $C^*\hat{C}$, $t^*\tau$ for N, C, and t respectively in equations (16a,b) and then exploit the fact that $\hat{N}$, $\hat{C}$, and τ are time-independent constants. We obtain

$$\frac{d(N^*\hat{N})}{d(t^*\tau)} = \left(\frac{K_{max}C^*\hat{C}}{K_n + C^*\hat{C}}\right)N^*\hat{N} - \frac{F}{V}(N^*\hat{N}), \tag{17a}$$

$$\frac{d(C^*\hat{C})}{d(t^*\tau)} = -\alpha\left(\frac{K_{max}C^*\hat{C}}{K_n + C^*\hat{C}}\right)N^*\hat{N} - \frac{FC^*\hat{C}}{V} + \frac{FC_0}{V}. \tag{17b}$$

Now multiply both sides by τ, divide by $\hat{N}$ or $\hat{C}$, and group constant terms together. The result is

$$\frac{dN^*}{dt^*} = \tau K_{max}\left(\frac{C^*}{K_n/\hat{C} + C^*}\right)N^* - \frac{\tau F}{V}N^*, \tag{18a}$$

$$\frac{dC^*}{dt^*} = \left(\frac{-\alpha\tau K_{max}\hat{N}}{\hat{C}}\right)\left(\frac{C^*}{K_n/\hat{C} + C^*}\right)N^* - \frac{\tau F}{V}C^* + \frac{\tau FC_0}{V\hat{C}}. \tag{18b}$$

By making judicious choices for the measuring scales $\hat{N}$, τ, and $\hat{C}$, which are as yet unspecified, we will be able to make the equations look much simpler and contain fewer parameters. Equations (18a,b) suggest a number of scales that are inherent to the chemostat problem. Notice what happens when we choose

$$\tau = \frac{V}{F}, \qquad \hat{C} = K_n, \qquad \hat{N} = \frac{K_n}{\alpha\tau K_{max}}.$$

The equations now can be written in the following form, in which we have dropped the stars for notational convenience.

$$\frac{dN}{dt} = \alpha_1\left(\frac{C}{1 + C}\right)N - N, \tag{19a}$$

$$\frac{dC}{dt} = -\left(\frac{C}{1 + C}\right)N - C + \alpha_2. \tag{19b}$$

The equations contain two dimensionless parameters, α_1 and α_2, in place of the original six (K_n, K_{max}, F, V, C_0, and α). These are related by the following equations:

$$\alpha_1 = (\tau K_{max}) = \frac{VK_{max}}{F},$$

$$\alpha_2 = \frac{\tau FC_0}{V\hat{C}} = \frac{C_0}{K_n}.$$

In problem 8 we discuss the physical meaning of the scales τ, $\hat{C}$, and $\hat{N}$ and of the new dimensionless quantities that appear here.

We have arrived at a dimensionless form of the chemostat model, given by equations (19a,b). Not only are these equations simpler; they are more revealing. By the above we see that only two parameters affect the chemostat. No other choice of τ, $\hat{C}$, and $\hat{N}$ yields less than two parameters (see problem 10). Thus the chemostat has two degrees of freedom.

Equations (19a,b) are nonlinear because of the term $NC/(1 + C)$. Generally this means that there is little hope of finding explicit analytic solutions for $N(t)$ and $C(t)$. However, we can still explore the nature of special classes of solutions, just as we did in the nonlinear difference-equation models. Since we are interested in maintaining a continuous culture in which bacteria and nutrients are present at some fixed densities, we will next determine whether equations (19a,b) admit a steady-state solution of this type.

4.6 STEADY-STATE SOLUTIONS

A steady state is a situation in which the system does not appear to undergo any change. To be more precise, the values of *state variables,* such as bacterial density and nutrient concentration within the chemostat, would be constant at steady state even though individual nutrient particles continue to enter, leave, or be consumed. Setting derivatives equal to zero,

$$\frac{dN}{dt} = 0, \tag{20a}$$

$$\frac{dC}{dt} = 0, \tag{20b}$$

we observe that the quantities on the RHS of equations (19a,b) must be zero at steady state:

$$F(\overline{N}, \overline{C}) = \alpha_1\left(\frac{\overline{C}}{1 + \overline{C}}\right)\overline{N} - \overline{N} = 0, \tag{21a}$$

$$G(\overline{N}, \overline{C}) = -\left(\frac{\overline{C}}{1 + \overline{C}}\right)\overline{N} - \overline{C} + \alpha_2 = 0. \tag{21b}$$

This condition gives two algebraic equations that are readily solved explicitly for $\overline{N}$ and $\overline{C}$.

From (21a) we see that

$$\text{either } \overline{N} = 0 \tag{22a}$$

$$\text{or } \frac{\overline{C}}{1 + \overline{C}} = \frac{1}{\alpha_1}. \tag{22b}$$

After some simplification, (22b) becomes $\overline{C} = 1/(\alpha_1 - 1)$. From equation (21b), if $\overline{N} = 0$ we get $\overline{C} = \alpha_2$; on the other hand, if $\overline{N} \neq 0$, we get

$$\left(\frac{\overline{C}}{1 + \overline{C}}\right)\overline{N} = (\alpha_2 - \overline{C}). \tag{23}$$

Using (22b), we get

$$\overline{N} = \frac{1 + \overline{C}}{\overline{C}}(\alpha_2 - \overline{C}) = \alpha_1(\alpha_2 - \overline{C}). \tag{24}$$

Combining the information in equations (23) and (24) leads to the conclusion that there are two steady states:

$$(\bar{N}_1, \bar{C}_1) = \left(\alpha_1\left(\alpha_2 - \frac{1}{\alpha_1 - 1}\right), \frac{1}{\alpha_1 - 1}\right). \tag{25a}$$

$$(\bar{N}_2, \bar{C}_2) = (0, \alpha_2). \tag{25b}$$

The second solution, $(\bar{N}_2, \bar{C}_2)$, represents a situation that is not of interest to the experimentalists: no bacteria are left, and the nutrient is at the same concentration as the stock solution (remember the meaning of α_2 and the concentration scale to which it refers). The first solution (25a) looks more inspiring, but note that it does not always exist biologically. This depends on the magnitudes of the terms α_1 and α_2. Clearly, if $\alpha_1 < 1$, we get negative values. Since population densities and concentrations must always be positive, negative values would be meaningless in the biological context. The conclusion is that α_1 and α_2 must be such that $\alpha_1 > 1$ and $\alpha_2 > 1/(\alpha_1 - 1)$. In problem 8 we reach certain conclusions about how to adjust the original parameters of the chemostat to satisfy these constraints.

4.7 STABILITY AND LINEARIZATION

Thus far we have arrived at two steady-state solutions that satisfy equations (19a, b). In realistic situations there are always small random disturbances. Thus it is of interest to determine whether such deviations from steady state will lead to drastic changes or will be damped out.

By posing these questions we return once more to stability, a concept that was intimately explored in the context of difference-equation models. In this section we retrace the steps that were carried out in Section 2.7 to reach essentially identical conclusions, namely that, *close to the steady state, the problem can be approximated by a linear one*.

Let us look at a more general setting and take our system of ordinary differential equations to be

$$\frac{dX}{dt} = F(X, Y), \tag{26a}$$

$$\frac{dY}{dt} = G(X, Y), \tag{26b}$$

where F and G are nonlinear functions. We assume that $\bar{X}$ and $\bar{Y}$ are steady-state solutions, i.e., they satisfy

$$F(\bar{X}, \bar{Y}) = G(\bar{X}, \bar{Y}) = 0. \tag{27}$$

Now consider the close-to-steady-state solutions

$$X(t) = \bar{X} + x(t), \tag{28a}$$

$$Y(t) = \bar{Y} + y(t). \tag{28b}$$

Frequently these are called *perturbations* of the steady state. Substituting, we arrive at

$$\frac{d}{dt}(\bar{X} + x) = F(\bar{X} + x, \bar{Y} + y), \tag{29a}$$

$$\frac{d}{dt}(\bar{Y} + y) = G(\bar{X} + x, \bar{Y} + y). \tag{29b}$$

On the left-hand side (LHS) we expand the derivatives and notice that by definition $d\bar{X}/dt = 0$ and $d\bar{Y}/dt = 0$. On the right-hand side (RHS) we now expand F and G in a Taylor series about the point $(\bar{X}, \bar{Y})$, remembering that these are functions of two variables (see Chapter 2 for a more detailed discussion). The result is

$$\frac{dx}{dt} = F(\bar{X}, \bar{Y}) + F_x(\bar{X}, \bar{Y})x + F_y(\bar{X}, \bar{Y})y$$

$$+ \text{ terms of order } x^2, y^2, xy, \text{ and higher}, \tag{30a}$$

$$\frac{dy}{dt} = G(\bar{X}, \bar{Y}) + G_x(\bar{X}, \bar{Y})x + G_y(\bar{X}, \bar{Y})y$$

$$+ \text{ terms of order } x^2, y^2, xy, \text{ and higher}. \tag{30b}$$

where $F_x(\bar{X}, \bar{Y})$ is $\partial F/\partial x$ evaluated at $(\bar{X}, \bar{Y})$, and similarly for F_y, G_x, G_y and other terms.

Again by definition, $F(\bar{X}, \bar{Y}) = 0 = G(\bar{X}, \bar{Y})$, so we are left with

$$\frac{dx}{dt} = a_{11}x + a_{12}y, \tag{31a}$$

$$\frac{dy}{dt} = a_{21}x + a_{22}y, \tag{31b}$$

where the matrix of coefficients

$$\mathbf{A} = \begin{pmatrix} a_{11} & a_{12} \\ a_{21} & a_{22} \end{pmatrix} = \begin{pmatrix} F_x & F_y \\ G_x & G_y \end{pmatrix}_{(\bar{X}, \bar{Y})}. \tag{32}$$

is the Jacobian of the system of equations (26a,b). See Section 2.7 for definition.

To ultimately determine the question of stability, we are thus led to the question of how solutions to equation (31a,b) behave. We shall spend some time on this topic in the next sections. The methods and conclusions bear a strong relation to those we use for systems of difference equations.

4.8 LINEAR ORDINARY DIFFERENTIAL EQUATIONS: A BRIEF REVIEW

In this section we rapidly survey the minimal mathematical background required for analysis of ordinary differential equations (ODEs) such as those encountered in this chapter. For a broader review this section could be supplemented with material from any standard text on ODEs. (See references for suggested sources.)

Terminology

Equivalent representations of the first derivative of a function $y = f(t)$ are dy/dt, y', and $\dot{y}$; also, $d^n y/dt^n = y^{(n)}$ is the nth derivative. An *ordinary differential equation* is any statement linking the values of a function to its derivatives and to a single independent variable; for example,

$$F(t, y, y', y'', \ldots, y^{(n)}) = 0.$$

The *order* of the equation is n, the degree of the highest derivative that appears in the equation. [See equation (33a) for an example of a *first-order* equation and (34) for an example of a *second-order* equation.]

The solution of an ODE is a function $y = f(t)$ that satisfies the equation for every value of the independent variable.

Linear equations have the special form

$$a_0 y^{(n)} + a_1 y^{(n-1)} + \cdots + a_{n-2} y'' + a_{n-1} y' + a_n y = g(t),$$

where no multiples or other nonlinearities in y or its derivatives occur. The coefficients $a_0, a_1, \ldots, a_n$ may be functions of the independent variable t. The case of *constant coefficients* (where $a_0, \ldots, a_n$ are all constants) is of particular importance to stability analysis and can in principle be solved completely.

The equation is called *homogeneous* when the term $g(t) = 0$.

Examples

1. A *second-order, nonhomogeneous, nonlinear* ODE:

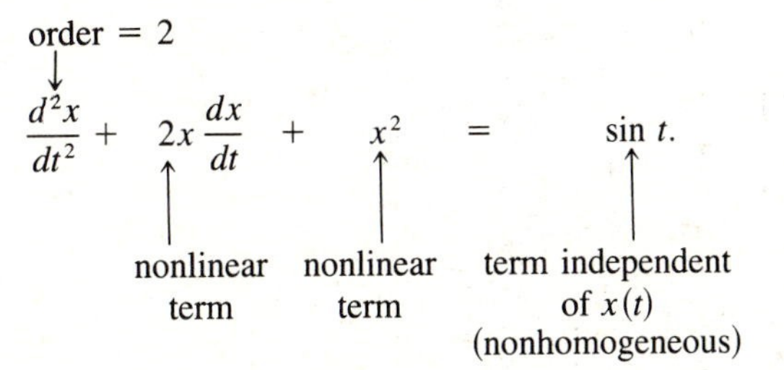

Independent variable = t; unknown function = $x(t)$.

2. A *third-order, linear, homogeneous* ODE with *nonconstant coefficients:*

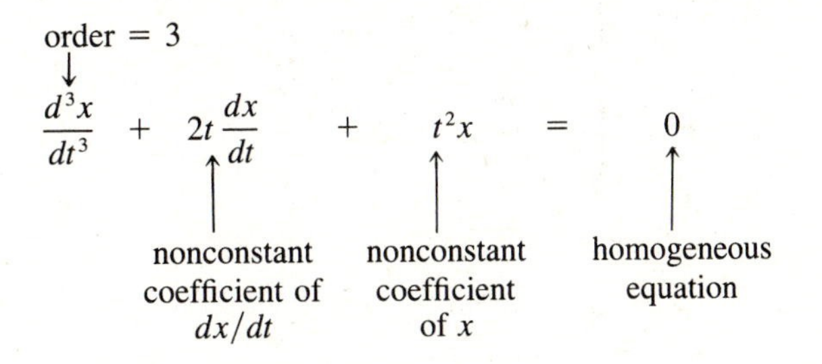

Independent variable = t; unknown function = $x(t)$.

3. An *nth order, linear, constant-coefficient* ODE:

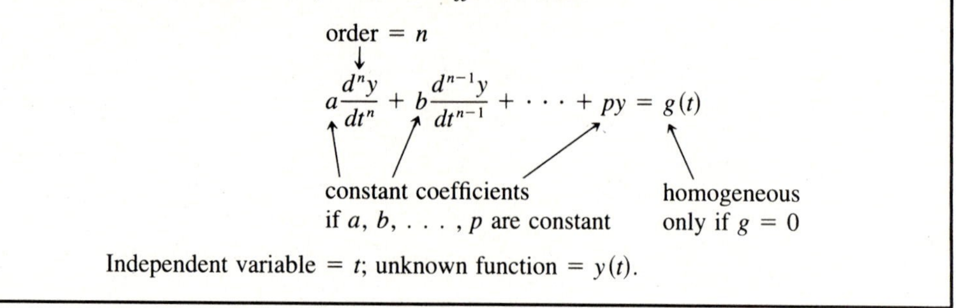

Now we consider only the case of *linear, homogeneous, constant-coefficient* ordinary differential equations. (See box for an explanation of terminology.)

First-Order ODEs

The simplest first-order ODE and its solution are

$$\frac{dx}{dt} = Kx, \tag{33a}$$

$$x(t) = x_0 e^{Kt}. \tag{33b}$$

(See Section 4.1.) The constant x_0 is the initial value of x at time $t = 0$. Below we see that the exponential function is useful in solving higher-order equations.

Second-Order ODEs

Consider the ODE

$$a\frac{d^2x}{dt^2} + b\frac{dx}{dt} + c = 0. \tag{34}$$

(The second derivative implies that the order is 2; a, b, and c are assumed to be constants.) Following a strategy similar to that of Section 1.3, we proceed with the assumption that solutions to equation (33) might work for equation (34). Thus, consider assuming that

$$x(t) = e^{\lambda t} \tag{35}$$

(where λ is a constant) solves equation (34). Then

$$x'(t) = \lambda e^{\lambda t}, \qquad x''(t) = \lambda^2 e^{\lambda t}$$

so by substitution and cancellation of a common factor, we get

$$a\lambda^2 e^{\lambda t} + b\lambda e^{\lambda t} + ce^{\lambda t} = 0,$$

$$a\lambda^2 + b\lambda + c = 0. \tag{36}$$

The latter characteristic equation has two roots called the eigenvalues:

$$\lambda_{1,2} = \frac{-b \pm \sqrt{b^2 - 4ac}}{2a} \tag{37}$$

so the two solutions to equation (34) are

$$x_1(t) = e^{\lambda_1 t}, \qquad x_2(t) = e^{\lambda_2 t}. \tag{38}$$

By the principle of linear superposition (see Chapter 1), if x_1 and x_2 are two solutions to a linear equation such as (34), then any linear combination is also a solution. The general solution is thus

$$x(t) = c_1 e^{\lambda_1 t} + c_2 e^{\lambda_2 t}, \tag{39}$$

where c_1 and c_2 are arbitrary constants determined from other information such as initial conditions. (See problems 17 and 18.)

Exceptional cases occur when

$$\lambda_1 = \lambda_2 = \lambda \qquad \text{(repeated eigenvalues)},$$
$$\lambda = a \pm bi \qquad \text{(complex conjugate eigenvalues)}.$$

The form of the solution is then amended as follows:

1. If $\lambda_1 = \lambda_2 = \lambda$, then

$$x(t) = c_1 e^{\lambda t} + c_2 t e^{\lambda t}.$$

(See problem 19.)

2. If $\lambda_{1,2} = a \pm bi$, then

$$x(t) = e^{at}\{c_1 \cos bt + c_2 \sin bt\}.$$

(See problem 20.)

(See problems or any standard text on differential equations.)

A System of Two First-Order Equations (Elimination Method)

Consider

$$\frac{dx}{dt} = a_{11}x + a_{12}y, \tag{40a}$$

$$\frac{dy}{dt} = a_{21}x + a_{22}y. \tag{40b}$$

This can be reduced by a procedure of elimination (see problem 21) to a single second-order equation in $x(t)$:

$$\frac{d^2x}{dt^2} - \beta\frac{dx}{dt} + \gamma x = 0, \tag{41}$$

where

$$\beta = a_{11} + a_{22},$$
$$\gamma = a_{11}a_{22} - a_{12}a_{21}.$$

By the procedure given in the subsection "Second-Order ODEs" we then find the general solution for $x(t)$ to be

$$x(t) = c_1 e^{\lambda_1 t} + c_2 e^{\lambda_2 t}, \tag{42a}$$

where

$$\lambda_{1,2} = \frac{\beta \pm \sqrt{\beta^2 - 4\gamma}}{2}. \tag{42b}$$

The quantity $\delta = \beta^2 - 4\gamma$ is called the *discriminant*. (When δ is negative, eigenvalues are complex.) $y(t)$ can be found by solving (40a); see problem 21b. (Again, in the cases of complex or multiple eigenvalues, the form of the general solution must be amended as before.)

A System of Two First-Order Equations (Eigenvalue-Eigenvector Method)

We write the system of equations in vector notation as

$$\frac{dx}{dt} = \mathbf{A}\mathbf{x}, \tag{43a}$$

where

$$\mathbf{x} = \begin{pmatrix} x \\ y \end{pmatrix}, \tag{43b}$$

$$\mathbf{A} = \begin{pmatrix} a_{11} & a_{12} \\ a_{21} & a_{22} \end{pmatrix}, \tag{43c}$$

for a 2×2 system. (More generally, $\mathbf{A}$ is a $n \times n$ matrix, and $\mathbf{x}$ is an n-vector when we are dealing with a system of n differential equations in n variables.) The notes matrix multiplication, and $\dfrac{d\mathbf{x}}{dt}$ stands for a vector whose entries are $d\mathbf{x}/dt$, dy/dt.

In the spirit of the example given in equations (33a, b) we shall assume solutions of the form

$$\mathbf{x}(t) = \mathbf{v}e^{\lambda t}. \tag{44}$$

Now $\mathbf{v}$ must be a vector whose entries are independent of time. Using this idea, we substitute (44) into (43a):

$$\frac{d\mathbf{x}}{dt} = \mathbf{v}\frac{d}{dt}e^{\lambda t} = \lambda\mathbf{v}e^{\lambda t} = \mathbf{A}\mathbf{v}e^{\lambda t}.$$

The last equality has to be satisfied if (44) is to be a solution of system (43). Cancelling the common factor $e^{\lambda t}$ results in

$$\mathbf{Av} = \lambda \mathbf{v}. \tag{45a}$$

Note: $\mathbf{v}$ cannot be "cancelled" since $\mathbf{Av}$ stands for matrix multiplication; however, we can rewrite (45a) as

$$\mathbf{Av} - \lambda \mathbf{Iv} = 0, \tag{45b}$$

where $\mathbf{I}$ is the identity matrix ($\mathbf{Iv} = \mathbf{vI} = \mathbf{v}$). Now $\lambda\mathbf{I}$ is also a matrix, namely

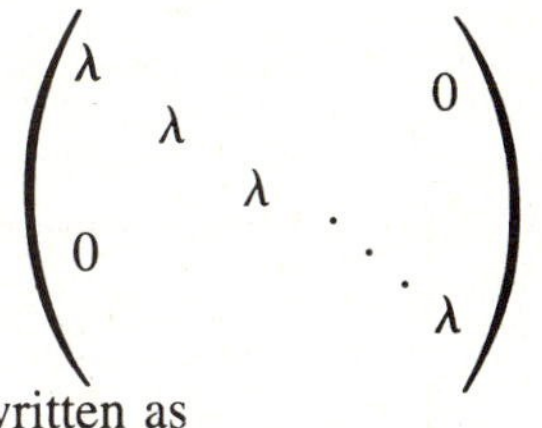

Thus equation (45b) can be written as

$$(\mathbf{A} - \lambda \mathbf{I})\mathbf{v} = 0. \tag{45c}$$

Readers familiar with linear algebra will recognize (45c) as an equation that is satisfied with eigenvalues λ and eigenvectors $\mathbf{v}$ of the matrix $\mathbf{A}$. For other than the trivial solution $\mathbf{v} = 0$, we must have

$$\det(\mathbf{A} - \lambda \mathbf{I}) = 0. \tag{46}$$

When λ satisfies this equation, vectors which satisfy (45c) can be found. These vectors will be nonunique; that is, they will depend on an arbitrary constant because the equations making up the algebraic system (45c) are redundant when (46) is true.

As in the subsection "First-Order ODEs" the eigenvalues of a 2×2 system will always be

$$\lambda_{1,2} = \frac{\beta \pm \sqrt{\delta}}{2},$$

where

$$\beta = \text{trace } \mathbf{A} = a_{11} + a_{22},$$
$$\gamma = \det \mathbf{A} = a_{11}a_{22} - a_{12}a_{21},$$
$$\delta = \text{disc } \mathbf{A} = \beta^2 - 4\gamma.$$

The eigenvectors will be

$$\mathbf{v}_i = \begin{pmatrix} 1 \\ \dfrac{\lambda_i - a_{11}}{a_{12}} \end{pmatrix}, \tag{47}$$

for $a_{12} \neq 0$. (See the examples 1 and 2 in the boxes at the end of this section.) Once the eigenvectors $\mathbf{v}_1$ and $\mathbf{v}_2$ are found, the general solution (provided $\lambda_1 \neq \lambda_2$ and both are real) is given by

$$\mathbf{x}(t) = c_1 \mathbf{v}_1 e^{\lambda_1 t} + c_2 \mathbf{v}_2 e^{\lambda_2 t}. \tag{48}$$

Recall that this is a shorthand version of the following:

$$x_1(t) = c_1 v_{11} e^{\lambda_1 t} + c_2 v_{21} e^{\lambda_2 t}, \tag{49a}$$
$$x_2(t) = c_1 v_{12} e^{\lambda_1 t} + c_2 v_{22} e^{\lambda_2 t}. \tag{49b}$$

Special cases

1. $\lambda_1 = \lambda_2 = \lambda$ (repeated eigenvalues; one eigenvector $\mathbf{v}$):
The form of the general solution must be amended (to allow for two distinct linearly independent parts). See any text on ODEs for details of the method.
2. $\lambda_{1,2} = r \pm ci$ (complex eigenvalues): *(50a)*
This case occurs when disc $\mathbf{A} = \beta^2 - 4\gamma < 0$; then $r = \beta/2$ = real part of λ, $c = \sqrt{\delta}/2$ = imaginary part of λ, and $i = \sqrt{-1}$. Note that complex eigenvalues always come in conjugate pairs.

Complex eigenvalues always have corresponding complex conjugate eigenvectors

$$\mathbf{v}_{1,2} = \mathbf{a} \pm \mathbf{b}i, \tag{50b}$$

where $\mathbf{a}$ and $\mathbf{b}$ are two real constant vectors. The general solution can be expressed in the complex notation

$$\mathbf{x}(t) = c_1(\mathbf{a} + \mathbf{b}i)e^{(r+ci)t} + c_2(\mathbf{a} - \mathbf{b}i)e^{(r-ci)t}. \tag{50c}$$

A real-valued solution can also be constructed by using the identity

$$e^{(r+ci)t} = e^{rt}(\cos ct + i \sin ct). \tag{51}$$

Define

$$\begin{aligned} \mathbf{u}(t) &= e^{rt}(\mathbf{a} \cos ct - \mathbf{b} \sin ct), \\ \mathbf{w}(t) &= e^{rt}(\mathbf{a} \sin ct - \mathbf{b} \cos ct), \end{aligned} \tag{52}$$

It can be shown that each of these parts is itself a real-valued solution, so that the general solution in the case of complex eigenvalues takes the form

$$\mathbf{x}(t) = c_1\mathbf{u}(t) + c_2\mathbf{w}(t). \tag{53}$$

As seen in the previous analysis, complex eigenvalues lead to oscillatory solutions. The imaginary part c governs the frequency of oscillation. The real part r governs the amplitude.

Summary: Solutions of a second-order linear equation such as (34) or (41), or, of a system of first-order equations such as (43), are made up of sums of exponentials or else products of oscillatory functions (sines and cosines) with exponentials. Figure 4.4 illustrates how the three basic ingredients contribute to the character of the solution. In the case of real eigenvalues, if one or both eigenvalues are positive, the solution grows with time. Only if both are negative, as in case 1 does the solution decrease. Furthermore, if complex eigenvalues are found, i.e., if $\lambda = r \pm ci$, their real part r determines whether the amplitude of the oscillation increases ($r > 0$), decreases ($r < 0$), or remains constant ($r = 0$). The complex part c determines the frequency of oscillation.

Figure 4.4 *The three basic ingredients of a solution to a linear system for* $r > 0$ *are (a) a decreasing exponential function* e^{-rt}, *(b) an oscillatory function such as* $\sin t$ *or* $\cos t$, *and (c) an increasing exponential function* e^{rt}. *When eigenvalues are real, the linear combination* $x(t) = c_1e^{\lambda_1 t} + c_2e^{\lambda_2 t}$ *is a decreasing function only when both* λ_1 *and* λ_2 *are negative (case 1). When eigenvalues are complex, the solution is oscillatory with either increasing or decreasing amplitude. (opposite page)*

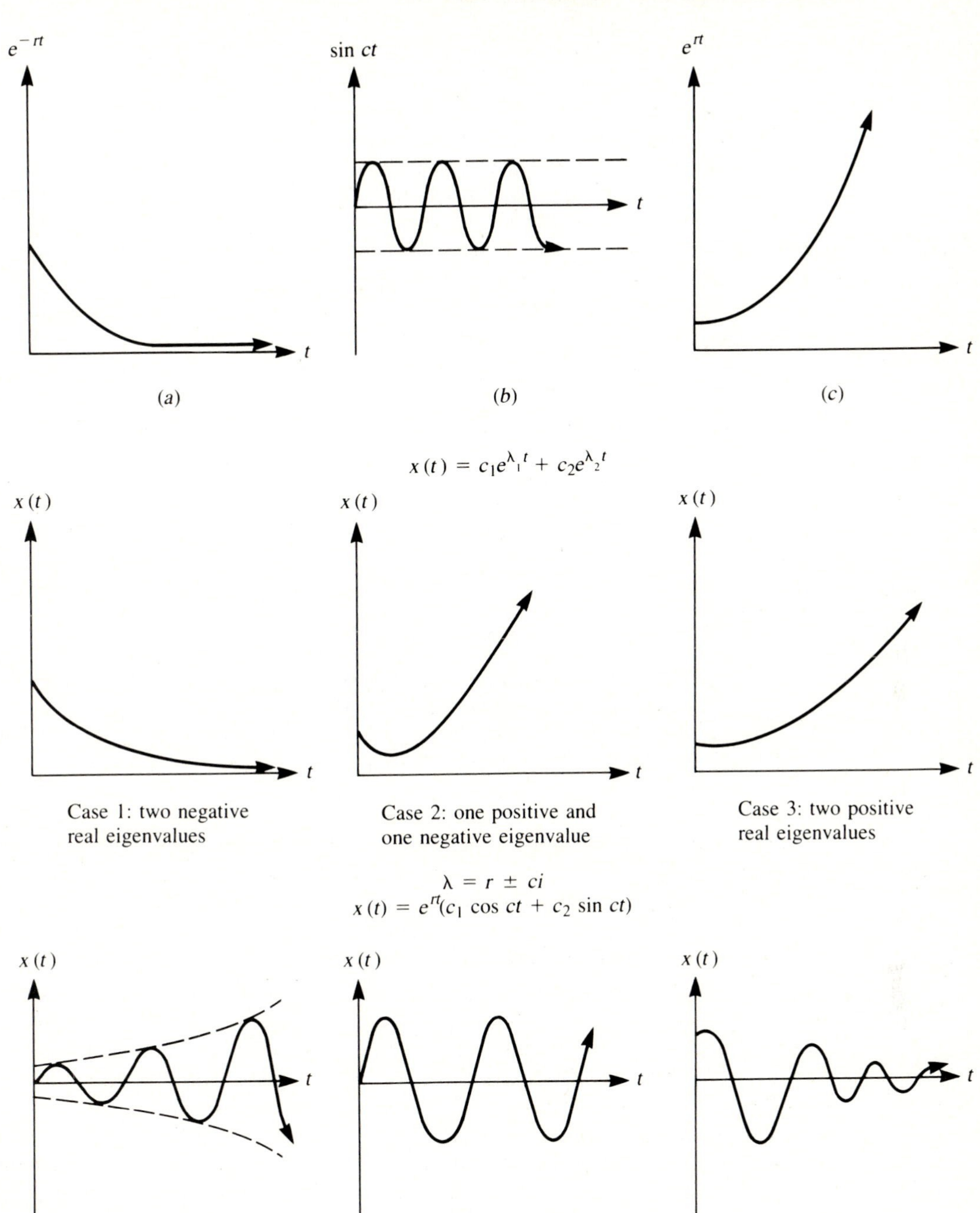

Case 4: complex conjugate eigenvalues, positive real part

Case 5: zero real part

Case 6: negative real part

Remarks

1. From equations (33a, b) it is apparent that a solution to a first-order differential equation involves one arbitrary constant (e.g., z_0) whose value depends on other information, such as the initial conditions of the problem. Similarly, we see that a second-order equation, such as equation (34), will have a general solution containing two arbitrary constants [for example, c_1 and c_2 in equation (39)]. These would again be determined from initial conditions. (See problem 18.)
2. Given a linear equation of nth order,

$$a_0\frac{d^n x}{dt^n} + a_1\frac{d^{n-1}x}{dt^{n-1}} + \cdots + a_n x = 0,$$

the procedure outlined in equations (35) and (36) would lead to a characteristic equation

$$a_0\lambda^n + a_1\lambda^{n-1} + \cdots + a_n = 0,$$

that is, an nth-order polynomial. Generally it is not an easy matter to find the roots of this polynomial (and thus determine what the eigenvalues are) when $n > 2$. We can at best hope to find out something about these eigenvalues. (See Section 6.4 for details.)

3. Nonlinear equations or equations with nonconstant coefficients are not generally solved in a straightforward way, unless they are of special form. Indeed, we cannot always be assured that a solution exists. The mathematical theory that deals with the question of existence and uniqueness of solutions to ODEs can be found in most advanced texts on ordinary differential equations.

Example 1

Solve the following system of equations:

$$\frac{dx_1}{dt} = 3x_1 - x_2, \qquad \frac{dx_2}{dt} = 6x_1 - 4x_2. \tag{54}$$

Solution

Rewrite the equations as

$$\frac{d\mathbf{x}}{dt} = \mathbf{A}\mathbf{x},$$

where
$$\mathbf{x} = \begin{pmatrix} x_1 \\ x_2 \end{pmatrix}, \quad \mathbf{A} = \begin{pmatrix} 3 & -1 \\ 6 & -4 \end{pmatrix}. \tag{55}$$

To find solutions $\mathbf{x}(t) = \mathbf{v}e^{\lambda t}$, we must find the eigenvalues λ and eigenvectors $\mathbf{v}$ of the matrix $\mathbf{A}$. The former are found by setting $\det(\mathbf{A} - \lambda\mathbf{I}) = 0$:

$$\begin{aligned} 0 &= \det(\mathbf{A} - \lambda\mathbf{I}) \\ &= \det\left[\begin{pmatrix} 3 & -1 \\ 6 & -4 \end{pmatrix} - \begin{pmatrix} \lambda & 0 \\ 0 & \lambda \end{pmatrix}\right] \\ &= \det\begin{pmatrix} 3-\lambda & -1 \\ 6 & -4-\lambda \end{pmatrix} \end{aligned}$$

$$= (3 - \lambda)(-4 - \lambda) - (-1)(6)$$
$$= \lambda^2 + \lambda - 12 + 6 = \lambda^2 + \lambda - 6$$
$$= (\lambda - 2)(\lambda + 3)$$

This quadratic equation has the two solutions

$$\lambda_1 = 2 \quad \text{and} \quad \lambda_2 = -3. \tag{56}$$

We now find eigenvectors associated with each eigenvalue by solving $(\mathbf{A} - \lambda\mathbf{I})\mathbf{v} = 0$. Corresponding to λ_1, $\mathbf{v}_1$ must satisfy

$$(\mathbf{A} - \lambda\mathbf{I})\mathbf{v}_1 = \begin{pmatrix} 3 - \lambda_1 & -1 \\ 6 & -4 - \lambda_1 \end{pmatrix}\begin{pmatrix} v_{11} \\ v_{12} \end{pmatrix} = \begin{pmatrix} 0 \\ 0 \end{pmatrix}. \tag{57}$$

Since $\lambda_1 = 2$, the system of equations is

$$\begin{pmatrix} 1 & -1 \\ 6 & -6 \end{pmatrix}\begin{pmatrix} v_{11} \\ v_{12} \end{pmatrix} = \begin{pmatrix} 0 \\ 0 \end{pmatrix},$$

that is,

$$\begin{aligned} v_{11} - v_{12} &= 0, \\ 6v_{11} - 6v_{12} &= 0. \end{aligned} \tag{58}$$

Notice that the equations are redundant. We use one of these, with an arbitrary value for one of the variables (for example, $v_{11} = 1$) to conclude that

$$\mathbf{v}_1 = \begin{pmatrix} 1 \\ 1 \end{pmatrix}, \tag{59}$$

or any constant multiple of (59).

To find $\mathbf{v}_2$, repeat the procedure with the second eigenvalue, $\lambda_2 = -3$. The system of equations is then

$$\begin{pmatrix} 6 & -1 \\ 6 & -1 \end{pmatrix}\begin{pmatrix} v_{21} \\ v_{22} \end{pmatrix} = \begin{pmatrix} 0 \\ 0 \end{pmatrix},$$

that is,

$$\begin{aligned} 6v_{21} - v_{22} &= 0, \\ 6v_{21} - v_{22} &= 0. \end{aligned} \tag{60}$$

Arbitrarily selecting $v_{21} = 1$, we obtain

$$\mathbf{v}_2 = \begin{pmatrix} 1 \\ 6 \end{pmatrix}. \tag{61}$$

It is worth remarking that the computations in equations 57–60, here done to reinforce a concept, can be omitted in practice by using equation (47). We conclude that two solutions to the system of equations are

$$\begin{pmatrix} 1 \\ 6 \end{pmatrix}e^{-3t} \quad \text{and} \quad \begin{pmatrix} 1 \\ 1 \end{pmatrix}e^{2t}. \tag{62}$$

The general solution is thus

$$\mathbf{x}(t) = c_1\begin{pmatrix} 1 \\ 6 \end{pmatrix}e^{-3t} + c_2\begin{pmatrix} 1 \\ 1 \end{pmatrix}e^{2t}. \tag{63}$$

Example 2
Solve the following system of equations:

$$\frac{dx_1}{dt} = x_1 - x_2, \qquad \frac{dx_2}{dt} = x_1 + x_2. \tag{64}$$

Solution
This system is equivalent to

$$\frac{d\mathbf{x}}{dt} = \mathbf{A}\mathbf{x} \qquad \text{where} \quad \mathbf{A} = \begin{pmatrix} 1 & -1 \\ 1 & 1 \end{pmatrix}. \tag{65}$$

The eigenvalues of **A** satisfy the relation

$$\begin{aligned} 0 = \det(\mathbf{A} - \lambda\mathbf{I}) &= \det\begin{pmatrix} 1-\lambda & -1 \\ 1 & 1-\lambda \end{pmatrix} \\ &= (1-\lambda)^2 + 1 \\ &= \lambda^2 - 2\lambda + 2. \end{aligned}$$

Thus

$$\lambda_{1,2} = \frac{2 \pm \sqrt{4-8}}{2} = 1 \pm i. \tag{66}$$

The eigenvector of **A** corresponding to $\lambda_1 = 1 + i$ satisfies

$$\begin{aligned} \begin{pmatrix} 0 \\ 0 \end{pmatrix} &= (\mathbf{A} - \lambda_1\mathbf{I})\mathbf{v}_1 \\ &= \begin{pmatrix} 1-(1+i) & -1 \\ 1 & 1-(1+i) \end{pmatrix}\begin{pmatrix} v_{11} \\ v_{12} \end{pmatrix} \\ &= \begin{pmatrix} -i & -1 \\ 1 & -i \end{pmatrix}\begin{pmatrix} v_{11} \\ v_{12} \end{pmatrix}. \end{aligned} \tag{67}$$

Thus

$$-iv_{11} - v_{12} = 0.$$

Taking arbitrarily $v_{11} = 1$ one obtains $v_{12} = -iv_{11} = -i$. Thus

$$\mathbf{v}_1 = \begin{pmatrix} 1 \\ -i \end{pmatrix} = \begin{pmatrix} 1 \\ 0 \end{pmatrix} - i\begin{pmatrix} 0 \\ 1 \end{pmatrix} = \mathbf{a} - i\mathbf{b}. \tag{68}$$

It follows that $\mathbf{v}_2$ is the complex conjugate of $\mathbf{v}_1$; that is,

$$\mathbf{v}_2 = \begin{pmatrix} 1 \\ i \end{pmatrix} = \begin{pmatrix} 1 \\ 0 \end{pmatrix} + i\begin{pmatrix} 0 \\ 1 \end{pmatrix} = \mathbf{a} + i\mathbf{b}. \tag{69}$$

Note that this can again be obtained directly from equation (47). The complex form of the general solution is

$$\mathbf{x}(t) = c_1\begin{pmatrix} 1 \\ -i \end{pmatrix}e^{(1+i)t} + c_2\begin{pmatrix} 1 \\ i \end{pmatrix}e^{(1-i)t}. \tag{70}$$

Defining

$$\mathbf{u}(t) = e^t\left[\begin{pmatrix} 1 \\ 0 \end{pmatrix}\cos t - \begin{pmatrix} 0 \\ 1 \end{pmatrix}\sin t\right] \tag{71a}$$

$$\mathbf{w}(t) = e^t\left[\begin{pmatrix}1\\0\end{pmatrix}\sin t + \begin{pmatrix}0\\1\end{pmatrix}\cos t\right] \tag{71b}$$

one obtains the real-valued general solution

$$\begin{aligned}\mathbf{x}(t) &= c_1\mathbf{u}(t) + c_2\mathbf{w}(t)\\ &= e^t\left[c_1\begin{pmatrix}\cos t\\ -\sin t\end{pmatrix} + c_2\begin{pmatrix}\sin t\\ \cos t\end{pmatrix}\right].\end{aligned} \tag{72}$$

4.9 WHEN IS A STEADY STATE STABLE?

In Section 4.8 we explored solutions to systems of linear equations such as (31) and concluded that the key quantities were the eigenvalues λ_i given by

$$\lambda_{1,2} = \frac{\beta \pm \sqrt{\beta^2 - 4\gamma}}{2},$$

where $\beta = a_{11} + a_{22}$ and $\gamma = a_{11}a_{22} - a_{12}a_{21}$. We then saw that when λ_1 and λ_2 are real numbers and not equal, the basic "building blocks" for solutions to the system (31) have the time-dependent parts

$$e^{\lambda_1 t} \quad \text{and} \quad e^{\lambda_2 t}. \tag{73}$$

We are now ready to address the main question regarding stability of a steady-state solution that was posed back in Section 4.7, namely whether the small deviations away from steady state (28) will grow larger (instability) or decay (stability). Since these small deviations satisfy a system of linear ordinary differential equations, the answer to this question depends on whether a linear combination of (73) will grow or decline with time.

Consider the following two cases:

1. λ_1, λ_2 are real eigenvalues.
2. λ_1, λ_2 are complex conjugates:

$$\lambda_{1,2} = r \pm ci, \qquad r = \frac{\beta}{2}, \qquad c = \frac{1}{2}(4\gamma - \beta^2)^{1/2}.$$

In case 2 we require that e^{rt} be *decreasing* [see equations (52) and (53)]; that is, r, the real part of λ, must be negative. In case 1 both $e^{\lambda_1 t}$ and $e^{\lambda_2 t}$ must be decreasing. Thus λ_1 and λ_2 should both be negative.

To summarize: in a continuous model, a steady state will be stable provided that eigenvalues of the characteristic equation (associated with the linearized problem) are both negative (if real) or have negative real parts (if complex). That is,

$$\text{Re}\ \lambda_i < 0 \text{ for all } i.$$

In case 2 we see that this criterion is satisfied whenever $\beta < 0$. In case 1 we use the following argument to derive necessary and sufficient conditions. Let

$$\lambda_1 = \frac{\beta + \sqrt{\beta^2 - 4\gamma}}{2},$$

$$\lambda_2 = \frac{\beta - \sqrt{\beta^2 - 4\gamma}}{2}.$$

We want both λ_1 and $\lambda_2 < 0$. For $\lambda_1 < 0$ it is essential that

$$\beta < 0.$$

Notice that this will always make $\lambda_2 < 0$. However, it is also necessary that

$$|\beta| > \sqrt{\beta^2 - 4\gamma}.$$

Otherwise λ_1 would be positive, since the radical would dominate over β. Squaring both sides and rewriting, we see that

$$\beta^2 > \beta^2 - 4\gamma, \qquad \text{or} \qquad 0 > -\gamma,$$

so that

$$\gamma > 0.$$

We conclude that the steady state will be stable provided that the following condition is satisfied

Stability Condition

$$\beta = a_{11} + a_{22} < 0 \tag{74a}$$

$$\gamma = a_{11}a_{22} - a_{12}a_{21} > 0 \tag{74b}$$

Now we rephrase this in the context of Section 4.7:

A steady state $(\bar{X}, \bar{Y})$ of a system of equations

$$\frac{dX}{dt} = F(X, Y), \qquad \frac{dY}{dt} = G(X, Y),$$

will be stable provided

$$F_x(\bar{X}, \bar{Y}) + G_y(\bar{X}, \bar{Y}) < 0, \tag{75a}$$

and

$$F_x(\bar{X}, \bar{Y})G_y(\bar{X}, \bar{Y}) - F_y(\bar{X}, \bar{Y})G_x(\bar{X}, \bar{Y}) > 0. \tag{75b}$$

where the terms are partial derivatives of F and G with respect to X and Y that are evaluated at the steady state.

4.10 STABILITY OF STEADY STATES IN THE CHEMOSTAT

Returning to the chemostat problem, we shall now determine whether $(\bar{N}_1, \bar{C}_1)$ and $(\bar{N}_2, \bar{C}_2)$ are stable steady-states. Define

$$F(N, C) = \alpha_1\left(\frac{C}{1 + C}\right)N - N, \tag{76a}$$

$$G(N, C) = -\left(\frac{C}{1 + C}\right)N - C + \alpha_2. \tag{76b}$$

Then we compute the partial derivatives of F and G and evaluate them at the steady states. In doing the evaluation step it is helpful to note the following:

1. At $(\bar{N}_1, \bar{C}_1)$ we know that $\bar{C}_1/(1 + \bar{C}_1) = 1/\alpha_1$.
2. The derivative of $x/(1 + x)$ is $1/(1 + x)^2$. (You should verify this.)
3. We define

$$A = \frac{\bar{N}_1}{(1 + \bar{C}_1)^2}$$

to avoid carrying this cumbersome expression.

4. We also define

$$B = \frac{\alpha_2}{1 + \alpha_2}$$

to simplify notation for $(\bar{N}_2, \bar{C}_2)$.

We see from Table 4.2 that for the steady state $(\bar{N}_1, \bar{C}_1)$ given by equation (25a)

$$\beta < 0 \qquad \text{and} \qquad \gamma > 0,$$

thus the steady state is stable whenever it exists, that is, whenever $\bar{N}_1$ and $\bar{C}_1$ in (25a) are positive. We also remark that

$$\beta^2 - 4\gamma = (A + 1)^2 - 4A = (A - 1)^2 > 0,$$

Table 4.2 ***Jacobian Coefficients for the Chemostat***

Coefficient in J	*Relevant Expressions*	*Evaluated at* $(\bar{N}_1, \bar{C}_1)$	*Evaluated at* $(\bar{N}_2, \bar{C}_2)$
a_{11}	$F_N = \alpha_1 C/(1 + C) - 1$	0	$\alpha_1 B - 1$
a_{12}	$F_C = \alpha_1 N/(1 + C)^2$	$\alpha_1 A$	0
a_{21}	$G_N = -C/(1 + C)$	$-1/\alpha_1$	$-B$
a_{22}	$G_C = -N/(1 + C)^2 - 1$	$-A - 1$	-1
$\beta = \text{Tr}(J)$	$a_{11} + a_{22}$	$-(A + 1)$	$\alpha_1 B - 2$
$\gamma = \det(J)$	$a_{11}a_{22} - a_{12}a_{21}$	A	$-(\alpha_1 B - 1)$

which means that the eigenvalues of the linearized equations for $(\bar{N}_1, \bar{C}_1)$ are always real. This means that no oscillatory solutions should be anticipated. Problem 10(d) demonstrates that the trivial steady state $(\bar{N}_2, \bar{C}_2)$ is only stable when $(\bar{N}_1, \bar{C}_1)$ is nonexistent.

As a conclusion to the chemostat model, we will interpret the various results so that useful information can be extracted from the mathematical analysis. To summarize our findings, we have determined that a sensibly operating chemostat will always have a stable steady-state solution (25a) with bacteria populating the growth chamber. Recall that this equilibrium can be biologically meaningful provided that α_1 and α_2 satisfy the inequalities

$$\alpha_1 > 1, \tag{77a}$$

$$\alpha_2 > \frac{1}{\alpha_1 - 1}, \tag{77b}$$

where these constraints must be satisfied to prevent negative values of the bacteria population $\bar{N}_1$ and nutrient concentration $\bar{C}_1$. In problem 8 it is shown that, in terms of original parameters appearing in the equations,

$$\alpha_1 = \frac{K_{\max} V}{F}, \tag{78a}$$

$$\alpha_2 = \frac{C_0}{K_n}. \tag{78b}$$

The first condition (77a) is thus equivalent to

$$K_{\max} > \frac{F}{V}. \tag{79}$$

We notice that both sides of this inequality have dimensions 1/time. It is more revealing to rewrite this as

$$\frac{1}{K_{\max}} < \frac{V}{F}. \tag{80}$$

To interpret this, observe that $K_{\max}$ is the maximal bacterial reproduction rate (in the presence of *unlimited* nutrient $dN/dt \simeq K_{\max}N$). Thus $1/K_{\max}$ is proportional to the doubling-time of the bacterial population. V/F is the time it takes to replace the whole volume of fluid in the growth chamber with fresh nutrient medium. Equation (80) reveals that if the bacterial doubling time τ_2 is smaller than the emptying time of the chamber ($\times$ 1/ln 2), the bacteria will be washed out in the efflux faster than they can be renewed by reproduction.

The second inequality (77b) can be rewritten in terms of the steady-state value $\bar{C}_1 = 1/(\alpha_1 - 1)$. When this is done the inequality becomes

$$\frac{C_0}{K_n} > C_1, \quad \text{or} \quad \frac{F}{V}\frac{K_n}{K_{\max} - F/V} < C_0. \tag{81}$$

Since $\hat{C} = K_n$ is the reference concentration used in rendering equations (16a,b) dimensionless, we see that

$$\overline{C} = \hat{C}\overline{C}_1 = K_n\overline{C}_1,$$

is the original dimension-carrying steady state (whose units are mass per unit volume). Thus (81) is equivalent to

$$C_0 > \overline{C}, \tag{82}$$

which summarizes an intuitively obvious result: that the nutrient concentration within the chamber cannot exceed the concentration of the stock solution of nutrients.

4.11 APPLICATIONS TO RELATED PROBLEMS

The ideas that we used in assembling the mathematical description of a chemostat can be applied to numerous related situations, some of which have important clinical implications. In this section we will outline a number of such examples and suggest similar techniques, mostly as problems for independent exploration.

Delivery of Drugs by Continuous Infusion

In many situations drugs that sustain the health of a patient cannot be administered orally but must be injected directly into the circulation. This can be done with serial injections, or in particular instances, using continuous infusion, which delivers some constant level of medication over a prolonged time interval. Recently there has even been an implantable infusion system (a thin disk-like device), which is surgically installed in patients who require long medication treatments. Apparently this reduces incidence of the infection that can arise from external infusion devices while permitting greater mobility for the individual. Two potential applications still in experimental stages are control of diabetes mellitus by insulin infusion and cancer chemotherapy. A team that developed this device, Blackshear et al. (1979), also suggests other applications, such as treatment of thromboembolic disease (a clotting disorder) by heparin, Parkinson's disease by dopamine, and other neurological disorders by hormones that could presumably be delivered directly to a particular site in the body.

Even though the internal infusion pump can be refilled nonsurgically, the fact that it must be implanted to begin with has its drawbacks. However, leaving aside these medical considerations we will now examine how the problem of adjusting and operating such an infusion pump can be clarified by mathematical models similar to one we have just examined.

In the application of cancer chemotherapy, one advantage over conventional methods is that local delivery of the drug permits high local concentrations at the tumor site with fewer systemic side effects. (For example, liver tumors have been treated by infusing via the hepatic artery.) Ideally one would like to be able to calculate in advance the most efficient delivery of drug to be administered (including concentration, flow rate, and so forth). This question can never be answered conclusively unless one has detailed information about the tumor growth rate, the extent to which the drug is effective at killing malignant cells, as well as a host of other complicating effects such as geometry, effect on healthy cells, and so on.

However, to gain some practice with continuous modeling, a reasonable first step is to extract the simplest essential features of this complicated system and think of an idealized caricature, such as that shown in Figure 4.5. For example, as a first step we could assume that the pump, liver, and hepatic artery together behave like a system of interconnected chambers or compartments through which the drug can flow. The tumor cells are restricted to the liver. In this idealization we might assume that (1) the blood bathing a tumor is perfectly mixed and (2) all tumor cells are equally exposed to the drug. This, of course, is a major oversimplification. However, it permits us to define and make statements about two variables:

N = the number of tumor cells per unit blood volume,

C = the number of drug units in circulation per unit blood volume.

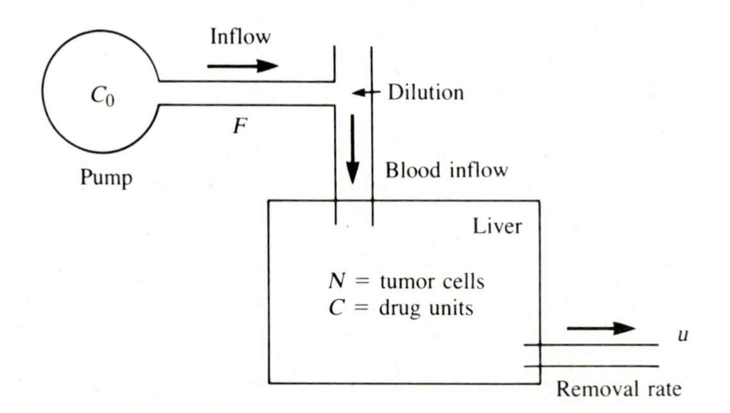

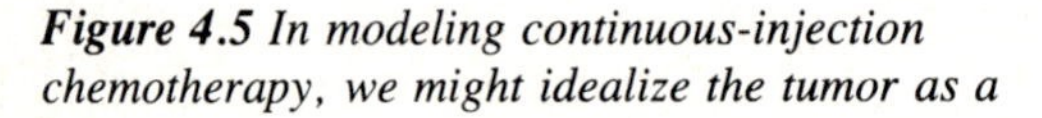

***Figure 4.5** In modeling continuous-injection chemotherapy, we might idealize the tumor as a collection of* N *identical cells that are all equally exposed to C units of drug.*

Several quantities might enter into the formulation of the equations. These include parameters that can be set or varied by the clinician, as well as those that are specific to the patient. We may also wish to define the following:

C_0 = the concentration of drug solution in the chamber (units/volume),

F = the pump flow rate (volume/unit time),

V = volume of the blood in direct contact with tumor area,

u = rate of blood flow away from tumor site,

a = reproductive rate of tumor cells.

Notice that these parameters are somewhat abstract quantities. In general, not all tumor cells will reproduce at the same rate, due to inherent variability and to differences in exposure rates to oxygen and nutrients in the blood. It may be difficult to estimate V and u. On the other hand, the parameters C_0 and F are theoretically known, since they are calibrated by the manufacturer of the pump; a typical value of F is in the range of 1.0–6.0 ml/day).

Keeping in mind the limitations of our assumptions, it is now possible to describe the course of chemotherapy as a system of equations involving the drug C and the tumor cells N. The equations might contain terms as follows:

$$\text{tumor} \qquad \frac{dN}{dt} = \begin{matrix}\text{growth rate}\\ \text{of cells}\end{matrix} - \begin{matrix}\text{drug-induced}\\ \text{death rate}\end{matrix}, \tag{83a}$$

$$\text{drug} \qquad \frac{dc}{dt} = \begin{matrix}\text{rate drug}\\ \text{infused}\end{matrix} - \begin{matrix}\text{rate of uptake}\\ \text{by cells}\end{matrix} - \begin{matrix}\text{rate of removal by}\\ \text{the circulation}\end{matrix}. \tag{83b}$$

Clearly this example is a somewhat transparent analog of the chemostat, because in abstracting the real situation, we made a caricature of the pump-tumor-circulatory system, as shown in Figure 4.5. A few remaining assumptions must still be incorporated in the model which in principle, can now be written fully. Some analysis of this example is suggested in problem 25.

It should be pointed out that this simple model for chemotherapy is somewhat unrealistic, as it treats all tumor cells identically. In most treatments, the drugs administered actually differentiate between cells in different stages of their cell cycle. Models that are of direct clinical applicability must take these features into account. Further reading on this subject in Swan (1984), Newton (1980), and Aroesty et al. (1973) is recommended. A more advanced approach based on the cell cycle will be outlined in a later chapter.

Modeling Glucose-Insulin Kinetics

A second area to which similar mathematical models can be applied is the physiological control of blood glucose by the pancreatic hormone *insulin*. Models that lead to a greater understanding of glucose-insulin dynamics are of potential clinical importance for treating the disorder known as *diabetes mellitus*.

There are two distinct forms of this disease, *juvenile-onset* (type I) and *adult-onset* (type II) diabetes. In the former, the pancreatic cells that produce insulin (islets of Langerhans) are destroyed and an insulin deficiency results. In the latter it appears that the fault lies with mechanisms governing the secretion or response to insulin when blood glucose levels are increased (for example, after a meal).

The chief role of insulin is to mediate the uptake of glucose into cells. When the hormone is deficient or defective, an imbalance of glucose results. Because glucose is a key metabolic substrate in many physiological processes, abnormally low or high levels result in severe problems. One way of treating juvenile-onset diabetes is by continually supplying the body with the insulin that it is incapable of making. There are clearly different ways of achieving this; currently the most widely used is a repeated schedule of daily injections. Other ways of delivering the drug are undergoing development. (Sources dating back several years are given in the references.)

Here we explore a simple model for the way insulin regulates blood glucose levels following a disturbance in the mean concentration. The model, due to Bolie (1960), contains four functions (whose exact forms are unspecified). These terms are meant to depict sources and removal rates of glucose, y, and of insulin, x, in the blood. The equations he gives are (see the definitions given in Table 4.3)

Table 4.3 ***Variables in Bolie's (1960) Model for Insulin-Glucose Regulation***

Symbol	*Definition*	*Dimensions*
V	Extracellular fluid volume	Volume
$\dot{I}$	Rate of insulin injection	Units/time
$\dot{G}$	Rate of glucose injection	Mass/time
$X(t)$	Extracellular insulin concentration	Units/volume
$Y(t)$	Extracellular glucose concentration	Units/volume
$F_1(X)$	Rate of degradation of insulin	(See problem 26)
$F_2(Y)$	Rate of production of insulin	"
$F_3(X, Y)$	Rate of liver accumulation of glucose	"
$F_4(X, Y)$	Rate of tissue utilization of glucose	"

$$\text{insulin:} \quad V\frac{dX}{dt} = \dot{I} - F_1(X) + F_2(Y), \tag{84a}$$

$$\text{glucose:} \quad V\frac{dY}{dt} = \dot{G} - F_3(X, Y) - F_4(X, Y). \tag{84b}$$

The model is a minimal one that omits many of the complicating features; see the original article for a discussion of the validity of the equations. Although the model's applicability is restricted, it serves well as an example on which to illustrate the ideas and techniques of this chapter. (See problem 26.)

An aspect of this model worth noting is that Bolie does not attempt to use experimental data to deduce the forms of the functions F_i, $i = 1, 2, 3, 4$, directly. Rather, he studies the behavior of the system close to the mean steady-state levels when no insulin or glucose is being administered [equations (84a, b) where $\dot{I} = \dot{G} = 0$ and $dX/dt = dY/dt = 0$]. He deduces values of the coefficients a_{11}, a_{12}, a_{21}, and a_{22} in a Jacobian of (84) by looking at empirical data for physiological responses to small disturbances. The approach is rather like that of the plant-herbivore model discussed in Section 3.5. His article is particularly suitable for independent reading and class presentation as it combines mathematical ideas with consideration of empirical results.

Equally accessible are contemporary articles by Ackerman et al. (1965, 1969), Gatewood et al. (1970), and Segre et al. (1973) in which linear models of the release of hormone and removal rate of both substances are presented and compared to data. An excellent recent summary of this literature and of the topic in general is given by Swan (1984, Chap. 3) whose approach to the problem is based on optimal control theory.

Aside from a multitude of large-scale simulation models that we will not dwell on here, more recent models have incorporated nonlinear kinetics (Bellomo et al, 1982) and much greater attention to the details of the physiology. Landahl and Grodsky (1982) give a model for insulin release in which they describe the spike-like pattern of insulin secretion in response to a stepped-up glucose concentration stimulus. Their model consists of four coupled ordinary differential equations. The same phenomenon has also been treated elsewhere using a partial differential equation model (for example, Grodsky, 1972; Hagander et al., 1978). These papers would be accessible to somewhat more advanced readers.

Compartmental Analysis

Physiologists are often interested in following the distribution of biological substances in the body. For clinical medicine the rate of uptake of drugs by different tissues or organs is of great importance in determining an optimal regime of medication. Other substances of natural origin, such as hormones, metabolic substrates, lipoproteins, and peptides, have complex patterns of distribution. These are also studied by related techniques that frequently involve radiolabeled tracers: the substance of interest is radioactively labeled and introduced into the blood (for example, by an injection at $t = 0$). Its concentration in the blood can then be ascertained by withdrawing successive samples at $t = t_1, t_2, \ldots, t_n$; these samples are analyzed for amount of radioactivity remaining. (Generally, it is *not* possible to measure concentrations in tissues other than blood.)

Questions of interest to a physiologist might be:

1. At what rate is the substance taken up and released by the tissues?
2. At what rate is the substance degraded or eliminated altogether from the circulation (for example, by the kidney) or from tissue (for example, by biochemical degradation)?

A common approach for modeling such processes is *compartmental analysis:* the body is described as a set of interconnected, well-mixed compartments (see Figure 4.6b) that exchange the substance and degrade it by simple linear kinetics. One of the most elementary models is that of a two-compartment system, where pool 1 is the circulatory system from which measurements are made and pool 2 consists of all other relevant tissues, not necessarily a single organ or physiological entity. The goal is then to use the data from pool 1 to make deductions about the magnitude of the exchange L_{ij} and degradation D_i from each pool.

To proceed, we define the following parameters:

m_1 = mass in pool 1,

m_2 = mass in pool 2,

V_1 = volume of pool 1,

V_2 = volume of pool 2,

x_1 = mass per unit volume in pool 1,

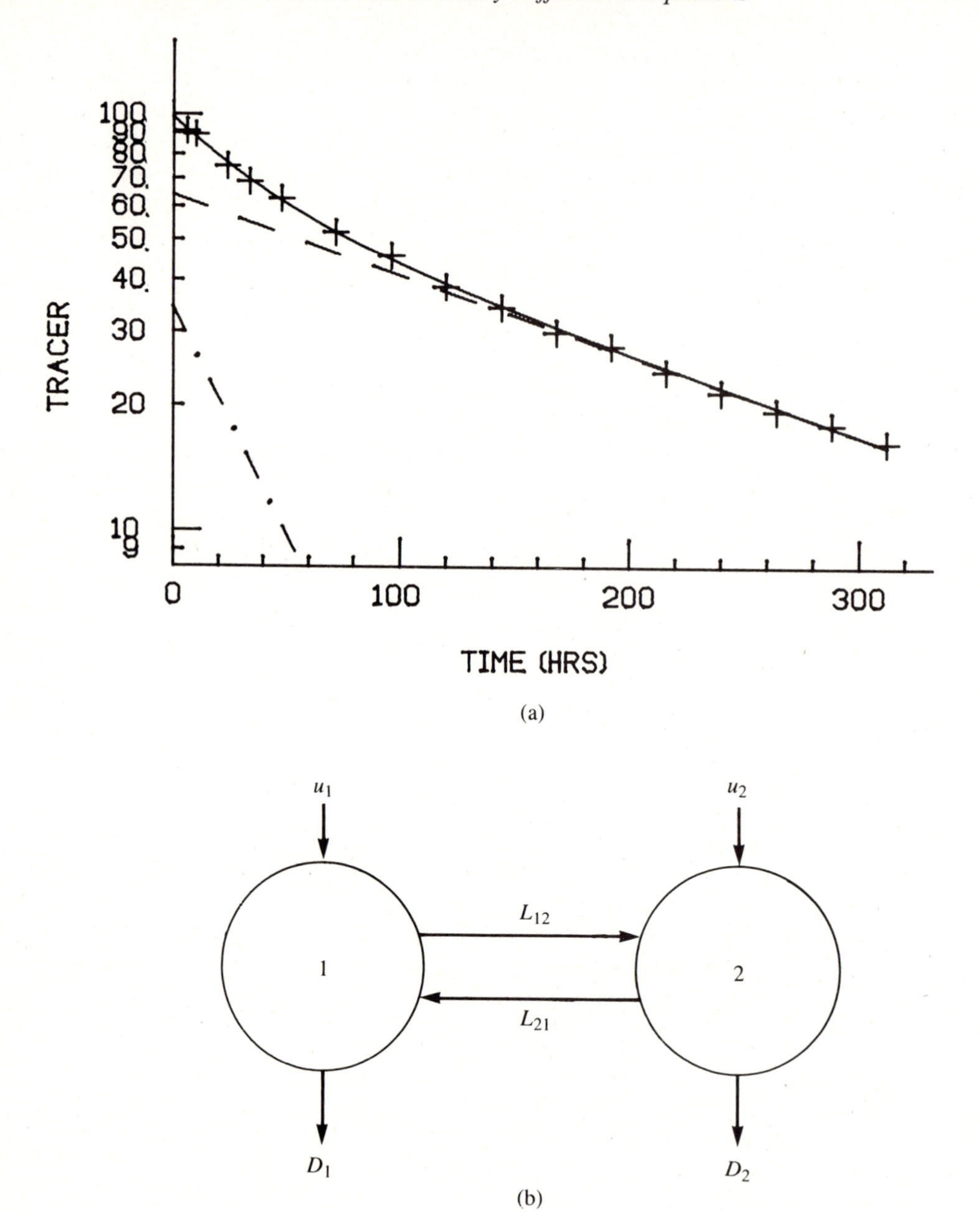

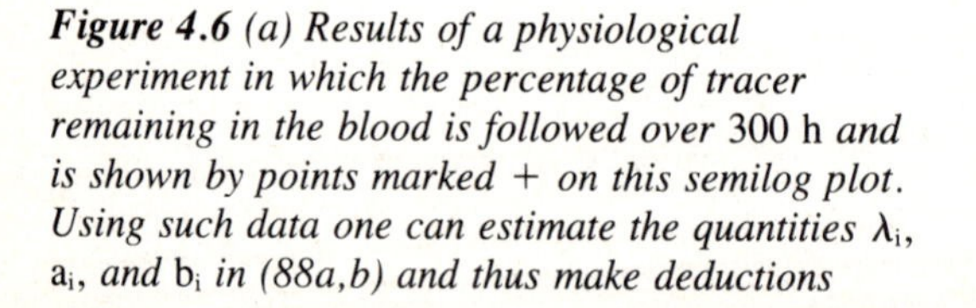
Figure 4.6 *(a) Results of a physiological experiment in which the percentage of tracer remaining in the blood is followed over 300 h and is shown by points marked + on this semilog plot. Using such data one can estimate the quantities λ_i, a_i, and b_i in (88a,b) and thus make deductions about the distribution of the substance in the body. Dotted-dashed line: $a_1e^{-\lambda_1 t}$; dashed line: $a_2e^{-\lambda_2 t}$; and solid line, $a_1e^{-\lambda_1 t} + a_2e^{-\lambda_2 t}$. Here $\lambda_2 < \lambda_1$. (b) A two-compartment model with exchange rates L_{ij}, degradation rates D_i and rates of injection u_1 and u_2. $u_1 = u_2 = 0$ (see text).*

x_2 = mass per unit volume in pool 2,

L_{ij} = exchange from pool i to pool j,

D_j = degradation from pool j,

u_j = rate of injection of substance into pool j.

Note that L_{ij} and D_j have units of 1/time, while u_j has units of mass/time. A linear model would then lead to the following mass balance equations:

$$\frac{dm_1}{dt} = -L_{12}m_1 + L_{21}m_2 - D_1m_1 + u_1, \tag{85a}$$

$$\frac{dm_2}{dt} = L_{12}m_1 - L_{21}m_2 - D_2m_2 + u_2. \tag{85b}$$

In problem 30 we show that this can be rewritten in the form

$$\frac{dx_1}{dt} = -K_1x_1 + K_{21}x_2 + w_1, \tag{86a}$$

$$\frac{dx_2}{dt} = K_{12}x_1 - K_2x_2 + w_2, \tag{86b}$$

where

$$K_1 = L_{12} + D_1, \qquad K_2 = L_{21} + D_2,$$

$$K_{21} = \frac{L_{21}V_2}{V_1}, \qquad K_{12} = \frac{L_{12}V_1}{V_2},$$

$$w_1 = \frac{u_1}{V_1}, \qquad w_2 = \frac{u_2}{V_2}.$$

Note that now coefficients are corrected by terms that account for effects of dilution since compartment sizes are not necessarily the same. This illustrates why equations should proceed from mass balance rather than from concentration balance.

Now suppose that a mass M_0 of substance is introduced into the bloodstream by a *bolus injection* (i.e. all at one time, say at $t = 0$). Assuming that it is rapidly mixed in the circulation, we may take

$$m_1(0) = m_0, \qquad m_2(0) = 0, \qquad u_1 = u_2 = 0. \tag{87}$$

Then equations (86a,b) are readily solved since they are linear and we obtain

$$\mathbf{x} = c_1\mathbf{v}_1e^{-\lambda_1 t} + c_2\mathbf{v}_2e^{-\lambda_2 t} \tag{88a}$$

where

$$\mathbf{x} = \begin{pmatrix} x_1(t) \\ x_2(t) \end{pmatrix} \quad \text{and} \quad c_i\mathbf{v}_i = \begin{pmatrix} a_i \\ b_i \end{pmatrix}. \tag{88b}$$

Note that the exponents are negative because substance is being removed. In problems 30 through 32 we discuss how, by fitting such solutions to data for $x_1(t)$ (concentration in the blood), we can gain appreciation for the rates of exchange and degradation in the body. A thorough treatment of this topic is to be found in Rubinow (1975).

PROBLEMS*

1. Discuss the deductions made by Malthus (1798) regarding the fate of humanity. Can the model in equation (1b) give long-range predictions of human population growth? Malthus assumed that food supplies increase at a linear rate at most, so that eventually their consumption would exceed their renewal. Comment on the validity of his model.

2. Determine whether Crichton's statement (The epigram at the beginning of this chapter. See p. 115) is true. Consider that the average mass of an *E. coli* bacterium is 10^{-12} gm and that the mass of the earth is 5.9763×10^{24} kg.

3. Show that for a decaying population

$$dN/dt = -KN \qquad (K > 0)$$

the time at which only half of the original population remains (the half-life) is

$$\tau_{1/2} = \frac{\ln 2}{K}.$$

4. Consider a bacterial population whose growth rate is $dN/dt = K(t)N$. Show that

$$N(t) = N_0 \exp\left(\int_0^t K\, ds\right).$$

5. Below we further analyze the nutrient-depletion model given in Section 4.1.

(a) Show that the equation

$$\frac{dN}{dt} = \kappa(C_0 - \alpha N)N$$

can be written in the form

$$\frac{dN}{dt} = r\left(1 - \frac{N}{B}\right)N,$$

where $r = C_0\kappa$ and $B = C_0/\alpha$. This is called a logistic equation. Interpret r and B.

(b) Show that the equation can be written

$$\frac{dN}{(1 - N/B)N} = r\, dt,$$

and integrate both sides.

(c) Rearrange the equation in (b) to show that the solution thereby obtained is

$$N(t) = \frac{N_0 B}{N_0 + (B - N_0)e^{-rt}}.$$

(d) Show that for $t \to \infty$ the population approaches the density B. Also show that if N_0 is very small, the population initially appears to grow exponentially at the rate r.

*Problems preceded by an asterisk are especially challenging.

(e) Interpret the results in terms of the original parameters of the bacterial model.
(f) Find the values of B, N_0, and r in the curve that Gause (1934) fit to the growth of the yeast *Schizosaccharomyces kephir* (see caption of Figure 4.1c).

In problems 6 through 12 we explore certain details in the chemostat model.

6. **(a)** Analyze the dimensions of terms in equation 12 and show that an inconsistency is corrected by changing the terms FC and FC_0.
(b) What are the physical dimensions of the constant α?

7. Michaelis-Menten kinetics were selected for the nutrient-dependent bacterial growth rate in Section 4.4.
(a) Show that if $K(C)$ is given by equation (15) a half-maximal growth rate is attained when the nutrient concentration is $C = K_n$.
(b) Suppose instead we assume that $K(C) = K_m C$, where K_m is a constant. How would this change the steady state $(\overline{N}_1, \overline{C}_1)$?
(c) Determine whether the steady state found in part (b) would be stable.

8. **(a)** By using dimensional analysis, we showed that equations (16a,b) can be rescaled into the dimensionless set of equations (19a,b). What are the physical meanings of the scales chosen and of the dimensionless parameters α_1 and α_2?
(b) Interpret the conditions on α_1, α_2 (given at the end of Section 4.6) in terms of the original chemostat parameters.

9. **(a)** Show that each term in equation (14b) has units of (number of bacteria) $(\text{time})^{-1}$.
(b) Similarly, show that each of the terms in equation (14a) has dimensions of (nutrient mass)$(\text{time})^{-1}$.

10. It is usually possible to render dimensionless a set of equations in more than one way. For example, consider the following choice of time unit and concentration unit:

$$\tau = \frac{1}{K_{\max}}, \qquad \hat{C} = \frac{\tau F C_0}{V},$$

where $\hat{N}$ is as before.
(a) Determine what would then be the dimensionless set of equations obtained from (18a, b).
(b) Interpret the meanings of the above quantities τ and $\hat{C}$ and of the new dimensionless parameters in your equations. How many such dimensionless parameters do you get, and how are they related to α_1 and α_2 in equations (19a, b)?
(c) Write the stability conditions for the chemostat in terms of new parameters. Determine whether or not this leads to the same conditions on $K_{\max}$, V, F, C_0, K_n, and so forth.
(d) Show that $(\overline{N}_2, \overline{C}_2)$ is stable only when $(\overline{N}_1, \overline{C}_1)$ is not.

11. In industrial applications, one wants not only to ensure that the steady state $(\overline{N}_1, \overline{C}_1)$ given by equations (25a, b) exists, but also to increase the yield of bacteria, $\overline{N}_1$. Given that one can in principle adjust such parameters as V, F, and C_0, how could this be done?

12. In this question we deal with a number of variants of the chemostat.
- **(a)** How would you expect the model to differ if there were two growth-limiting nutrients?
- **(b)** Suppose that at high densities bacteria start secreting a chemical that inhibits their own growth. How would you model this situation?
- **(c)** In certain cases two (or more) bacterial species are kept in the same chemostat and compete for a common nutrient. Suggest a model for such competition experiments.

13. Consider equation (8) for the growth of a microorganism in a nutrient-limiting environment.
- **(a)** By making appropriate choices for units of measurement $\hat{N}$ and τ (for time), bring the equation to dimensionless form.
- **(b)** What are the steady states of the equation?
- **(c)** Determine the stability of these steady states by linearizing the equation about the steady states obtained in (b).
- **(d)** Verify that your results agree with the exact solution given by equation (10).

14. In the hypothetical growth chamber shown here, the microorganisms (density $N(t)$) and their food supply are kept in a chamber separated by a semipermeable membrane from a reservoir containing the stock nutrient whose concentration $[C_0 > C(t)]$ is assumed to be fixed. Nutrient can pass across the membrane by a process of diffusion at a rate proportional to the concentration difference. The microorganisms have mortality μ.

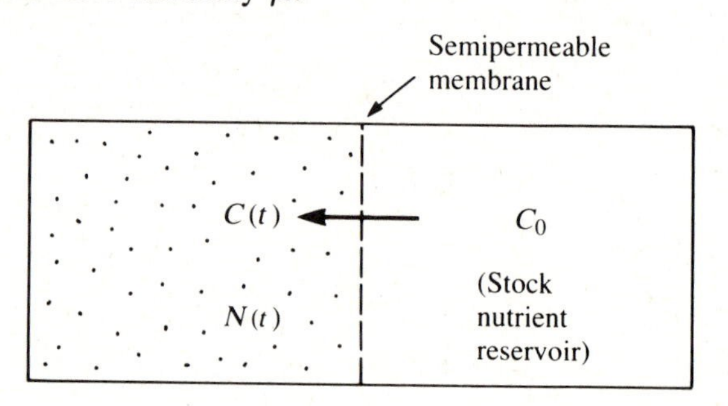

Figure for problem 14.

- **(a)** Explain the following equations:

$$\frac{dN}{dt} = N\left(\frac{K_{\max} C}{K_n + C}\right) - \mu N,$$

$$\frac{dC}{dt} = D[C_0 - C(t)] - \alpha N \frac{K_{\max} C}{K_n + C}.$$

- **(b)** Determine the dimensions of all the quantities in (a).

(c) Bring the equations to a dimensionless form.
(d) Find all steady states.
(e) Carry out a stability analysis and find the constraints that parameters must satisfy to ensure stability of the nontrivial steady state.

Problems 15 through 24 deal with ODEs and techniques discussed in Section 4.8.

15. Classify the following ordinary differential equations by determining whether they are linear, what their order is, whether they are homogeneous, and whether their coefficients are constant.

(a) $(\sin x)y'' + \cos x = 0.$

(b) $y'' + y^2 = 2y'.$

(c) $\dfrac{d^3y}{dt^3} + \dfrac{2dy}{dt} = \sin y.$

(d) $\dfrac{d}{dt}(y^2 + 2y) = y.$

(e) $\dfrac{d^2y}{dt^2} + 2\dfrac{dy}{dt} + 3y = e^t + e^{-t}.$

(f) $\dfrac{dy}{dt} = \dfrac{1}{1+y}.$

(g) $\dfrac{dy}{dx} = \dfrac{1}{1+x}.$

(h) $\dfrac{d^5y}{dx^5} = x^6 + 5x + 6.$

(i) $t\dfrac{dy}{dt} + ty = 1.$

16. Find the steady states of the following systems of equations, and determine the Jacobian of the system for these steady states:

(a) $\dfrac{dx}{dt} = x^2 - y^2,$ $\dfrac{dy}{dt} = x(1 - y).$

(b) $\dfrac{dx}{dt} = y - xy,$ $\dfrac{dy}{dt} = xy.$

(c) $\dfrac{dx}{dt} = x - x^2 - xy,$ $\dfrac{dy}{dt} = y(1 - y).$

(d) $\dfrac{dx}{dt} = x - xy,$ $\dfrac{dy}{dt} = xy - y.$

17. Consider the equation

$$\frac{d^2x}{dt^2} - 2\frac{dx}{dt} - 3x = 0.$$

(a) Show that $x_1(t) = e^{3t}$ and $x_2(t) = e^{-t}$ are two solutions.
(b) Show that $x(t) = c_1x_1(t) + c_2x_2(t)$ is also a solution.

18. The differential equation

$$\frac{d^2x}{dt^2} + 3\frac{dx}{dt} + 2x = 0$$

has the general solution

$$x(t) = c_1e^{-t} + c_2e^{-2t}.$$

If we are told that, when $t = 0$, $x(0) = 1$ and its derivative $x'(0) = 1$, we can determine c_1 and c_2 by solving the equations

$$x(0) = c_1 e^{-0} + c_2 e^{(-2)(0)} = c_1(1) + c_2(1) = 1,$$
$$x'(0) = c_1(-1)e^{-0} + c_2(-2)e^{-0} = -c_1 - 2c_2 = 1.$$

(a) Find the values of c_1 and c_2 by solving the above.

For questions (b) through (e) find the solution of the differential equation subject to the specified initial condition.

(b) $y' = 10y$; $y(0) = 0.001$.
(c) $y'' - 3y' - 4y = 0$; $y(0) = 0$, $y'(0) = 1$.
(d) $y'' - 9y = 0$; $y(0) = 5$, $y'(0) = 0$.
(e) $y'' - 5y' = 0$; $y(0) = 1$, $y'(0) = 2$.

19. When $\lambda_1 = \lambda_2 = \lambda$, one solution of equation (34) is

$$x_1(t) = e^{\lambda t}.$$

To find another solution (that is not just a constant multiple but is *linearly independent* of the above solution), consider the assumption

$$x_2(t) = v(t)e^{\lambda t}.$$

Find first and second derivatives, substitute into equation (34), and show that $v(t) = t$. Conclude that the general solution is

$$x(t) = c_1 e^{\lambda t} + c_2 t e^{\lambda t}.$$

20. If $\lambda_{1,2} = a \pm bi$ are complex conjugate eigenvalues, the complex form of the solutions is

$$x_1(t) = e^{(a+bi)t}, \qquad x_2(t) = e^{(a-bi)t}.$$

Use the identity

$$e^{(a+bi)t} = e^{at}(\cos bt + i \sin bt)$$

and consider

$$u(t) = \frac{x_1(t) + x_2(t)}{2}, \qquad w(t) = \frac{x_1(t) - x_2(t)}{2i}.$$

Reason that these are also solutions by linear superposition and thus show that a real-valued solution (as in Section 4.8) can be defined.

21. **(a)** Show that system (40a, b) can be reduced to equation (41) by eliminating one variable. [*Hint:* differentiate both sides of (40a) first, and then make two other substitutions.]

(b) Once $x(t)$ is found [see equation (42a)], $y(t)$ can be found from (40a) by setting

$$y(t) = \frac{1}{a_{12}}\left(\frac{dx}{dt} - a_{11}x\right) \qquad (a_{12} \neq 0).$$

Determine $y(t)$ in terms of the expressions in (42a).

(c) Conclude that in vector form the solution can be written as

$$\begin{pmatrix} x(t) \\ y(t) \end{pmatrix} = c_1 \begin{pmatrix} 1 \\ \dfrac{\lambda_1 - a_{11}}{a_{12}} \end{pmatrix} e^{\lambda_1 t} + c_2 \begin{pmatrix} 1 \\ \dfrac{\lambda_2 - a_{11}}{a_{12}} \end{pmatrix} e^{\lambda_2 t}.$$

22. In the following exercises find the general solution to the system of equations $d\mathbf{x}/dt = \mathbf{A}\mathbf{x}$, where the matrix $\mathbf{A}$ is as follows:

(a) $\mathbf{A} = \begin{pmatrix} -1 & 0 \\ 0 & 1 \end{pmatrix}$. **(d)** $\mathbf{A} = \begin{pmatrix} -1 & 4 \\ -2 & 5 \end{pmatrix}$.

(b) $\mathbf{A} = \begin{pmatrix} 3 & 1 \\ 1 & 3 \end{pmatrix}$. **(e)** $\mathbf{A} = \begin{pmatrix} 2 & -3 \\ 1 & -2 \end{pmatrix}$.

(c) $\mathbf{A} = \begin{pmatrix} -2 & 7 \\ 2 & 3 \end{pmatrix}$. **(f)** $\mathbf{A} = \begin{pmatrix} -4 & 1 \\ 3 & 0 \end{pmatrix}$.

23. For problem 22(a–f) write out the system of equations

$$\frac{dx_1}{dt} = a_{11}x_1 + a_{12}x_2, \qquad \frac{dx_2}{dt} = a_{21}x_1 + a_{22}x_2.$$

Then eliminate one variable to arrive at a single second-order equation, using the method given in problem 21. Find the characteristic equation and solve for the eigenvalues of the equation. Find solutions for $x_1(t)$ and $x_2(t)$.

24. Consider the nth-order linear ordinary differential equation

$$y^{(n)} + a_1 y^{(n-1)} + \cdots + a_{n-1} y' + a_n y = 0.$$

Show that by assuming solutions of the form

$$y(t) = Ce^{\lambda t}$$

one obtains a characteristic equation that is an nth-order polynomial.

25. In this problem we write equations to model the continuous chemotherapy described in Section 4.11.

(a) Assume that the effect of the drug on mortality of tumor cells is given by Michaelis-Menten kinetics [as in equation (15)] and that the drug is removed from the site at the rate u. Suggest equations for the tumor cell population N and the drug concentration C, assuming that tumor cells grow exponentially.

(b) Carry out dimensional analysis of your equations and indicate which dimensionless combinations of parameters are important.

(c) Determine whether the system admits steady-state solutions, and if so, what their stability properties are.

(d) Interpret your results in the biological context.

(e) A solid tumor usually grows at a declining rate because its interior has no access to oxygen and other necessary substances that the circulation supplies. This has been modeled empirically by the Gompertz growth law,

$$\frac{dN}{dt} = \gamma N \qquad \text{where} \qquad \frac{d\gamma}{dt} = -\alpha\gamma.$$

γ is the effective tumor growth rate, which will decrease exponentially by this assumption. Show that equivalent ways of writing this are

$$\frac{dN}{dt} = \gamma_0 e^{-\alpha t} N = (-\alpha \ln N)N.$$

[*Hint:* use the fact that

$$\frac{1}{N}\frac{dN}{dt} = \frac{d}{dt}(\ln N)\Bigg].$$

(f) Use the Gompertz law to make the model equations more realistic. Assume that the drug causes an increase in α as well as greater tumor mortality.

26. *Insulin-glucose regulation*. Equations (84a,b) due to Bolie (1960) are a simple model for insulin-mediated glucose homeostasis. The following questions are a guide to investigating this model.

(a) When neither component is injected in normal, healthy individuals, the blood glucose and insulin levels are regulated to within fairly restricted concentration ranges. What does this imply about equations (84a, b)?

(b) Explain the appearance of the factor V on the LHS of equations (84a, b). What are the dimensions of the functions F_1, F_2, F_3, and F_4?

(c) Bolie defines X_0 and Y_0 as the mean equilibrium levels of insulin and glucose when none is being injected into the body. What equations do X_0 and Y_0 satisfy?

(d) Consider the following four parameters:

$$\alpha = \frac{1}{V}\left(\frac{\partial F_1}{\partial X}\right), \qquad \beta = \frac{1}{V}\left(\frac{\partial F_2}{\partial Y}\right),$$

$$\gamma = \frac{1}{V}\left(\frac{\partial F_3}{\partial X} + \frac{\partial F_4}{\partial X}\right), \qquad \delta = \frac{1}{V}\left(\frac{\partial F_3}{\partial Y} + \frac{\partial F_4}{\partial Y}\right)$$

[Partial derivatives are evaluated at (X_0, Y_0).]

Interpret what these represent and comment on the fact that these are assumed to be positive constants.

(e) Suppose that $\dot{I} = \dot{G} = 0$, but that at time $t = 0$ a rapid ingestion of glucose followed by a single insulin injection changes the internal concentrations to

$$X = X_0 + x', \qquad Y = Y_0 + y',$$

where x', y' are small compared to X_0, Y_0. Discuss what you expect to happen and how it depends on the parameters α, β, γ, and δ.

(f) By extrapolating empirical data for canines to the body mass of a human, Bolie suggests the values

$$\alpha = 0.8\ \text{hr}^{-1}, \qquad \gamma = 4.8\ \text{g hr}^{-1}\ \text{unit}^{-1},$$

$$\beta = 0.3\ \text{unit hr g}^{-1}, \qquad \delta = 3.2\ \text{hr}^{-1}.$$

Are these values consistent with a stable equilibrium?

27. The following equations were suggested by Bellomo et al. (1982) as a model for the glucose-insulin (g, i) hormonal system.

$$\frac{di}{dt} = -K_i i + K_g(g - g_d) + K_s i_r,$$

$$\frac{dg}{dt} = K_h g - K_0 gi - K_s K_f .$$

The coefficients K_i, K_g, K_s, K_h, K_0 are constants whose exact definitions may be ignored in this problem.

(a) Suggest an interpretation of the terms in these equations.

(b) Explore the steady state(s) and stability properties of this system of equations.

Note: For a more extended project, the problem can be extended to a full report or a class presentation based on this model.

28. A model due to Landahl and Grodsky (1982) for insulin secretion is based on the assumption that there are separate storage and labile compartments, a provisionary factor, and a signal for release. They define the following variables:

G = glucose concentration,

X = moiety formed from glucose in presence of calcium ions Ca^{++},

P = provisionary quantity required for insulin production,

Q = total amount of insulin available for release = CV, where C is its concentration and V is the volume,

S = secretion rate of insulin,

I = concentration of an inhibitory quantity.

(a) The following equation describes the amount of insulin available for release:

$$V\frac{dC}{dt} = k_+ C_s - k_- C + \gamma VP - S,$$

where k_+, k_- and γ are constant and C_s is the concentration in the storage compartment (of volume V_s), assumed constant. Explain the terms and assumptions made in deriving the equation.

(b) Show that another way of expressing part (a) is

$$\frac{dQ}{dt} = H(Q_0 - Q) + \gamma' P - S.$$

How do H, γ', and Q_0 relate to parameters which appear above?

(c) An equation for the provisional factor P is given as follows:

$$\frac{dP}{dt} = \alpha[P_\infty(G) - P],$$

where $P_\infty(G)$ is just some function of G (e.g., $P_\infty(G) = G$). Explain what has been assumed about P.

(d) The inhibitory entity I is assumed to be produced at the rate

$$\frac{dI}{dt} = B(NX - I),$$

where B is a rate constant and N is a proportionality constant. Explain this equation.

(e) Secretion of insulin S is assumed to be determined by two processes, and governed by the equation

$$S = [M_1 Y(G) + M_2(X - I)]Q,$$

where M_1 is constant, Y is a function of the glucose concentration, and M_2 is a step function; that is,

$$M_2(X - I) = \begin{cases} 0 & \text{if } X < I, \\ M_2 & \text{if } X \geq I. \end{cases}$$

Explain the equation for S.

29. A semitoxic chemical is ingested by an animal and enters its bloodstream at the constant rate D. It distributes within the body, passing from blood to tissue to bones with rate constants indicated in the figure. It is excreted in urine and sweat at rates u and s respectively. Let x_1, x_2, and x_3 represent concentrations of the chemical in the three pools. The equation for x_1 is

$$\frac{dx_1}{dt} = D - ux_1 - k_{12}x_1 + k_{21}x_2.$$

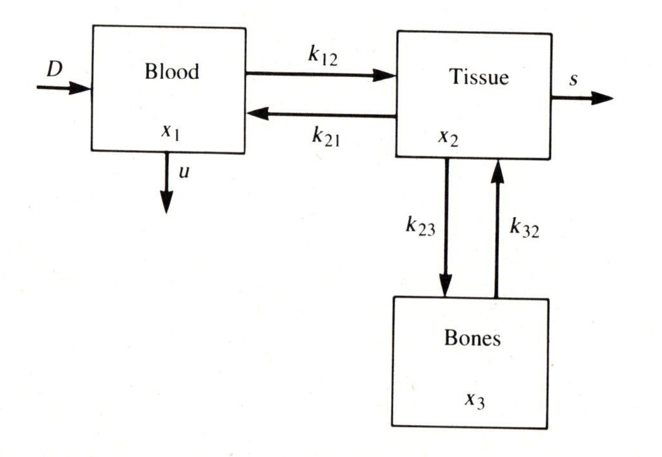

Figure for problem 29.

(a) Assuming *linear* exchange between the three compartments, write equations for x_2 and x_3.

(b) Find the steady-state values x_1, x_2, and x_3. Simplify notation by using your own symbols for ratios of rate constants that appear in the expressions.

(c) How would you investigate whether this steady state is stable?

[For an example of this type of model with realistic parameter values see Batschelet, E.; Brand, L.; and Steiner, A. (1979), On the kinetics of lead in the human body, *J. Math. Biol.*, *8*, 15 – 23.]

30. Verify and explain equations (85a, b) and (86a, b).

31. **(a)** Consider the set of measurements in Figure 4.6(a) indicated by (+). Assume that in equations 88 $\lambda_2 < \lambda_1$ and that both are positive. Then for large t it is approximately true that

$$x_1(t) \stackrel{\text{def}}{=} a_1 e^{-\lambda_1 t} + a_2 e^{-\lambda_2 t}.$$

Why? Use Figure 4.6(a) to indicate how λ_2 and a_2 can be approximated using this information.

(b) Now define

$$y(t) = x_1(t) - a_2 e^{-\lambda_2 t}.$$

This curve is shown in Figure 4.6(a) as the dotted-dashed line. Reason that

$$y(t) = a_1 e^{-\lambda_1 t},$$

and use Figure 4.6(a) to estimate a_1 and λ_1. This procedure is known as *exponential peeling*. It can be used to give a rough estimate of the eigenvalues of equations (86a, b). More sophisticated statistical and computational techniques are used when a more reliable estimate is desired.

32. **(a)** Show that the quantities in equations (86) and (88) are related as follows:

$$\begin{aligned}
\lambda_1 a_1 &= -K_1 a_1 + K_{21} b_1, \\
\lambda_1 b_1 &= K_{12} a_1 - K_2 b_1, \\
\lambda_2 a_2 &= -K_1 a_2 + K_{21} b_2, \\
\lambda_2 b_2 &= K_{12} a_2 - K_2 b_2.
\end{aligned}$$

(*Hint:* Use the fact that

$$\mathbf{x}_1 = \begin{pmatrix} a_1 \\ b_1 \end{pmatrix} e^{\lambda_1 t} \quad \text{and} \quad \mathbf{x}_2 = \begin{pmatrix} a_2 \\ b_2 \end{pmatrix} e^{\lambda_2 t}$$

are both solutions.)

(b) Another useful relation is

$$b_1 + b_2 = 0.$$

Why is this true if substance is injected only into pool 1?

(c) Assuming that the mass injected at $t = 0$ is m, what is the volume of pool 1?

***(d)** Show that

$$K_1 = \frac{a_1\lambda_1 + a_2\lambda_2}{a_1 + a_2}, \qquad K_2 = \frac{a_1\lambda_2 + a_2\lambda_1}{a_1 + a_2}.$$

(e) Find the product $K_{21}K_{12}$.

(f) Can any of the other parameters be determined (e.g., V_2, L_{12}, L_{21}, D_1, or D_2)?

(g) Discuss what sort of conclusions could be drawn from such determinations.

[A good source for further details on this topic is Rubinow (1975).]

REFERENCES

Ordinary Differential Equations

Boyce, W. E., and DiPrima, R. C. (1977). *Elementary Differential Equations*. 3rd ed. Wiley, New York.

Braun, M. (1979). *Differential Equations and Their Applications*. 3d ed. Springer-Verlag, New York.

Braun, M.; Coleman, C. S.; and Drew, D. A., eds. (1983). *Differential Equation Models*. Springer-Verlag, New York.

Henderson West, B. (1983). Setting up first order differential equations from word problems. Chap. 1 in Braun et al. (1983).

Spiegel, M. R. (1981). *Applied Differential Equations*. 3d ed. Prentice-Hall, Englewood Cliffs, N.J.

The Chemostat and Growth of Microorganisms

Biles, C. (1982). Industrial Microbiology, *UMAP Journal, 3*(1), 31–38.

Gause, G. F. (1969). *The Struggle for Existence*. Hafner Publishing, New York.

Malthus, T. R. (1970). *An essay on the principle of population*. Penguin, Harmondsworth, England.

Rubinow, S. I. (1975). *Introduction to Mathematical Biology*. Wiley, New York.

Segel, L. A. (1984). *Modeling Dynamic Phenomena in Molecular and Cellular Biology*. Cambridge University Press, Cambridge.

Chemotherapy and Continuous Infusion

Aroesty, J.; Lincoln, T.; Shapiro, N.; and Boccia, G. (1973). Tumor growth and chemotherapy: Mathematical methods, computer simulations, and experimental foundations. *Math Biosci., 17,* 243–300.

Blackshear, P. J.; Rohde, T. D.; Prosl, F.; and Buchwald, H. (1979). The implantable infusion pump: A new concept in drug delivery. *Med. Progr. Technol., 6,* 149–161.

Newton, C. M. (1980). Biomathematics in oncology: Modeling of cellular systems. *Ann. Rev. Biophys. Bioeng., 9,* 541–579.

Swan, G. W. (1984). *Applications of Optimal Control Theory in Biomedicine*. Marcel Dekker, New York, chap. 6.

Diabetes, Insulin, and Blood Glucose

Ackerman, E.; Gatewood, L. C.; Rosevear, J. W.; and Molnar, G. D. (1965). Model studies of blood-glucose regulation. *Bull. Math. Biophys., 27,* 21–37.

Ackerman, E. L.; Gatewood, L. C.; Rosevear, J. W.; and Molnar, G. (1969). Blood glucose regulation and diabetes. In F. Heinmets, ed., *Concepts and Models of Biomathematics*. Marcel Dekker, New York.

Albisser, A. M.; Leibel, B. S.; Ewart, T. G.; Davidovac, Z., Botz, C. K.; Zingg, W.; Schipper, H.; and Gander, R. (1974). Clinical control of diabetes by the artificial pancreas. *Diabetes, 23,* 397–404.

Bellomo, J.; Brunetti, P.; Calabrese, G.; Mazotti, D.; Sarti, E.; and Vincenzi, A. (1982). Optimal feedback glycaemia regulation in diabetics. *Med. Biol. Eng. Comp., 20,* 329–335.

Bolie, V. W. (1960). Coefficients of normal blood glucose regulation. *J. Appl. Physiol., 16,* 783–788.

Gatewood, L. C.; Ackerman, E.; Rosevear, J. W.; and Molnar, G. D. (1970). Modeling blood glucose dynamics. *Behav. Sci., 15,* 72–87.

Grodsky, G. M. (1972). A threshold distribution hypothesis for packet storage of insulin and its mathematical modeling. *J. Clin. Invest., 51,* 2047–2059.

Hagander, P.; Tranberg, K. G.; Thorell, J.; and Distefano, J. (1978). Models for the insulin response to intravenous glucose. *Math. Biosci., 42,* 15–29.

Landahl, H. D., Grodsky, G. M. (1982). "Comparison of Models of Insulin Release," *Bull. Math. Biol., 44,* 399–409.

Santiago, J. V.; Clemens, A. H.; Clarke, W. L.; and Kipnis, D. M. (1979). Closed-loop and open-loop devices for blood glucose control in normal and diabetic subjects. *Diabetes, 28,* 71–81.

Segre, G.; Turco, G. L.; and Vercellone, G. (1973). Modeling blood glucose and insulin kinetics in normal, diabetic, and obese subjects. *Diabetes, 22,* 94–103.

Swan, G. W. (1982). An optimal control model of diabetes mellitus. *Bull. Math. Biol., 44,* 793–808.

Swan, G. W. (1984). *Applications of Optimal Control Theory in Biomedicine.* Marcel Dekker, New York, chap. 3.

General References on Tracer Kinetics

Berman, M. (1979). Kinetic analysis of turnover data. (Intravascular metabolism of lipoproteins). *Prog. Biochem. Pharmacol., 15,* 67–108.

Matthews, C. (1957). The theory of tracer experiments with ^{131}I-Labelled Plasma Proteins. *Phys. Med. Biol., 2,* 36–53.

Rubinow, S. I. (1975). *Introduction to Mathematical Biology.* Wiley, New York.

5 Phase-Plane Methods and Qualitative Solutions

Nothing is permanent but change.

Heraclitus (500 B.C.)

Nonlinear phenomena are woven into the fabric of biological systems. Interactions between individuals, species, or populations lead to relationships that depend on the variables (such as densities) in ways more complicated than that of simple proportionality. Among other things, this means that models proporting to describe such phenomena contain nonlinear equations that are often difficult if not impossible to solve explicitly in closed analytic form.

To give a rather elementary example, consider the following two superficially similar differential equations:

$$\text{Linear:} \qquad \frac{dy}{dt} = t^2 - y, \tag{1a}$$

$$\text{Nonlinear:} \qquad \frac{dy}{dt} = y^2 - t. \tag{1b}$$

The first is linear (in the dependent variable y) and can be solved by a rather standard method (see problem 15). The second is nonlinear since it contains the term y^2; equation (1b) is not solvable in terms of elementary functions such as those encountered in calculus. While the equations both look simple, the nonlinearity in (1b) means that special methods must be applied in analyzing the nature of its solutions. Several qualitative approaches to understanding ordinary differential equations (ODEs) or systems of such equations will make up the subject of this chapter.

Our aim is to circumvent the necessity for calculating explicit solutions to ODEs; we shall be concerned with determining qualitative features of these solutions. The flavor of this approach is in large measure graphical and geometric. By blending certain geometric insights with some intuition, we will describe the behavior of solutions and thus understand the phenomena captured in a model in a pictorial form. These pictures are generally more informative than mathematical expressions and lead to a much more direct comprehension of the way that parameters and constants that appear in the equations affect the behavior of the system.

This introduction to the subject of qualitative solutions and phase-plane methods is meant to be intuitive rather than formal. While the mathematical theory underlying these methods is a rich one, the techniques we speak of can be mastered rather easily by nonmathematicians and applied to a host of problems arising from the natural sciences. Collectively these methods are an important tool that is equally accessible to the nonspecialist as to the more experienced modeler.

Reading through Sections 5.4–5.5, 5.7–5.9, and 5.11 and then working through the detailed example in Section 5.10 leads to a working familiarity with the topic. A more gradual introduction, with some background in the geometry of curves in the plane, can be acquired by working through the material in its fuller form.

Alternative treatments of this topic can be found in numerous sources. Among these, Odell's (1980) is one of the best, clearest, and most informative. Other versions are to be found in Chapter 4 of Braun (1979) and Chapter 9 of Boyce and DiPrima (1977). For the more mathematically inclined, Arnold (1973) gives an appealing and rigorous exposition in his delightful book.

5.1 FIRST-ORDER ODEs: A GEOMETRIC MEANING

To begin on relatively familiar ground we start with a single first-order ODE and introduce the concept of qualitative solutions. Here we shall assume only an acquaintance with the meaning of a derivative and with the graph of a function.

Consider the equation

$$\frac{dy}{dt} = f(y, t), \tag{2a}$$

and suppose that with this differential equation comes an initial condition that specifies some starting value of y:

$$y(0) = y_0. \tag{2b}$$

[To ensure that a unique solution to (2a) exists, we assume from here on that $f(y, t)$ is continuous and has a continuous partial derivative with respect to y.]

A solution to equation (2a) is some function that we shall call $\phi(t)$. Given a formula for this function, we might graph $y = \phi(t)$ as a function of t to display its time behavior. This graph would be a curve in the ty plane, as follows. According to equation (2b) the curve starts at the point $t = 0$, $\phi(0) = y_0$. The equation (2a) tells us that at time t, the slope of any tangent to the curve must be $f(t, \phi(t))$. (Recall that the derivative of a function is interpreted in calculus as the *slope of the tangent to its graph*.)

Let us now drop the assumption that a formula for the solution $\phi(t)$ is known and resort to some intuitive reasoning. Suppose we make a sketch of the ty plane and use only the information in equation (2a): at every point (t, y) we could draw a small line segment of slope $f(t, y)$. This can be done repeatedly for many points, resulting in a picture aptly termed a *direction field* [Figure 5.1(*a*)]. The solution curves shown in Figure 5.1(*b*), must be tangent to the directions of the line segments in Figure 5.1(*a*). Now we reconstruct an approximate graph of the solution by beginning at $(0, y_0)$ and sketching a curve that winds its way through the plane in the general direction depicted by the field. (The more line segments we have drawn, the better our approximation will be.) Starting at many different initial points one can generate a whole family of solution curves that summarize the qualitative behavior specified by the differential equation. See example 1.

Example 1

Here we explore the nature of solutions to equation (1b). We tabulate several values as follows:

Location y	t	*Slope of Tangent Line* $f(t, y) = y^2 - t$
0	0	0
1	1	0
1	2	−1
2	1	3
⋮	⋮	⋮

In a somewhat more systematic approach, we notice that $f(t, y) = K$ is the locus of points $K = y^2 - t$. (This is a parabola about the t axis, displaced from the origin by an amount $-K$.) Along each of these loci, tangent lines are parallel and of slope $= K$, as in Figure 5.1(*a*). Figure 5.1(*b*) is an approximate sketch of solution curves for several initial values. We have made no attempt to depict exact solutions in this picture, but rather to describe a general behavior pattern.

Example 2

The equation

$$\frac{dy}{dt} = y(1 - y)(2 - y) \tag{3}$$

is autonomous. Its solutions have zero slope whenever $y = 0$, 1, or 2. The slopes are positive for $0 < y < 1$ and $y > 2$ and negative for $1 < y < 2$. (The exact values of these slopes could be tabulated but are not important since the sketch is meant to be only approximate.) From the sketch in Figure 5.2b it is clear that for y initially smaller than 2, the solution approaches the value $y = 1$. For y initially larger than 2, the solution grows without bound. The values $y = 0$, 1, and 2 are steady states ($dy/dt = 0$). $y = 1$ is stable; the others are unstable.

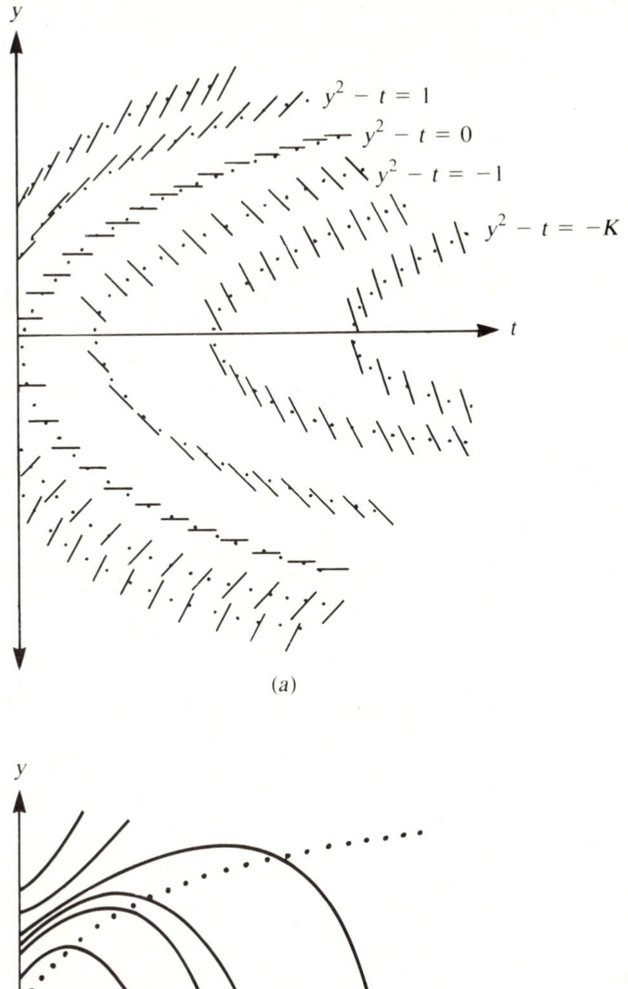

(a)

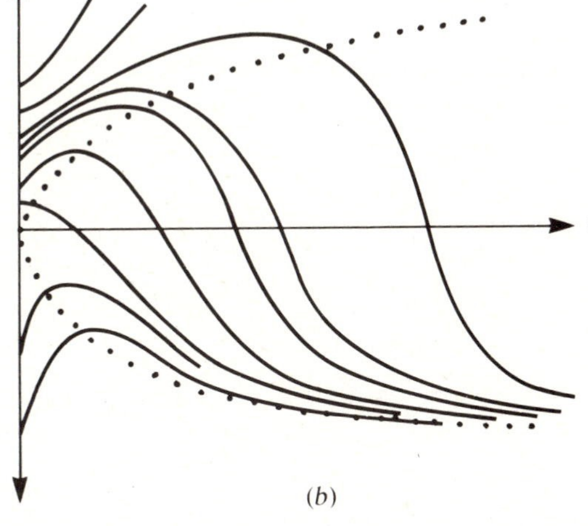

(b)

***Figure 5.1** Solutions to* $y' = y^2 - t$. *(a) For each pair of values* (t, y), *line segments whose slope is* $f(t, y) = y^2 - t$ *are shown. (Note that slopes are constant along parabolic curves for which* $K = y^2 - t$, *where* K *is any constant.) (b) Solution curves are constructed by maintaining tangency to the directions shown in (a).*

In example 2, the function appearing on the RHS of equation (3) depends explicitly only on y, not on t. A system described by such an equation would be unfolding at some inherent rate independent of the clock time or the time at which the process began. The differential equation is said to be *autonomous*, and solutions to it can be represented in an especially convenient way, as will presently be shown.

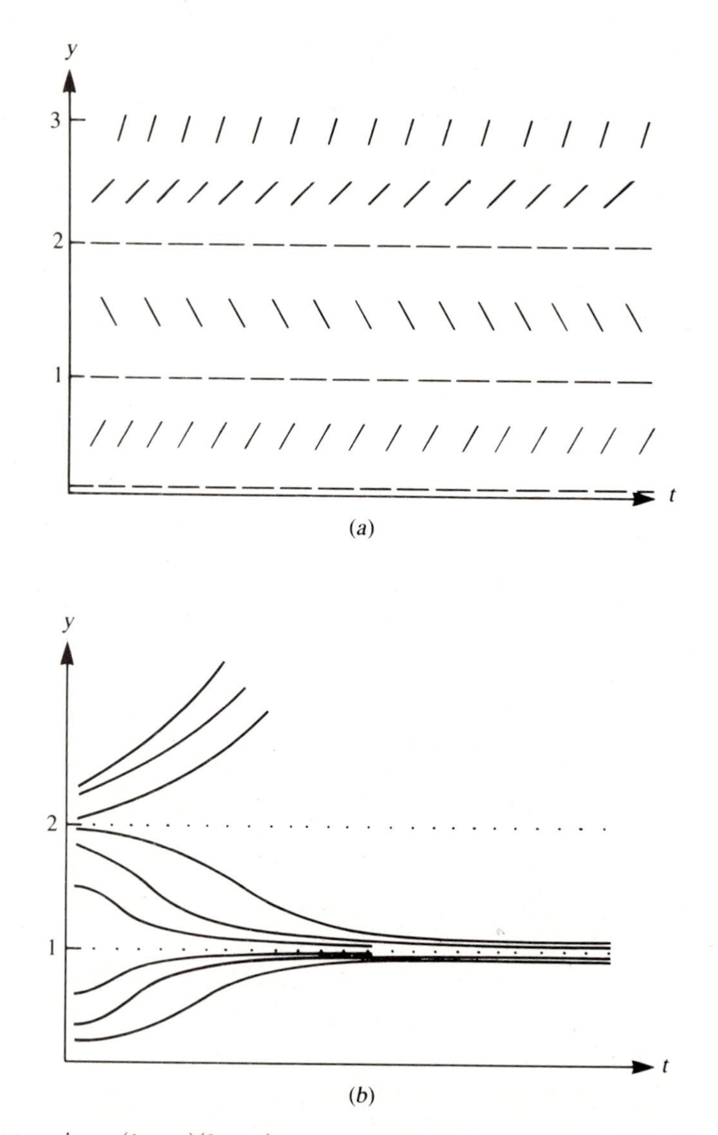

Figure 5.2 *Solutions to* $y' = y(1 - y)(2 - y)$. *(a) For each value of* y, *line segments bearing the slope* $f(y) = y(1 - y)(2 - y)$ *have been drawn. The slopes are zero when* $y = 0, 1,$ *or* 2, *and positive for* $0 < y < 1$ *or* $y > 3$. *(b) Solution curves are constructed by maintaining tangency to the line segments drawn in (a).*

The fact that a differential equation is autonomous means, pictorially, that the tangent line segments do not "wobble" along the time axis. This can be used to represent the same qualitative information in a more condensed form. Let us suppress the time dependence and instead plot dy/dt as a function of y. See Figure 5.3(*a*). Whenever $f(y)$ is positive (that is, for $0 < y < 1$ or $y > 2$), y must be increasing. Whenever $f(y)$ is negative, y must be decreasing. This can be represented by drawing arrows pointing to the left or to the right directly along the y axis, as shown in Figure 5.3(*b*). This abbreviated representation is called a *one-dimensional phase portrait*, or a *phase flow on a line*. Figure 5.3(*b*) conveys roughly the same qualitative information as does Figure 5.2, with the omission of the time course, or speed with which the solution $y(t)$ changes.

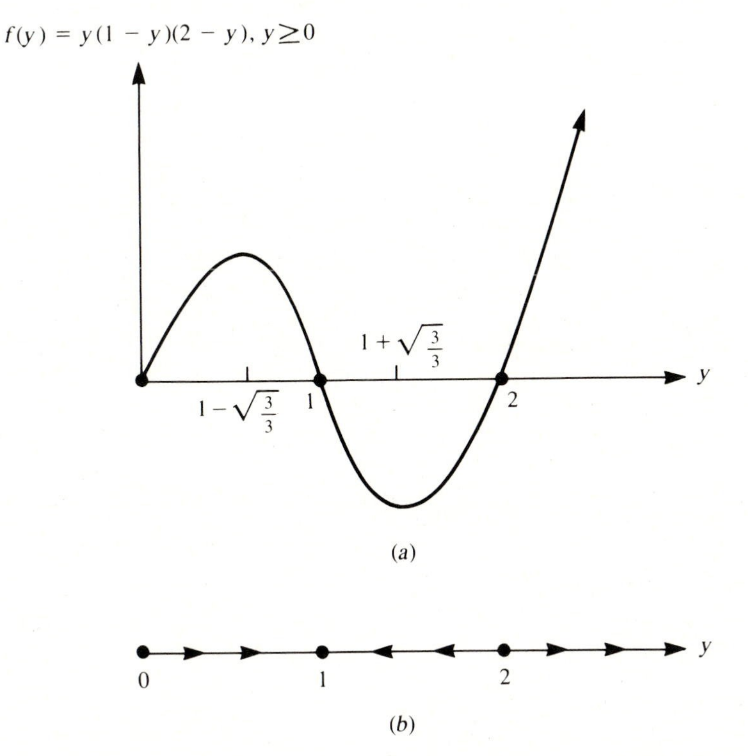

Figure 5.3 *(a) Graph of* f(y) *versus* y *for equation (3). Since* y′ = f(y), y *is increasing when* f *is positive, decreasing when* f *is negative, and stationary when* f = 0. *(b) The qualitative features described in (a) can be summarized by drawing the directions of motion along the* y *axis.*

Example 3 again illustrates the procedure of extracting information from the equation and depicting the solution as a one-dimensional flow.

As mentioned previously, when a differential equation is autonomous, the qualitative behavior of its solutions can be characterized even when time dependence is suppressed. Think of a qualitative solution as a *trajectory:* a flow that begins

Example 3
The differential equation

$$\frac{dy}{dt} = \sin y \tag{4}$$

can be treated in the same way, as shown in Figure 5.4. The following are some convenient values to tabulate:

y	$dy/dt = \sin y$
0	0
$n\pi$	0
$-n\pi$	0
$\pi/2 \pm 2n\pi$	1
$-\pi/2 \pm 2n\pi$	-1

Solution curves and directions of flow are given in Figure 5.4.

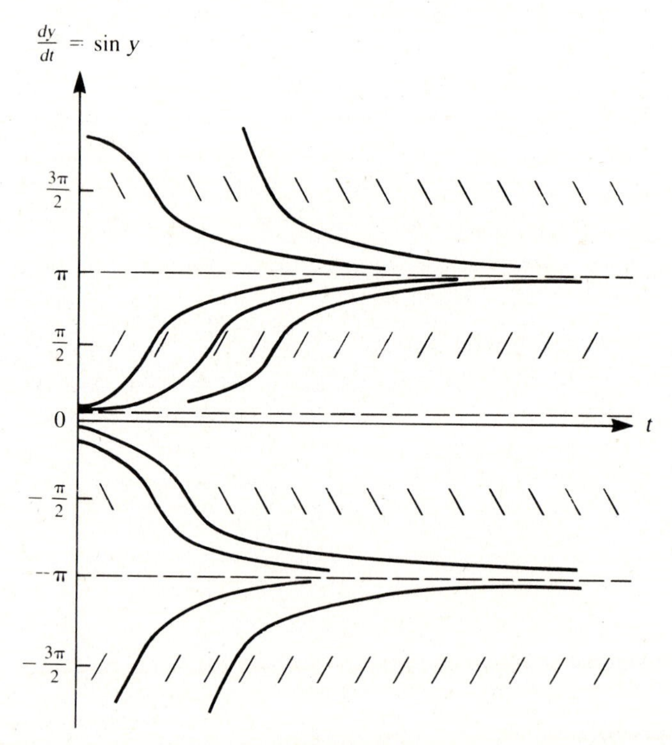

(a)

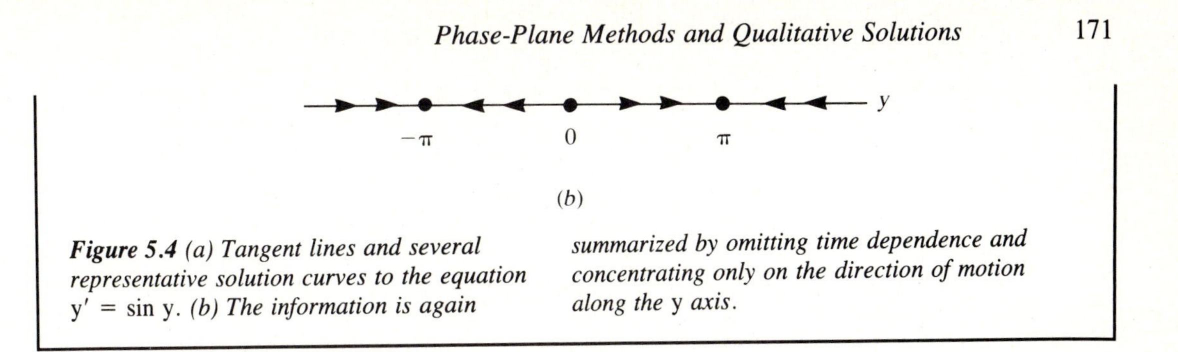

Figure 5.4 (a) Tangent lines and several representative solution curves to the equation y′ = sin y. (b) The information is again summarized by omitting time dependence and concentrating only on the direction of motion along the y axis.

somewhere (at an initial point) and has an orientation consistent with increasing values of time. We shall presently see that these ideas have a natural and important generalization to systems of differential equations.

5.2 SYSTEMS OF TWO FIRST-ORDER ODEs

In modeling biological systems, which are generally composed of several *interacting variables,* we are frequently confronted with systems of nonlinear ODEs. The ideas of Section 5.1 can be extended to encompass such systems; in the present section we deal in great detail with systems of two equations that describe the interaction of two species. The reason for dealing almost exclusively with these will emerge after some preliminary familiarity is established.

Let us therefore turn attention to a system of two autonomous first-order equations, a prototype of which follows:

$$\frac{dx}{dt} = f_1(x, y), \tag{5a}$$

$$\frac{dy}{dt} = f_2(x, y). \tag{5b}$$

Technically, we assume that f_1 and f_2 are continuous functions having partial derivatives with respect to x and y; this ensures existence of a unique solution given an initial value for x and y. A solution to system (5) would be two functions, $x(t)$, and $y(t)$, that satisfy the equations together with the initial conditions, if any.

As a preliminary to understanding the equations, let us consider an approximate form of these equations, whereby derivatives are replaced by finite differences, as follows:

$$\frac{\Delta x}{\Delta t} = f_1(x, y), \tag{6a}$$

$$\frac{\Delta y}{\Delta t} = f_2(x, y). \tag{6b}$$

The changes Δx and Δy in the two independent variables are thus specified whenever x and y are known, since

$$\Delta x = f_1(x, y)\,\Delta t, \tag{7a}$$

$$\Delta y = f_2(x, y)\,\Delta t. \tag{7b}$$

These equations can be interpreted as follows: Given a value of x and y, after some small increment of time Δt, x will change by an amount Δx and y by an amount Δy. This is represented pictorially in Figure 5.5, where a point (x, y) is assigned a vector with components $(\Delta x, \Delta y)$ that describe changes in the two variables simultaneously. We see that *equations (6) and (7) are mathematical statements that assign a vector (representing a change) to every pair of values* (x, y).

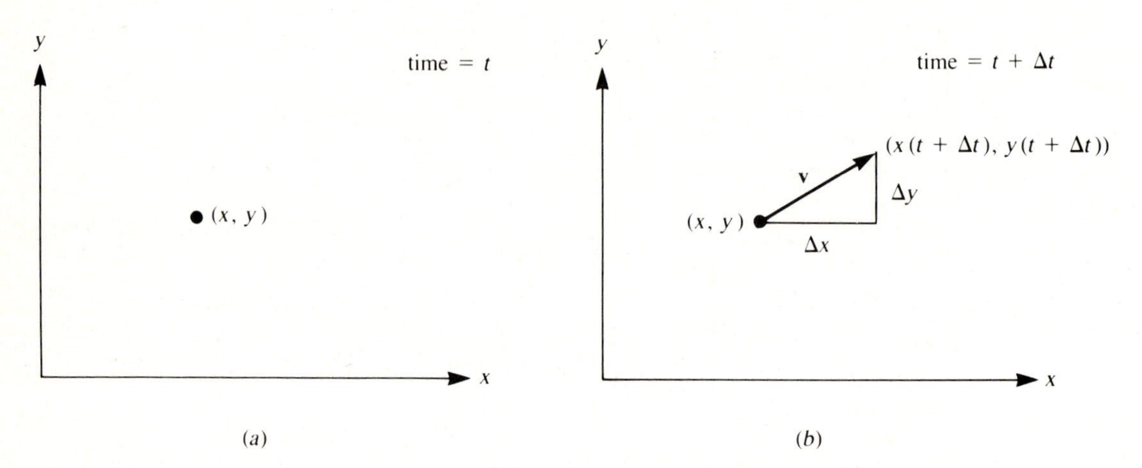

Figure 5.5 *(a) Given a point* (x, y), *(b) a change in its location can be represented by a vector* **v**.

In calculus such concepts are made more precise. Indeed, we know that derivatives are just limits of expressions such as $\Delta x/\Delta t$ when ever-smaller time increments are considered. Using calculus, we can understand equations (5a,b) directly without resorting to their approximated version. (A review of these ideas is presented in Section 5.3, which may be skipped if desired.)

5.3 CURVES IN THE PLANE

In calculus we learn that the concepts *point* and *vector* are essentially interchangeable. The pair of numbers (x, y) can be thought of as a point in the cartesian plane with coordinates x and y [as in Figure 5.6(a)] or as an arrow strung out between the origin (0, 0) and (x, y) that points to the location of this point [Figure 5.6(b)]. When the coordinates x and y vary with time or with some other parameter, the point (x, y) moves over the plane tracing a curve as it moves. Equivalently, the arrow twirls and stretches as its head tracks the position of the point $(x(t), y(t))$. For this reason, it is often called a *position vector,* symbolized by $\mathbf{x}(t)$.

As previously remarked, since the solution of a system of equations such as (5a,b) is a pair $(x(t), y(t))$, the idea that a solution corresponds geometrically to a curve carries through from the one-dimensional case. To be precise, the *graph* of a solution would be a curve $(t, x(t), y(t))$ in the three-dimensional space, depicting the time evolution of the values of x and y. We shall use the fact that equations (5a,b) are autonomous to suppress time dependence as before, that is, to depict solutions by trajectories in the plane. Such trajectories, each representing a solution, together make up a phase-plane portrait of the system of equations under consideration.

We observed in Section 5.2 that $(\Delta x, \Delta y)$ given by equations (7a,b) is a vector that depicts both the magnitude and the direction of changes in the two variables. A limiting value of this vector,

$$\left(\frac{dx}{dt}, \frac{dy}{dt}\right), \tag{8a}$$

is obtained when the time increment Δt gets vanishingly small in $(\Delta x/\Delta t, \Delta y/\Delta t)$. The latter, often symbolized

$$\frac{d\mathbf{x}}{dt} \tag{8b}$$

represents the *instantaneous change* in x and y, and can also be depicted as an arrow attached to the point $(x(t), y(t))$ and tangent to the curve. This vector is often called the *velocity vector,* since its magnitude indicates how quickly changes are occurring.

A summary of all these facts is collected here:

A Summary of Facts about Vector Functions (from Calculus)

1. The pair $(x(t), y(t))$ represents a curve in the xy plane with t as a parameter.
2. $\mathbf{x}(t) = (x(t), y(t))$ also represents a position vector: a vector attached to $(0, 0)$ that points to the position along the curve, that is, the location corresponding to the value t.
3. The vector $d\mathbf{x}/dt$, which is just the pair $(dx/dt, dy/dt)$ has a well-defined geometric meaning. It is a vector that is tangent to the curve at $\mathbf{x}(t)$. Its magnitude, written $|d\mathbf{x}/dt|$ represents the speed of motion of the point $(x(t), y(t))$ along the curve.
4. The set of equations (5a,b) can be written in vector form,

$$\frac{d\mathbf{x}}{dt} = \mathbf{F}(\mathbf{x}).$$

Here the vector function $\mathbf{F} = (f_1, f_2)$ assigns a vector to every location $\mathbf{x}$ in the plane; $\mathbf{x}$ is the position vector (x, y), and $d\mathbf{x}/dt$ is the velocity vector $(dx/dt, dy/dt)$.

Figure 5.6 *(a) Point and (b) vector representations of a pair* (x, y). *(c) A curve* (x(t), y(t)) *can also be represented by moving vector* **x**(t), *as in (d).*

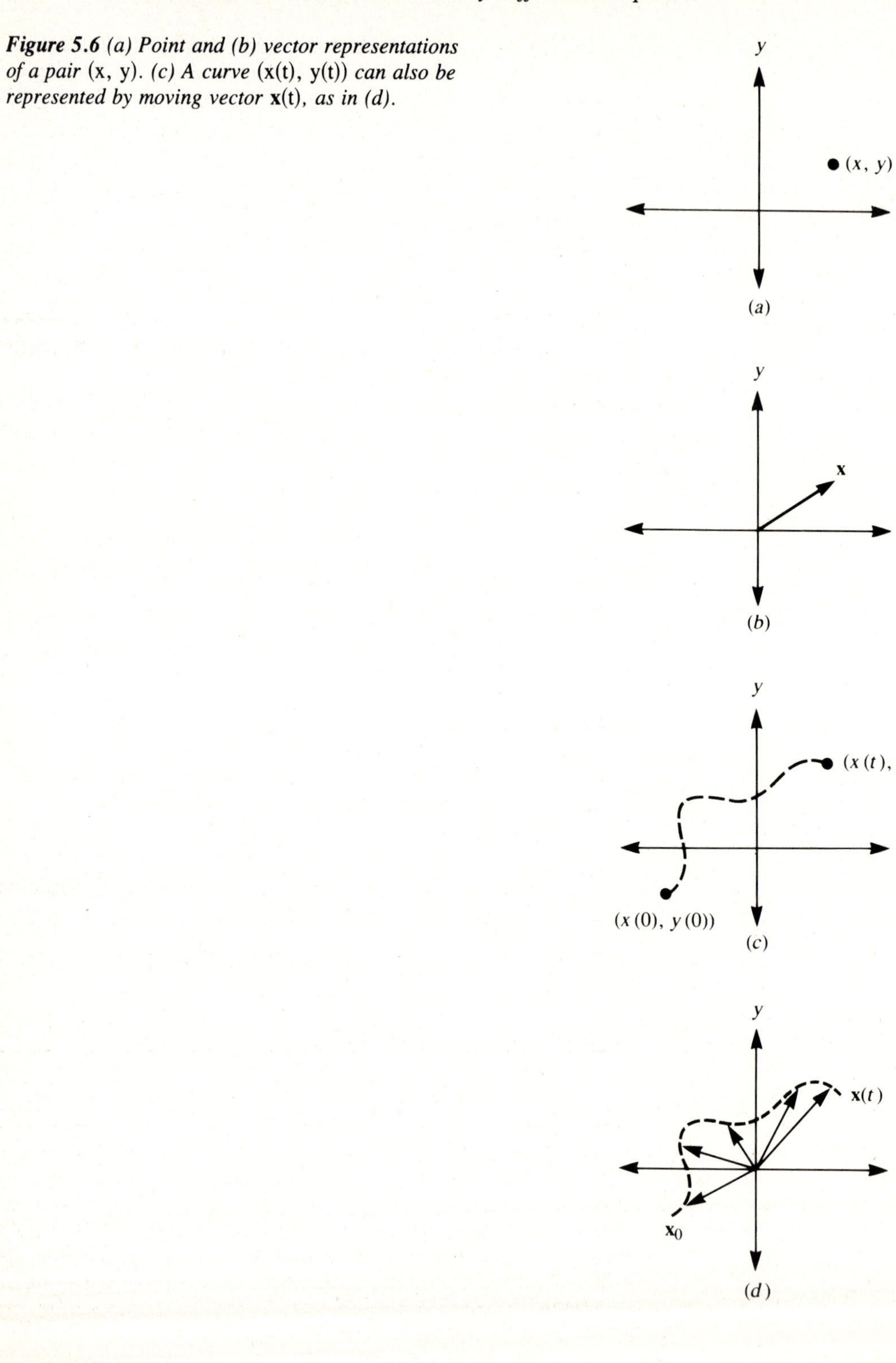

5.4 THE DIRECTION FIELD

From concepts that arise in calculus we surmise that solutions to ODEs, whether in one dimension or higher, correspond to curves, and differential equations are "recipes" for tangent vectors to these curves. This insight will now be applied to reconstructing a qualitative picture of solutions to a system of two equations such as (5). For such autonomous systems each point (x, y) in the plane is assigned a unique vector $(f_1(x, y), f_2(x, y))$ that does not change with time. A solution curve passing through (x, y) must have these vectors as its tangents. Thus a collection of such vectors defines a direction field, which can be used as a visual guide in sketching a family of solution curves, collectively a *phase-plane portrait*. Example 4 clarifies how this is done in practice.

Example 4
Let

$$\frac{dx}{dt} = xy - y, \qquad (9a)$$

$$\frac{dy}{dt} = xy - x, \qquad (9b)$$

and let $f_1(x, y) = xy - y, f_2(x, y) = xy - x$. In the following table the values of f_1 and f_2 are listed for several values of (x, y).

x	y	$f_1(x, y)$	$f_2(x, y)$
0	0	0	0
0	1	−1	0
1	0	0	−1
−1	0	0	1
0	−1	1	0
1	1	0	0
1	−1	2	−2
−2	−1	3	4

After tabulating arbitrarily many values of (x, y) and the corresponding values of $f_1(x, y)$ and $f_2(x, y)$, we are ready to construct the direction field. To each point (x, y) we attach a small line segment in the direction of the vector $(f_1(x, y), f_2(x, y))$.

See Figure 5.7. The slope $\Delta y/\Delta x$ of the line segment is to have the ratio $f_2(x, y)/f_1(x, y)$. Notice that a vector $(f_1(x, y), f_2(x, y))$ has the magnitude $[f_1(x, y)^2 + f_2(x, y)^2]^{1/2}$, which we shall not attempt to portray accurately. This magnitude represents a rate of motion, the speed with which a trajectory is traced. A cluttered picture emerges should we attempt to draw the vectors (f_1, f_2) in their true sizes. Since we are interested in establishing only the direction field, making all tangent vectors some uniform small size proves most convenient.

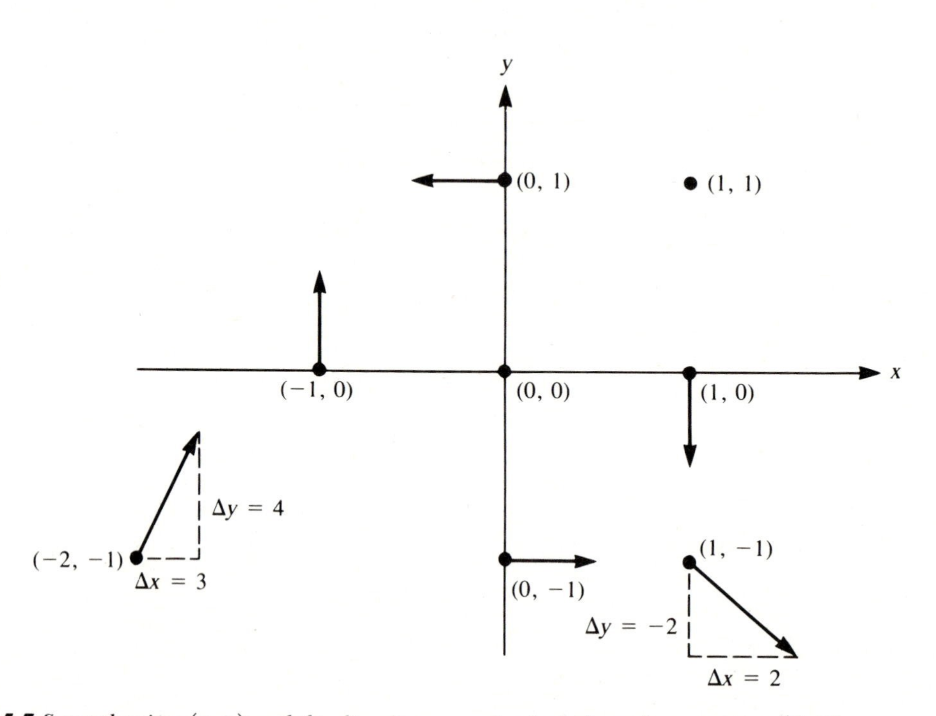

Figure 5.7 *Several points* (x, y) *and the direction vectors* (f_1, f_2) *associated with them have been sketched above for equations (9a,b).*

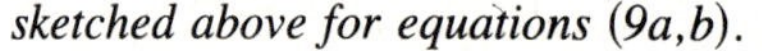

Two notable locations in example 4 are the points (0, 0) and (1, 1), at both of which $f_1 = 0$ and $f_2 = 0$. Neither x nor y changes given these initial values; the terms *steady state, equilibrium point,* or *singular point* are synonymously used to denote such locations. Presently we will see that such points play a central role in determining global phase-plane behavior.

The chore of tabulating and sketching direction fields is in principle straightforward but tedious. Rather than belabor the process we might consign the job to a computer, as we have done in Figures 5.8 (*a,b*). A simple BASIC program run on an IBM personal computer produced these results.

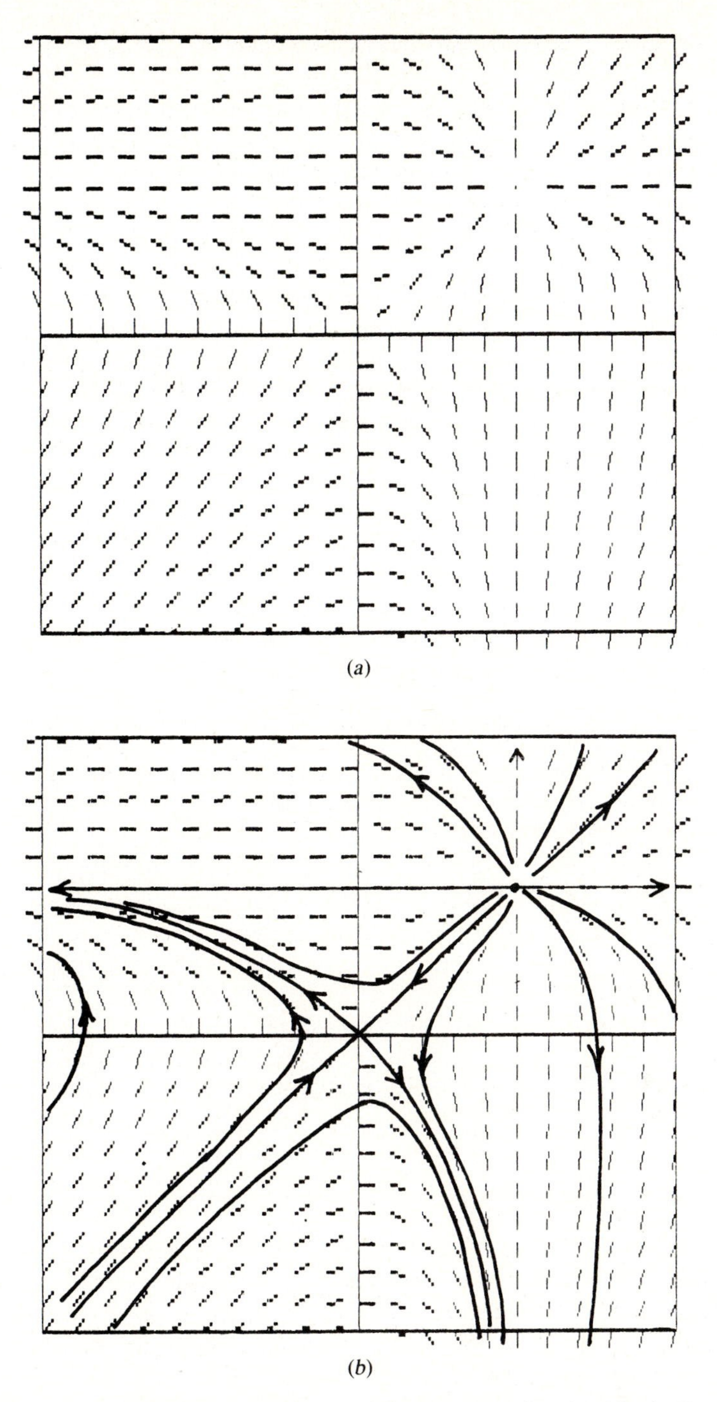

Figure 5.8 *(a) Computer-generated vector field for example 4. The vectors point* away *from the points to which they are attached. For example, along the positive* x *axis, they point down. (b) Hand-sketched solution curves for example 4. The directions are ascertained by noting whether vectors point into or out of the region at the boundary of the square. (Computer plot by Yehoshua Keshet.)*

From the direction field thus generated one gets a good general idea of solution curves consistent with the flow. Through every point in the plane there is a curve (by existence of a solution) and only one curve (by uniqueness). Thus curves may not intersect or touch each other, except at the steady states designated by heavy dots in Figures 5.7 and 5.8. Rules governing the possible pattern of curves will be outlined in a subsequent section.

As a word of caution, note that a phase-plane diagram is not a quantitatively accurate graph. In practice, because only a finite number of tangent vectors can be drawn in the plane, there will always be some small error in the curve that we inscribe. Such initially small mistakes could propagate if they result in an improper choice of tangent vectors along the way. For this reason, solution curves drawn in this way are approximate. There may be cases where ambiguity arises close to a steady state and where it is difficult to distinguish between several alternatives. Such situations call for a more rigorous technique. Before turning to these matters, we investigate a more systematic way of establishing the direction field in a computationally efficient way.

5.5 NULLCLINES: A MORE SYSTEMATIC APPROACH

Rather than arbitrarily plotting tabulated values, we prepare the way by noticing what happens along the locus of points for which one of the two functions, either $f_1(x, y)$ or $f_2(x, y)$ is zero. We observe that

1. If $f_1(x, y) = 0$, then $dx/dt = 0$, so x does not change. This means that the direction vector must be parallel to the y axis, since its Δx component is zero.
2. Similarly, if $f_2(x, y) = 0$, then $dy/dt = 0$, so y does not change. Thus the direction vector is parallel to the x axis, since its Δy component is zero.

The locus of points satisfying one of these two conditions is called a *nullcline*. The x *nullcline* is the set of points satisfying condition 1; similarly, the y *nullcline* is the set of points satisfying condition 2. Because the arrows are parallel to the y and x axis respectively on these loci, it proves helpful to sketch these as a first step. Example 5 illustrates the procedure.

Example 5

For equations (9a,b) the nullclines are loci for which

1. $\dot{x} = 0$ (the x nullcline); that is, $xy - y = 0$. This is satisfied when $x = 1$ or $y = 0$. See dotted lines in Figure 5.9(a). On these lines, direction vectors are vertical.
2. $\dot{y} = 0$ (the y nullcline); that is, $xy - x = 0$. This is satisfied when $x = 0$ or $y = 1$. See the dotted-dashed line in Figure 5.9(a). On these lines direction vectors are horizontal.

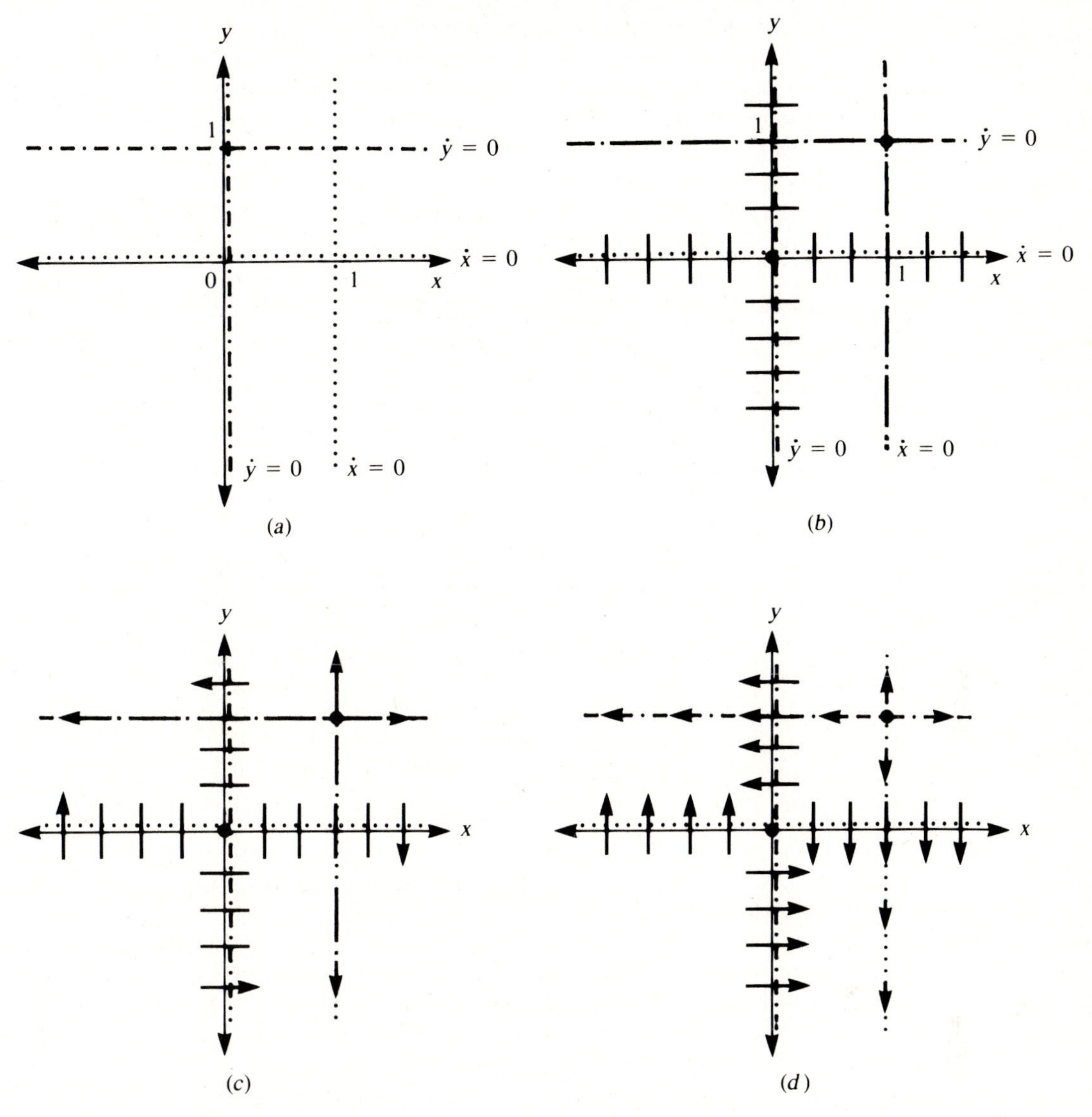

Figure 5.9 *Nullclines and flow directions for example 5. (a) Nullclines, which happen to be straight lines here, are sketched in the* xy*-plane and assigned vertical or horizontal line segments in (b). (c) Directions are determined by tabulating several values and inscribing arrowheads. (d) Neighboring arrows are deduced by preserving a continuous flow.*

Points of intersection of nullclines satisfy both $\dot{x} = 0$ and $\dot{y} = 0$ and thus represent steady states. To identify these and determine the directions of flow, several guidelines are useful.

Rules for determining steady states and direction vectors on nullclines

1. Steady states are located at intersections of an x nullcline with a y nullcline.
2. At steady states there is no change in either x or y values; that is, the vectors have zero length.
3. Direction vectors must vary continuously from one point to the next on the nullclines. Thus a change in the orientation (for example, from pointing up to pointing down) can take place only at steady states.

We note that (0, 0) and (1, 1) are the only two steady states in example 5. It is important to avoid confusing these with other intersections, for example (1, 0) and (0, 1), for which only one of the two nullcline conditions is satisfied. Generally it is a good idea to distinguish between the x and y nullclines by using different symbols or colors for each type.

It should be remarked that in affixing orientations to the arrows along nullclines we can economize on algebra by being aware of certain geometric properties. For instance, in example 5 we observe the following patterns of signs:

x	y	$f_1(x, y)$	$f_2(x, y)$
−	1	−	0
0	−	+	0
1	−	0	−
0	+	0	−
+, > 1	1	+	0
1	+, > 1	0	+
0	+	−	0
−	0	0	+

It is evident that on opposite sides of a steady-state point (along a given nullcline) the orientation of arrows is reversed. This is a property shared by most systems of equations with the exception of certain singular cases. (We shall be able to distinguish these exceptions by calculating the Jacobian **J** and evaluating it at the steady state in question. If det **J** $\neq 0$, the property of arrow reversal holds.) In most cases where we encounter det **J** $\neq 0$, it suffices to determine the direction vectors at one or two select places and deduce the rest by preserving continuity and switching orientation as a steady state is crossed. Thus the arrow-nullcline method can reveal a fairly complete picture with relatively little calculation (see example 6).

Example 6
Consider the equations

$$\frac{dx}{dt} = x + y^2, \tag{10a}$$

$$\frac{dy}{dt} = x + y. \tag{10b}$$

The x nullcline is the curve $0 = x + y^2$; the y nullcline is the line $0 = x + y$. Steady states are thus (0, 0) and (−1, 1). The Jacobian of system (10) is

$$\mathbf{J}(x_0, y_0) = \begin{pmatrix} 1 & 2y \\ 1 & 1 \end{pmatrix}_{(x_0, y_0)}.$$

Thus det $\mathbf{J}(0, 0) = 1 \neq 0$, det $\mathbf{J}(-1, 1) = -1 \neq 0$, so the property of arrow reversal holds. It suffices to tabulate two values, for example, as follows:

x	y	$\dot{x} = x + y^2$	$\dot{y} = x + y$
+	$y = -x$	+	0
−	$x = -y^2$	0	−

After drawing these two arrows, all others follow by the above method. (See Figure 5.10.)

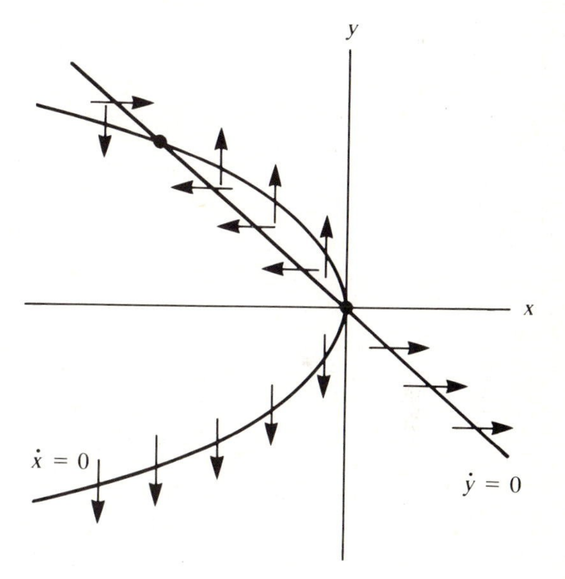

Figure 5.10 *Nullclines and arrows for example 6, equations (10a,b).*

5.6 CLOSE TO THE STEADY STATES

The examples we have seen give evidence to the notion that dramatic local changes in the flow pattern can only take place in the vicinity of steady-state points. We now invoke a metaphorical magnifying glass to scrutinize the behavior close to these locations. In the discussions of Chapter 4, we established that close to a steady state

$(\bar{x}_0, \bar{y}_0)$ [defined by $f_1(\bar{x}_0, \bar{y}_0) = f_2(\bar{x}_0, \bar{y}_0) = 0$] the nonlinear system (5) behaves very nearly like a linear one,

$$\frac{dx}{dt} = a_{11}x + a_{12}y, \tag{11a}$$

$$\frac{dy}{dt} = a_{21}x + a_{22}y, \tag{11b}$$

where a_{ij}, related to partial derivatives of f_1 and f_2, make up the coefficient of the Jacobian matrix $\mathbf{J}(\bar{x}_0, \bar{y}_0)$ as follows:

$$\mathbf{J}(\bar{x}_0, \bar{y}_0) = \begin{pmatrix} a_{11} & a_{12} \\ a_{21} & a_{22} \end{pmatrix} = \begin{pmatrix} \dfrac{\partial f_1}{\partial x} & \dfrac{\partial f_1}{\partial y} \\ \dfrac{\partial f_2}{\partial x} & \dfrac{\partial f_2}{\partial y} \end{pmatrix}_{(\bar{x}_0, \bar{y}_0)}. \tag{12}$$

This result is important, as it reduces the problem to one we understand well. It remains to interpret the phase-plane equivalents of solutions to systems of linear ODEs (described in Chapter 4). This will give us the local picture of the flow pattern about the steady states.

Example 7

Equations (9a,b) can be linearized about the steady states (0, 0) and (1, 1). The Jacobian is

$$\mathbf{J}(\bar{x}_0, \bar{y}_0) = \begin{pmatrix} y & x-1 \\ y-1 & x \end{pmatrix}_{(\bar{x}_0, \bar{y}_0)}.$$

One obtains

$$\mathbf{J}(0, 0) = \begin{pmatrix} 0 & -1 \\ -1 & 0 \end{pmatrix}, \quad \mathbf{J}(1, 1) = \begin{pmatrix} 1 & 0 \\ 0 & 1 \end{pmatrix}.$$

Thus close to (0, 0) the system behaves much like the linearized version,

$$\frac{dx}{dt} = -y, \quad \frac{dy}{dt} = -x.$$

Similarly, close to (1, 1) the linearized equations are

$$\frac{dx}{dt} = x, \quad \frac{dy}{dt} = y.$$

A summary of properties of linear systems (of two ordinary differential equations) is given in Table 5.1, in which we consider only the real, distinct eigenvalues case.

Table 5.1 ***Linear Systems of two ODEs***

	Full algebraic notation	*Equivalent Vector–Matrix Notation*
Equations	$\dfrac{dx}{dt} = a_{11}x + a_{12}y$ $\dfrac{dy}{dt} = a_{21}x + a_{22}y$	$\dfrac{d\mathbf{x}}{dt} = \mathbf{A}\mathbf{x}, \qquad \mathbf{A} = \begin{pmatrix} a_{11} & a_{12} \\ a_{21} & a_{22} \end{pmatrix}$
Significant quantities	$\beta = a_{11} + a_{22},$ $\gamma = a_{11}a_{22} - a_{12}a_{21},$ $\delta = \beta^2 - 4\gamma$	Tr $\mathbf{A}$, det $\mathbf{A}$, disc $\mathbf{A}$
Characteristic equation	$\lambda^2 - \beta\lambda + \gamma = 0$	$\det(\mathbf{A} - \lambda\mathbf{I}) = 0$
Eigenvalues	$\lambda_{1,2} = \dfrac{\beta \pm \sqrt{\delta}}{2}$	$\lambda_{1,2} = \dfrac{\text{Tr } \mathbf{A} \pm \sqrt{\text{disc } \mathbf{A}}}{2}$
Identities	$\lambda_1 + \lambda_2 = \beta, \qquad \lambda_1\lambda_2 = \beta$	$\lambda_1 + \lambda_2 = \text{Tr } \mathbf{A}, \qquad \lambda_1\lambda_2 = \det \mathbf{A}$
Eigenvectors	$\begin{pmatrix} a_{12} \\ \lambda_1 - a_{11} \end{pmatrix}, \begin{pmatrix} a_{12} \\ \lambda_2 - a_{11} \end{pmatrix}$	$\mathbf{v}_1, \mathbf{v}_2$ such that $(\mathbf{A} - \lambda\mathbf{I})\mathbf{v}_i = 0$
Solutions	$x = c_1 a_{12} e^{\lambda_1 t} + c_2 a_{12} e^{\lambda_2 t},$ $y = d_1 e^{\lambda_1 t} + d_2 e^{\lambda_2 t},$ where $d_1 = c_1(\lambda_1 - a_{11})$, $d_2 = c_2(\lambda_2 - a_{11})$.	$\mathbf{x} = c_1 \mathbf{v}_1 e^{\lambda_1 t} + c_2 \mathbf{v}_2 e^{\lambda_2 t}.$

5.7 PHASE-PLANE DIAGRAMS OF LINEAR SYSTEMS

We observe that a linear system can have at most one steady state, at (0, 0) provided $\gamma = \det \mathbf{A} \neq 0$. In the particular case of real eigenvalues there is a rather distinct geometric meaning for eigenvectors and eigenvalues:

1. For real λ_i the *eigenvectors* $\mathbf{v}_i$ *are directions on which solutions travel along straight lines towards or away from* (0, 0).
2. If λ_i is positive, the direction of flow along $\mathbf{v}_i$ is away from (0, 0), whereas if λ_i is negative, the flow along $\mathbf{v}_i$ is towards (0, 0).

Proof of these two statements is given below.

An Interpretation of Eigenvectors

Solutions to a linear system are of the form

$$\mathbf{x}(t) = c_1\mathbf{v}_1 e^{\lambda_1 t} + c_2\mathbf{v}_2 e^{\lambda_2 t}. \tag{13}$$

Recall that c_1 and c_2 are arbitrary constants. If initial conditions are such that $c_1 = 0$ and $c_2 = 1$, the corresponding solution is

$$\mathbf{x}(t) = \mathbf{v}_2 e^{\lambda_2 t}. \tag{14}$$

For any value of t, $\mathbf{x}(t)$ is a scalar multiple of $\mathbf{v}_2$. (This means that $\mathbf{x}(t)$ is always parallel to the direction specified by the vector $\mathbf{v}_2$.) If λ is negative, then for very large values of t $\mathbf{x}(t)$ is small. In the limit as t approaches $+\infty$, $\mathbf{x}(t)$ approaches the steady state (0, 0). Thus $\mathbf{x}(t)$ describes a straight-line trajectory moving parallel to the direction $\mathbf{v}_2$ and towards the origin.

A similar result is obtained when $c_1 = 1$ and $c_2 = 0$. Then we arrive at

$$\mathbf{x}(t) = \mathbf{v}_1 e^{\lambda_1 t}. \tag{15}$$

The solution is a straight-line trajectory parallel to $\mathbf{v}_1$.

It follows that any solution curve that starts on a straight line through (0, 0) in either direction $\pm\mathbf{v}_1$ or $\pm\mathbf{v}_2$ will stay on that line for all t, $-\infty < t < \infty$ either approaching or receding from the origin. Note also from the above that a steady state can only be attained as a limit, when t gets infinitely large, because time dependence of solutions is exponential. This tells us that the rate of motion gets progressively slower as one approaches a steady state.

Solution curves that begin along directions different from those of eigenvectors tend to be curved (because when both c_1 and c_2 are nonzero, the solution is a linear superposition of the two fundamental parts, $\mathbf{v}_1 e^{\lambda_1 t}$ and $\mathbf{v}_2 e^{\lambda_2 t}$, whose relative contributions change with time). There is a tendency for the "fast" eigenvectors (those associated with largest eigenvalues) to have the strongest influence on the solutions. Thus trajectories curve towards these directions, as shown in Figure 5.11.

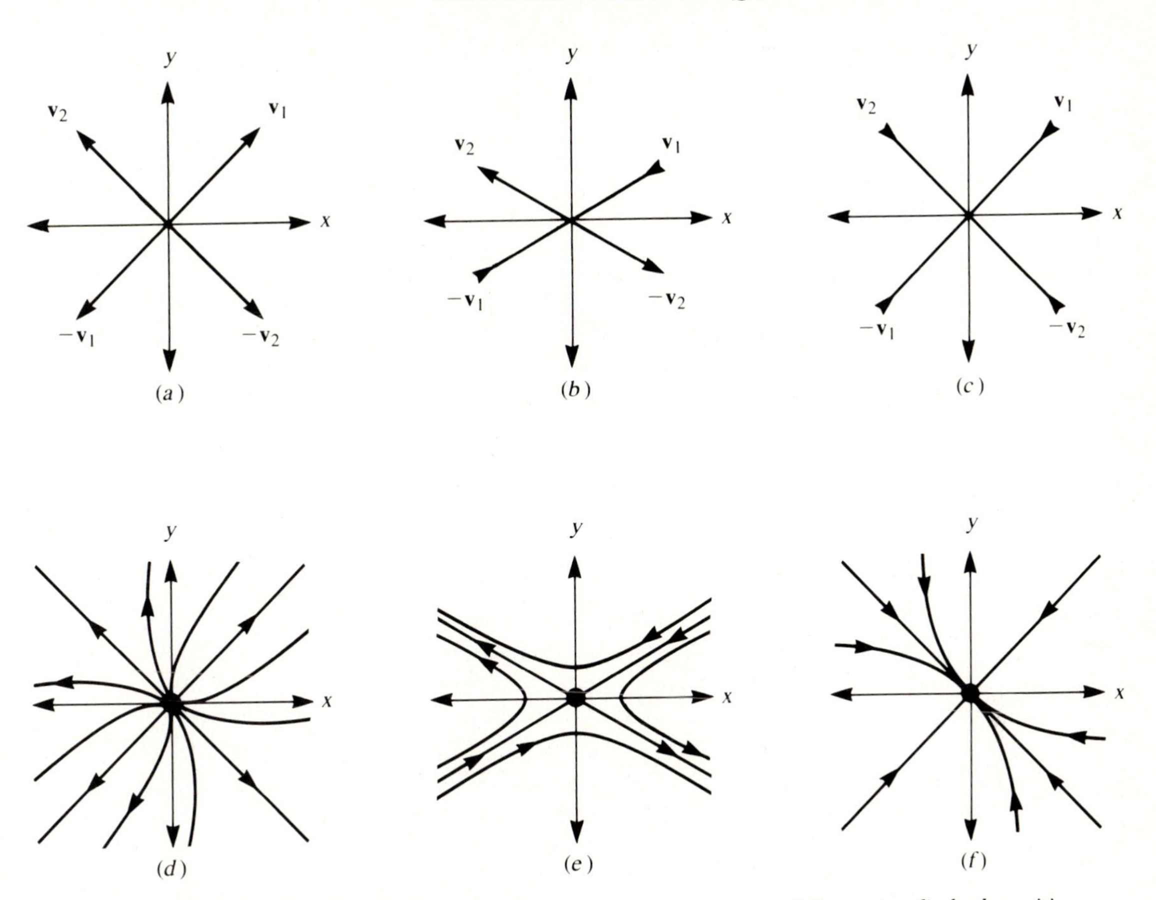

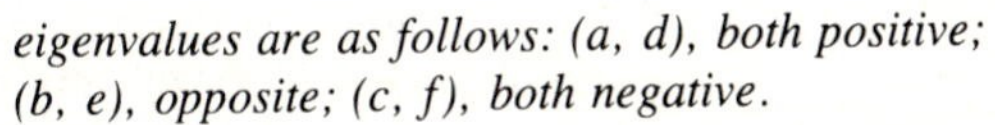
Figure 5.11 Sketches of the eigenvectors (a–c) and solution curves (d–f) of the linear equations (11a,b) for real eigenvalues. The signs of the two eigenvalues are as follows: (a, d), both positive; (b, e), opposite; (c, f), both negative.

Real Eigenvalues

Assuming that eigenvalues are real and distinct (with $\gamma \neq 0$, $\beta^2 - 4\gamma > 0$ where β, γ are as defined in Table 5.1 and equation (16), the behavior of solutions can be classified into one of the three possible categories:

1. Both eigenvalues are positive: $\lambda_1 > 0$, $\lambda_2 > 0$.
2. Eigenvalues are of opposite signs: e.g., $\lambda_1 > 0$, $\lambda_2 < 0$.
3. Both eigenvalues are negative: $\lambda_1 < 0$, $\lambda_2 < 0$.

In these three cases the eigenvectors also are real. Both vectors point away from the origin in case 1 and towards it in case 3. In case 2 they are of opposite orientations, with the one pointing outwards associated with the positive eigenvalue. Figure 5.11(a–c) illustrates this point.

All solutions grow with time in case 1 and decay with time in case 3; hence in each case the point (0, 0) is an *unstable* or a *stable node,* respectively. Case 2 is somewhat different in that solutions approach (0, 0) along one direction and recede from it along another. This unstable behavior is descriptively termed a *saddle point* (see Figure 5.11(*e*)).

Complex Eigenvalues

For $\lambda = a \pm bi$, we distinguish between the following cases:

4. Eigenvalues have a positive real part ($a > 0$).
5. Eigenvalues are pure imaginary ($a = 0$).
6. Eigenvalues have a negative real part ($a < 0$).

Note that when the linear equations have real coefficients, complex eigenvalues can occur only in conjugate pairs since they are roots of the quadratic characteristic equation.

The eigenvectors are then also complex and have no direct geometric significance. In building up real-valued solutions, recall that the expressions we obtained in Section 4.8 were products of exponential and sinusoidal terms. We remarked on the property that these solutions are oscillatory, with amplitudes that depend on the real part a of the eigenvalues $\lambda = a \pm bi$. In the xy plane, oscillations are depicted by trajectories that wind around the origin. When a is positive, the amplitude of oscillation grows, so the pair (x, y) spirals away from (0, 0); whereas if a is negative, it spirals towards it. The case where $a = 0$ is somewhat special. Here $e^{at} = 1$, and the amplitude of such solutions does not change. These trajectories are disjoint closed curves encircling the origin, which is then termed a *neutral center*. In this case a somewhat precarious balance exists between the forces that lead to increasing and decreasing oscillations. It is recognized that small changes in a system that oscillates in this way may disrupt the balance, and hence a neutral center is said to be *structurally unstable*. Cases 4, 5, and 6 are illustrated in Figure 5.12.

5.8 CLASSIFYING STABILITY CHARACTERISTICS

From certain combinations of the coefficients appearing in the linear equations, we can deduce criteria for each of the six classifications described in the previous section. We shall catalog the nature of the eigenvalues and thus the stability properties of a steady state using three quantities,

$$\beta = a_{11} + a_{22} = \text{Tr } \mathbf{A}, \tag{16a}$$

$$\gamma = a_{11}a_{22} - a_{12}a_{21} = \det \mathbf{A}, \tag{16b}$$

$$\delta = \beta^2 - 4\gamma = \text{disc } \mathbf{A}, \tag{16c}$$

where $\mathbf{A}$ is the 2×2 matrix of coefficients (a_{ij}) and $\mathbf{A} = \mathbf{J}(x_0, y_0)$. [See equation (12)] and Tr ($\mathbf{A}$) = trace, det ($\mathbf{A}$) = determinant, and disc ($\mathbf{A}$) = discriminant of $\mathbf{A}$.

Figure 5.12 *Solution curves for linear equations (11a,b) when eigenvalues are complex with (a) positive, (b) zero, and (c) negative real parts.*

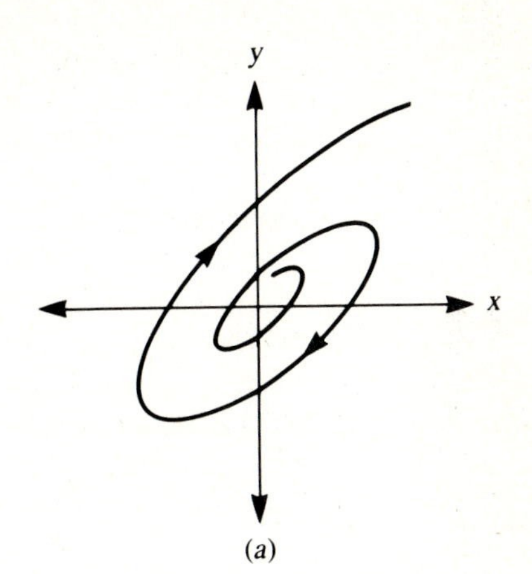

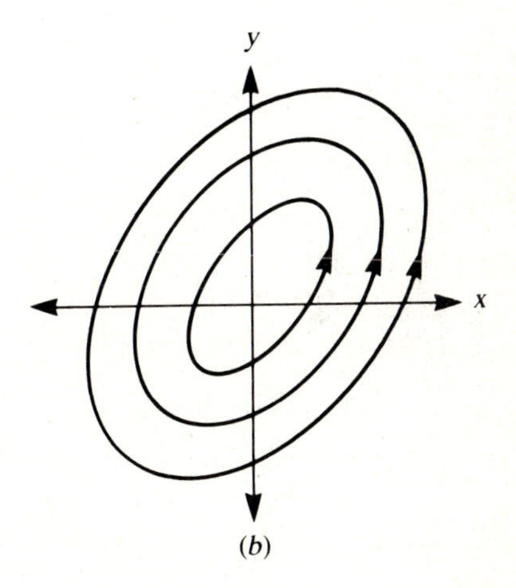

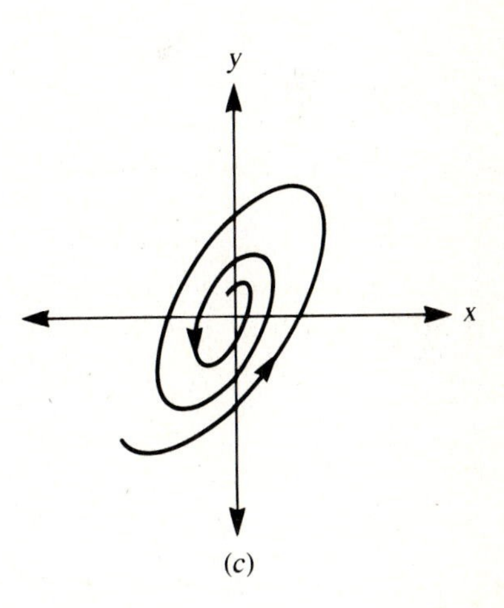

Criteria stem from the fact that eigenvalues are related to these by

$$\lambda_{1,2} = \frac{\beta \pm \sqrt{\delta}}{2}. \tag{17}$$

Consult Figure 5.13 for a graphical interpretation of the arguments that follow.

For real eigenvalues, δ must be a positive number. Now if γ is positive, $\delta = \beta^2 - 4\gamma$ will be smaller than β^2 so that $\sqrt{\delta} < \beta$. In that case, $\beta + \sqrt{\delta}$ *and* $\beta - \sqrt{\delta}$ will have the same sign [see Figures 5.13(*a*) and 5.13(*c*)]. In other words, the eigenvalues will then be positive if $\beta > 0$ [case 1, Figure 5.13(*a*)] and negative if $\beta < 0$ [case 3, Figure 5.13(*c*)]. On the other hand, if γ is negative, we arrive at the conclusion that $\sqrt{\delta}$ is bigger than β. Thus $\beta + \sqrt{\delta}$ and $\beta - \sqrt{\delta}$ will have opposite signs whether β is positive or negative [case 2, Figure 5.13(*b*)].

Example 8

In Section 5.6 we saw that the Jacobian of equations (9a,b) for the two steady states (0, 0) and (1, 1) are

$$\mathbf{J}(0,0) = \begin{pmatrix} 0 & -1 \\ -1 & 0 \end{pmatrix}, \qquad \mathbf{J}(1,1) = \begin{pmatrix} 1 & 0 \\ 0 & 1 \end{pmatrix}.$$

Thus for (0, 0), $\beta = 0$ and $\gamma = -1$; so (0, 0) is a saddle point. For (1, 1), $\beta = 2$ and $\gamma = 1$; so (1, 1) is an unstable node.

Example 9

Consider the system of equations

$$\frac{dx}{dt} = 2x - y, \qquad \frac{dy}{dt} = 3x + 2y.$$

Then

$$\beta = (2 + 2) = 4, \qquad \gamma = (2)(2) + (1)(3) = 7,$$
$$\delta = \beta^2 - 4\gamma = 16 - 28 = -12.$$

Since $\beta^2 < 4\gamma$, the eigenvalues will be complex. Since $\beta = 4 > 0$, the behavior is that of an unstable spiral.

Example 10

Consider the system

$$\frac{dx}{dt} = -4x + y, \qquad \frac{dy}{dt} = x - 2y.$$

Then

$$\beta = (-4 - 2) = -6, \qquad \gamma = (-4)(-2) - (1)(1) = 7,$$
$$\delta = \beta^2 - 4\gamma = 36 - 28 = 12.$$

Since $\beta < 0$ and $\gamma > 0$, the system is a stable node.

Figure 5.13 *Eigenvalues are those values λ at which the parabola $y = \lambda^2 - \beta\lambda + \gamma$ crosses the λ axis. Signs of these values depend on β and on the ratio of $\sqrt{\delta}$ to β where $\delta = \beta^2 - 4\gamma$. When $\gamma > 0$, both eigenvalues have the same sign as β. If $\delta < 0$, the parabola does not intersect the λ axis, so both eigenvalues are complex.*

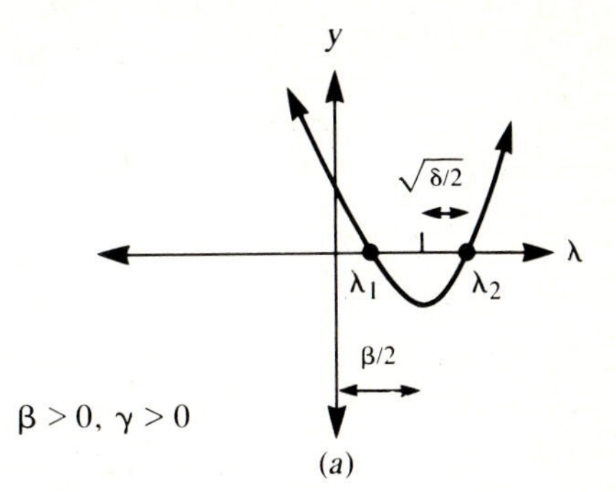

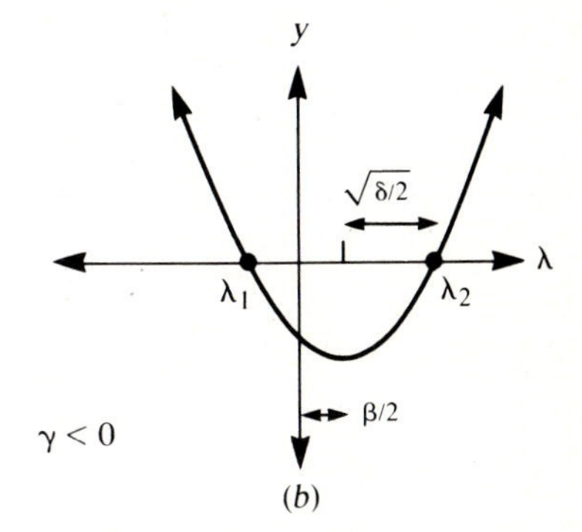

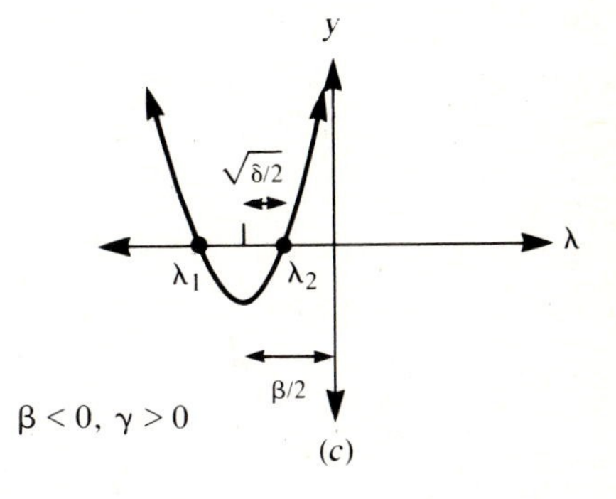

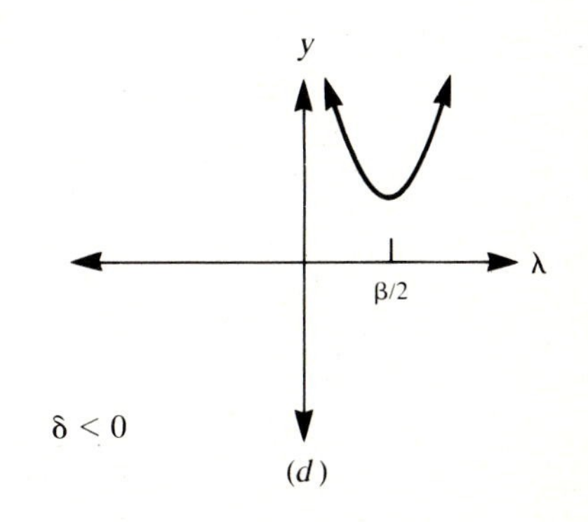

For eigenvalues to be complex (and not real) it is necessary and sufficient that $\delta = \beta^2 - 4\gamma$ be negative. Then

$$\lambda = \frac{\beta \pm i\sqrt{-\delta}}{2}.$$

Cases 4, 5, and 6 then follow for positive, zero, or negative β respectively.

To summarize, the steady state can be classified into six cases as follows:

1. Unstable node: $\beta > 0$ and $\gamma > 0$.
2. Saddle point: $\gamma < 0$.
3. Stable node: $\beta < 0$ and $\gamma > 0$.
4. Unstable spiral: $\beta^2 < 4\gamma$ and $\beta > 0$.
5. Neutral center: $\beta^2 < 4\gamma$ and $\beta = 0$.
6. Stable spiral: $\beta^2 < 4\gamma$ and $\beta < 0$.

The $\beta\gamma$ *parameter plane,* shown in Figure 5.14, consists of six regions in which one of the above qualitative behaviors obtains. This figure captures in a com-

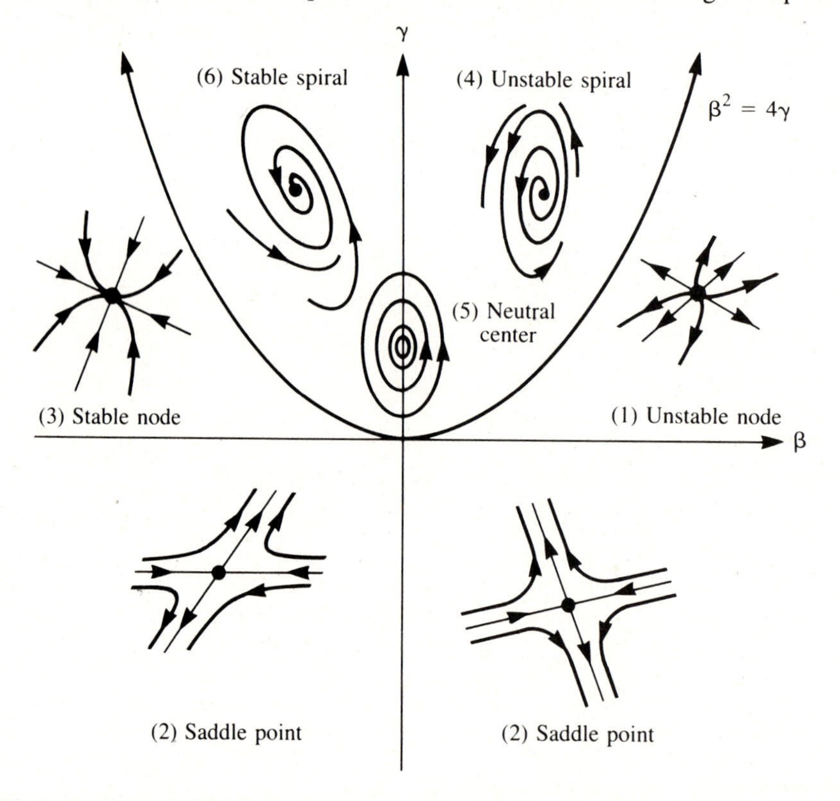

Figure 5.14 *To get a general idea of what happens in a linear system such as*

$$\dot{x} = a_{11}x + a_{12}y, \qquad \dot{y} = a_{21}x + a_{22}y,$$

we need only compute the quantities

$$\beta = a_{11} + a_{22}, \qquad \gamma = a_{11}a_{22} - a_{12}a_{21}.$$

The above parameter plane can then be consulted to determine whether the steady state (0, 0) *is a node, a spiral point, a center, or a saddle point.*

prehensive way the fundamental characteristics of a linear system. Notice that the region associated with a neutral center occupies a small part of parameter space, namely the positive γ axis.

The stability and behavior of a linear system, or the properties of a steady state of a nonlinear system can in practice be ascertained by determining β and γ and noting the region of the parameter plane in which these values occur. See examples 8, 9, and 10.

5.9 GLOBAL BEHAVIOR FROM LOCAL INFORMATION

Systems of nonlinear ODEs may have multiple steady states (see examples 5 and 6). Close to the steady states, behavior is approximated by the linearized equations, a fact that does not depend on the degree of the system; that is, it holds true in general for $n \times n$ systems.

An attribute of 2×2 systems that is *not* shared by those of higher dimensions is that local behavior at steady states can be used to reconstruct global behavior. By this we mean that stability properties of steady states and various gross features of the direction field determine a flow in the plane in an unambiguous way. The reason bigger systems of equations cannot be treated in the same way is that curves in higher dimensions are far less constrained by imposing a continuity requirement. A result that holds in the plane but not in higher dimensions is that a simple closed curve (for example, an ellipse or a circle) separates the plane into two disjoint regions, the "inside" and the "outside." It can be shown in a mathematically rigorous way that this limits the ways in which curves can form a smooth flow pattern in a planar region. Problem 16 gives some intuitive feeling for why this fact plays such a central role in establishing the qualitative behavior of 2×2 systems.

The terminology commonly used in the theory of ODEs reflects an underlying analogy between abstract mathematical equations and physical flows. We tend to associate the behavior of solutions to a 2×2 system with the motion of a two-dimensional fluid that emanates or vanishes at steady-state points. This at least imparts the idea of what a smooth phase-plane picture should look like. (We note a slight exception since saddle points have no readily apparent fluid analogy.) By smooth, or continuous flow we understand that a small displacement from a position (x_1, y_1) to one close to it (x_2, y_2) should not cause a drastic change in the direction of the flow.

There are a limited number of ways that trajectories can be combined to create a flow pattern that accommodates the local (steady-state) properties with the global property of continuity. A partial list follows:

1. Solution curves can only intersect at steady-state points.
2. If a solution curve is a closed loop, it must encircle at least one steady state that cannot be a saddle point (see Chapter 8).

Trajectories can have any one of several *asymptotic behaviors* (limiting behavior for $t \to +\infty$ or $t \to -\infty$). It is customary to refer to the *α-limit set* and *ω-limit set*, which are simply the sets of points that are approached along a trajectory for

$t \rightarrow -\infty$ and $t \rightarrow +\infty$ respectively. Limit sets may include any of the following (see Figure 5.15):

1. A steady-state point.
2. Infinity. (Trajectories emanating from or approaching infinitely large values in phase space are said to be unbounded.)
3. A *closed-loop trajectory*. (A trajectory may itself be a closed curve or else may approach or recede from one. Such solution curves represent oscillating systems; see Chapters 6 and 8.)
4. A *cycle graph* (a set containing a finite number of steady states connected by an equal number of trajectories).

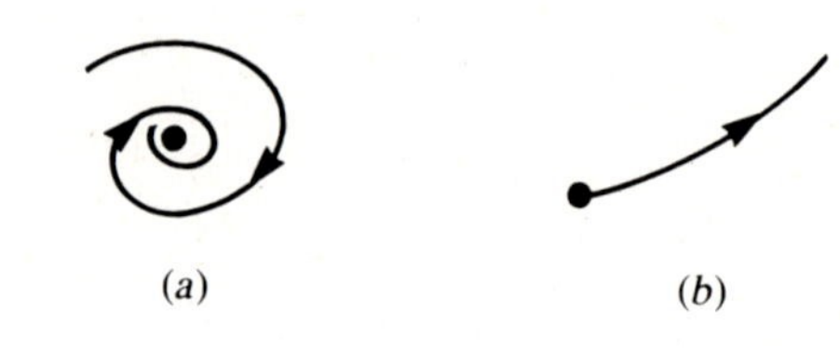

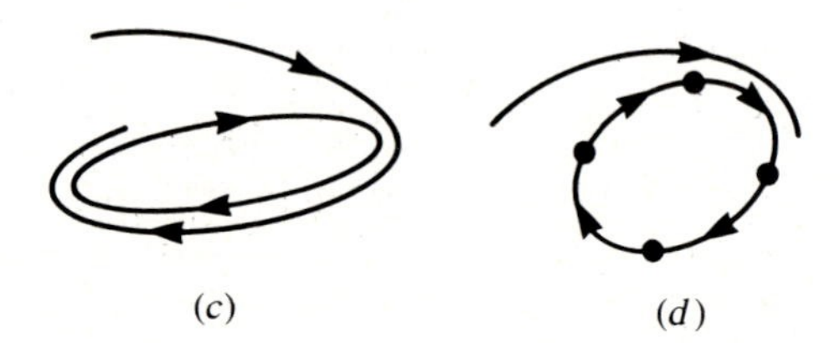

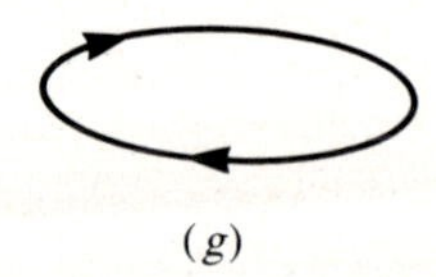

Figure 5.15 *Limit sets described in text: (a) steady-state point, (b) infinity, (c) closed-loop trajectory, (d) cycle graph, (e) heteroclinic trajectory, (f) homoclinic trajectory, and (g) limit cycle.*

Certain types of trajectories are further distinguished by name since they represent interesting or important properties. Three of these are listed here:

5. A *heteroclinic trajectory* connects two (different) steady states. (The term *connects* is often used loosely to convey that an orbit tends to each of the steady states for $t \to \pm\infty$.)

6. A *homoclinic trajectory* returns to the same steady state from whence it originates.

7. A *limit cycle* is a closed orbit that is the α or ω limit set of neighboring orbits (see Figure 5.15 and Chapter 8).

It has been shown that by linearizing a set of (nonlinear) equations about a given steady state, we can understand local behavior rather thoroughly. Indeed, this behavior falls into a small number of possible cases, six of which were described in Figure 5.14. (We did not go into details of several other singular cases, for example, if det $\mathbf{A} = 0$ or disc $\mathbf{A} = 0$. These are discussed in several sources in the references.)

Suppose we arrive at a prediction that some steady state is a spiral, a node, or a saddle point according to linear theory. The nonlinearity of the equations might distort that local behavior somewhat, but its basic features would not change. An exception to this occurs when linearization predicts a neutral center. In that case, somewhat more advanced analysis is necessary to establish whether this prediction holds true. A hint for why this prediction is not trustworthy has been given previously and involves the concept of structural stability. Briefly, even though the effect of nonlinearities is small near a steady state, it may suffice to disrupt the delicate balance of a neutral center. What happens when the delicate rings of a neutral center are broken? We postpone discussion of this to a later chapter.

5.10 CONSTRUCTING A PHASE-PLANE DIAGRAM FOR THE CHEMOSTAT

To demonstrate how to apply the theory given in Chapters 4 and 5 to a given situation, we return to the example of the chemostat. In Section 4.5 we discovered the following set of dimensionless equations depicting bacterial density N and nutrient concentration C:

$$\frac{dN}{dt} = \alpha_1\left(\frac{C}{1 + C}\right)N - N, \tag{18a}$$

$$\frac{dC}{dt} = -\left(\frac{C}{1 + C}\right)N - C + \alpha_2. \tag{18b}$$

As we saw, these nonlinear equations have two steady states, one of which represents a stable level of nutrient and cells. We now apply the method of phase-plane analysis to this example. Because only positive values of N and C are biologically meaningful, we shall restrict attention to the positive quadrant of the NC plane.

Step 1: Nullclines

The N nullcline

$\dot{N} = 0$ represents all the points such that

$$\alpha_1\left(\frac{C}{1 + C}\right)N - N = 0,$$

which are $N = 0$, or $\alpha_1(C/1 + C) = 1$. After rearranging, the latter leads to

$$C = \frac{1}{\alpha_1 - 1}. \tag{19}$$

This horizontal line crosses the C axis at $1/(\alpha_1 - 1)$. On this line and on the line $N = 0$, the value of N cannot change, so arrows are parallel to the C axis.

The C nullcline

$\dot{C} = 0$ represents all the points satisfying

$$-\left(\frac{C}{1 + C}\right)N - C + \alpha_2 = 0.$$

For a better way of expressing this implicit equation of a nullcline, we solve for N to get

$$N = (\alpha_2 - C)\frac{1 + C}{C}. \tag{20}$$

This is a single curve with the following properties:

1. It passes through $(\alpha_2, 0)$.
2. It is asymptotic to $C = 0$ and tends to $+\infty$ there.
3. Arrows along this nullcline are parallel to the N axis.

The curves corresponding to the N nullclines and C nullcline are shown on Figure 5.16. Notice that we have drawn the two curves intersecting in the first quadrant; in other words, we assume that

$$\frac{1}{\alpha_1 - 1} < \alpha_2. \tag{21}$$

When this fails to be true, the picture will be quite different, as shown in problem 11. Direction of arrows will be determined by tabulating several judiciously chosen values. Having calculated the Jacobian of equations (18a,b) previously, we observe that det $\mathbf{J} \neq 0$ at either steady state. This means that arrows along nullclines have opposite orientations on opposite sides of a steady state.

In determining the signs of dC/dt and dN/dt, it is sometimes helpful to prepare the ground by rewriting the equations in a more transparent form. For example, after rearranging equation (18a) we get

$$\frac{dN}{dt} = \frac{(\alpha_1 - 1)C - 1}{(1 + C)}N. \tag{22}$$

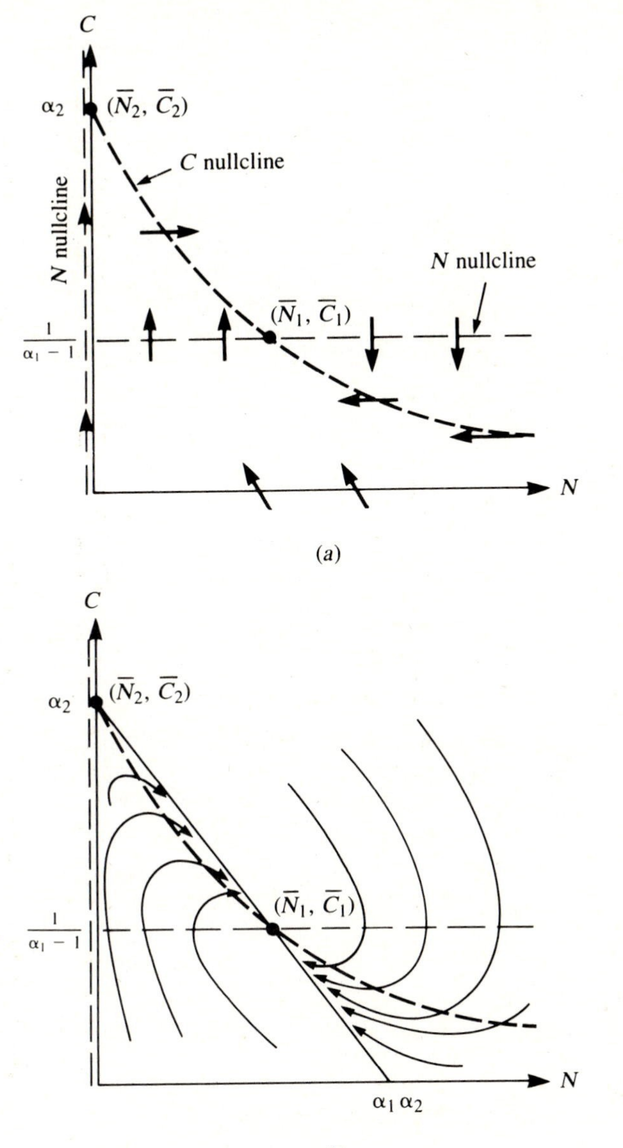

Figure 5.16 *Phase-plane portrait of the chemostat model based on equations (18a,b) showing nullclines (dashed and dotted lines) and steady-state points (heavy dots). (a) The directions of flow as given in Table 5.2. (b) The trajectories based on flow directions, steady state stability, and all other analysis.*

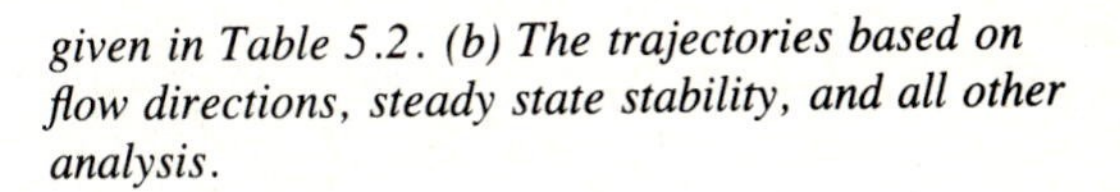

This allows us to conclude in a more direct way that dN/dt is negative whenever $C < 1/(\alpha_1 - 1)$ and positive when $C > 1/(\alpha_1 - 1)$. A similar procedure can be used for equation (18b). Table 5.2 summarizes these conclusions.

Table 5.2 Directions of Flow in the NC Plane (Fig. 5.16)

Case	C	N	dC/dt	dN/dt
1	small, > 0	on C nullcline	0	$\approx$ small term $\times N - N$; < 0; N must be decreasing
2	large, $> \alpha_2$	0	$\simeq -C + \alpha_2 < 0$; C must be decreasing	0
3	$\dfrac{1}{\alpha_1 - 1}$	$N >$ steady-state value $\overline{N}_1$	$\simeq -$large term + small term < 0; C must be decreasing	0
4	0	0	$\alpha_2 > 0$; C must be increasing	0

Step 2: Steady States

The two steady states of the chemostat, $(\overline{N}_1, \overline{C}_1)$ and $(\overline{N}_2, \overline{C}_2)$, are given by the expressions

$$\left(\alpha_1\left(\alpha_2 - \frac{1}{\alpha_1 - 1}\right), \frac{1}{\alpha_1 - 1}\right) \quad \text{and} \quad (0, \alpha_2). \tag{23}$$

These are the two points of intersection of $\dot{N} = 0$ and $\dot{C} = 0$. [Note that $(0, 1/(\alpha_1 - 1))$ is not such an intersection since it satisfies only the condition $\dot{N} = 0$.] The nullclines always intersect at two places, but the first of these intersections is in the positive NC quadrant only when $\alpha_2 > 1/(\alpha_1 - 1)$ and $\alpha_1 > 1$. We have already noted that these inequalities must be satisfied in order to apply to biological systems.

Step 3: Close to Steady States

We now summarize the calculations of stability characteristics of the two steady states:

1. In the steady state $(\overline{N}_1, \overline{C}_1)$ the Jacobian is

$$\mathbf{J} = \begin{pmatrix} 0 & \alpha_1 A \\ -1/\alpha_1 & -(A + 1) \end{pmatrix}, \tag{24}$$

where $A = \overline{N}_1/(1 + \overline{C}_1)^2$. Since

$$\beta = -(A + 1) < 0, \qquad \gamma = A > 0,$$
$$\beta^2 - 4\gamma = (A - 1)^2 > 0,$$

this steady state is always a stable node.

2. In the second steady state $(\overline{N}_2, \overline{C}_2)$, the Jacobian is

$$\mathbf{J} = \begin{pmatrix} \alpha_1 B - 1 & 0 \\ -B & -1 \end{pmatrix}, \tag{25}$$

where $B = \alpha_2/(1 + \alpha_2)$. Thus

$$\beta = \alpha_1 B - 2 \qquad \text{and} \qquad \gamma = 1 - \alpha_1 B.$$

This steady state will be a saddle point whenever $1 - \alpha_1 B < 0$, that is, when

$$\alpha_1 > \frac{1}{B}. \tag{26}$$

In problem 11(a) it is shown that this is satisfied precisely when

$$\alpha_2 > \frac{1}{\alpha_1 - 1}. \tag{27}$$

This condition ensures that the nonzero steady state $(\overline{N}_1, \overline{C}_1)$ exists. Thus, when $(\overline{N}_1, \overline{C}_1)$ is a biologically meaningful steady state, $(\overline{N}_2, \overline{C}_2) = (0, \alpha_2)$ is a saddle point.

The Shape of Trajectories Close to $(\overline{N}_1, \overline{C}_1)$

Problem 12 demonstrates that eigenvalues and corresponding eigenvectors of the Jacobian in equation (24) are as follows:

$$\lambda_1 = -A, \qquad \lambda_2 = -1; \tag{28}$$

$$\mathbf{v}_1 = \begin{pmatrix} \alpha_1 \\ -1 \end{pmatrix}, \qquad \mathbf{v}_2 = \begin{pmatrix} \alpha_2 A \\ -1 \end{pmatrix}. \tag{29}$$

In problem 12(d) we show that $\mathbf{v}_1$ defines a straight line through the steady state $(\overline{N}_1, \overline{C}_1)$ and two other points $(\alpha_1\alpha_2, 0)$ and $(0, \alpha_2)$. In problem 13 it is also shown that all trajectories approach this line as t approaches infinity.

It is worth remarking that several steps carried out in the chemostat example simplify the analysis. The first was that of reducing equations to dimensionless form; this eliminated many parameters that would complicate the expressions appearing in the Jacobian. The second step was recognizing certain recurring expressions, such as $\overline{N}_1/(1 + \overline{C}_1)^2$, and representing these by suitably defined constants. Such steps are recommended as an aid to organization when analyzing the behavior of a model.

We can complete a phase-plane portrait of the chemostat by combining the nullcline-and-arrow method with knowledge of the steady-state behavior ascertained above. [See also box on the shape of trajectories near the steady state $(\overline{N}_1, \overline{C}_1)$.] Figure 5.16(b) shows a smooth flow pattern consistent with both local and global clues. Other details of the flow are worked out in problems 10 through 13. We see that no matter what the initial values of C and N, solution curves eventually approach the steady state $(\overline{N}_1, \overline{C}_1)$.

Step 4: Interpreting the Solutions

Three hypothetical ways of starting a chemostat culture are described below. Figure 5.16 is used to deduce what happens in each situation.

First, suppose that the growth chamber in the chemostat initially has no bacteria or nutrient. As the stock solution of nutrient flows into the chamber, it causes the nutrient level there to increase. From Figure 5.16 we see that after starting at $(0, 0)$ we gradually approach the steady state $(0, \alpha_2)$. Thus C is building up to a level equivalent to that of the stock solution (recall the definition of α_2). N never increases, because bacteria are not present and thus cannot reproduce.

Now consider inoculating the chamber with a small bacterial population, $N = \epsilon$, and again starting with $C = 0$. Note that the solution curves through the N axis (for N small) sweep into the positive quadrant. N initially decreases, because until a nutrient level is established, bacteria cannot reproduce fast enough to replace those that are lost in the effluent. Once excess nutrient is available, bacterial densities rise dramatically, so that the solution curve has a nearly vertical "kink." At this point, rapid consumption causes decline in the nutrient and N and C approach their steady-state values. (In theory, the steady state is only attained at $t = \pm\infty$. In practice it may take only a finite time such as a few hours to be close enough to steady state as to be indistinguishable from it.)

As a third example, starting with $N > \overline{N}_1$, $C > \overline{C}_1$ we find that N initially increases, thereby causing nutrient depletion. (C drops below its steady-state value.) The bacterial population declines so that nutrient consumption is less rapid. Again, after these transients, the steady state is once more established.

In problem 14 we return once again to the original parameters of the chemostat. There it is shown that the following relevant conclusions are reached:

Summary of the Chemostat Model

1. If either

$$\frac{F}{V} \leq K_{\max}, \qquad \text{or}$$

$$\frac{F}{V} > K_{\max} \qquad \text{and} \qquad C_0 \leq \frac{(F/V)K_n}{K_{\max} - (F/V)},$$

$(\overline{N}_2, \overline{C}_2) = (0, C_0)$ is the only steady-state point and it is stable. This situation is called a *washout* since the microbe will be washed out of the chemostat.

2. If both

$$\frac{F}{V} > K_{\max} \qquad \text{and}$$

$$\frac{(F/V)K_n}{K_{\max} - (F/V)} < C_0,$$

then $(\overline{N}_1, \overline{C}_1)$ is a stable steady-state point. Provided $N(0)$ is initially nonzero and $C_0 > 0$, the bacterial density and nutrient concentration will converge to $\overline{N}_1$ and $\overline{C}_1$ respectively.

5.11 HIGHER-ORDER EQUATIONS

So far we have dealt only with systems of first-order equations. However, the geometric theory used here can also be applied to problems consisting of higher-order equations, such as

$$\frac{d^n y}{dt^n} = F\left(\frac{d^{n-1}y}{dt^{n-1}}, \frac{d^{n-2}y}{dt^{n-2}}, \ldots, y', y\right). \tag{30}$$

The problem will be reduced to one that is familiar by converting this nth-order equation to a set of n first-order equations. To do so, define $y_0 = y$ and $n - 1$ new variables, each of which represents the derivative of the preceding variable:

$$\begin{aligned} y_0 &= y, \\ y_1 &= \frac{dy_0}{dt} = \frac{dy}{dt}, \\ &\vdots \\ y_{n-1} &= \frac{dy_{n-2}}{dt} = \cdots = \frac{d^{n-1}y}{dt^{n-1}}. \end{aligned}$$

Now rewrite this as a "system" of equations in the variables $y_0, \ldots, y_{n-1}$, using equation (30) in the final equation:

$$\begin{aligned} \frac{dy_0}{dt} &= y_1, \\ \frac{dy_1}{dt} &= y_2, \\ &\vdots \\ \frac{dy_{n-2}}{dt} &= y_{n-1}, \\ \frac{dy_{n-1}}{dt} &= F(y_{n-1}, y_{n-2}, \ldots, y_1, y_0). \end{aligned} \tag{31}$$

The system (31) can be summarized by a vector equation,

$$\frac{d\mathbf{y}}{dt} = \mathbf{f}(\mathbf{y}) = (f_0, f_1, \ldots, f_{n-1}), \tag{32}$$

where $f_0 = y_1, f_1 = y_2, \ldots, f_{n-1} = F$, and so on.

A solution to (32), $\mathbf{y}(t)$ is a curve in n-dimensional space, parameterized by t. While $\mathbf{f}(\mathbf{y})$ again represents a direction field, it is now much more difficult to visualize. Nullclines are hyperplanes or hypersurfaces of dimension $n - 1$; in the examples given here, the subspaces are $y_1 = 0$, $y_2 = 0, \ldots,$ and $F(y_{n-1}, y_{n-2}, \ldots, y_1, y_0) = 0$. It is clear that while the geometric interpretation underlying the equations can be thus generalized, we must abandon the idea of visualizing qualitative behavior in all but the simplest cases.

In theory, steady states can be determined analytically (when equations such as $\dot{y}_0 = 0, \ldots, \dot{y}_{n-1} = 0$ can be solved). The stability of these steady states is ascertained by linearizing the equations, but technical difficulties ensue (see Section 6.4). Even given complete local information about steady states, the global qualitative behavior is generally unknown, with few exceptions. So while in theory the scope of the analysis of the 2 × 2 case can be extended, in practice we obtain valuable insights in the general case only rarely.

PROBLEMS*

1. For the following first-order ordinary differential equations, sketch solution curves $y(t)$ by first plotting the tangent vectors specified by the differential equations:

(a) $\dfrac{dy}{dt} = y^2.$

(b) $\dfrac{dy}{dt} = 1 - \dfrac{y}{1+y}.$

(c) $\dfrac{dy}{dt} = y(y-2).$

(d) $\dfrac{dy}{dt} = ye^{(y-1)}$

(e) $\dfrac{dy}{dt} = \sin y \cos y.$

2. For problem 1(a–e) above, graph dy/dt as a function of y. Use this graph to summarize the behavior of solutions to the equations on the y axis by drawing arrows to indicate when y increases or decreases.

3. Prove that solution curves of equation (3) have inflection points at

$$y = 1 \pm \frac{\sqrt{3}}{3}.$$

(*Hint:* Consider $f'(y)$ and see Figure 5.3(*a*).)

4. *Curves in the plane*. The locus of points for which $x = y^3$ can be written in the form $(x(t), y(t))$ by choosing some parameter t. For example,

$$x(t) = t, \qquad y(t) = t^3.$$

Other choices are possible, for example,

$$x = s^{1/3}, \qquad y = s.$$

These would depict the same curve but a different rate of motion along the curve. Then, using this parameterized form we can depict any position on the curve by the vector

$$\mathbf{x}(t) = (t, t^3)$$

and any tangent vector to the curve by the vector

$$\mathbf{v}(t) = \frac{d\mathbf{x}}{dt} = \left(\frac{dx(t)}{dt}, \frac{dy(t)}{dt}\right) = (1, 3t^2).$$

* Problems preceded by an asterisk are especially challenging.

For example, at $t = 1$, $\mathbf{x} = (1, 1)$ and $\mathbf{v} = (1, 3)$.

(a) Using the parameterized form given here, sketch the curve, and compute the tangent vectors at points $(0, 0)$, $(2, 8)$, and $(-1, -1)$.

(b) Find a way of parameterizing the following curves, and determine the form of the tangent vector to the curve:

(1) $y = x(x - 1)$.
(2) $y^2 = \sin x$.
(3) $x = 1/y$.
(4) $x^2 + y^2 = 1$.
(5) $y = ax + b$.
(6) $y = 4x^2$.

5. Sketch the nullclines in the xy phase plane, identify steady states, and draw directions of arrows on the nullclines for the following systems of first-order equations:

(a) $$\frac{dx}{dt} = y^2 - x^2, \qquad \frac{dy}{dt} = x - 1.$$

(b) $$\frac{dx}{dt} = x(y^2 - y), \qquad \frac{dy}{dt} = x - y.$$

(c) $$\frac{dx}{dt} = x^2 + y, \qquad \frac{dy}{dt} = -y.$$

(d) $$\frac{dx}{dt} = -xy, \qquad \frac{dy}{dt} = (1 + x)(1 - y).$$

(e) $$\frac{dx}{dt} = x^2 - y, \qquad \frac{dy}{dt} = y^2 - x.$$

(f) $$\frac{dx}{dt} = \frac{-xy}{1 + x} + x, \qquad \frac{dy}{dt} = \frac{xy}{1 + x} - y.$$

(g) $$\frac{dx}{dt} = xy(1 - x) + C, \qquad \frac{dy}{dt} = y\left(1 - \frac{y}{x}\right).$$

6. For problem 5(a–g) find the Jacobian of each system of equations and determine stability properties of each steady state.

7. Sketch the phase-plane behavior of the following systems of linear equations and classify the stability characteristic of the steady state at $(0, 0)$:

(a) $$\frac{dx}{dt} = -2y, \qquad \frac{dy}{dt} = x.$$

(b) $$\frac{dx}{dt} = 3x + 2y, \qquad \frac{dy}{dt} = 4x + y.$$

(d) $$\frac{dx}{dt} = 5x + 8y, \qquad \frac{dy}{dt} = -3x - 5y.$$

(e) $$\frac{dx}{dt} = -4 - 2y, \qquad \frac{dy}{dt} = 3x - y.$$

(c) $$\frac{dx}{dt} = 2x + y,\qquad \frac{dy}{dt} = x + 2y.$$

(f) $$\frac{dx}{dt} = x - 4y,\qquad \frac{dy}{dt} = x + y.$$

8. Write a system of linear first-order ODEs whose solutions have the following qualitative behaviors:
(a) $(0,0)$ is a stable node with eigenvalues $\lambda_1 = -1$ and $\lambda_2 = -2$.
(b) $(0,0)$ is a saddle point with eigenvalues $\lambda_1 = -1$ and $\lambda_2 = 3$.
(c) $(0,0)$ is a center with eigenvalues $\lambda = \pm 2i$.
(d) $(0,0)$ is an unstable node with eigenvalues $\lambda_1 = 2$ and $\lambda_2 = 3$.

Hint: Use the fact that λ_1 and λ_2 are eigenvalues of a matrix **A**, then

$$\lambda_1 + \lambda_2 = \text{Tr}\,\mathbf{A} = a_{11} + a_{22},$$
$$\lambda_1\lambda_2 = \det \mathbf{A} = a_{11}a_{22} - a_{12}a_{21}.$$

Note that there will be many possible choices for each of the above.

9. Consider the system of equations

$$\dot{x} = y - x^2, \qquad \dot{y} = y - 2x^2.$$

(a) Show that the only steady state is $(0,0)$.
(b) Draw nullclines and determine the directions of arrows on the nullcline. Note that $(0,0)$ is a point of tangency of the two nullclines, which intersect but do not cross.
(c) Find the Jacobian at $(0,0)$, show that its determinant is equal to zero, and conclude that two eigenvalues are $\lambda_1 = 0$ and $\lambda_2 = 1$.
(d) Sketch solution curves in the *xy* plane.

10. By examining Figure 5.16 describe in words what would happen if we set up the chemostat to contain the following:
(a) A small number of bacteria with excess nutrient in the growth chamber.
(b) A large number of bacteria with very little nutrient in the growth chamber.

11. In drawing the phase-plane diagram of the chemostat, we assumed that $\alpha_2 > 1/(\alpha_1 - 1)$.
(a) Show that $(\overline{N}_2, \overline{C}_2)$ is a saddle point whenever this inequality is satisfied.
(b) Now suppose this inequality is not satisfied. Sketch the resulting phase-plane diagram and interpret the biological meaning.

12. **(a)** In the chemostat model find the quantity

$$A = \frac{\overline{N}_1}{(1 + \overline{C}_1)^2}$$

in terms of α_1 and α_2 [where $(\overline{N}_1, \overline{C}_1)$ is given by (23)].
(b) Show that the two eigenvalues of the Jacobian given by (24) are

$$\lambda_1 = -A, \qquad \text{and} \qquad \lambda_2 = -1.$$

(c) Show that the corresponding eigenvectors are

$$\mathbf{v}_1 = \begin{pmatrix} \alpha_1 \\ -1 \end{pmatrix}, \qquad \mathbf{v}_2 = \begin{pmatrix} \alpha_1 A \\ -1 \end{pmatrix}.$$

***(d)** Show that the eigenvector $\mathbf{v}_1$ and the steady state $(\overline{N}_1, \overline{C}_1)$ define a straight line whose equation is

$$N - \alpha_1\alpha_2 = -\alpha_1 C.$$

[*Hint:* Use the fact that the slope is given by the ratio $\alpha_1/(-1) = -\alpha_1$ of the components of $\mathbf{v}_1$.]

(e) Show that this line passes through the points $(\alpha_1\alpha_2, 0)$ and $(0, \alpha_2)$.

13. In this problem we establish that, for the chemostat all trajectories approach the line

$$N - \alpha_1\alpha_2 = \alpha_1 C.$$

(a) Multiply equation (18b) by α_1 and add to equation (18a). Show that this leads to

$$\frac{d}{dt}(N + \alpha_1 C) = \alpha_1\alpha_2 - (N + \alpha_1 C).$$

(b) Let $x = N + \alpha_1 C$ and integrate the equation in part (a). Show that

$$x(t) = Ke^{-t} + \alpha_1\alpha_2$$

is a solution (K = a constant of integration).

(c) Show that in the limit for $t \to \infty$ one obtains $x(t) \to \alpha_1\alpha_2$; that is,

$$N + \alpha_1 C = \alpha_1\alpha_2.$$

Conclude that as t approaches infinity, all points $(N(t), C(t))$ approach this line.[1]

14. (a) Verify that conclusions outlined in the summary of the chemostat model at the end of Section 5.10 are correct.

(b) Sketch the phase-plane behavior of the original dimension-carrying variables of the problem. (Your sketch should be similar to Figure 5.16 but with relabeled axes.)

15. Equation (1a) is linear but nonhomogeneous. To solve this problem consider first the corresponding homogeneous problem

$$\frac{dy}{dt} + y = 0.$$

Find the solution $y = \Phi(t)$ of this equation and look for solutions of the equation (1a) of the form

$$y = \Phi(t)C(t)$$

where $C(t)$ is an unknown function. Solve for $C(t)$. This procedure is known as the method of *variation of parameters*.

1. This problem was kindly suggested by C. M. Biles.

16. In the accompanying figure, locations and stability properties of steady states have been indicated by arrows. Fill in the global flow pattern using the fact that continuity of the flow must be preserved (that is, no sharp transitions at neighboring points except in the vicinity of steady states). In some cases more than one qualitative flow pattern is possible. Can you determine which of the following gives ambiguous clues?

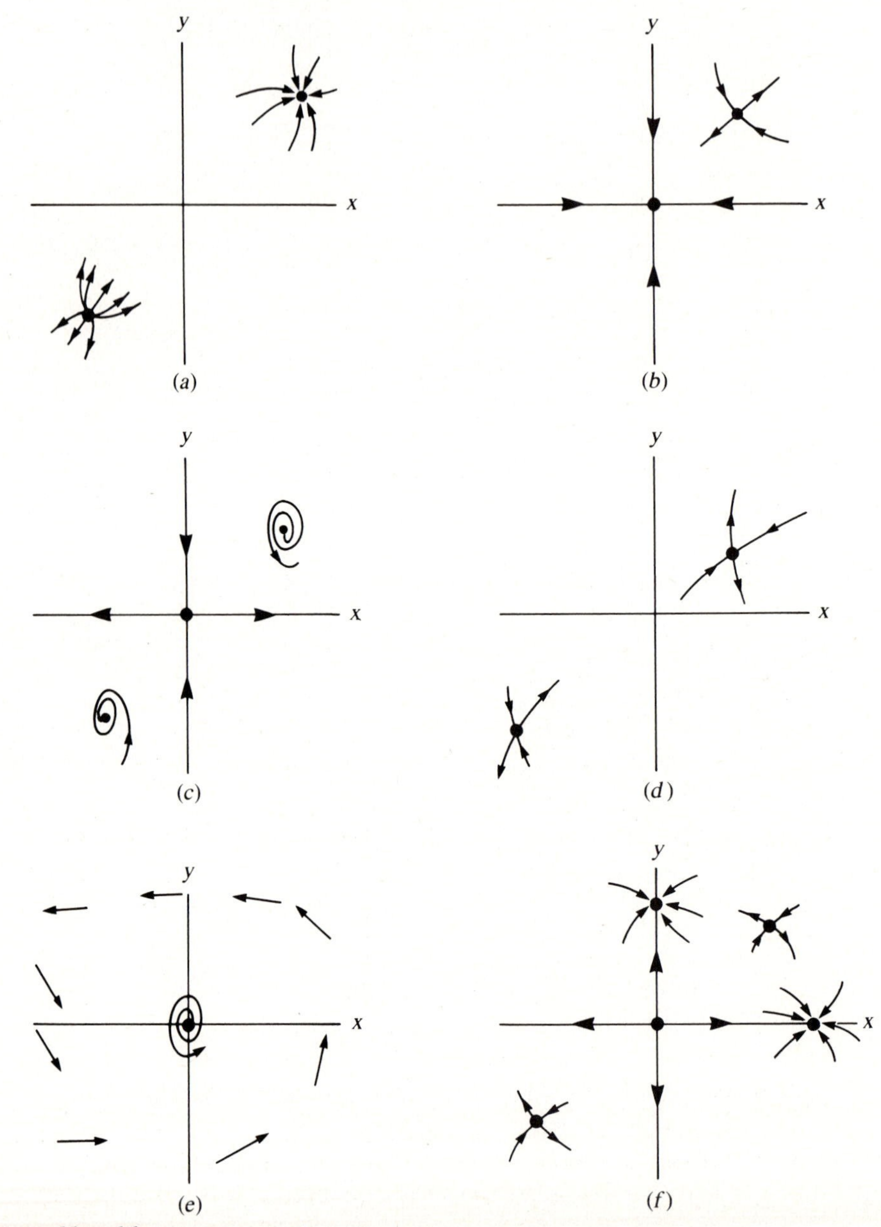

Figure for problem 16.

17. In this problem we explore one of the major distinctions between 2×2 systems of equations, which are represented by flows in the plane, and those of higher dimensionality.

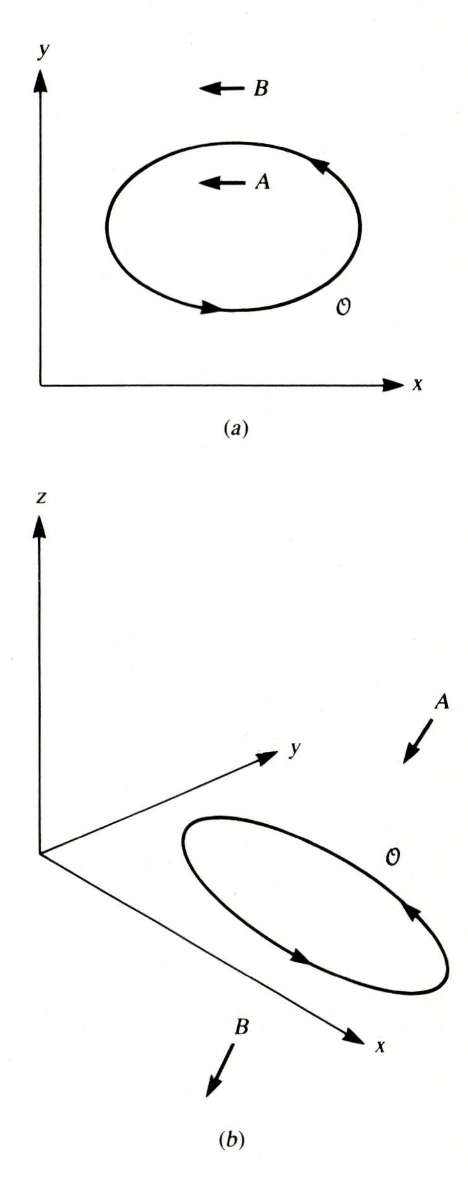

Figure for problem 17.

(a) In diagram (*a*) of a 2×2 system, a closed orbit ($\mathcal{O}$) has been drawn in the *xy* plane. The arrows *A* and *B* represent the local directions of motion at two points on the inside and outside of the closed curve.

(1) By preserving a continuous flow, sketch several *different* qualitative flow patterns consistent with the diagram.

(2) Section 5.11 tells us there must be a steady state somewhere in the diagram. In which region must it be, and why?

(b) A similar diagram in three dimensions (for a system *xyz* of three equations) leads to some ambiguity. Is it possible to define inside and outside regions for the orbit? Give some sketches or verbal descriptions of flow patterns consistent with this orbit. Show that it is not necessary to assume that a steady state is associated with the closed orbit.

18. Use phase-plane methods to find qualitative solutions to the model for the glucose-insulin system due to Bellomo et al. (1982). (See problem 27 in Chapter 4.) Draw nullclines, identify steady states, and sketch trajectories in the *ig* plane. Interpret your graph and discuss how parameters might influence the nature of the solutions.

19. Use methods similar to those mentioned in problem 17 to explore the model for continuous chemotherapy that was suggested in problem 25 of Chapter 4.

20. *Extended problem or project.* Using plausible assumptions or sources in the literature, suggest appropriate forms for the functions $F_1(X)$, $F_2(Y)$, $F_3(X, Y)$, and $F_4(X, Y)$ in the model for insulin and glucose proposed by Bolie (1960) (see equations (84a) and (84b) in Chapter 4). Use these functions to treat the problem by phase-plane methods and interpret your solutions.

21. In this problem we examine a continuous plant-herbivore model. We shall define q as the chemical state of the plant. Low values of q mean that the plant is toxic; higher values mean that the herbivores derive some nutritious value from it. Consider a situation in which plant quality is enhanced when the vegetation is subjected to a low to moderate level of herbivory, and *declines* when herbivory is extensive. Assume that herbivores whose density is I are small insects (such as scale bugs) that attach themselves to one plant for long periods of time. Further assume that their growth rate depends on the quality of the vegetation they consume. Typical equations that have been suggested for such a system are

$$\frac{dq}{dt} = K_1 - K_2 qI(I - I_0),$$

$$\frac{dI}{dt} = K_3 I\left(1 - \frac{K_4 I}{q}\right).$$

(a) Explain the equations, and suggest possible meanings for K_1, K_2, I_0, K_3, and K_4.

(b) Show that the equations can be written in the following dimensionless form:

$$\frac{dq}{dt} = 1 - KqI(I - 1),$$

$$\frac{dI}{dt} = \alpha I\left(1 - \frac{I}{q}\right).$$

Determine K and α in terms of original parameters.

(c) Find qualitative solutions using phase-plane methods. Is there a steady state? What are its stability properties?

(d) Interpret your solutions in part (c).

22. *A continuous ventilation-volume model*. In Chapters 1 and 2 we considered a simple model for irregular patterns of breathing. It was assumed that the sensitivity of the CO_2 chemoreceptor controls the depth (volume) of breathing. Over the time scale of 10 to 100 breaths the discontinuous nature of breathing and the delay in CO_2 sensitivity play significant roles. However, suppose we now view the process over a much longer time length (t = several hours). Define

$C(t)$ = blood CO_2 concentration at time t

$V(t)$ = magnitude of ventilation volume at time t.

(a) By making the approximations

$$\frac{C_{n+1} - C_n}{\Delta t} \simeq \frac{dC}{dt}, \qquad \frac{V_{n+1} - V_n}{\Delta t} \simeq \frac{dV}{dt}$$

reason that the continuous equations for the CO_2 ventilation volume systems based on equations (49) of Chapter 1 take the form

$$\frac{dC}{dt} = -\mathcal{L}(V, C) + m,$$

$$\frac{dV}{dt} = \mathcal{S}(C) - \epsilon V,$$

where $\mathcal{L}$ = CO_2 loss rate per unit time and $\mathcal{S}$ = CO_2-induced ventilation change per unit time. What is ϵ in terms of Δt?

(b) Now investigate the problem using the following steps: First assume that $\mathcal{L}$ and $\mathcal{S}$ are linear functions (as in problem 18 in Chapter 1); that is,

$$\mathcal{L}(V, C) = \beta V, \qquad \mathcal{S}(C) = \alpha C.$$

(1) Write out the system of equations, find their steady state, and determine the eigenvalues of the equations. Show that decaying oscillations may occur if $\epsilon^2 < 4\alpha\beta$.

(2) Sketch the phase-plane diagram of the system. Interpret your results biologically.

(c) Now consider the situation where

$$\mathcal{L}(V, C) = \beta VC, \qquad \mathcal{S}(C) = \alpha C.$$

(1) Explain the biological significance of this system. When is this a more valid assumption?

(2) Repeat the analysis requested in part (b).

(d) Finally, suppose that

$$\mathcal{L}(V, C) = \beta V, \qquad \mathcal{S}(C) = \frac{V_{\max} C}{K + C}.$$

How does this model and its predictions differ from that of part (b)?

23. The following equations were given by J. S. Griffith (1971, pp. 118–122), as a model for the interactions of messenger RNA M and protein E:

$$\dot{M} = \frac{aKE^m}{1 + KE^m} - bM, \qquad \dot{E} = cM - dE.$$

(See problem 25 in Chapter 7 for an interpretation.)

(a) Show that by changing units one can rewrite these in terms of dimensionless variables, as follows

$$\dot{M} = \frac{E^m}{1 + E^m} - \alpha M, \qquad \dot{E} = M - \beta E.$$

Find α and β in terms of the original parameters.

(b) Show that one steady state is $E = M = 0$ and that others satisfy $E^{m-1} = \alpha\beta(1 + E^m)$. For $m = 1$ show that this steady state exists only if $\alpha\beta \leq 1$.

(c) *Case 1*. Show that for $m = 1$ and $\alpha\beta > 1$, the only steady state $E = M = 0$ is stable. Draw a phase-plane diagram of the system.

(d) *Case 2*. Show that for $m = 2$, at steady state

$$E = \frac{1 \pm (1 + 4\alpha^2\beta^2)^{1/2}}{2\alpha\beta}.$$

Conclude that there are two solutions if $2\alpha\beta < 1$, one if $2\alpha\beta = 1$, and none if $2\alpha\beta > 1$.

(e) *Case 2 continued*. For $m = 2$ and $2\alpha\beta < 1$, show that there are two stable steady states (one of which is at $E = M = 0$) and one saddle point. Draw a phase-plane diagram of this system.

24. In modeling the effect of spruce budworm on forest, Ludwig et al. (1978) defined the following set of variables for the condition of the forest:

$S(t)$ = total surface area of trees

$E(t)$ = energy reserve of trees.

They considered the following set of equations for these variables in the presence of a constant budworm population B:

$$\frac{dS}{dt} = r_S S\left(1 - \frac{S}{K_S}\frac{K_E}{E}\right),$$

$$\frac{dE}{dt} = r_E E\left(1 - \frac{E}{K_E}\right) - P\frac{B}{S}.$$

The factors r, K, and P are to be considered constant.

***(a)** Interpret possible meanings of these equations.

(b) Sketch nullclines and determine how many steady states exist.

(c) Draw a phase-plane portrait of the system. Show that the outcomes differ qualitatively depending on whether B is small or large.

***(d)** Interpret what this might imply biologically.

[*Note:* You may wish to consult Ludwig et al. (1978) or to return to parts (a) and (d) after reading Chapter 6.]

REFERENCES

Arnold, V. I. (1973). *Ordinary Differential Equations*. MIT Press, Cambridge, Mass.

Bellomo, J.; Brunetti, P.; Calabrese, G.; Mazotti, D.; Sarti, E.; and Vincenzi, A. (1982). Optimal feedback glaecemia regulation in diabetics. *Med. & Biol. Eng. & Comp., 20,* 329–335.

Bolie, V. W. (1960). Coefficients of normal blood glucose regulation. *J. Appl. Physiol., 16,* 783–788.

Boyce, W. E., and Diprima, R. C. (1977). *Elementary Differential Equations*. 3rd ed. Wiley, New York.

Braun, M. (1979). *Differential equations and their applications: An introduction to applied mathematics*. 3rd ed. Springer-Verlag, New York.

Griffith, J. S. (1971). *Mathematical Neurobiology*. Academic Press, New York.

Hale, J. K. (1980). *Ordinary Differential Equations*. Krieger Publishing Company, Huntington, N.Y.

Ludwig, D.; Jones, D. D.; and Holling, C. S. (1978). Qualitative analysis of insect outbreak systems: the spruce budworm and forest. *J. Anim. Ecol., 47,* 315–332.

Odell, G. M. (1980). Qualitative theory of systems of ordinary differential equations, including phase plane analysis and the use of the Hopf bifurcation theorem. In L. A. Siegel, ed., *Mathematical Models in Molecular and Cellular Biology*. Cambridge University Press, Cambridge.

Ross, S. L. (1984). *Differential Equations*. 3d ed. Wiley, New York, chap. 13.

6 Applications of Continuous Models to Population Dynamics

Each organic being is striving to increase in a geometrical ratio . . . each at some period of its life, during some season of the year, during each generation or at intervals has to struggle for life and to suffer great destruction. . . . The vigorous, the healthy, and the happy survive and multiply.

Charles R. Darwin. (1860). *On the Origin of Species by Means of Natural Selection,* D. Appleton and Company, New York, chap. 3.

The growth and decline of populations in nature and the struggle of species to predominate over one another has been a subject of interest dating back through the ages. Applications of simple mathematical concepts to such phenomena were noted centuries ago. Among the founders of mathematical population models were Malthus (1798), Verhulst (1838), Pearl and Reed (1908), and then Lotka and Volterra, whose works were published primarily in the 1920s and 1930s.

The work of Lotka and Volterra, who arrived independently at several models including those for predator-prey interactions and two-species competition, had a profound effect on the field now known as *population biology*. They were among the first to study the phenomena of interacting species by making a number of simplifying assumptions that led to nontrivial but tractable mathematical problems. Since their pioneering work, many other notable contributions were made. Among these is the work of Kermack and McKendrick (1927), who addressed the problem of outbreaks of epidemics in a population.

Today, students of ecology and population biology are commonly taught such classical models as part of their regular biology curriculum. Critics of these historical models often argue that certain biological features, such as environmental effects, chance random events, and spatial heterogeneity to mention a few, were ignored. However, the importance of these models stems not from realism or the accuracy of their predictions but rather from the simple and fundamental principles that they set forth; the propensity of predator-prey systems to oscillate, the tendency of competing species to exclude one another, the threshold dependence of epidemics on population size are examples.

While appreciation of the Lotka-Volterra models in the biological community is mixed, it is nevertheless interesting to note that in subtle yet important ways they have helped to shape certain research directions in current biology. As demonstrated by the Nicholson-Bailey model of Chapter 3, a model does not have to be accurate to serve as a helpful diagnostic tool. We shall later discuss more specific ways in which the unrealistic predictions of simple models have led to new empirical as well as theoretical progress.

The classic population biology models serve several purposes in this text. Aside from being interesting in their own right, models of two interacting species or of epidemics in a fixed population are ideal illustrations of the techniques and concepts outlined in Chapters 4 and 5. The models also demonstrate how the predictions of a model change when slight alterations are made in the equations or in values of the critical quantities that appear in them. Finally, the fact that these models are fairly simple allows us to assess critically the various assumptions and their consequences.

As in previous discussions, we set the stage by a brief discussion of models for single-species populations. (Section 4.1 introduced this topic; here we somewhat broaden the context.) In Sections 6.2 and 6.3 the Lotka-Volterra predator-prey and species competition models are described and then analyzed. The story of Volterra's initiation to this biological area is well known. This Italian mathematician became interested in the area of population biology through conversations with a colleague, U. d'Ancona, who had observed a puzzling biological trend. During World War I, commercial fishing in the Adriatic Sea fell to rather low levels. It was anticipated that this would cause a rise in the availability of fish for harvest. Instead, the population of commercially valuable fish declined on average while the number of sharks, which are their predators, increased. The two populations were also perceived to fluctuate.

Volterra suggested a somewhat naive model to describe the predator-prey interactions in the fish populations and was thereby able to explain the trends d'Ancona had observed. As we shall see, the model's basic prediction is that predators tend to overrespond to increases in the population of their prey. This can give rise to oscillations in the populations of both species.

Because natural communities are composed of numerous interacting species no two of which can be entirely isolated from the rest, theoretical tools for dealing with larger systems are often required. The Routh-Hurwitz criteria and the methods of qualitative stability are thus briefly outlined in Sections 6.4 and 6.6. For rapid coverage of this chapter, these sections may be omitted without loss of continuity. In Sections 6.6 and 6.7 we study models for the spread of an epidemic in a population and then explore certain consequences of the policy of vaccinating against disease-causing agents.

Since the scope of this material is vast, a thorough documentation of sources is impossible. There are numerous recent reviews (for example, May, 1973). An excellent companion to this chapter is Van der Vaart (1983), which contains historical, biological, and mathematical details on certain topics and which uses an instructive and guided approach. [See also Braun (1979, 1983).] All of these sources have been used repeatedly in putting together the material for this chapter.

6.1 MODELS FOR SINGLE-SPECIES POPULATIONS

Two examples of ODEs modeling continuous single-species populations have already been encountered and analyzed in Section 4.1. To summarize, these are

1. *Exponential growth (Malthus, 1798):*

$$\frac{dN}{dt} = rN, \tag{1a}$$

Solution: $$N(t) = N_0 e^{rt}. \tag{1b}$$

2. *Logistic growth (Verhulst, 1838):*

$$\frac{dN}{dt} = r\left(1 - \frac{N}{K}\right)N, \tag{2a}$$

Solution: $$N(t) = \frac{N_0 K}{N_0 + (K - N_0)e^{-rt}}. \tag{2b}$$

$N_0 = N(0) =$ the initial population. (See Figure 6.1.)

To place both of the above into a somewhat broader context we proceed from a more general assumption, namely that for an isolated population (no migration) the rate of growth depends on population density. Therefore

$$\frac{dN}{dt} = f(N). \tag{3}$$

This approach is based on an instructive summary by Lamberson and Biles (1981), which should be consulted for further details.

Observe that for equations (1a) and (2a) the function f is the polynomial

$$f(N) = a_0 + a_1 N + a_2 N^2$$

where $a_0 = 0$; for equation (1a) $a_1 = r$ and $a_2 = 0$; for equation (2a) $a_1 = r$ and $a_2 = -r/K$. More generally, it is possible to write an infinite power (Taylor) series for f if it is sufficiently smooth:

$$f(N) = \sum_{n=0}^{\infty} a_n N^n = a_0 + a_1 N + a_2 N^2 + a_3 N^3 + \cdots.$$

Thus any growth function may be written as a (possibly infinite) polynomial (see Lamberson and Biles, 1981).

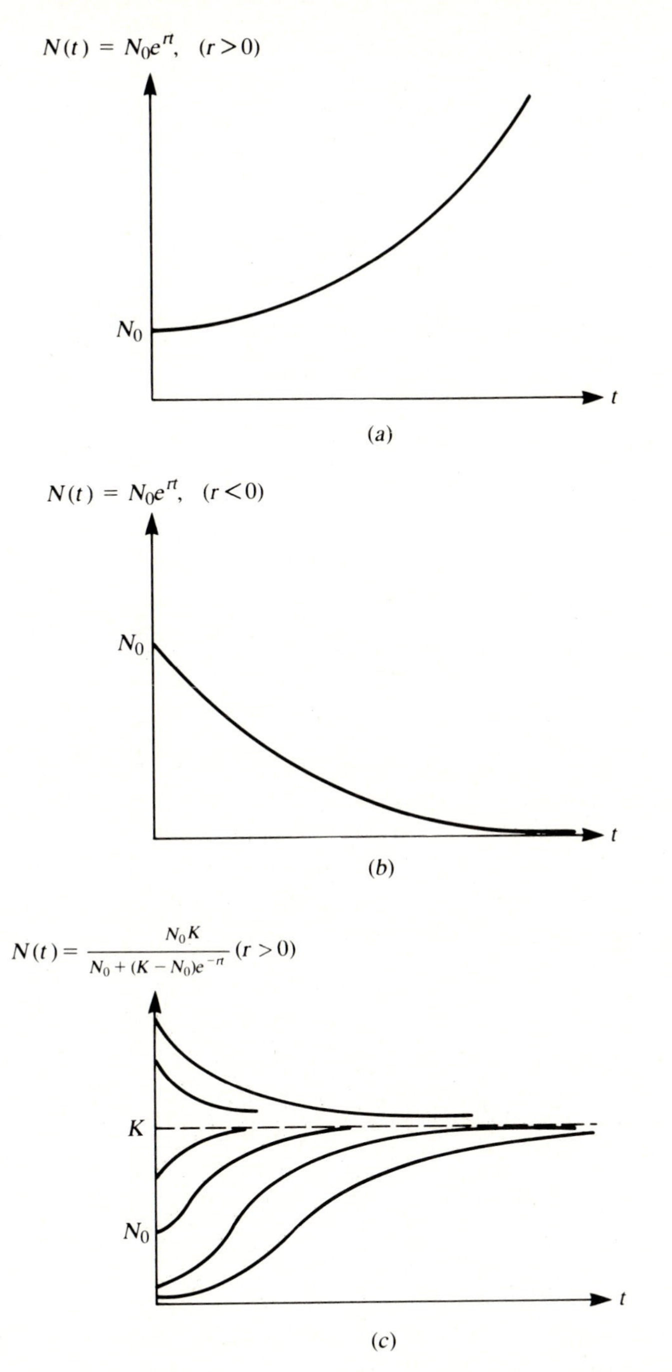

Figure 6.1 *Changes in population size* N(t) *predicted by two models for single-species growth: Exponential growth with (a)* $r > 0$, *(b)* $r < 0$, *and (c) logistic growth. See equations (1a) and (2a).*

About (3) we require that $f(0) = 0$ to dismiss the possibility of *spontaneous generation,* the production of living organisms from inanimate matter. (See also Hutchinson, 1978 for this *Axiom of Parenthood:* every organism must have parents.) In any growth law this is equivalent to

$$\left.\frac{dN}{dt}\right|_{N=0} = f(0) = 0,$$

so that we may assume that

$$\begin{aligned} a_0 &= 0, \\ \frac{dN}{dt} &= a_1 N + a_2 N^2 + a_3 N^3 + \cdots \\ &= N(a_1 + a_2 N + a_3 N^2 + \cdots), \\ &= N g(N). \end{aligned} \tag{4}$$

The polynomial $g(N)$ is called the *intrinsic growth rate* of the population.

Now we examine more closely several specific growth models, including those given in equations (1a) and (2a).

Malthus Model

This can be viewed as the simplest form of equation (4) in which the coefficients of $g(N)$ are $a_1 = r$ and $a_2 = a_3 = \cdots = 0$. As noted before, this model predicts exponential growth if $r > 0$ and exponential decline if $r < 0$.

Logistic Growth

To correct the prediction that a population can grow indefinitely at an exponential rate, consider a nonconstant intrinsic growth rate $g(N)$. The logistic growth model is perhaps the simplest extension of equation (1a). It can be explained by any of the following comments.

Formal mathematical justification

Equation (2a) makes use of more terms in the (possibly infinite) series for $f(N)$ and is thus more faithful to the true population growth rate.

Density-dependent growth rate

Equation (2a) takes the form of equation (4), where

$$g(N) = r\left(1 - \frac{N}{K}\right).$$

It is essentially the simplest rule in which the intrinsic growth rate g depends on the population density (in a *linear decreasing* relationship). It thus accounts for a decreasing per capita growth rate as population size increases.

Carrying capacity

From equation (2a) we observe that

$$\frac{dN}{dt} = 0 \qquad (N = K).$$

Thus $N = K$ is a steady state of the logistic equation. It is easy to establish that this steady state is stable; note in particular that

$$\frac{dN}{dt} > 0 \qquad (N < K),$$

$$\frac{dN}{dt} < 0 \qquad (N > K).$$

The constant K can represent the carrying capacity of the environment for the species. See also Section 4.1 for a derivation of (2a) based on nutrient consumption.

Intraspecific competition

The fact that individuals compete for food, habitat, and other limited resources means that such an increase in the net population mortality may be observed under crowded conditions. Such effects are most pronounced when there are frequent *encounters* between individuals. Equation (2a) can be written in the form

$$\frac{dN}{dt} = rN - \frac{r}{K}N^2.$$

The second term thus depicts a mortality proportional to the rate of paired encounters.

The solution of equation (2a) given by (2b) can be obtained in a relatively straightforward calculation (see problem 5 of Chapter 4). Aside from Gause's work on yeast cultures (Section 4.1), such models have been applied to a variety of populations including humans (Pearl and Reed, 1920), microorganisms (Slobodkin, 1954), and other species. See Lamberson and Biles (1981) for examples and references.

Allee Effect

A further direct extension of equations (1) and (2) is an assumption of the form

$$g(N) = a_1 + a_2N + a_3N^2.$$

Provided $a_2 > 0$, and $a_3 < 0$, one obtains the *Allee effect,* which represents a population that has a maximal intrinsic growth rate at intermediate density. This effect may stem from the difficulty of finding mates at very low densities.

Figure 6.2 is an example of a density-dependent form of $g(N)$ that depicts the

Allee effect. Its general character can be summarized by the inequalities

$$g'(N) > 0 \qquad (N < \eta),$$
$$g'(N) < 0 \qquad (N > \eta),$$

where η is the density for optimal reproduction.

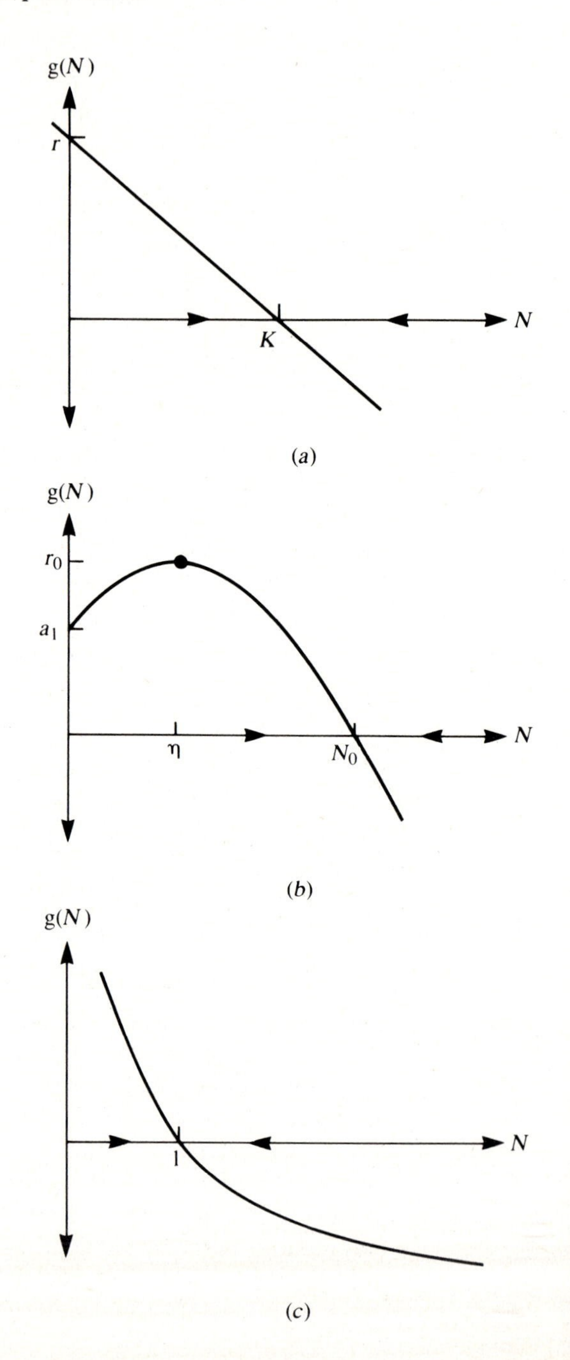

Figure 6.2 *A comparison between three types of density-dependent intrinsic growth rates* g(N). *(a) Logistic growth decreases linearly with density (or population size). (b) In the Allee effect the rate of reproduction is maximal at intermediate densities. (c) The Gompertz law shows a negative logarithmic dependence of growth rate on population size. See text for details.*

The simplest example of an Allee effect would be

$$g(N) = r_0 - \alpha(N - \eta)^2, \qquad (\eta < \sqrt{r_0/\alpha}). \tag{5}$$

Notice that this inverted parabola, shown in Figure 6.2*(b)*, intersects the axis at $r_0 - \alpha\eta^2$, has a maximum of r_0 when $N = \eta$, and drops below 0 when

$$N > N_0 = \eta + \sqrt{r_0/\alpha}.$$

Thus for densities above N_0, the population begins to decline. From the curve in Figure 6.2*(a)* we can deduce that $N = N_0$ is a stable equilibrium for the population. (N_0 is an equilibrium point because $g(N_0) = 0$; it is stable because $g'(N_0) < 0$.)

In equation (5) we assumed that $a_1 = r_0 - \alpha\eta^2$, $a_2 = 2\alpha\eta$, and $a_3 = -\alpha$.

Other Assumptions; Gompertz Growth in Tumors

Yet a fourth growth law that frequently appears in models of single-species growth is the *Gompertz law* (introduced in Chapter 4), which is used mainly for depicting the growth of solid tumors. The problems of dealing with a complicated geometry and with the fact that cells in the interior of a tumor may not have ready access to nutrients and oxygen are simplified by assuming that the growth rate declines as the cell mass grows. Three equivalent versions of this growth rate are as follows:

$$\frac{dN}{dt} = \lambda e^{-\alpha t} N, \tag{6a}$$

$$\frac{dN}{dt} = \gamma N, \qquad \frac{d\gamma}{dt} = -\alpha\gamma, \tag{6b}$$

$$\frac{dN}{dt} = -\kappa N \ln N. \tag{6c}$$

See Figure 6.2(*c*). In (6c) we can identify the intrinsic growth rate as

$$g(N) = -\kappa \ln N.$$

Since $\ln N$ is undefined at $N = 0$, this relation is not valid for very small populations and cannot be considered a direct extension of any of the previous growth laws. It is, however, a popular model in clinical oncology. (See Braun, 1979, sec. 1.8; Newton, 1980; Aroesty et al., 1973.) Biological interpretations for these equations are discussed in problem 7. Considering their relatively simple form, the predictions of any of the Gompertz equations agree remarkably well with the data for tumor growth. (See Aroesty et al., 1973, or Newton, 1980, for examples.)

A valid remark about most of the models for population growth is that they are at best gross simplifications of true events and often are used simply as an expedient fit to the data. To be more realistic one needs a greater mathematical sophistication. For example, in Chapter 13 we will see that partial differential equations provide a more powerful way to deal with age-dependent growth, fecundity, or mortality rates. Equations such as (3) or (6) are frequently used by modelers as a convenient first approach to complicated situations and thus are quite useful provided their limitations are not ignored.

6.2 PREDATOR-PREY SYSTEMS AND THE LOTKA-VOLTERRA EQUATIONS

The fact that predator-prey systems have a tendency to oscillate has been observed for well over a century. The Hudson Bay Company, which traded in animal furs in Canada, kept records dating back to 1840. In these records, oscillations in the populations of lynx and its prey the snowshoe hare are remarkably regular (see Figure 6.3).

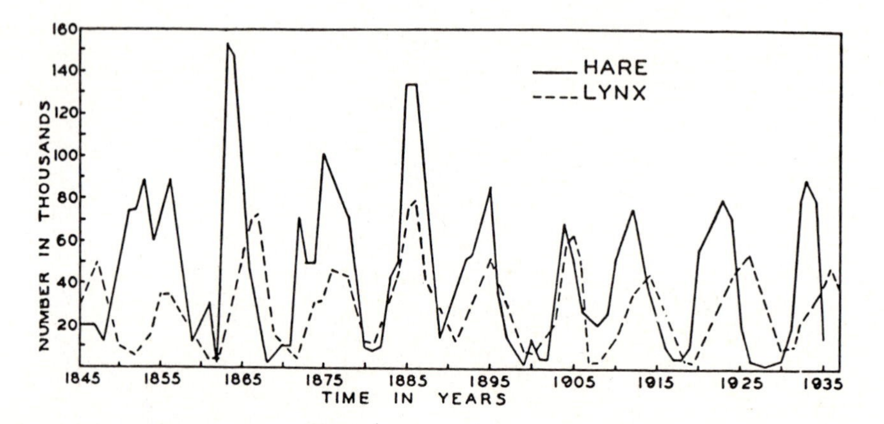

Figure 6.3 Records dating back to the 1840s kept by the Hudson Bay Company. Their trade in pelts of the snowshoe hare and its predator the lynx reveals that the relative abundance of the two species undergoes dramatic cycles. The period of these cycles is roughly 10 years. [From E. P. Odum (1953), fig. 39.]

In this section we explore a model for predator-prey interactions that Volterra proposed to explain oscillations in fish populations in the Mediterranean. To reconstruct his line of reasoning and arrive at the equations independently, let us list some of the simplifying assumptions he made:

1. Prey grow in an unlimited way when predators do not keep them under control.
2. Predators depend on the presence of their prey to survive.
3. The rate of predation depends on the likelihood that a victim is encountered by a predator.
4. The growth rate of the predator population is proportional to food intake (rate of predation).

Taking the simplest set of equations consistent with these assumptions, Volterra wrote down the following model:

$$\frac{dx}{dt} = ax - bxy, \tag{7a}$$

$$\frac{dy}{dt} = -cy + dxy, \tag{7b}$$

where x and y represent prey and predator populations respectively; the variables can represent, for example, biomass or population densities of the species. To acquaint ourselves with this model we proceed by answering several questions. First let us consider the meaning of parameters a, b, c, and d and of each of the four terms on the RHS of the equations.

The net growth rate a of the prey population when predators are absent is a positive quantity (with dimensions of 1/time) in accordance with assumption 1. The net death rate c of the predators in the absence of prey follows from assumption 2. The term xy approximates the likelihood that an encounter will take place between predators and prey given that both species move about randomly and are uniformly distributed over their habitat.

The form of this encounter rate is derived from the *law of mass action* that, in its original context, states that the rate of molecular collisions of two chemical species in a dilute gas or solution is proportional to the product of the two concentrations (see Chapter 7). We should bear in mind that this simple relationship may be inaccurate in describing the subtle interactions and motion of organisms. An encounter is assumed to decrease the prey population and increase the predator population by contributing to their growth. The ratio b/d is analogous to the efficiency of predation, that is, the efficiency of converting a unit of prey into a unit of predator mass.

Further practice in linear stability techniques given in Chapter 5 can be revealing:

It is clear that two possible steady states of equation (7) exist:

$$(\bar{x}_1, \bar{y}_1) = (0, 0) \qquad \text{and} \qquad (\bar{x}_2, \bar{y}_2) = \left(\frac{c}{d}, \frac{a}{b}\right).$$

Their stability properties are determined by the methods given in Chapter 5.

The Jacobian of this system is

$$\mathbf{J} = \begin{pmatrix} a - by & -bx \\ dy & dx - c \end{pmatrix}_{(\bar{x}, \bar{y})}.$$

for steady state 1

$$\mathbf{J} = \begin{pmatrix} a & 0 \\ 0 & -c \end{pmatrix},$$

for steady state 2

$$\mathbf{J} = \begin{pmatrix} 0 & \dfrac{-bc}{d} \\ \dfrac{da}{b} & 0 \end{pmatrix},$$

eigenvalues are	
$\lambda_1 = a, \quad \lambda_2 = -c.$	$\lambda_{1,2} = \pm\sqrt{ca}\,i.$
Thus $(\bar{x}_1, \bar{y}_1)$ is a saddle.	Thus $(\bar{x}_2, \bar{y}_2)$ is a center[1].

From the analysis of this model we arrive at a number of somewhat counterintuitive results. First, notice that the steady-state level of prey is independent of its own growth rate or mortality; rather, it depends on parameters associated with the predator ($x_2 = c/d$). A similar result holds for steady-state levels of the predator ($y_2 = a/b$). It is the particular *coupling* of the variables that leads to this effect. To paraphrase, the presence of predator ($y \neq 0$) means that the available prey has to just suffice to make growth rate due to predation, dx, equal predator mortality c for a steady predator population to persist. Similarly, when prey are present ($x \neq 0$), predators can only keep them under control when prey growth rate a and mortality due to predation, by, are equal. This helps us to understand the steady-state equations.

A second result (see problem 10) is that the steady state $(\bar{x}_2, \bar{y}_2)$ is neutrally stable (a center). The eigenvalues of $\mathbf{J}(\bar{x}_2, \bar{y}_2)$ are pure imaginary and the steady state is not a spiral point. See problem 10. Note that the off-diagonal terms, $-bc/d$ and da/b, are of opposite sign (since the influence of each species on the other is opposite) and that the diagonal terms evaluated at $(\bar{x}_2, \bar{y}_2)$ are *zero*. Stability analysis predicts oscillations about the steady state $(\bar{x}_2, \bar{y}_2)$. The factor $\sqrt{(ca)}$ governs the frequency of these oscillations, so that larger prey reproduction or predator mortality (which means a greater turnover rate) result in more rapid cycles. A complete phase-plane diagram of the predator-prey system (7) can be arrived at with minimal further work. See Figure 6.4(*a*).

To gain deeper understanding of the neutral stability of system (7) we will examine a slight variant in which prey populations have the property of self-regulation. Assuming logistic prey growth, equations (7a, b) become

$$\frac{dx}{dt} = \frac{ax\,(K - x)}{K} - bxy, \tag{8a}$$

$$\frac{dy}{dt} = -cy + dxy. \tag{8b}$$

This leads to steady-state values

$$(\bar{x}_2, \bar{y}_2) = \left(\frac{c}{d}, \frac{a}{b} - \frac{ca}{dbK}\right),$$

1. To be more accurate we must include the possibility that this steady state could be a spiral point since the system is nonlinear. (See Section 5.10 for comments.) In problem 10 we will demonstrate that this option can be dismissed for the predator-prey equations.

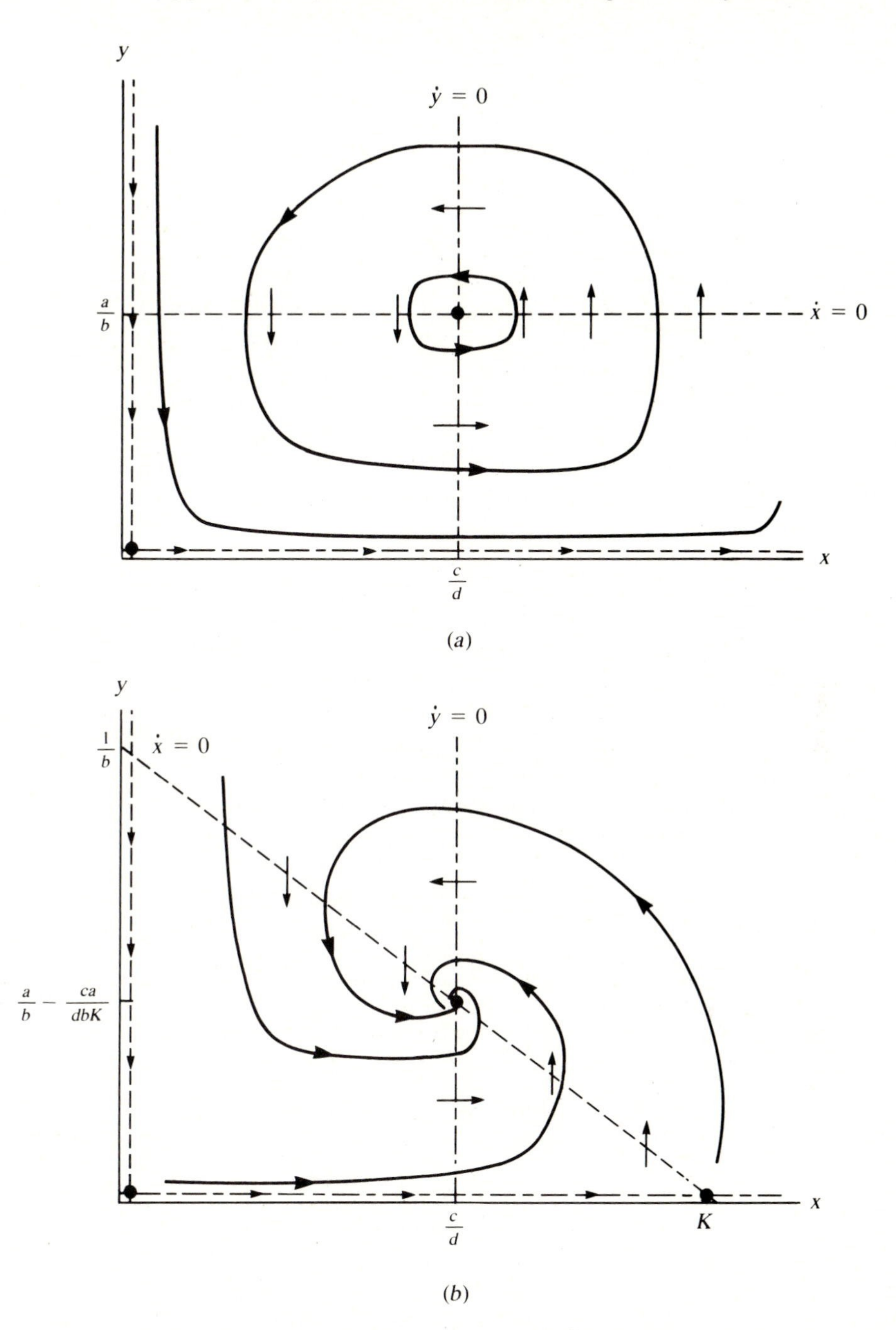

Figure 6.4 *(a) The Lotka-Volterra equations (7a,b) predict neutral stability at the steady state (c/d, b/a). (b) When the prey grow logistically [as in equation (8a)], the steady state becomes a stable spiral with somewhat depressed predator population levels.*

and the Jacobian is then

$$\mathbf{J} = \begin{pmatrix} \dfrac{-ac}{dK} & \dfrac{-bc}{d} \\ d\left(\dfrac{a}{b} - \dfrac{ca}{dbK}\right) & 0 \end{pmatrix}.$$

(The condition $1 > c/dK$ must be satisfied so that the steady-state predator level y is positive.) Now $\text{Tr}\,\mathbf{J} = -ac/dK$ is always negative, and $\det \mathbf{J} = bc\bar{y}_2$ is positive, so that the steady state is always stable. In other words, its neutral stability has been lost. In problem 14 you are asked to investigate whether oscillations accompany the return to steady state after a perturbation.

The lesson to be learned from this example is that a relatively minor change in equations (7a, b) has a major influence on the predictions. In particular, this means that neutral stability, and thus also the oscillations that accompany a neutrally stable steady state, tend to be somewhat ephemeral. This is a serious criticism of the realism of the Lotka-Volterra model.

Taking a somewhat more philosophical approach, we could argue that the Lotka-Volterra model serves a useful purpose precisely because it is so delicately balanced between stability and instability. We could use this model together with minor variants to test out a set of assumptions and so identify stabilizing and destabilizing influences. Following are some of the frequently suggested alterations. It is a relatively easy task to understand what effects such changes have on the stability of the equilibrium. More theoretical results on stable cycles due to Kolmogorov (1936) and others (briefly mentioned below) are recommended for further independent exploration and will be discussed in detail in Chapter 8.

For Further Study

Stable cycles in predator-prey systems

The main objection to the Lotka-Volterra model is that its cycles are only neutrally stable. What additional features are necessary to yield stable oscillations? As we shall see in Chapter 8, stable oscillations (usually called limit cycles) are closed trajectories that attract nearby flow in the phase plane. Kolmogorov (1936) investigated conditions on the general predator-prey system

$$\frac{dx}{dt} = xf(x, y),$$

$$\frac{dy}{dt} = yg(x, y),$$

that would lead to such solutions. The functions f and g are assumed to satisfy several relations consistent with the nature of predator-prey systems:

$$\partial f/\partial x < 0 \text{ (for large } x), \qquad \partial g/\partial x > 0,$$
$$\partial f/\partial y < 0, \qquad \partial g/\partial y < 0.$$

An interpretation of these is left as an exercise. Additional conditions (for example, Coleman, 1978; May 1973) are equivalent to the nullcline geometry shown in Figure 6.5 (Rosenzweig, 1969). It can be proved that when the steady state S is unstable, any trajectory winding out of its vicinity approaches a stable limit cycle that is trapped somewhere inside the rectangular region. See Chapter 8 for further details.

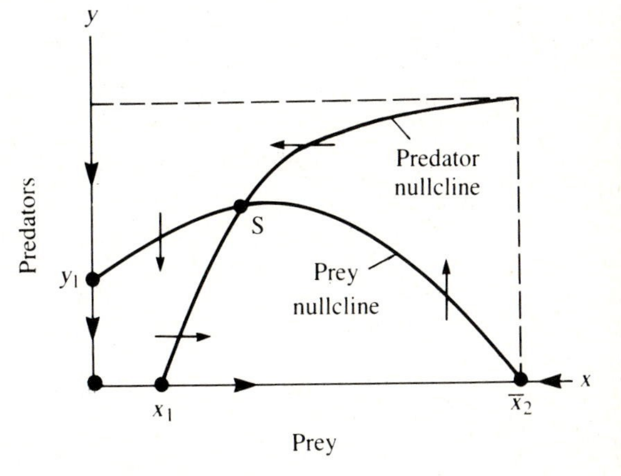

Figure 6.5 *With a set of conditions given by Kolmogorov (1936), the phase plane for a predator-prey system has a nullcline geometry that gives rise to stable (limit cycle) oscillations. See Chapter 8.*

Other modifications of Volterra's equations

Other assumptions which have been made over the years to modify Volterra's equations are listed below. Details about their effect can be found in May (1973) and in the references as follows.

1. *Density dependence:* More realistic prey growth-rate assumptions in which a is replaced by a density-dependent function f:

$$f(x) = r\left(1 - \frac{x}{K}\right) \qquad \text{Pielou (1969, pp. 19–21)},$$

$$f(x) = r\left[\left(\frac{K}{x}\right)^{-g} - 1\right] \qquad (1 \geq g > 0) \qquad \text{Rosenzweig (1971)},$$

$$f(x) = r\left(\frac{K}{x} - 1\right) \qquad \text{Schoener (1973)}.$$

2. *Attack rate:* More realistic rates of predation where the term bxy is replaced by a term in which the attack capacity of predators is a limited one. Terms replacing bxy in equation (7a) are:

$$ky(1 - e^{-cx}) \qquad \text{Ivlev (1961)},$$

$$\frac{kxy}{x + D} \qquad \text{Holling (1965)},$$

$$kyx^{g} \qquad (1 \geq g > 0) \qquad \text{Rosenzweig (1971)},$$

$$\frac{kyx^2}{x^2 + D^2} \qquad \text{Takahashi (1964)}.$$

6.3 POPULATIONS IN COMPETITION

When two or more species live in proximity and share the same basic requirements, they usually compete for resources, habitat, or territory. Sometimes only the strongest prevails, driving the weaker competitor to extinction. (This is the *principle of competitive exclusion,* a longstanding concept in population biology.) One species wins because its members are more efficient at finding or exploiting resources, which leads to an increase in population. Indirectly this means that a population of competitors finds less of the same resources and cannot grow at its maximal capacity.

In the following model, proposed by Lotka and Volterra and later studied empirically by Gause (1934), the competition between two species is depicted without direct reference to the resources they share. Rather, it is assumed that the presence of each population leads to a depression of its competitor's growth rate. We first give the equations and then examine their meanings and predictions systematically. See also Braun (1979, sec. 4.10) and Pielou (1969, sec. 5.2) for further discussion of this model.

The Lotka-Volterra model for species competition is given by the equations

$$\frac{dN_1}{dt} = r_1 N_1 \frac{\kappa_1 - N_1 - \beta_{12} N_2}{\kappa_1}, \tag{9a}$$

$$\frac{dN_2}{dt} = r_2 N_2 \frac{\kappa_2 - N_2 - \beta_{21} N_1}{\kappa_2}, \tag{9b}$$

where N_1 and N_2 are the population densities of species 1 and 2. Again we proceed to understand the equations by addressing several questions:

1. Suppose only species 1 is present. What has been assumed about its growth? What are the meanings of the parameters r_1, κ_1, r_2, and κ_2?
2. What kind of assumption has been made about the effect of competition on the growth rate of each species? What are the parameters β_{12} and β_{21}?

To answer these questions observe the following:

1. In the absence of a competitor ($N_2 = 0$) the first equation reduces to the logistic equation (2a). This means that the population of species 1 will stabilize at the value $N_1 = \kappa_1$ (its carrying capacity), as we have already seen in Section 6.1.
2. The term $\beta_{21} N_2$ in equation (9a) can be thought of as the contribution made by species 2 to a *decline in the growth rate* of species 1. β_{12} is the per capita decline (caused by individuals of species 2 on the population of species 1).

The next step will be to study the behavior of the system of equations. The task will again be divided into a number of steps, including (1) identifying steady states, (2) drawing nullclines, and (3) determining stability properties as necessary in putting together a complete phase-plane representation of equation (9) using the

Nullclines are just all point sets that satisfy one of the following equations:

$$\frac{dN_1}{dt} = 0 \quad \text{or} \quad \frac{dN_2}{dt} = 0.$$

1. From equation (9a) we arrive at the N_1 nullclines:

$$N_1 = 0 \quad \text{and} \quad \kappa_1 - N_1 - \beta_{12}N_2 = 0,$$

2. Whereas equation (9b) leads to the N_2 nullclines:

$$N_2 = 0 \quad \text{and} \quad \kappa_2 - N_2 - \beta_{21}N_1 = 0.$$

To simplify the notation slightly, we shall refer to these lines as L_{1a}, L_{1b}, L_{2a}, and L_{2b} respectively. Notice that L_{1a} and L_{2a} are just the N_2 and N_1 axes respectively, whereas L_{1b} and L_{2b} intersect the axes as follows:

L_{1b} goes through $(0, \kappa_1/\beta_{12})$ and $(\kappa_1, 0)$.
L_{2b} goes through $(0, \kappa_2)$ and $(\kappa_2/\beta_{21}, 0)$.

methods given in Chapter 5. (For practice, it is advisable to attempt this independently before continuing to the procedure in the box.)

It follows that the points $(0, 0)$, $(\kappa_1, 0)$, and $(0, \kappa_2)$ are always steady states. These correspond to three distinct situations:

$(0, 0)$ = both species absent,
$(\kappa_1, 0)$ = species 2 absent and species 1 at its carrying capacity κ_1,
$(0, \kappa_2)$ = species 1 absent and species 2 at its carrying capacity.

There is a fourth possible steady-state value that corresponds to coexistence of the two species. (We leave the computation of this steady state as an exercise.)

Proceeding to the second stage, we sketch the nullcline curves on a phase plane. If you have already attempted this independently, you may have hesitated slightly because numerous situations are possible. Figure 6.6 illustrates four distinct possibilities, all of them correct. In order to choose any one of the four cases we must make some assumptions about the relative magnitudes of κ_2 and κ_1/β_{12}, and of κ_1 and κ_2/β_{21}. The cases shown in Figure 6.6 correspond to the following situations:

$$\text{case 1: } \frac{\kappa_2}{\beta_{21}} > \kappa_1 \quad \text{and} \quad \kappa_2 > \frac{\kappa_1}{\beta_{12}},$$

$$\text{case 2: } \kappa_1 > \frac{\kappa_2}{\beta_{21}} \quad \text{and} \quad \frac{\kappa_1}{\beta_{12}} > \kappa_2,$$

$$\text{case 3: } \kappa_1 > \frac{\kappa_2}{\beta_{21}} \quad \text{and} \quad \kappa_2 > \frac{\kappa_1}{\beta_{12}},$$

$$\text{case 4: } \frac{\kappa_2}{\beta_{21}} > \kappa_1 \quad \text{and} \quad \frac{\kappa_1}{\beta_{12}} > \kappa_2.$$

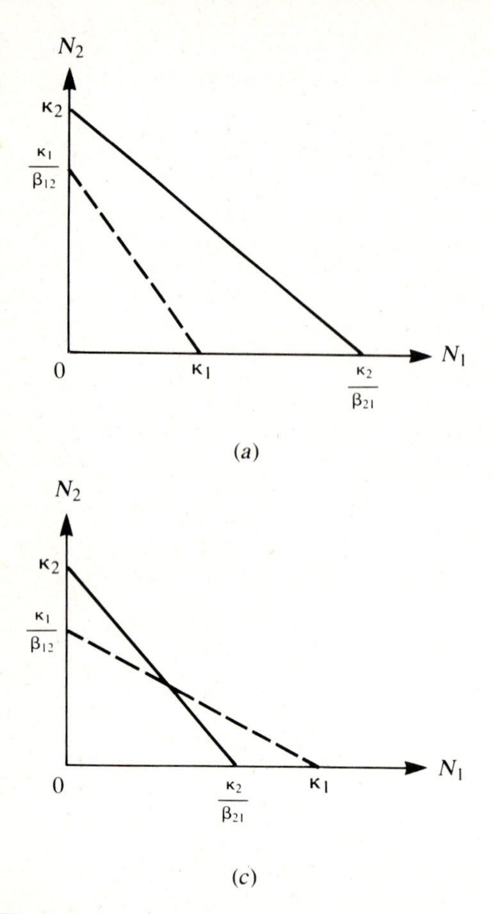

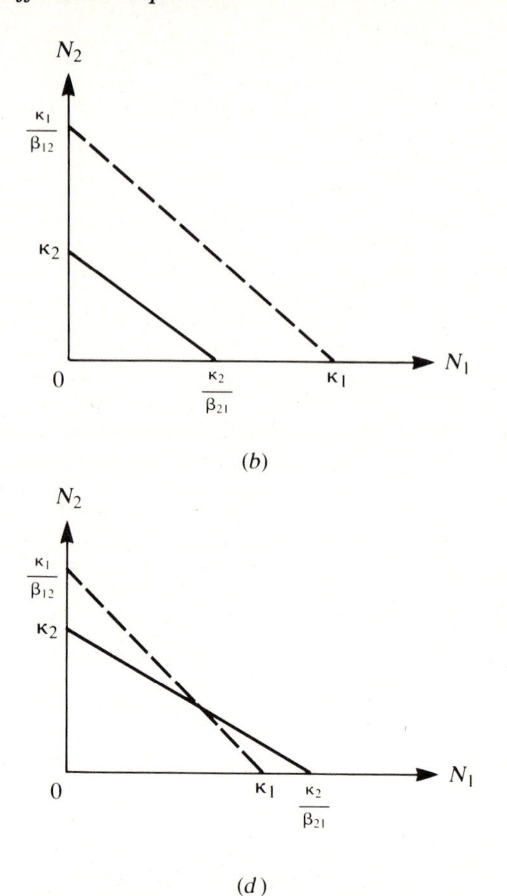

***Figure 6.6** Four possible cases corresponding to four choices in the relative positions of nullclines of equations (9): (a) $\kappa_2/\beta_{21} > \kappa_1$ and $\kappa_2 > \kappa_1/\beta_{12}$; (b) $\kappa_1 > \kappa_2/\beta_{21}$ and $\kappa_1/\beta_{12} > \kappa_2$; (c) $\kappa_1 > \kappa_2/\beta_{21}$ and $\kappa_2 > \kappa_1/\beta_{12}$; (d) $\kappa_1/\beta_{12} > \kappa_2$ and $\kappa_2/\beta_{21} > \kappa_1$.*

In problem 15 the reader is asked to interpret these inequalities within the biological context of the problem.

Our next step will be to identify the steady states of equations (9a,b) in Figure 6.6(a–d). By drawing arrows on the nullclines we will also indicate the directions of flow in the N_1N_2 plane for each of the four cases shown. To do this, one can combine geometric reasoning with results of analysis. We recall that steady states are located at the intersections of two nullclines (which must be of opposite types). It helps to remember that L_{1b} (the line $N_1 = 0$) is an N_1 nullcline; it is simply the N_2 axis. Thus the point at which any N_2 nullcline meets the N_2 axis will be a steady state. It is evident that this happens at (0, 0) as well as at (0, κ_2). By similar reasoning we find that (κ_1, 0) is at the intersection of two (opposite type) nullclines. A fourth steady state occurs only when L_{1b} and L_{2b} intersect, as is true in (c) and (d) of Figure 6.6.

To sketch arrows on the N_1 and N_2 nullclines, recall that the directions of flow on these are parallel to the N_2 and N_1 axes respectively. Arrows have been put in for cases 1 and 4 in Figure 6.7, with case 2 and 3 left as an exercise. Notice that once the flow along the N_1 and N_2 axes is drawn the rest of the picture can be completed by preserving the continuity of flow. (See remarks in Section 5.5.) For a more pedestrian approach, we can use equations (9a,b) to tabulate the directions associated with several points in the plane.

At this stage the problem is practically solved; with the directions of flow determined on the nullclines, we can draw sensible phase-plane pictures in only one distinct way for each case. For example, it should be evident in case 1 that for any starting value of (N_1, N_2) provided $N_2 \neq 0$, the populations eventually converge to the steady state on the N_2 axis. (To see this, notice that there is no other exit from the region bounded by the two slanted lines L_{1b} and L_{2b} in case 1; moreover, all flows pass through this region.) In case 4, any point within the two triangular regions must eventually converge to the steady state at the intersection of L_{1b} and L_{2b}. (What can be said about other regions of the plane in case 4?)

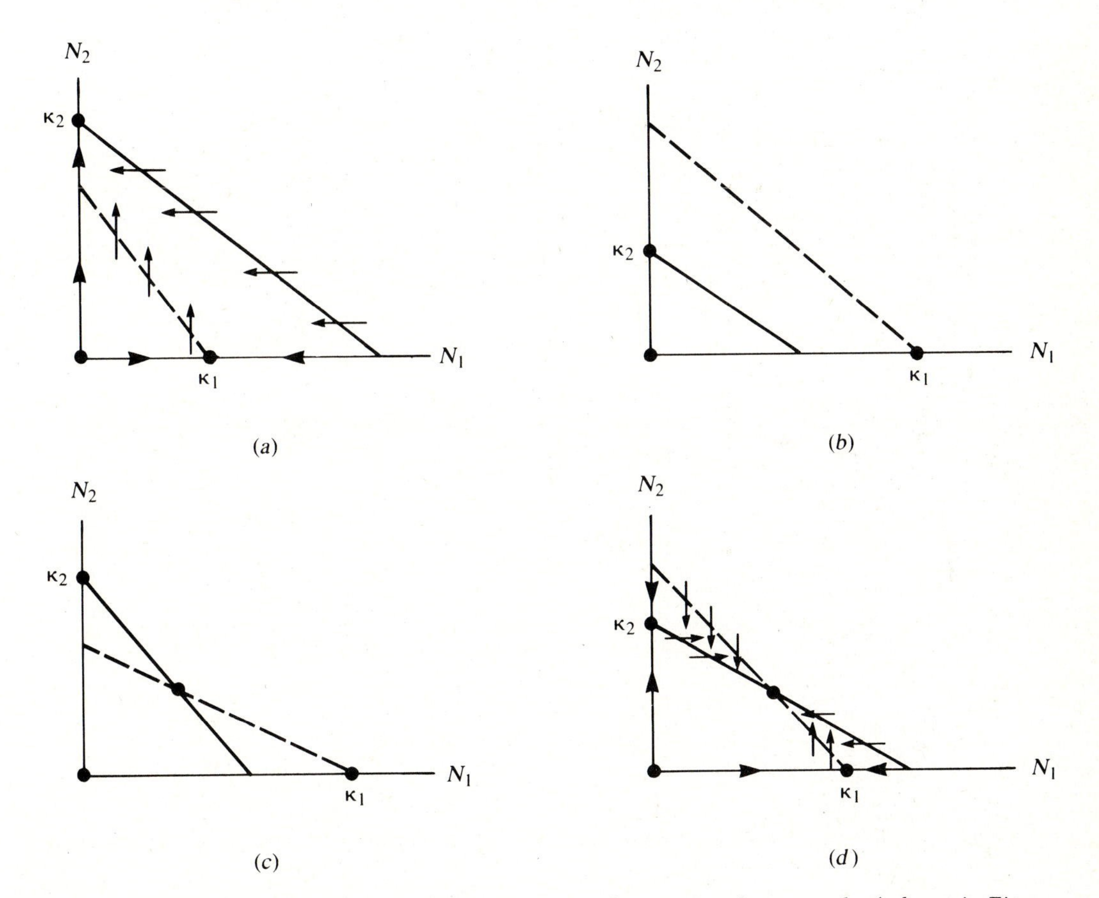

Figure 6.7 *Steady states of equation (9) shown as heavy dots at the intersections of the* N_1 *nullclines (dashed lines) and the* N_2 *nullclines (solid lines). (a–d) correspond to cases 1–4 shown in Figure 6.6 and described in text.*

As a somewhat optional final step, we can confirm the conjectured flow by determining what happens close to steady-state values, using the linearization procedures outlined in Chapter 5 [see problem 15(c)]. By carrying out this analysis it can be shown that the outcome of competition is as follows:

case 1: only $(0, \kappa_2)$ is stable,
case 2: only $(\kappa_1, 0)$ is stable,
case 3: both $(0, \kappa_1)$ and $(0, \kappa_2)$ are stable,
case 4: only the steady state given by the expression in problem 15(b) is stable.

With the combined information above, the qualitative pictures in Figure 6.8 can be confirmed, and the mathematical steps in understanding the model are complete. It is now necessary to make a biological interpretation of the result. Part of this is left as a problem for the reader. A rather clear prediction is that in three out of the four cases, competition will lead to extinction of one species. Only in case 4 does the interaction result in coexistence, and then at population levels *below* the normal carrying capacities.

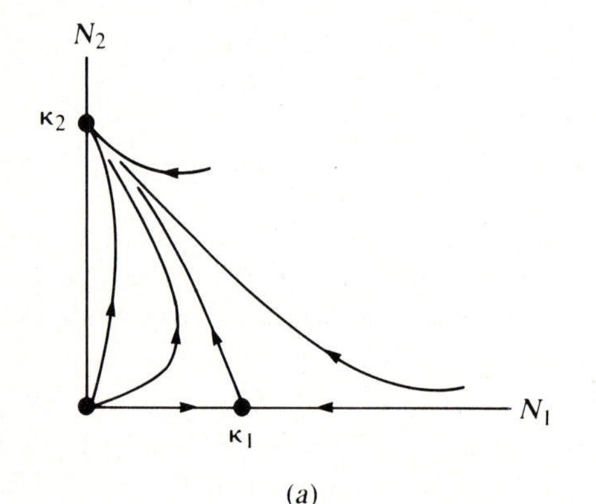

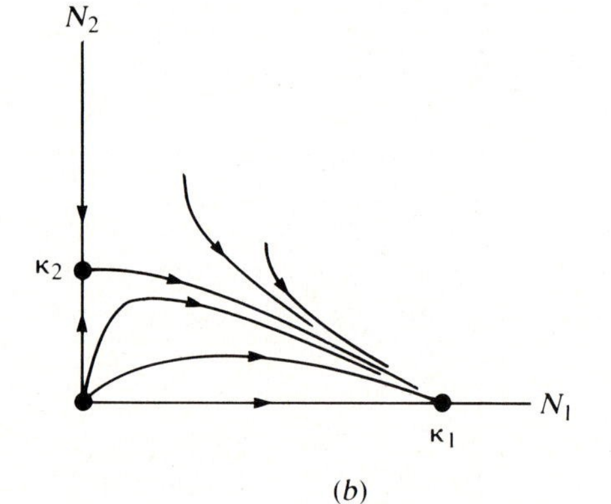

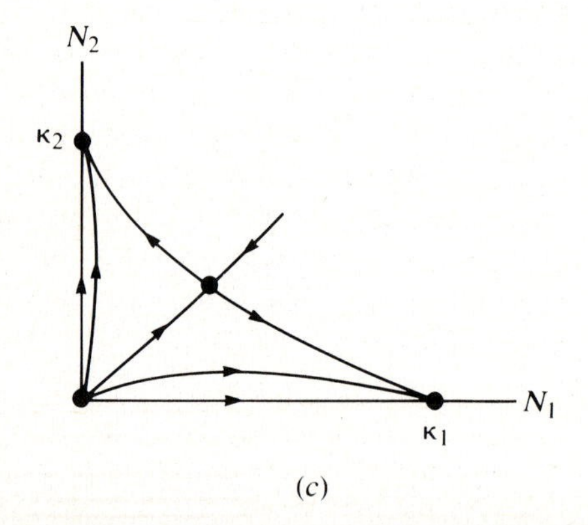

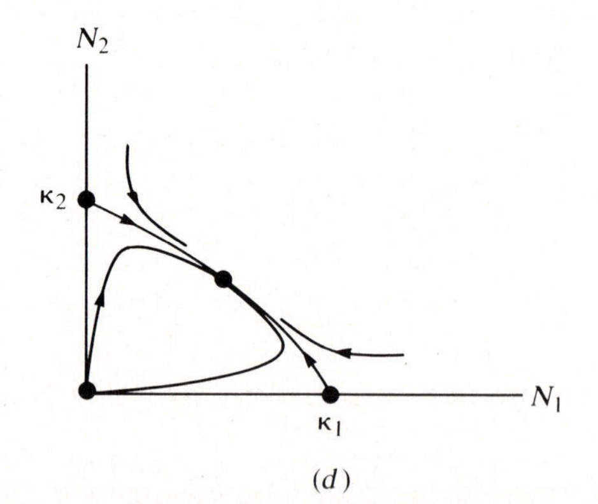

Figure 6.8 *The final phase-plane behavior of solutions to equations (9a,b). (a–d) correspond to cases 1–4 in Figure 6.6. See text for details.*

A small change in the format of the inequalities for cases 1 through 4 will reveal how the *intensity of competition,* which is represented by the β parameters, influences the outcome. To make things more transparent, suppose the carrying capacities are equal ($\kappa_1 = \kappa_2$). Conditions 1 to 4 can be written as follows:

1. $\beta_{21} < 1$ and $\beta_{12} > 1$.
2. $\beta_{21} > 1$ and $\beta_{12} < 1$.
3. $\beta_{21} > 1$ and $\beta_{12} > 1$.
4. $\beta_{21} < 1$ and $\beta_{12} < 1$.

From this, observe that in cases 1, 2, and 3, one or both species are aggressive in competing with their adversary (that is, at least one β is large). In case 4, for which coexistence is obtained, β_{21} and β_{12} are both small, indicating that competition is less intense.

An accepted biological fact is that species very similar in habits, size, and/or feeding preferences tend to compete more strongly for resources when confined to the same habitat (Roughgarden, 1979). For example, species of fish that have similar mouth parts and thus seek the same type of food would overlap in their resource utilization and, thus be more aggressive competitors than those that feed differently. With this observation, a prediction of the model is that similar species in the same habitat will not coexist. (This is a popular version of the principle of competitive exclusion.)

Recent research directions in population biology have focused on questions raised by this principle. Because ecosystems frequently consist of many competitors that appear to vie for common resources, the predictions of this simple model have reshaped some preconceptions about coexistence and species interactions. It has become more challenging to discover the numerous ways competitive exclusion can be foiled.

The model ignores spatial distributions of species and variations in both space and time of the significant quantities as well as many other subtle influences (such as the effects of predation on one of the species). This points to numerous possible effects that could come into play in permitting species to live and share a common habitat. In fact, it is now recognized that species are distributed in a patchy way, rather than uniformly partitioning their habitat so that competition tends to diminish somewhat. A time-sharing arrangement with succession of species or seasonal variability can effect a similar result. Other factors include gradual evolution of differing traits (*character displacement*) to minimize competition, and more complex multispecies interactions in which predation mediates competition. Observations of such special cases are abundant in the current biological literature. Sources for additional readings are Whitaker and Levin (1975) and a forthcoming monograph on theoretical ecology by Simon Levin (Cornell University). Chapter 21 of Roughgarden (1979) also makes for good reading on the competition model and its implications.

There are recent extensions of the competition model to handle n species. Luenberger (1979, sec. 9.5) gives an excellent presentation. A good discussion of the principle of competitive exclusion is given in Armstrong and McGehee (1980). A number of other contributors have included T. G. Hallam, T. C. Gard, R. M. May, H. I. Freedman, P. Waltman, and J. Hofbauer.

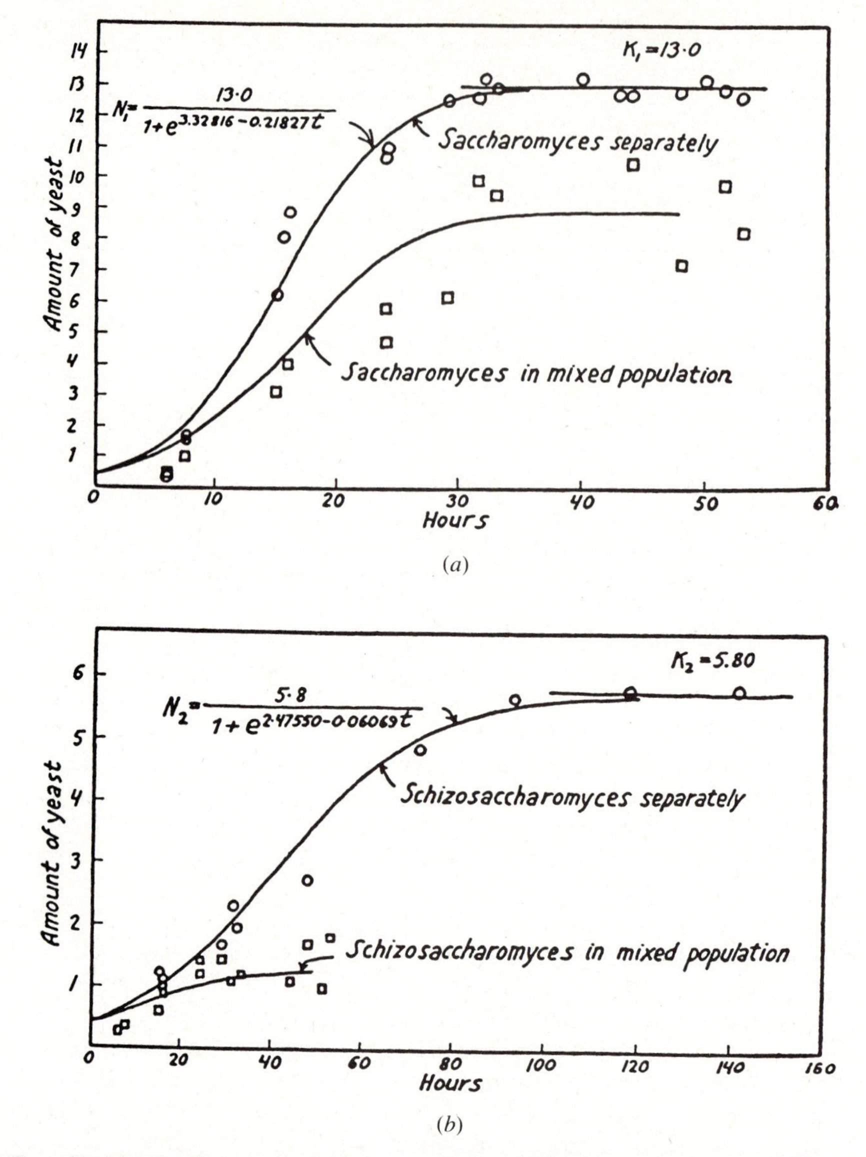

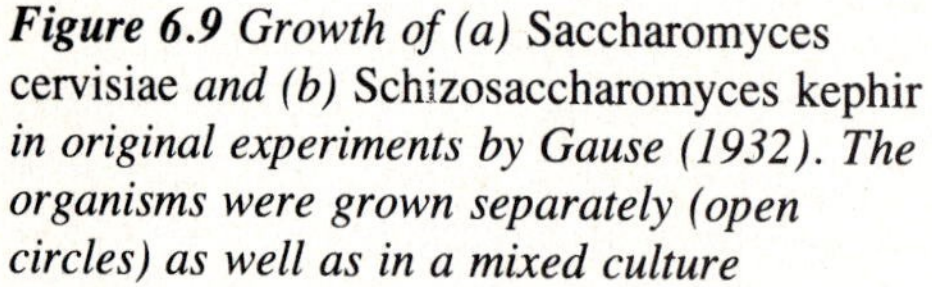

***Figure 6.9** Growth of (a)* Saccharomyces cervisiae *and (b)* Schizosaccharomyces kephir *in original experiments by Gause (1932). The organisms were grown separately (open circles) as well as in a mixed culture containing both (open rectangles). From Gause, G. F. (1932), Experimental studies on the struggle for existence. I. Mixed population of two species of yeast.* J. Exp. Biol., 9, *Figures 2 and 3.*

Gause: Empirical Tests of the Species-Competition Model

In his book, *The Struggle for Existence,* Gause (1934) describes a series of laboratory experiments in which two yeast species, *Saccharomyces cerevisiae* and *Schizosaccharomyces kephir* were grown separately and then paired in a mixed population (Figure

6.9). Using results shown in Figure 6.9 (*a* and *b*), he was able to estimate the following values for parameters in equation (9):

$$r_1 = 0.21827, \quad K_1 = 13.0, \quad \beta_{12} = 3.15,$$
$$r_2 = 0.06069, \quad K_2 = 5.8, \quad \beta_{21} = 0.439.$$

See problems 33 and 34 for some details and analysis, and Gause for a very readable summary of these and other experiments.

6.4 MULTIPLE-SPECIES COMMUNITIES AND THE ROUTH-HURWITZ CRITERIA

Now that we have spent some time mastering the techniques of linear stability theory, it seems discouraging to realize that the elegance and simplicity of phase-plane methods apply only to two-species systems. In this section we briefly touch on methods for gaining insight into models for k species interacting in a community, where $k > 2$.

The models we have seen thus far take the form

$$\frac{dN_1}{dt} = f(N_1, N_2), \tag{10a}$$

$$\frac{dN_2}{dt} = g(N_1, N_2). \tag{10b}$$

More generally a system comprised of k species with populations $N_1, N_2, \ldots, N_k$ would be governed by k equations:

$$\begin{aligned} \frac{dN_1}{dt} &= f_1(N_1, N_2, \ldots, N_k), \\ \frac{dN_2}{dt} &= f_2(N_1, N_2, \ldots, N_k), \\ &\vdots \\ \frac{dN_k}{dt} &= f_k(N_1, N_2, \ldots, N_k). \end{aligned} \tag{11}$$

Since it is cumbersome to carry this longhand version, one often sees the shorthand notation

$$\frac{dN_i}{dt} = f_i(N_1, N_2, \ldots, N_k) \qquad (i = 1, 2, \ldots, k), \tag{12}$$

or, better still, the vector notation

$$\frac{d\mathbf{N}}{dt} = \mathbf{F}(\mathbf{N}), \tag{13}$$

for $\mathbf{N} = (N_1, N_2, \ldots, N_k)$, $\mathbf{F} = (f_1, f_2, \ldots, f_k)$, where each of the functions f_1, $f_2, \ldots, f_k$ may depend on all or some of the species populations $N_1, N_2, \ldots, N_k$.

We shall now suppose that it is possible to solve the equation (or set of equations)

$$\mathbf{F}(\mathbf{N}) = 0, \tag{14}$$

so as to identify one (or possibly several) steady-state points, $\overline{\mathbf{N}} = (N_1, N_2, \ldots, N_k)$, satisfying $\mathbf{F}(\overline{\mathbf{N}}) = 0$. The next step, as a diligent reader might have guessed, would be to determine stability properties of this steady solution. While the idea is essentially identical to previous linear stability analysis, a slightly greater sophistication may be necessary to extract an answer. Let us see why this is true.

In linearizing equation (13) we find, as before, the Jacobian of $\mathbf{F}(\mathbf{N})$. This is often symbolized

$$\mathbf{J} = \frac{\partial \mathbf{F}}{\partial \mathbf{N}}(\overline{\mathbf{N}}). \tag{15}$$

Recall that this really means

$$\mathbf{J} = \begin{pmatrix} \dfrac{\partial f_1}{\partial N_1} & \dfrac{\partial f_2}{\partial N_2} & \cdots & \dfrac{\partial f_k}{\partial N_k} \\ \vdots & & & \\ \dfrac{\partial f_k}{\partial N_1} & \dfrac{\partial f_k}{\partial N_2} & \cdots & \dfrac{\partial f_k}{\partial N_k} \end{pmatrix}_{\overline{\mathbf{N}}} \tag{16}$$

so that $\mathbf{J}$ is now a $k \times k$ matrix. Population biologists frequently refer to $\mathbf{J}$ as the *community matrix* (see Levins, 1968). Eigenvalues λ of this matrix now satisfy

$$\det(\mathbf{J} - \lambda \mathbf{I}) = 0. \tag{17}$$

Thinking of what this means, you should arrive at the conclusion that λ must satisfy a characteristic equation of the form

$$\lambda^k + a_1\lambda^{k-1} + a_2\lambda^{k-2} + \cdots + a_k = 0. \tag{18}$$

If you find this baffling, you may wish to verify this with a 3×3 matrix, that is, by evaluating

$$\det \begin{pmatrix} a-\lambda & b & c \\ d & e-\lambda & f \\ g & h & i-\lambda \end{pmatrix}.$$

(The result is a cubic polynomial.) In general, the characteristic equation is a *polynomial* whose degree k is equal to the number of species interacting. Although for $k = 2$ the quadratic characteristic equation is easily solved, for $k > 2$ this is no longer true.

While we are unable in principle to *find* all eigenvalues, we can still obtain information about their magnitudes. Suppose $\lambda_1, \lambda_2, \ldots, \lambda_k$ are all (known) eigenvalues of the linearized system

$$\frac{d\mathbf{N}}{dt} = \mathbf{J} \cdot \mathbf{N}. \tag{19}$$

What must be true about these eigenvalues so that the steady state $\overline{\mathbf{N}}$ would be stable? Recall that they must *all* have negative real parts since close to the steady states each of the species populations can be represented by a sum of exponentials in $\lambda_i t$ as follows:

$$N_i = \overline{N}_i + a_1 e^{\lambda_1 t} + a_2 e^{\lambda_2 t} + \cdots + a_k e^{\lambda_k t}. \tag{20}$$

(This is a direct generalization of Section 5.6.) If one or more eigenvalues have positive real parts, $N_i - \overline{N}_i$ will be an increasing function of t, meaning that N_i will not return to its equilibrium value $\overline{N}_i$. Thus the question of stability of a steady state can be settled if it can be determined whether or not all eigenvalues $\lambda_1, \ldots, \lambda_k$ have negative real parts. (Contrast this with stability conditions for difference equations.) This can be done without actually solving for these eigenvalues by checking certain criteria. Recall that in the two-species case we derived conditions on quantities β and γ (which were, respectively, the trace and the determinant of the Jacobian) that ensured eigenvalues with negative real parts. For $k > 2$ these conditions are known as the *Routh-Hurwitz criteria* and are summarized in the box.

The Routh-Hurwitz Criteria

Given the characteristic equation (18), define k matrices as follows:

$$\mathbf{H}_1 = (a_1), \qquad \mathbf{H}_2 = \begin{pmatrix} a_1 & 1 \\ a_3 & a_2 \end{pmatrix}, \qquad \mathbf{H}_3 = \begin{pmatrix} a_1 & 1 & 0 \\ a_3 & a_2 & a_1 \\ a_5 & a_4 & a_3 \end{pmatrix}, \ldots$$

$$\mathbf{H}_j = \begin{pmatrix} a_1 & 1 & 0 & 0 & \cdots 0 \\ a_3 & a_2 & a_1 & 1 & \cdots 0 \\ a_5 & a_4 & a_3 & a_2 & \cdots 0 \\ a_{2j-1} & a_{2j-2} & a_{2j-3} & a_{2j-4} & \cdots a_j \end{pmatrix} \ldots \mathbf{H}_k = \begin{pmatrix} a_1 & 1 & 0 \ldots & 0 \\ a_3 & a_2 & a_1 \ldots & 0 \\ \vdots & \vdots & \vdots & \vdots \\ 0 & 0 & \cdots & a_k \end{pmatrix},$$

where the (l, m) term in the matrix $\mathbf{H}_j$ is

a_{2l-m} for $0 < 2l - m < k$,

1 for $2l = m$,

0 for $2l < m$ or $2l > k + m$.

Then all eigenvalues have negative real parts; that is, *the steady-state* $\overline{\mathbf{N}}$ *is stable* if and only if the determinants of all Hurwitz matrices are positive:

$$\det \mathbf{H}_j > 0 \qquad (j = 1, 2, \ldots, k).$$

See Pielou (1969, chap. 6) for a treatment of the above and further references.

May (1973) summarizes these stability conditions in the cases $k = 2, \ldots, 5$. Some of these are given in the small box.

Example 1 illustrates how the Routh-Hurwitz criteria might be applied in a situation in which three species interact.

Routh-Hurwitz Criteria for k = 2, 3, 4

$k = 2$: $a_1 > 0$, $a_2 > 0$.

$k = 3$: $a_1 > 0$, $a_3 > 0$; $a_1 a_2 > a_3$.

$k = 4$: $a_1 > 0$, $a_3 > 0$; $a_4 > 0$; $a_1 a_2 a_3 > a_3^2 + a_1^2 a_4$.

Example 1

Suppose x is a predator and y and z are both its prey. z grows logistically in the absence of its predator. x dies out in the absence of prey, and y grows at an exponential rate in the absence of predator. We shall use the Routh-Hurwitz techniques to discover whether these species can coexist in a stable equilibrium.

Step 1

Writing equations for this system we get

$$\frac{dx}{dt} = \alpha(xz) + \beta(xy) - \gamma x, \tag{21a}$$

growth from eating z — from eating y — mortality

$$\frac{dy}{dt} = \delta y - \epsilon(xy), \tag{21b}$$

growth when no predator present — predation mortality

$$\frac{dz}{dt} = \mu z(\nu - z) - \chi(xz). \tag{21c}$$

logistic growth — predation mortality

Step 2

Solving for steady state values we get

$$\alpha xz + \beta xy - \gamma x = 0 \Rightarrow \alpha \bar{z} + \beta \bar{y} = \gamma, \text{ or } \bar{x} = 0, \tag{22a}$$

$$\delta y - \epsilon(xy) = 0 \Rightarrow \bar{x} = \frac{\delta}{\epsilon}, \text{ or } \bar{y} = 0, \tag{22b}$$

$$\mu z(\nu - z) - \chi xz = 0 \Rightarrow \mu\nu - \mu\bar{z} - \chi\bar{x} = 0. \tag{22c}$$

From the above we arrive at the nontrivial steady state

$$\bar{x} = \frac{\delta}{\epsilon}, \qquad \bar{y} = \gamma - \alpha\bar{z}, \qquad \bar{z} = \nu - \frac{\chi}{\mu}\bar{x}.$$

This equilibrium makes sense biologically whenever $\gamma > \alpha\bar{z}$ and $\nu > \chi/\mu\,\bar{x}$.

Step 3

Calculating the Jacobian of the system, we get

$$\mathbf{J} = \begin{pmatrix} \alpha\bar{z} + \beta\bar{y} - \gamma & \beta\bar{x} & \alpha\bar{x} \\ -\epsilon\bar{y} & \delta - \epsilon\bar{x} & 0 \\ -\chi\bar{z} & 0 & \mu\nu - 2\mu\bar{z} - \chi\bar{x} \end{pmatrix}. \tag{23}$$

Using the conclusions of step 2, we notice that terms on the diagonal of **J** evaluated at steady state lead to particularly simple forms, so that **J** is

$$\mathbf{J} = \begin{pmatrix} 0 & \beta\bar{x} & \alpha\bar{x} \\ -\epsilon\bar{y} & 0 & 0 \\ -\chi\bar{z} & 0 & -\mu\bar{z} \end{pmatrix} \tag{24}$$

Step 4
To find eigenvalues we must set

$$\det(\mathbf{J} - \lambda\mathbf{I}) = 0.$$

Thus we must evaluate

$$\det \begin{pmatrix} 0 - \lambda & \beta\bar{x} & \alpha\bar{x} \\ -\epsilon\bar{y} & 0 - \lambda & 0 \\ -\chi\bar{z} & 0 & -\mu\bar{z} - \lambda \end{pmatrix}, \tag{25}$$

and set the result equal to zero to get the characteristic equation. Expanding, we get

$$-\lambda\begin{pmatrix} -\lambda & 0 \\ 0 & -\mu\bar{z} - \lambda \end{pmatrix} - (-\epsilon\bar{y}) \det \begin{pmatrix} \beta\bar{x} & \alpha\bar{x} \\ 0 & -\mu\bar{z} - \lambda \end{pmatrix} + (-\chi\bar{z}) \det \begin{pmatrix} \beta\bar{x} & \alpha\bar{x} \\ -\lambda & 0 \end{pmatrix}$$

$$= (-\lambda)(-\lambda)(-\mu\bar{z} - \lambda) - (-\epsilon\bar{y})(\beta\bar{x})(-\mu\bar{z} - \lambda) + (-\chi\bar{z})(-\alpha\bar{x})(-\lambda)$$

$$= -\lambda^3 - \lambda^2\mu\bar{z} + \lambda(-\epsilon\bar{y}\beta\bar{x} - \chi\bar{z}\alpha\bar{x}) + (-\mu\bar{z}\epsilon\bar{y}\beta\bar{x}) = 0.$$

Cancelling a factor of -1 we obtain

$$\lambda^3 + a_1\lambda^2 + a_2\lambda + a_3 = 0, \tag{26a}$$

where

$$a_1 = \mu\bar{z}, \tag{26b}$$

$$a_2 = \epsilon\beta\bar{x}\bar{y} + \chi\alpha\bar{x}\bar{z}, \tag{26c}$$

$$a_3 = \mu\epsilon\beta\bar{x}\bar{y}\bar{z}. \tag{26d}$$

Step 5
Now we check the three conditions using the Routh-Hurwitz criteria for the case $k = 3$ (three species): The three conditions are

1. $a_1 > 0$,
2. $a_3 > 0$,
3. $a_1 a_2 > a_3$.

Condition 1 is true since $a_1 = \mu\bar{z}$ is a positive quantity. Condition 2 is true for the same reason. Looking at condition 3, we note that

$$a_1 a_2 = \mu\bar{z}(\epsilon\beta\bar{x}\bar{y} + \chi\alpha\bar{x}\bar{z}).$$

This is clearly bigger than $a_3 = \mu\epsilon\beta\bar{x}\bar{y}\bar{z}$ since the quantity $\chi\alpha\mu\bar{x}\bar{z}^2$ is positive. Thus condition 3 is also satisfied.

We conclude that the steady state is a stable one.

Remark 1

Because calculations of 3×3 (and higher-order) determinants can be particularly cumbersome, it is advisable to express the Jacobian in the simplest possible notation. We do this by leaving entries in terms of $\overline{x}$, $\overline{y}$, and $\overline{z}$ except where further simplification can be made (such as along the diagonal of $\mathbf{J}$). In step 5 we then use the fact that the quantities $\overline{x}$, $\overline{y}$, and $\overline{z}$ are positive.

Remark 2

In some situations the *magnitudes* of the steady-state values also enter into the stability conditions. We will see in a later section why this is not the case here in example 1.

6.5 QUALITATIVE STABILITY

The Routh-Hurwitz criteria outlined in Section 6.4 are an exact but cumbersome method for determining stability of a large system. For communities of five or more species, the technique proves so computationally involved that it is of diminishing practical value. Shortcuts, when available, can be quite useful.

In this section we explore a shortcut method for investigating large systems that needs little if any computation. Because this method is not universally applicable, its importance is viewed as secondary. Furthermore, to understand exactly why the method works requires knowledge of matrices beyond elementary linear algebra. Nevertheless, what makes the technique of *qualitative stability* appealing is that it is easy to explain, easy to test, and thus a refreshing change from intensive computations.

The technique of qualitative stability analysis applies ideally to large complicated systems in which there is no quantitative information about the interrelationship of species or subsystems. Motivation for this method actually came from economics. A paper by the economists Quirk and Ruppert (1965) was followed later by further work and application to ecology by May (1973), Levins (1974), and Jeffries (1974).

In a complex community composed of many species, numerous interactions take place. The magnitudes of the mutual effects of species on each other are seldom accurately known, but one can establish with greater certainty whether predation, competition, or other influences are present. This means that technically the functions appearing in equations that describe the system [such as equation (11)] are not known. What is known instead is the pattern of signs of partial derivatives of these functions [contained, for example, in the Jacobian of equation (16)]. We encountered a similar problem in the context of a plant-herbivore system (Chapter 3) and of a glucose-insulin model (Chapter 4). Here the problem consists of larger systems in a continuous setting, and even the magnitudes of partial derivatives may not be known.

There are two equivalent ways of representing qualitative information. A more obvious one is to assign the symbols $+$, 0, and $-$ to the (i, j)th entry of a matrix if the species j has respectively a positive influence, no influence, or a negative influence on species i. An alternate, visual representation captures the same ideas in a *directed graph* (also called *digraph*) in which nodes represent species and arrows between them represent the mutual interactions, as shown in Figures 6.10 and 6.11. The question is then whether it can be concluded, *from this graph or sign pattern only,* that the system is stable. If so, the system is called *qualitatively stable*.

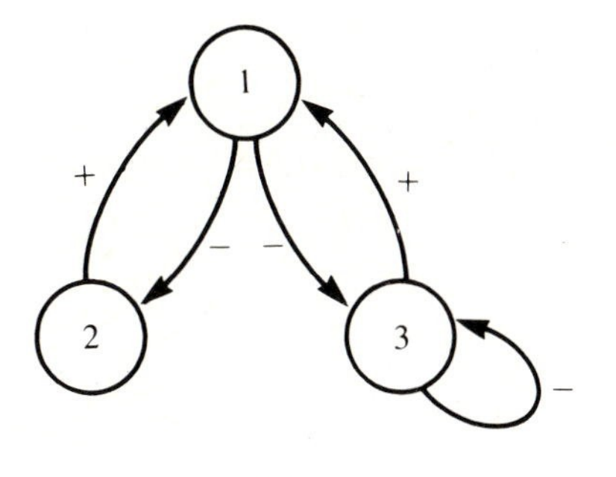

(*a*)

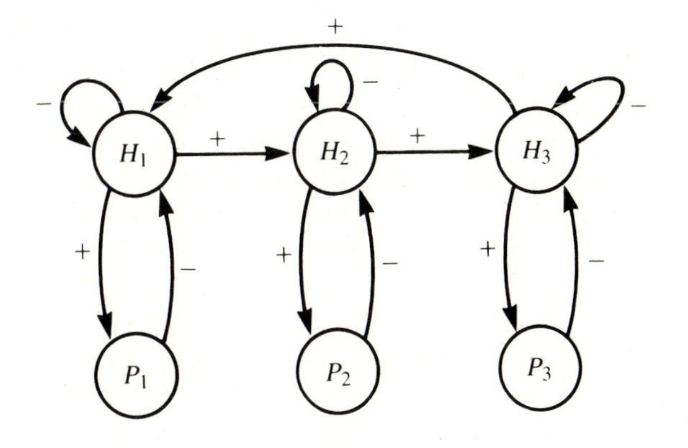

(*b*)

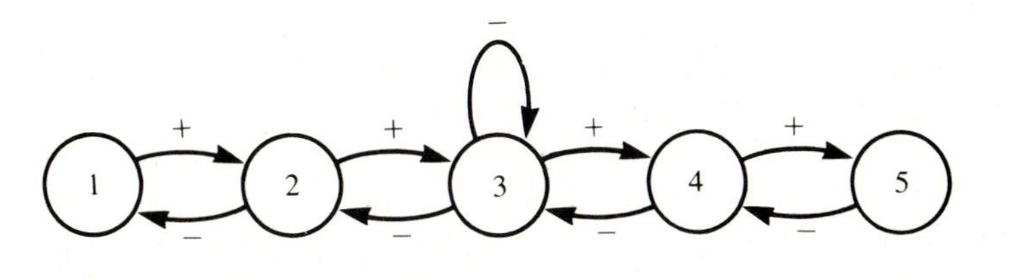

(c)

Figure 6.10 *Signed directed graphs (digraphs) can be used to represent species interactions in a complex ecosystem. The graphs shown here are equivalent to the matrix representation of sign patterns given in the text (a) example 2, (b) example 3, and (c) example 4.*

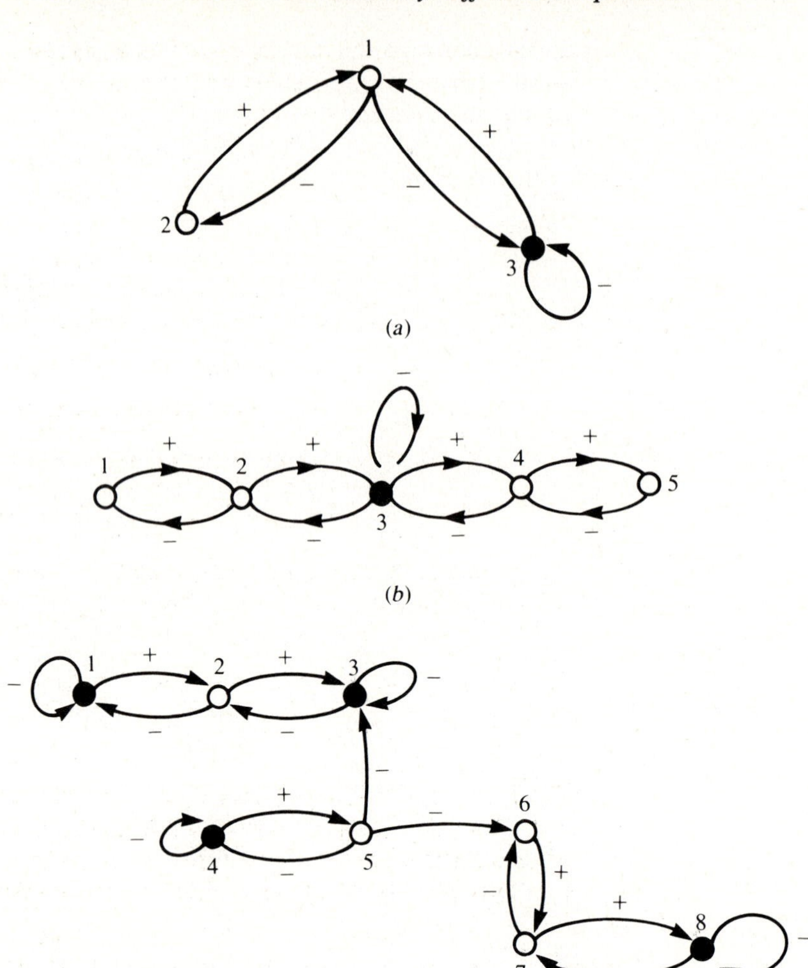

Figure 6.11 *Properties of signed directed graphs can be used to deduce whether the system is* qualitatively stable *(stable regardless of the magnitudes of mutual effects). The Jeffries color test and the Quirk-Ruppert conditions are applied to these graphs to conclude that (a), which corresponds to example 2, and (c) are stable communities, whereas (b), which corresponds to example 4, is not.*

Systems that are qualitatively stable are also stable in the ordinary sense. (The converse is not true.) Systems that are *not* qualitatively stable can still be stable under certain conditions (for example, if the magnitudes of interactions are appropriately balanced.)

Following Quirk and Ruppert (1965), May (1973) outlined five conditions for

Example 2

Here we study the sign pattern of the community described in equations (21a, b, c) of Section 6.4. From Jacobian (24) of the system we obtain the qualitative matrix

$$\mathbf{Q} = \text{sign } \mathbf{J} = \begin{bmatrix} 0 & + & + \\ - & 0 & 0 \\ - & 0 & - \end{bmatrix}.$$

This means that close to equilibrium, the community can also be represented by the graph in Figure 6.10. Reading entries in **Q** from left to right, top to bottom:

Species 1 gets positive feedback from species 2 and 3.
Species 2 gets negative feedback from species 1.
Species 3 gets negative feedback from species 1 and from itself.

Example 3 (Levins, 1977)

In a closed community, three predators or parasitoides, labeled P_1, P_2, and P_3, attack three different stages in the life cycle of a host, H_1, H_2, and H_3. The presence of hosts is a positive influence for their predators but predators have a negative influence on their prey. Figure 6.10(*b*) and the following matrix summarize the interactions:

$$\begin{array}{c} \\ P_1 \\ P_2 \\ P_3 \\ H_1 \\ H_2 \\ H_3 \end{array} \begin{array}{c} \begin{array}{cccccc} P_1 & P_2 & P_3 & H_1 & H_2 & H_3 \end{array} \\ \begin{bmatrix} 0 & 0 & 0 & + & 0 & 0 \\ 0 & 0 & 0 & 0 & + & 0 \\ 0 & 0 & 0 & 0 & 0 & + \\ - & 0 & 0 & - & 0 & + \\ 0 & - & 0 & + & - & 0 \\ 0 & 0 & - & 0 & + & - \end{bmatrix} \end{array}.$$

Note that H_1, H_2, and H_3 each exert negative feedback on themselves.

Example 4 (Jeffries, 1974)

In a five-species ecosystem, species 2 preys on species 1, species 3 on species 2, and so on in a food chain up to species 5. Species 3 is also self-regulating. A qualitative matrix for this community is

$$\mathbf{Q} = \begin{bmatrix} 0 & - & 0 & 0 & 0 \\ + & 0 & - & 0 & 0 \\ 0 & + & - & - & 0 \\ 0 & 0 & + & 0 & - \\ 0 & 0 & 0 & + & 0 \end{bmatrix}.$$

See Figure 6.10(*c*).

qualitative stability. Suppose a_{ij} is the ijth element of the matrix of signs $\mathbf{Q}$. Then it is necessary for all of the following conditions to hold:

1. $a_{ii} \leq 0$ for all i.
2. $a_{ii} < 0$ for at least one i.
3. $a_{ij}a_{ji} \leq 0$ for all $i \neq j$.
4. $a_{ij}a_{jk} \cdots a_{qr}a_{ri} = 0$ for any sequences of three or more distinct indices $i, j, k, \ldots, q, r$.
5. $\det \mathbf{Q} \neq 0$.

These conditions can be interpreted in the following way:

1. No species exerts positive feedback on itself.
2. At least one species is self-regulating.
3. The members of any given pair of interacting species must have opposite effects on each other.
4. There are no closed chains of interactions among three or more species.
5. There is no species that is unaffected by interactions with itself or with other species.

For mathematical proof of these five necessary conditions, consult Quirk and Ruppert (1965). May (1973) and Pielou (1969) comment on the biological significance, particularly of conditions 3 and 4. The conditions can be tested by looking at graphs representing the communities. One must check that these graphs have all the following properties:

1. No + loops on any single species (that is, no positive feedback).
2. At least one − loop on some species in the graph.
3. No pair of like arrows connecting a pair of species.
4. No cycles connecting three or more species.
5. No node devoid of input arrows.

These five conditions are equivalent to the original algebraic statement.

Example 5
For examples 2 to 4 we check off the five conditions given earlier:

Condition Number	*Example 2*	*Example 3*	*Example 4*
1	√	√	√
2	√	√	√
3	√	√	√
4	√	No	√
5	√	√	√

It was shown by Jeffries (1974) that these five conditions alone cannot distinguish between *neutral stability* (as in the Lotka-Volterra cycles) and *asymptotic stability,* wherein the steady state is a stable node or spiral. (In other words, the conditions are *necessary but not sufficient* to guarantee that the species will coexist in a constant steady state). In example 4 Jeffries notes that pure imaginary eigenvalues can occur, so that even though the five conditions are met, the system will oscillate. To weed out such marginal cases, Jeffries devised an auxiliary set of conditions, which he called the "color test," that *replaces* condition 2. Before describing the color test, it is necessary to define the following:

A *predation link* is a pair of species connected by one + line and one − line.
A *predation community* is a subgraph consisting of all interconnected predation links.

If one defines a species not connected to any other by a predation link as a *trivial* predation community, then it is possible to decompose any graph into a set of distinct predation communities. The systems shown in Figure 6.10 have predation communities as follows: (*a*) {2, 1, 3}; (*b*) $\{H_1, P_1\}$, $\{H_2, P_2\}$, $\{H_3, P_3\}$; and (*c*) {1, 2, 3, 4, 5}. In Figure 6.11(*c*) there are three predation communities: {1, 2, 3}, {4, 5}, and {6, 7, 8}.

The following color scheme constitutes the test to be made. A predation community is said to *fail the color test* if it is not possible to color each node in the subgraph black or white in such a way that

1. Each self-regulating node is black.
2. There is at least one white point.
3. Each white point is connected by a predation link to at least one other white point.
4. Each black point connected by a predation link to one white node is also connected by a predation link to one other white node.

Jeffries (1974) proved that for asymptotic stability, a community must satisfy the original Quirk-Ruppert conditions 1, 3, 4, and 5, and in addition must have only predation communities that fail the color test.

Example 5 (continued)

Examples 2 and 4 satisfy the original conditions. In Figure 6.11 the color test is applied to their communities. We see that example 2 consists of a single predation community that *fails* part 4 of the color test. Example 4 *satisfies* the test. A final example shown in Figure 6.11(*c*) has three predation communities, and each one fails the test. We conclude that Figure 6.11(*a*) and (*c*) represent systems that have the property of asymptotic stability; that is, these ecosystems consist of species that coexist at a stable fixed steady state without sustained oscillations.

A proof and discussion of the revised conditions is to be found in Jeffries (1974). For other applications and properties of graphs, you are encouraged to peruse Roberts (1976).

6.6 THE POPULATION BIOLOGY OF INFECTIOUS DISEASES

Infectious diseases can be classified into two broad categories: those caused by viruses and bacteria are *microparasitic* diseases, and those due to worms (more commonly found in third-world countries) are *macroparasitic*. Other than the relative sizes of the infecting agents, the main distinction is that microparasites reproduce within their host and are transmitted directly from one host to another. Most macroparasites, on the other hand, have somewhat more complicated life cycles, often with a secondary host or carrier implicated. (Examples of these include malaria and schistosomiasis; see Anderson, 1982 for a review.)

This section briefly summarizes some of the classical models for microparasitic infections. The mathematical techniques required for analyzing the models parallel the techniques applied in Sections 6.2 and 6.3. However, as a general remark, it should be said that the *flavor* of the models differs somewhat from the species-interactions models introduced in this chapter.

With no *a priori* knowledge, suppose we are asked to model the process of infection of a viral disease such as measles or smallpox. In keeping with the style of population models for predation or competition, it would be tempting to start by defining variables for population densities of the host x and infecting agent y. Here is how such a model might proceed:

Primitive Model for a Viral Infection

This model is for illustrative purposes only. Let

x = population of human hosts,

y = viral population.

The assumptions are that

1. There is a constant human birth rate α.
2. Viral infection causes an increased mortality due to disease, so $g(y) > 0$.
3. Reproduction of viral particles depends on human presence.
4. In the absence of human hosts, virus particles "die" or become nonviable at rate γ.

The equations then read:

$$\frac{dx}{dt} = [\alpha - g(y)]x,$$

$$\frac{dy}{dt} = \beta xy - \gamma y.$$

The approach leads us to a modified Lotka-Volterra predation model. This view, to put it simply, is that viruses y are predatory organisms searching for human prey x to consume. The conclusions given in Section 6.2 follow with minor modification.

The philosophical view of disease as a process of predation is an unfortunate and somewhat misleading analogy on several counts. First, no one can reasonably suppose it possible to measure or even estimate total viral population, which may range over several orders of magnitude in individual hosts. Second, a knowledge of this number is at best uninteresting and trivial since it is the distribution of viruses over hosts that determines what percentage of people will actually suffer from the disease. To put it another way, some hosts will harbor the infecting agent while others will not. Finally, in the "primitive" model an underlying hidden assumption is that viruses roam freely in the environment, randomly encountering new hosts.This is rarely true of microparasitic diseases. Rather, diseases are spread by contact or close proximity between infected and healthy individuals. How the disease is spread in the population is an interesting question. This crucial point is omitted and is thus a serious criticism of the model.

A new approach is necessary. At the very least it seems sensible to make a distinction between sick individuals who harbor the disease and those who are as yet healthy. This forms the basis of all microparasitic epidemiological models, which, as we see presently, virtually omit the population of parasites from direct consideration.

Instead, the host population is subdivided into distinct classes according to the health of its members. A typical subdivision consists of susceptibles S, infectives I, and a third, removed class R of individuals who can no longer contract the disease because they have recovered with immunity, have been placed in isolation, or have died. If the disease confers a *temporary immunity* on its victims, individuals can also move from the third class to the first.

Time scales of epidemics can vary greatly from weeks to years. *Vital dynamics* of a population (the normal rates of birth and mortalities in the absence of disease) can have a large influence on the course of an outbreak. Whether or not immunity is conferred on individuals can also have an important impact. Many models using the general approach with variations on the assumptions have been studied. An excellent summary of several is given by Hethcote (1976) and Anderson and May (1979), although different terminology is unfortunately used in each source.

Some of the earliest classic work on the theory of epidemics is due to Kermack and McKendrick (1927). One of the special cases they studied is shown in Figure 6.12(a). The diagram summarizes transition rates between the three classes with the parameter β, the rate of transmission of the disease, and the rate of removal ν. It is assumed that each compartment consists of identically healthy or sick individuals and that no births or deaths occur in the population. (In more current terminology, the situation shown in Figure 6.12(a) would be called an *SIR model without vital dynamics* because the transitions are from class S to I and then to R; see for example, Hethcote, 1976.)

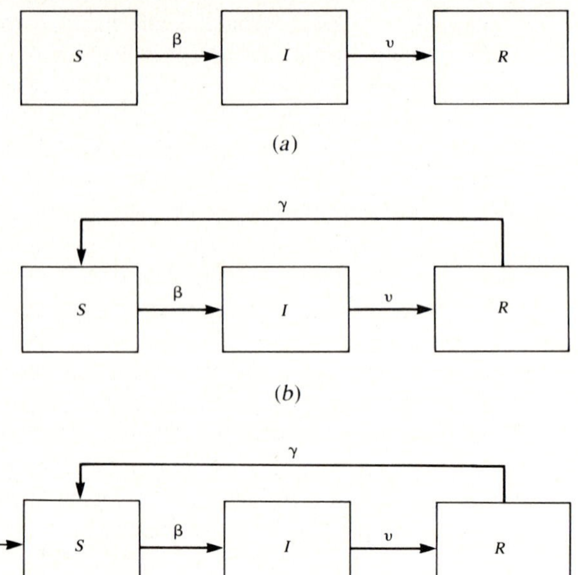

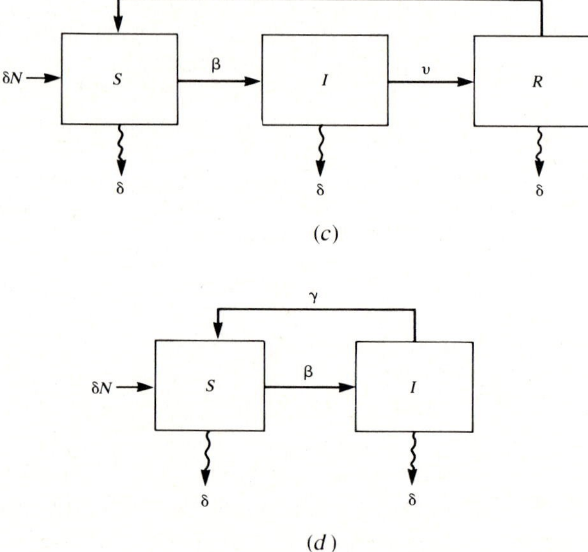

Figure 6.12 *A number of epidemic models that have been studied. The total population* N *is subdivided into susceptible* (S), *infective* (I) *and removed* (R) *classes. Transitions between compartments depict the course of transmission, recovery, and loss of immunity with rate constants* β, ν, *and* γ. *A population with vital dynamics is assumed to be producing new susceptibles at rate* δ *which is identical to the mortality rate. (a)* SIR *model; (b, c)* SIRS *models; and (d)* SIS *model.*

Figure 6.12(a) and those following it are somewhat reminiscent of models we have already studied for the physical flows between well-mixed compartments (for example, the chemostat). A subtle distinction must be made though, since the passage of individuals from the susceptible to the infective class generally occurs as a result of close proximity or contact between healthy and infective individuals. Thus the rate of exchange between S and I has a special character summarized by the following assumption:

> ***Assumption***
> The rate of transmission of a microparasitic disease is proportional to the rate of encounter of susceptible and infective individuals modelled by the product (βSI).

The equations due to Kermack and MacKendrick for the disease shown in Figure 6.12(a) are thus

$$\frac{dS}{dt} = -\beta IS, \tag{27a}$$

$$\frac{dI}{dt} = \beta IS - \nu I, \tag{27b}$$

$$\frac{dR}{dt} = \nu I. \tag{27c}$$

It is easily verified that the total population $N = S + I + R$ does not change. Though these equations are nonlinear, Kermack and MacKendrick derived an approximate expression for the rate of removal dR/dt (in their paper called dz/dt) as a function of time. The result is a rather messy expression involving hyperbolic secants; when plotted with the appropriate values given to the parameters it compares rather well with data for death by plague in Bombay during an epidemic in 1906 (see Figure 6.13).

A more instructive approach is to treat the problem by qualitative methods. Now we shall carry out this procedure on a slightly more general case, allowing for a loss of immunity that causes recovered individuals to become susceptible again [Figured 6.12(b)]. It will be assumed that this takes place at a rate proportional to the population in class R, with proportionality constant γ. Thus the equations become

$$\frac{dS}{dt} = -\beta SI + \gamma R, \tag{28a}$$

$$\frac{dI}{dt} = \beta SI - \nu I, \tag{28b}$$

$$\frac{dR}{dt} = \nu I - \gamma R. \tag{28c}$$

This model is called an *SIRS* model since removed individuals can return to class $S (\gamma = 0$ is the special case studied by Kermack and McKendrick). It is readily shown that these equations have two steady states:

$$\bar{S}_1 = N, \qquad \bar{I}_1 = 0, \qquad \bar{R}_1 = 0; \tag{29a}$$

$$\bar{S}_2 = \frac{\nu}{\beta}, \qquad \bar{I}_2 = \gamma\frac{N - \bar{S}_2}{\nu + \gamma}, \qquad \bar{R}_2 = \frac{\nu\bar{I}_2}{\gamma}. \tag{29b}$$

In (29a) the whole population is healthy (but susceptible) and disease is eradicated. In (29b) the community consists of some constant proportions of each type provided $(\bar{S}_2, \bar{I}_2, \bar{R}_2)$ are all positive quantities. For $\bar{I}_2$ to be positive, N must be larger than $\bar{S}_2$. Since $\bar{S}_2 = \nu/\beta$, this leads to the following conclusion:

The disease will be established in the population provided the total population N exceeds the level ν/β, that is,

$$\frac{N\beta}{\nu} > 1.$$

This important *threshold effect* was discovered by Kermack and McKendrick; the population must be "large enough" for a disease to become endemic.

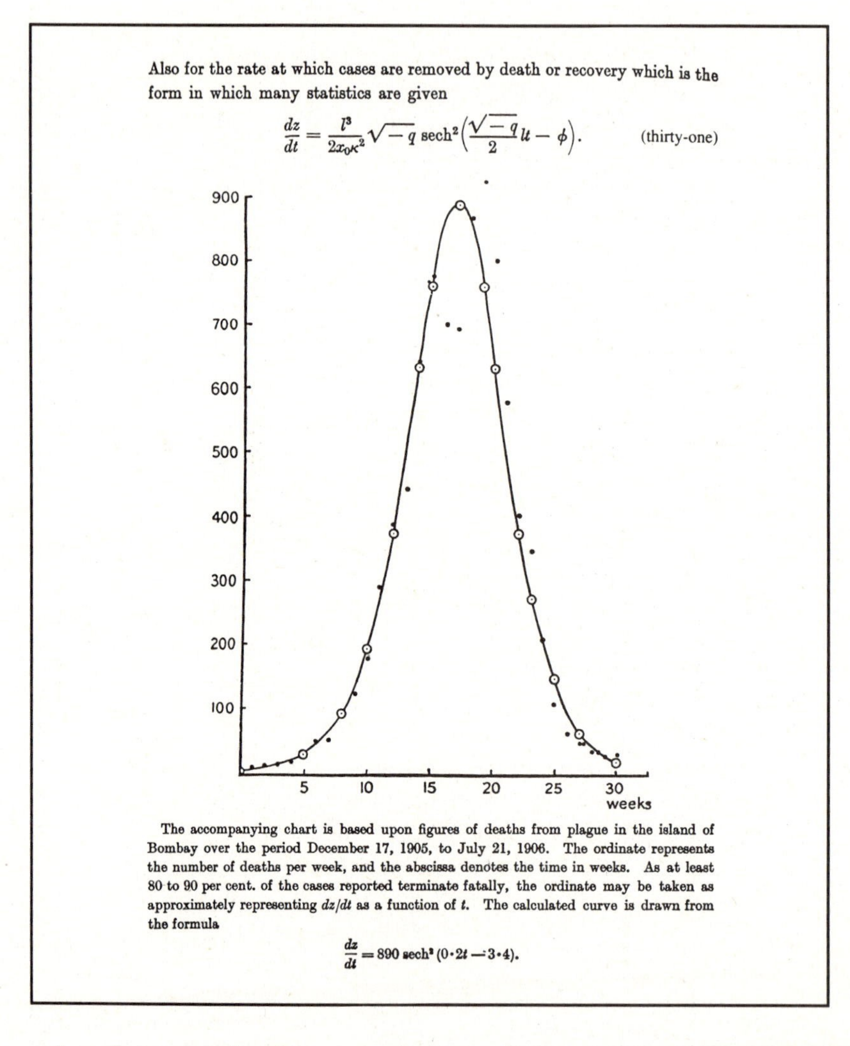

Also for the rate at which cases are removed by death or recovery which is the form in which many statistics are given

$$\frac{dz}{dt} = \frac{l^3}{2x_0\kappa^2}\sqrt{-q}\,\text{sech}^2\left(\frac{\sqrt{-q}}{2}lt - \phi\right). \qquad \text{(thirty-one)}$$

The accompanying chart is based upon figures of deaths from plague in the island of Bombay over the period December 17, 1905, to July 21, 1906. The ordinate represents the number of deaths per week, and the abscissa denotes the time in weeks. As at least 80 to 90 per cent. of the cases reported terminate fatally, the ordinate may be taken as approximately representing dz/dt as a function of t. The calculated curve is drawn from the formula

$$\frac{dz}{dt} = 890\,\text{sech}^2(0\cdot 2t - 3\cdot 4).$$

Figure 6.13 *On a page from their original article, Kermack and McKendrick compare predictions of the model given by equations (27a,b,c) with data for the rate of removal by death. Note:* dz/dt *is equivalent to* dR/dt *in equations (27).[Kermack, W. O., and McKendrick, A. G. (1927). A contribution to mathematical theory of epidemics.* Roy Stat. Soc. J., 115, *714.]*

The ratio of parameters β/ν has a rather meaningful interpretation. Since removal rate from the infective class is ν (in units of 1/time), the average period of infectivity is $1/\nu$. Thus β/ν is the fraction of the population that comes into contact with an infective individual during the period of infectiousness. The quantity $R_0 = N\beta/\nu$ has been called the *infectious contact number,* σ (Hethcote, 1976) and the *intrinsic reproductive rate of the disease* (May, 1983). R_0 represents the average number of secondary infections caused by introducing a single infected individual into a host population of N susceptibles. (In papers by May and Anderson, the threshold result is usually written $R_0 > 1$.)

In further analyzing the model we can take into account the particularly convenient fact that the total population

$$N = S + I + R$$

does not change (see problem 25 for verification). This means that one variable, say R, can always be eliminated so that the model can be given in terms of two equations in two unknowns. In the following analysis this fact is exploited in applying phase-plane methods to the problem.

Qualitative Analysis of a* SIRS *Model: Epidemic with Temporary Immunity and No Vital Dynamics

Since the total population is constant, we eliminate R from equations (28) by substituting

$$R = N - S - I. \tag{30}$$

The equations for S and I are then

$$\frac{dS}{dt} = -\beta SI + \gamma(N - S - I) \overset{\text{def}}{\equiv} F(S, I), \tag{31a}$$

$$\frac{dI}{dt} = \beta SI - \nu I \equiv G(S, I). \tag{31b}$$

Nullclines

$$I' = 0 \quad \text{for} \quad I = 0 \quad \text{and} \quad S = \nu/\beta,$$
$$S' = 0 \quad \text{for} \quad \beta SI = \gamma(N - S - I).$$

After rearranging,

$$I = \frac{\gamma(N - S)}{(\beta S + \gamma)}.$$

This curve intersects the axes at $(N, 0)$ and $(0, N)$.

Steady states

$$(\bar{S}_1, \bar{I}_1) = (N, 0); (\bar{S}_2, \bar{I}_2) = \left(\frac{\nu}{\beta}, \frac{N - (\nu/\beta)}{\nu + \gamma}\right).$$

Jacobian

$$\mathbf{J} = \begin{pmatrix} F_S & F_I \\ G_S & G_I \end{pmatrix}_{ss} = \begin{pmatrix} -(\beta\bar{I} + \gamma) & -(\beta\bar{S} + \gamma) \\ \beta\bar{I} & \beta\bar{S} - \nu \end{pmatrix}.$$

Stability
For $(\bar{S}_2, \bar{I}_2)$,

$$\text{Tr}\,\mathbf{J} = -(\beta\bar{I}_2 + \gamma) \text{ is always negative,}$$
$$\det\mathbf{J} = \beta\bar{I}_2(\nu + \gamma) \text{ is always positive.}$$

Thus this steady state is always stable when it exists, namely when the threshold condition is satisfied. It is evident from Figures 6.14 and 6.15(b) and from further analysis that the approach to this steady state can be oscillatory.

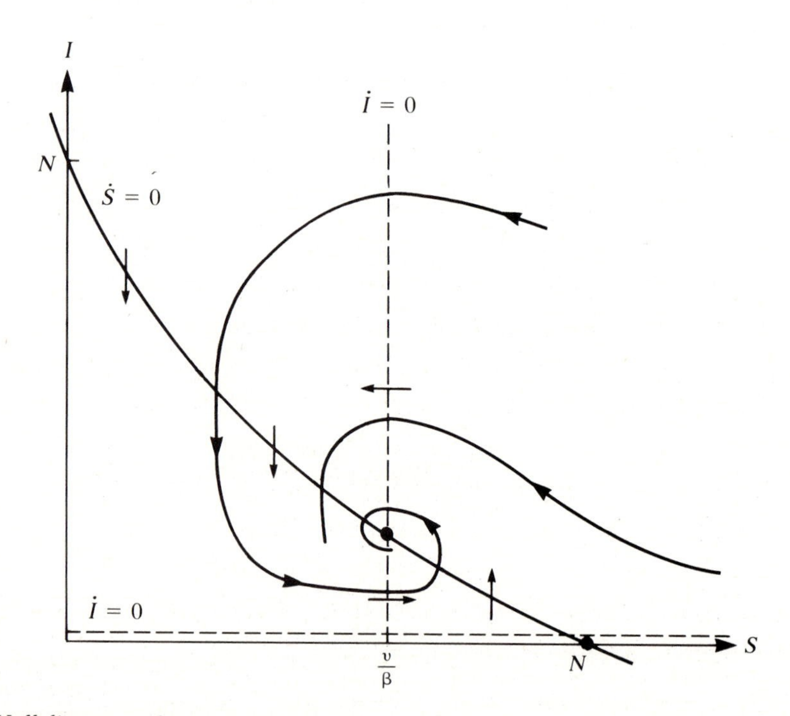

Figure 6.14 *Nullclines, steady states, and several trajectories for the* SIRS *model given by equations (31a,b), which are equivalent to (28a,b,c).*

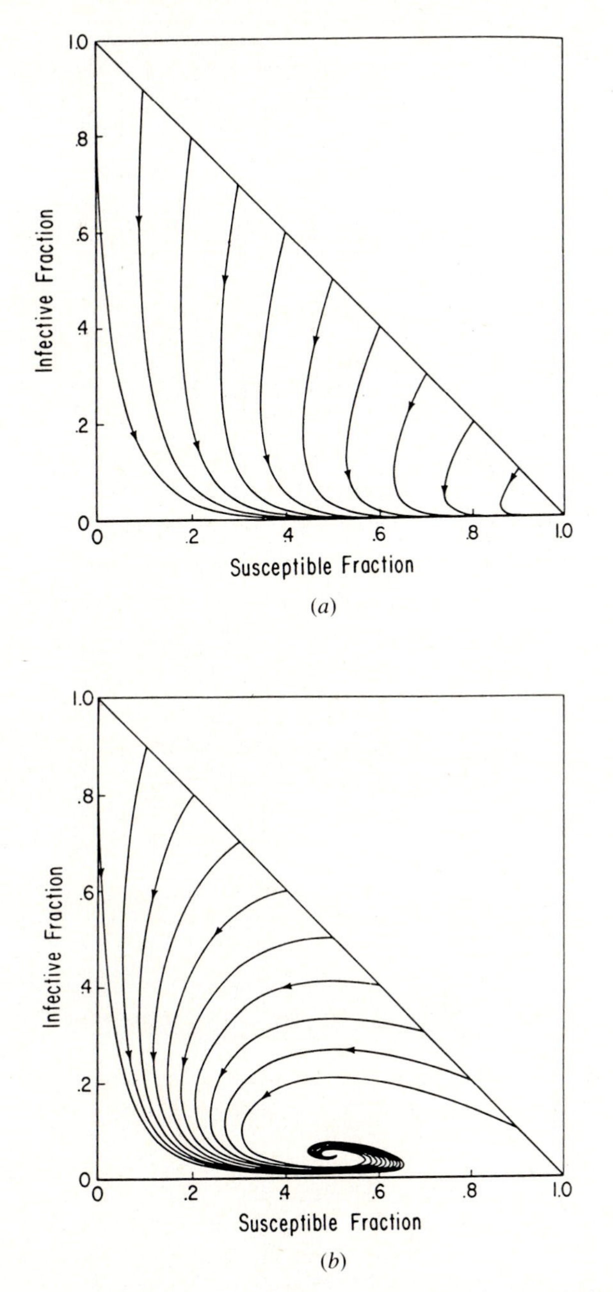

Figure 6.15 *Epidemic models are characterized by the magnitude of an infectious contact number* σ. *(a) When* $\sigma < 1$, *the infective class will disappear. (b) When* $\sigma > 1$, *there is some stable steady state in which both susceptibles and infectives are present. Shown here is an* SIRS *model with vital dynamics.) [Reprinted by permission of the publisher from Hethcote, H. W. (1976). Qualitative analyses of communicable disease models.* Math. Biosci., *28, 344 and 345. Copyright 1976 by Elsevier Science Publishing Co., Inc.]*

MEMENTO MORI

LONDON'S *Dreadful Viſitation:*

Or, A COLLECTION of All the

Bills of Mortality

For this Preſent Year:

Beginning the 27th of *December* 1664. and
ending the 19th. of *December* following:

As alſo, *The GENERAL or whole years BILL:*

According to the Report made to the
KING'S Moſt Excellent Majeſty,

By the Company of Pariſh Clerks of London. &c

LONDON:
Printed and are to be ſold by *E. Cotes* living in *Alderſgate-ſtreet,*
Printer to the ſaid Company 1665.

Mortality from a variety of afflictions, only some of which were caused by disease, were systematically recorded as early as the 1600s in the Bills of Mortality published in London. Reproduced here is the title page of the London Bills of Mortality for 1665, the year of the great plague. The people of the city followed with anxiety the rise and fall in the number of deaths from the plague, hoping always to see the sharp decline which they knew from past experience indicated that the epidemic was nearing its end. When the decline came the refugees, mostly from the nobility and wealthy merchants, returned to the city, and then for a time the mortality rose again as the disease attacked these new arrivals. The plague of 1665 started in June; its peak came in September and its decline in October. The secondary rise occurred in November and cases of the disease were reported as late as March of the following year. *[From H. W. Haggard (1957),* Devils, Drugs, Doctors, *Harper & Row, New York.]*

The Diseases, and Casualties this year being 1632.

Disease	Number	Disease	Number
Abortive, and Stilborn	445	Grief	11
Affrighted	1	Jaundies	43
Aged	628	Jawfaln	8
Ague	43	Impostume	74
Apoplex, and Meagrom	17	Kil'd by several accidents	46
Bit with a mad dog	1	King's Evil	38
Bleeding	3	Lethargie	2
Bloody flux, scowring, and flux	348	Livergrown	87
Brused, Issues, sores, and ulcers,	28	Lunatique	5
Burnt, and Scalded	5	Made away themselves	15
Burst, and Rupture	9	Measles	80
Cancer, and Wolf	10	Murthered	7
Canker	1	Over-laid, and starved at nurse	7
Childbed	171	Palsie	25
Chrisomes, and Infants	2268	Piles	1
Cold, and Cough	55	Plague	8
Colick, Stone, and Strangury	56	Planet	13
Consumption	1797	Pleurisie, and Spleen	36
Convulsion	241	Purples, and spotted Feaver	38
Cut of the Stone	5	Quinsie	7
Dead in the street, and starved	6	Rising of the Lights	98
Dropsie, and Swelling	267	Sciatica	1
Drowned	34	Scurvey, and Itch	9
Executed, and prest to death	18	Suddenly	62
Falling Sickness	7	Surfet	86
Fever	1108	Swine Pox	6
Fistula	13	Teeth	470
Flocks, and small Pox	531	Thrush, and Sore mouth	40
French Pox	12	Tympany	13
Gangrene	5	Tissick	34
Gout	4	Vomiting	1
		Worms	27

Christened	Males....4994 Females..4590 In all....9584	Buried	Males....4932 Females..4603 In all....9535

Whereof, of the Plague.8

Increased in the Burials in the 122 Parishes, and at the Pest-house this year.......... 993

Decreased of the Plague in the 122 Parishes, and at the Pest-house this year.......... 266 [10]

Numerous other cases have been analyzed in detail. Perhaps the best summary is given by Hethcote (1976), in which theoretical results are followed by *biocorollaries* that spell out the biological predictions. His paper was used in drawing up Table 6.1, a composite that describes a number of cases.

One point worth mentioning is the essential difference between models in which the susceptible class is *renewed* (by recovery or loss of immunity) and those in which it is not. [The *SIRS* and *SIS* examples shown in Figure 6.12($b-d$) belong to the former category.] These distinct types behave differently when the normal turnover of births and deaths is superimposed on the dynamics of the disease. In the *SIR* type without births, the continual decrease of the susceptible class results in a decline in the effective reproductive rate of the disease. The epidemic stops for want of infectives, not, as it might seem, for want of susceptibles (Hethcote, 1976). On the other hand, if the susceptible class is replenished by births or recoveries, the subpopulation that participates in the disease is maintained, and the disease can persist.

From Table 6.1 we see that *SIR* models are subdivided into those with and without births and deaths. In other models the chief effect of normal birth and mortality at rates δ is to decrease the infectious contact number σ. This means that a smaller population can sustain an endemic disease. Note that the total population N is taken to be constant in all of these models since the number of deaths from all classes is assumed to exactly balance the births of new susceptibles. Among other things, this permits all such models to be analyzed by methods similar to the method used here since one variable can always be eliminated.

A somewhat different philosophical approach was taken by Anderson and May (1979), who were less interested in the dynamics of the disease itself. By analyzing a model in which a disease-free population grows exponentially, rather than being maintained at a constant level, they demonstrated that epidemics increasing host mortality have the potential to regulate population levels (see problem 30). This adds yet another interesting possiblity to the list of causes of decelerating growth rates in natural populations. Aside from inter- and intraspecies competition and predation, disease-causing agents (much like parasites) can control the population dynamics of their hosts.

The theory of epidemics has numerous ramifications, some of which are mathematical and some practical. In recent years more advanced mathematical models have been studied to determine the effects of delay factors (such as a waiting time in the infectious class), age structure, migration, and spatial distributions. Many of these models require sophisticated mathematical methods of analysis (see, for example, Busenberg and Cooke, 1978). An excellent survey with detailed references is given by Hethcote et al. (1981).

One theoretical question such papers often address is whether models with particular structures lead to stable (limit-cycle) oscillations. This question is of interest since some diseases are associated with periodic outbreaks with very low endemic periods followed by peak epidemic cycles. In some cases the forces driving such cyclic behavior are related to seasonality and to changes in contact rates. (A good example is childhood diseases, which invariably peak during the school year when contact between their potential hosts is greatest.) However, even in the absence of externally imposed periodicity, models similar to *SIRS* can have an inherent ten-

Table 6.1 A Summary of Several Epidemic Models

Type	*Immunity*	*Birth/Death*	*Significant quantity*	*Results*	*Figures*
SIS	None	Rate = δ	$\sigma = \dfrac{\beta}{\gamma + \delta}$	(1) $\sigma > 1$: constant endemic infection (2) $\sigma < 1$: infection disappears	6.9(*d*)
		Additional disease fatality rate η	σ as above, and $\epsilon = \dfrac{\eta}{\gamma + \delta}$	Disease always eventually disappears leaving some susceptibles.	
SIR	Yes, recovery gives immunity.	None	$\sigma = \dfrac{S_0 \beta}{\gamma}$ (S_0 = initial S)	(1) $\sigma > 1$: infection peaks and then disappears (2) $\sigma < 1$: infection disappears	6.9(*a*)
		Yes, rate = δ	$\sigma = \dfrac{\beta}{\nu + \delta}$	(1) $\sigma < 1$: susceptibles and infectives approach constant levels (2) $\sigma < 1$: infectives disappear; only S remains	6.9(*c*) with $\gamma = 0$
(*SIR* with carriers)	Yes	Yes		Disease always remains endemic.	
SIRS	Temporary, lost at rate γ	Rate = δ	$\sigma = \dfrac{\beta}{\nu + \delta}$	(1) $\sigma > 1$: same as *SIR* (1) but higher levels of infectives (2) $\sigma < 1$: same as *SIR* (2)	6.9(*c*)

dency to give rise to oscillations. This is particularly true of models with long periods of immunity or some other delaying factor. Hethcote notes that a sequence of at least three removed classes will also achieve the result (for example, $SIR_1R_2R_3S$).

The implications of many aspects of applying mathematical theory to natural populations are eloquently described in numerous papers by May and Anderson. Some questions are of a basic scientific nature. For example, the extent to which diseases and hosts have coevolved is a fascinating topic; a second controversial question is whether or not diseases are in fact a predominant factor in controlling natural populations. Other questions have more immediate medical ramifications. Anderson and May suggest that theory has an important place in illuminating the impact of disease on human populations and the ability to eradicate or control disease. Two applications of the theory to vaccination programs are briefly highlighted in the following section.

6.7 FOR FURTHER STUDY: VACCINATION POLICIES

Models for infectious diseases lead to a better understanding of how vaccination programs affect the control or eradication of the disease. Several popular articles by Anderson and May (1982) and a more detailed mathematical version (1983) are thought-provoking and informative. The full theory that takes into account age structure of the population uses a partial differential equation model (which would be understood more fully after covering Chapter 11). However, a number of rather interesting consequences of the theory can be understood with no further preparation.

Eradicating a Disease

Immunization can reduce or eliminate the incidence of infection, even when only part of the population receives the treatment. Those individuals who have been vaccinated will be protected from acquiring infection (this is the obvious direct effect). A secondary effect is that since vaccinated individuals are essentially removed from participating in transmission of the disease, there will be fewer infectious individuals and thus a decreased likelihood that an unvaccinated susceptible will come in contact with the disease. This indirect effect is known as *herd immunity*.

Administering vaccinations to an entire population can be costly. Some vaccines (for example, some measles and whooping cough vaccines) also carry the risk, though rare, of causing various reactions or neurological damage. Thus, if disease eradication can be achieved by partially vaccinating some fraction p of the population, an advantage is gained.

The fraction to be immunized must be such that the remaining population, $(1 - p)N$, will no longer exceed the threshold level necessary to perpetuate the disease. In the terminology of Anderson and May, the reproductive factor R_0 of the infection is to be reduced below 1. Since $R_0 = N\beta/\nu$, for a given disease this factor can be estimated from epidemiological and population records. Table 6.2 lists several common diseases with their corresponding R_0 factors.

Table 6.2 ***Estimates of the intrinsic reproductive rate R_0 for human diseases and the corresponding percentage of the population p that must be protected by immunization to achieve eradication. [Reprinted by permission, American Scientist,* journal of Sigma Xi, *"Parasitic Infections as Regulator of Animal Populations," by Robert M. May, 71:36–45 (1983).]***

Infection	*Location and Time*	R_0	*Approximate Value of p (%)*
Smallpox	Developing countries, before global campaign	3–5	70–80
Measles	England and Wales, 1956–68;	13	92
	U.S., various places, 1910–30	12–13	92
Whooping cough	England and Wales, 1942–50;	17	94
	Maryland, U.S., 1908–17	13	92
German measles	England and Wales, 1979;	6	83
	West Germany, 1972	7	86
Chicken pox	U.S., various places, 1913–21 and 1943	9–10	90
Diphtheria	U.S., various places, 1910–47	4–6	~80
Scarlet fever	U S., various places, 1910–20	5–7	~80
Mumps	U.S., various places, 1912–16 and 1943	4–7	~80
Poliomyelitis	Holland, 1960: U.S., 1955	6	83

The fraction p to be immunized is then deduced from the following simple calculation:

Intrinsic reproductive rate of disease = R_0, Fraction immunized = p, Fraction not immunized = $1 - p$, Population participating in disease = $N(1 - p)$, Effective intrinsic reproduction rate of disease (after immunization) = $R_0' = (1 - p)R_0$.

Thus

$$R_0' < 1 \Rightarrow (1 - p)R_0 < 1 \Rightarrow p > 1 - \frac{1}{R_0}$$

The percentage of the population to be vaccinated thus depends strongly on the infectiousness of the disease. It is noteworthy that smallpox, a disease essentially eradicated by vaccination, has one of the lowest R_0 values and a correspondingly low required vaccination fraction. By contrast, measles and whooping cough require a much higher percentage of immunization and would be harder to eradicate.

Average Age of Acquiring a Disease

Is it always wise to vaccinate at least some people, even if the disease will not be eradicated? When a disease has different impacts on individuals of different ages, vaccination at a young age can have a somewhat surprising deleterious effect. A case in point is German measles (Rubella). Normally a mild short-lasting infection, Rubella can be particularly devastating when contracted by a pregnant woman, as it results in birth defects to the fetus during the first trimester of pregnancy. Vaccinating *against* Rubella raises the average age at which the disease is first acquired (see box). Thus, rather than incurring the disease on average at age 12, it may be more prevalent at ages 20 to 30, precisely the most dangerous period for women of childbearing ages.

How Vaccinations Raise the Age of First Acquiring a Disease

Define $\lambda = \beta I$. Called the *per capita force of infection,* λ has units of $1/\text{time}$ and is the rate of acquiring the disease given a population containing I infectives and a transmission constant β.

Let $A = 1/\lambda$. A is the *average age of first infection,* or average waiting time in the susceptible compartment before acquiring the disease. A has units of time.

Now note that a vaccination program tends to reduce the number of infectives I, thus reducing λ and *raising* A.

For other aspects of the topic of vaccinations, epidemiology, and population dynamics, the many excellent sources quoted in the references are highly recommended.

PROBLEMS*[1]

1. (a) Assuming that $N(0) = N_0$, integrate equation (2a) and show that its solution is given by equation (2b). (See problem 5 of Chapter 4 or Braun, 1979; sec. 1.5.)
   (b) Show that the solution given by (2b) has the following properties:
   (1) $N \to K$ as $t \to \infty$.
   (2) The graph is concave up for $N_0 < N < K/2$.
   (3) The graph is concave down for $K/2 < N < K$.
   (4) If $N_0 > K$, the graph is concave up.

2. Consider the model

$$\frac{dN}{dt} = rN(K - N)(N - M) \qquad \text{where } r > 0 \text{ and } 0 < M < K$$

   (a) Express the intrinsic growth rate $g(N)$ as a polynomial in N and find the coefficients a_1, a_2, a_3.
   (b) Show that $N = 0$, $N = K$, and $N = M$ are steady states and determine their stability.
   (c) Solve the equation and graph the solution.

3. Models that are commonly used in fisheries are

$$\frac{dN}{dt} = Ng(N),$$

where $g(N)$ is given by

$$\text{Ricker model:} \qquad g(N) = re^{-\beta N}.$$

$$\text{Beverton-Holt:} \qquad g(N) = \frac{r}{\alpha + N}.$$

Analyze the behavior of the solutions to these questions. (Assume α, β, $r > 0$).

4. For single-species populations, which of the following density-dependent growth rates would lead to a decelerating rate of growth as the population increases? Which would result in a stable population size?
   (a) $g(N) = \dfrac{\beta}{1 + N}$, $\beta > 0$.
   (b) $g(N) = \beta - N$, $\beta > 0$.
   (c) $g(N) = N - e^{\alpha N}$, $\alpha > 0$.
   (d) $g(N) = \log N$.

5. List the assumptions that underlie the logistic equation (2a). You may wish to think about such factors as environmental or individual variability, reproductive ages, and the effects of the spatial distribution of the population. Which assumptions are not generally valid?

*Problems preceded by an asterisk are especially challenging.
1. Several problems were kindly suggested by C. Biles.

6. The factor $g(N) = r(1 - N/K)$ in equation (2a) is a per capita growth rate. Smith (1963) observed that in cultures of the unicellular alga *Daphnia magna* g decreases at a nonlinear rate as N increases. To account for this fact, Smith suggested that the growth rate depends on the rate at which food is utilized:

$$g(N) = r\frac{T - F}{T}$$

where F is the rate of utilization when the population size is N, and T is the maximal rate, when the population has reached a saturated level. He further assumed that

$$F = c_1N + c_2\frac{dN}{dt}, \qquad (c_1, c_2 > 0)$$

as long as $dN/dt > 0$.

(a) Explain this assumption for F.

(b) Show that the modified logistic equation is then

$$\frac{dN}{dt} = rN\left[\frac{K - N}{K + (\gamma N)}\right],$$

where $\gamma = rc_2/c_1$ and $K = T/c_1$.

(c) Sketch the expression in square brackets as a function of N.

(d) What would be the qualitative behavior of this population growth? (For a deeper analysis of this problem see Pielou, 1977.)

7. (a) Show that equations (6a,b,c) for the Gompertz growth law are equivalent. Find κ in terms of α.

(b) In a tumor the cells without access to nutrients and oxygen stop reproducing and generally die, leaving a *necrotic center*. The Gompertz growth law can be interpreted as a description of this necrosis. Discuss this point and give alternative interpretations of equations (6a,b).

8. Suppose that prey have a refuge from predators into which they can retreat. Assume the refuge can hold a fixed number of prey. How would you model this situation, and what predictions can you make?

9. (a) Suppose a one-time fishing expedition reduced the prey population by 10% of its current level. What does the Lotka-Volterra model predict about the subsequent behavior of the system? (*Note:* this prediction is one of the most objectionable features of the model and will be dealt with in a later chapter.)

(b) Now consider the situation in which there is a constant level of fishing in which both prey and predatory fish are caught and removed at rates proportional to their densities, ϕx and ϕy. Compare this to the situation in the absence of fishing, and show what Volterra concluded about d'Ancona's observation. (For one treatment of this problem see Braun, 1979; a more advanced mathematical treatment can be found in Brauer and Soudack, 1979.)

***10.** In Section 6.2 we showed that the steady state $(\bar{x}_2, \bar{y}_2) = (c/d, a/b)$ is associated with pure imaginary eigenvalues. Since the equations are nonlinear, it is necessary to consider the possibility that the steady state is a spiral point.

(a) Write the system of equations in the form

$$\frac{dy}{dx} = \frac{-cy + dxy}{ax - bxy}.$$

Separate variables and integrate both sides to obtain

$$y^a e^{-by} = Kx^c e^{-dx}$$

(where K is an arbitrary constant).

(b) On the nullcline $x = c/d$ observe that

$$y^a e^{-by} = \text{constant}.$$

Graph the function $f(y) = y^a e^{-by}$ and use your graph to demonstrate that $f(y) = \text{constant}$ can have at most two solutions for any given constant.

(c) Conclude that the trajectory cannot be a spiral. (*Hint:* Consider how many times it intersects the line $x = c/d$.)

11. Interpret the assumptions 1 to 4 made in the Kolmogorov equations for a predator-prey system.

12. **(a)** At the end of Section 6.2 a number of modifications of the Lotka-Volterra equations are described. Explain these modifications, paying particular attention to their predictions for low and high values of the prey population x.

(b) For each modification you discuss in (a), assume it is the only change made in equations (7a,b) and determine the effect on steady states and their stability properties.

13. Show it is possible, by introducing dimensionless variables, to rewrite the Lotka-Volterra equations as

$$\frac{dv}{dt} = v(1 - e), \qquad \frac{de}{dt} = \alpha e(v - 1),$$

where v = victims and e = exploiters.

14. Determine whether the nontrivial steady state of equations (8a,b) is a stable node or a stable spiral.

15. In this problem we attend to several details that arise in the species competition model [equations (9a,b)].

(a) The four cases shown in Figure 6.6 correspond to four possible sets of inequalities satisfied by the parameters κ_1, κ_2, β_{12}, and β_{21}. Give a biological interpretation of these relations.

(b) Show that a fourth steady state to equations (9a,b) is

$$(\bar{x}, \bar{y}) = \left(\frac{\beta_{21}\kappa_1 - \kappa_2}{\beta_{21}\beta_{12} - 1}, \frac{\beta_{12}\kappa_2 - \kappa_1}{\beta_{21}\beta_{12} - 1}\right),$$

and demonstrate that this steady state has biological relevance only in cases 3 and 4.

(c) The Jacobian of (9a,b) has off-diagonal elements as follows:

$$\mathbf{J} = \begin{pmatrix} ? & \dfrac{-r_1\beta_{12}N_1}{\kappa_1} \\ \dfrac{-r_2\beta_{21}N_2}{\kappa_2} & ? \end{pmatrix}_{(\bar{N}_1, \bar{N}_2)}.$$

Verify this fact, find the remaining entries, and then compute **J** for each of the steady states of (9a,b). (*Note:* one of these involves rather cumbersome expressions.)

(d) Verify the stability properties of steady states to the species competition model [equations (9a,b)] in the four cases discussed in Section 6.3.

(e) Give a biological interpretation of cases 1 through 4 in Figure 6.8.

(1) What is the outcome of competition in case 1? How does this differ from case 3?

(2) In cases 1, 2, and 4 the final outcome does not depend on the initial levels of the competing populations. This is not true in case 3. Give a rough rule of thumb for determining which species wins in case 3.

***16.** In this problem you are asked to generalize the competition model of equations (9a,b) to the case of k species, with particular emphasis on $k = 3$.

(a) Write a set of equations for the populations of species 1, 2, . . . , k that compete pairwise as in equations (9a,b).

(b) For $k = 3$ how many steady states are possible? (*Hint:* in the absence of a third species, each pair behaves according to the original two-species competition model.)

(c) The accompanying figure sketches the case roughly corresponding to case 4. Now suppose that the populations are such that (1) species 2 wins

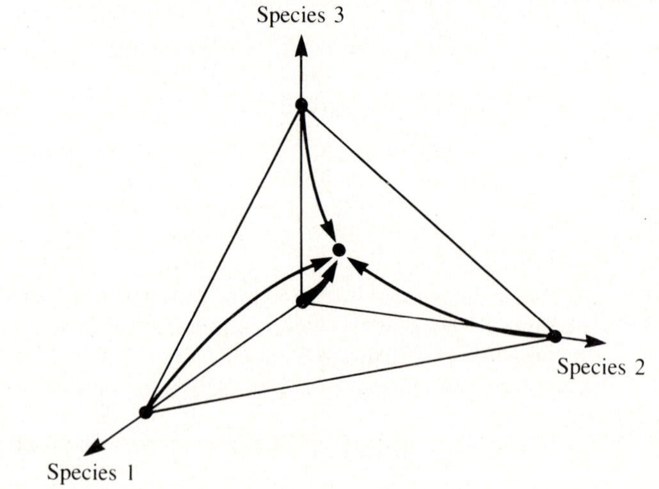

Figure for problem 16(c).

when 3 is absent; (2) species 3 wins when 1 is absent; and (3) species 1 wins when 2 is absent. Sketch the expected dynamics.

17. Species may derive mutual benefit from their association; this type of interaction is known as *mutualism*. May (1976) suggests the following set of equations to describe a possible pair of mutualists:

$$\frac{dN_1}{dt} = rN_1 \frac{1 - N_1}{\kappa_1 + \alpha N_2}, \qquad \frac{dN_2}{dt} = rN_2 \frac{1 - N_2}{\kappa_2 + \beta N_1},$$

where N_i is the population of the ith species, and $\alpha\beta < 1$.

(a) Explain why the equations describe a mutualistic interaction.

(b) Determine the qualitative behavior of this model by phase-plane and linearization methods.

(c) Why is it necessary to assume that $\alpha\beta < 1$?

18. Write down all Routh-Hurwitz matrices $\mathbf{H}_1$, $\mathbf{H}_2$, and $\mathbf{H}_3$ for the case of three species. Show that May's conditions are equivalent to the original Routh-Hurwitz criteria by evaluating the determinants of these matrices.

19. Suppose that in the three-species model discussed in example 1 (Section 6.4), species x and z are competitors. How would the model and its conclusions change?

20. Suppose that in the same example species y is also a prey of species z. How would the model and its conclusions change?

21. In the accompanying directional graph, arrows represent positive and negative effects that each of three species exert on each other when they are at equilibrium. For example, the effect of y on x is negative. Use the Routh-Hurwitz criteria to show that this system must be unstable.

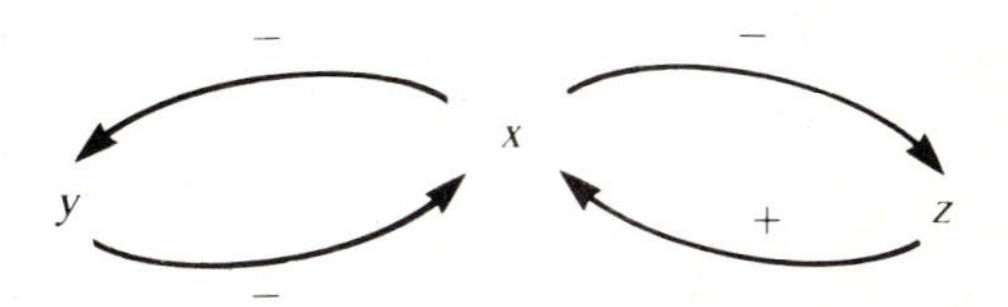

Figure for problem 21.

22. Use the Routh-Hurwitz criteria to investigate stability in problem 29 of Chapter 4.

23. Analyze the community structures shown in the accompanying figures in the following way:

(1) Give the patterns of signs in the qualitative matrix $\mathbf{Q}$ that describes the system.

(2) Identify predation communities.

(3) Determine whether or not the system is qualitatively stable. If not, identify which condition(s) it does not satisfy.

(4) Suggest what kind of community might be represented by the graph.

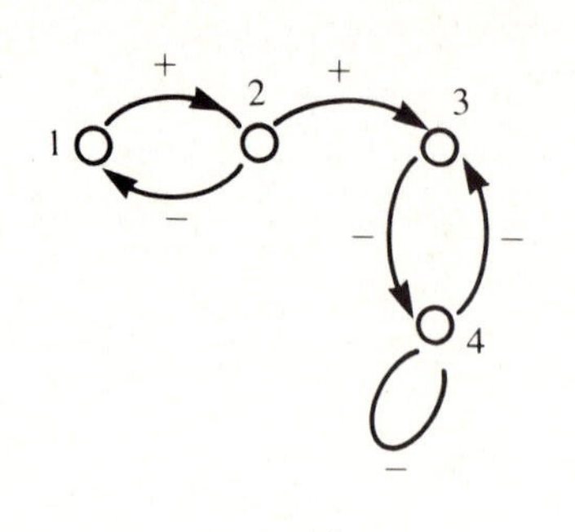

(a)

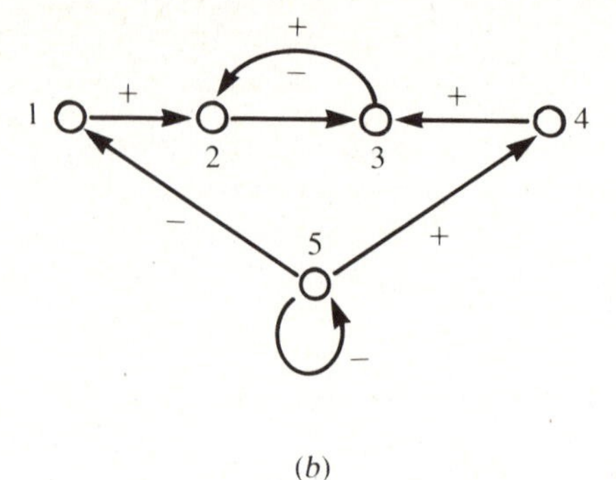

(b)

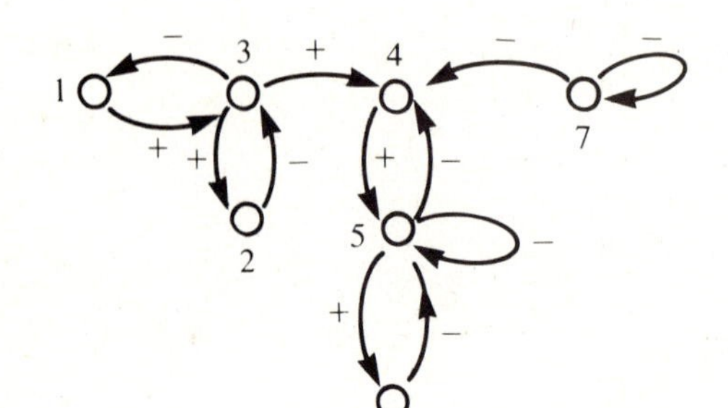

(c)

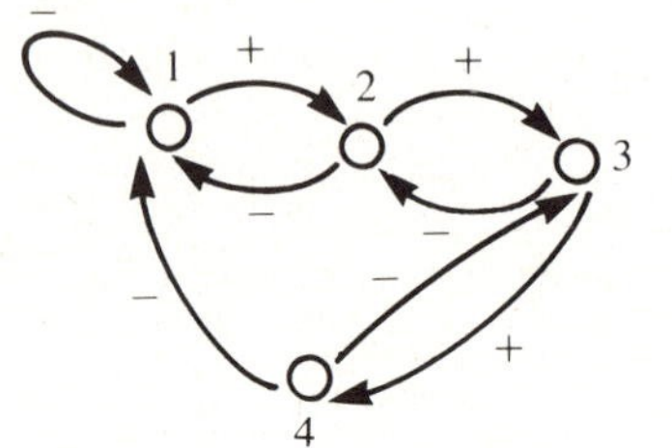

(d)

(e)

Figures (a–e) for problem 23.

24. Draw directed graphs, identify predation communities, and determine whether the systems depicted in the following qualitative matrices are stable or not.

(a) $\begin{bmatrix} 0 & + & 0 & 0 \\ - & + & 0 & - \\ 0 & 0 & 0 & + \\ 0 & + & - & 0 \end{bmatrix}$ (Levins, 1977: one of two competitors for a common resource is itself preyed on: the other is resistant.)

(b) $\begin{bmatrix} 0 & 0 & 0 & + \\ - & 0 & + & + \\ + & - & - & 0 \\ - & - & 0 & - \end{bmatrix}$ (Levins, 1974: nutrients and organisms in a lake. Two species are algae; the others represent nutrients, one of which is produced by an alga.)

(c) $\begin{bmatrix} - & - & 0 & 0 & + \\ + & - & - & 0 & 0 \\ 0 & + & - & - & 0 \\ 0 & 0 & + & - & - \\ - & 0 & 0 & + & 0 \end{bmatrix}$ Unlikely food chain.

(d) $\begin{bmatrix} 0 & - & 0 & 0 \\ + & - & 0 & 0 \\ + & 0 & - & + \\ 0 & + & - & 0 \end{bmatrix}$ A larval prey and predator whose roles reverse in adulthood.

(e) $\begin{bmatrix} - & 0 & 0 & 0 & 0 \\ + & - & 0 & 0 & 0 \\ + & 0 & - & 0 & 0 \\ 0 & + & 0 & 0 & - \\ 0 & 0 & + & - & 0 \end{bmatrix}$ A species that can exist in two types that compete in adulthood.

25. In this problem you are asked to verify several results quoted in Section 6.6.
(a) Show that $N = S + I + R$ is constant for the *SIRS* model given by equations (28a,b,c).
(b) Verify that the two steady states of (28) are given by (29a,b).
(c) Suppose that all members of a population give birth to susceptibles (at rate δ) and die (at rate δ). How would the equations change?
(d) Find steady states for part (c), and determine what the infectious contact number is in terms of the parameters.
(e) Compare the model with and without the above vital dynamics.

***26.** Analyze an *SIR* model with and without vital dynamics. Verify the results summarized in Table 6.1.

***27.** Show that in an *SIR* model with disease fatality at rate η the disease will always eventually disappear.

***28.** Show that in an *SIR* model with carriers who show no symptoms of the disease, the disease always remains endemic.

29. Capasso and Serio (1978) considered the following model with emigration of susceptibles:

$$\frac{dS}{dt} = -g(I)S - \lambda S,$$

$$\frac{dI}{dt} = g(I)S - \gamma I,$$

$$\frac{dR}{dt} = \lambda S + \gamma I.$$

The function $g(I)$, shown in the accompanying graph, takes into account "psychological" effects. Explain the equations and show that the epidemic will always tend to extinction with respect to both infectives and susceptibles.

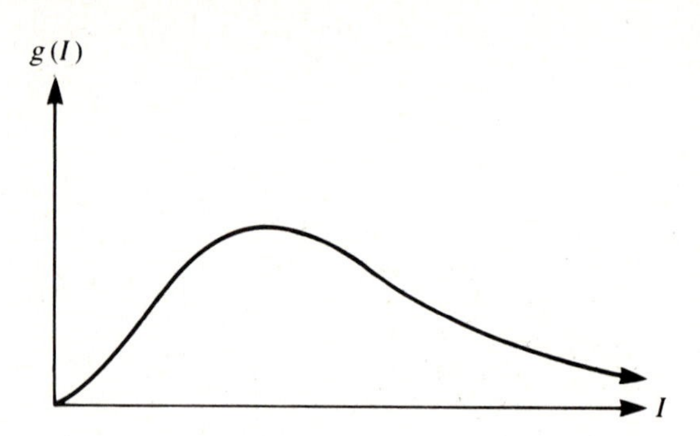

Figure for problem 29.

30. Anderson and May (1979) describe a model in which the natural birth and death rates, a and b, are not necessarily equal so that the disease-free population may grow exponentially:

$$\frac{dN}{dt} = (a - b)N.$$

The disease increases mortality of infected individuals (additional rate of death by infection $= \alpha$), as shown in the accompanying figure. In their terminology the population consists of the following:

X = susceptible class (= S),
Y = infectious class (= I),
Z = temporarily immune class (= R).

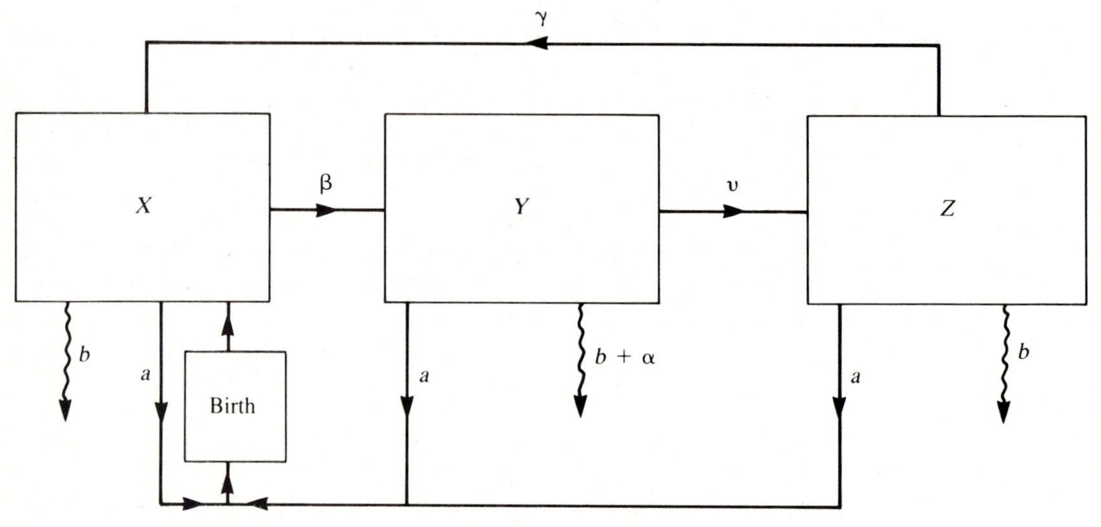

Figure for problem 30.

(a) Write equations describing the disease.
(b) Show that the steady-state solution representing the presence of disease is given by

$$\bar{X}_2 = \frac{\alpha + b + \nu}{\beta}, \qquad \bar{Y}_2 = \frac{r}{\alpha}N, \qquad \bar{Z}_2 = \frac{r}{\alpha}\left(\frac{\nu}{b + \gamma}\right)N_2,$$

where

$$N_2 = \frac{\alpha(\alpha + b + \nu)}{\beta(\alpha - r[1 + \nu/(b + \gamma)])}, \qquad \text{and} \qquad r = a - b.$$

(c) The steady state in part (b) makes biological sense whenever

$$\alpha > r\left(1 + \frac{\nu}{b + \gamma}\right).$$

Interpret this inequality.
(*Note:* Anderson and May conclude that disease can be a regulating influence on the population size.)

31. Consider a lake with some fish attractive to fishermen. We wish to model the fish-fishermen interaction.

Fish Assumptions:

i. Fish grow logistically in the absence of fishing.

ii. The presence of fishermen depresses fish growth at a rate jointly proportional to the fish and fishermen populations.

Fishermen Assumptions:

i. Fishermen are attracted to the lake at a rate directly proportional to the amount of fish in the lake.

ii. Fishermen are discouraged from the lake at a rate directly proportional to the number of fishermen already there.

(a) Formulate, analyze, and interpret a mathematical model for this situation.

(b) Suppose the department of fish and game decides to stock the lake with fish at a constant rate. Formulate, analyze, and interpret a mathematical model for the situation with stocking included. What effect does stocking have on the fishery?

32. Beddington and May (1982) have proposed the following model to study the interactions between baleen whales and their main food source, *krill* (a small shrimp-like animal), in the southern ocean:

$$\text{krill:} \qquad \dot{x} = rx\frac{1 - x}{K} - axy$$

$$\text{whales:} \qquad \dot{y} = sy\frac{1 - y}{bx}$$

Here the whale carrying capacity is not constant but is a function of the krill population:

$$K_{\text{whales}} = bx$$

Analyze this model by determining the steady states and their stability; include a phase-plane diagram.

33. *Estimating parameters in the logistic equation:* We would like to address the curve-fitting problem of how to find the logistic equation of best fit when given appropriate data. First, rearrange equation (2b) and show that

$$N(t) = \frac{K}{1 + [(K - N_0)/N_0]e^{-rt}}.$$

Then conclude that

$$N(t) = \frac{K}{1 + \exp\{-rt + \ln[(K - N_0)/N_0]\}}.$$

Now rearrange the above to demonstrate that

$$\frac{K - N}{N} = \exp\left(-rt + \ln\frac{K - N_0}{N_0}\right).$$

Thus

$$\ln\frac{K - N}{N} = -rt + \ln\frac{K - N_0}{N_0}.$$

Now the quantity $\ln[(K - N)/N]$ is a linear function of time with slope r. The following data is given in Gause (1969) for the growth in volume of each of the yeasts *Saccharomyces* and *Schizosaccharomyces* growing separately. (See Figure 6.9, top curves in each graph.)

Age (h)	*(1) Saccharomyces*	*(2) Schizosaccharomyces*
6	0.37	—
16	8.87	1.00
24	10.66	—
29	12.50	1.70
40	13.27	—
48	12.87	2.73
53	12.70	—
72	—	4.87
93	—	5.67
141	—	5.83

(a) Use the maximal population levels to estimate K_1 and K_2, the carrying capacities for each of the two species.

(b) Plot $(K_i - N_i)/N_i$ on a log scale and use this to determine r_1 and r_2. Do your values agree with those given by Gause in Figure 6.9?

34. Estimating the species-competition parameters.

(a) Use equations (9a,b) to show that

$$\beta_{12} = \frac{K_1 - (dN_1/dt)K_1 - N_1}{r_1 N_1 N_2},$$

$$\beta_{21} = \frac{K_2 - (dN_2/dt)K_2 - N_2}{r_2 N_1 N_2}.$$

(b) Suggest how such parameters might be estimated given empirical observations of N_1 and N_2 growing in a mixed population. (See Gause for experimental data.)

(c) The values of all parameters determined by Gause are given in Section 6.3, Figure 6.9. Determine whether the two species can coexist in a stable mixed population or if one wins over the other.

35. Populations of lemmings, voles, and other small rodents are known to fluctuate from year to year. Early Scandinavians believed the lemmings to fall down from heaven during stormy weather. Later in history, the legend developed that they migrate periodically into the sea for suicide in order to reduce their numbers. . . . None of these theories, however, was supported by any accurate observations. (H. Dekker, 1975)

An alternate hypothesis was suggested by Dekker to account for rodent population cycles. His theory is based on the idea that the rodents fall into two genotypic classes (Myers and Krebs, 1971) that interact. Type 1 reproduces rapidly, but migrates in response to overcrowding; type 2 is less sensitive to high densities but has a lower reproductive capacity.

The following simple mathematical model was given by Dekker to demonstrate that oscillations could be produced when types 1 and 2 rodents were both present in the population:

$$\frac{dn_1}{dt} = n_1[a_1 - (b_1 - c_1)n_2 - c_1(n_1 + n_2)],$$

$$\frac{dn_2}{dt} = n_2[-a_2 + b_2 n_1],$$

where

$$n_1 = \text{density per acre of type 1,}$$

$$n_2 = \text{density per acre of type 2.}$$

The term $b_1 - c_1$ was chosen for convenience in the mathematical calculations rather than for particular biological reasons.

(a) On the basis of the information, give an interpretation of the individual terms in the equations.

(b) Using phase-plane methods, determine the qualitative behavior of solutions to Dekker's equations. If there is more than one case, pay particular attention to the case in which oscillations are present. Give conditions on the parameters a_j, b_j, and c_j for which oscillatory behavior will be seen.

(c) Give a short critique of Dekker's model, indicating whether you would change his assumptions and/or equations. Dekker's article has received a somewhat critical peer review by Nichols et. al. (1979). You may wish to comment on their specific points of contention.

REFERENCES

Single Species Growth

Lamberson, R., and Biles, C. (1981). Polynomial models of biological growth, *UMAP Journal, 2*(2), 9–25.

Malthus, T. R. (1798). An essay on the principle of population, and A summary view of the principle of population. Penguin, Harmondsworth, England.

Pearl, R., and Reed, L. J. (1920). On the rate of growth of the population of the United States since 1790 and its mathematical representation. *Proc. Natl. Acad. Sci. USA, 6,* 275–288.

Slobodkin, L. B. (1954). Population dynamics in *Daphnia obtusa* Kurz. *Ecol. Monogr., 24,* 69–88.

Smith, F. E. (1963). Population dynamics in *Daphnia magna*. *Ecology, 44,* 651–663.

Verhulst, P. F. (1838). Notice sur la loi que la population suit dans son accroissement. *Correspondance Mathematique et Physique, 10,* 113–121.

Predator-Prey Interactions and Species Competition

Armstrong, R. A., and McGehee, R. (1980). Competitive exclusion. *Am. Nat., 115*(2), 151–170.

Brauer, F., and Soudack, A. C. (1979). Stability regions and transition phenomena for harvested predator-prey systems. *J. Math. Biol., 7,* 319–337.

Braun, M. (1979). *Differential Equations and Their Applications*. 3d ed., Springer-Verlag, New York.

Braun, M. (1983). Chaps. 4, 5, 15, 17 in M. Braun, C. S. Coleman, and D. A. Drew, eds. *Differential Equation Models*. Springer-Verlag, New York.

Coleman, C. S. (1978). Biological cycles and the fivefold way. In M. Braun, C. S. Coleman, and D. A. Drew, eds., *Differential Equation Models,* Springer-Verlag, New York.

Gause, G. F. (1932). Experimental studies on the struggle for existence. I. Mixed population of two species of yeast. *J. Exp. Biol., 9,* 389–402.

Gause, G. F. (1934). *The Struggle for Existence*. Hafner Publishing, New York. (Reprinted 1964, 1969).

Holling, C. S. (1965). The functional response of predators to prey density and its role in mimicry and population regulation. *Mem. Entomol. Soc. Can., 45,* 1–60.

Ivlev, V. S. (1961). *Experimental Ecology of the Feeding of Fishes*. Yale University Press, New Haven, Conn.

Kolmogorov, A. (1936). Sulla Teoria di Volterra della Lotta per l'Esistenza. *G. Ist. Ital. Attuari, 7,* 74–80.

Lotka, A. J. (1925). *Elements of Physical Biology*. Williams & Wilkins, Baltimore.

May, R. (1973). *Stability and Complexity in Model Ecosystems*. Princeton University Press, Princeton, N. J.

May, R. (1976). *Theoretical Ecology, Principles and Applications*. Saunders, Philadelphia.

Odum, E. P. (1953). *Fundamentals in Ecology*. Saunders, Philadelphia.

Pielou, E. C. (1969). *An Introduction to Mathematical Ecology*. Wiley-Interscience, New York. (Reprinted 1977).

Rosenzweig, M. L. (1969). Why the prey curve has a hump. *Am. Nat. 103,* 81–87.

Rosenzweig, M. L. (1971). Paradox of enrichment: Destabilization of exploitation ecosystems in ecological time. *Science, 171,* 385–387.

Roughgarden, J. (1979). *Theory of Population Genetics and Evolutionary Ecology: An Introduction*. Macmillan, New York.

Schoener, T. W. (1973). Population growth regulated by intraspecific competition for energy or time. *Theor. Pop. Biol., 4,* 56–84.

Takahashi, F. (1964). Reproduction curve with two equilibrium points: A consideration on the fluctuation of insect population. *Res. Pop. Ecol., 6,* 28–36.

Van der Vaart, H. R. (1983). Some examples of mathematical models for the dynamics of several-species ecosystems. Chap. 4 in H. Marcus-Roberts and M. Thompson, eds. *Life Science Models*. Springer-Verlag, New York.

Volterra, V. (1926). Variazioni e fluttuazioni del numero d'individui in specie animal conviventi. *Mem. Acad. Lincei., 2,* 31–113. Translated as an appendix to Chapman, R. N. (1931). *Animal Ecology*. McGraw-Hill, New York.

Volterra, V. (1931). *Leçons sur la theorie mathematique de la lutte pour la vie*. Gauthier-Villars, Paris.

Volterra, V. (1937). Principe de biologie mathematique. *Acta Biotheor., 3,* 1–36.

Whitaker, R. H., and Levin, S. A., eds. (1975). *Niche: Theory and Application*. Dowden, Hutchinson & Ross, New York.

Qualitative Stability

Jeffries, C. (1974). Qualitative stability and digraphs in model ecosystems. *Ecology, 55,* 1415–1419.

Levins, R. (1974). The qualitative analysis of partially specified systems. *Ann. N. Y. Acad. Sci., 231,* 123–138.

Levins, R. (1977). Qualitative analysis of complex systems, in D. E. Matthews, ed., *Mathematics and the Life Sciences. Lecture Notes in Biomathematics,* vol. 18, 153–199. Springer-Verlag, New York.

Roberts, F. S. (1976). *Discrete Mathematical Models, with Applications to Social, Biological and Environmental Problems*. Prentice-Hall, Englewood Cliffs, N. J.

Quirk, J., and Ruppert, R. (1965). Qualitative economics and the stability of equilibrium. *Rev. Econ. Stud., 32,* 311–326.

Mathematical Theory of Epidemics

Anderson, R. M., ed. (1982). *Population Dynamics of Infectious Diseases, Theory and Applications*. Chapman & Hall, New York.

Anderson, R. M., and May, R. M. (1979). Population biology of infectious diseases, part I. *Nature, 280,* 361–367; part II, *Nature, 280,* 455–461.

Busenberg, S. N., and Cooke, K. L. (1978). Periodic solutions of delay differential equations arising in some models of epidemics. In *Proceedings of the Applied Nonlinear Analysis Conference, University of Texas*. Academic Press, New York.

Capasso, V., and Serio, G. (1978). A generalization of the Kermack-McKendrick deterministic epidemic model. *Math. Biosci., 42,* 43–61.

Hethcote, H. W.; Stech, H. W.; and van den Driessche, P. (1981). Periodicity and stability in epidemic models: A Survey. In *Differential Equations and Applications in Ecology, Epidemics, and Population Models*. Academic Press, New York, pp. 65–82.

Hethcote, H. W. (1976). Qualitative analyses of communicable disease models. *Math. Biosci., 28,* 335–356.

Kermack, W. O., and McKendrick, A. G. (1927). Contributions to the mathematical theory of epidemics. *Roy. Stat. Soc. J., 115*, 700–721.

May, R. M., (1983). Parasitic infections as regulators of animal populations. *Am. Sci., 71*, 36–45.

Vaccination Programs

Anderson, R. M., and May, R. M. (1982). The logic of vaccination. *New Scientist*, (November 18, 1982 issue) pp. 410–415.

Anderson, R. M., and May, R. M. (1983). Vaccination against rubella and measles: Quantitative investigations of different policies. *J. Hyg. Camb., 90*, 259–325.

May, R. M. (1982). Vaccination programmes and herd immunity. *Nature, 300*, 481–483.

Miscellaneous

Aroesty, J., Lincoln, T., Shapiro, N., and Boccia, G. (1973). Tumor growth and chemotherapy: mathematical methods, computer simulations, and experimental foundations. *Math. Biosci., 17*, 243–300.

Beddington, J. R., and May, R. M. (1982). The harvesting of interacting species in a natural ecosystem. *Sci. Am.*, November 1982, pp. 62–69.

Dekker, H. (1975). A simple mathematical model of rodent population cycles. *J. Math. Biol., 2*, 57–67.

Freedman, H. I. (1980). *Deterministic Mathematical Models in Population Ecology*. Marcel Dekker, New York.

Greenwell, R. (1982). *Whales and Krill: A Mathematical Model*. UMAP module unit 610. COMAP, Lexington, Mass.

Hofbauer, J. (in press). Permanence and persistence of Lotka-Volterra systems.

Hutchinson, G. E. (1978). *Introduction to Population Ecology*. Yale University Press, New Haven, Conn.

Kingsland, S. E. (1985). *Modelling Nature*. University of Chicago Press, Chicago.

Levins, R. (1968). *Evolution in Changing Environments*. Princeton University Press, Princeton.

Luenberger, D. G. (1979). *Introduction to Dynamic Systems*. Wiley, New York, pp. 328–331.

May, R. M., Beddington, J. R., Clark, C. W., Holt, S. J., and Laws, R. M. (1979). Management of multispecies fisheries. *Science, 205*, 267–277.

Myers, J. H., and Krebs, C. J. (1971). Genetic, behavioral, and reproductive attributes of dispersing field voles *Microtus pennsylvanicus* and *Microtus Ochrogaster*. *Ecol. Monogr., 41*, 53–78.

Newton, C. M. (1980). Biomathematics in oncology: modelling of cellular systems. *Ann. Rev. Biophys. Bioeng., 9*, 541–579.

Nichols, J. D.; Hestbeck, J. B.; and Conley, W. (1979). Mathematical models and population cycles: A critical evaluation of a recent modelling effort. *J. Math. Biol., 8*, 259–263.

Nisbet, R. M., and Gurney, W. S. C. (1982). *Modelling Fluctuating Populations*. Wiley, New York.

7 Models for Molecular Events

All the effects of nature are only the mathematical consequences of a small number of immutable laws.

-P. S. Laplace (1749–1827) quoted in E. T. Bell (1937) *Men of Mathematics* p. 172 Simon & Schuster, N.Y.

The realm of molecular biology lies at the outermost limits of resolution of our best microscopes. Just beyond is a world that is both fascinating and mysterious. This is the world of *macromolecules:* versatile entities that give cells their structure, store or transmit information, recognize and respond to other macromolecules, build or synthesize each other, and regulate all the other chemical events in the living cell.

It is sometimes easy to forget that our familiarity with the subcellular realm is very recent. Until 1838, when Schleiden and Schwann proposed their *cell theory,* scientists scarcely appreciated that living things were made up of smaller units called cells. Only by the mid 1900s had the electron microscope extended our visual frontier into the finer structures of the cell. Numerous contemporary techniques eventually led to the important discovery by J. Watson and F. Crick in 1953 of the structure of DNA (deoxyribonucleic acid), the fundamental genetic material. Since that time, and especially within the last two decades, the field of molecular biology has undergone rapid, explosive growth. It is now universally recognized as a field of nearly unlimited promise in medicine, industry, agriculture, and many other areas of application.

Despite this recent surge of knowledge and fervent ongoing exploration of the molecular world, much is still unknown about the microcosm inside the living cell. How do all of these complicated entities work so well as a unit? How are a multitude of simultaneous processes controlled, each with split-second accuracy? What makes it all a cohesive living unit that can respond to its environment while maintaining an identity despite potentially destructive influences? Answers to these questions are not within our immediate grasp and must await further discoveries on the experimental frontier. They are, broadly speaking, related to similar questions one could pose about any complex system that consists of myriad interrelated units working in concert.

While molecular biology owes much, in its advances, to the technological breakthroughs that stem from the "hard" sciences, it may be considered presumptuous to overplay the role of mathematics. Nevertheless, mathematics does indeed have a contribution to make, at least in helping us understand the very basic building blocks of behavior exhibited by the cell, the gene, and the enzyme. A traditionally mathematical approach underlies *enzymology,* the study of dynamic interactions between *enzymes* (large protein molecules that catalyze reactions in the living cell) and their *substrates*. It is not always clear whether mathematical modeling can help illuminate fundamental questions about "how things work." Yet, it is our gradual perception that such approaches have borne fruit in disciplines of parallel complexity such as ecology and physiology; it is to be hoped that similar combinations of theory and experiment will lead to progress in molecular biology as well.

This chapter could be viewed as an analog of Chapter 6 dealing with the "population dynamics" of molecules rather than whole organisms. Perhaps the key ideas presented here are that some parallels exist between such disparate realms and that certain *dynamic properties* are shared by unrelated entities. To begin this excursion, we delve into a familiar macroscopic observation and study its molecular foundation. Looking more closely at events on a bacterial cell's membrane, we show that Michaelis-Menten kinetics (used in Section 4.4) correspond to saturating nutrient-conveying carrier molecules. Mathematical methods applied to this problem are used again for studying two related situations (Sections 7.3 and 7.4) in which sigmoidally saturating kinetics are implicated.

As a departure from the somewhat technical enzyme kinetics, we explore how two simple molecular events lead to an aspect of behavior that mimics certain properties of the cell. Here a somewhat abstract mathematical approach leads to insights that, although oversimplified, are nevertheless useful.

An extension of the geometric and graphical analysis of Chapter 5 is applied to two chemical situations (Sections 7.7 and 7.8) simply for further practice in abstract reasoning. Finally, we conclude with a brief exposé of mathematical theories that could be applied to certain biochemical and molecular systems too complicated to be analyzed by standard modeling techniques.

For a condensed coverage of this chapter, the following material could be covered: Section 7.1 and part of Section 7.2, Sections 7.3, 7.5, and 7.6; the material in Sections 7.7 and 7.8 can be assigned to more advanced students or omitted.

7.1 MICHAELIS-MENTEN KINETICS

In describing bacterial growth within a chemostat (Chapter 4) we assumed an expression for the nutrient-dependent growth rate that had the property of saturation; for low levels of the nutrient concentration c, bacterial growth rate given by equation

(15) in Chapter 4 is roughly proportional to c. At high c levels, though, this rate approaches a constant value, K_{max}. (See Figure 4.3.) Numerous biological phenomena exhibit saturating kinetics. The expression

$$K(C) = \frac{K_{max}\, C}{K_n + C}, \tag{1}$$

which depicts such a property, is called the *Michaelis-Menten kinetics*. This expression actually stems from a particular set of assumptions about what may be occurring at the molecular level on the surface of the bacterial cell membrane. We explore this mechanism in some detail in this and the following section.[1]

Figure 7.1 gives a schematic version of how bacteria consume organic substances such as glucose. Most water-soluble molecules are unable to pass through the hydrophobic environment of the cell membrane directly and must be carried across by special means. Typically, molecular receptors embedded in the bacterial cell membrane are involved in "capturing" these polar molecules in a loose complex, conveying them across the membrane barrier, and releasing them to the interior of the cell. *Saturation* results from the limited number of receptors and the limited rate at which their "conveyor-belt" mechanism can operate.

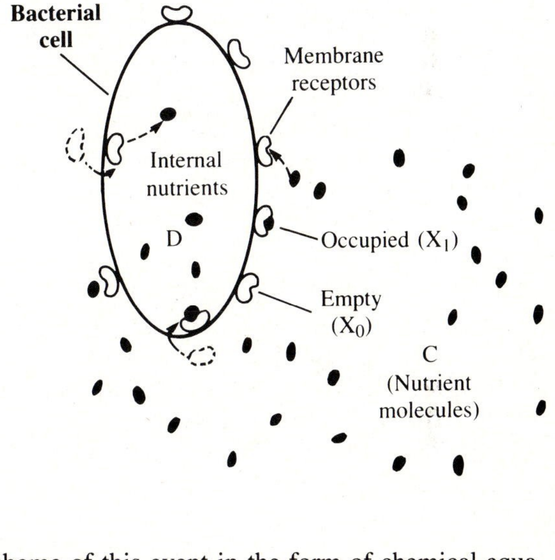

Figure 7.1 *The passage of nutrient molecules into a cell may be mediated by membrane-bound receptors. This saturating mechanism for nutrient uptake can be described by Michaelis-Menten kinetics.*

Let us write the molecular scheme of this event in the form of chemical equations. We will denote an external nutrient molecule by C, an unoccupied receptor by X_0, a nutrient-receptor complex by X_1, and a nutrient molecule successfully captured by the cell by P. The constants k_1, k_{-1}, and k_2 depict the various rates with which these reactions proceed. The following equations summarize the directions and rates of reactions:

1. The approach in this material is partly based on a lecture given by L. A. Segel at the Weizmann Institute, where the author was a graduate student.

$$C + X_0 \underset{k_{-1}}{\overset{k_1}{\rightleftharpoons}} X_1, \tag{2a}$$

$$X_1 \xrightarrow{k_2} P + X_0. \tag{2b}$$

That is, C and X_0 can combine to form the complex X_1, which either breaks down into the former constituents, or else produces P and X_0. The fact that reaction (2a) is reversible means that the carrier receptor sometimes fails to transport its nutrient load into the cell interior, dumping it instead outside the cell.

A reaction diagram such as the one given by equations (2a,b) can be translated into a set of differential equations that describe rates of change of concentrations of the participating reactants. The diagram encodes both the sequence of steps and the rates with which these steps occur. To write corresponding equations, we must use the law of mass action (encountered previously), which states that when two or more reactants are involved in a reaction step, the rate of reaction is proportional to the product of their concentrations. By convention, the rate constants k_i in the reaction diagram are the proportionality constants. We define the *volumetric receptor concentration* by averaging over a population of bacteria, as follows:

$$\begin{matrix}\text{number of receptors} \\ \text{per unit volume}\end{matrix} = \begin{matrix}\text{average number of} \\ \text{receptors per cell}\end{matrix} \times \begin{matrix}\text{number of cells} \\ \text{per unit volume}\end{matrix}.$$

The lower-case letters c, x_0, x_1, and p will be used to denote concentrations of C, X_0, X_1, and P. Keeping track of each chemical participant allows us to derive the set of equations (3). (It is a good idea to attempt to write these equations independently before proceeding.)

$$\frac{dc}{dt} = -k_1 c x_0 + k_{-1} x_1, \tag{3a}$$

$$\frac{dx_0}{dt} = -k_1 c x_0 + k_{-1} x_1 + k_2 x_1, \tag{3b}$$

$$\frac{dx_1}{dt} = k_1 c x_0 - k_{-1} x_1 - k_2 x_1, \tag{3c}$$

$$\frac{dp}{dt} = k_2 x_1. \tag{3d}$$

Adding equations (3b,c) reveals a feature common to many *enzymatic reactions:*

$$\frac{dx_0}{dt} + \frac{dx_1}{dt} = 0, \tag{4}$$

which simply means that $x_0 + x_1$, the total of occupied and unoccupied receptors, is a constant. This is not at all surprising, since receptors are neither formed nor destroyed in the process of conveying their cargo into the cell. Suppose we started out with an initial concentration of receptors r. Then the total concentration would remain r:

$$x_1 + x_0 = r. \tag{5}$$

It is generally true that the enzymes are conserved in the reactions in which they participate.

Equation (5) permits us to simplify the system of equations (3a,b,c) by eliminating either x_1 or x_0. We arbitrarily choose to eliminate x_0. Furthermore, because equations (3a,b,c) are not dependent on concentration p, we shall set aside (3d), which can always be solved independently once solutions for the other variables are known. In problem 1 you are asked to verify that these steps lead to the following:

$$\frac{dc}{dt} = -k_1 rc + (k_{-1} + k_1 c)x_1, \tag{6a}$$

$$\frac{dx_1}{dt} = k_1 rc - (k_{-1} + k_2 + k_1 c)x_1. \tag{6b}$$

7.2 THE QUASI-STEADY-STATE ASSUMPTION

At this point we could proceed with a full analysis of equations (6a,b) using the phase-plane methods described in Chapter 6 (see problem 7). However, in this section we concentrate on an assumption that leads directly to the Michaelis-Menten rate law and then examine the restrictions to which this approximation is subject.

Typically, small molecules such as glucose or other nutrients are found in concentrations much higher than those of the receptors (in the sense of our previous definition of receptor concentration). We could argue, therefore, that receptors are always working at maximal capacity, so that their occupancy rate is virtually constant. This assumption, which leads us to write

$$\frac{dx_1}{dt} \simeq 0, \tag{7a}$$

or equivalently,

$$k_1 rc - (k_{-1} + k_2 + k_1 c)x_1 = 0 \tag{7b}$$

is called the *quasi-steady-state* hypothesis and permits further simplification by allowing x_1 to be eliminated from the system of equations (6a,b).

In problem 2 you are asked to detail the algebraic steps showing that this assumption results in

$$\frac{dc}{dt} = -\frac{K_{\max} c}{k_n + c}, \tag{8}$$

where

$$K_{\max} = k_2 r, \qquad k_n = \frac{k_{-1} + k_2}{k_1}.$$

There is one serious problem with this reasoning. Setting $dx_1/dt = 0$ in equation (6b) changes the character of the mathematical problem from a system of two ODEs to a simple ODE coupled with the algebraic equation (7b). This change in the problem can have drastic consequences and should not be taken lightly. In order to more fully understand when this approximation can be used, it seems prudent to make a more careful comparison of magnitudes of terms in these equations.

Because these magnitudes really depend on the system of units adopted to measure concentrations and time, a prerequisite step in making such comparisons is to reduce the equations to a dimensionless form. (Detailed steps are left as an exercise in problem 3.) First we choose the following scales:

$$\tau = 1/k_1 r, \text{ for units of time,} \tag{9a}$$

$$\hat{x}_1 = r = \text{initial receptor concentration, for units of } x, \tag{9b}$$

$$\hat{c} = c_0 = \text{initial nutrient concentration, for units of } c. \tag{9c}$$

The equations can then be written in the dimensionless form,

$$\frac{dc^*}{dt^*} = -c^* + \left(\frac{k_{-1}}{k_1 c_0} + c^*\right) x_1^*, \tag{10a}$$

$$\frac{dx_1^*}{dt^*} = \frac{c_0}{r} c^* - \frac{c_0}{r}\left(\frac{k_{-1} + k_2}{k_1 c_0} + c^*\right) x_1^*. \tag{10b}$$

Next, drop the asterisks and define

$$\epsilon = \frac{r}{c_0}, \qquad K = \frac{k_{-1} + k_2}{k_1 c_0}, \qquad \lambda = \frac{k_2}{k_1 c_0}.$$

Notice that ϵ is the ratio of concentrations of receptors and nutrient molecules. The equations become

$$\frac{dc}{dt} = -c + (K - \lambda + c) x_1, \tag{11a}$$

$$\epsilon \frac{dx_1}{dt} = c - (K + c) x_1. \tag{11b}$$

We now realize that neglecting the LHS of equation (6b) is equivalent to assuming that $\epsilon\, dx_1/dt$ is small. This would be fine provided ϵ is small, that is, the receptor concentration is lower than nutrient concentration. Notice that dimensional analysis has given a much more precise meaning to the quasi-steady-state assumption.

To summarize results, it has been concluded that for time scales on the order of $\tau = 1/(k_1 r)$ the process of receptor-mediated nutrient uptake is, to first-order approximation, given by the equations

$$\frac{dc}{dt} = -c + (K - \lambda + c) x_1,$$

$$0 = c - (K + c) x_1.$$

More simply stated, this means that

$$\frac{dc}{dt} = \frac{-\lambda c}{K + c}, \tag{12a}$$

$$x_1 = \frac{c}{K + c}. \tag{12b}$$

We recognize this as another version of the Michaelis-Menten rate law. From equation (12a) observe that whenever $c > 0$, $dc/dt < 0$ so that c is a decreasing function of time. [This equation can be integrated to obtain an implicit solution for $c(t)$:

See problem 4a.] From (12b) we also observe that x_1 decreases as c decreases. [See problem 4b.] Thus on the time scale τ, the concentrations of both the nutrient and the nutrient-receptor complex will be decreasing with time. This is one approximation of the nutrient-receptor kinetics.[2]

We now repeat a step in our analysis by examining the same process again but with a different choice of time scale. Let us now choose

$$\tilde{\tau} = \frac{1}{k_1 c_0}, \tag{13}$$

and retain the previous choices $\hat{x}_1 = r$ and $\hat{c} = c_0$. In problem 5 it is shown that resulting dimensionless equations are

$$\frac{dc^*}{dt} = -\epsilon c^* + \epsilon(K - \lambda + c^*)x_1, \tag{14a}$$

$$\frac{dx_1^*}{dt} = c^* - (K + c^*)x_1^*, \tag{14b}$$

where ϵ, K, and λ have their previous meaning.

How do the two time scales τ and $\tilde{\tau}$ compare? According to our assumption, ϵ is small; that is, $c_0 \gg r$. Thus

$$\tau \gg \tilde{\tau}.$$

With our second choice of time scales we are studying the behavior for *short times*, close to $t = 0$. For example, in a situation in which substrate at concentration c_0 is abruptly added to the solution at $t = 0$, this second time scale would be appropriate for understanding the way initially free receptor sites fill up with their ligands.

Again exploiting the fact that ϵ is small now leads to the conclusion that the RHS of equation (14a) can be neglected to first-order approximation, so that for time scales on the order of $\tilde{\tau}$ we can say that

$$\frac{dc^*}{dt^*} \simeq 0 \qquad (c^* = 1,\ c = c_0), \tag{15a}$$

$$\frac{dx_1^*}{dt^*} \simeq 1 - (K + 1)x_1^*. \tag{15b}$$

The equation for x_1^* can then be integrated (this is left as an exercise), and we then observe that the receptors that at $t = 0$ are unoccupied $[x_1(0) = 0]$ quickly fill up, approaching a fixed fractional occupancy rate. By our previous analysis, x_1 eventually decreases as c is depleted from the environment of the cells.

2. To achieve greater accuracy, we can refine this approximation by assuming that the functions x_1 and c are made up of sums of terms that are proportional to $\epsilon^0, \epsilon^1, \epsilon^2, \ldots, \epsilon^n$. These are called *asymptotic expansions,* and the procedure for then getting successive approximations for x_1 and c is called a *singular perturbation method*. This important method has rather broad application to problems in applied mathematics. However, the numerous technical details involved are beyond the scope of this book.

A thorough exposition of the method of asymptotic expansions and its application to enzyme kinetics is given by Murray (1977) and Lin and Segel (1974).

Summary of Steps Leading to the Michaelis-Menten Rate Equation

1. Draw the following reaction diagram:
$$C + X_0 \underset{k_{-1}}{\overset{k_1}{\rightleftharpoons}} X_1 \xrightarrow{k_2} X_0 + P.$$
2. Write equations for changes in concentrations c, x_0, x_1, and p using the law of mass action.
3. Use the fact that the total number of receptor molecules $x_0 + x_1 = r$ is fixed and eliminate one variable.
4. Assume receptors are at quasi steady state so that $dx_1/dt = 0$ to get a relationship between x_1 and c.
5. Eliminate x_1 from the equation for dc/dt, and obtain[3]
$$\frac{dc}{dt} = -K_{max}\frac{c}{k_n + c},$$
where
$$K_{max} = k_2 r, \qquad k_n = \frac{k_{-1} + k_2}{k_1}.$$

We have seen previously that on two different time scales one can ascertain the behavior by solving different approximate versions of the equations. The final step, that of matching these short and long time solutions, is accomplished by the technique of *matched asymptotic analysis* and will not be discussed here. However, the results enable us to establish a complete time sequence of events, as illustrated in Figure 7.2.

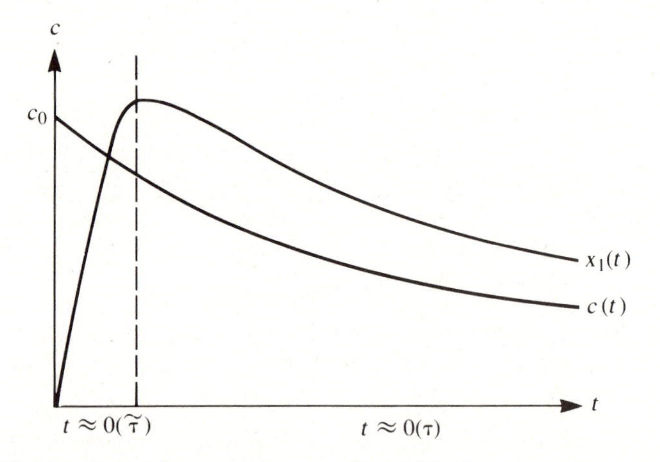

***Figure 7.2** A reversible reaction such as equation (2a) has distinctly different kinetics at different time scales: On a short time scale ($t \simeq 0(\tilde{\tau})$) receptors fill up quickly, and $c \simeq c_0$. On longer time scales ($t \simeq 0(\tau)$) the receptor occupancy x_1 decreases as c is progressively depleted. (Most of the nutrient has been transported into the cell interior.)*

3. dc/dt is negative since the concentration of substance outside of the cell is decreasing.

In problem 7 it is shown that these results are consistent with a phase-plane analysis of equations (6a,b) in which no quasi-steady-state approximation is made.

7.3 A QUICK, EASY DERIVATION OF SIGMOIDAL KINETICS

One of the features of Michaelis-Menten kinetics not shared by more complex molecular pathways is that for low precursor concentrations it yields an approximately linear rate of reaction. (The graph of $-dc/dt$ versus c is almost a straight line close to $c = 0$.) Stated another way, for moderately low precursor levels (those that do not oversaturate the receptors), increasing the precursor concentration by 50% tends to increase the reaction rate by 50% simply because the chances of encounter between receptors and precursors increase proportionately.

We shall see that this simple proportionality changes when more than one precursor molecule is implicated in forming a complex. Instead we typically observe a sigmoidally saturating graph of the rate kinetics. A simple but naive way of demonstrating this is to consider the following double-substrate complexing reaction:

$$2C + X_0 \underset{k_{-1}}{\overset{k_1}{\rightleftharpoons}} X_2 \xrightarrow{k_2} X_0 + 2P. \tag{16}$$

Here two molecules of the substance C are required for forming the complex X_2, which then yields products X_0 and P. With the preparation given in Sections 7.1 and 7.2, it is a straightforward matter to draw the necessary conclusions. Indeed, we need only insert a single change in the previous steps to see the result. According to the law of mass action the reaction that feeds on two molecules of C and one of X_0 proceeds at a rate $k_1c^2x_0$. Thus the first two equations describing the reaction are

$$\frac{dc}{dt} = -k_1c^2x_0 + k_{-1}x_2, \tag{17a}$$

$$\frac{dx_0}{dt} = -k_1c^2x_0 + k_{-1}x_2 + k_2x_2, \tag{17b}$$

Others in the series are equally easy to write down. We recognize (17) as a thinly disguised version of the previous rate laws but with the following changes:

c is replaced by c^2,

x_1 is replaced by x_2.

For practice, you may want to reconstruct the steps (identical to those of the previous sections) that lead to the rate equation for c given a quasi-steady-state assumption for x_2. Because the only substantive change is replacing c with c^2, it should come as no surprise that the result is

$$\frac{dc}{dt} = -\frac{K_{\max}c^2}{k_n + c^2}, \tag{18}$$

where $K_{\max} = k_2r$ and $k_n = (k_{-1} + k_2)/k_1$ as before. The graph of this function, shown in Figure 7.3, is sigmoidal, where $\sqrt{k_n}$ the concentration required for half-maximal response. $c = \sqrt{k_n}$ gives $dc/dt = K_{\max}/2$. For small c the graph is approximately quadratic; that is,

$$\frac{dc}{dt} \simeq -K_{\max}\frac{c^2}{k_n},$$

where $c^2 \ll k_n$.

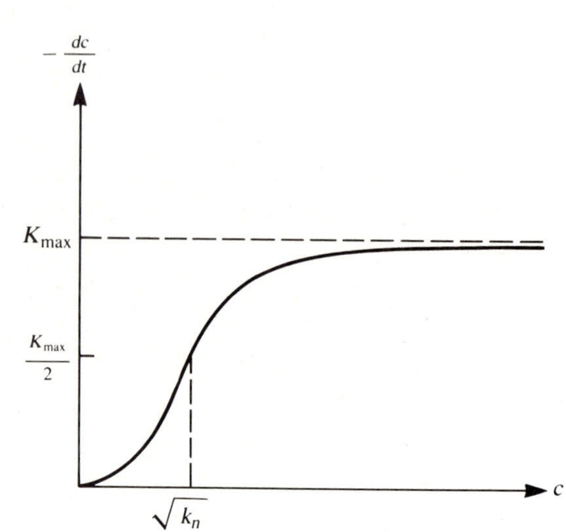

Figure 7.3 A sigmoidal reaction rate for a reaction involving two molecules of precursor.

7.4 COOPERATIVE REACTIONS AND THE SIGMOIDAL RESPONSE

In most biochemical systems, trimolecular reactions are considered unlikely, as they involve a collision between three molecules. For this reason, simultaneous complexing of two molecules with a receptor is generally unrealistic. In this section we examine a more plausible version that involves only sequential bimolecular steps. The connection between these distinct mechanisms will then be established by comparing results.

Let us suppose that a double complex forms in the following way. First a single molecule attaches to the receptor, then a second. Products might be formed at an intermediate stage (with rate κ_1) or at the end (rate κ_2). A diagram typical for this reaction would be as follows:

$$C + X_0 \underset{k_{-1}}{\overset{k_1}{\rightleftharpoons}} X_1 \xrightarrow{\kappa_1} X_0 + P, \tag{19a}$$

$$C + X_1 \underset{k_{-2}}{\overset{k_2}{\rightleftharpoons}} X_2 \xrightarrow{\kappa_2} X_1 + P. \tag{19b}$$

Corresponding equations for c and x_1 must now include all reactions in which these appear as products or reactants, as follows:

$$\frac{dc}{dt} = -k_1 c x_0 + x_1(k_{-1} - k_2 c) + k_{-2} x_2, \tag{20a}$$

$$\frac{dx_0}{dt} = -k_1 c x_0 + x_1(\kappa_1 + k_{-1}), \tag{20b}$$

$$\frac{dx_1}{dt} = k_1 c x_0 + x_2(\kappa_2 + k_{-2}) - x_1(\kappa_1 + k_{-1} + k_2 c), \tag{20c}$$

$$\frac{dx_2}{dt} = k_2 c x_1 - x_2(k_{-2} + \kappa_2), \tag{20d}$$

$$\frac{dp}{dt} = \kappa_1 x_1 + \kappa_2 x_2. \tag{20e}$$

The conservation of receptors in the reaction now implies that

$$x_0 + x_1 + x_2 = r \tag{20f}$$

where r is a fixed constant. This equation can be used to eliminate any x from equations (20a–e) for example, x_0.

We now make a quasi-steady-state assumption for each receptor occupancy state and set $dx_1/dt = dx_2/dt = 0$ after eliminating x_0). The resulting relations involve certain ratios of rate constants that we shall define, following Rubinow (1975), as follows:

$$K_m = \frac{\kappa_1 + k_{-1}}{k_1}, \tag{21a}$$

$$K'_m = \frac{\kappa_2 + k_{-2}}{k_2}. \tag{21b}$$

In problem 9 you are asked to demonstrate that the quasi-steady-state assumptions lead to the relations

$$K_m x_1 = c x_0, \tag{22a}$$

$$K_m K'_m x_2 = c^2 x_0. \tag{22b}$$

Consequently, using equation (20f), it is possible to express x_0 in terms of c; when this is done, the following relation is obtained:

$$x_0 = \frac{r}{K_m K'_m}(K_m K'_m + K'_m c + c^2). \tag{23}$$

As a last step, we rewrite the equation for dc/dt using equations (21) to (23). The result is

$$\frac{dc}{dt} = \frac{-rc(\kappa_1 K'_m + \kappa_2 c)}{K_m K'_m + K'_m c + c^2}. \tag{24}$$

This equation bears an apparent connection with the simpler sigmoidal kinetics in equation (18), but it contains several terms that were absent before. It is somewhat revealing to examine when these terms can be ignored so as to establish a connection between the two mechanisms shown in equations (16) and (19).

We notice in the numerator that the term linear in c vanishes if $\kappa_1 = 0$, that is, if products are not formed in the intermediate steps of the reaction. When is the term $K'_m c$ in the denominator of equation (24) small enough to neglect? This term is small relative to c^2 and to $K_m K'_m$ provided

$$K_m K'_m \gg K'_m c \qquad \text{and} \qquad c^2 \gg K'_m c.$$

Combining inequalities leads to

$$K_m \gg c \gg K'_m, \tag{25}$$

which indicates that the constant K_m must be larger than K'_m. Furthermore, the term $K'_m c$ can only be neglected at intermediate levels of concentrations c; that is, at lower or higher levels the presence of this term tends to distort somewhat the graph of the function shown in Figure 7.3 (see problem 11).

Rewriting the above inequalities in terms of original parameters leads to the following:

$$\frac{1}{K'_m} > \frac{1}{K_m},$$

$$\frac{k_2}{\kappa_2 + k_{-2}} > \frac{k_1}{\kappa_1 + k_{-1}}. \tag{26}$$

This indicates that the tendency of the first reaction step in (19b) to proceed in the forward direction is greater than that of the first step in (19a). Stated another way, once a single molecule of C has complexed with a receptor, a second molecule complexes more readily. Thus the intermediate complex X_1 is short-lived and can almost be neglected, as it has been in the simplified scheme of the trimolecular mechanism, equation (16).

Many biologically important reactions have the characteristic that once a first step is complete, others follow rapidly. A notable example is that of *hemoglobin* (a macromolecule in red blood cells which conveys oxygen). Hemoglobin has four *polypeptide* (protein) components, each of which contains a heme group that can bind with an oxygen molecule. After a single oxygen molecule is attached, the binding of others is enhanced. This reaction and others like it are termed *positively cooperative*. Mechanisms for cooperativity may include *conformational changes* (changes in shape) of the macromolecule that enhance exposure of active sites. Further details about these fascinating topics can be obtained from any current text on biochemistry or molecular biology.

Based on the investigation in this section we conclude that for highly cooperative bimolecular reactions involving a complex between one macromolecule and two substrate molecules, equation (18) is a reasonable approximation for the reaction rate (subject to all the appropriate conditions outlined earlier). A generalization of this rate law to *n-substrate complexes* is

$$\frac{dc}{dt} = \frac{-K_{\max} c^n}{(k_n + c^n)}. \tag{27}$$

7.5 A MOLECULAR MODEL FOR THRESHOLD-GOVERNED CELLULAR DEVELOPMENT

Over the years our understanding of molecular processes within the cell has increased. This knowledge has led to much greater insight into the way cells acquire a commitment to specific developmental pathways. The quest to probe these complex processes further is at the forefront of science, technology, and perhaps even our philosophical view of living things.

All cells in our body have one ancestral cell: the fertilized egg created at conception. Since the advent of molecular biology of the gene we know that all these progeny cells, no matter how diverse their functions, have identical "blueprints" encoded in genetic material in the nucleus. Somehow during the life history of the cell these blueprints are selectively transcribed and used in building the unique character of the cell, be it neuron, epithelial cell, hepatic cell, or one of thousands of other cell types in our bodies. How this developmental process occurs is still largely a mystery.

Mathematics has played an admittedly modest role in solving the mysteries of molecular biology. Nevertheless, mathematical reasoning can illuminate specific questions that may then be clues to a tremendously complex puzzle. We shall see two examples in this and the following sections.

We first discuss a model by Lewis et al. (1977) that illustrates the idea that chemical reactions can act as logical elements, helping to make decisions about the developmental processes that occur in a cell. To discover how this works, a rather simple idealized example serves as the focal point of our discussion.

Consider a row of cells connected to each other in a one-dimensional filament. Originally the cells are identical; after a process of differentiation the row consists of two distinct cell types (say, pigmented and unpigmented cells). A well-defined border between these types appears in a predictable and controlled position. How is this achieved?

One theory, by no means the only one, is that cells have *positional information,* cues by which they assess their locations relative to particular points of demarcation. (The ends of a filament and the boundary of a two-dimensional tissue are examples of such demarcation points.) These cues, which may be carried by chemical messages, then have to be interpreted by the cell to arrive at a set of instructions that determine the course of differentiation.

We can imagine how positional information might be created and maintained. For example, in our example of filament of cells, a chemical source at one end (say the "head") and a sink at the other (the "tail" end) could result in a permanent and continuous gradient of a chemical signal S across the tissue. Cells closest to the head would be exposed to high levels of S; those closest to the tail would sense low S concentrations; and those at intermediate positions would detect moderate levels [see Figure 7.4(b)]. Each cell could then "feel its position" by assessing the concentrations of S about it. S could be called a *morphogenetic* substance since it controls the differentiation and development of form in the tissue.

It still remains to determine how a continuous spatially varying signal is inter-

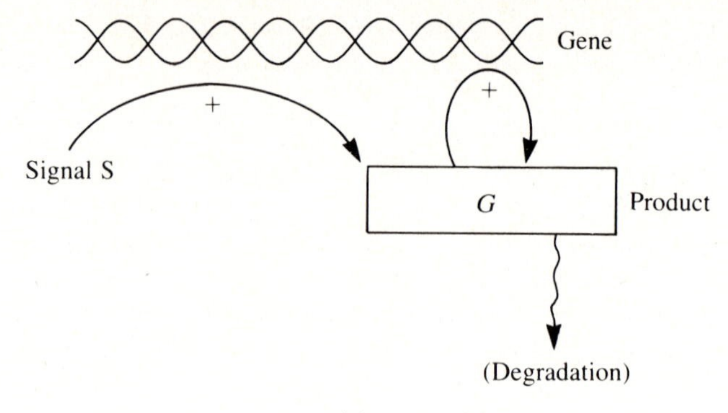

(a)

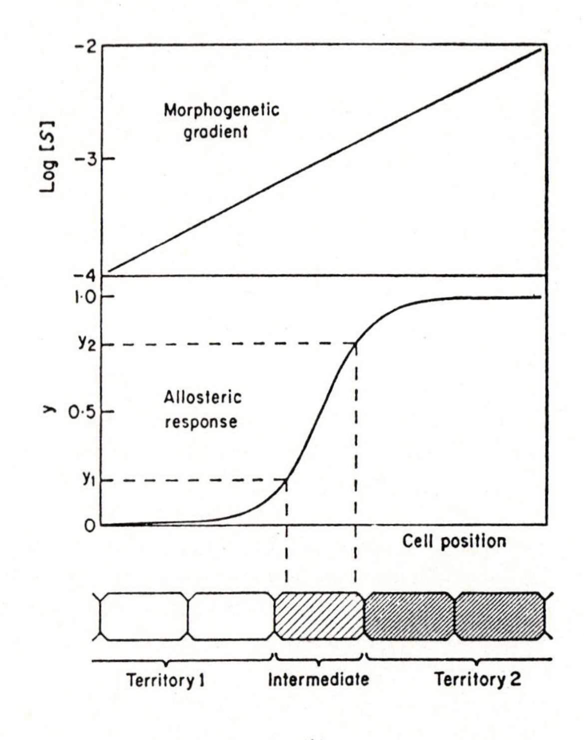

(b)

Figure 7.4 *(a) In this model,* S *is the signal for gene activation and* G *is the product synthesized by the gene that contributes to further activation. (b) A continuous morphogenetic gradient of* S *thus results in a response that undergoes a rapid transition from* 0 *to* 1. *[y represents the magnitude of the sigmoidal term in equation (28).] This means that sharp transitions in developmental processes can occur in a row of cells that experience a continuous signal gradient. (Reprinted with permission of the authors and publisher from Lewis, J., Slack, J.M.W., and Wolpert, L. (1977). Thresholds in development. J. Theor. Biol., 65, 579–590.)*

preted in a discontinuous developmental response. Clearly what is needed is a cellular on-off switch, a mechanism that governs whether pigment is synthesized or not, depending on the concentration of S. It happens that simple biochemical processes can provide such decision elements.

While many possible hypothetical molecular systems could work in principle, we examine now a simple example due to Lewis et al. (1977). What is attractive about this example is that it is relatively elementary, uses chemical kinetics discussed in a previous section, and has rather interesting dynamical behavior that we will ascertain by methods given in Section 5.1.

It is an acceptable assumption that production of a pigment (or, for that matter, any other protein product of the cell) requires activation of a gene that may normally be quiescent. Suppose gene G produces its product (whose concentration is g) at a rate that depends linearly on the level s of signal S. Furthermore, suppose G exerts a sigmoidal positive feedback effect on its formation and is degraded at a rate proportional to its concentration [see Figure 7.4(*a*)]. The level of the product of G within the cell could then be governed by the equation

$$\frac{dg}{dt} = k_1 s - k_2 g + \frac{Kg^2}{k_n + g^2}. \tag{28}$$

To understand the implications of (28), we study a graph of dg/dt versus g. A simple way of arriving at such a graph is to superimpose graphs of the two functions $k_1 s - k_2 g$ and $Kg^2/(k_n + g^2)$ and add these. The first is a straight line of slope $-k_2$ that intercepts the g axis at $k_1 s$. The second is a sigmoidal function like the one in Figure 7.3. The sum of the two leads to one of the graphs shown in Figure 7.5 depending on the size of $k_1 s$.

This model consists of a single nonlinear differential equation (28), whose form is $dg/dt = f(g)$. (s is assumed to be a known parameter, so g is the only variable.) We have encountered such equations in Section 5.1. Figure 7.5 identifies the steady states of (28) as points of intersection of $y = f(g)$ with the g axis. It can be seen that the number of such intersections can vary from one to three, depending on the height of the dip in the curve. Moreover, the stability properties of these states can also change in the transition from Figure 7.5(*a*) to (*c*). Recall that stability of the steady states can be inferred from the direction in which changes take place close to these points. This in turn depends only on the sign of dg/dt, which can be read directly from the graphs.

For example, in Figures 7.5(*a,b*), dg/dt is positive for all g values to the left of $\bar{g}_1$. In Figure 7.5(*c*) the fact that the valley in the graph dips below the g axis means that dg/dt is negative for $\bar{g}_3 < g < \bar{g}_2$. As shown in the graphs, these observations lead us to deduce a "flow" along the g axis. In cases (*a*) and (*b*) this flow is towards $\bar{g}_1$ (which is then a *globally stable steady state*). In case (*c*) both $\bar{g}_3$ and $\bar{g}_1$ are stable; if g is initially smaller than $\bar{g}_2$ it will approach $\bar{g}_3$ with time, whereas if g is initially higher than $\bar{g}_2$, it will be attracted to $\bar{g}_1$. Using the methods given in Chapter 5 we have determined qualitative properties of the chemical kinetics depicted by equation (28) without explicitly calculating anything.

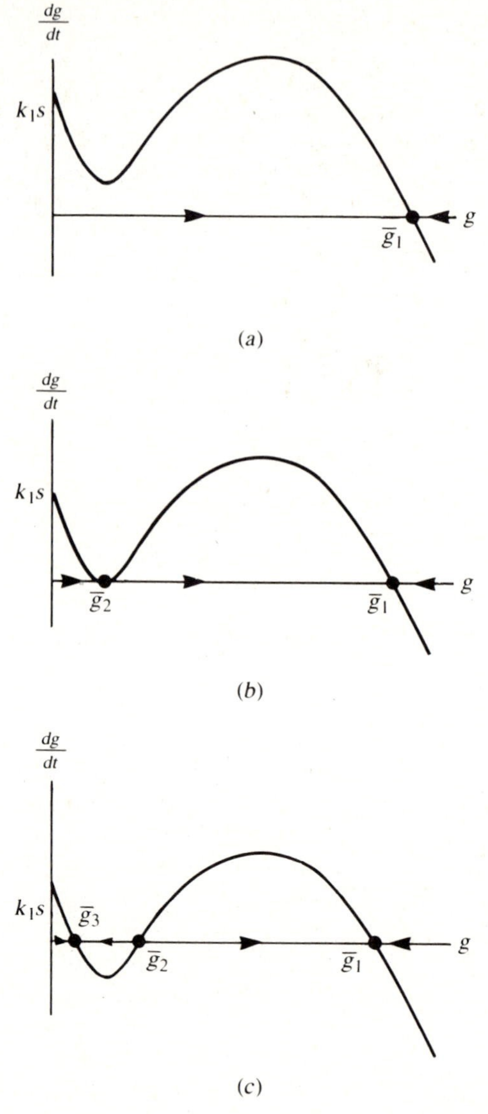

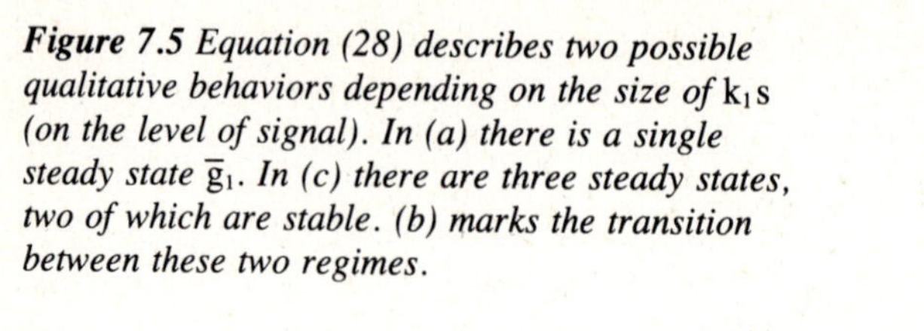
Figure 7.5 *Equation (28) describes two possible qualitative behaviors depending on the size of* $k_1 s$ *(on the level of signal). In (a) there is a single steady state* $\overline{g}_1$. *In (c) there are three steady states, two of which are stable. (b) marks the transition between these two regimes.*

Now we consider what this model implies about cellular differentiation. Suppose that initially all cells have no gene product, so $g = 0$. As the chemical gradient of S is established in the row of cells, there will be a continuous spatial transition from cases (*a*) to (*b*) to (*c*) (and all intermediate cases) across the length of the tissue. Close to the tail end of the row of cells, where s is very low, the kinetics of the gene product is given by Figure 7.5(*c*): the intercept $k_1 s$ is close to zero. Thus, since initially $g < \overline{g}_2$, the cell is led into steady state $\overline{g}_3$; that is, very low levels of gene product are formed. Somewhere along the row of cells a threshold level of s is present; there Figure 7.5(*b*) applies. The steady state $\overline{g}_3$ has disappeared. $\overline{g}_2$ and $\overline{g}_1$

are the only remaining steady states; the former is an ephemeral one, vanishing when s increases by the slightest amount. Thenceforth, in all cells beyond the transition point, gene product is synthesized up to a concentration $g = \bar{g}_1$, which is the unique steady state.

We see that the mechanism indeed captures the essence of a threshold switch. Another interesting feature noted by Lewis et al. (1977) is the memory built into the scheme: Once s is raised above threshold, the state of the cell changes permanently to $\bar{g}_1$. Even if s subsequently decreases, so that Figure 7.5(c) is obtained, the cell is "trapped" in $\bar{g}_1$ and will not return to its former state. The authors point out that a transient signal can thus be used to control discontinuous transitions in this developmental model as well as in other cellular processes.

7.6 SPECIES COMPETITION IN A CHEMICAL SETTING

A second example of biochemical control is discussed in this section. The model is presented here more to provoke your imagination and amuse you than to make a serious claim about cellular development. Perhaps as important as the message is the approach that differs from previous modeling in that a mechanism is *inferred* from an abstract set of equations.

Consider the following hypothetical situation. A cell can produce two types of chemical products X and Y. Under some circumstances it is advantageous to produce only one of the two products, while under other circumstances it is requisite to form both in a predetermined ratio. How is this to be achieved?

Suppose X and Y are components of some structural macromolecules. Figure 7.6 illustrates an artist's conception of how the *axis of polarity* of a two-dimensional cell could be determined by combining monomers of two types into longer structures. The scheme only works if the cell is "glued" into position with a permanently affixed immobile cornerstone on which macromolecular structures are built like scaffolds. The idea is then to combine appropriate ratios of X-type and Y-type bricks, thereby obtaining a structure whose orientation is given approximately by the vector (percent X type, percent Y type). Thus a way of controlling the polarity of the cell (whether it elongates horizontally, vertically, or in some other direction) is by controlling the relative percentages of X- and Y-type monomers that are synthesized inside the cell.

How cellular polarity is actually controlled is still an unsolved problem. Beautiful and exciting examples of the effects of polarity differentiation are often demonstrated in plant cells (which, incidentally are largely immobile, unlike the fluid-like cells of animals). A particularly striking pattern of horizontal-versus-vertical cellular orientation is exhibited in tree trunks and other woody parts of plants in the tissue called the *xylem* (which consists of cells that have died after differentiating). There one sees "islands" of horizontal radially directed cells (called *ray cells*) embedded in a sea of vertical cells (*tracheids*). (See Figure 7.7.) The remarkable point is that both cell types arise from rather similar precursors in the actively growing tissue called the *cambium*. There are virtually no cells that do not fall into this dichotomy, indicating that the control mechanism governing polarity commitment is rather strict.

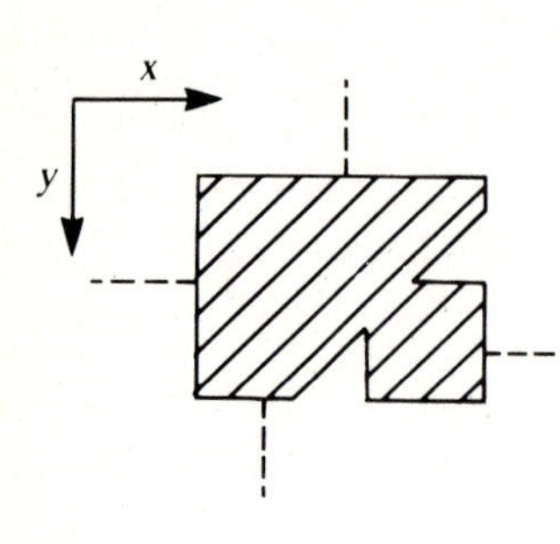

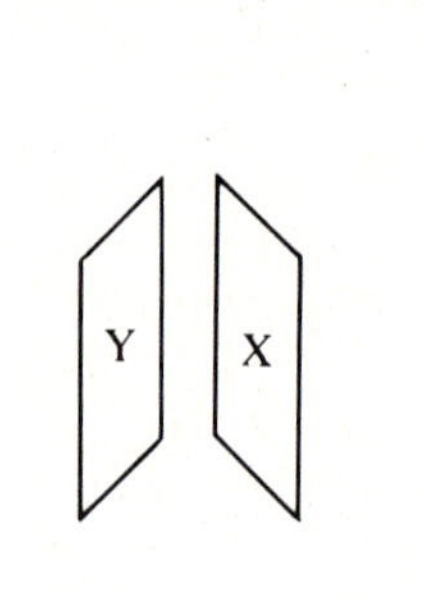

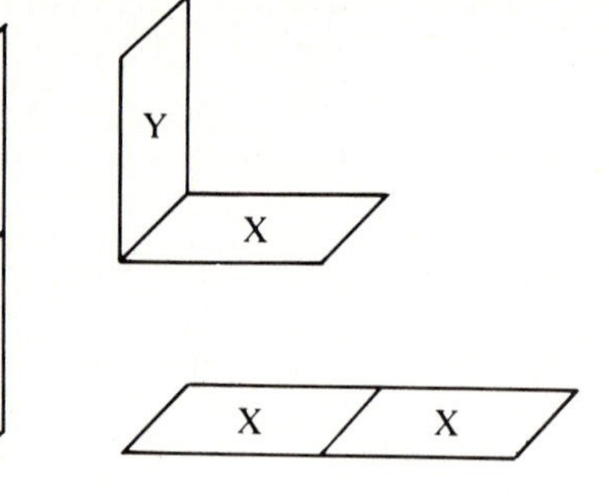

(a) (b) (c)

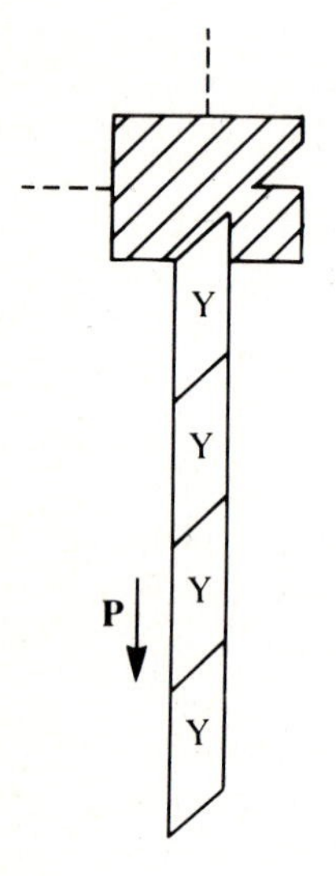

(d)

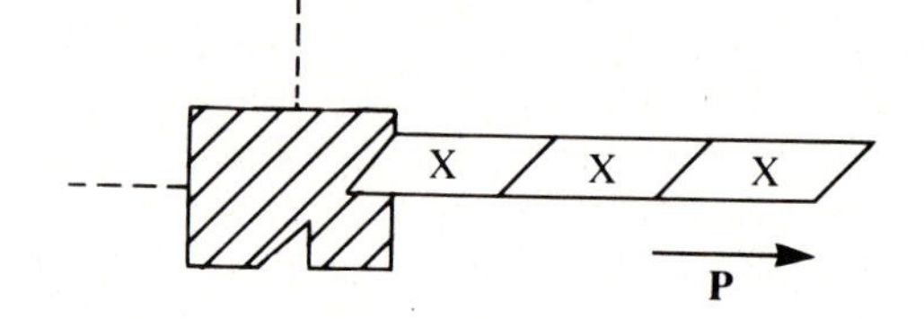

(e)

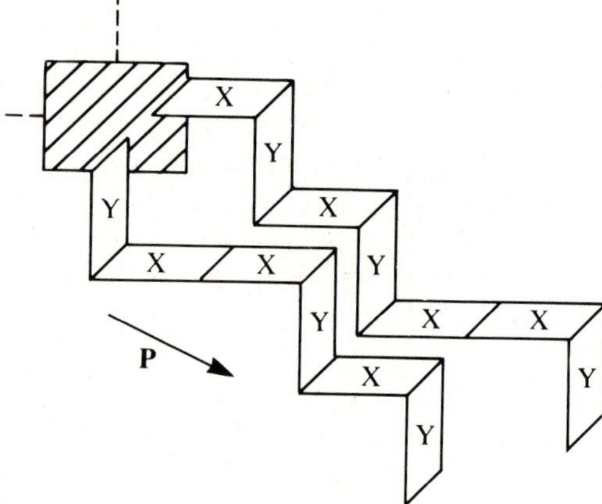

(f)

Figure 7.6 *The polarity of structures within the cell and hence the orientation of the cell could perhaps be governed by the relative proportions of monomers* X *and* Y *that are made. The cell must have (a) a fixed "cornerstone" and (b) synthesize* X *and* Y, *which can form complexes such as (c) dimers or larger polymers. (d) If only* Y *is made or (e) only* X, *the cell will have a horizontal or vertical polarity. (f) Intermediate orientations occur if both products are made in some relative proportion.*

Looking at examples like that of *rays* and *tracheids* in the xylem of plants leads us to feel that some sort of competition takes place within the precursor cell between the tendencies to promote horizontal and vertical characteristics. This intuition leads to the model that follows, though clearly many other approaches to the problem are possible.[4]

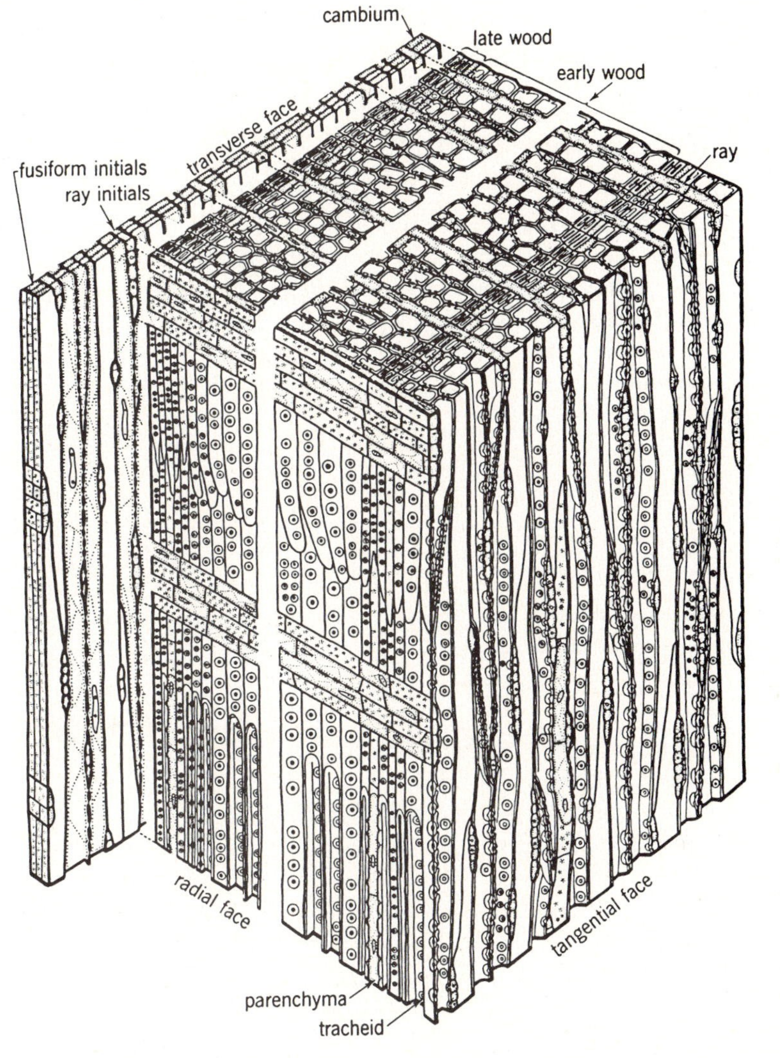

Figure 7.7 *Cells that are oriented precisely in the horizontal or vertical directions can be observed in the woody parts of higher plants. The question of how such polarity is determined led to the model given by equations (29a,b). This diagram shows the xylem of white cedar with its ray cells and tracheids. [From Esau, K. (1965). Plant Anatomy, Wiley, New York. Copyright © 1965 by K. Esau. Reprinted by permission of John Wiley & Sons, Inc.]*

4. The models of plant cell polarity were inspired through conversations with Tsvi Sachs of the Hebrew University, Israel.

At this point we pause for a pitch on the advantages of modeling. We are about to witness the fact that a mathematical abstraction permits us to make a connection between two seemingly unrelated phenomena that share similar dynamical properties. In this case the suspicion that competition between two forces or two chemical species might be involved leads us to recollect that a competition model in another context is already known to us. From Section 6.3 we borrow equations (9a,b):

$$\frac{dN_1}{dt} = r_1 N_1 \frac{\kappa_1 - N_1 - \beta_{12} N_2}{\kappa_1}, \qquad \textit{(9a, Chap. 6)}$$

$$\frac{dN_2}{dt} = r_2 N_2 \frac{\kappa_2 - N_2 - \beta_{21} N_1}{\kappa_2}. \qquad \textit{(9b, Chap. 6)}$$

These equations describe the populations of two animal species. One wonders to what extent they could pertain to chemical species, where reproduction, survivorship, or mortality have vague meanings if any. If we are to proceed with a reinterpretation, we must use a more general restatement of the equations. Accordingly, we replace variable names by x and y (concentrations of the two molecular species X and Y), use new parameters μ, α, γ and multiply the RHS expressions to get the following rejuvenated model:

$$\frac{dx}{dt} = \mu_1 x - \alpha_1 x^2 - \gamma_{12} xy, \qquad (29a)$$

$$\frac{dy}{dt} = \mu_2 y - \alpha_2 y^2 - \gamma_{21} xy. \qquad (29b)$$

Note that the new parameters are related to the old as follows: $\kappa_i = \mu_i/\alpha_i$ and $\kappa_i/\beta_{ij} = \mu_i/\gamma_{ij}$. The behavior of solutions to these equations falls into one of the four categories shown in Figure 6.6:

1. species Y always predominates.
2. species X always predominates.
3. X or Y predominates depending on initial conditions.
4. Stable coexistence of both species at some ratio.

Which case is obtained depends on the relative magnitudes of certain combinations of the parameters:

1. $\mu_2/\mu_1 > \gamma_{21}/\alpha_1$ and $\mu_2/\mu_1 > \alpha_2/\gamma_{12}$.
2. $\mu_2/\mu_1 < \gamma_{21}/\alpha_1$ and $\mu_2/\mu_1 < \alpha_2/\gamma_{12}$.
3. $\mu_2/\mu_1 < \gamma_{21}/\alpha_1$ and $\mu_2/\mu_1 > \alpha_2/\gamma_{12}$.
4. $\mu_2/\mu_1 > \gamma_{21}/\alpha_1$ and $\mu_2/\mu_1 < \alpha_2/\gamma_{12}$.

On the face of it, the equations can potentially describe the very phenomenon that we are attempting to understand in this section, namely a mechanism of controlling synthesis of species at some relative proportions. However, in order to reap some benefit from this conclusion we might wish for some kind of molecular interpretation for terms in equations (29a,b). Since the control of synthesis of large molecules ultimately resides within the genome, a suitable interpretation would be to view these terms as effects on the genetic material that codes for species X and Y.

A positive feature of equations (29a,b) that was not readily apparent in their previous form is that the quadratic terms they contain are rather familiar mass-action terms for interactions of pairs of molecules (X with X, Y with Y, or X with Y). This suggests the following intriguing hypothesis.

Suppose the X and Y molecules can form dimers such as X—X, X—Y, and Y—Y in some rapidly equilibrating reversible reaction. In this case, the cellular concentration of such dimers would be approximately proportional to the products of concentrations of the participating monomers (see problem 12). Precisely such terms appear in equations (29a,b) accompanied by minus signs. This suggests that these dimers tend to inhibit the production of the substances X and/or Y (or possibly activate enzymes that degrade these chemicals).

We can go further in putting together the puzzle by interpreting a complete molecular mechanism as follows:

1. Each monomer activates its own gene. (Witness the positive contributions of terms $\mu_1 x$ and $\mu_2 y$ in the equations.)
2. Dimers made up of identical monomers (X—X and Y—Y) repress only the gene that codes for that particular molecule.
3. Mixed dimers (X—Y) repress both the X gene and the Y gene.

See Figure 7.8(*a*) for a schematic view of these events.

We have seen earlier that with the appropriate relations between the various rate constants this regulatory mechanism would select for the synthesis of a single product or some proportion of both products. What do such rate constants represent? Previous analysis in this chapter demonstrates that rate constants are often ratios or more complicated combinations of parameters that depict forward or reverse reaction rates. Loosely speaking, the constants appearing in equations (29a,b) may depict affinities of molecules for each other (as for dimers) or for regions of the genome that control synthesis of the products X and Y.

Since it is known that slight changes in molecular conformations can alter such affinities, it is reasonable to think of cells as having a whole range of permissible values of $(\mu_i, \alpha_i, \gamma_{ij})$. Some values would lead to all-or-none behavior, while others would govern the relative frequency of X and Y synthesis. What makes this fact intriguing is that we can envision a *developmental pathway,* in which a cell changes its character throughout various stages of its cycle to meet various needs. This could be accomplished by a gradual variation of one or several rate constants. (See problem 13.)

For example, if $\mu_2/\mu_1 < \gamma_{21}/\alpha_1$ and $\mu_2/\mu_1 > \alpha_2/\gamma_{12}$ (case 3) the cell would have the dynamical behavior shown in Figure 7.8(*b*): given any initial concentrations (x_0, y_0) the final outcome would be synthesis of only X or only Y depending on which gene gets more strongly repressed. This in turn depends on whether (x_0, y_0) falls above or below a *separatrix* in the *xy* plane, a curve that subdivides the positive

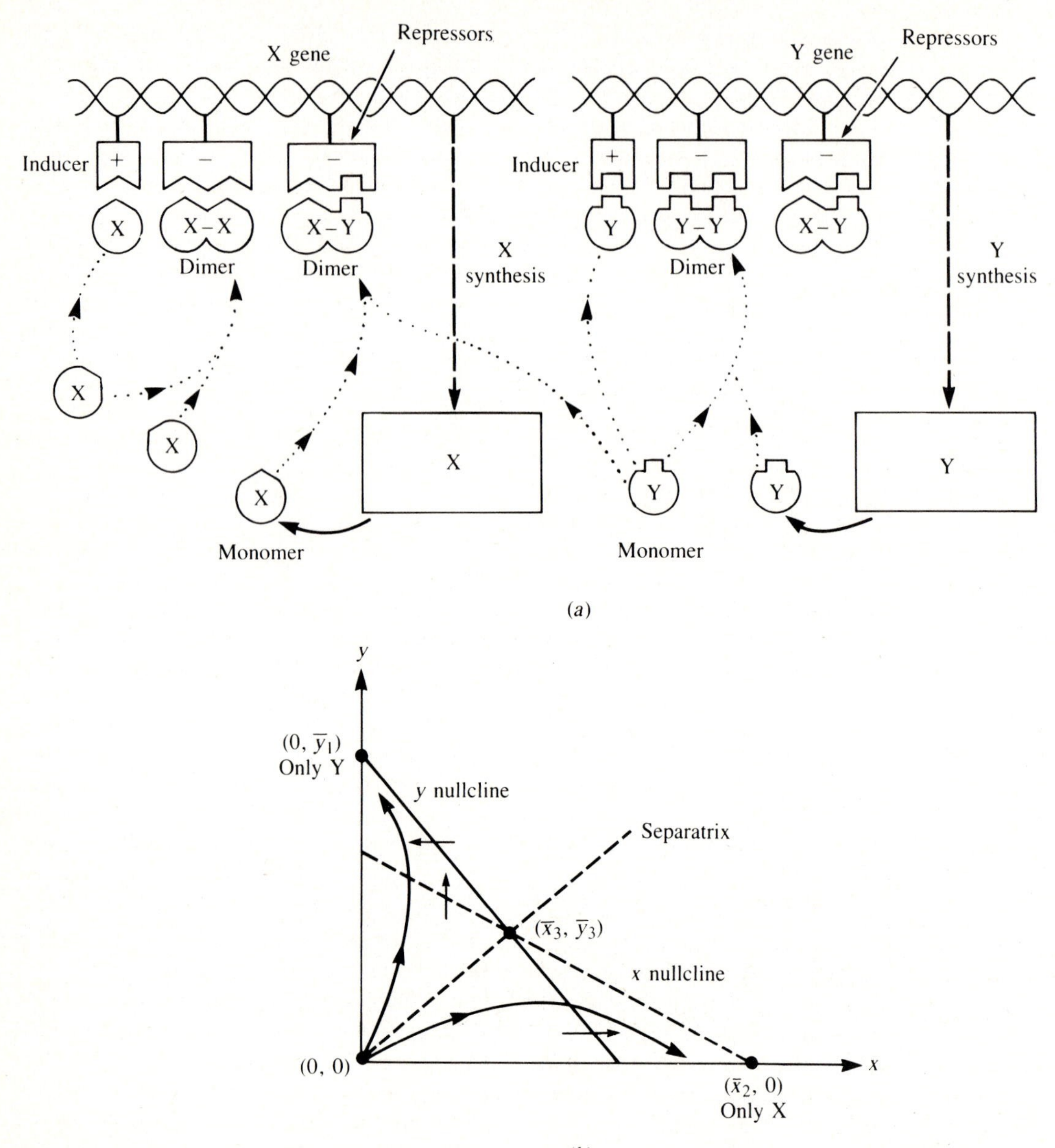

Figure 7.8 *(a) The model given by equations (29a,b) could be interpreted in terms of a molecular mechanism for genetic control. Shown are the two genes coding for* X *and* Y *molecules. Repressors on these genes are sensitive to concentrations of the dimers* X—X, Y—Y, *and* X—Y. *Inducers are sensitive to monomer concentrations of* X *or* Y. *(b) Provided condition (3) is satisfied by the parameters (see text), the dynamic behavior of equations (29a,b) resembles that of a switch. There are two possible outcomes (only* X *or only* Y *synthesized), depending on the initial concentrations* (x, y). *Note that the phase plane is divided into two domains by a curve called a separatrix. Points in a given domain are attracted to one of the two steady states on the axes. The steady state in the positive xy quadrant is a saddle point.*

xy quadrant into two separate *basins of attraction* [see Figure 7.8(b)]. Should the parameters change so that one or both of the above inequalities is reversed, the outcome would be independent of the initial concentrations.

Within the context of cellular polarity, we see that depending on molecular affinities and initial conditions, the (percent X type, percent Y type) structural molecules in the cell could so evolve that the cell is eventually polarized entirely in the x or in the y direction. (Alternately, in case 4 the cell could attain some intermediate orientation governed by the coordinates $(\bar{x}_3, \bar{y}_3)$ of the steady state in the positive quadrant.)

The problem of control of relative proportions of biosynthesis is clearly of much wider applicability. To give but one other related example, consider the events underlying synthesis of an enzyme, lactate dehydrogenase (LDH). It is known that a single enzyme molecule is comprised of four subunits. Subunits come in two varieties, A and B, and any of the following five combinations can occur:

LDH-1: BBBB, LDH-4: AAAB,
LDH-2: ABBB, LDH-5: AAAA.
LDH-3: AABB,

It is interesting to learn that in certain cells in the body (for example, mouse kidney cells) there is a progressive shift from production of LDH-5 to LDH-1 from birth to adulthood. This may imply that biosynthesis gradually changes from production of all B subunits to certain proportions of B to A and finally to all A subunits. There is no indication that the genetic mechanism is related to the simple scheme discussed in this section. However, we nevertheless observe that such developmental transitions could stem from gradual parameter changes in some underlying system of molecular interactions.

What do we gain from this modeling exercise? First and foremost is an appreciation of the fact that mathematical equations are abstract statements that may have applications to more than one area. But can we really believe that the simple mechanism proposed on the basis of the species-competition equations could be at work inside a cell? Here we must take some care, for even appealing analogies such as the one used here may be false or inaccurate. Ultimately the test lies in empirical evidence for or against a given hypothesis. You may wish to consult Edelstein (1982) for indications of possible empirical tests of the model just discussed.

The two models examined in these sections were obtained in a way different from previous examples; here a dynamical behavior was known *a priori*. The behavior was reminiscent of solutions to previous equations that derived from quite different contexts. These equations were then rewritten and reinterpreted, leading to new suggestions for underlying mechanisms. This approach cannot be expected to work in every case, but it is of surprising value when it does. One is led to feel that some control mechanisms are universal, reappearing in the contexts of ecology, engineering, molecular biology, and other systems. This observation underscores the power and generality of mathematical modeling that directly illuminates the connection between such diverse problems.

In the next section we regard one of the four cases discussed here somewhat more abstractly and identify the particular attributes that generate switch-like behavior. For further examples of this abstract modeling approach, a good source is Rosen (1970, 1972). For more about the versatility of biochemical control, consult Savageau (1976).

7.7 A BIMOLECULAR SWITCH

In this section we examine case 3 of equations (29a,b) and explore what general assumptions suffice to produce similar dynamic behavior in other systems. Recall in case 3 that whether x or y predominates depends only on initial concentrations of X and Y. In the xy-plane shown in Figure 7.8(b), the first quadrant is divided into two regimes of influence; in one, all starting points head towards $(0, \bar{y}_1)$, whereas in the other the attractor is $(\bar{x}_2, 0)$. ($\bar{x}_2$ and $\bar{y}_1$ stand for the steady state values on the two axes). A system that behaves in this way can be described as a *switch*, a means for sorting an initial situation into one of two possible outcomes.

Since such mechanisms can have potential aplications to other physical phenomena, we shall extract the features of equations (29a,b) that lead to this behavior.

From Figure 7.8(b) it is evident that a rather general property of the switch is that the positions of steady states meet the following conditions:

1. There are four steady states: one at (0, 0), two on the axes, and one in the first quadrant at $(\bar{x}_3, \bar{y}_3)$.
2. (0, 0) is an unstable node.
3. $(\bar{x}_3, \bar{y}_3)$ is a saddle point.

Below we restrict attention to equations of the form

$$\dot{x} = xf(x, y) \tag{30a}$$

$$\dot{y} = yg(x, y). \tag{30b}$$

These will satisfy condition 1 provided there are values $\bar{x}_3$, and $\bar{y}_3$ such that

$$f(\bar{x}_3, \bar{y}_3) = g(\bar{x}_3, \bar{y}_3) = 0 \qquad (x^*, y^* > 0). \tag{31}$$

The proof of this assertion is left as a problem. We note that the fact that $(\bar{x}_3, \bar{y}_3)$ is in the first quadrant is equivalent to assuming that the nullclines $f = 0$ and $g = 0$ intersect in this quadrant.

To satisfy condition 2 we need to assume that $f(0, 0)$ and $g(0, 0)$ are both positive. (See problem 14.)

To satisfy condition 3 it is necessary that $(f_x/f_y)|_{ss} < (g_x/g_y)|_{ss}$ where f_x, f_y, g_x, and g_y are partial derivatives of f and g evaluated at $(\bar{x}_3, \bar{y}_3)$. This condition is rather interesting. It can be obtained by either one of two reasoning processes. A straightforward calculation of the Jacobian of (30) at $(\bar{x}_3, \bar{y}_3)$ reveals that

$$\det \mathbf{J} = \overline{xy}(f_x g_y - f_y g_x). \tag{32}$$

(See problem 14.) Requiring that $\det \mathbf{J} < 0$ leads to the above inequality.

A more novel approach is to use geometric reasoning. To obtain case 3 the nullclines have to be so situated that the y nullcline $g(x, y) = 0$ is more steeply inclined than the x nullcline $f(x, y) = 0$ at their intersection $(\bar{x}_3, \bar{y}_3)$ such that both have *negative* slopes.

Using *implicit differentiation* below, we arrive at the following results:

$$x \text{ nullcline:} \qquad f(x, y) = 0, \tag{33a}$$

$$f_x + f_y \frac{dy}{dx} = 0, \tag{33b}$$

$$\text{slope of } x \text{ nullcline:} \qquad \frac{dy}{dx} = -\frac{f_x}{f_y} < 0, \tag{33c}$$

$$y \text{ nullcline:} \qquad g(x, y) = 0, \tag{34a}$$

$$g_x + g_y \frac{dy}{dx} = 0, \tag{34b}$$

$$\text{slope of } y \text{ nullcline:} \qquad \frac{dy}{dx} = -\frac{g_x}{g_y} < 0. \tag{34c}$$

The slopes must satisfy the relation $f_x/f_y < g_x/g_y$, establishing the result. In the alternate form $(f_x g_y < f_y g_x)$, we can interpret this result as follows. The product of the effect of each species on its own growth rate should be smaller than the product of the cross effects of species i on species j.

We see now that more general kinetic expressions can also be used to construct a switching mechanism. [See problem 14(d).] Another rather nice example of a switching mechanism is given by Thornley (1976) for the biochemistry of flower initiation.

The abstract way in which we studied properties of the nullclines of equations (30) can be used for other general questions. In the next section a similar approach will be used to derive geometric conditions for stability of interacting chemical species.

7.8 STABILITY IN ACTIVATOR-INHIBITOR AND POSITIVE FEEDBACK SYSTEMS

Systems in which only qualitative properties of the interactions are known were discussed in Section 6.5. We now investigate a pair of chemical systems that have particular qualitative sign patterns (and associated graphs shown in Figure 7.9). The analysis is of interest for two reasons. First, it illustrates a method for understanding stability in a geometric way. Second, the results will be of relevance to material discussed in Chapter 11, where activator-inhibitor and positive-feedback systems are of special importance.

The systems considered here each consist of two chemicals, with mutual effects depicted by the following sign patterns:

1. Activator-inhibitor system: $\mathbf{Q}_1 = \begin{bmatrix} + & - \\ + & - \end{bmatrix}$, *(35a)*

2. Positive feedback system: $\mathbf{Q}_2 = \begin{bmatrix} - & - \\ + & + \end{bmatrix}$. *(35b)*

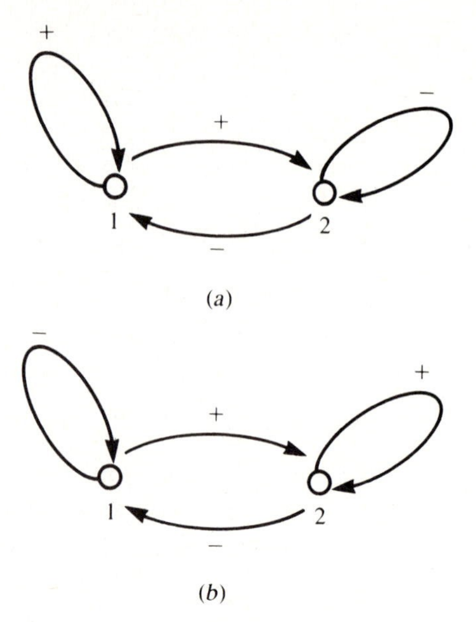

Figure 7.9 *Signed directed graphs for (a) the activator-inhibitor and (b) positive feedback systems discussed in Section 7.8.*

To be more explicit, $\mathbf{Q}_1$ and $\mathbf{Q}_2$ are sign patterns of elements in the Jacobian of each system, evaluated at some steady state. In equation (35a) the distribution of signs implies that chemical 1 has a positive effect on its own synthesis and on the synthesis of chemical 2, whereas chemical 2 inhibits the formation of both substances. For this reason, chemical 1 is termed the *activator* and chemical 2 the *inhibitor*. (Notice that this is a molecular analog of an ecological predator-prey pair.) In equation (35b) either participant promotes increase in the second chemical and decrease in the first. The term *positive feedback system* has a historical source and should not be taken too literally since, in fact, both (35a) and (35b) have positive as well as negative feedback loops.

Indeed from Figure 7.9 it is evident that neither system is qualitatively stable in the sense discussed in Section 6.5 because of the positive feedback loops on one of the participating species. Stability thus depends on other constraints, to be reviewed presently. Rather than merely restating these in terms of the coefficients in the Jacobian, we will derive analogous conditions on the intersection properties of the nullclines. Thus, let us momentarily suppose that we have functional expressions for the kinetic terms and work backwards. We deal first with (35a) and then with (35b).

The Activator-Inhibitor System

Suppose the kinetics of chemical interactions are governed by the rate laws

$$\frac{dx}{dt} = f_1(x, y), \tag{36a}$$

$$\frac{dy}{dt} = f_2(x, y), \tag{36b}$$

and that $(\bar{x}, \bar{y})$ is a nontrivial steady state of this system. In the xy phase plane, nullclines for x and y would then be those curves for which

$$f_1(x, y) = 0 \qquad (x \text{ nullcline}), \tag{37a}$$

$$f_2(x, y) = 0 \qquad (y \text{ nullcline}), \tag{37b}$$

and $(\bar{x}, \bar{y})$ would be a point of intersection of these curves. We now resort to implicit differentiation to draw conclusions about the slope of the nullclines at $(\bar{x}, \bar{y})$.

First note that, after differentiating both sides of the equations with respect to x, we arrive at

$$\frac{\partial f_1}{\partial x} + \frac{\partial f_1}{\partial y}\left(\frac{dy}{dx}\right)_1 = 0, \tag{38a}$$

$$\frac{\partial f_2}{\partial x} + \frac{\partial f_2}{\partial y}\left(\frac{dy}{dx}\right)_2 = 0. \tag{38b}$$

Here $(dy/dx)_1$ means "slope of the nullcline $f_1 = 0$ at some point P." Similarly $(dy/dx)_2$ is the slope of $f_2 = 0$ at some point P.

We must now use information specified in the problem, namely that the sign pattern in equation (35a) determines the signs of elements of the Jacobian. Since these elements are precisely partial derivatives evaluated at $(\bar{x}, \bar{y})$, we use this fact in deducing that

$$\left.\frac{\partial f_1}{\partial x}\right|_{ss} = a, \qquad \left.\frac{\partial f_1}{\partial y}\right|_{ss} = -b, \qquad \left.\frac{\partial f_2}{\partial x}\right|_{ss} = c. \qquad \left.\frac{\partial f_2}{\partial y}\right|_{ss} = -d \tag{39}$$

where a, b, c, and d are some positive constants.

Define the quantities

$$s_1 = \left.\left(\frac{dy}{dx}\right)_1\right|_{(\bar{x},\bar{y})}, \tag{40a}$$

$$s_2 = \left.\left(\frac{dy}{dx}\right)_2\right|_{(\bar{x},\bar{y})}. \tag{40b}$$

Then, as mentioned above, s_1 and s_2 are slopes of the two nullclines at their intersection, $(\bar{x}, \bar{y})$. Equations (38a,b) can now be written in terms of the new quantities as follows:

$$a - bs_1 = 0 \tag{41a}$$

$$c - ds_2 = 0 \tag{41b}$$

This implies that $a = bs_1$ and $c = ds_2$. Thus the Jacobian of equations (36a,b) can be written in two ways:

$$\mathbf{J} = \begin{pmatrix} a & -b \\ c & -d \end{pmatrix} = \begin{pmatrix} bs_1 & -b \\ ds_2 & -d \end{pmatrix}. \tag{42}$$

Now we may determine when $(\bar{x}, \bar{y})$ is stable. The conditions are that

$$\beta = \operatorname{Tr} \mathbf{J} = bs_1 - d < 0 \qquad \Rightarrow s_1 < d/b, \tag{43a}$$

and

$$\gamma = \det \mathbf{J} = -dbs_1 + dbs_2 > 0 \Rightarrow s_2 > s_1. \tag{43b}$$

We note from equations (41a,b) that s_1 and s_2 must both be positive since a, b, c, and d are assumed to be positive. Thus stability implies that at $(\bar{x}, \bar{y})$ the y nullcine will be more steeply sloped than the x nullcline. Figure 7.10 illustrates how these conclusions affect the *local* geometry of the phase plane. Further discussion of this case is suggested in problem 15.

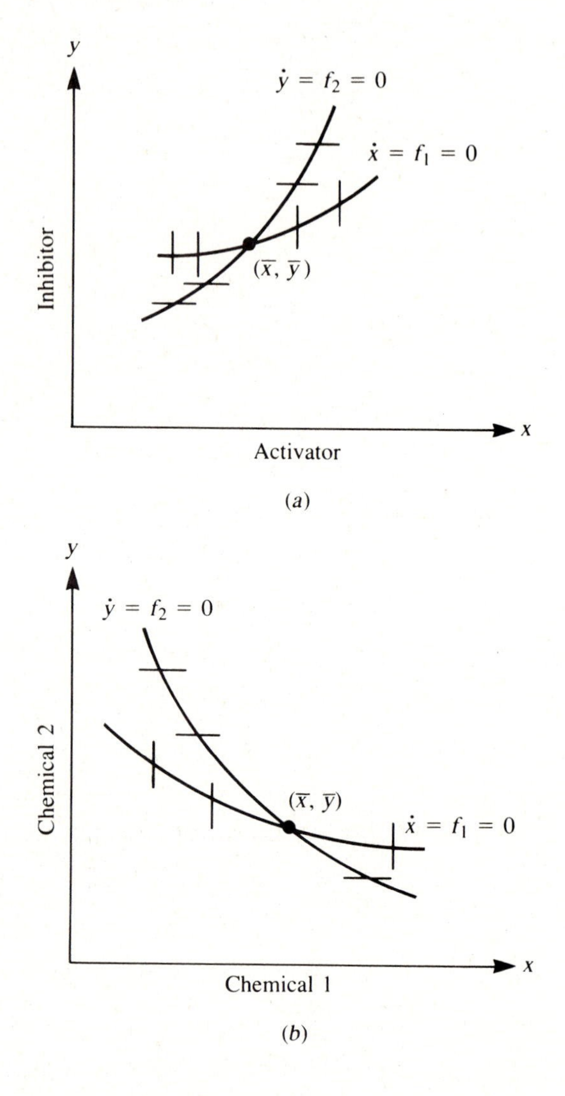

Figure 7.10 *(a) In an activator-inhibitor system, a steady state $(\bar{x}, \bar{y})$ is stable only if the inhibitor nullcline ($\dot{y} = 0$) is steeper than the activator nullcline ($\dot{x} = 0$) with both curves having positive slopes. It is further necessary that the slope of $\dot{x} = 0$ at the steady state be not too steep, that is, less than d/b; see equation (43a). (b) In a positive-feedback system the two nullclines must have negative slopes such that $\dot{y} = 0$ is steeper than $\dot{x} = 0$ for stability of $(\bar{x}, \bar{y})$.*

Positive Feedback

We proceed in a similar way in dealing with the second case, which is left largely as an exercise. Now, however, we define

$$\left.\frac{\partial f_1}{\partial x}\right|_{ss} = -a, \qquad \left.\frac{\partial f_1}{\partial y}\right|_{ss} = -b, \qquad \left.\frac{\partial f_2}{\partial x}\right|_{ss} = c, \qquad \left.\frac{\partial f_2}{\partial y}\right|_{ss} = \mathrm{d}, \qquad (44)$$

where a, b, c and d are again positive constants. In problem 16 you are asked to verify that this method, with s_1, and s_2 defined as before, leads to the conclusion that both s_1 and s_2 are negative, such that

$$s_2 < s_1 < \frac{-d}{b}. \tag{45}$$

That is, s_2 is "more negative" than s_1, so that the y nullcline is again steeper than the x nullcline, but now both have negative slopes. Figure 7.10(b) shows what this implies about the local geometry of the nullclines in the positive feedback case.

Based on these properties, it was proved by Kadas (1982) that *a two-species reaction mechanism with a monotonic nullcline cannot have steady states of both types* (*35a*) and (*35b*). This allows one to classify the mechanisms as activator-inhibitor or positive-feedback in a broader sense, even though their properties are only known locally (close to their steady-state levels).

7.9 SOME EXTENSIONS AND SUGGESTIONS FOR FURTHER STUDY

In this section we outline several alternate approaches to molecular systems, some of which are longstanding and others more recent. References for independent exploration are suggested.

1. Summaries of enzyme action and of the pertinent mathematical methods appear in the encyclopedic book by Dixon and Webb (1979). Reiner (1969), and Boyer (1970) are much shorter. Numerous special cases, such as *multivalent enzymes, product inhibition, allosteric effects,* and *endogenous activators* are discussed and accompanied by standard kinetic analysis. The chief visual device used in studying such systems is graphs of the reaction rate v plotted against concentration c of one of the chemical participants. (The reaction rate is a measure of the disappearance of substrate or appearance of product; for example, $v = dc/dt$; see Figure 7.3 for example.) Alternative graphical constructions include the Lineweaver-Burk plot, which is simply a graph of $1/v$ versus $1/c$. Such graphs have conventionally been used to identify rate constants in chemical kinetic studies and as a convenient way of summarizing and comparing enzyme systems.
2. *Understanding large chemical systems from network structure*. Most biochemical pathways contain a large number of intermediates that react and affect each other in complex chemical networks. Mathematical analysis of such chemical systems by standard methods is impractical, yet one is often interested in addressing fairly general and important questions, such as:
   - **(a)** Does the system admit steady-state solutions? (Are there *multiple* steady states?)
   - **(b)** Are such steady states stable?
   - **(c)** Can such systems admit temporal oscillations of chemical concentrations?

   In a series of papers, Feinberg (1977, 1980, in press) has addressed such questions by methods that utilize the structure of the chemical network rather than the differential equations that correspond to the chemical kinetics. The approach consists of defining three integers, n, l, and s, which represent respectively the number of entities appearing in the network, the extent of linkage of the network, and the span of a vectorspace defined by assigning a vector to each of the reactions. The integer $\delta = n - \ell - s$, called the *deficiency* of the network, is then indicative of the expected dynamics. For example, in a network in which all reactions are reversible, if $\delta = 0$, then there is always a single (strictly positive) steady state that is stable and *no* oscillatory solutions exist. (See references for details of the definitions and stronger statements of these results.) The Feinberg network method is unfortunately not yet general enough to lead to strong conclusions in every case. However, where applicable it is a valuable and computationally inexpensive technique.

Papers given in the references are expository and would be accessible to students who have a minimal background in linear algebra. (It is, for example, necessary to be able to find the *rank* of a matrix in computing the integer s.)

3. A brief but thorough summary of enzyme kinetics that gives most of the technical highlights is to be found in Rubinow (1975). This source demonstrates a somewhat different application of graph theory (based on Volkenshtein, 1969); here the goal is to derive expressions for the reaction velocity of a chemical system. Using the quasi-steady-state assumption, this problem is essentially one of solving a system of linear algebraic equations. When the network is large, the corresponding system of algebraic equations can be rather cumbersome. However, by invoking certain rules, it is possible to simplify the network (for example, by adding parallel branches and merging nodes in a particular way) and so deduce a relation between the reaction velocity v and the concentrations and kinetic rate constants in the pathway without solving a complicated system of algebraic equations. You should recognize that this method does not address questions regarding the dynamic behavior of a chemical reaction scheme under general conditions since the quasi-steady-state assumption underlies the method. Rubinow (1975) gives details and several worked-out examples.

4. A contemporary approach to the analysis of biochemical systems has been described by Rosen (1970, 1972) and Savageau (1976) and comes under the general heading "biochemical systems analysis." An important point their work addresses is that enzymes are not only catalysts of reactions but also the control elements that can be modulated by a variety of influences. It is commonplace in biochemical pathways to encounter examples of feedback control. End products of the reaction may directly affect the catalysis of a key enzyme by attaching to it and changing its physical conformation. This leads to changes in the biological function of the enzyme and may result in total inhibition of further product synthesis.

Savageau (1976) compares the design of a number of biochemical and genetic control pathways, summarized by reaction diagrams that convey the sequence of products and their feedback control. (This systems approach appears in Rosen, 1970, 1972.) An interesting feature Savageau then explores is comparison of alternative control patterns, in which control is exerted at a variety of nodes and by a variety of intermediates. In the cases where the number of intermediates is small (for example, $n = 3$), explicit stability analysis is carried out. Certain networks lead to more stable interactions than others. (For example, a simple end-product inhibition in which the last product inhibits the first reaction step has a steady state that can more readily be destabilized by parameter variations than can a system of the same size with another pattern of feedback interactions.) Using such analysis, Savageau addresses the question of optimality of design in assessing whether real biochemical networks have advantages over other possible networks less commonly encountered. Detailed discussions of stability, Routh-Hurwitz tests, and many interesting examples make Savageau's book a good source for further study.

Rapp (1979) gives a control-theory approach to metabolic regulation. His paper, suitable for advanced students, contains numerous interesting examples, a clear discussion of techniques, and a thorough bibliography.

PROBLEMS*

The following questions pertain to Michaelis-Menten kinetics.

1. Verify that equations (6a,b) are obtained by eliminating x_0 from equations (3a,c).

2. Show that when the quasi-steady-state assumption is made for x_1 in equations (6a,b) one can algebraically simplify the model with the following procedure:
   (a) First write x_1 in terms of an expression involving only c.
   (b) Substitute this expression into equation (6a) and simplify to obtain equation (8).

3. Show that equations (6a,b) can be reduced to the dimensionless equations (10a,b) by the choice of reference scales given by equations (9a–c). Interpret the meanings of τ, ϵ, κ, and λ. Verify that the equations can be written in the form (11a,b).

4. **(a)** Integrate equation (12a) to obtain an implicit solution (an expression linking the variables c and t.) Use the fact that at $t = 0$, $c = c_0$ to eliminate the constant of integration.
   (b) Use equation (12b) and the fact that [by (12a)] $c(t) \to 0$ to reason that $x(t) \to 0$. (*Hint:* Consider $\lim_{c \to 0} c/[k + c]$.)

5. Show that equations (6a,b) can be reduced to the second dimensionless equations (14a,b) by choosing the reference time scale of equation (13).

*Problems preceded by an asterisk are especially challenging.

6. Integrate equation (15b). Is $x_1(t)$ an increasing or a decreasing function of time on this time scale? (Note: You should use the initial conditions described in the text to get a meaningful solution.)

***7.** Equations (6a,b) can be studied by phase-plane methods.

(a) Show that c and x_1 nullclines have the form

$$\dot{c} = 0 \quad \text{when} \quad x_1 = \frac{Kc}{a + c},$$

$$\dot{x}_1 = 0 \quad \text{when} \quad x_1 = \frac{Kc}{b + c}.$$

Identify K, a, and b in terms of the original parameters, r_0, k_1, k_{-1}, and k_2. Which is larger, a or b?

(b) Sketch the nullclines in an x_1c phase plane. One point of intersection is $x_1 = c = 0$. Is there another? Determine directions of flow along these curves and along the c and x_1 axes.

(c) Draw a trajectory beginning at the state in which all receptors are unoccupied and the initial nutrient concentration is c_0. What is the eventual outcome? Explain your result in terms of the original cellular process.

(d) Which portions of your trajectory correspond to the initial fast transition and which to the gradual slow decline shown in Figure 7.2?

In problems 8–11 we discuss details of the derivation and implications of sigmoidal kinetics.

8. Write down a complete set of equations for the reaction diagram (16). [The first two equations are given in (17a,b).] Show that equation (18) is obtained by making a quasi-steady-state assumption.

9. **(a)** Verify that the relations (22a,b) are obtained when we assume that $dx_2/dt \simeq 0$ and $dx_0/dt \simeq 0$.

(b) Demonstrate that equation (23) is obtained by eliminating all variables except c and x_0 in equation (20f).

(c) Show that this leads to equation (24) for c.

10. Find examples of other positively cooperative biochemical reactions and describe their kinetics.

11. Determine how a graph of the kinetics given by equation (24) compares with that for equation (18).

12. Suppose A and B are monomers that undergo dimerization in a rapidly equilibrating reaction:

$$A + B \underset{k_{-1}}{\overset{k_1}{\rightleftharpoons}} [A{-}B]$$

Show that the concentration of dimers is proportional to the product of the monomer concentrations.

13. Discuss what gradual changes in the rate constants appearing in equations (29a,b) would lead to the following developmental process:

(a) A cell that initially produces only product x will eventually produce only y.

(b) A cell that initially produces some fixed ratio of product x to product y eventually produces either x or y but not both.

14. **(a)** Suggest why it is reasonable to assume equations of the form (30) in Section 7.7.

(b) Show that to satisfy condition 2 we must assume that $f(0, 0)$ and $g(0, 0)$ are both positive.

(c) Show that the determinant of the Jacobian of equations (30a,b) at the steady state $(\bar{x}, \bar{y})$ is given by equation (32).

(d) Suggest other reaction mechanisms that would give dynamic behavior like that of the biochemical switch discussed in Section 7.6. Interpret your model(s) biochemically.

***15.** *Stability in an activator-inhibitor system.* From the information given in Section 7.8 can one deduce the directions of arrows on the nullclines shown in Figure (10a,b)? Is the result unique, or are there several possibilities?

16. *Stability in a positive-feedback system*

(a) With the definitions given in equation (44), use implicit differentiation along the nullclines to show that

$$a = -bs_1 \qquad \text{and} \qquad c = -ds_2.$$

(b) What is the Jacobian matrix in terms of b, d, s_1, and s_2?

(c) Use stability conditions to verify equation (45).

17. The following chemical reaction mechanism was studied by Lotka in 1920 and later in 1956:

$$\mathrm{A} + \mathrm{X} \xrightarrow{k_1} 2\mathrm{X}$$
$$\mathrm{X} + \mathrm{Y} \xrightarrow{k_2} 2\mathrm{Y}$$
$$\mathrm{Y} \xrightarrow{k_3} \mathrm{B}$$

Assume that A and B are kept at a constant concentration.

(a) Write a set of equations for the concentrations of X and Y using the law of mass action. Suggest a dimensionless form of the equations.

(b) Show that there are two steady states, and use the methods of Chapter 5 to demonstrate that Lotka's system has oscillatory solutions. Compare with the Lotka-Volterra predator-prey system.

18. A system of three chemical species has a steady state $\bar{x}, \bar{y}, \bar{z}$. The Jacobian of the system at steady state is

$$\mathbf{J} = \begin{pmatrix} a_{11} & a_{12} & a_{13} \\ a_{21} & 0 & 0 \\ a_{31} & 0 & a_{33} \end{pmatrix}.$$

Magnitudes of a_{ij} are not known. It is known that a_{11}, a_{33}, a_{21}, and a_{13} are negative, while a_{12}, and a_{31} are positive. Is the steady state stable or unstable?

19. *The glycolytic oscillator.* A biochemical reaction that is ubiquitous in metabolic systems contains the following sequence of steps:

$$\text{glucose} \rightarrow \text{GGP} \rightarrow \text{FGP} \rightarrow \xrightarrow[\text{ATP} \curvearrowright \text{ADP}]{\text{PFK}} \text{FDP} \rightarrow \text{products},$$

where

GGP = glucose-6-phosphate,
FGP = fructose-6-phosphate,
FDP = fructose-1,6-diphosphate,
ATP = adenosine triphosphate,
ADP = adenosine diphosphate,
PFK = phosphofructokinase.

An assumption generally made is that the enzyme phosphofructokinase has two states, one of which has a higher activity. ADP stimulates this allosteric regulatory enzyme and produces the more active form. Thus a product of the reaction step mediated by PFK *enhances* the rate of reaction. A schematic version of the kinetics is

$$\text{substrates} \rightarrow \text{FGP} \dashrightarrow \overset{+}{\frown} \text{ADP} \rightarrow \text{products}$$

Equations for this system, where x stands for FGP and y for ADP, are as follows:

$$\frac{dx}{dt} = \delta - kx - xy^2,$$

$$\frac{dy}{dt} = kx + xy^2 - y.$$

These equations, derived in many sources (see references), are known to have stable oscillations as well as other interesting properties.

(a) Show that the steady state of these equations is

$$(\bar{x}, \bar{y}) = \left(\frac{\delta}{(k + \delta^2)}, \delta\right).$$

(b) Find the Jacobian of the glycolytic oscillator equations at the above steady state.

(c) Find conditions on the parameters so that the system is a positive-feedback system as described in Section 7.8.

20. *The Brusselator.* This hypothetical system was first proposed by a group working in Brussels [see Prigogine and Lefever (1968)] in connection with spatially nonuniform chemical patterns. Because certain steps involve trimolecular reactions, it is not a model of any real chemical system but rather a prototype that has been studied extensively. The reaction steps are

$$\begin{aligned} A &\rightarrow X, \\ B + X &\rightarrow Y + D, \\ 2X + Y &\rightarrow 3X, \\ X &\rightarrow E. \end{aligned}$$

It is assumed that concentrations of A, B, D, and E are kept artificially constant so that only X and Y vary with time.

(a) Show that if all rate constants are chosen appropriately, the equations describing a Brusselator are:

$$\frac{dx}{dt} = A - (B + 1)x + x^2y,$$

$$\frac{dy}{dt} = Bx - x^2y.$$

(b) Find the steady state.

(c) Calculate the Jacobian and show that if $B > 1$, the Brusselator is a positive-feedback system as described in Section 7.8.

21. *An activator-inhibitor system (Gierer-Meinhardt).* A system also studied in connection with spatial patterns (see Chapter 10) consists of two substances. The activator enhances its own synthesis as well as that of the inhibitor. The inhibitor causes the formation of both substances to decline. Several versions have been studied, among them is the following system:

$$\frac{dx}{dt} = \rho + \frac{x^2}{y} - x,$$

$$\frac{dy}{dt} = x^2 - \gamma y.$$

(a) Find the steady state for this system.

(b) Show that if the input of activator is sufficiently small compared to the decay rate of inhibitor, the system is an activator-inhibitor system as described in Section 7.8.

(c) In a modified version of these equations the term x^2/y is replaced by $x^2/y(1 + kx^2)$. Suggest what sort of chemical interactions may be occurring in this and in the original system.

(d) Why is it not possible to solve for the steady state of the modified Gierer-Meinhardt equations?

(e) Find the Jacobian of the modified system. When does the Jacobian have the sign pattern of an activator-inhibitor system?

22. Consider the following hypothetical chemical system:

$$M_0(\text{a catalyst}) + X \underset{k_{-1}}{\overset{k_1}{\rightleftharpoons}} M_1 \qquad \text{(active complex)},$$

$$M_1 + X \underset{k_{-2}}{\overset{k_2}{\rightleftharpoons}} M_2 \qquad \text{(inactive complex)},$$

$$M_1 + Y \xrightarrow{k_3} P + Q + M_0 \qquad \text{(products plus catalyst)}.$$

This system is called a *substrate-inhibited reaction* since the chemical X can deactivate the complex M_1 which is required in forming the products.

(a) Write equations for the chemical components.

(b) Assume $M_0 + M_1 + M_2 = C$ (where m_0, m_1, and m_2 are concentrations of M_0, M_1, and M_2, and C is a constant). Make a quasi-steady-state assumption for M_0, M_1, and M_2 and show that

$$m_0 = m_1 \frac{k_{-1} + k_3 y}{k_1 x}, \qquad m_2 = k_2 x \frac{m_1}{k_{-2}}.$$

(c) Use these results to show that

$$m_1 = \frac{C k_1 x}{k_{-1} + k_3 y + k_1 x[1 + (k_2/k_{-2})x]}$$

(d) Now show that x will satisfy an equation whose dimensionless form is

$$\frac{dx}{dt} = \gamma - x - \frac{\beta x y}{1 + x + y + (\alpha/\delta)x^2}$$

Identify the various combinations of parameters. When can the term y in the denominator be neglected?

23. The following model was proposed by Othmer and Aldridge (1978): A cell can produce two chemical species x and y from a substrate according to the reaction

$$\text{substrate} \rightarrow x \rightarrow y \rightarrow \text{products}.$$

Species x can diffuse across the cell membrane at a rate that depends linearly on its concentration gradient. The ratio of the volume of cells to the volume of external medium is given by a parameter ϵ; x and y are intracellular concentrations of X and Y and x^0 is the extracellular concentration of X. The equations they studied were

$$\frac{dx}{dt} = \delta - F(x, y) + P(x^0 - x)m,$$

$$\frac{dy}{dt} = \alpha[F(x, y) - G(y)],$$

$$\frac{dx^0}{dt} = \epsilon P(x - x^0).$$

(a) Explain the equations. Determine the values of $\bar{x}$, $F(\bar{x}, \bar{y})$ and $G(\bar{y})$ at the steady state $(\bar{x}, \bar{y}, \bar{x}^0)$.

(b) The matrix of linearization of these equations about this steady state is

$$\mathbf{J} = \begin{pmatrix} k_{11} - P & k_{12} & P \\ k_{21} & k_{22} & 0 \\ \epsilon P & 0 & -\epsilon P \end{pmatrix}.$$

What are the constants k_{ij}?

(c) For the characteristic equation

$$\lambda^3 + a_1\lambda^2 + a_2\lambda + a_3 = 0,$$

find a_1, a_2, and a_3 in terms of k_{ij} and in terms of partial derivatives of F and G.

24. A model for control of synthesis of a gene product that activates mitosis was suggested by Tyson and Sachsenmaier (1979), based on repression and derepression (the reversal of repression) of a genetic *operon* (a sequence of genes that are controlled as a unit by a single gene called the *operator*.) They assumed that the gene consists of two portions, one replicating earlier than the second, with the following control system.

Protein R (coded by 1) binds to the operator region O of gene 2, repressing transcription of genes G_P and G_A. Protein P (product of G_P) inactivates the repressor and thus has a positive influence on its own synthesis, as well as on synthesis of A (see figure).

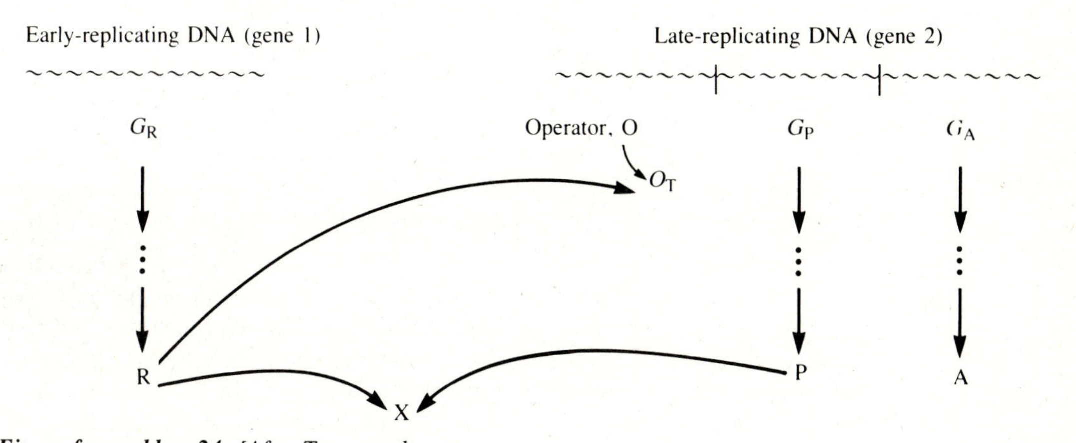

Figure for problem 24. *[After Tyson and Sachsenmaier (1979).]*

Assume that R, P, and A have removal rates ℓ_1, ℓ_2, and ℓ_3. Let G_R, G_P, and G_A be the *number* of genes coding for R, P, and A, at rates K_1, K_2, and K_3 (when actively transcribing). Let f be the fraction of operons of the late-replicating DNA that are active at a given time.

(a) Give equations governing the concentrations of R, P, and A in terms of R_is, G_is, and f.

(b) Following are the operator and repressor binding reactions:

$$P + R \underset{k_{-1}}{\overset{k_1}{\rightleftharpoons}} X$$

$$R + O \underset{k_{-2}}{\overset{k_2}{\rightleftharpoons}} O_R$$

where X is an inducer-repressor complex and O_R is a repressed operator. Write down equations for P, R, and O based on these kinetics.

(c) Now assume that these reactions are always in equilibrium and that the total number of operator molecules is O_T, a constant. Let

$$K_2 = \frac{k_2}{k_{-2}} \quad \text{and} \quad K_1 = \frac{k_1}{k_{-1}}.$$

Find an expression relating the fraction f of repressed operators ($f = O/O_T$) (1) to these rate constants and to the concentration of R.

(d) For the specific situation in which more than one repressor molecule can bind the operator,

$$n\text{R} + \text{O} \rightleftharpoons \text{OR}_n,$$

(again where $K_2 = k_2/k_{-2}$ is the equilibrium constant), Tyson and Sachsenmaier showed that

$$f = (\epsilon + x)^n[1 + (\epsilon + x)^n]^{-1}, \qquad \text{for} \qquad x = \frac{K_1 P}{K_2 X}$$

$$\epsilon = (K_2 R_T)^{-1}.$$

provided ($O_T \ll X$).

Sketch f as a function of x for $n = 2$.

25. *Positive feedback to one gene*. The following model and analysis appears in Griffith (1971, pp 118 – 122). Parts d–f of this problem depend on the solution to problem 22 in Chapter 5. Consider a gene that is directly induced by m copies of the protein E for which it codes. Suppose

$$\text{G} + m\text{E} \rightarrow \text{X}$$

where G is the gene and X is a complex composed of m molecules of E and one molecule of G. Let M be the concentration of *messenger RNA* (mRNA) that conveys the code for synthesis of the protein to the *ribosomes* (where the protein is assembled from amino acids).

(a) Griffith assumes that the fraction of time p for which the gene G is active given by

$$p = \frac{KE^m}{1 + KE^m},$$

where E is the concentration of E and m the number of E molecules that participate in forming a complex X. Explain this assumption.

(b) Explain the following equations:

$$\dot{M} = \frac{aKE^m}{1 + KE^m} - bM$$

$$\dot{E} = cM - dE.$$

(c) What are the meanings of the constants K, b, c, and d?

(d) In problem 22 of Chapter 5 it was shown that the behavior of this system depends on m and on a dimensionless multiple where

$$\alpha = b\tau, \qquad \beta = d\tau, \qquad \tau = \frac{K^{-m}}{c}.$$

Interpret the meanings of each of these quantities.

(e) Use the results of phase-plane analysis in each of the cases given in problem 22 of Chapter 5 to draw biological conclusions about this system.

(f) Suggest situations in which the dynamic behavior may be of biological relevance.

REFERENCES

Enzyme and Biochemical Kinetics (General)

Boyer, P. D. (1970). *Kinetics and Mechanisms*. (*The Enzymes*, vol. 2.) Academic Press, New York.

Dixon, M., and Webb, E. C. (1979). *Enzymes*. 3d ed. Academic Press, New York.

Reiner, J. M. (1969). *Behavior of Enzyme Systems*. 2d ed. Van Nostrand, New York.

Rubinow, S. I. (1975). *Introduction to Mathematical Biology*. Wiley, New York.

Savageau, M. A. (1976). *Biochemical Systems Analysis: A Study of Function and Design in Molecular Biology*. Addison-Wesley, London.

Volkenshtein, M. V. (1969). *Enzyme Physics*. Plenum Press, New York.

Molecular Biology and Biochemistry (Nonmathematical References)

Lehninger, A. L. (1982). *Principles of Biochemistry*. Worth Publishing, New York.

Price, F. N. (1979). *Basic Molecular Biology*. Wiley, New York.

Yudkin, M., and Offord, R. (1980). *A Guidebook to Biochemistry*. 4th ed. Cambridge University Press, Cambridge.

Genetic Control

Edelstein, L. (1982). A molecular switching mechanism in differentiation: Induction of polarity in cambial cells. *Differentiation*, *23*, 1–9.

Griffith, J. S. (1971). *Mathematical Neurobiology*. Academic Press, New York.

Lewis, J.; Slack, J. M. W.; and Wolpert, L. (1977). Thresholds in development. *J. Theor. Biol.*, *65*, 579–590.

Rosen, R. (1970). *Dynamical System Theory in Biology*. Wiley-Interscience, New York.

Rosen, R. (1972). Mechanics of Epigenetic Control. In R. Rosen, ed. *Foundations of Mathematical Biology, II*. Academic Press, New York, pp. 79–140.

Thornley, J. H. M. (1976). *Mathematical Models in Plant Physiology*. Academic Press, New York.

Network Analysis

Feinberg, M. (1977). Mathematical aspects of mass action kinetics. In L. Lapidus and N. Amundson, eds., *Chemical Reactor Theory: A Review*. Prentice-Hall, Englewood Cliffs, N.J., pp. 59–130.

Feinberg, M. (1980). Chemical oscillations, multiple equilibria, and reaction network structure. In W. E. Stewart, W. H. Ray, and C. C. Conley, eds., *Dynamics and Modelling of Reactive Systems*. Academic Press, New York.

Feinberg, M. (in press). Chemical reaction network structure and the stability of complex isothermal reactors. I. The deficiency zero and deficiency one theorems; II. Multiple steady states for networks of deficiency one. *Chemical Engineering Science*.

Other References:

Banks, H. T., and Mahaffy, J. M. (1979). *Mathematical Models for Protein Biosynthesis*. LCDS Report. Brown University, Providence, Rhode Island.

Banks, H. T.; Miech, R. P.; and Zinberg, D. S. (1975). Nonlinear systems in models for enzyme cascades. In A. Ruberti and R. Mohler, eds. *Variable Structure Systems with Applications to Economy and Biology*. (*Lecture Notes in Economy and Mathematical Systems*, vol. 3.) Springer-Verlag, New York.

Kadas, Z. M. (1982). A piecewise linear activator-inhibitor model. (preprint of unpublished paper).

Lin, C. C., and Segel, L. A. (1974). *Mathematics Applied to Deterministic Problems in the Natural Sciences*. Macmillan, New York.

Murray, J. D. (1977). *Lectures on Nonlinear Differential Equation Models in Biology*. Clarendon Press, Oxford.

Othmer, H., and Aldridge, J. (1978). The effects of cell density and metabolite flux on cellular dynamics. *J. Math. Biol.*, *5*, 169–200.

Prigogine, I. and Lefever, R. (1968). Symmetry-breaking instabilities in dissipative systems. II *J. Chem. Phys.*, *48*, 1695–1700.

Prigogine, I., and Nicolis, G. (1967). On symmetry-breaking instabilities in dissipative systems. *J. Chem. Phys.*, *46*, 3542–3550.

Rapp, P. E. (1979): Bifurcation theory, control theory, and metabolic regulation. in D. A. Linkens, ed. *Biological Systems, Modelling and Control*. Peregrinus, New York.

Segel, L. A., ed. (1980). *Mathematical Models in Molecular and Cellular Biology*. Cambridge University Press, Cambridge.

Tyson, J., and Sachsenmaier, W. (1979). Derepression as a model for control of the DNA-division cycle in eukaryotes. *J. Theor. Biol.*, *79*, 275–280.

8 Limit Cycles, Oscillations, and Excitable Systems

One ring to rule them all, one ring to find them,
One ring to bring them all, and in the darkness bind them
J.R.R. Tolkien (1954) *The Lord of the Rings*, Part 1 Ballantine Books, NY (1965)

Periodicity is an inherent phenomenon in living things. From the cell cycle, which governs the rate and timing of *mitosis* (cell division), to the *diurnal* (circadian) cycle that results in sleep-wake patterns, to the ebb and flow of populations in their natural environment—life proceeds in a rhythmic and periodic style. The stability of these periodic phenomena, the fact that they are not easily disrupted or changed by a noisy and random environment, leads us to believe that the pattern is a ubiquitous part of the process of growth, of biochemical and metabolic control systems, and of population fluctuations.

Oscillations are easily found in such physical examples as spring-mass systems and electrical circuits, which ideally are linear in behavior. Nonlinear equations such as the Lotka-Volterra predation model yield oscillations too. However, as previously suggested, the Lotka-Volterra equations are not sufficiently descriptive of the oscillations encountered in natural population cycles. For one thing, the Lotka-Volterra cycles are neutrally stable. This means that the amplitude of oscillations depends on the initial population level. (We shall see in Section 8.8 that all *conservative systems* share this property.) Another unrealistic feature is that the Lotka-Volterra model is structurally unstable: very slight modifications of the equations will disrupt the cycling behavior (see comments in Chapter 6).

For this combination of reasons, we are led to consider more suitable descriptions of the stable biological cycles. It transpires that the notion of a limit cycle answers this need. In phase-plane plots a *limit cycle* is any simple oriented closed curve trajectory that does not contain singular points (points we have been calling steady states and at which the phase flow is stagnant). The curve must be *closed* so that a point moving along the cycle will return to its starting position at fixed time intervals and thus execute periodic motion. It must be *simple* (cannot cross itself) by the uniqueness property of differential equations [see Figure 8.1(a,b) and problem 1].

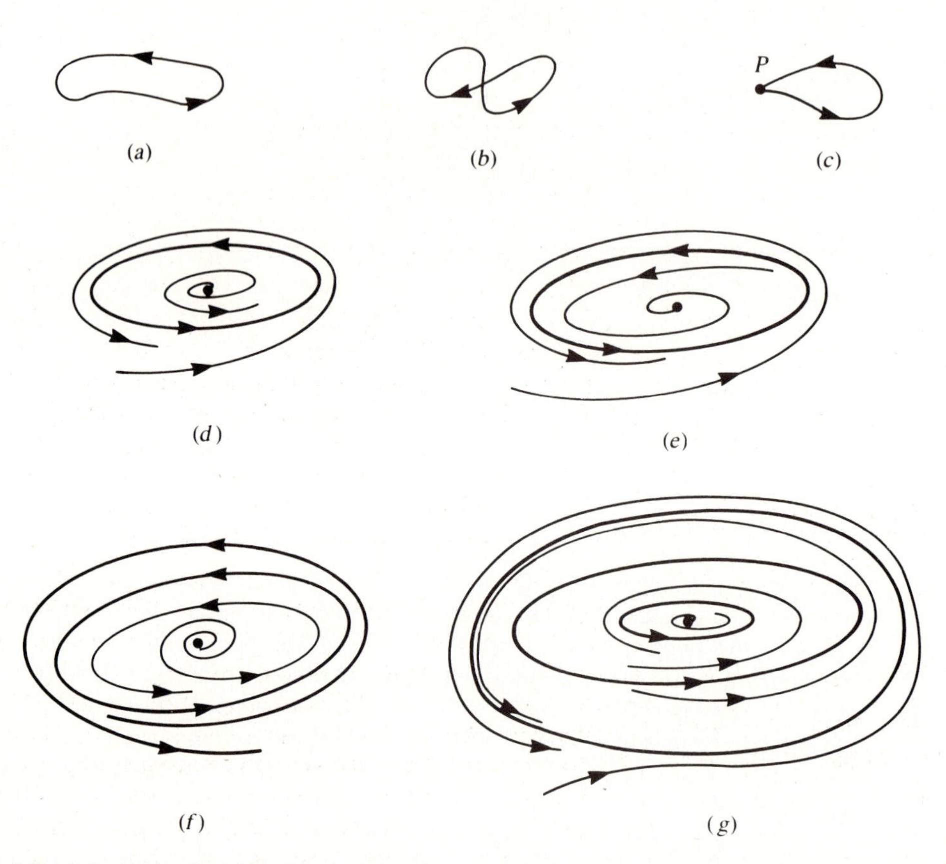

Figure 8.1 *(a) A simple closed, oriented curve. (b) This curve is not simple since it crosses itself. (c) If P is a steady state, this curve cannot be a limit cycle. Limit cycles come in several varieties. (d) This limit cycle is stable since all neighboring points are contained in trajectories that approach it as* $t \to +\infty$. *(e,f) These are unstable. (g) A multiplicity of limit cycles, some stable and some not, are shown.*

What distinguishes a limit cycle from the cycles that surround a neutral center (see Chapter 5) is the fact that it represents the limiting behavior of adjacent trajectories; points nearby will approach the limit cycle either for $t \to +\infty$ or for $t \to -\infty$. If the former case holds (for all adjacent trajectories) the limit cycle is *stable*. Otherwise it is *unstable*. Figure 8.1 provides several examples of stable and unstable limit cycles in a two-dimensional phase plane.

In this chapter we are concerned with identifying criteria that point to the existence (or nonexistence) of limit cycles. We shall address this issue primarily within the context of the system of the two equations

$$\frac{dx}{dt} = F(x, y), \tag{1a}$$

$$\frac{dy}{dt} = G(x, y). \tag{1b}$$

About F and G we shall assume that these are continuous functions with continuous partial derivatives with respect to x and y so that a unique solution to (1a,b) will exist for a given set of initial values (x_0, y_0).

Before launching into the mathematical techniques that are useful in determining whether limit cycles to a system such as equations (1a,b) exist, we discuss neural excitation by way of a motivating example. Some of the basic concepts and physiological detail are described in Section 8.1. A model due to Hodgkin and Huxley is then derived and partially analyzed in Section 8.2, to be followed later by a simpler set of equations due to Fitzhugh and Nagumo in Section 8.5. We find that both models have the potential for exhibiting sustained oscillations typical of a limit cycle solution, as well as *excitable behavior:* by this we mean that a stimulus larger than some threshold will provoke a very large response. (This type of behavior is depicted by a large excursion away from some steady state and then back to it as excitation subsides.) While phase-plane analysis (described in Chapter 5) suffices for understanding Sections 8.1 and 8.2, we find it useful to draw on new results in Section 8.5.

The Poincaré-Bendixson theory is a cornerstone on which much of the theory of limit cycles rests. Section 8.3 provides an informal introduction, and the results are then applied in Section 8.4 to a system of equations known as the van der Pol oscillator. This prototype serves to illustrate how a nullcline configuration in which one or both nullclines is S-shaped tends to produce oscillatory or excitable behavior. Many examples drawn from the literature are based on similar principles. Among these is Fitzhugh's model for neural excitation described in Section 8.5.

Another mathematical method commonly encountered in the quest for limit-cycle solutions is the Hopf bifurcation theorem (Section 8.6). Here we are more specific about the spectrum of possible effects that are encountered close to a steady state of a nonlinear system at which linear stability calculations predict a transition from a stable focus through a neutral center to an unstable focus as some parameter is varied. (The comments at the end of Section 5.9 were deliberately vague in anticipation of the upcoming discussion.)

In Sections 8.7 and 8.8 we apply the new mathematical techniques to problems stemming from population fluctuations and oscillations in chemical systems. These sections are extensions of material covered in Chapters 6 and 7 respectively.

For a shorter course, any one of the following sequences is suitable: (1) Sections 8.1 and 8.2 (based only on previous material); (2) Sections 8.1 to 8.5

(physiological emphasis); (3) Sections 8.3 and 8.7 (population biology); (4) Sections 8.3, 8.4, 8.6, and 8.8 (mathematical techniques with examples drawn from molecular models).

8.1 NERVE CONDUCTION, THE ACTION POTENTIAL, AND THE HODGKIN-HUXLEY EQUATIONS

One of the leading frontiers of biophysics is the study of neurophysiology, which only several decades ago spawned an understanding of the basic processes underlying the unique electrochemical communication system that constitutes our nervous system. Our brains and every other subsystem in the nervous system are composed of cells called *neurons*. While these vary greatly in size, shape, and properties, such cells commonly share certain typical features (see Figure 8.2). Anatomically, the cell body (*soma*) is the site at which the nucleus and major subcellular structures are located and is the central point from which synthesis and metabolism are coordinated.

A more prominent feature is a long tube-like structure called the *axon* whose length can exceed 1 meter (that is, $\sim 10^5$ times the dimension of the cell body). It is known that the propagation of a nerve signal is electrical in nature; after being initiated at a site called the *axon hillock* (see Figure 8.2) propagates down the length of the axon to terminal branches, which form loose connections (*synapses*) with neighboring neurons. A propagated signal is called an *action potential* (see Figure 8.3).

A neuron has a collection of *dendrites* (branched, "root-like" appendages), which receive incoming signals by way of the synapses and convey them to the soma.

How the detailed electrochemical mechanism operates is a fascinating story that, broadly speaking, is now well understood. It is known that neuronal signals travel *along the cell membrane* of the axon in the form of a local voltage difference across the membrane. A word of explanation is necessary. In the *resting state* the cytoplasm (cellular fluid) inside the axon contains an ionic composition that makes the cell interior slightly negative in potential (-50 mV difference) with respect to the outside (see Figure 8.4). Such a potential difference is maintained at a metabolic expense to the cell by *active pumps* located on the membrane. These continually transport sodium ions (Na^+) to the outside of the cell and convey potassium ions (K^+) inwards so that concentration gradients in both species are maintained. The differences in these and other ionic concentrations across the membrane result in the net electric potential that is maintained across the membrane of the living cell. In this section we take the convention that the voltage v is the potential difference (inside minus outside) for the membrane.

Thinking of the axon as a long electrical cable is a vivid but somewhat erroneous conception of its electrical properties. First, while a current is implicated, it is predominantly made up of ionic flow (not electrons), and its direction is not longitudinal but transverse (into the cell) as shown in Figure 8.5. Second, while a passive cable has fixed resistance per unit length, an axon has an *excitable membrane* whose resistance to the penetration of ions changes as the potential difference v is raised.

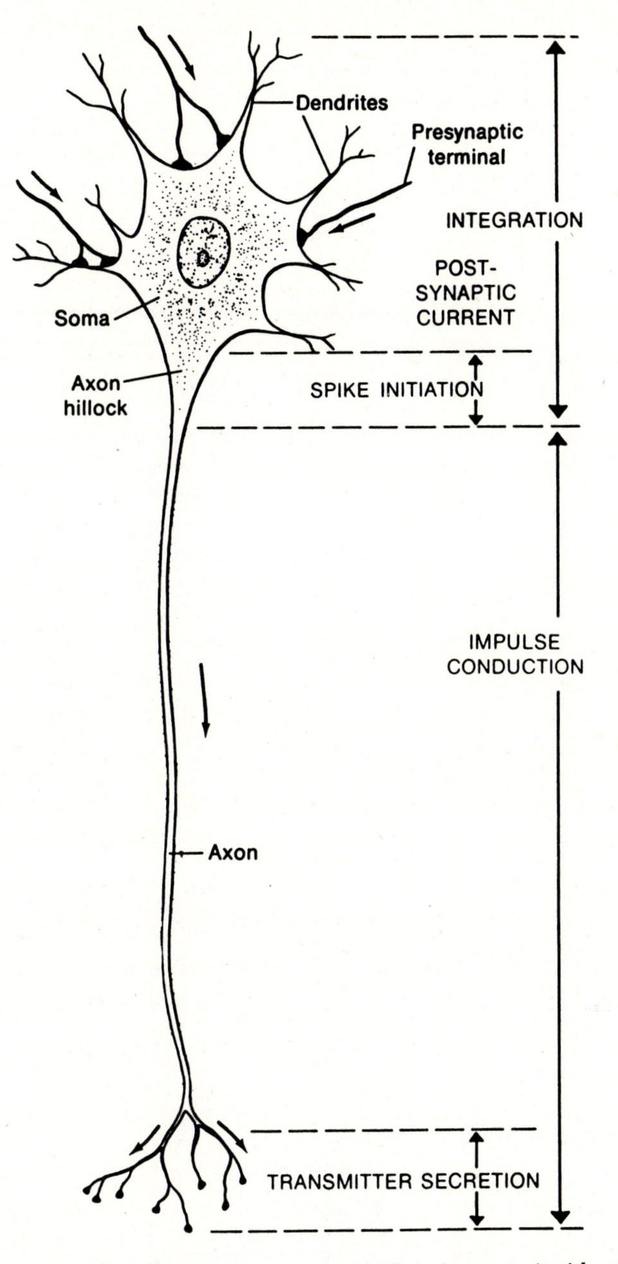

Figure 8.2 *Schematic representation of a neuron showing the cell body (soma) which receives stimuli via the dendrites, the axon along which impulses are conducted, and the terminal branches that form connections (synapses) with other neurons. [From Eckert, R., and Randall, D.* Animal Physiology, *2d edition. W. H. Freeman and Company. Copyright © 1983, p. 179.]*

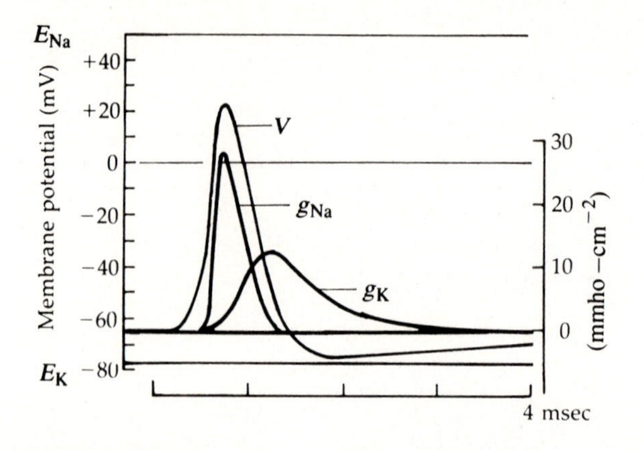

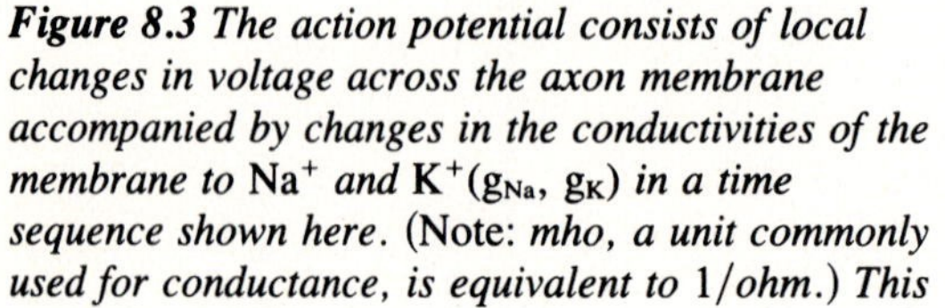

***Figure 8.3** The action potential consists of local changes in voltage across the axon membrane accompanied by changes in the conductivities of the membrane to* Na^+ *and* K^+(g_{Na}, g_K) *in a time sequence shown here.* (Note: *mho, a unit commonly used for conductance, is equivalent to 1/ohm.) This signal is generally propagated along the neuronal axon from soma to terminal branches. [After Hodgkin and Huxley (1952), from Kuffler, Nicholls, and Martin (1984) p. 151, fig, 13A,* From Neuron to Brain, *2nd edition, by permission of Sinauer Associates Inc.]*

The flow of charged ions across a cell membrane is restricted to specific molecular sites called *pores,* which are sprinkled liberally along the membrane surface. It is now known that many different kinds of pores (each specific to a given ion) are present and that these open and close in response to local conditions including the electrical potential across the membrane. This can be broadly understood in terms of changes in the conformation of the proteins making up these pores, although the biophysical details are not entirely known.

To understand the process by which an action potential signal is propagated, we must look closely at events happening in the immediate vicinity of the membrane. Starting the process requires a *threshold voltage:* the potential difference must be raised to about −30 to −20 mV at some site on the membrane. Experimentally

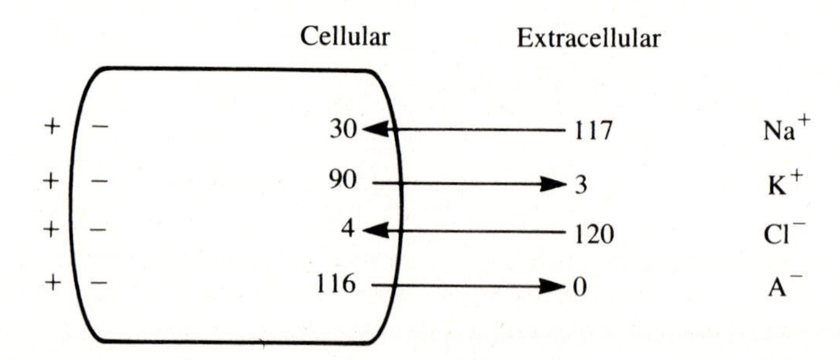

***Figure 8.4** In the resting state, cells have an ionic composition (given here in millimolar units) that differ from that of their environment. Active transport maintains a lower sodium* (Na^+) *and a higher potassium* (K^+) *concentration inside the cell.* Cl^- *and* A^- *represent respectively chlorine ions and other ionic species such as proteins.*

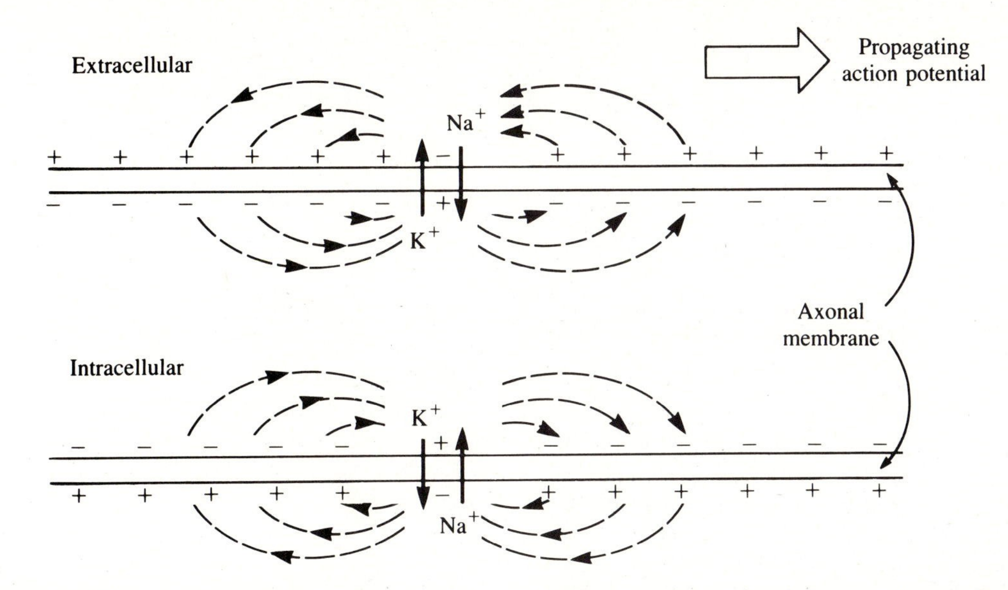

Figure 8.5 *When the axon of a neuron receives a sufficiently large stimulus at some point along its length, the conductivities of the membrane for* Na^+ *and* K^+ *changes. This permits ions to cross the membrane, creating local currents. Since adjacent portions of the membrane are thereby stimulated, the wave of activity known as the action potential can be propagated.*

this can be done by a stimulating electrode that pierces a single neuron. Biologically this happens at the axon hillock in response to an integrated appraisal of excitatory inputs impinging on the soma. As a result of reaching this threshold voltage, the following sequence of events occurs (see Figure 8.6):

1. Sodium channels open, letting a flood of Na^+ ions enter the cell interior. This causes the membrane potential to *depolarize* further; that is, the inside becomes positive with respect to the outside, the reverse of resting-state polarization.
2. After a slight delay, the potassium channels open, letting K^+ *leave* the cell. This restores the original polarization of the membrane, and further causes an overshoot of the negative rest potential.
3. The sodium channels then close in response to a decrease in the potential difference.
4. Adjacent to a site that has experienced these events the potential difference exceeds the threshold level necessary to set in motion step 1. The process repeats, leading to spatial conduction of the spike-like signal. The action potential can thus be transported down the length of the axon without attenuation or change in shape. Mathematically, this makes it a *traveling wave*.

The finer details of this somewhat impressionistic description were uncovered in 1952 in a series of brilliant but painstaking experiments due to Hodgkin, Huxley, and Katz on the giant squid axon, a cell whose axonal diameter is large enough to

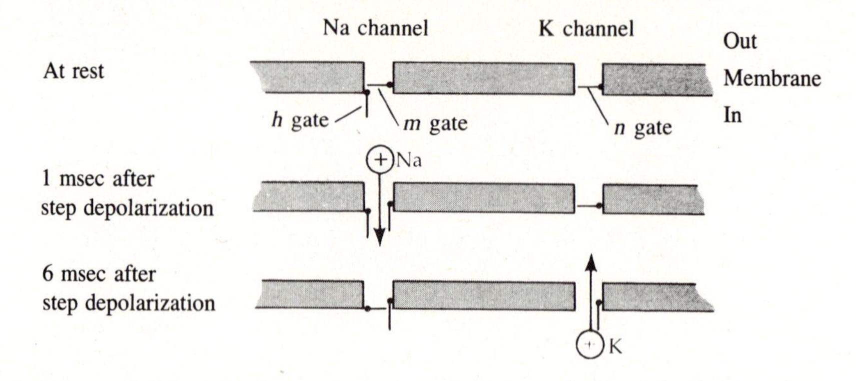

Figure 8.6 *Schematic time sequence depicting the membrane of the axon. Shown are separate pores, governed by gates, for the ions* Na^+ *and* K^+. *At rest the* m *and* n *gates are closed. When a threshold voltage is applied, the* m *gates open rapidly, followed by changes in the other gates. [From Kuffler, Nicholls, and Martin (1984), p. 149, fig. 12.* From Neuron to Brain, *2nd edition, by permission of Sinauer Associates Inc.]*

permit intracellular recording of the voltage by microelectrodes. One technique particularly useful in elucidating the time sequence of ionic conductivities is the *voltage-clamp* experiments. In these, an axon is excised from its cell and its contents are emptied (in a manner akin to squeezing a tube of toothpaste). A thin wire inserted into the hollow axon replaces its cytoplasm and permits an artificially constant voltage to be applied simultaneously along its length. Provided the preparation is kept physiologically active, one can observe a spatially constant but time-varying voltage across the membrane. This voltage has typical action-potential characteristics.

It is further possible to follow the time behavior of the ionic conductivities by a variety of techniques. These include *patch-clamp* experiments, in which single pores are isolated on bits of membrane by suction using fine micropipettes or by selective ionic blocking using agents, such as tetrodotoxin, that bind to ionic pores in a specific way. The detailed structure of some ionic pores is beginning to emerge by a combination of techniques including electron microscopy. (See Figure 8.7.)

With this physiological description we can now discuss the mathematical model that has played a significant role in the advances in neurophysiology. First, a brief review of terminology and properties of an electrical circuit is provided in the box.

The example given in the box, when somewhat modified, can be used to depict electrical properties of the axonal membrane. The idea underlying the approach on which the Hodgkin-Huxley model is based is to use an electric-circuit analog in which physical properties such as ionic conductivities are represented as circuit elements (resistors). The voltage across the membrane thus corresponds to voltage across a collection of resistors, each one depicting a set of ionic pores that selectively permit a limited current of ions. By previous discussions, the resistance (or equivalently, the *conductivities* of such pores) depends on voltage.

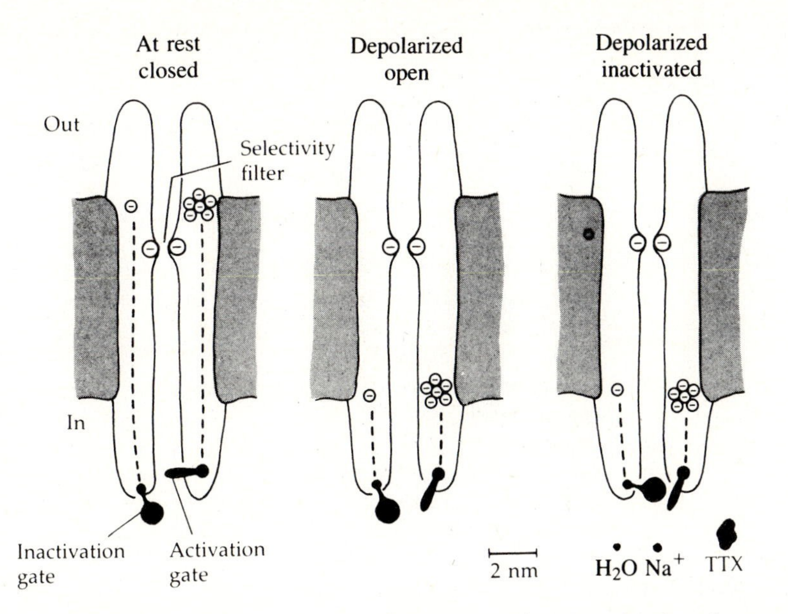

Figure 8.7 *Recent biochemical, electron microscope, and electrophysiological information leads to this schematic sketch of a voltage-sensitive sodium channel. Shown are the activation gates and a constriction that permits selectivity to* Na^+. *Also shown to scale are molecules of water, sodium, and the blocking agent tetrodotoxin (TTX). [From Kuffler, Nicholls, and Martin (1984), p. 156, fig. 15.* From Neuron to Brain, *2nd edition, by permission of Sinauer Associates Inc.]*

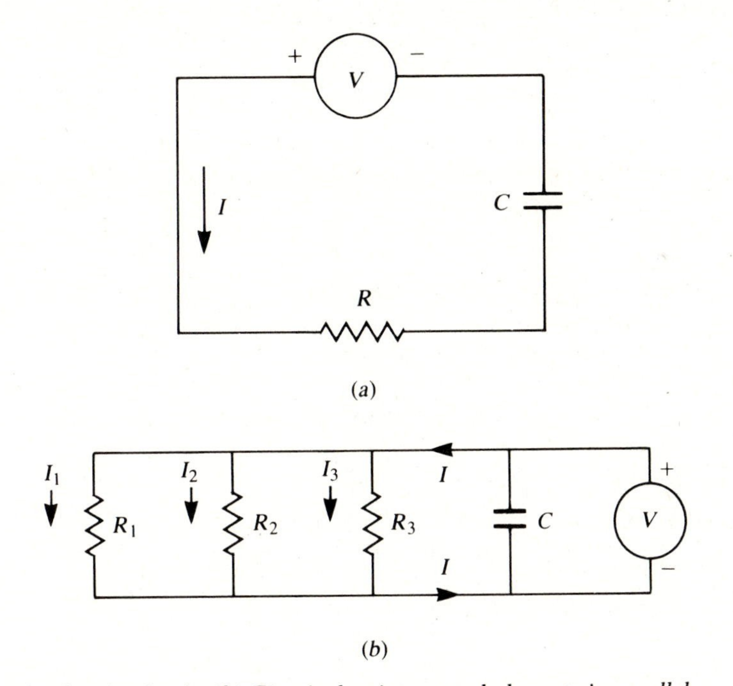

Figure 8.8 *(a) Simple electric circuit. (b) Circuit showing several elements in parallel.*

A Simple Electric Circuit

The following terms are applied in describing a circuit, such as the one shown in Figure 8.8(*a*).

$q(t)$ = the charge (net positive or negative charge carried by particles in the circuit at time t),
$I(t)$ = the current (rate of flow of charge in the circuit) = dq/dt,
$\mathrm{V}(t)$ = the voltage (electromotive force that causes motion of charge; also a measure of the difference in the electrical potential across a given element or set of elements),
R = resistance (property of a material that tends to impede the flow of charged particles),
g = conductance = $1/R$,
C = capacitance (a property of any element that tends to separate physically one group of charged particles from another; this causes a difference in electric potential across the element, called a capacitor).

The following physical relationships hold in a circuit:

1. *Ohm's law:* the voltage drop across a resistor is proportional to the current through the resistor; R or $1/g$ is the factor of proportionality:

$$\mathrm{V}_R(t) = I(t)R = \frac{I(t)}{g}. \tag{2}$$

2. *Faraday's law:* the voltage drop across a capacitor is proportional to the electric charge; $1/C$ is the factor of proportionality:

$$\mathrm{V}_C(t) = \frac{q(t)}{C}. \tag{3}$$

3. *Kirchhoff's law:* the voltage supplied is equal to the total voltage drops in the circuit. For example:

$$\mathrm{V}(t) = \mathrm{V}_R(t) + \mathrm{V}_C(t).$$

4. For several elements in parallel, the total current is equal to the sum of currents in each branch; the voltage across each branch is then the same. In the example shown in Figure 8.8(*b*) the current is

$$I(t) = I_1(t) + I_2(t) + I_3(t) = \frac{\mathrm{V}}{R_1} + \frac{\mathrm{V}}{R_2} + \frac{\mathrm{V}}{R_3}$$
$$= \mathrm{V}(g_1 + g_2 + g_3). \tag{4a}$$

Also,

$$\mathrm{V}(t) = q(t)C. \tag{4b}$$

Differentiating (4b) leads to

$$\frac{d\mathrm{V}}{dt} = \frac{1}{C}\frac{dq}{dt} = \frac{I(t)}{C}. \tag{4c}$$

Thus

$$\frac{dV}{dt} = \frac{V(t)}{C}(g_1 + g_2 + g_3). \tag{4d}$$

In the circuit analog of an axon shown in Figure 8.9, resistance to ionic flow across the membrane is depicted by the conductivities g_K, g_{Na}, and g_L. Resistance to ionic motion inside the axon in an axial direction is represented by longitudinal elements whose resistance per unit length is fixed and much higher than that of the medium surrounding the outer membrane. (The axon has a small radius, which implies greater resistance to flow.) The finite thickness of the membrane is associated with its property of *capacitance,* that is, a separation of charge. We must remember, in looking at this schematic representation, that the axon is cylindrical. Therefore the following modified definitions prove convenient:

$q(x, t)$ = charge density inside the axon at location x and time t (units of charge per unit length),
C = capacitance of the membrane per unit area,
a = radius of the axon,
$I_i(x, t)$ = net rate of exit of positive ions from the exterior to the interior of the axon per unit membrane area at (x, t),
$v(x, t)$ = departure from the resting voltage of the membrane at (x, t).

Then by previous remarks the following relationship is satisfied:

$$q(x, t) = 2\pi a C v(x, t). \tag{5}$$

We shall now assume that a voltage clamp is applied to the axon so that $q = q(t)$, $I_i = I_i(t)$, and $v = v(t)$ and thus all points on the inside of the axon are at the same voltage at any instant. This means that charge will not move longitudinally (there is no force leading to its motion). It can only change by currents that convey ions across the cell membrane. In this case the rate of change of internal charge can be written

$$\frac{dq}{dt} = -2\pi a I_i. \tag{6}$$

The current I_i can be further expressed as a sum of the three currents I_{Na}, I_K, and I_L (sodium, potassium, and all other ions) and related to the potential difference that causes these, as follows:

$$\frac{dq}{dt} = -2\pi a(I_{Na} + I_K + I_L), \tag{7}$$

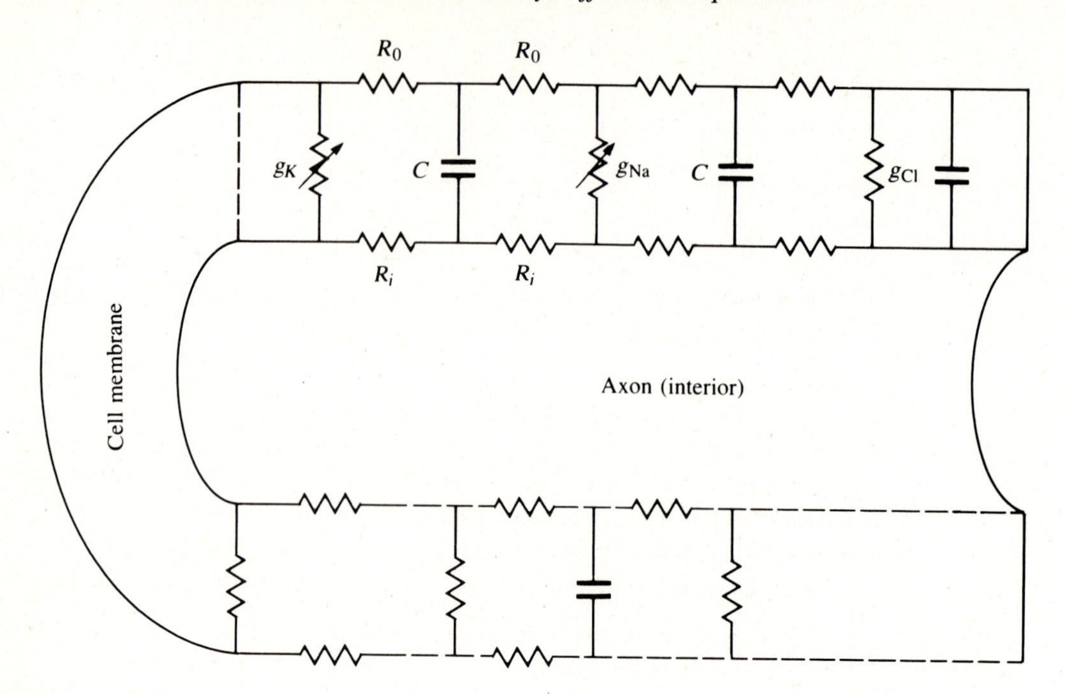

***Figure 8.9** A schematic version of the electric wiring diagram roughly equivalent to the axonal membrane.* g_K, g_{Na}, *and* g_{Cl} *are the voltage-dependent conductivites to* K^+, Na^+, *and* Cl^-; R_i *and* R_0 *represent the resistance of inside and outside environments;* C *depicts the membrane capacitance.* (Note: g_{Cl} *is assumed to be constant.)*

$$I_{Na} = g_{Na}(v - v_{Na}), \tag{8a}$$
$$I_K = g_K(v - v_K), \tag{8b}$$
$$I_L = g_L(v - v_L). \tag{8c}$$

Here v_{Na}, v_K, and v_L represent that part of the resting membrane potential that is due to the contributions of the ions Na^+, K^+, and L (all other mobile species). Furthermore, equation (7) in its entirety may be written in terms of voltage by using equation (5), with the result that

$$\frac{dv}{dt} = -\frac{1}{C}[g_{Na}(v)(v - v_{Na}) + g_K(v)(v - v_K) + g_L(v - v_L)]. \tag{9}$$

It is generally assumed that g_L is independent of v (is constant). At this point Hodgkin, Huxley, and Katz departed somewhat from a straightforward electrical analysis and went on to speculate on a possible mechanism governing the ionic conductivities g_{Na} and g_K. After numerous trial-and-error models, laboriously solved on mechanical calculators, they found it necessary to introduce three variables n, m, and h in the dynamics of the ionic pores. These hypothetical quantities could perhaps be interpreted as concentrations of proteins that must act in concert to open or close a pore. (See Figure 8.6.) However, the equations were chosen to fit the data, not from a more fundamental knowledge of molecular mechanisms.

They defined

$$g_{\mathrm{Na}} = \bar{g}_{\mathrm{Na}} m^3 h, \tag{10}$$

$$g_{\mathrm{K}} = \bar{g}_{\mathrm{K}} n^4, \tag{11}$$

where $\bar{g}$'s are constant conductivity parameters. They suggested that n, m, and h are voltage-sensitive gate proteins (see Figures 8.6 and 8.7), that obey differential equations in which voltage dependence is described:

$$\frac{dn}{dt} = \alpha_n(v)(1 - n) - \beta_n(v)n, \tag{12a}$$

$$\frac{dm}{dt} = \alpha_m(v)(1 - m) - \beta_m(v)m, \tag{12b}$$

$$\frac{dh}{dt} = \alpha_h(v)(1 - h) - \beta_h(v)h. \tag{12c}$$

In addition, the quantities α_n, α_m, α_h, β_n, β_m, and β_h are assumed to be voltage-dependent as follows:

$$\alpha_m(v) = 0.1(v + 25)(e^{(v+25)/10} - 1)^{-1}, \qquad \beta_m(v) = 4\, e^{v/18}, \tag{13a,b}$$

$$\alpha_h(v) = 0.07\, e^{v/20}, \qquad \beta_h(v) = (e^{(v+30)/10} + 1)^{-1}, \tag{13c,d}$$

$$\alpha_n(v) = 0.01(v + 10)(e^{(v+10)/10} - 1)^{-1}, \qquad \beta_n(v) = 0.125 e^{v/80}. \tag{13e,f}$$

The values of other constants appearing in the equations are $\bar{g}_{\mathrm{Na}} = 120$, $\bar{g}_{\mathrm{K}} = 36$, and $g_{\mathrm{L}} = 0.3$ mmho cm^{-2}; $v_{\mathrm{Na}} = -115$, $v_{\mathrm{K}} = 12$, and $v_{\mathrm{L}} = -10.5989$ mV.

With a physiological system as intricate as the neural axon, it is reasonable to expect rather complicated interactions between variables. In assessing the Hodgkin-Huxley model, we should keep in mind that all but one of its equations were tailored to fit experimental observations. Part of the surprisingly great success of the model lies in its ability to predict fairly accurately the results of many other observations *not* used in formulating the equations. A valid criticism of the model is that the internal variables m, n, and h do not clearly relate to underlying molecular mechanisms; these were, of course, unknown at the time).

The Hodgkin-Huxley equations consist of four coupled ODEs with highly nonlinear terms. For this reason they are quite difficult to understand in an analytic mathematical way. In the next section we explore this model in the elegant way suggested by Fitzhugh. After drawing certain conclusions about the behavior of these equations, we will go on to a much simpler model that captures essential features of the dynamics.

8.2 FITZHUGH'S ANALYSIS OF THE HODGKIN-HUXLEY EQUATIONS

In an elegant paper written in 1960 Fitzhugh set out "to expose to view part of the inner working mechanism of the Hodgkin-Huxley equations." In the year this paper appeared, the three most advanced techniques applied to analysis of nerve conduction models were (1) calculations on a desk calculator (Hodgkin and Huxley's method), (2) Runge-Kutta integration on the digital computers of the late 1950s, (3) use of the analog computer. Fitzhugh (1960) notes that on the digital computer the solution was very slow, "involving a week or more for a solution, including time for relaying instructions to the operating personnel . . . ." The analog computer used by Fitzhugh was an electronic device consisting of 40 operational amplifiers, six diode function generators, and five servo multipliers. Being much faster than the digital machines then in use, it permitted greater flexibility in experimenting with the equations, but special precautions were necessary to overcome inaccuracies that would have drastically changed the results for reasons that will become clear presently.

Fitzhugh was the first investigator to apply qualitative phase-plane methods to understanding the Hodgkin-Huxley equations. Since this is a system of four coupled equations (in the variables V, m, n, and h), the phase space resides in R^4. To make headway in gaining analytic insight, Fitzhugh first considered the variables that change most rapidly, viewing all others as slowly varying parameters of the system. In this way he derived a reduced two-dimensional system that could be viewed as a phase plane. We follow his method here.

The voltage convention adopted by Fitzhugh $V = v_{out} - v_{in}$ is unfortunately opposite to what subsequently became entrenched in the scientific literature. Thus the first and second quadrants of his phase plane appear reversed relative to the Hodgkin-Huxley model. To avoid possible confusion in conventions we use capital V when referring to Fitzhugh's analysis.

From the Hodgkin-Huxley equations Fitzhugh noticed that the variables V and m change more rapidly than h and n, at least during certain time intervals. By arbitrarily setting h and n to be constant we can isolate a set of two equations which describe a two-dimensional (V, m) phase plane. Plotting the functional relationships representing the nullcline equations ($\dot{m} = 0$ and $\dot{V} = 0$) we obtain Figure 8.10. On this figure the three intersections A, B, and C are steady states; A and C are stable nodes, and B is a saddle point.

The directions of flow in this plane prescribe the following: A small displacement from the "rest state" at A causes a return to this stable node, but a slightly larger deviation (for example, $m = 0$ and $V = -20$ mV) will lead to a large excursion whose final destination is the attracting point C. Note, however, that the nullclines intersect at very small angles at the points A and B. [See enlarged view, Figure 8.11(a).] Consequently there is great sensitivity to any parameter variations that tend to produce small displacements in these curves. This essentially is the effect of incorporating the modulating influence of the other variables. As some critical parameter changes, a displacement is produced, resulting in the following sequence of dynamic behavior:

1. The points A and B approach each other and coalesce.
2. These both vanish as the nullclines separate, leaving a single stable steady state at C. When this transition has occurred, any initial state in the Vm plane is drawn towards C.

Adding a third variable, we might next consider the (V, m, h) equations. A heuristic interpretation given by Fitzhugh is to imagine that as a point traces out a trajectory of Figure 8.10(a) the orbits themselves are "wiggling," so that the phase plane is changing with time. By considering the equation for h, Fitzhugh remarks

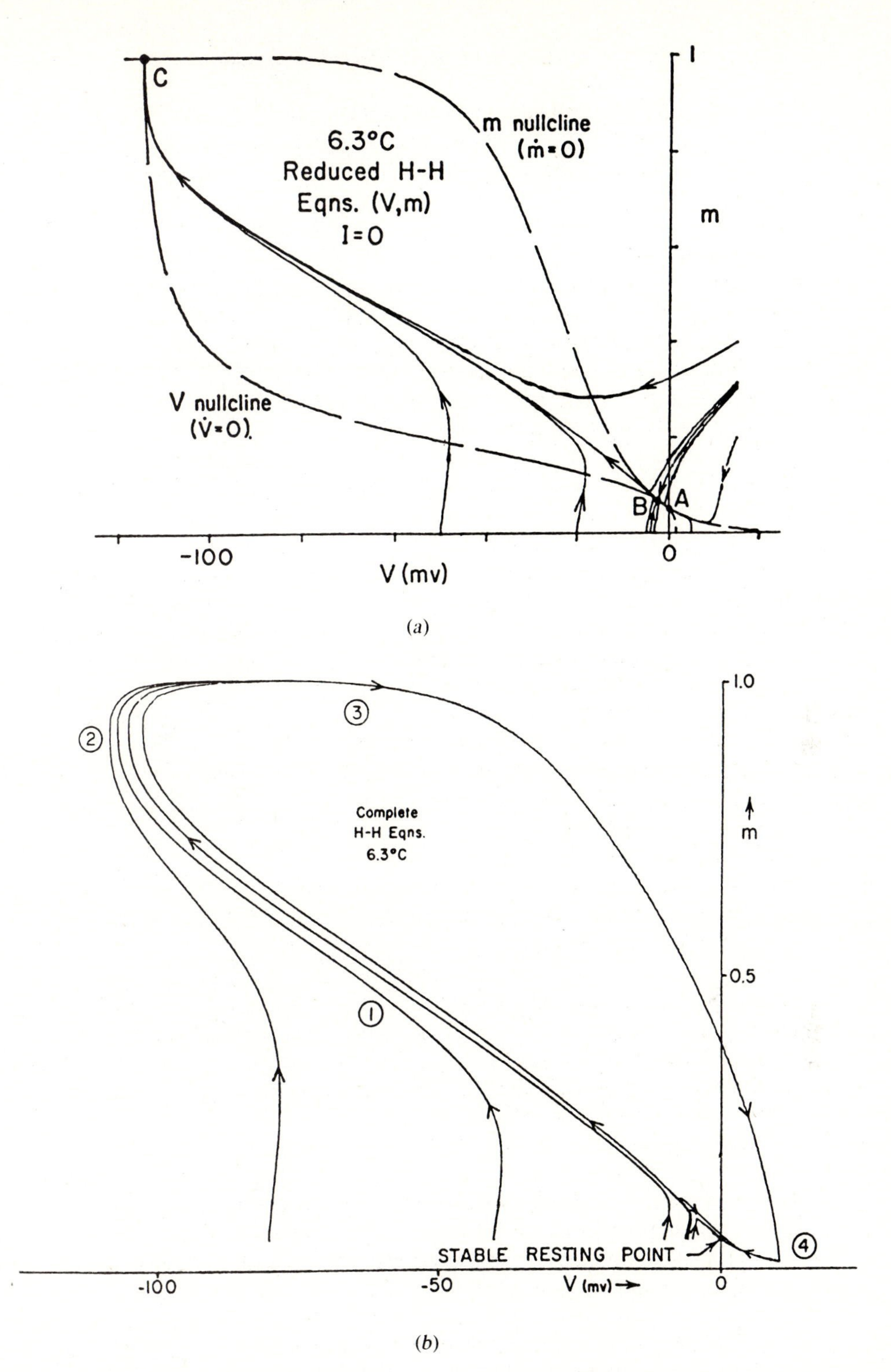

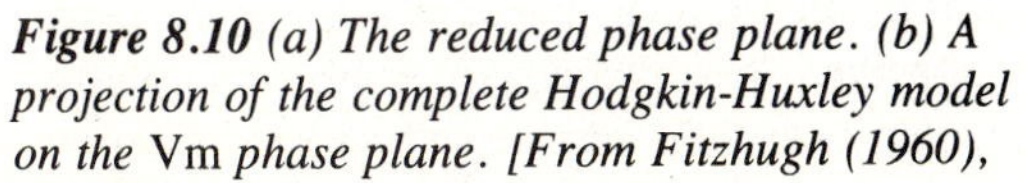

Figure 8.10 *(a) The reduced phase plane. (b) A projection of the complete Hodgkin-Huxley model on the* Vm *phase plane. [From Fitzhugh (1960), figs. 2 and 9. Reproduced from the* J. Gen. Physiol. *1960, vol. 43, pp. 867–896 by copyright permission of the Biophysical Society.]*

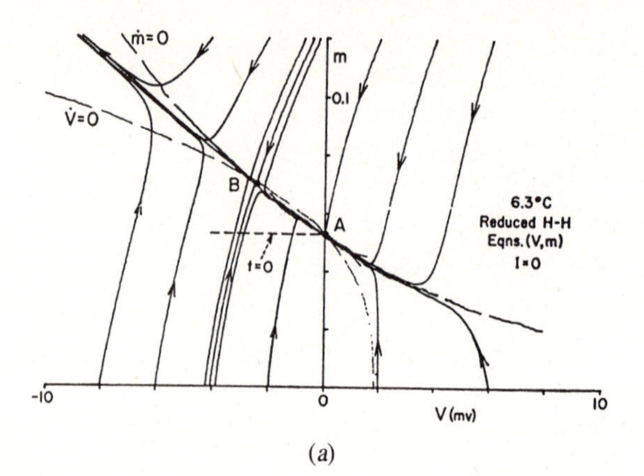

(a)

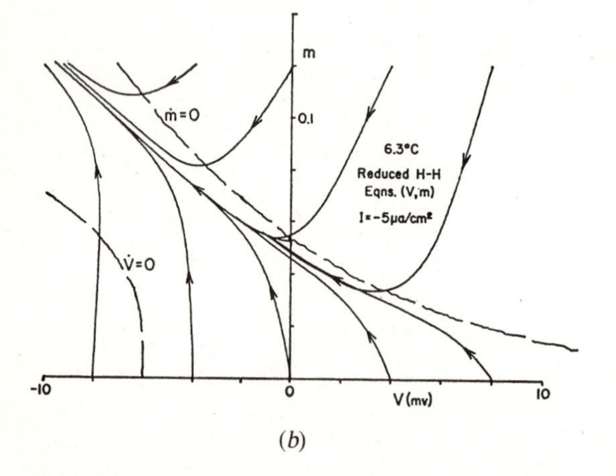

(b)

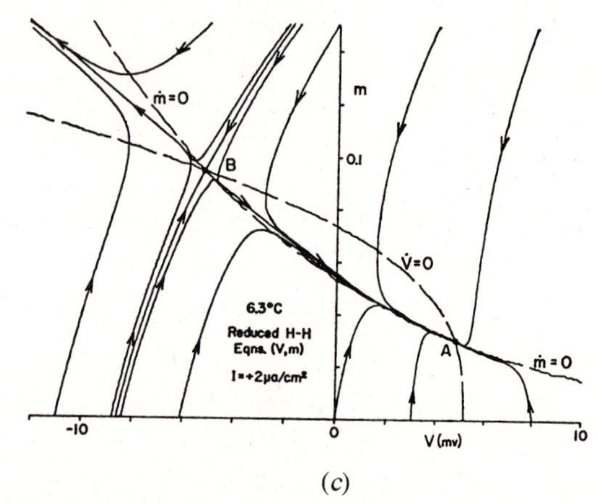

(c)

Figure 8.11 *Expanded view of the region of the* Vm *phase plane near the steady states* B *and* A. *(a) In the absence of stimulating current,* A *is a stable node and* B *a saddle point. (b) For super-threshold stimulation,* A *and* B *coalesce and disappear; only* C *will be a steady state. (c) An inhibitory stimulus can cause a greater separation of* A *and* B, *making it less likely that threshold will be exceeded. [From Fitzhugh (1960), figs. 3, 5, and 6. Reproduced from the* J. Gen. Physiol. *1960, vol. 43, pp. 867–896 by copyright permission of the* Biophysical Society.*]*

that as V becomes negative, h decreases, causing an upwards movement in the V nullcline. For small displacements from rest state A this means that B moves to the left, escaping from the moving phase point, which would then return to A. Larger displacements may initially lie to the left of B, in a region that is initially attracted towards C. However, as the geometry shifts, these points may be overtaken by the moving nullclines and forced back to the rest state of A. Fitzhugh gives more details, described in greater subtlety, in a good expository way.

The entire *Vmhn* phase space was reconstructed and represented by a schematic diagram, a projection into the *Vm* plane. In the complete system there is no saddle point. However, for the variables h and n close to their resting values the system is essentially equivalent to the reduced (V, m) system in which the point B is a saddle point. As seen in Figure 8.10(b) small deviations from the stable resting point do not lead to excitation, but rather to a gradual return to rest. Larger, above-threshold deviations result in a large excursion through phase space, in which V first increases and finally returns with overshoot to the resting state. Such superthreshold trajectories are the phase-space representations of an action potential. The regions marked on these curves with circled numbers correspond to parts of the physiological response which have been called the (1) regenerative, (2) active, (3) absolutely refractory, and (4) relatively refractory phases.

A familiarity with the Hodgkin-Huxley equations underscores the following:

1. *Excitability:* Above-threshold initial voltage leads to rapid response with large changes in the state of the system.
2. *Stable oscillations:* While not described earlier, the presence of an applied input current represented by an additional term, $I(t)$, on the RHS of equation (9) (e.g. a step function with $I = -10\ \mu\text{A cm}^{-2}$) can lead to the formation of a stable limit cycle in the full model (see Fitzhugh, 1961).

Working with these basic characteristics of the Hodgkin-Huxley model led Fitzhugh to propose a simpler model that gives a descriptive portrait of the neural excitation without direct reference to known or conjectured physiological variables. In preparation for an analysis of his much simpler model we take a mathematical detour to become acquainted with several valuable techniques that will prove useful in a number of upcoming results.

8.3. THE POINCARÉ-BENDIXSON THEORY

As previously mentioned, two-dimensional vector fields and thus also two-dimensional phase planes have attributes quite unlike those of their n-dimensional counterparts. One important feature, on which much of the following theory depends, is the fact that a simple closed curve (for example, a circle) subdivides a plane into two disjoint open regions (the "inside" and the "outside"). This result, known as the *Jordan curve theorem* implies (through a chain of reasoning we shall briefly highlight in Appendix 2 for this chapter) that there are restrictions on the trajectories of a smooth two-dimensional phase flow. As discussed in Chapter 5, a trajectory can approach as its limiting value only one of the following: (1) a critical point, (2) a periodic orbit, (3) a cycle graph (see Figure 8.12), and (4) infinite xy values. A trajectory contained in a *bounded* region of the plane can only fall into cases 1 to 3.

The following result is particularly useful for establishing the existence of periodic orbits.

Theorem 1: The Poincaré-Bendixson Theorem
If for $t \geq t_0$ a trajectory is bounded and does not approach any singular point, then it is either a closed periodic orbit or approaches a closed periodic orbit for $t \to \infty$.

Comment: The theorem still holds if we replace $t \geq t_0$ with $t \leq t_0$ and $t \to \infty$ with $t \to -\infty$.

The boxed material outlines properties of a phase plane that are essentially equivalent to the Poincaré-Bendixson theorem and that serve equally well for discovering periodic orbits.

Theorem 2

Suppose the direction field of the system of equations (1a,b) has the following properties:

1. There is a bounded region D in the plane that contains a single repelling steady state and into which flow enters but from which it does not exit. Then the system (1a,b) possesses a periodic solution (represented by a closed orbit lying entirely inside A or D.
2. There is a bounded annular region A in the plane into which flow enters but from which flow does not exit, and A contains no steady states of equations (1a,b).

It is shown in the appendix that the steady state in part 1 can only be an unstable node or focus.

Two other statements outline the stability properties of periodic solutions.

Theorem 3

1. If either of the regions described in theorem 2 contains only a *single* periodic solution, that solution is a stable limit cycle.
2. If Γ_1 and Γ_2 are *two* periodic orbits such that Γ_2 is in the interior of the region bounded by Γ_1 and no periodic orbits or critical points lie between Γ_1 and Γ_2, then one of the orbits must be unstable on the side facing the other orbit.

Much of the theoretical work on proving the existence of oscillatory solutions to nonlinear equations such as (1a,b) rests on identifying regions in the phase plane that have the properties described in theorem 2. We now summarize the Poincaré-Bendixson limit-cycle recipe.

Existence of Periodic Solutions

If you can find a region in the xy phase plane containing a single repelling steady state (i.e. unstable node or spiral) and show that the arrows along the boundary of the region never point outwards, you may conclude that there must be at least one closed periodic trajectory inside the region.

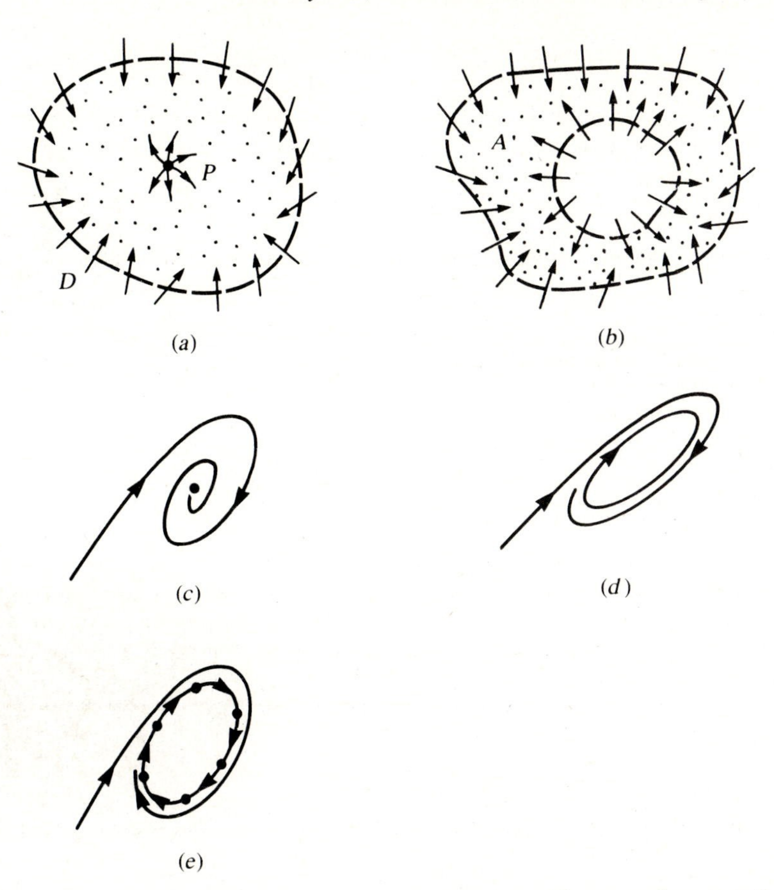

Figure 8.12 *The Poincaré-Bendixson theory prescribes the existence of a limit cycle in two equivalent cases: (a) Flow cannot leave some region* D *that contains an unstable node or focus. (b) Flow is trapped inside an annular region* A *in the* xy *plane. There are three possible fates of a bounded semiorbit: (c) approach to a steady state, (d) approach to a periodic orbit, and (e) approach to a cycle graph.*

An analogous statement corresponding to Figure 8.12(b) can be made. We shall see several examples of the usefulness of the Poincaré-Bendixson theory in this chapter.

The two following criteria are sometimes useful in ruling out the presence of a limit cycle, and for this reason have been called the *negative criteria:*

1. *Bendixson's criterion.* Suppose D is a simply connected region of the plane (that is, D is a region without holes). If the expression $\partial F/\partial x + \partial G/\partial y$ is not identically zero (i.e. is not zero for all (x, y) in D) and does not change sign in D, then there are no closed orbits in this region.
2. *Dulac's criterion:* Suppose D is a simply connected region in the plane, and suppose there exists a function $B(x, y)$, continuously differentiable on D, such that the expression

$$\frac{\partial(BF)}{\partial x} + \frac{\partial(BG)}{\partial y}$$

is not identically zero and does not change sign in D. Then there are no closed orbits in this region.

The proof of Bendixson's criterion is based on Green's theorem and is accessible to students who have had advanced calculus (see appendix to this chapter). Dulac's criterion is an extension that results by substituting BF for F and BG for G in the proof of Bendixson's criterion. For an interesting example of the utility of Dulac's criterion, consider a two-species competition model with carrying capacity κ_i:

$$\frac{dx}{dt} = r_1 x \frac{\kappa_1 - x - \beta_{12} y}{\kappa_1} \tag{14a}$$

$$\frac{dy}{dt} = r_2 y \frac{\kappa_2 - y - \beta_{21} x}{\kappa_2}. \tag{14b}$$

For eliminating limit cycles, Bendixson's criterion fails, but Dulac's criterion succeeds by choosing $B(x, y) = 1/xy$. (See problem 5.) Based on Bendixson's criterion, the following result is readily established.

Corollary of Bendixson's Criterion:

If equations (1a,b) are linear in x and y, then the only possible oscillations are the neutrally stable ones. (Limit cycles can only be obtained with nonlinear equations.)

To understand why this is true, consider the system (1a,b) where $f(x, y) = ax + by$, $G(x, y) = cx + dy$; then $F_x + G_y = a + d$. This is a constant and has a fixed sign. Thus the criterion is only satisfied trivially if $a + d = 0$ in which case the equations would be $dx/dt = by$ and $dy/dt = dx$. Such equations have neutral cycles (not limit cycles), provided b and d have opposite signs.

Comments: Bendixson's negative criterion does not say what happens if the expression $\partial F/\partial x + \partial G/\partial y$ does change sign. (No conclusions can then be drawn about the existence of limit cycles.) In other words, the theorem gives a necessary but *not* a sufficient condition to test.

8.4 THE CASE OF THE CUBIC NULLCLINES

As one application of the Poincaré-Bendixson theorem we examine a rather classical phase-plane geometry that almost invariably leads to the properties of oscillation or excitability. We first discuss a prototype in which one of the nullclines is a simple cubic curve [equation (16)]. As the qualitative analysis will illustrate, this configuration creates the geometry to which the Poincaré-Bendixson theorem applies. An extension to more general S-shaped nullclines will easily follow.

Consider the system of equations

$$\dot{u} = v - G(u), \tag{15a}$$

$$\dot{v} = -u, \tag{15b}$$

where for our prototype we take $G(u)$ to be

$$G(u) = \frac{u^3}{3} - u. \tag{16}$$

We shall postpone a discussion of the motivation underlying these equations and concentrate first on understanding their behavior. One important feature of the function to be exploited presently is that

$$G(u) = -G(-u),$$

that is, G is an *odd* function. Nullclines of this system are the loci of points

$$v = G(u) \qquad \text{(the } u \text{ nullcline)}, \tag{17}$$

$$u = 0 \qquad \text{(the } v \text{ nullcline)}. \tag{18}$$

The term *cubic nullcline* now becomes somewhat more transparent. The shape of the loci given by equation (17) is that of a cubic curve, symmetric about the origin. The two humps to the left and right of the origin also play an important role in the properties of the system. (For this reason the function $G(u) = u^3$ would not be satisfactory; see problem 10 and Figure 8.13.)

Now consider the pattern of flow along these nullclines. The following points can be deduced from the equations:

1. The direction must be "vertical" on the u nullcline and "horizontal" on the v nullcline (since $\dot{u} = 0$ or $\dot{v} = 0$, respectively).
2. Whenever u is positive, v decreases.
3. On the v nullcline, u is zero so that $G(u)$ is also zero. Thus by equation (15a) $\dot{u} = v$, and u will increase when v is positive and decrease when v is negative.

These conclusions are depicted in Figure 8.13(a).

Now consider a trajectory emanating from some arbitrary point $P_0(x_0, y_0)$ in the stippled annulus in Figure 8.13(b). The flow in proximity to the u nullcline will carry it across and towards decreasing u values. After arriving at P_1, the flow drifts horizontally across the v axis and over to the left branch of the cubic curve (P_2). Here a current in the positive v direction conveys the point to P_3 and then back across the top of the hump and into the positive quadrant. From the construction in the diagram it is further evident that the direction of flow is everywhere *into* or parallel to the boundary of the annular region, indicating that once a trajectory has entered the region, it is forever trapped. There are no steady-state points in A, and A is bounded. By the Poincaré-Bendixson theorem we can conclude that there is a limit-cycle trajectory inside this region.

Furthermore, it is possible to shrink the thickness of A to an arbitrarily fine region and draw similar conclusions. In particular, this means that we can dismiss the

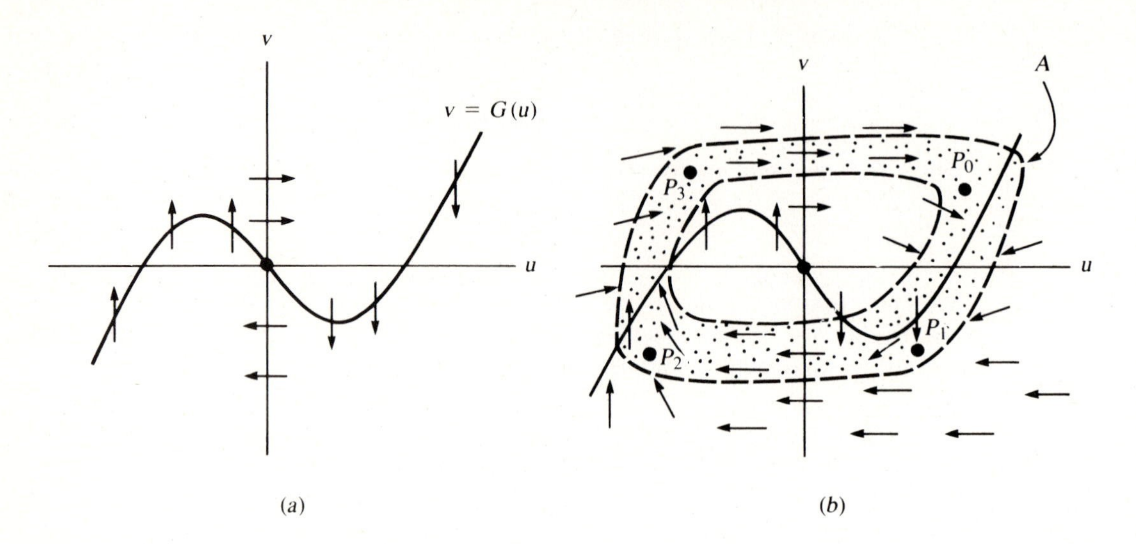

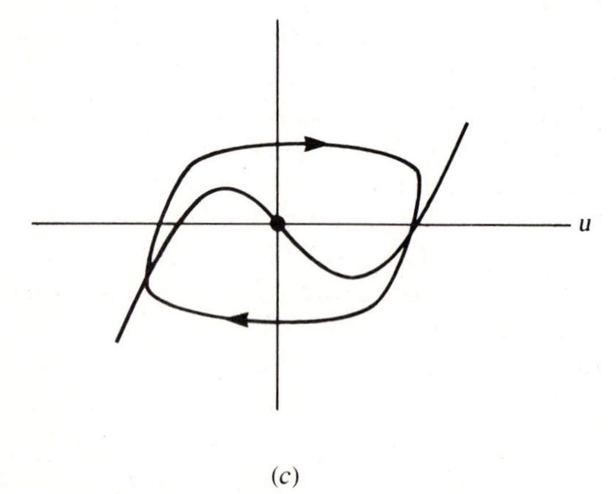

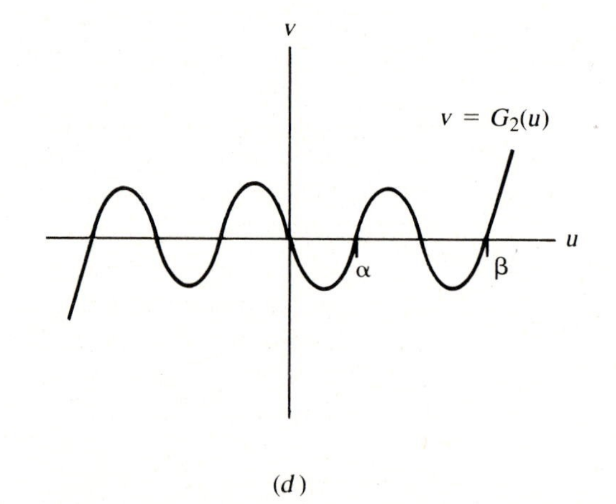

Figure 8.13 *Flow along cubic nullclines described by equations (15) and (16); (b) an annular region* A *that traps flow in the* uv *plane; (c) a limit cycle oscillation contained in the region* A*; (d) more general shapes of the function* G *that lead to similar results.*

possibility that there is more than one limit cycle in the dynamical system. (See Problem 10.)

We have chosen a particular example in which the form of the equations leads to certain specific features: (steady state at (0, 0), flow symmetric with respect to the origin, and one nullcline along the y axis). These features can be changed somewhat without losing the main dynamic features of the system. More generally, a broader class known as the *Lienard equations* exhibit similar behavior. Sometimes written as the following single equation,

$$\frac{d^2u}{dt^2} + g(u)\frac{du}{dt} + u = 0, \tag{19}$$

it can be shown to be equivalent to the system of (15a,b), where

$$G(u) = \int_0^u g(s)\, ds. \tag{20}$$

Then the following properties of $G(u)$ lead to a generalized cubic that results in essentially identical conditions:

1. $G(u) = -G(-u)$ (thus $G(u)$ is an odd function).
2. $G(u) \to \infty$ for $u \to \infty$ (the right and left branches of G extend to $+\infty$ and $-\infty$) and for some positive β, $G(u) > 0$ and $dG/du > 0$ whenever $u > \beta$; (G is eventually positive and monotonically increasing).
3. For some positive α, $G(\alpha) = 0$ and $G(u) < 0$ whenever $u < \alpha$. (G is negative for small positive u values).

Condition 1 means that all trajectories will be symmetric about the origin. Condition 2 is necessary to cause the flow to be trapped or confined to the given annulus. Condition 3 means that the steady state at (0, 0) is unstable. (Details are investigated in problem 11.) An example of a "bumpy" function satisfying these conditions is shown in Figure 8.13(*d*). It can be rigorously established (with reasoning similar to that used earlier) that such conditions guarantee that the system of equations (15) (or the single equation 19) admits a nontrivial periodic solution, that is, a limit cycle. In the particular case where $\alpha = \beta$, as in the example we have analyzed, there is indeed a single periodic orbit that is asymptotically stable. Rigorous proof and further details may be found in Hale (1980), (p. 57–63).

The example used as a prototype in this section is called the *Van der Pol oscillator,* sometimes written in the form

$$\ddot{u} - k(1 - u^2)\dot{u} + u = 0 \qquad (k > 0). \tag{21}$$

(See problem 12.) Van der Pol first used it in 1927 to represent an electric circuit containing a nonlinear element (a triode valve whose resistance depends on the applied current). Even then, van der Pol realized the parallel between this circuit and certain biological oscillations such as the heart beat. For large values of the constant k, the corresponding system of equations

$$\epsilon\dot{u} = v - G(u), \tag{22a}$$

$$\dot{v} = -u\epsilon, \tag{22b}$$

contains a small parameter $\epsilon = 1/k$. (Recall that this can be exploited in calculating approximate solutions using techniques of asymptotic expansions. See problem 12 for a taste of the idea.) The solutions to this small-parameter system are called *relaxation oscillations* for the following reasons: as long as v is close to $G(u)$ (that is, in vicinity of the cubic curve), $\dot{u}$ and $\dot{v}$ both change rather slowly. When the trajectory departs from this curve, $\dot{u} = [v - G(u)]/\epsilon$ is quite large. The horizontal progression across from P_1 to P_2 is thus rapid. A plot of $u(t)$ reveals a succession of time intervals in which u changes slowly followed by ones in which it changes more rapidly.

Models related to van der Pol's oscillator have been important in many physical settings and, as we shall see, have also been valuable in describing oscillating biological systems. (A example of an application to the heart-beat cycle is discussed in Jones and Sleeman, 1983.) More generally, the idea underlying s-shaped nullclines has been exploited in a variety of models for excitable and oscillatory phenomena. Figures 8.14 to 8.17 are a sampling drawn from the literature, and Section 8.5 deals in greater detail with one particular application in the study of neural signals.

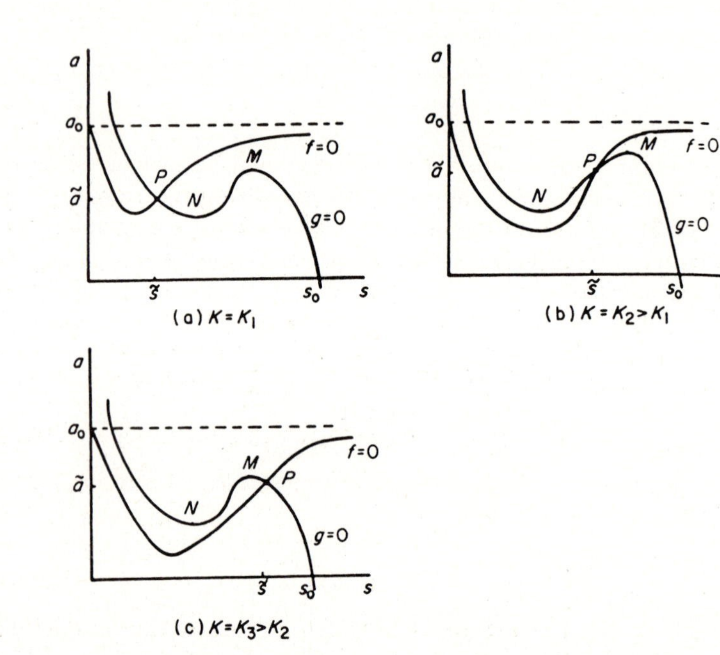

Figure 8.14 *Several regimes of behavior in a model for substrate inhibition with S-shaped nullclines.* s *and* a *represent substrate and cosubstrate. Their chemical kinetics are represented by the equations*

$$\frac{ds}{dt} = g(s, a), \qquad \frac{da}{dt} = f(s, a)$$

where

$$g = s_0 - s - \frac{\rho s a}{1 + s + Ks^2},$$

$$f = \alpha(a_0 - a) - \frac{\rho s a}{1 + s + Ks^2}.$$

K *is the inhibition parameter,* α, ρ, a_0 *and* s_0 *are constants (see details in the original reference.)* M *and* N *represent the maximum and minimum points along the nullcline* $g = 0$, *and* P *is the steady state. The transition* a) → b) → c) *is for decreasing* K-*values. [From Murray (1981), fig. 2, p. 168.* J. Theor. Biol., *88, 161–199. Reprinted by permission of Academic Press Inc. (London)]*

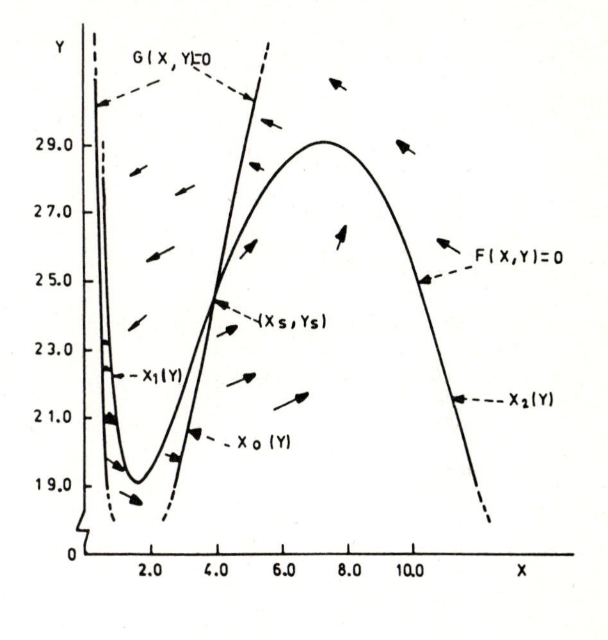

Figure 8.15 *A vector field and cubic nullclines in a model for respiration in a bacterial culture. The nutrient* x *and oxygen* y *are assumed to satisfy the equations*

$$\frac{dx}{dt} = B - x - \frac{xy}{1 + qx^2} = F(x, y),$$

$$\frac{dy}{dt} = A - \frac{xy}{1 + qx^2} = G(x, y).$$

[From Fairen and Velarde (1979). Fig. 3, p. 152, J. Math. Biol., *8, 147–157, by permission of Springer-Verlag.]*

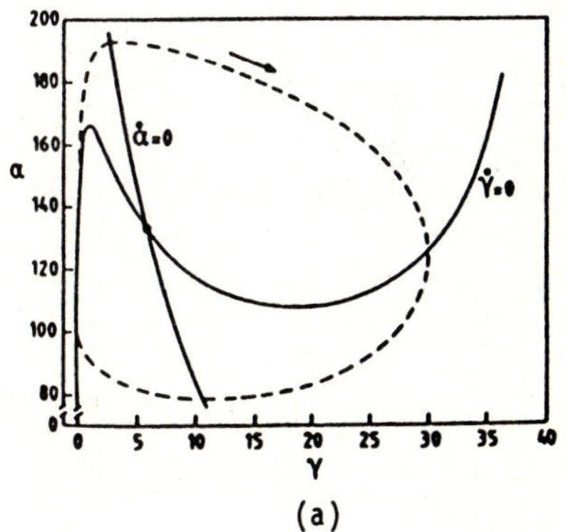

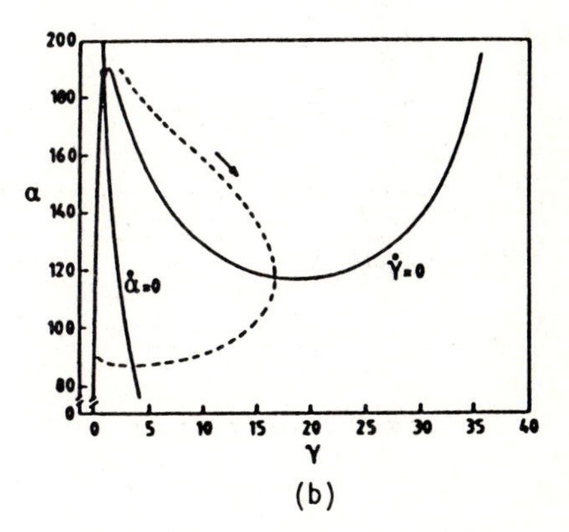

Figure 8.16 *(a) Limit-cycle oscillations, and (b) excitability on suprathreshold perturbation, in a two-variable system derived from equations for a cyclic-AMP signaling system devised by Goldbeter and Segel (1977). The original equations are*

$$\frac{d\alpha}{dt} = v - \sigma\Phi \qquad \text{(intracellular ATP)},$$

$$\frac{d\beta}{dt} = q\sigma\Phi - k_t\beta \qquad \text{(intracellular CAMP)},$$

$$\frac{d\gamma}{dt} = \frac{k_t\beta}{h} - k_s\gamma \qquad \text{(extracellular CAMP)},$$

α, β *and* γ *represent dimensionless metabolic concentrations and*

$$\Phi = \frac{\alpha(1 + \alpha)(1 + \gamma)^2}{L + (1 + \alpha)^2(1 + \gamma)^2} \qquad \text{(allosteric kinetics of adenylate cyclase)}$$

is a chemical kinetic term depicting enzyme—receptor interactions. See Segel (1984) for a simplified model and definitions of parameters. [From Goldbeter and Martiel (1983). A critical discussion of plausible models for relay and oscillation of cyclic AMP in dictyostelium cells, in Rhythms in Biology and other Fields of Application Lecture Notes in Biomath, *vol. 49, pp 173–178. Reprinted with author's permission.]*

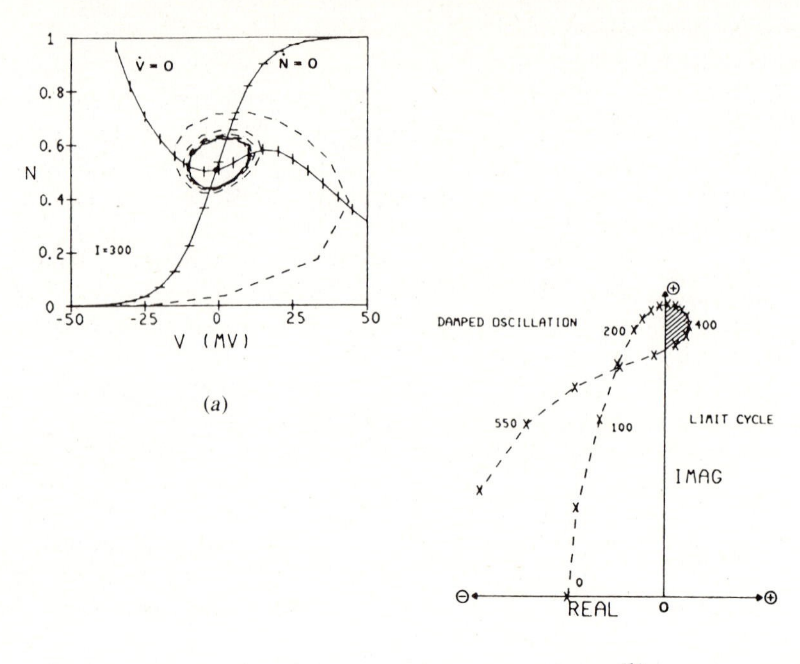

Figure 8.17 *(a) S-shaped nullclines in a model for voltage oscillations in the barnacle giant muscle fiber. Shown is a limit cycle in a reduced* V, N *system (*V = *voltage and* N = *fraction of open* K^+ *channels);* V *satisfies an equation like (19), and* N *is given by*

$$\frac{d\dot{N}}{dt} = \lambda_N(V)(N_\infty(V) - N),$$

where

$$N_\infty = \frac{1}{2}\left(1 + \tanh\frac{v - v_3}{v_4}\right).$$

(b) The eigenvalues of the linearized system change as the current is increased and cross the imaginary axis twice. [From Morris and Lecar (1981), figs. 7 and 8. Reproduced from the Biophysical Journal *(1981) vol 35, 193–213, by copyright permission of the Biophysical Society.]*

Conclusions

A system of equations (1a,b) in which one of the nullclines $F(x, y) = 0$ or $G(x, y) = 0$ is an S-shaped curve can give rise to *oscillations* provided the steady state on this curve is unstable.

When the steady state is stable, the system may exhibit a somewhat different behavior called *excitability*. We discuss this property further in Section 8.5.

8.5 THE FITZHUGH-NAGUMO MODEL FOR NEURAL IMPULSES

In Section 8.2 we followed an analysis of the Hodgkin-Huxley model published by Fitzhugh in the biophysical literature in 1960. The elegance of applying phase-plane methods and reduced systems of equations to this rather complicated problem should not be underestimated. (Similar ideas are used in more recent examples; see Morris and Lecar, 1981.)

Using such analysis, Fitzhugh was able to explain the occurrence of thresholds in the Hodgkin-Huxley model of neural excitation. The analysis was, however, less instructive in demonstrating causes of *repetitive impulses* (a sequence of action potentials in the neuron) because here the interactions of all four variables—V, m, h, and n—were important [see equations (9) to (12).] A greater reliance on numerical, rather than analytic techniques was thus necessary.

In a succeeding paper published in 1961, Fitzhugh proposed to demonstrate that the Hodgkin-Huxley model belongs to a more general class of systems that exhibit the properties of excitability and oscillations. As a fundamental prototype, the van der Pol oscillator was an example of this class, and Fitzhugh therefore used it (after suitable modification). A similar approach was developed independently by Nagumo et al. (1962) so the following model has subsequently been called the Fitzhugh-Nagumo equations.

To avoid misunderstanding, it should be emphasized that the main purpose of the model is not to portray accurately quantitative properties of impulses in the axon. Indeed, the variables in the equations have somewhat imprecise meanings, and their interrelationship does not correspond to exact physiological facts or conjectures. Rather, the system is meant as a simpler paradigm in which one can exhibit the sorts of interactions between variables that lead to properties such as excitability and oscillations (repetitive impulses).

Fitzhugh proposed the following equations:

$$\frac{dx}{dt} = c\left[y + x - \frac{x^3}{3} + z(t)\right], \tag{23a}$$

$$\frac{dy}{dt} = -\frac{x - a + by}{c}. \tag{23b}$$

In these equations the variable x represents the excitability of the system and could be identified with voltage (membrane potential in the axon); y is a recovery variable, representing combined forces that tend to return the state of the axonal membrane to rest. Finally, $z(t)$ is the applied stimulus that leads to excitation (such as input current). In typical physiological situations, such stimuli might be impulses, step functions, or rectangular pulses. It is thus of interest to explore how equations (23a,b) behave when various functions $z(t)$ are used as inputs. Before addressing this question we first take $z = 0$ and analyze the behavior of the system in the xy phase plane.

In order to obtain suitable behavior, Fitzhugh made the following assumptions about the constants a, b, and c:

$$1 - \frac{2b}{3} < a < 1, \qquad 0 < b < 1, \qquad b < c^2, \tag{24}$$

Inspection reveals that when $z = 0$, nullclines of equations (23a,b) are given by the following loci:

$$y = \frac{x^3}{3} - x \qquad (\dot{x} = 0, \text{ the } x \text{ nullcline}), \tag{25}$$

$$y = \frac{a - x}{b} \qquad (\dot{y} = 0, \text{ the } y \text{ nullcline}). \tag{26}$$

The humps on the cubic curve are located at $x = \pm 1$.

We shall not explicitly solve for the steady state of this system, which satisfies the cubic equation

$$\frac{x^3}{3} + x\left(\frac{1}{b} - 1\right) = \frac{a}{b}. \tag{27}$$

However, the conditions given by equation (24) on the parameters guarantee that there will be a *single* (real) steady-state value $(\overline{x}, \overline{y})$ located just beyond the negative hump on the cubic x nullcline, at its intersection with the skewed y nullcline, as shown in Figure 8.18.

Calculating the Jacobian of equations (23) leads to

$$\mathbf{J} = \begin{pmatrix} (1 - \overline{x}^2)c & c \\ \dfrac{-1}{c} & \dfrac{-b}{c} \end{pmatrix} \tag{28}$$

(problem g). Thus, by writing the characteristic equation in terms of $\overline{x}$, we obtain the quadratic equation

$$\lambda^2 + \left[\frac{b}{c} - (1 - \overline{x}^2)c\right]\lambda + [1 - (1 - \overline{x}^2)b] = 0. \tag{29}$$

The steady state will therefore be stable provided that

$$-\left[\frac{b}{c} - (1 - \overline{x}^2)c\right] < 0, \tag{30a}$$

$$1 - (1 - \overline{x}^2)b > 0. \tag{30b}$$

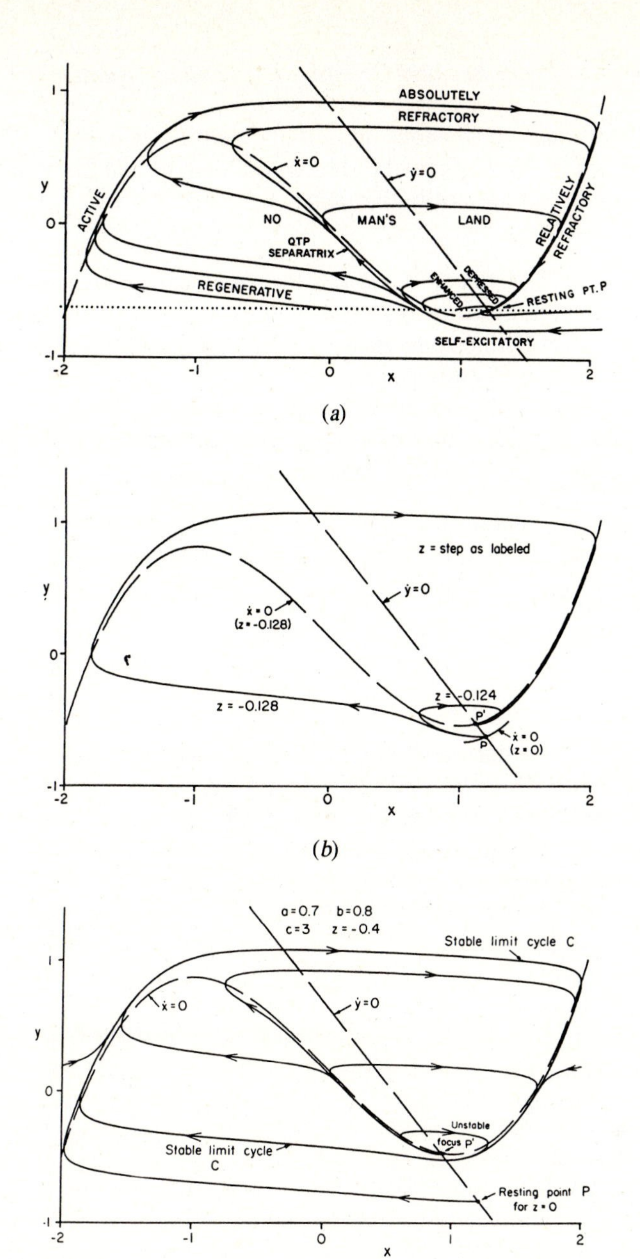

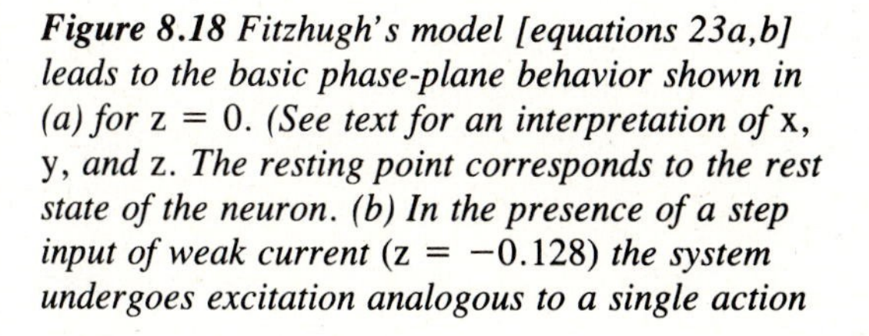
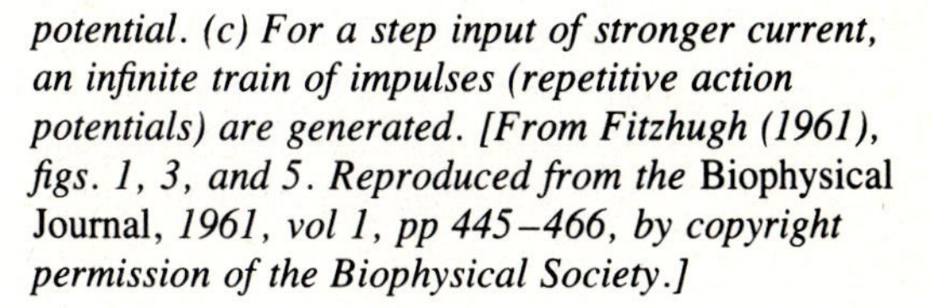

***Figure 8.18** Fitzhugh's model [equations 23a,b] leads to the basic phase-plane behavior shown in (a) for* z = 0. *(See text for an interpretation of* x, y, *and* z. *The resting point corresponds to the rest state of the neuron. (b) In the presence of a step input of weak current* (z = −0.128) *the system undergoes excitation analogous to a single action potential. (c) For a step input of stronger current, an infinite train of impulses (repetitive action potentials) are generated. [From Fitzhugh (1961), figs. 1, 3, and 5. Reproduced from the* Biophysical Journal, *1961, vol 1, pp 445–466, by copyright permission of the Biophysical Society.]*

Note that these conclusions are not affected if z in equation (23a) is nonzero. Since $b < 1$ and $b < c^2$, it may be shown (problem *g*) that the steady state is stable for all values of $\bar{x}$ that are not in the range

$$-\gamma \leq \bar{x} \leq \gamma, \tag{31}$$

where $\gamma = (1 - b/c^2)^{1/2}$ is a positive number whose magnitude is smaller than 1. The geometry shown in Figure 8.18 tells us that if the y nullcline were to intersect the cubic x nullcline somewhere on the middle branch between the two humps, the steady state would be unstable.

We next consider the behavior of the *stimulated* neuron, described by equations (23a,b) with a nonzero stimulating current $z(t)$. A particularly simple set of stimuli might consist of the following:

1. A step function $z = 0$ for $t < 0$; $z = i_0$ for $t \geq 0$).
2. A pulse ($z = i_0$ for $0 \leq t \leq t_0$; $z = 0$ otherwise.
3. A constant current $z = i_0$.

For as long as $z = i_0$, the configuration of the x nullcline is given by the equation

$$y = \frac{x^3}{3} - x + i_0. \tag{32}$$

This cubic curve has been shifted in the positive y direction if i_0 is positive and in the opposite direction if i_0 is negative. Let S^* represent the intersection of (32) with the y nullcline, S^0 the intersection of (25) with the y nullcline, and S the instantaneous state of the system. Thus S^* and S^0 are the steady states of the stimulated and unstimulated system, respectively and $S = (x(t), y(t))$.

At the instant a stimulus is applied, $S = S^0$ is no longer a rest state, since the steady state has shifted to S^*. This means that S will change, tracing out some trajectory in the xy plane. The following possibilities arise:

1. If i_0 is very small, S^* will be close to S^0, on the region of the cubic curve for which steady states are stable and $S = (x(t), y(t))$ will be attracted to S^* without undergoing a large displacement.
2. If i_0 is somewhat larger, S^* may still be in the stable regime, but a more abrupt dynamics could ensue: in particular, if S falls beyond the separatrix shown in Figure 8.18(*a*), the state of the system will undergo a large excursion in the xy plane before settling into the attracting steady state S^*.

Such cases represent a single action-potential response that occurs for superthreshold stimuli. (See Figure 8.3.) This type of behavior, in which a steady state is attained only after a long detour in phase space, is typical of *excitable systems*. As we have seen, it is also a property intimately associated with systems in which one or both of the nullclines have the S shape (also referred to sometimes as N shape or z shape) similar to that of the cubic curve.

3. For yet larger i_0, S^* will fall into the middle branch of the cubic curve so that it is no longer stable. In this case the situation discussed in Section 8.5 occurs and a stable, closed, periodic trajectory is created. All points, and in particular

S, will undergo cyclic dynamics, approaching ever closer to the stable limit cycle. This behavior corresponds to *repetitive firing* of the axon, which results when a step stimulus of sufficiently high intensity is applied.

In all of these cases, when the applied stimulus is removed (at $z = 0$), S^0 becomes an attracting rest state once more; the repetitive firing ceases, and the excited state eventually returns to rest.

With a few masterful strokes, Fitzhugh has painted a *caricature* of the behavior of neural excitation. His model is not meant to accurately portray the physiological mechanisms operating inside the axonal membrane. Rather, it is a behavioral paradigm, phrased in terms of equations that are mathematically tractable. As such it has played an important role in leading to an understanding of the nature of excitable systems and in studying more complicated models of the action potential that include the effect of spatial propagation in the native (nonclamped) axon.

8.6 THE HOPF BIFURCATION

A further diagnostic tool that can help in establishing the existence of a limit-cycle trajectory is the *Hopf bifurcation theorem*. It is quoted widely, applied in numerous contexts, and for this reason merits discussion.

Subject to certain restrictions, this theorem predicts the appearance of a limit cycle about any steady state that undergoes a transition from *a stable to an unstable focus* as some parameter is varied. The result is local in the following sense: (1) The theorem only holds for parameter values close to the *bifurcation value* (the value at which the just-mentioned transition occurs). (2) The predicted limit cycle is close to the steady state (has a small diameter). (The Hopf bifurcation does not specify what happens as the *bifurcation parameter* is further varied beyond the immediate vicinity of its critical bifurcation value.)

In the following box the theorem is stated for the case $n = 2$. A key requirement corresponding to our informal description is that the given steady state be associated with *complex* eigenvalues whose real part changes sign (from $-$ to $+$). In popular phrasing, such eigenvalues are said to "cross the real axis." To recall the connection between complex eigenvalues and oscillatory trajectories, consider the discussion in Section 5.7 (and in particular, Figure 5.12 of Chapter 5).

Advanced mathematical statements of this theorem and its applications can be found in Marsden and McCracken (1976). Odell (1980) and Rapp (1979) give good informal descriptions of this result that are suitable for nonmathematical readers.

One of the attractive features of the Hopf bifurcation theorem is that it applies to larger systems of equations, again subject to certain restrictions. This makes it somewhat more applicable than the Poincaré-Bendixson theorem, which holds only for the case $n = 2$.

The Hopf Bifurcation Theorem for the Case n = 2

Consider a system of two equations that contains a parameter γ.

$$\frac{dx}{dt} = f(x, y; \gamma), \tag{33a}$$

$$\frac{dy}{dt} = g(x, y; \gamma). \tag{33b}$$

The usual differentiability and continuity assumptions are made about f and g as functions of x, y, and γ (see Chapter 5). Suppose that for each value of γ the equations admit a steady state whose value may depend on γ, that is, $(\bar{x}(\gamma), \bar{y}(\gamma))$. Consider the Jacobian matrix evaluated at the parameter-dependent steady state:

$$\mathbf{J}(\gamma) = \begin{pmatrix} \frac{\partial f}{\partial x} & \frac{\partial f}{\partial y} \\ \frac{\partial g}{\partial x} & \frac{\partial g}{\partial y} \end{pmatrix}_{(\bar{x}, \bar{y})} \tag{34}$$

Suppose eigenvalues of this matrix are $\lambda(\gamma) = a(\gamma) \pm b(\gamma)i$. Also suppose that there is a value γ^*, called the bifurcation value, such that $a(\gamma^*) = 0$, $b(\gamma^*) \neq 0$, and as γ is varied through γ^*, the real parts of the eigenvalues change signs ($da/d\gamma \neq 0$ at $\gamma = \gamma^*$).

Given these hypotheses, the following possibilities arise:

1. At the value $\gamma = \gamma^*$ a center is created at the steady state, and thus infinitely many neturally stable concentric closed orbits surround the point $(\bar{x}, \bar{y})$.
2. There is a range of γ values such that $\gamma^* < \gamma < c$ for which a *single* closed orbit (a limit cycle) surrounds $(\bar{x}, \bar{y})$. As γ is varied, the diameter of the limit cycle changes in proportion to $|\gamma - \gamma^*|^{1/2}$. There are no other closed orbits near $(\bar{x}, \bar{y})$. Since the limit cycle exists for γ values *above* γ^*, this phenomenon is known as a *supercritical bifurcation*.
3. There is a range of values such that $d < \gamma < \gamma^*$ for which a conclusion similar to case 2 holds. (The limit cycle exists for values below γ^*, and this is termed a *subcritical bifurcation*.)

Figure 8.19 illustrates what can happen close to a bifurcation value γ^* (in the case $n = 2$). Here the xy phase plane has been drawn for several values of the parameter γ. In Figure 8.19(a) observe that as γ increases from small values through γ^* and up to $\gamma = c$ the steady state changes from a stable focus to an unstable focus. A stable limit cycle appears at γ^*, and its diameter grows as indicated by the parabolic envelope. This succession, called a supercritical bifurcation, is often represented schematically by the bifurcation diagram of Figure 8.19(c).

In Figure 8.19(b) an unstable limit cycle accompanies the stable focus until γ increases beyond γ^* and disappears. This type of transition, called a *subcritical* bifurcation, is represented by Figure 8.19(d). (Recall that in bifurcation diagrams solid

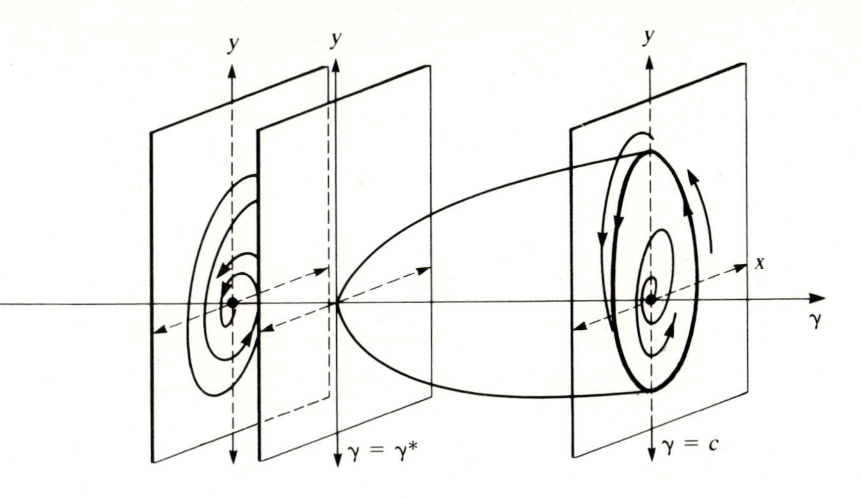

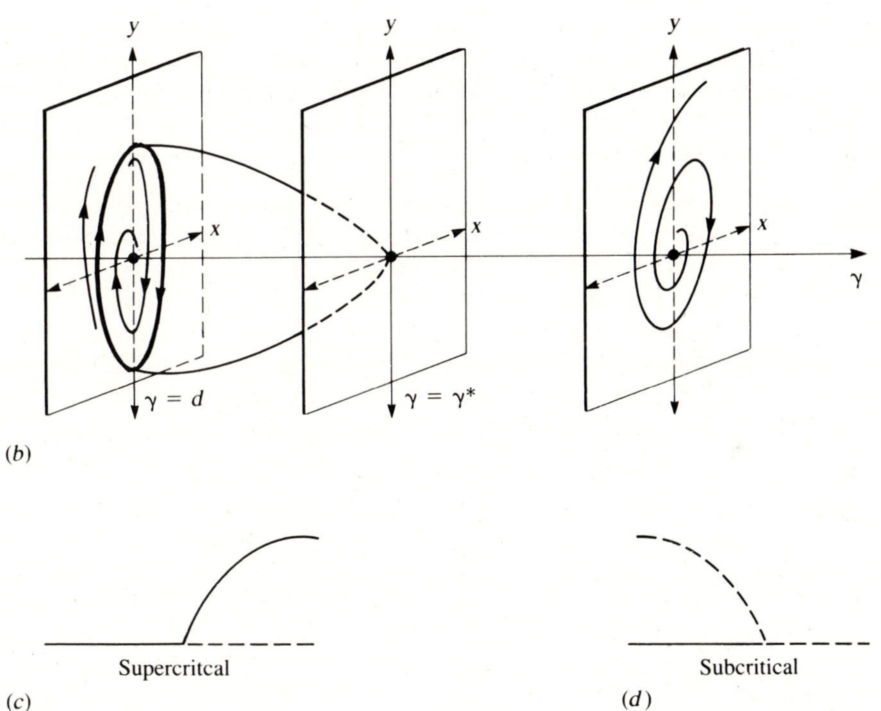

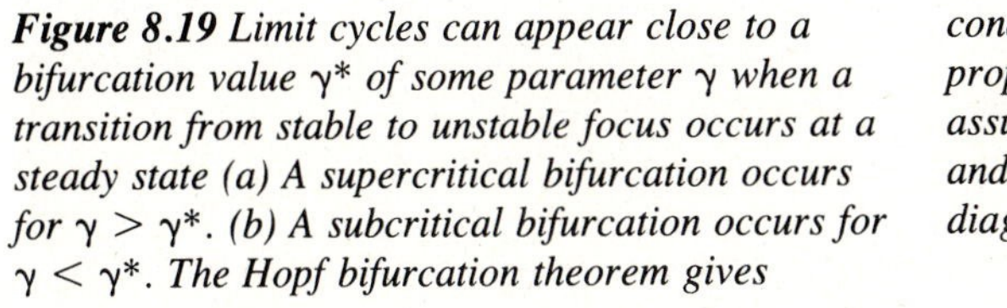

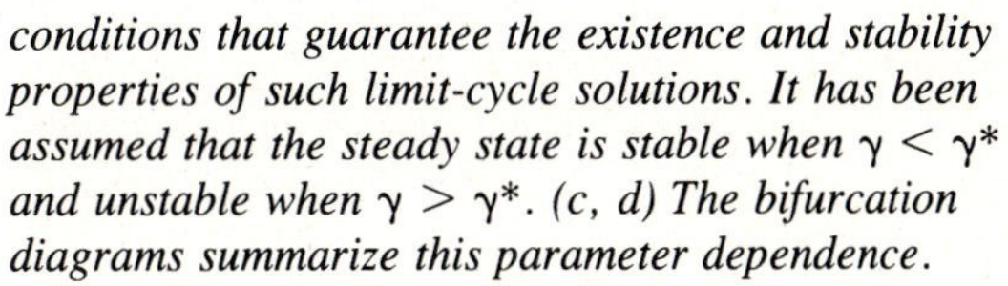

Figure 8.19 *Limit cycles can appear close to a bifurcation value γ^* of some parameter γ when a transition from stable to unstable focus occurs at a steady state (a) A supercritical bifurcation occurs for $\gamma > \gamma^*$. (b) A subcritical bifurcation occurs for $\gamma < \gamma^*$. The Hopf bifurcation theorem gives conditions that guarantee the existence and stability properties of such limit-cycle solutions. It has been assumed that the steady state is stable when $\gamma < \gamma^*$ and unstable when $\gamma > \gamma^*$. (c, d) The bifurcation diagrams summarize this parameter dependence.*

The Hopf Bifurcation Theorem for the Case n > 2

Consider a system of n equations in n variables,

$$\frac{d\mathbf{x}}{dt} = \mathbf{F}(\mathbf{x}, \gamma),$$

where

$$\mathbf{x} = (x_1, x_2, \ldots, x_n),$$
$$\mathbf{F} = (f_1(\mathbf{x}; \gamma), f_2(\mathbf{x}; \gamma), \ldots, f_n(\mathbf{x}; \gamma)),$$

with the appropriate smoothness assumptions on f_i which are functions of the variables and a parameter γ. If $\overline{\mathbf{x}}$ is a steady state of this system and linearization about this point yields n eigenvalues

$$\lambda_1, \lambda_2, \ldots, \lambda_{n-2}, a + bi, a - bi,$$

where eigenvalues λ_1 through λ_{n-2} have negative real parts and λ_{n-1}, λ_n (precisely these two) are complex conjugates that cross the imaginary axis when γ varies through some critical value, then the theorem still predicts closed (limit-cycle) trajectories. The extension of this theorem to arbitrary dimensions depends in large part on another important theorem (called the *center manifold theorem*), which ensures that close to the steady state the interesting events take place on a plane-like subset of R^n (mathematically called a *two-dimensional manifold*).

curves represent stable steady states or periodic solutions, whereas dotted curves represent unstable states.)

While in Figure 8.19(*a,b*) the limit cycle is shown originating at γ^*, the theorem actually does not specify at what exact value the closed orbit appears. Furthermore, to determine whether or not the limit cycle is stable a rather complicated calculation must be done. A simple case is given as an example in the following box. Other examples can be found in Marsden and McCracken (1976, sections 4a,b), Odell (1980), Guckenheimer and Holmes (1983) and Rand et al. (1981).

Calculations of Stability of the Limit Cycle in Hopf Bifurcation for the Case n = 2

For the system of equations (33a,b) it will be assumed that for $\gamma = \gamma^*$ the Jacobian matrix is of the form

$$\mathbf{J} = (\overline{x}, \overline{y})\,\big|_{\gamma=\gamma^*} = \begin{pmatrix} 0 & 1 \\ -b & 0 \end{pmatrix}. \tag{35}$$

In this case the eigenvalues (at the critical parameter value) are

$$\lambda_1 = bi, \qquad \lambda_2 = -bi.$$

Marsden and McCracken (1976) derive a rather elaborate stability criterion.

The following expression must be calculated and evaluated at the steady state when $\gamma = \gamma^*$:

$$V''' = \frac{3\pi}{4b}(f_{xxx} + f_{xyy} + g_{xxy} + g_{yyy}) + \frac{3\pi}{4b^2}[f_{xy}(f_{xx} + f_{yy}) + g_{xy}(g_{xx} + g_{yy}) + f_{xx}g_{xx} - f_{yy}g_{yy}]. \quad (36)$$

Then the conclusions are as follows:

1. If $V''' < 0$, then the limit cycle occurs for $\gamma > \gamma^*$ (supercritical) and is stable.
2. If $V''' > 0$, the limit cycle occurs for $\gamma < \gamma^*$ (subcritical) and is repelling (unstable).
3. If $V''' = 0$, the test is inconclusive.

Note: If the Jacobian matrix obtained by linearization of equations (33a,b) has diagonal terms but the eigenvalues are still complex and on the imaginary axis when $\gamma = \gamma^*$, then it is necessary to *transform variables* so that the Jacobian appears as in (35) before the stability receipe can be applied. (See Marsden and McCracken, 1976, for several examples.)

Example 1

Consider the system of equations

$$\frac{dx}{dt} = y = f(x, y; \gamma), \quad (37a)$$

$$\frac{dy}{dt} = -y^3 + \gamma y - x = g(x, y; \gamma). \quad (37b)$$

(This system comes from Marsden and McCracken, 1976, pp. 136, and references therein.) We consider these equations, where γ plays the role of a bifurcation parameter.

The only steady state occurs when $x = y = 0$; at that point the Jacobian is

$$\mathbf{J}(0, 0) = \begin{pmatrix} f_x & f_y \\ g_x & g_y \end{pmatrix} = \begin{pmatrix} 0 & 1 \\ -1 & \dfrac{-y^2}{3} + \gamma \end{pmatrix}_{(0,0)} = \begin{pmatrix} 0 & 1 \\ -1 & \gamma \end{pmatrix} \quad (38)$$

Eigenvalues of this system are given by

$$\lambda = \frac{\gamma \pm \sqrt{\gamma^2 - 4}}{2}. \quad (39)$$

In the range $|\gamma| < 2$, the eigenvalues are complex $\lambda = \{a \pm bi\}$, where

$$a = \text{Re}\,\lambda = \frac{\gamma}{2} \quad \text{and} \quad b = \text{Im}\,\lambda = \frac{(\sqrt{4 - \gamma^2})}{2}.$$

Note that as γ increases from negative values through 0 to positive values, the eigenvalues cross the imaginary axis. In particular, at the bifurcation value $\gamma = 0$ we have

$$\lambda = \pm i$$

(pure imaginary eigenvalues) and $da/d\gamma \neq 0$.

The Hopf bifurcation theorem therefore applies, predicting the existence of periodic orbits for equations (37a,b). We next calculate the expression V''' and determine stability. For equations (37a,b) we have

$$\begin{aligned} f_{xxx} &= 0, & f_{xy} &= 0, \\ f_{xyy} &= 0, & g_{xy} &= 0, \\ g_{xxy} &= 0, & f_{xx} &= 0, \\ g_{yyy} &= -6, & f_{yy} &= 0. \end{aligned} \tag{40}$$

This means that

$$\begin{aligned} V''' &= \frac{3\pi}{4b}(0 + 0 + 0 - 6) + \frac{3\pi}{4b^2}(0) \\ &= \frac{3\pi}{4}(-6) < 0. \end{aligned} \tag{41}$$

This falls into case 1 (see previous box), and so the limit cycle is stable.

8.7 OSCILLATIONS IN POPULATION-BASED MODELS

A number of natural populations are known to exhibit long-term oscillations. Although conjectures about the underlying mechanisms vary, it is commonly recognized that the relationship between prey and predators can lead to such fluctuations in species abundance.

One model that predicts the tendency of predator-prey systems to oscillate is the Lotka-Volterra model, which we discussed in some detail in Chapter 6, Section 6.2. As previously remarked, however, this model is not sufficiently faithful to the real dynamical behavior of populations. Part of the problem stems from the fact that the Lotka-Volterra model is structurally unstable, that is, subject to drastic changes when relatively minor changes are made in the equations. A second problem is that the cycles in species abundance are sensitive to initial population levels. For example, if we start with large populations, the fluctuations too will be very large (see Figure 6.4a). (In the next section we shall discover that this property stems from the fact that the Lotka-Volterra system is *conservative*.)

Since oscillations in natural populations are more regular and stable than those of the simple Lotka-Volterra model, we explore the possibility that the underlying dynamical behavior suitable for depicting these is the limit cycle. Our main goal in this section is to start with a more general set of equations, retain some of the properties of the predator-prey model, and make minimal further assumptions in order to ensure that stable limit-cycle oscillations exist. (This approach is to some extent similar to problems discussed in Sections 3.5, 4.11, and 7.7.) With the theory of Poincaré, Bendixson, and Hopf, we are in a position to reason geometrically and analytically about stable oscillations.

As a starting set of equations we shall consider

$$\frac{dx}{dt} = xf(x, y), \tag{42a}$$

$$\frac{dy}{dt} = yg(x, y), \tag{42b}$$

where x is the prey density, y is the predator density, and the functions f and g represent the species-dependent per capita growth and death rates. Furthermore, we consider, as before, spatially uniform populations comprised of identical individuals.

Based on this form we may conclude that system (42) has nullclines that satisfy

$$\dot{x} = 0: \quad x = 0 \quad \text{or} \quad f(x, y) = 0, \tag{43a}$$

$$\dot{y} = 0: \quad y = 0 \quad \text{or} \quad g(x, y) = 0. \tag{43b}$$

Thus the x and y axes are nullclines, as are the other loci described by equations (43a,b). We shall assume that the loci corresponding to $f(x, y) = 0$ and $g(x, y) = 0$ are single curves and that these intersect once in the positive xy quadrant at $(\bar{x}, \bar{y})$; that is, that there is a single, strictly positive steady state.

In connection with these assumptions, recall that in the Lotka-Volterra model, $f(x, y) = 0$, $g(x, y) = 0$ are straight lines, parallel to the x or y axes respectively, and that the steady state at their intersection is a neutral center for all parameter values. This presents two features to be corrected if we are to discover stable limit-cycle solutions. First, as evident from Figure 6.4, there is the possibility of flow escaping to infinitely large x values, whereas, to use the Poincaré-Bendixson theorem we are required to exhibit a *bounded* region that traps flow. Second, if we were to exploit the Hopf bifurcation theorem, the steady state should not be forever poised at the brink of neutral stability; rather, it should undergo some transition (from stable to unstable focus) as parameters are changed.

We now list four assumptions that will be made in view of the minimal information consistent with the biological system.

1. $\partial f/\partial y < 0$ (Predators adversely affect the prey; that is, net growth rate of the prey population is smaller when they are being exploited by more predators.)

2. $\partial g/\partial x > 0$ (The availability of more prey enhances the predator's growth rate.)

3. There is a threshold level of predators, y_1, that reduces the per capita prey growth rate to zero (even when the prey population is very small); that is, y_1 is defined by

$$f(0, y_1) = 0.$$

4. There is a level of prey x_1 that is minimally required to sustain the predator population; x_1 is defined by the relationship

$$g(x_1, 0) = 0.$$

Assumptions 3 and 4 simply mean that we retain the property that prey (x) and predator (y) nullclines intersect the predator and prey axes as in Figure 8.20(a). Further, assumptions 1 and 2 imply that along the y and x axes the flow is respectively toward and away from (0, 0) (see problem 17). Moreover, this also determines the directions of flow along the nullclines $f(x, y) = 0$ and $g(x, y) = 0$ since continuity must be preserved.

We now consider what other minimal assumptions could produce a geometrical situation to which the Poincaré-Bendixson theorem might apply, that is, a region in the xy plane that confines flow. According to our previous remark, it is necessary to prevent the escape of trajectories along the x axis to infinite x values. This could be

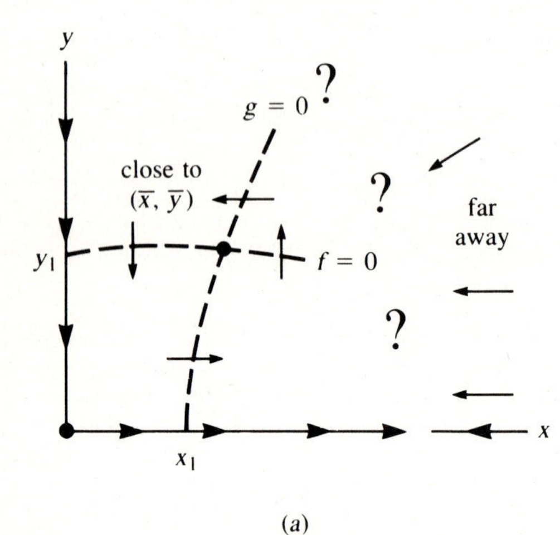

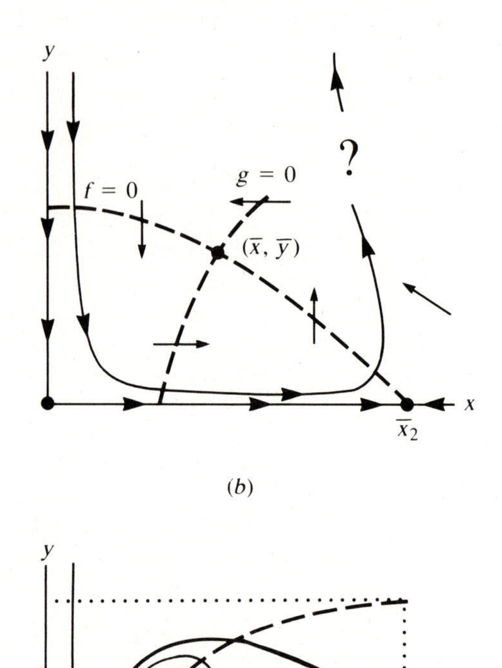

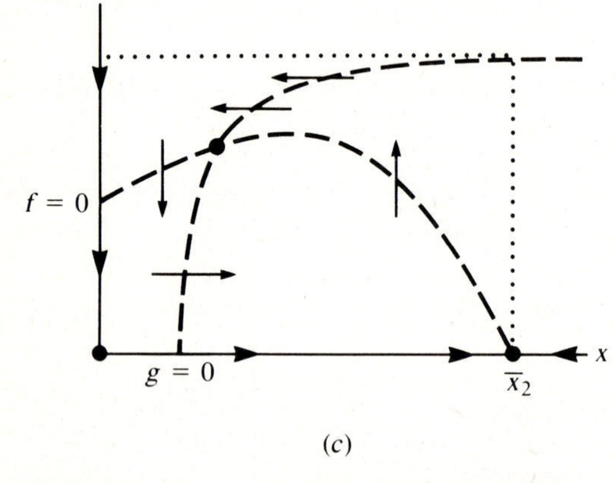

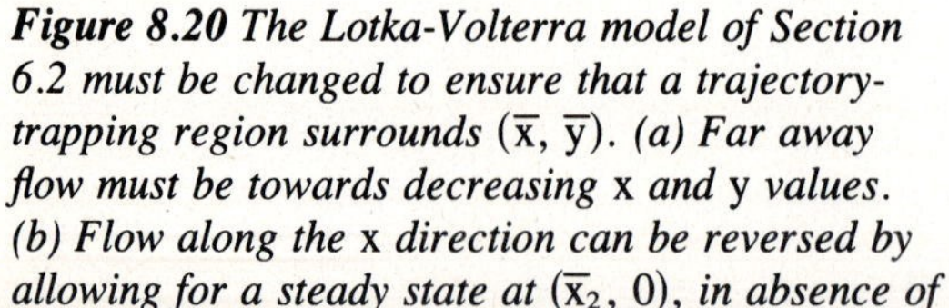

Figure 8.20 *The Lotka-Volterra model of Section 6.2 must be changed to ensure that a trajectory-trapping region surrounds ($\bar{x}$, $\bar{y}$). (a) Far away flow must be towards decreasing* x *and* y *values. (b) Flow along the* x *direction can be reversed by allowing for a steady state at ($\bar{x}_2$, 0), in absence of predators. (c) Flow along the* y *axis can be controlled by bending the* y *nullcline (g = 0). To ensure that ($\bar{x}$, $\bar{y}$) is unstable, the curve f = 0 must have positive slope at its intersection with g = 0 (hence f = 0 has a "hump"). These conditions lead to a stable limit cycle, shown in (d).*

done by assuming that there is a second steady state somewhere along the positive x axis, at $(\bar{x}_2, 0)$. At such a point the flow will be halted and reversed. The appropriate assumption is that

5. There is a value of x, say $\bar{x}_2$, such that $(\bar{x}_2, 0)$ is also a steady state.

In other words, in the absence of predators, the prey population will eventually attain a constant level given by its natural carrying capacity, here denoted as $\bar{x}_2$. The prey population does not continue to grow ad infinitum.

An easy way of achieving this predation-free steady state geometrically is to bend the curve $f(x, y) = 0$ so that it intersects the x axis at $\bar{x}_2$, in other words, so that $f(\bar{x}_2, 0) = 0$ is exactly as shown in Figure 8.20(*b*). This then implies that

6. The x-nullcline corresponding to $f(x, y) = 0$ has *a negative slope* for prey levels that are sufficiently large (so that it eventually comes down and intersects the x axis).

The condition can be made precise by implicit differentiation: Everywhere on the curve

$$f(x, y) = 0 \tag{44}$$

it is true that

$$\frac{\partial f}{\partial x}\frac{dx}{dx} + \frac{\partial f}{\partial x}\frac{dy}{dx} = 0, \tag{45}$$

that is,

$$\frac{\partial f}{\partial x} + \frac{\partial f}{\partial y} s = 0, \tag{46}$$

where $s = dy/dx$ is the slope of the curve. Requiring a negative slope means that

$$s = -\frac{\partial f/\partial x}{\partial x/\partial y} < 0 \qquad \text{(for large } x\text{)}.$$

By assumption 1 this can only be true if

$$\frac{\partial f}{\partial x} < 0 \qquad \text{(for large } x\text{)}. \tag{47}$$

This means that at large population levels the prey's growth rate diminishes as density increases. The qualifier "large x" simply means that we wish the intersection $\bar{x}_2$ to occur at prey values bigger that x_1 (see problem 17).

The above six conditions lead to the sketch shown in Figure 8.20(*b*). This restricts flow along the x axis but is not sufficient to rule out narrow but infinitely long flow regimes down the y axis, around the steady state, and back up to $y \to \infty$. The problem is alleviated by suitably positioning the "tail end" of the y nullcline $g(x, y) = 0$. Figure 8.20(*c*) illustrates the way to proceed. We see that the slope of this curve is positive for all values of x. By implicit differentiation it can be established (as previously) that this leads to a condition on g, namely

7. $$\partial g/\partial y < 0. \tag{48}$$

We see that the predator population also experiences density-dependent growth regulation and cannot grow explosively, even in an abundance of prey.

Conditions 1 to 7 result in the phase-plane behavior shown in Figure 8.20(c). To summarize, we have now obtained a subset of the xy plane, shown as a dotted rectangle in Figure 8.20(c) from which flow cannot depart. In order to obtain a limit cycle it is now necessary to ensure that the steady state $(\bar{x}, \bar{y})$ is unstable (repellent) so that the conditions of the Poincaré-Bendixson theorem will be met. To determine the constraints that this requirement imposes, we define the Jacobian of equations (42):

$$\mathbf{J} = \begin{pmatrix} \dfrac{\partial(xf)}{\partial x} & \dfrac{\partial(xf)}{\partial y} \\ \dfrac{\partial(yg)}{\partial x} & \dfrac{\partial(yg)}{\partial y} \end{pmatrix}_{(\bar{x}, \bar{y})} \equiv \begin{pmatrix} a & b \\ c & d \end{pmatrix}. \tag{49}$$

Then, since $f(\bar{x}, \bar{y}) = g(\bar{x}, \bar{y}) = 0$, we have

$$a = xf_x|_{ss}, \tag{50a}$$
$$b = xf_y|_{ss}, \tag{50b}$$
$$c = yg_x|_{ss}, \tag{50c}$$
$$d = yg_y|_{ss}. \tag{50d}$$

[where ss denotes that the quantities are to be evaluated at $(\bar{x}, \bar{y})$]. Furthermore, requiring $(\bar{x}, \bar{y})$ to be an unstable node or focus is equivalent to the conditions for two positive eigenvalues of (49):

$$a + d > 0, \tag{51a}$$
$$ad - bc > 0. \tag{51b}$$

Details leading to this result and further comments are given in problem 18. It transpires that the appropriate instability at $(\bar{x}, \bar{y})$ can only be obtained if one assumes that at $(\bar{x}, \bar{y})$ the slope of the nullcline $f(x, y) = 0$ is positive but smaller than that of the nullcline $g(x, y) = 0$; that is,

$$0 < s_f(\bar{x}, \bar{y}) < s_g(\bar{x}, \bar{y}). \tag{52}$$

where

$$s_f(x, y) = \left.\frac{dy}{dx}\right|_{(\bar{x}, \bar{y})}$$

is the slope of the nullcline $f(x, y) = 0$, and similarly s_g is the slope of $g(x, y) = 0$ at $(\bar{x}, \bar{y})$. Therefore, the curve $f(x, y) = 0$ must have curvature as shown in Figure 8.20(c), with $(\bar{x}, \bar{y})$ to the left of the hump.

With this informal reasoning we have essentially arrived at a rather famous theorem due to Kolmogorov, summarized here:

Kolmogorov's Theorem

For a predator-prey system given by equations (42a,b) let the functions f and g satisfy the following conditions:

1. $\partial f/\partial y < 0$;
2. $\partial g/\partial x > 0$;
3. For some $y_1 > 0$, $f(0, y_1) = 0$.
4. For some $x_1 > 0$, $g(x_1, 0) = 0$.
5. There exists $\bar{x}_2 > 0$ such that $f(\bar{x}_2, 0) = 0$.
6. $\partial f/\partial x < 0$ for large x (equivalently, $\bar{x}_2 > x_1$), but $\partial f/\partial x > 0$ for small x.
7. $\partial g/\partial y < 0$.
8. There is a positive steady state $(\bar{x}, \bar{y})$ that is unstable; that is

$$(xf_x + yg_y)_{ss} > 0 \qquad \text{and} \qquad (f_x g_y - f_y g_x)_{ss} > 0.$$

9. Moreover, $(\bar{x}, \bar{y})$ is located on the section of the curve $f(x, y) = 0$ whose slope is positive.

Then there is a (strictly positive) limit cycle, and the populations will undergo sustained periodic oscillations.

Figure 8.20(d) depicts the nature of these oscillations in the xy plane.

A Summary of the Biological Conditions Leading to Predator-Prey Oscillations

1. An increase in the predator population causes a net decrease in the per capita growth rate of the prey population.
2. An increase in the prey density results in an increase in the per capita growth rate of the predator population.
3. There is a predator density y_1 that will prevent a small prey population from growing.
4. There is a prey density x_1 that is minimally required to maintain growth in the predator population.

5, 6. Growth rate of the prey is density-dependent, so that there is a density $\bar{x}_2$ at which the trend reverses from net growth to net decline. At small densities an increase in the population leads to increased net rate of growth (for example, by enhancing reproduction or survivorship). The opposite is true for densities greater than $\bar{x}_2$. (This is the so-called Allee effect, discussed in Section 6.1.)

7. The predator population undergoes density-dependent growth. As density increases, competition for food or similar effects causes a net decline in the growth rate (for example, by decreasing the rate of reproduction).
8. The species coexistence at some constant levels $(\bar{x}, \bar{y})$ is unstable, so that small fluctuations lead to bigger fluctuations. This is what creates the limit-cycle oscillations.

These eight conditions are sufficient to guarantee stable population cycles in any system in which one of the species exploits the other (sometimes called an *exploiter-victim system;* see Odell, 1980, for example). Of course it should be remarked that other population interactions may also lead to stable oscillations and that we have by no means exhausted the list of reasonable interactions leading to the appropriate dynamics.

For an example of oscillations in a plant-herbivore system, see problem 21. You may wish to consult Freedman (1980) for more details and an extensive bibliography, or Coleman (1983) for a good summary of this material.

8.8 OSCILLATIONS IN CHEMICAL SYSTEMS

The discovery of oscillations in chemical reactions dates back to 1828, when A.T.H. Fechner first reported such behavior in an electrochemical cell. About seventy years later, in the late 1890s, J. Liesegang also discovered periodic precipitation patterns in space and time. The first theoretical analysis, dating back to 1910 was due to Lotka (whose models are also used in an ecological contex; see Chapter 6). However, misconceptions and old ideas were not easily changed. The scientific community held the unshakable notion that chemical reactions always proceed to an equilibrium state. The arguments used to support such intuition were based on thermodynamics of *closed* systems (those that do not exchange material or energy with their environment). It was only later recognized that many biological and chemical systems are *open* and thus not subject to the same thermodynamic principles.

Over the years other oscillating reactions have been found (see, for example, Bray, 1921). One of the most spectacular of these was discovered in 1959 by a Russian chemist, B. P. Belousov. He noticed that a chemical mixture containing citric acid, acid bromate, and a cerium ion catalyst (in the presence of a dye indicator) would undergo striking periodic changes in color, right in the reaction beaker. His results were greeted with some skepticism and disdain, although his reaction (later studied also by A. M. Zhabotinsky) has since received widespread acclaim and detailed theoretical treatment. It is now a common classroom demonstration of the effects of nonlinear interactions in chemistry. (See Winfree, 1980 for the recipe and Tyson, 1979, for a review.)

The bona fide acceptance of chemical oscillations is quite recent, in part owing to the discovery of oscillations in biochemical systems (for example, Ghosh and Chance, 1964; Pye and Chance, 1966). After much renewed interest since the early 1970s, the field has blossomed with the appearance of hundreds of empirical and theoretical publications. Good summaries and references can be found in Degn (1972), Nicolis and Portnow (1973), Berridge and Rapp (1979), and Rapp (1979).

The first theoretical work on the subject by Lotka (discussed presently) led to a reaction mechanism which, like the Lotka-Volterra model, suffered from the defect of structural instability. That is, his equations predict neutral cycles that are easily disrupted when minor changes are made in the dynamics. An important observation is that this property is shared by all *conservative systems* of which the Lotka system is an example. A simple mechanical example of a conservative system is the ideal linear pendulum: the amplitude of oscillation depends only on its initial configuration since some quantity (namely the total energy in this mechanical system) is conserved. Such systems cannot lead to the structurally stable limit-cycle oscillations. It can be shown (see box) that in the Lotka system there is a quantity that remains constant along solution curves. Each value of this constant corresponds to a distinct periodic solution. (Thus there are uncountably many closed orbits, each within an infinitesimal distance of one another.) This leads to structural instability.

Lotka's Reaction

The following mechanism was studied by Lotka (1920):

$$\begin{aligned} A + x_1 &\xrightarrow{k_1} 2x_1, \\ x_1 + x_2 &\xrightarrow{k_2} 2x_2, \\ x_2 &\xrightarrow{k_3} B. \end{aligned} \tag{53}$$

Corresponding equations are

$$\frac{dx_1}{dt} = -kAx_1 + 2k_1Ax_1 - k_2x_1x_2 = k_1Ax_1 - k_2x_1x_2 \tag{54a}$$

$$\frac{dx_2}{dt} = -k_2x_1x_2 + 2k_2x_1x_2 - k_3x_2 = k_2x_1x_2 - k_3x_2 \tag{54b}$$

When the substance A is maintained at a constant concentration, the above equations (after renaming constants) are identical with the predator-prey model analyzed in some detail in Chapter 6. The presence of a steady state with neutral cycles is thus clear.

It can also be shown that the following expression is constant along trajectories:

$$v = x_1 + x_2 - \frac{k_3}{k_2}\ln x_1 - \frac{k_1A}{k_2}\ln x_2. \tag{55}$$

To see this it is necessary to rewrite the system of equations (54) as follows:

$$\frac{dx_1}{dx_2} = \frac{k_1Ax_1 - k_2x_1x_2}{k_2x_1x_2 - k_3x_2}, \tag{56}$$

or equivalently as

$$\left(1 - \frac{k_3}{k_2x_1}\right)dx_1 + \left(1 - \frac{k_1A}{k_2x_2}\right)dx_2 = 0. \tag{57}$$

Now we shall discuss the properties of this so called *exact* equation.

Let us inquire whether the LHS of equation (57) can be written as a total differential of some scalar function $v(x_1, x_2)$. If so then

$$dv = 0, \tag{58}$$

(which means that v is a constant). This implies that

$$\frac{\partial v}{\partial x_1} dx_1 + \frac{\partial v}{\partial x_2} dx_2 = 0. \tag{59}$$

Let us define the following, taken from expressions in equation (57):

$$M = 1 - \frac{k_3}{k_2 x_1}, 1 \qquad N = 1 - \frac{k_1 A}{k_2 x_2}. \tag{60}$$

These should satisfy the relations

$$M dx_1 + N dx_2 = 0$$

where

$$M = \frac{\partial v}{\partial x_1}, \qquad N = \frac{\partial v}{\partial x_2}. \tag{61}$$

As will be emphasized in Chapter 10 in connection with partial differentiation, this can only be true when the following relation between M and N holds (provided v has continuous second partial derivatives):

$$\frac{\partial M}{\partial x_2} = \frac{\partial v}{\partial x_2 \partial x_1} = \frac{\partial v}{\partial x_1 \partial x_2} = \frac{\partial N}{\partial x_1}. \tag{62}$$

Checking equations (60) for this condition leads to

$$\frac{\partial M}{\partial x_2} = \frac{\partial}{\partial x_2}\left(1 - \frac{k_3}{k_3 x_1}\right) = 0 = \frac{\partial}{\partial x_1}\left(1 - \frac{k_1 A}{k_2 x_2}\right) = \frac{\partial N}{\partial x_1}. \tag{63}$$

Equations that satisfy (61), (62) are called *exact* equations. Once identified as such, they can be solved in a standard way (see problem 23). Geometrically the solution curves of such equations can be viewed as the *level curves* of a potential function (v in this case) or, in a more informal description, as the constant-altitude contours on a topographical map of a mountain range. (See Figure 8.21.) Each curve corresponds to a fixed "total energy", a fixed potential v, or a fixed height on the mountain, depending on the way one interprets the conserved quantity. Since there is an infinity of possible closed curves, they are neutrally but not asymptotically stable.

Just as it was possible to modify the Lotka-Volterra model in population dynamics, so too we can write modified Lotka reactions in which limit-cycle trajectories exist. (See problem 22.) This has been the basis of much recent work in the topic of oscillating reactions.

More generally, a number of guidelines and requirements for chemical oscillations has been established. A partial list follows.

Criteria for Oscillations in a Chemical System

1. The theory of Poincaré and Bendixson (in two dimensions) and Hopf bifurcations (in n dimensions) applies as before to a given system.
2. Some sort of feedback is required to obtain oscillations; following are

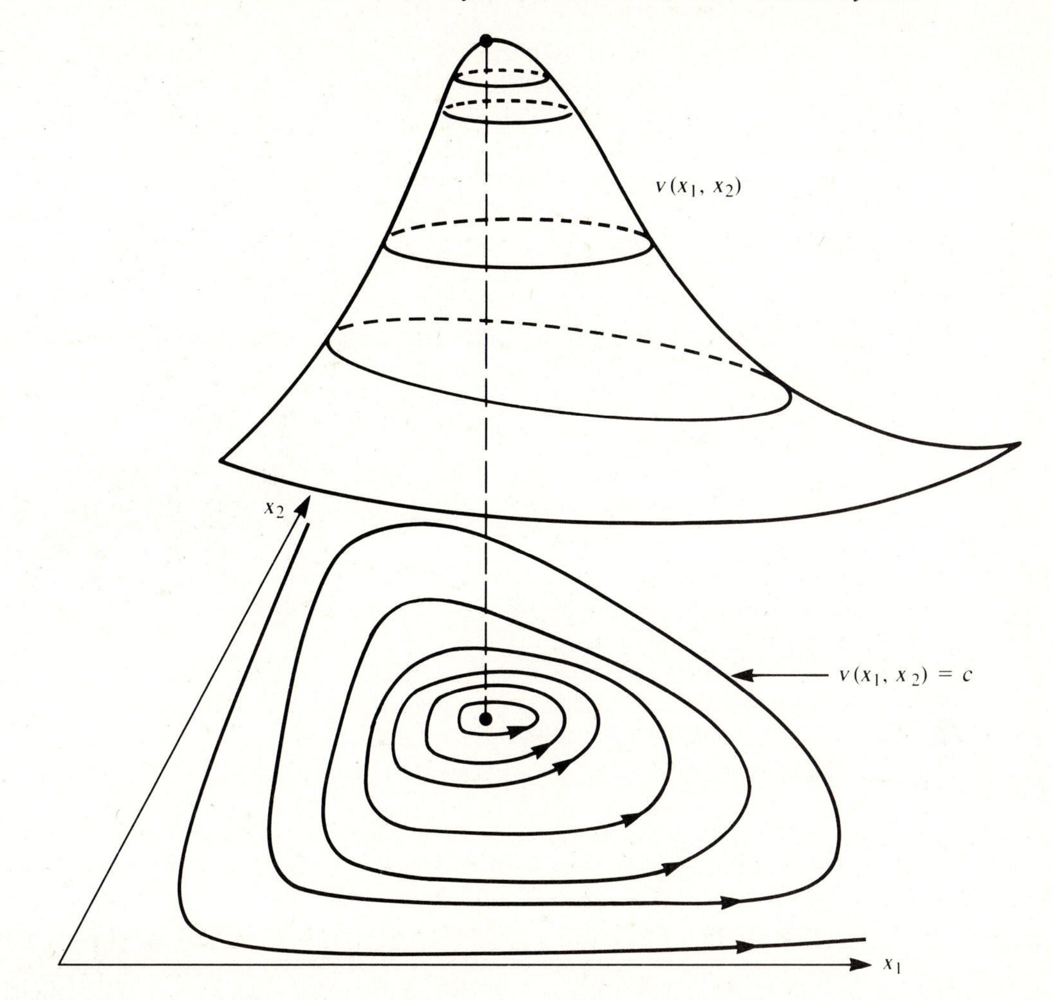

Figure 8.21 *A conservative system is characterized by the property that some quantity (for example, potential, total energy, or height on a hill) is constant along solution curves. Such systems can have neutrally stable cycles but* not *limit-cycle trajectories.*

examples of *activation* and *inhibition,* which are particular types of feedback:

activation

$$\longrightarrow a_1 \rightleftharpoons \hat{a}_2 \cdots a_{n-1} \overset{+}{\rightleftharpoons} a_n \longrightarrow$$

activation

$$a_1 \overset{+}{\longrightarrow} a_2 \longrightarrow a_3 \longrightarrow \cdots \longrightarrow a_n \longrightarrow$$

inhibition

$$a_1 \longrightarrow a_2 \overset{-}{\longrightarrow} \cdots \longrightarrow a_n \longrightarrow$$

inhibition

$$\longrightarrow a_1 \rightleftharpoons a_2 \cdots a_{n-1} \overset{-}{\rightleftharpoons} a_2 \;\; a_n \longrightarrow$$

In these examples one of the chemicals affects the rate of a reaction step in the network. (Activation implies enhancing a reaction rate, whereas inhibition implies the opposite effect.) Since most biological reactions are mediated by enzymes, a common mechanism through which such effects could occur is *allosteric modification;* for example, the substrate attaches to the enzyme, thereby causing a conformational change that reduces the activity of the enzyme. It is common to refer to a *feedback* influence if the chemical has an effect on its *precursors* (substances required for its own formation). Similarly, *feedforward* refers to a chemical's influence on its *products* (those chemicals for which it is a precursor).

Nicolis and Portnow (1973) comment particularly on chemical systems that can be described by a pair of differential equations such as

$$\frac{dx_1}{dt} = v_1(x_1, x_2), \tag{64a}$$

$$\frac{dx_2}{dt} = v_2(x_1, x_2), \tag{64b}$$

The Lotka chemical system and equations (54a,b) would be an example of this type. Equations (64a,b) indicate that to obtain limit-cycle oscillations, one of the following two situations should hold:

2a. At least one intermediate, x_1 or x_2 is *autocatalytic,* (catalyses its own production or activates a substance that produces it).

2b. One substance participates in *cross catalysis* (x_1 activates x_2 or vice versa).

These two conditions are based on the mathematical prerequisites of the Poincaré-Bendixson theory (see Nicolis and Portnow, 1973). For more than two variables it has been shown the inhibition without additional catalysis can lead to oscillations.)

3. Thermodynamic considerations dictate that *closed chemical systems* (which receive no input and cannot exchange material or heat with their environment) cannot undergo sustained oscillations because as reactants are used up the system settles into a steady state.

4. Oscillations cannot occur close to a thermodynamic equilibrium (see Nicolis and Portnow, 1973, for definition and discussion).

Schnakenberg (1979), who also considers in generality the case of chemical reactions involving *two* chemicals, concludes that when each reaction has a rate that depends monotonically on the concentrations (meaning that increasing x_1 will always increase the rate of reactions in which it participates) trajectories in the x_1x_2 phase space are bounded. Thus, when the steady state of the network is an unstable focus, conditions for limit-cycle oscillations are produced.

The following additional observations were made by Schnakenberg:

1. Chemical systems with two variables and less than three reactions cannot have limit cycles.

2. Limit cycles can only occur if one of the three reactions is autocatalytic.
3. Furthermore, in such systems limit cycles will only occur if the autocatalytic step involves the reaction of at least three molecules. When exactly three molecules are involved, the reaction is said to be *trimolecular;* for example, the reaction (65a).

Details and further references may be found in Schnakenberg (1979).

To demonstrate oscillations in a simple chemical network, we consider a system described by Schnakenberg (1979):

$$2X + Y \rightleftharpoons 3X, \tag{65a}$$

$$\mathrm{A} \longrightarrow Y, \tag{65b}$$

$$X \longrightarrow \mathrm{B}. \tag{65c}$$

The network bears similarity to the Brusselator (see problem 20 of Chapter 7). It is also related to a model for the glycolytic oscillator due to Sel'kov (1968). The reaction is said to be trimolecular since there is a reaction step in which an encounter between three molecules is required. Presently we shall use this example for the purposes of illustrating techniques rather than as a prototype of real chemical oscillators, which tend to be far more complicated.

Before beginning the calculations, it is worth remarking that the network has an autocatalytic step (X makes more X), consists of three reaction steps, and has a trimolecular step. It is thus a prime candidate for oscillations.

Equations for the Schnakenberg system (65a,b,c) are as follows:

$$\frac{dx}{dt} = x^2y - x, \tag{66a}$$

$$\frac{dy}{dt} = a - x^2y, \tag{66b}$$

where x, y, and a are the concentrations of X, Y, and A. These have a steady state at $(\bar{x}, \bar{y}) = (a, 1/a)$ and a Jacobian

$$\mathbf{J} = \begin{pmatrix} 2xy - 1 & x^2 \\ -2xy & -x^2 \end{pmatrix}_{ss} = \begin{pmatrix} 1 & a^2 \\ -2 & -a^2 \end{pmatrix}; \tag{67}$$

since

$$\beta = \operatorname{Tr} \mathbf{J} = 1 - a^2, \tag{68a}$$

$$\gamma = \det \mathbf{J} = a^2(-1 + 2) = a^2, \tag{68b}$$

we observe that eigenvalues are

$$\lambda_{1,2} = \frac{(1 - a^2) \pm \sqrt{(1 - a^2)^2 - 4a}}{2}. \tag{69}$$

If $a^2 = 1$, these are pure imaginary ($\lambda = \pm ai$), and for $a^2 < 1$ eigenvalues have a positive real part, so that the steady state is an unstable focus. The Hopf bifurcation indicates the presence of a small-amplitude limit cycle as a is decreased from $a > 1$ to $a < 1$. However, the stability of the limit cycle is uncertain. Notice that for $a = 1$ the Jacobian is not in the "normal" form necessary for direct determination of the stability expression V''' discussed in Section 8.6. While it is possible to transform variables to get this normal form, the calculations are formidably messy (and preferably avoided).

Looking at a phase-plane diagram of equations (66a,b) in Figure 8.22(a) reveals a problem: There is the possibility that close to the y axis trajectories may sweep off to $y = \infty$ so that it is impossible to define a Poincaré-Bendixson region that traps flow.

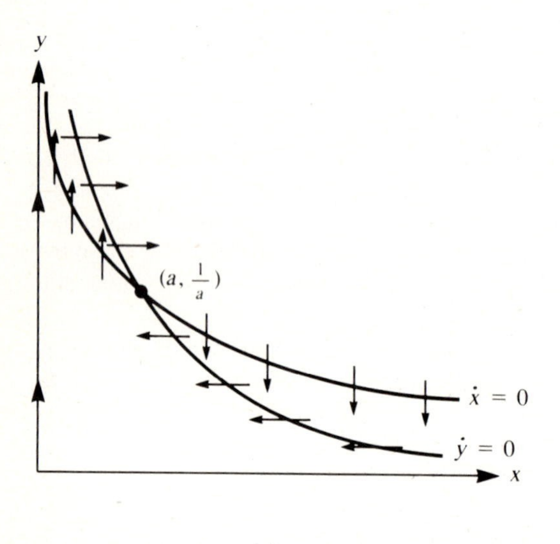

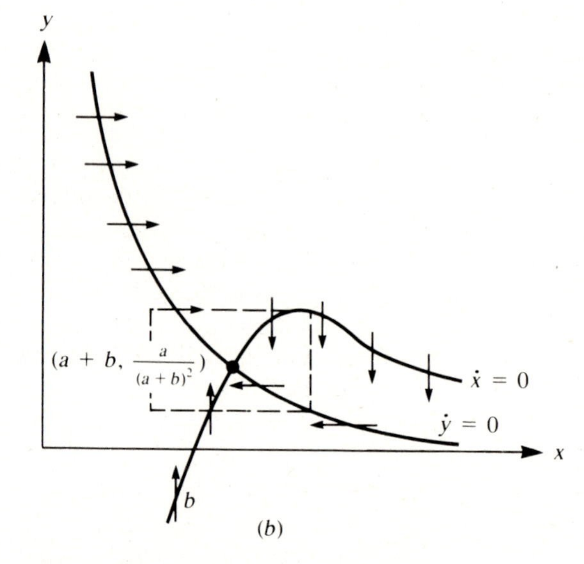

Figure 8.22 *(a) A simple chemical network taken from equations (66a,b) (suggested by Schnakenberg, 1979) has this qualitative phase plane. (b) (By modifying the system slightly to equations (70a,b), we can use the Poincaré-Bendixson theorem to conclude that a limit cycle exists somewhere in the dotted box.*

To circumvent this problem, Schnakenberg changed the last reaction step to a reversible one,

$$x \rightleftharpoons \text{B}. \qquad (65c')$$

This changes the equations into the system

$$\frac{dx}{dt} = x^2y - x + b, \qquad (70a)$$

$$\frac{dy}{dt} = -x^2y + a. \qquad (70b)$$

The new steady state is $(a + b, a/(a + b)^2)$, and the new Jacobian is

$$\mathbf{J} = \begin{pmatrix} \dfrac{a-b}{a+b} & (a+b)^2 \\ \dfrac{-2a}{a+b} & -(a+b)^2 \end{pmatrix}. \tag{71}$$

Letting

$$\beta = \text{Tr}\,\mathbf{J} = \frac{a-b}{a+b} - (a+b)^2, \tag{72a}$$

$$\gamma = \det \mathbf{J} = -(a-b)(a+b) + 2a(a+b) = (a+b)^2, \tag{72b}$$

we see that γ is always a positive quantity and that a bifurcation occurs when $\beta = 0$, that is, when

$$\frac{a-b}{a+b} = (a+b)^2. \tag{73}$$

The eigenvalues are then

$$\lambda_{1,2} = \frac{\pm\sqrt{-4\gamma}}{2} = \pm(a+b)i. \tag{74}$$

Suppose b is much smaller than a. An approximate result is that

$$\frac{a-b}{a+b} \simeq \frac{a}{a} = 1, \qquad (a+b)^2 \simeq a^2, \tag{75}$$

so that from equation (73) we have

$$a^2 \simeq 1.$$

Thus the bifurcation occurs close to $a = 1$. (For larger a values, the steady state is stable; for smaller values it is an unstable focus.) Again, the Hopf bifurcation theorem can be applied to conclude that closed periodic trajectories will be found.

Summary of Biological Factors Necessary for Limit-Cycle Dynamics

1. *Structural stability:* infinitesimal changes in parameters or terms appearing in the model description of the system should not totally disrupt the dynamic behavior (as in the Lotka-Volterra model).
2. *Open systems:* systems should be open in the thermodynamic sense. Systems that are not open have finite, nonrenewable components that are consumed. Such *closed* systems will eventually be dissipated and thus cannot undergo undamped oscillations.
3. *Feedback:* some form of autocatalysis or feedback is generally required to maintain oscillations. For example, a species may indirectly influence its growth rate by affecting a secondary species upon which it depends.
4. *Steady state:* the system must have at least one steady state. In the case of populations or chemical concentrations, this steady state must be strictly positive. For oscillations to occur, the steady state must be unstable.

5. *Limited growth:* there should be limitations on the growth rates of all intermediates in the system to ensure that *bounded* oscillations can occur.

(See Murray, 1977 for detailed discussion.)

Summary of Mathematical Criteria for the Existence of Limit Cycles

1. There must exist at least one steady state that loses stability to become an unstable node or focus. (See Appendix 1 to this chapter on *topological index theory*.)
2. At such a steady state, there must be complex eigenvalues $\lambda = a \pm bi$ whose real part a changes sign as some parameter in the equations is tuned (Hopf bifurcation).
3. A bounded annular region in xy phase space admits flow inwards but not outwards (Poincaré-Bendixson theory).
4. A region in the xy phase space must have the quantity $\partial F/\partial x + \partial G/\partial y$ change sign (Bendixson's negative criterion). This is a necessary but not a sufficient condition.
5. A region in the xy phase space must have the quantity $\partial(BF)/\partial x + \partial(BG)/\partial y$ change sign where B is any continuously differentiable function (Dulac's criterion). This is a necessary but not a sufficient condition.
6. One or more S-shaped nullclines, suitably positioned, often lead to oscillations or excitability. Examples are the Lienard equation or van der Pol oscillator.

(For more advanced methods see also Cronin, 1977.)

Now examining the phase-plane behavior of the modified system, we remark that the nullclines, given by curves

$$y = \frac{x - b}{x^2} \qquad \left(\frac{dx}{dt} = 0\right), \tag{76a}$$

$$y = \frac{a}{x^2} \qquad \left(\frac{dy}{dt} = 0\right), \tag{76b}$$

prevent flow from leaving a finite region in the first quadrant. Thus, by the Poincaré-Bendixson theory, it can be conclusively established that stable periodic behavior results whenever the steady state is unstable.

8.9 FOR FURTHER STUDY: PHYSIOLOGICAL AND CIRCADIAN RHYTHMS

One of the earliest recorded observations of biological rhythms was made by an officer in the army of Alexander the Great who noted (in 350 BC) that the leaves of certain plants were open during daytime and closed at night. Until the 1700s such rhythms were viewed as passive responses to a periodic environment, that is, to the succession of light and dark cycles due to the natural day length (see Figure 8.23).

Figure 8.23 *In 1751 Linnaeus designed a "flower clock" based on the times at which petals for various plant species opened and closed. Thus, a botanist should be able to estimate the time of day without a watch merely by noticing which flowers were open. [Drawing by Ursula Schleicher-Benz from* Lindauer Bilderbogen *no. 5, edited by Friedrich Boer. Copyright by Jan Thorbecke Verlag, eds., Sigmaringen, West Germany.]*

In 1729 the astronomer Jean Jacques d'Ortous De Mairan conducted experiments with a plant and reported that its periodic behavior persisted in a total absence of light cues. Although his results were disputed at first, further demonstrations and experiments by others (such as Wilhelm Pfeffer, in 1875, 1915) gave clear evidence in support of the observations that many physiological rhythms are *endogeneous* (independent of any external environment influences).

We now know that most organisms have innate "clocks" that govern peaks of activity, times devoted to sleep, and a variety of physiological states that fluctuate over the course of time. Rhythms close in period to the day length have been called *circadian*. Recent experimental work on a variety of organisms, including numerous plants, the fruitfly *Drosophila,* a variety of fungi, bees, birds, and mammals have been carried out (C. S. Pittendrigh, E. Bunning, J. Aschoff, R. A. Wever, A. T. Winfree). The results are often intriguing. For example, the circadian clocks of migratory birds allow them to navigate by compensating for the constantly shifting position of the sun.

As yet, our knowledge of the basic mechanisms underlying circadian rhythm, as well as other longer or shorter physiological cycles, remains uncertain. It appears that there are often numerous distinct physiological or cellular oscillators that are *coupled* (influence one another) but that maintain certain mutually independent characteristics. A brief summary of theoretical questions related to such problems is given by Winfree (1979). For fascinating historical summaries see Reinberg and Smolensky (1984) and Moore-Ede et al. (1982). Other references are provided as a starting point for further independent research on this topic. An excellent up-to-date synopsis of the human sleep-wake cycle is given by Strogatz (1986).

PROBLEMS*

1. Discuss what is meant by the statement that a limit cycle must be a *simple* closed curve. Why is Figure 8.1(*b*) not permissible? Why is Figure 8.1(*c*) not a limit cycle?

2. Consider the following system of equations (Lefschetz, 1977):

$$\frac{dx}{dt} = y = f(x, y),$$

$$\frac{dy}{dt} = x(a^2 - x^2) + by = g(x, y) \qquad (a \neq 0, b \neq 0).$$

(a) Show that the critical points are $(0, 0)$, $(a, 0)$, and $(-a, 0)$.
(b) Evaluate the expression $\partial f/\partial x + \partial g/\partial y$ at these steady states.
(c) Use Bendixson's negative criterion to rule out the presence of limit-cycle solutions about each one of these steady states.
(d) Sketch the phase-plane behavior.

***3.** Use Bendixson's negative criterion to show that if P is an isolated saddle point there cannot be a limit cycle in the neighborhood of P that contains only P.

4. Use Bendixson's negative criterion to indicate whether or not limit cycles can be ruled out in the following systems of equations in the indicated regions of the plane:

* Problems preceded by an asterisk (*) are especially challenging.

(a) $\dfrac{dx}{dt} = 2x - 3y,$

$\dfrac{dy}{dt} = 10x + y,$

(in R^2).

(b) $\dfrac{dx}{dt} = x + 2y(\sin^2 y) + y^{1/2},$

$\dfrac{dy}{dt} = xe^x + \dfrac{x^2}{2} + y,$

(in R^2).

(c) $\dfrac{dx}{dt} = a_1x + b_1y + c_1x^2,$

$\dfrac{dy}{dt} = d_2x + b_2y + c_2y^2,$

$(x > 0,\ y > 0)$.

(d) $\dfrac{dx}{dt} = ax - bxy,$

$\dfrac{dy}{dt} = -cy + dxy,$

$(x > 0,\ y > 0)$.

5. Show that Dulac's criterion but not Bendixson's criterion can be used to establish the fact that no limit cycles exist in the two-species competition model given in Section 6.3. (Hint: Let $B(x, y) = 1/xy$.)

6. Consider the following system of equations (Odell, 1980), which are said to describe a predator-prey system:

$$\frac{dx}{dt} = x[x(1 - x) - y],$$

$$\frac{dy}{dt} = k\left(x - \frac{1}{\mu}\right)y.$$

(a) Interpret the terms in these equations.

(b) Sketch the nullclines in the xy plane and determine whether the Poincaré-Bendixson theory can be applied.

(c) Show that the steady states are located at

$$(0, 0), \qquad (1, 0), \qquad \left(\frac{1}{\mu}, \frac{1 - 1/\mu}{\mu}\right).$$

(d) Show that at the last of these steady states the linearized system is characterized by the matrix

$$\begin{pmatrix} \dfrac{1}{\mu}\left(1 - \dfrac{2}{\mu}\right) & -\dfrac{1}{\mu} \\ \dfrac{k}{\mu}\left(1 - \dfrac{1}{\mu}\right) & 0 \end{pmatrix}$$

(e) Can the Bendixson negative criterion be used to rule out limit-cycle oscillations?

(f) If your results so far are not definitive, consider applying the Hopf bifurcation theorem. What is the stability of the strictly positive steady state? What is the bifurcation parameter, and at what value does the bifurcation occur?

***(g)** Show that at bifurcation the matrix in part (d) is not in the "normal" form required for Hopf stability calculations. Also show that if you transform the variables by defining

$$\tilde{x} = x, \qquad \tilde{y} = \frac{2}{k^{1/2}}y,$$

you obtain a new system that is in normal form.

***(h)** Find the transformed system, calculate V''' and show that it is negative.

(i) What conclusions can be drawn about the system?

7. If all steady states (indicated by large dots) in Figures (a) through (d) are unstable, can the Poincaré-Bendixson theorem be applied to prove existence of limit cycles?

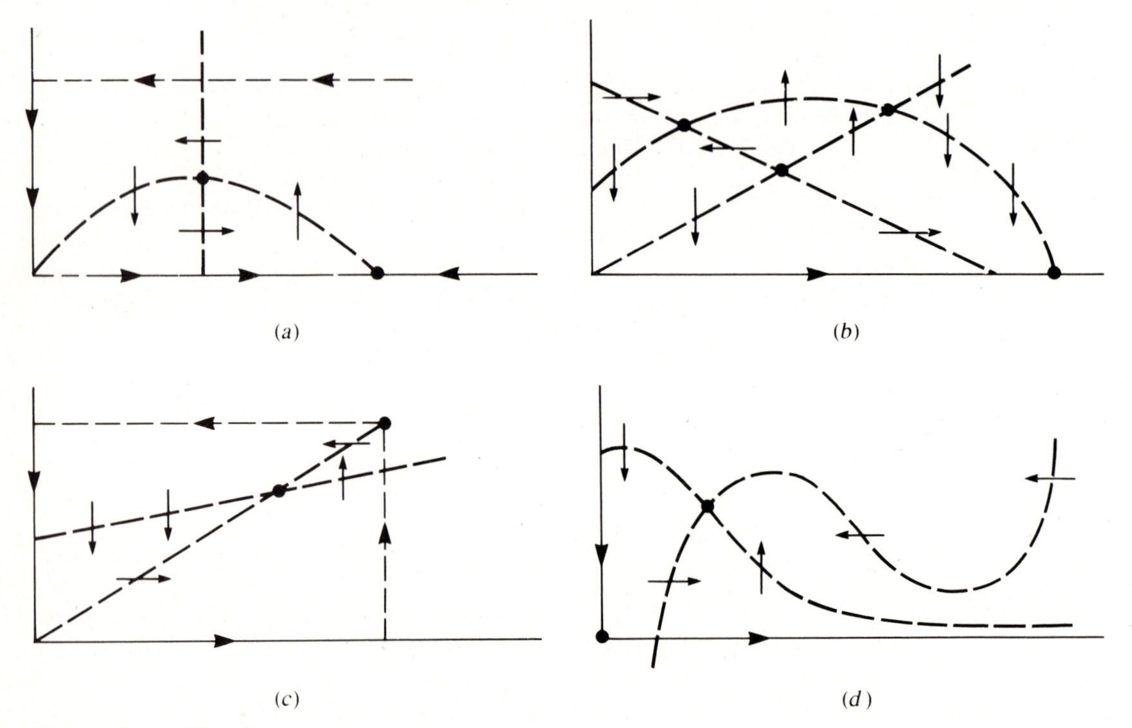

Figures for problem 7.

8. *The Hodgkin-Huxley model*

(a) What might be an interpretation of equations (10) and (11)? Of (12a–c)?

(b) Sketch the functions given in equations (13a–c). You may do this by hand or by computer.

(c) Discuss Fitzhugh's assumption that V and m change more rapidly than h and n.

(d) Demonstrate that the nullclines of Fitzhugh's reduced system are as shown in Figure 8.10(a).

9. *The Fitzhugh model*

(a) Verify the Jacobian given by (28) and the stability conditions of equations (30a,b).

(b) Show that the steady state of Fitzhugh's model is stable if it falls in the range

$$-\gamma \leq \bar{x} \leq \gamma$$

provided $b < 1$, $b < c^2$ and $\gamma = (1 - b/c^2)^{1/2}$.

(c) Show that this constraint places $\bar{x}$ on the portion of the cubic nullcline between the two humps.

10. (a) Graph the functions $G_1(u) = u^3$ and $G_2(u) = u^3/3 + 1$. In what way do these differ from the function $v = G(u)$ shown in Figure 8.13(*a*)?

(b) Draw nullclines and sketch the phase-plane flow behavior for the system of equations (15) where $G = G_1(u)$ and $G = G_2(u)$ given in part (a).

11. (a) Show that the Lienard equation (19) is equivalent to the system of equations (15a,b) where $G(u)$ is given by equation (20).

(b) Give justification for the claim that condition 1 of Section 8.4 guarantees that all trajectories will be symmetric about the origin.

(c) Show that condition 3 implies that (0, 0) is an unstable steady state of (15) where $G(u)$ is given by equation (20).

(d) If $G(u)$ satisfies conditions 1 to 3 and $\alpha = \beta$, give a reason for the assertion that there can be only one limit cycle in the Lienard system.

***12. (a)** *The van der Pol oscillator.* Show that equations (15a,b) and (16) are equivalent to equation (21).

(b) Suppose ϵ in equations (22a,b) is a small quantity and that the solutions to (22) can be expressed as

$$u(t) = f_0(t) + \epsilon f_1(t) + \epsilon^2 f_2(t) + \cdots + \epsilon^n f_n(t) + \cdots$$
$$v(t) = g_0(t) + \epsilon g_1(t) + \epsilon^2 f_2(t) + \cdots + \epsilon^n g_n(t) + \cdots.$$

What equations do the functions f_0, g_0, f_1, g_1, f_2, and g_2 satisfy?

13. (a) Suggest a possible molecular mechanism that might lead to the equations derived by Murray (1981) and thus the nullcline formation shown in Figure 8.14. (*Hint:* refer to problem 22 of Chapter 7.)

(b) Of the various configurations in Figure 8.14, which would you expect to lead to a limit-cycle oscillation?

(c) Is it possible to determine the steady state and its stability directly from Murray's equations?

14. (a) Suggest a possible mechanism that would lead to the equations given by Fairen and Velarde (1979) for bacterial respiration (see Figure 8.15).

(b) How does this model compare with that of Murray?

15. (a) Demonstrate that the Segel-Goldbeter model for oscillations in the cyclic AMP signaling system leads to the phase-plane configuration shown in Figure 8.16. What has been assumed to obtain the reduced $\gamma\alpha$ system?

(b) Investigate the mechanism underlying the assumption for Φ that supposedly depicts allosteric kinetics of adenylate cyclase. (You may wish to consult original papers by Segel and Goldbeter or Segel, 1984.)

16. Morris and Lecar (1981) describe a semiqualitative model for voltage oscillations in the giant muscle fiber of the barnacle. In this system the important ions are potassium and calcium (not sodium). The equations they suggest are the following:

$$I = C\frac{dv}{dt} + g_{\mathrm{L}}(v - v_{\mathrm{L}}) + g_{\mathrm{Ca}}M(v - v_{\mathrm{Ca}}) + g_{\mathrm{K}}N(v - v_{\mathrm{K}}),$$

$$\frac{dM}{dt} = \lambda_M(v)[M_\infty(v) - M],$$

$$\frac{dN}{dt} = \lambda_N(v)[N_\infty(v) - N],$$

where

v = voltage,

M = fraction of open Ca^{2+} channels,

N = fraction of open K^+ channels.

(a) Interpret these three equations. For reasons detailed in their papers Morris and Lecar define the functions M_∞, λ_M, N_∞, and λ_N as follows:

$$M_\infty(v) = \frac{1}{2}\left(1 + \tanh\frac{v - v_1}{v_2}\right), \qquad N_\infty(v) = \frac{1}{2}\left(1 + \tanh\frac{v - v_3}{v_4}\right),$$

$$\lambda_M(v) = \overline{\lambda}_M \cosh\frac{v - v_1}{2v_2}, \qquad \lambda_N(v) = \overline{\lambda}_N \cosh\frac{v - v_3}{2v_4}.$$

(b) Sketch or describe the voltage dependence of these functions.

Note: $\cosh x = \dfrac{e^x + e^{-x}}{2}, \qquad \sinh x = \dfrac{e^x - e^{-x}}{2},$

$$\tanh x = \frac{\sinh x}{\cosh x}.$$

(c) Morris and Lecar consider the reduced vN system to be an approximation to the whole model. What assumption underlies this approximation?

(d) Show that in the reduced vN system the variables are constrained to satisfy the following inequalities:

$$0 < N < 1,$$

$$\frac{g_{\mathrm{L}}v_{\mathrm{L}} + g_{\mathrm{K}}v_{\mathrm{K}} + I}{g_{\mathrm{L}} + g_{\mathrm{K}}} < v < \frac{g_{\mathrm{L}}v_{\mathrm{L}} + g_{\mathrm{Ca}}v_{\mathrm{Ca}} + I}{g_{\mathrm{L}} + g_{\mathrm{Ca}}}.$$

(You may need to use the fact that $0 < M < 1$.)

(e) Give support for the configuration of nullclines shown in Figure 8.17(a). Notice that there is an intersection within the region described by the inequalities in part (d) of this problem.

(f) Suppose you are told that at the steady state the linearized system has a pair of complex eigenvalues $\lambda = a + bi$ such that $a + bi$ behaves as in Figure 8.17(b) as the current I changes. Use your results from parts (a) to (e) to make a statement about the existence and stability of a limit-cycle solution.

(g) Interpret this in the biological context of electrical signal propagation in the barnacle giant muscle fiber.

(h) Compare the assumptions and results obtained from this model to the Hodgkin-Huxley and Fitzhugh models.

(This problem could be extended to an independent project with in-class discussion. See Morris and Lecar, 1981, for other details.)

17. *Limit cycles in predator-prey systems*. In this problem we investigate details that arise in Section 8.7.

(a) Justify the particular form of equations (42a,b) used in discussing a predator-prey system.

***(b)** Suppose the orientations of the nullclines is as shown in Figure 8.20(a). Use conditions 1 and 2 in Section 8.7, along with the fact that *on* these curves $f(x, y) = 0$ and $g(x, y) = 0$ respectively, to reason that the direction of flow along nullclines must conform to that shown in Figure 8.20(a). (*Hint:* Use the inequalities to determine in which regions f, or g, must be positive and in which negative.)

(c) A condition for a limit cycle is that $\bar{x}_2 > x_1$. Interpret this biologically and describe why this inequality is necessary.

(d) Verify condition 7 [equation (48)] by implicit differentiation of $g(x, y) = 0$.

(e) A steady state $(\bar{x}, \bar{y})$ is unstable if any *one* of the eigenvalues of the equations [linearized about $(\bar{x}, \bar{y})$] has a positive real part. Why then is it necessary to have *two* positive eigenvalues to ensure that a limit cycle exists?

18. In this problem we compute the stability properties of the nonzero steady state of the predator-prey equations (42a,b).

(a) Find the Jacobian of equations (42a,b).

(b) Show that conditions for an unstable node or spiral at $(\bar{x}, \bar{y})$ are as given in equations (51a–b).

(c) Verify the following relationship between the slope s_g of the nullcline $g = 0$ and the partial derivatives of g:

$$\text{slope of } g \text{ nullcline:} \quad s_g = -\frac{g_x}{g_y}.$$

(d) Use the inequality $a + d > 0$ and the inequalities $f_y < 0$, $g_x > 0$, and $g_y < 0$ to establish the following result: for $(\bar{x}, \bar{y})$ to be an unstable node or spiral, it is necessary and sufficient that

$$\bar{x} f_x(\bar{x}, \bar{y}) > \bar{y} \, | g_y(\bar{x}, \bar{y}) | .$$

Why does this imply that

$$f_x(\bar{x}, \bar{y}) > 0?$$

Why does it follow that

$$s_f(\bar{x}, \bar{y}) > 0,$$

in other words, that the slope of the curve $f(x, y) = 0$ must be positive at the steady state?

(e) Use the inequality $ad - bc > 0$ to show that at the steady state

$$\bar{x}\,\bar{y} f_y g_x \left(\frac{s_f}{s_g} - 1 \right) > 0.$$

Note: all quantities are evaluated at the steady state, as in part (i). Reason that this inequality implies that

$$s_f(\bar{x}, \bar{y}) < s_g(\bar{x}, \bar{y}).$$

19. The following predator-prey system is discussed by May (1974):

$$\frac{dH}{dt} = rH\left(1 - \frac{H}{K}\right) - \frac{kPH}{H + D} \qquad \text{(host)},$$

$$\frac{dP}{dt} = sP\left(1 - \frac{P}{\gamma H}\right) \qquad \text{(parasite)}.$$

(a) Interpret the terms appearing in these equations and suggest what the various parameters might represent.
(b) Sketch the H and P nullclines on an HP phase plane.
(c) Apply Bendixson's criterion and Dulac's criterion (for $B = 1/HP$).

20. A number of modifications of the Lotka-Volterra predator-prey model that have been suggested over the years are given in the last box in Section 6.2. By considering several combinations of prey density-dependent growth and predator density-dependent attack rate, determine whether such modifications might lead to limit-cycle oscillations. You may wish to do the following:
(a) Check to see whether the Kolmogorov conditions are satisfied.
(b) Plot nullclines by hand (or by writing a simple computer program and check to see whether the Poincaré-Bendixson conditions are satisfied.
(c) Determine the stability properties of steady states.

21. This problem arises in a model of a plant-herbivore system: (Edelstein-Keshet, 1986). Assume that a population of herbivores of density y causes changes in the vegetation on which it preys. An internal variable x reflects some physical or chemical property of the plants which undergo changes in response to herbivory. We refer to this attribute as the *plant quality* of the vegetation and assume that it may in turn affect the fitness or survivorship of the herbivores. If this happens in a graded, continuous interaction, plant quality may be modeled by a pair of ODEs such as

$$\frac{dx}{dt} = f(x, y) \qquad \text{(rate of change of vegetation quality)},$$

$$\frac{dy}{dt} = yg(x, y) \qquad \text{(herbivore density)}.$$

(a) In one case the function $f(x, y)$ is assumed to be

$$f(x, y) = x(1 - x)[\alpha(1 - y) + x] \qquad (0 \leq x \leq 1).$$

Sketch this as a function of x and reason that the plant quality x always remains within the interval $(0, 1)$ if $x(0)$ is in this range. Show that plant quality may either decrease or increase depending on (1) initial value of x

and (2) population of herbivores. For a given herbivore population density $\hat{y}$, what is the "breakeven" point (the level of x for which $dx/dt = 0$)?

(b) It is assumed that the herbivore population undergoes logistic growth (see Section 6.1) with a carrying capacity that is directly proportional to current plant quality and reproductive rate β. What is the function g?

(c) With a suitable definition of constants in this problem your equations should have the following nullclines:

$$\begin{aligned} x \text{ nullclines:} \quad & x = 0, \\ & x = 1, \\ & x = \alpha(y - 1), \\ y \text{ nullclines:} \quad & y = 0, \\ & y = Kx. \end{aligned}$$

Draw these curves in the xy plane. (There is more than one possible configuration, depending on the parameter values.)

(d) Now find the direction of motion along all nullclines in part (c). Show that under a particular configuration there is a set in the xy plane that "traps" trajectories.

(e) Define $\gamma = \alpha/(\alpha K - 1)$. Interpret the meaning of this parameter. Show that $(\gamma, K\gamma)$ is a steady state of your equations and locate it on your phase plot. Find the other steady state.

(f) Using stability analysis, show that for $\gamma > 1$, $(\gamma, K\gamma)$ is a saddle point whereas for $\gamma < 1$ it is a focus.

(g) Now show that as β decreases from large to small values, the steady state ($\gamma < 1$) undergoes the transition from a stable to an unstable focus.

(h) Use your results in the preceding parts to comment on the existence of periodic solutions. What would be the biological interpretation of your answer?

22. Lotka's chemical model given in Section 8.8 is dynamically equivalent to the Lotka-Volterra predator-prey model. Thus modifications in the chemical kinetics should result in system(s) that exhibit limit-cycle oscillations. Explore whether the Kolmogorov conditions (see box in Section 8.8) apply to a chemical system; if so, interpret the inequalities (points 1 through 8 in the box) in the context of chemical (rather than animal) species.

23. *Exact equations and Lotka's model.* Let M and N be the functions defined by equations (60), and let V be some function satisfying

$$\frac{\partial V}{\partial x_1} = M, \qquad \frac{\partial V}{\partial x_2} = N.$$

Define

$$V_1 = \int M\, dx_1 + h(x_2),$$

$$V_2 = \int N\, dx_2 + g(x_1).$$

The integrals are to be performed with respect to one of the variables only, the other one being held fixed. Show that

$$V_1 = x_1 - \frac{k}{k_2} \ln x_1 + h(x_2),$$

$$V_2 = x_2 - \frac{k_1 A}{k_2} \ln x_2 + g(x_1).$$

If $V_1 = V_2 = V$, show that V must be given by equation (55).

24. The following equations were developed by Goodwin (1963) as a model of protein-*m*RNA interactions:

$$\frac{dM}{dt} = \frac{1}{1 + E} - \alpha, \qquad \frac{dE}{dt} = M - \beta.$$

Show that this system is conservative and has oscillatory solutions.

In the remaining problems we consider systems in which it is possible to explicitly solve for limit cycles and deduce their stability by transforming variables to polar form.

25. Consider a system of equations such as (1a,b). Transform variables by defining

$$x = r \cos \theta, \qquad y = r \sin \theta.$$

(a) Show that

$$r^2 = x^2 + y^2, \qquad \theta = \arctan \frac{y}{x}.$$

(b) Show that

$$\frac{1}{2}\frac{d(r^2)}{dt} = x\frac{dx}{dt} + y\frac{dy}{dt}.$$

(c) Verify that

$$r^2 \frac{d\theta}{dt} = x\frac{dy}{dt} - y\frac{dx}{dt}.$$

26. Consider the equations

$$\frac{dx}{dt} = y, \qquad \frac{dy}{dt} = -x.$$

(b) By transforming variables, obtain

$$\frac{dr^2}{dt} = 0, \qquad \frac{d\theta}{dt} = -1.$$

(b) Conclude that there are circular solutions. What is the direction of rotation? Are these cycles stable?

27. Show that the system

$$\frac{dx}{dt} = -2y, \qquad \frac{dy}{dt} = x - 5,$$

has closed elliptical orbits. (*Hint:* Consider first transforming x to $x - 5$ and y to $y/2$ and then using a polar transformation.)

28. Consider the nonlinear system of equations

$$\frac{dx}{dt} = -y + x(x^2 + y^2 - 1) = f(x, y),$$

$$\frac{dy}{dt} = x + y(x^2 + y^2 - 1) = g(x, y).$$

Show that $r = 1$ is an unstable limit cycle of the equations.

29. Lefschetz (1977) discusses the following system of equations

$$\frac{dx}{dt} = -y + xf(x^2 + y^2),$$

$$\frac{dy}{dt} = x + yf(x^2 + y^2).$$

Show that this system is equivalent to the polar equation

$$\frac{dr}{d\theta} = rf(r^2).$$

30. Find the polar form of the following equations and determine whether periodic trajectories exist. If so, find their stability.

(a) $$\frac{dx}{dt} = \frac{2\pi y}{1 + (x^2 + y^2)^{1/2}}, \qquad \frac{dy}{dt} = \frac{2\pi x}{1 + (x^2 + y^2)^{1/2}}.$$

(b) $$\frac{dx}{dt} = y + \frac{x}{(x^2 + y^2)^{1/2}}[1 - (x^2 + y^2)],$$

$$\frac{dy}{dt} = -x + \frac{y}{(x^2 + y^2)^{1/2}}[1 - (x^2 + y^2)].$$

Problems 31 and 32 follow Appendix 1 for Chapter 8.

REFERENCES

General Mathematical Sources

Arnold, V. I. (1981). *Ordinary Differential Equations.* MIT Press, Cambridge, Mass.

Cronin, J. (1977). Some Mathematics of Biological Oscillations, *SIAM Rev., 19,* 100–138.

Golubitsky, M., and Schaeffer, D. G. (1984). *Singularity and Groups in Bifurcation Theory.* (*Applied Mathematical Sciences,* vol. 51.) Springer-Verlag, New York.

Guckenheimer, J. and Holmes, P. (1983). *Nonlinear Oscillations, Dynamical Systems, and Bifurcations of Vector Fields.* Springer-Verlag, New York.

Hale, J. K. (1980). *Ordinary Differential Equations.* Krieger, Huntington, New York.

Lefschetz, S. (1977). *Differential Equations: Geometric Theory.* 2d ed. Dover, New York.

Minorsky, N. (1962). *Nonlinear Oscillations.* Van Nostrand, New York.

Odell, G. M. (1980). Appendix A3 in L. A. Segel, ed., *Mathematical Models in Molecular and Cellular Biology*. Cambridge University Press, Cambridge.

Ross, S. (1984). *Differential Equations*. 3d ed. Wiley, New York, chap. 13 and pp. 693–695.

Oscillations (Summaries and Reviews)

Berridge, M. J., and Rapp, P. E. (1979). A comparative survey of the function, mechanism and control of cellular oscillators. *J. Exp. Biol., 81,* 217–279.

Chance, B.; Pye, E. K.; Ghosh, A. K.; and Hess, B. (1973). *Biological and Biochemical Oscillators*. Academic Press, New York.

Goodwin, B. C. (1963). *Temporal Organization in Cells*. Academic Press, London.

Hoppensteadt, F. C., ed. (1979). *Nonlinear Oscillations in Biology*. (*AMS Lectures in Applied Mathematics*, vol. 17). American Mathematical Society, Providence, R.I.

Murray, J. D. (1977). *Lectures on Nonlinear Differential Equation Models in Biology*. Clarendon Press, Oxford, chap. 4.

Pavlidis, T. (1973). *Biological Oscillators: Their mathematical analysis*. Academic Press, New York.

Rapp, P. E. (1979). Bifurcation theory, control theory and metabolic regulation. In D. A. Linkens, ed., *Biological Systems, Modelling and Control*. Peregrinus, New York.

Winfree, A. T. (1980). *The Geometry of Biological Time*. Springer-Verlag, New York.

Nerves and Neuronal Excitation

Eckert, R., and Randall, D. (1978). *Animal Physiology*. W. H. Freeman, San Francisco.

Fitzhugh, R. (1960). Thresholds and plateaus in the Hodgkin-Huxley nerve equations. *J. Gen. Physiol., 43,* 867–896.

Fitzhugh, R. (1961). Impulses and physiological states in theoretical models of nerve membrane. *Biophys. J., 1,* 445–466.

Hodgkin, A. L. (1964). *The Conduction of the Nervous Impulse*. Liverpool University Press, Liverpool.

Hodgkin, A. L., and Huxley, A. F. (1952). A quantitative description of membrane current and its application to conduction and excitation in nerve. *J. Physiol., 117,* 500–544.

Kuffler, S. W.; Nicholls, J. G.; and Martin, A. R. (1984). *From Neuron to Brain*. Sinauer Associates, Sunderland, Mass.

Morris, C. and Lecar, H. (1981). Voltage oscillations in the barnacle giant muscle fiber. *Biophys. J., 35,* 193–213.

Nagumo, J., Arimoto, S., and Yoshizawa, S. (1962). An active pulse transmission line simulating nerve axon. *Proc. IRE. 50,* 2061–2070.

Models with S-shaped ("Cubic") Nullclines

Fairen, V., and Velarde, M. G. (1979). Time-periodic oscillations in a model for the respiratory process of a bacterial culture. *J. Math. Biol., 8,* 147–157.

Goldbeter, A., and Martiel, J. L. (1983). A critical discussion of plausible models for relay and oscillation of cyclic AMP in *dictyostelium* cells. In M. Cosnard, J. Demongeot, and A. L. Breton, eds., *Rhythms in Biology and Other Fields of Application*. Springer-Verlag, New York, pp. 173–188.

Goldbeter, A., and Segel, L. A. (1977). Unified mechanism for relay and oscillations of cyclic AMP in *Dictyostelium discoideum*. *Proc. Natl. Acad. Sci. USA, 74,* 1543–1547.

Murray, J. D. (1981). A Pre-pattern formation mechanism for animal coat markings. *J. Theor. Biol., 88,* 161–199.

Segel, L. A. (1984). *Modeling dynamic phenomena in molecular and cellular biology*. Cambridge University Press, Cambridge, chap. 6.

The Hopf Bifurcation and Its Applications

See previous references and also the following:

Marsden, J. E., and McCracken, M. (1976). *The Hopf Bifurcation and Its Applications*. (*Applied Mathematical Sciences,* vol. 19.) Springer-Verlag, New York.

Pham Dinh, T.; Demongeot, J.; Baconnier, P.; and Benchetrit, G. (1983). Simulation of a biological oscillator: The respiratory system. *J. Theor. Biol., 103,* 113–132.

Rand, R. H.; Upadhyaya, S. K.; Cooke, J. R.; and Storti, D. W. (1981). Hopf bifurcation in a stomatal oscillator. *J. Math. Biol., 12,* 1–11.

The van der Pol Oscillator

See Hale (1980), Minorsky (1962), Ross (1984) and also the following:

Jones, D. S., and Sleeman, B. D. (1983). *Differential Equations and Mathematical Biology*. Allen & Unwin, London.

van der Pol, B. (1927). Forced oscillations in a circuit with nonlinear resistance (receptance with reactive triode). *Phil. Mag.* (London, Edinburgh, and Dublin), *3,* 65–80.

van der Pol, B., and van der Mark, J. (1928). The heart beat considered as a relaxation oscillation, and an electrical model of the heart. *Phil. Mag.* (7th ser.), *6,* 763–775.

Limit Cycles in Population Dynamics

Coleman, C. S. (1983). Biological cycles and the five-fold way. Chap. 18 in M. Braun, C. S. Coleman, and D. Drew, eds., *Differential Equation Models*. Springer-Verlag, New York.

Edelstein-Keshet, L. (1986). Mathematical theory for plant-herbivore systems. *J. Math. Biol., 24,* 25–58.

Freedman, H. I. (1980). *Deterministic Mathematical Models in Population Ecology*. Dekker, New York.

Kolmogorov, A. (1936). Sulla Teoria di Volterra della Lotta per l'Esistenza, *G. Ist. Ital. Attuari, 7,* 74–80.

May, R. M. (1974). *Stability and Complexity in Model Ecosystems*. 2d ed. Princeton University Press, Princeton, N.J.

Rescigno, A., and Richardson, I. W. (1967). The struggle for life I: Two species, *Bull. Math. Biophys., 29,* 377–388.

Oscillations in Chemical Systems

Belousov, B. P. (1959). An oscillating reaction and its mechanism. *Sb. Ref. Radiats. Med.*, Medgiz, Moscow. p. 145. (In Russian.)
Bray, W. C. (1921). A periodic reaction in homogeneous solution and its relation to catalysis. *J. Amer. Chem. Soc.*, *43*, 1262–1267.
Degn, H. (1972). Oscillating chemical reactions in homogeneous phase. *J. Chem. Ed.*, *49*, 302–307.
Escher, C. (1979). Models of chemical reaction systems with exactly evaluable limit cycle oscillations. *Z. Physik B*, *35*, 351–361.
Ghosh, A., and Chance, B. (1964). Oscillations of glycolytic intermediates in yeast cells. *Biochem. Biophys. Res. Commun.*, *16*, 174–181.
Hyver, C. (1984). Local and global limit cycles in biochemical systems. *J. Theor. Biol.*, *107*, 203–209.
Lotka, A. J. (1920). Undamped oscillations derived from the law of mass action. *J. Amer. Chem. Soc.*, *42*, 1595–1599.
Nicolis, G., and Portnow, J. (1973). Chemical oscillations. *Chem. Rev.*, *73*, 365–384.
Pye, K., and Chance, B. (1966). Sustained sinusoidal oscillations of reduced pyridine nucleotide in a cell-free extract of *Saccharomyces carlsbergensis*. *Proc. Natl. Acad. Sci.*, *55*, 888–894.
Schnakenberg, J. (1979). Simple chemical reaction systems with limit cycle behavior, *J. Theor. Biol.*, *81*, 389–400.
Sel'kov, E. E. (1968). Self-oscillations in glycolysis. A simple kinetic model. *Eur. J. Biochem.*, *4*, 79–86.
Tyson, J. J. (1979). Oscillations, bistability, and echo waves in models of the Belousov-Zhabotinskii reaction. In O. Gurel and O. E. Rossler, eds., *Bifurcation Theory and Applications in Scientific Disciplines. Ann. N.Y. Acad. Sci.*, *316*, pp. 279–295.
Zhabotinskii, A. M. (1964). Periodic process of the oxidation of malonic acid in solution. *Biofizika*, *9*, 306–311. (In Russian.)

Circadian and Physiological Oscillations

See Winfree (1980) as well as the following:

Brown, F. A.; Hastings, J. W.; and Palmer, J. D. (1970). *The Biological Clock: Two Views*. Academic Press, New York.
Bunning, E. (1967). *The Physiological Clock*. 2d ed. Springer-Verlag, New York.
Edmunds, L. N., ed. (1984). *Cell Cycle Clocks*. Marcel Dekker, New York.
Kronauer, R. E.; Czeisler, C. A.; Pilato, S. F.; and Moore-Ede, M. C. and Weitzman, E. D. (1982). Mathematical model of the human circadian system with two interacting oscillators. *Am. J. Physiol.*, *242*, R3–R17.
Moore-Ede, M. C.; Czeisler, C. A., eds. (1984). *Mathematical Models of the Circadian Sleep-Wake Cycle*. Raven Press, New York.
Moore-Ede, M. C.; Sulzman, F. M.; and Fuller, C. A. (1982). *The Clocks that Time Us*. Harvard University Press, Cambridge, Mass.
Reinberg, A., and Smolensky, M. H. (1984). *Biological Rhythms and Medicine*. Springer-Verlag, New York.
Strogatz, S. (1986). *The Mathematical Structure of the Human Sleep-Wake Cycle, Lecture Notes in Biomathematics*, *69*, Springer-Verlag, New York.
Ward, R. R. (1971). *The Living Clocks*. Knopf, New York.

APPENDIX 1 TO CHAPTER 8: SOME BASIC TOPOLOGICAL NOTIONS

Many of the theorems that apply to systems of two ordinary differential equations such as (1a,b) are based on what are called *topological properties* of curves in the plane. While the topological theory itself is abstract and exceptionally beautiful, we shall avoid formal details, giving instead a descriptive outline of some key concepts. (See Arnold, 1981, for an introductory summary.)

Imagine a vector field in the xy plane. The vector field may have been generated by the set of equations (1a,b). Figure 8.24 serves as an example. In this vector field we place a *simple oriented closed curve;* (we may think of this curve as a deformed circle along which some point moves in a particular direction). We first consider arbitrary curves as "test" objects that are used to understand the vector field. (Only later will we turn to closed curves generated by the vector field itself.) Tracing the path of a single point around the test curve we may follow the rotation of a field vector attached to the point. If the curve does not go through any singular point [point at which $F = G = 0$ in equations (1)], the rotation will be continuous as the curve is traversed, and the field vector will have returned to its original orientation when the point has returned to its initial position. The *number of revolutions* and the *direction of rotation* executed by the field vector depend on details of the flow patterns. (Several possibilities are shown in Figure 8.24.) The *index of a curve* is the number of revolutions, and the sign of the index reflects whether the rotation is in the same sense or in an opposite sense to that of the curve.

A number of properties make the concept of the index important:

1. The index does not change if the curve is distorted, twisted, or enlarged, provided that in the process of change the curve does not go through any singular points.

2. Similarly, if the vector field is warped, distorted, or rearranged, the index of the curve will not change if no singular points cross the curve or are on the curve during the process.

Using such properties it can be shown (see Problem 31a) that the following fact holds true:

3. The index of a simple closed curve is zero unless there is at least one singular point in the region bounded by the curve.

We now define the *index of a singular point P* as the index of any circle of *small* radius centered at P. (See Figure 8.24.) According to property 1 the radius of this circle is immaterial as long as the curve encloses only *one* singular point and does not itself go through any singular points.

4. The indices of common isolated singular points are as follows [see Figure 8.24(f–h)]:
  (i) The index of a node is $+1$.
  (ii) The index of a focus is $+1$.
  (iii) The index of a center is $+1$.
  (iv) The index of a saddle point is -1.

(These are independent of stability in cases 1 and 2 and independent of direction of rotation in cases 3 and 4. See problem 31b.)

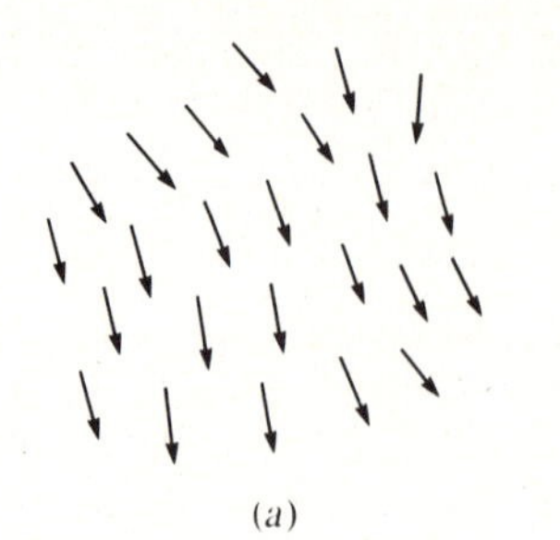

(a)

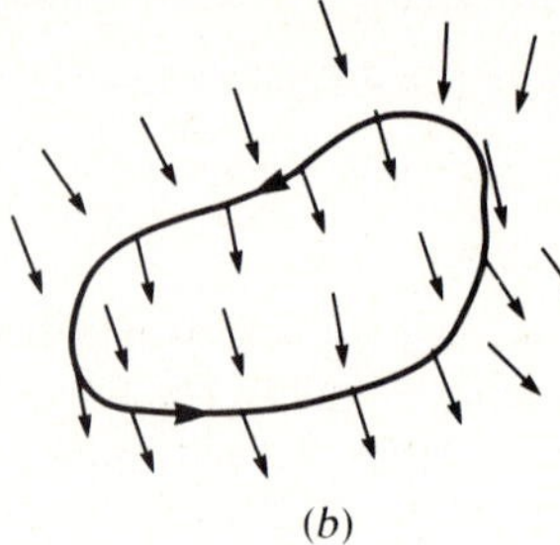

(b)

(c)

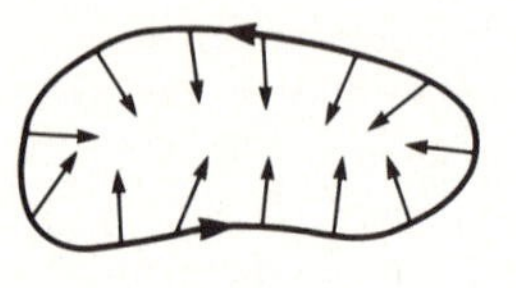

(d)

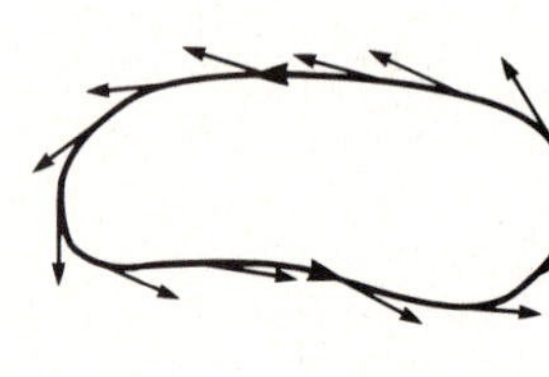

(e)

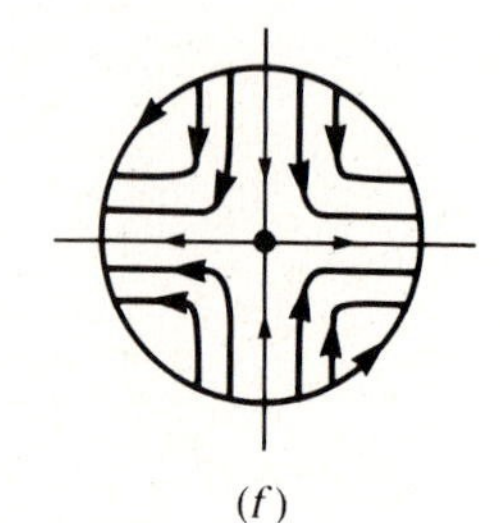

(f)

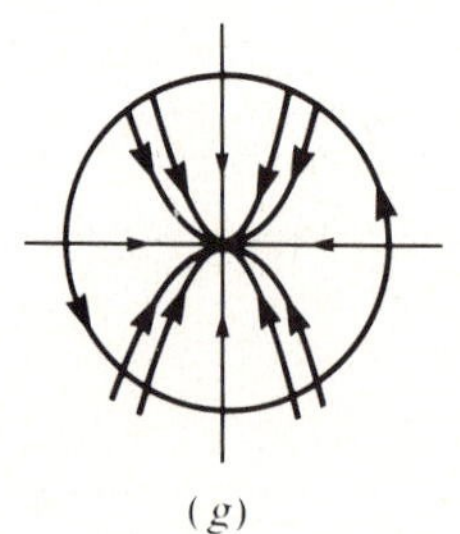

(g)

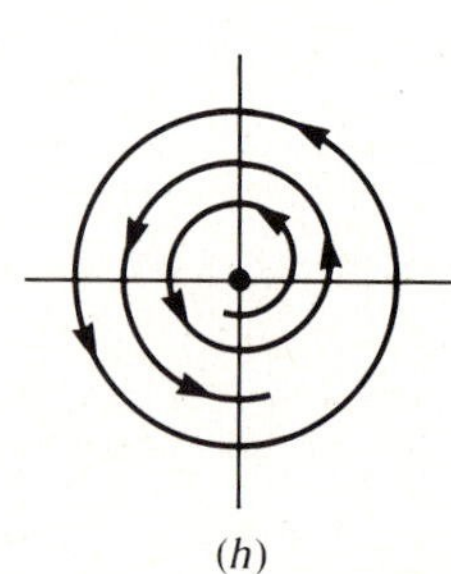

(h)

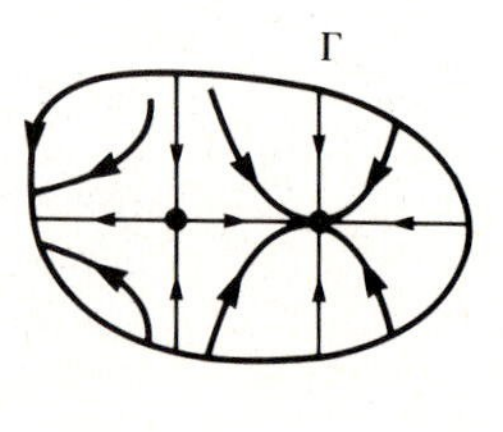

(i)

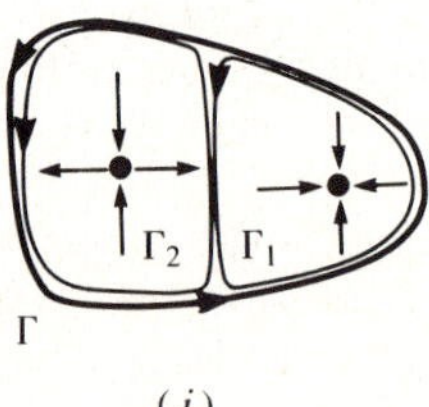

(j)

Figure 8.24 *(a) A vector field. (b) A test curve Γ; the vector field does not complete a revolution as curve is traversed (index = 0). (c) The field vector rotates by one revolution in the direction opposite to the curve (index = −1). (d) The field vector rotates by one revolution in the same direction (index = +1). (e) Limit cycle or any closed periodic trajectory (index = +1). (f) Saddle point (index = −1). (g) Node (index = +1). (h) Focus/center (index = +1). (The index of a singular point = index of a small circle encircling the point.) (i) The index of a test curve encircling several points equals (j) the sum of indices of the individual points. [After Arnold (1981).]*

Now consider a curve Γ that surrounds a region containing several singular points. Then the following can be proved:

5. The index of a curve is equal to *the sum of the indices of singular points* in the region D which it surrounds.

This result can be established by considering a picture similar to Figure 8.24(*i, j*). Here two artificial extensions of the original curve have been so added that their net contribution to the total rotation "cancels." The index of Γ must therefore be the same as that of Γ_1 plus Γ_2. However, according to property 1, these can be distorted to small circles about their respective singular points without change of index. This verifies the claim. (The argument is made more formal by giving a rigorous definition of index in terms of a *line integral* and demonstrating that the sum of line integrals around Γ_1 and Γ_2 equals the line integral of Γ. Students who have had advanced calculus may recognize Figure 8.24(*i, j*) as a familiar trick used in Green's theorem.)

These notions are useful in establishing the relation between a periodic solution of equations (1a,b) and singular points (or in our popular phrasing, steady states) of the flow $(F(x, y), G(x, y))$. According to previous remarks, a periodic solution corresponds to a solution curve that is itself simple and closed; the flow causes the point $(x(t), y(t))$ to rotate around this curve. In particular, the field vectors are tangent to the curve itself. By referring to Figure 8.24(*e*) we can clearly see that the field vector must therefore execute *one complete rotation* in the positive sense as the curve is traversed. We have thus observed the following:

6. The index of a closed (periodic-orbit) solution curve is $+1$.

Collecting all of our remarks and observations, we conclude the following:

7. **(a)** A limit cycle (or any closed periodic orbit) must contain at least one singular point.
(b) If it contains *exactly* one, that singular point must be a node, a focus, or a center.
(c) If it contains *more than* one, the number of saddle points must be one less than the total number of foci, nodes, and centers. (Note that only these four types of singular points are permitted).

More discussion of these observations is given in the problems. We now consider their applicability. First, an important comment is that the properties we have described derive from the basic topology of the plane. It is possible to generalize ideas to other "locally flat" objects called *manifolds* (such as the surface of a sphere, a torus, and so forth.) We shall leave the abstract development at this point and remark merely that conclusions of this section are specific to two-dimensional systems. In three dimensions, for example, more complicated flow patterns may accompany closed periodic trajectories, so that the number and locations of singular points may have no bearing on the presence of a limit cycle.

While the index properties do not predict whether limit cycles occur, they do shed light on restrictions that apply. Using these we can rule out, for example, any closed periodic trajectories about regions that contain exactly two nodes or exactly one node and one saddle point.

PROBLEMS FOR APPENDIX 1*

31. *Index of a curve*

(a) Give reasons to support the assertion that the index of a simple closed curve is zero unless there is at least one singular point in the region bounded by the curve.

(b) Show that the index of a singular point is independent of its stability.

(c) A rigorous definition of the index of a curve is as follows: for the vector field $\mathbf{V} = (F(x, y), G(x, y))$ the quantity

$$\frac{G(x, y)}{F(x, y)}$$

is the slope of a field vector in the xy plane. Define

$$d\phi = d\left(\arctan \frac{G}{F}\right).$$

Show that

$$d\phi = \frac{GdF - FdG}{F^2 + G^2} \qquad (G \neq 0).$$

The rest of this problem requires familiarity with line integrals.

***(d)** Now define the index of a closed curve ind γ by the following line integral

$$\text{ind } \gamma = \frac{1}{2\pi} \int_\gamma d\phi.$$

Show that this definition corresponds to the concept of index previously described.

***(e)** Use this definition together with Green's theorem to verify the claim that the index of a curve is equal to the sum of the indices of all singular points inside the region bounded by the curve. [*Hint:* Consult figure (*a*) for problem 32.]

32. **(a)** Show or give reason for the assertion that the index of a node, a focus, or a center is $+1$ and that of a saddle point is -1.

(b) Which of the cases shown in the accompanying figures is possible?

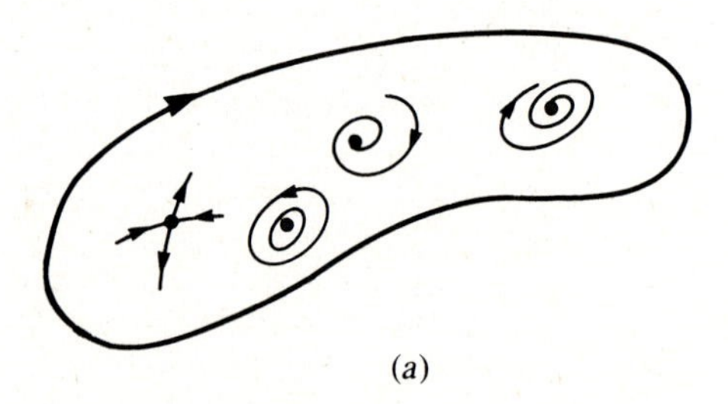

(*a*)

* Problems preceded by asterisks (*) are especially challenging.

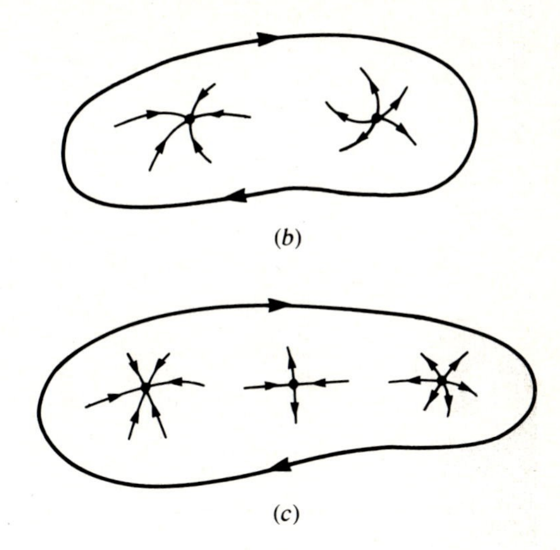

Figure for problem 32.

APPENDIX 2 TO CHAPTER 8: MORE ABOUT THE POINCARÉ-BENDIXSON THEORY

In this appendix we collect some mathematical terminology commonly encountered in the Poincaré-Bendixson theory. Using these definitions we then give a more precise statement of the Poincaré-Bendixson theorem. Also included is a proof of Bendixson's negative criterion.

Definitions

1. An orbit Γ through the point P is the curve

$$\{\mathbf{x}_P(t)\,|\,-\infty < t < \infty \text{ with } \mathbf{x}_P(0) = P\}.$$

2. A positive semiorbit Γ^+ through P is the curve

$$\{\mathbf{x}_P(t)\,|\,0 \le t < \infty \text{ with } \mathbf{x}_P(0) = P\}.$$

Similarly the negative semiorbit Γ^- is defined for $-\infty < t \le 0$.

3. The ω limit set of Γ is the set of points in R^2 that are approached along Γ with increasing time. Similarly, the α limit set of Γ is defined as the set of points approached with decreasing time.

4. A *limit cycle* is a periodic orbit Γ_0 that is the ω limit set or the α limit set for all other orbits in some neighborhood of Γ_0.

To paraphrase, we draw a distinction between solutions of equations (1a,b) for all time (which are represented by orbits Γ in the plane) and those for $t \ge t_0$ or $t \le t_0$ (represented by semiorbits Γ^+ and Γ^-). We have also introduced above the important notion of *limiting sets;* these come in two varieties (ω and α) depending on whether the limit is taken for $t \to \infty$ or $t \to -\infty$ respectively. They are thus the point sets that are approached along a trajectory in the forward (ω) or reverse (α) time direction.

A limit cycle is a special periodic solution of the autonomous dynamic system (1a,b) that is also simultaneously a limiting set for nearby trajectories. Physically this means that for $t \to \infty$ (or $t \to -\infty$, depending on stability) a solution that starts out close to the periodic solution will eventually be indistinguishable from it. (Of course, from the mathematical standpoint the two will never be exactly equal during finite time.)

We now state the Poincaré-Bendixson theorem, whose proof is to be found in numerous advanced books on ODEs (for example, see Hale, 1980):

Theorem 1: The Poincaré-Bendixson Theorem:

A bounded semiorbit that does not approach any singular point is either a closed periodic orbit or approaches a closed periodic orbit.

Finally, we prove Bendixson's criterion (stated in Section 8.3) using Green's theorem.

A Proof of Bendixson's Criterion

Suppose C is a closed-curve trajectory in the simply connected region D. Then by Green's theorem

$$\int_C F(x, y)\, dy - G(x, y)\, dx = \iint_S \left(\frac{\partial F}{\partial x} + \frac{\partial G}{\partial y}\right) dx\, dy, \tag{77}$$

where S is the region contained within the curve C.

For the system of equations (1a,b) we have the following relations:

$$\frac{dx}{dy} = \frac{dx/dt}{dy/dt} = \frac{F(x, y)}{G(x, y)}, \tag{78}$$

so that $G(x, y)\, dx = F(x, y)\, dy$.

The integral of the LHS above must therefore be zero, forcing the conclusion that

$$\iint_S \left(\frac{\partial F}{\partial x} + \frac{\partial G}{\partial y}\right) dx\, dy = 0. \tag{79}$$

The quantity $\partial F/\partial x + \partial G/\partial y$ will not have a vanishing integral over S unless it is (1) always zero or (2) alternately positive *and* negative in S. This proves the theorem.

III Spatially Distributed Systems and Partial Differential Equation Models

9 An Introduction to Partial Differential Equations and Diffusion in Biological Settings

I do not know what I may appear to the world; but to myself I seem to have been only like a boy playing on the seashore, and diverting myself in now and then finding a smoother pebble or prettier shell than ordinary, whilst the great ocean of truth lay all undiscovered before me.

Isacc Newton (1642–1727) p 90 E. T. Bell (1937) Men of Mathematics
Simon & Schuster, N.Y.

Part of our admiration for nature stems from the fact that it continually surprises us with its infinite variation, regardless of the scale of observation. This holds true of microscopic worlds; the surface of a cell for example, consists of myriad buoyant macromolecules distributed haphazardly in a viscous lipid sea. On the broad scale, that of continents or ecosystems, the fabric of habitats is like a patchwork quilt with a wide variety of local conditions, some favoring one species, some favoring another.

For this reason, real natural systems behave in a way that reflects an underlying spatial variation. Despite our idealizations, no species actually consists of identical individuals, since not all individuals are equally exposed to a constant environment. Similarly, on the molecular level, rarely do reactions take place in a homogeneous soup of chemicals. Somehow the effect of spatial organization does influence the way individual particles or molecules interact.

In the three chapters to follow, our purpose is to expose how spatial variation influences the motion, distribution, and persistence of species. We shall see that in the fine balance that exists between interdependent species, the spatial diversity of the system can have subtle but important effects. Conversely, the interactions of unlike species can result in spatial heterogeneity and lead to the appearance of patterns out of a uniform state. Our initial goal is to introduce the concepts underlying spatially dependent processes and the partial differential equations (PDEs) that describe these. The discussion is somewhat general, with examples drawn from molecular, cellular, and population levels. Later we will apply the ideas to more specific cases with the aim of gaining an understanding of phenomena.

In this chapter we discover primarily how partial differential equations arise and by what procedures they can be assembled into statements that are reasonable mathematically as well as physically. We see that under appropriate assumptions the motion of groups of particles (whether molecules, cells, or organisms) can be represented by statements of mass or particle conservation that involve partial derivatives. Such statements, often called *conservation* or *balance equations*, are universal in mathematical descriptions of the natural sciences. Indeed practically every PDE that depicts a physical process is ultimately based on principles of conservation—of matter, momentum, or energy.

Before undertaking the derivation of balance equations, we devote Section 9.1 to a review of the material that forms much of the structural underpinning of the mathematical framework. Students well versed in advanced calculus may skim through this section. One of the key observations we make is that the spatial variation in a distribution can lead to directional information. This proves conceptually useful in later discussions.

With this preparation we then proceed with the derivation of statements of conservation. This is accomplished in two stages. First, a simple argument for one-dimensional settings is given in Section 9.2. This is followed by more rigorous derivations and a generalization to other geometries and higher dimensions. We then consider several specific phenomena—including convection, diffusion, and attraction—that result in the motion of particles. Each phenomenon leads to special cases of the conservation equation. Such equations are derived in Section 9.4 and explored more fully later.

One example of applying such ideas to a universal process—that of diffusion—is illustrated in Sections 9.5 to 9.9. Derivation of the equation governing diffusion is rather straightforward if one accepts an assumption known as Fick's law. A more fundamental approach based on random-walk models is rather more sophisticated. Okubo (1980) and references therein should be consulted for finer details. Less straightforward is the process of actually solving the diffusion equation (or any other) PDE. Exploring the host of powerful techniques commonly applied by mathematicians in analyzing PDEs is beyond our scope. However, even before attempting to find a full solution, the form of the equation leads to an appreciation for the role of diffusion as a biological transport mechanism. A ubiquitous and metabolically free process on the subcellular level, diffusion proves inefficient or totally useless on somewhat larger distance scales. Some of these observations and their implications are presented in Sections 9.5 to 9.7.

Section 9.8 and the Appendix give some guidance on ways of solving the diffusion equation. We limit ourselves to separation of variables, a technique that is readily applied given a familiarity with ordinary differential equations (ODEs). Several basic solutions are derived, and others are given without formal justification in order to circumvent a lengthy mathematical excursion into the relevant techniques.

Section 9.9 describes an application of the diffusion equation to bioassay for mutation-inducing substances.

For a rapid coverage of the key ideas in this chapter, the following sequence is recommended: Section 9.1 should be included or covered briefly in the interests of review. Sections 9.2 and 9.4 are essential for later material. Sections 9.3 and 9.5 can be assigned as independent reading or further research. Some highlights of the material in Section 9.8 or in the Appendix should be given, with particular emphasis on the role of boundary conditions in solutions of the diffusion equation. Familiarity with the examples may prove helpful but is not essential for mastering the material in Chapter 11.

9.1 FUNCTIONS OF SEVERAL VARIABLES: A REVIEW

We begin this chapter by briefly reviewing the theory of functions of several variables with emphasis on the geometric concepts behind the mathematical ideas. Those of you who have had advanced calculus can skim quickly through this section or go directly to the next one.

First consider a real-valued function of two variables x and y. In this chapter x and y will symbolize spatial coordinates of a point (x, y), and

$$z = f(x, y), \tag{1}$$

the *value* assigned to (x, y) by the function f, will generally represent some spatially distributed quantity. Examples include

1. The density of a population at (x, y).
2. The concentration of a substance at (x, y).
3. The temperature at (x, y).

A *graph* of the function f is the set of points $(x, y, f(x, y))$ in R^3. The value $z = f(x, y)$ can be visualized as the height [assigned by the function f to each point in the plane, (x, y)]. Equation (1) thus describes a surface, as shown in Figure 9.1(*a*).

Similarly, a function of three variables

$$z = g(x, y, w) \tag{2}$$

has a graph consisting of all points $(x, y, w, g(x, y, w))$. This is not as easy to draw, but the idea is analogous. (Every point in space is assigned a value by the function.) Sometimes it is more convenient to depict functions in other ways, some of which are shown in Figure 9.1 for functions of two and three variables. It is common to represent the behavior of a function of two variables by a set of contours for which

$$f(x, y) = \text{constant}. \tag{3}$$

In R^2 these are called *level curves* and are simply loci for which a constant concentration or a constant density (or height) is maintained. As we shall see, they play an important role in the geometry of gradients and gradient fields.

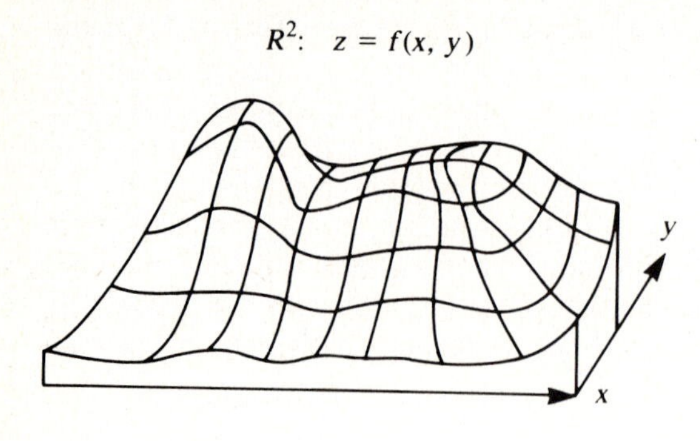

(a)

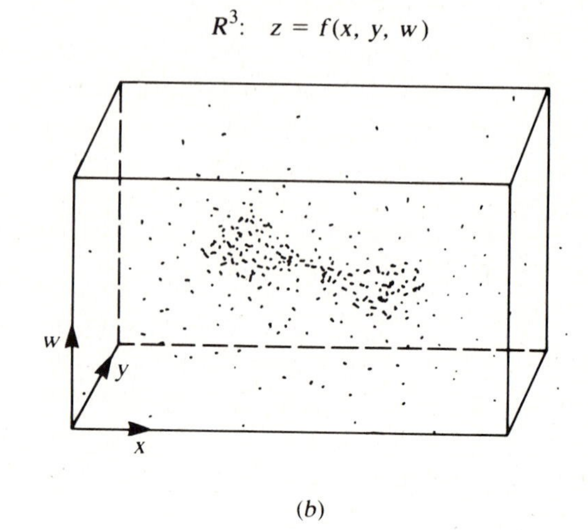

(b)

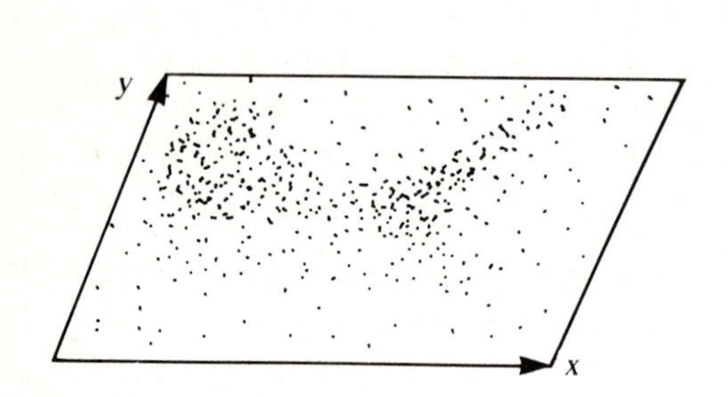

(c)

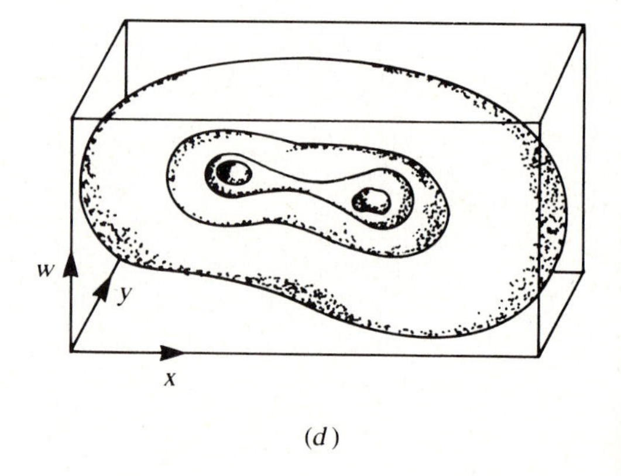

(d)

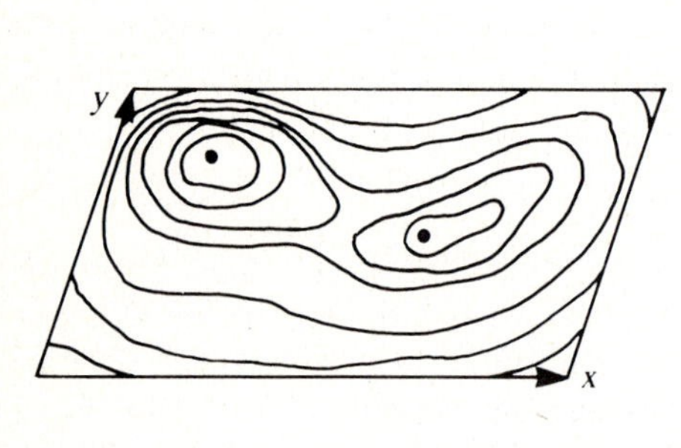

(e)

Figure 9.1 *Functions of two or three variables can be represented graphically in several ways: (a) as a surface (two variables only); (b, c) by density distributions in* R^3 *or* R^2*; or (d, e) by loci representing constant* f. *The latter are called (d) level surfaces or (e) level curves.*

In R^3 the function of three variables given by (2) can similarly be represented by sets of points for which

$$f(x, y, w) = \text{constant}. \tag{4}$$

Such loci are a generalized version of level curves, but for obvious reasons these are called *level surfaces* [see Figure 9.1(*d*)]. To clarify with an example, consider a temperature field in three dimensions. The *equitherms* are then surfaces at which some given constant temperature is maintained. If heat sources are located at two points, the resulting equithermal surfaces might look something like those shown in Figure 9.1(*d*).

In the context of this chapter, level curves or surfaces might represent the loci on which (1) population density is constant or (2) chemical concentration is constant.

We shall be concerned primarily with statements about how spatial distributions change with time; frequently it will be clear that the movement of one substance or population is closely linked to the distribution of another.

Consider the following simple example. An organism crawling on a flat surface may adapt its motion to the search for food particles. Imagine then that the dots in Figure 9.1(*c*) represent nutrient particle concentration. The observed path should ideally lead to the site of greatest concentration. To sense an increase in the ambient concentration level, an organism must continually *cross* level curves of the particle distribution. Per unit distance traveled, this crossing can be done most efficiently by maintaining a path *orthogonal* to the level curves; in other words, a tangent vector to the path should be perpendicular to a tangent to a level curve through a given point. This assertion can be verified using rather elementary calculus of several variables. It can also be shown that the destination will be a *critical point* of the function (in this case a local maximum) but not necessarily the *global* maximum.

In calculus a commonplace analogy is often drawn in explaining these ideas. Hikers often use topographical maps which are two-dimensional representations of the height of the terrain. The curves on such maps are level curves for the function $f(x, y)$ = height above sea level at latitude x and longitude y. Mountain peaks, valleys, and mountain passes are critical points of $f(x, y)$ that correspond to local maxima, minima, and saddle points respectively. Using local information only (for example, walking uphill with no information other than the local slope), one can attain a local maximum, but this may or may not be the highest possible peak.

These two-dimensional examples can be extended to higher dimensions. A motile organism that swims in a droplet of water might also use local cues in orienting itself and moving towards sites that have higher nutrient levels. This type of motion, called *chemotaxis,* will be discussed at greater length in Section 10.2. In R^3 the path of a highly efficient chemotactic organism would be orthogonal to the level surfaces of the nutrient distribution $c(x, y, z)$.

The descriptive statements in this section can be made more rigorous by introducing partial derivatives and gradients which are reviewed in the boxed material. Several examples follow the general discussion and definitions.

Partial Derivatives (A Review)

For a function of two variables $f(x, y)$ we define

$$\frac{\partial f}{\partial x} = \lim_{\Delta x \to 0} \frac{f(x + \Delta x, y) - f(x, y)}{\Delta x}. \tag{5}$$

A similar definition holds for $\partial f/\partial y$. Shorthand notation for partial derivatives is f_x and f_y.

To understand the geometrical meaning of these derivatives, imagine standing at a point (x_0, y_0) on a plane. Suppose $f(x_0, y_0)$ is the height of a surface above this location. The expression following "lim" in equation (5) (and similarly for $\partial f/\partial y$) represents the changes in the height of the surface per unit distance as we take a step in the x (or the y) direction. A *partial derivative* is the limit of this quantity as the length of the step shrinks to an infinitesimal size. It is therefore analogous to an ordinary derivative and also represents a *slope*.

To clarify, suppose we slice away part of the surface $z = f(x, y)$ along a direction parallel to the x (or y) axis. (See Figure 9.2.) In such cutaway drawings the partial derivative is the slope of a tangent to the curve forming the surface edge. The idea of a partial derivative is a special case of the somewhat more general concept of *directional derivative*. We shall not deal in more depth with this but rather refer the reader to any standard calculus text for a definition and explanation.

The following properties of partial derivatives follow from their basic definition:

$$\frac{\partial(cf)}{\partial x} = c\frac{\partial f}{\partial x}, \tag{6a}$$

$$\frac{\partial(f + g)}{\partial x} = \frac{\partial f}{\partial x} + \frac{\partial g}{\partial x}. \tag{6b}$$

for functions f and g and constant c. Similar equations ensue for partial differentiation with respect to y.

For functions that are continuously differentiable sufficiently many times, it is also true that mixed partial differentiation in any order produces the same result. For example

$$f_{yx} = \frac{\partial}{\partial x}\left(\frac{\partial f}{\partial y}\right) = \frac{\partial^2 f}{\partial x\,\partial y} = \frac{\partial^2 f}{\partial y\,\partial x} = \frac{\partial}{\partial y}\left(\frac{\partial f}{\partial x}\right) = f_{xy}. \tag{6c}$$

Note: Equation (6c) also defines the equivalent notation used for multiple partial differentiation.

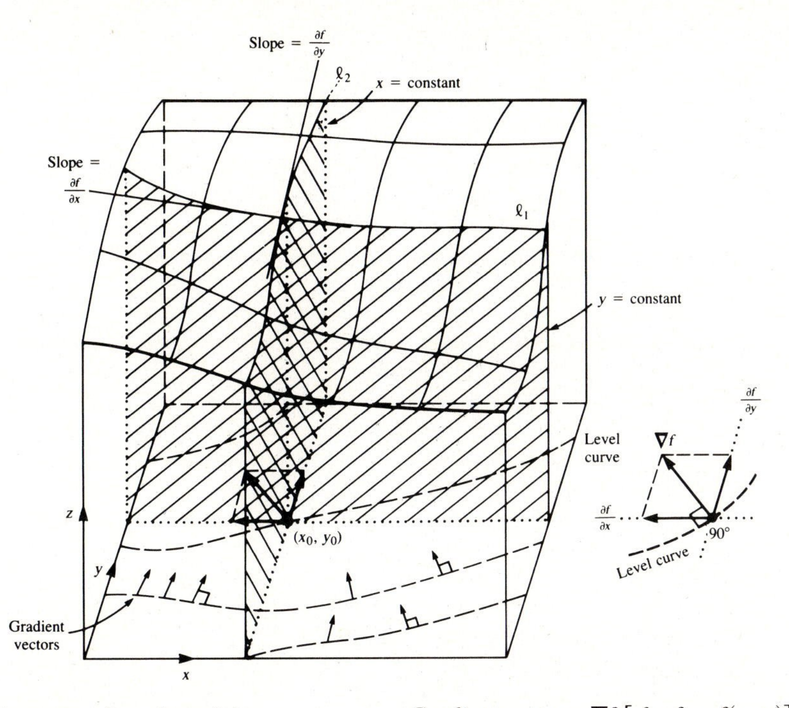

Figure 9.2 An interpretation of partial derivatives and gradient vectors. The surface z = f(x, y) *intersects planes for which* y = *constant or* x = *constant along curves* ℓ_1 *and* ℓ_2. *The slope of a tangent to* ℓ_1 *is* ∂f/∂x, *and the slope of a tangent to* ℓ_2 *is* ∂f/∂y.

Gradient vectors, **∇**f [*for* f = f(x, y)] *live in the* xy *plane and have* (∂f/∂x, ∂f/∂y) *as components. These vectors are always orthogonal to level curves of* z = f(x, y), *shown here by dashed lines (see inset).*

Gradients

For a function f of several variables the *gradient,* symbolized ∇f, is a vector consisting of the partial derivatives of f. For example, if $f = f(x, y)$, then

$$\nabla f = \left(\frac{\partial f}{\partial x}, \frac{\partial f}{\partial y}\right). \tag{7}$$

If $f = f(x, y, z)$, then

$$\nabla f = \left(\frac{\partial f}{\partial x}, \frac{\partial f}{\partial y}, \frac{\partial f}{\partial z}\right), \tag{8}$$

and so on for functions of n variables. The symbol ∇ is called the *del operator,* and is discussed in greater detail in Section 9.3.

The gradient vector has the following properties:

1. The magnitude of ∇f, $|\nabla f|$, represents the steepness of the local variations in the function f. For example,

$$|\nabla f| = (f_x^2 + f_y^2)^{1/2}. \tag{9}$$

2. (a) The direction of ∇f,

$$\mathbf{u} = \frac{\nabla f}{|\nabla f|},$$

is a unit vector in the direction of steepest (increasing) slope, in the sense that a step in this direction leads to the greatest increase in f per unit distance.

(b) The gradient vector at a point (x_0, y_0) is perpendicular to a level curve $f(x, y) = c$ that goes through (x_0, y_0) as long as (x_0, y_0) is not a local maximum, minimum, or saddle point.

For every point in R^2 (analogously, R^3 or R^n) at which the function f is defined, is continuous, and has partial derivatives, there will be a gradient vector. The vector will have all zero components at the critical points of f.

It is common to visualize a whole collection of these vectors, one at each point in space, as a *vector field*. A vector field that arises thus is called a *gradient field* and has certain special properties. (Note that a gradient field is a vector field, but the converse is not necessarily true.)

Gradient fields can always be paired (up to an arbitrary constant) with differentiable multivariate functions and vice versa (see examples in the following boxes). We see that the variations in a spatial distribution lead to orientation cues that are represented by the geometry of the gradient field.

The proof of the statements in this box are based on the chain rule of functions of several variables and on the properties of curves and vector dot products. These can be found in any text dealing with the calculus of several variables.

Example 1

Consider the function

$$f(x, y) = x^2 - 2x + y^2 + 4y + 5.$$

Level curves of this function have the equation

$$\begin{aligned} c &= x^2 - 2x + y^2 + 4y + 5 \\ &= (x - 1)^2 + (y + 2)^2. \end{aligned}$$

These are circles of radius $c^{1/2}$ centered at the point $(1, -2)$. Partial derivatives of $f(x, y)$ are the following:

$$\frac{\partial f}{\partial x} = 2x - 2, \qquad \frac{\partial f}{\partial y} = 2y + 4,$$

$$\frac{\partial^2 f}{\partial x\,\partial y} = 0, \qquad \frac{\partial^2 f}{\partial y\,\partial x} = 0,$$

$$\frac{\partial^2 f}{\partial x^2} = 2, \qquad \frac{\partial^2 f}{\partial y^2} = 2.$$

The gradient vector at a point (x, y) is

$$\nabla f = \left(\frac{\partial f}{\partial x}, \frac{\partial f}{\partial y}\right) = (2x - 2, 2y + 4).$$

This vector is perpendicular to a level curve going through the point (x, y). A critical point of f occurs at $(1, -2)$, where $\nabla f = (0, 0)$. At this point, $f(1, -2) = 0$. At any other point f is *greater*. For example, $f(1, 1) = 1 - 2 + 1 + 4 + 5 = 9$. Therefore $(1, -2)$ is a local minimum. (A more rigorous *second-derivative test* to distinguish between local minima, maxima, and saddle points is given in most calculus books).

Example 2

Consider the vector field

$$\begin{aligned} \mathbf{F} &= (M(x, y), N(x, y)) \\ &= (2x + 2y + y \cos xy, 2x + x \cos xy). \end{aligned} \tag{10}$$

We would like to determine whether $\mathbf{F}$ is a gradient field, that is, whether there is a function $f(x, y)$ such that

$$\mathbf{F} = \nabla f. \tag{11a}$$

If so, then

$$\mathbf{F} = (M, N) = \left(\frac{\partial f}{\partial x}, \frac{\partial f}{\partial y}\right), \tag{11b}$$

where $M(x, y) = \partial f/\partial x$ and $N = \partial f/\partial y$.

By a previous observation we must have

$$\frac{\partial M}{\partial y} = f_{xy} = f_{yx} = \frac{\partial N}{\partial x}. \tag{12}$$

Checking this, we note that

$$\frac{\partial M}{\partial y} = 2 + \cos xy - xy \sin xy = \frac{\partial N}{\partial x},$$

so that no contradiction results by assuming that (11a) holds. The condition given by (12) in fact guarantees that $\mathbf{F}$ is a gradient field (a fact whose proof will be omitted here).

To find f we need a function whose partial derivatives satisfy the following:

$$\frac{\partial f}{\partial x} = 2x + 2y + y \cos xy, \tag{13a}$$

$$\frac{\partial f}{\partial y} = 2x + x \cos xy. \tag{13b}$$

Integrating each expression with respect to a single variable while holding the other variable constant leads to these results:

$$f(x, y) = \int_{(y=\text{const})} (2x + 2y + y \cos xy)\, dx \tag{14}$$
$$= x^2 + 2xy + \sin xy + H,$$

and

$$f(x, y) = \int_{(x=\text{const})} (2x + x \cos xy)\, dy \tag{15}$$
$$= 2xy + \sin xy + G.$$

In ordinary one-variable calculus, a single integration introduces a single arbitrary constant. However, in the *partial integration* of (14) and (15) one must account for the distinct possibility that the integration "constants" H and G may depend on the values given to the fixed variables (to $y =$ const and to $x =$ const). For this reason it is necessary to presuppose that

$$H = h(y) \qquad \text{and} \qquad G = g(x) \tag{16}$$

are functions. Indeed, the only possibility for matching the two different expressions, (14) and (15), both of which equal the same function $f(x, y)$, would be to take

$$G(x) = x^2 + c, \qquad H(y) = c,$$

for constant c.

The conclusion then is that

$$f(x, y) = x^2 + 2xy + \sin xy + c. \tag{17}$$

To check this result, observe that

$$\nabla f = (2x + 2y - y \cos xy, 2x - x \cos xy) = \mathbf{F},$$

which confirms the calculation. Note that adding any constant to $f(x, y)$ results in the same gradient. Thus $f(x, y)$ is defined only up to some arbitrary additive constant.

Example 3

The concentration of nutrient particles suspended in a pond is given by the expression

$$c(x, y, z) = C_0 \exp -\alpha(x^2 + y^2 + z^2). \tag{18}$$

An organism located at $(x, y, z) = (1, -1, 1)$ moves in the direction of increasing concentration. In which direction should it move? Where is the maximum concentration?

Answer

To find the direction of greatest increase per unit distance, compute the gradient vector. Since c is a function of three variables, ∇c is a vector in R^3:

$$\nabla c = \left(\frac{\partial c}{\partial x}, \frac{\partial c}{\partial y}, \frac{\partial c}{\partial z}\right) \tag{19a}$$

$$= (-2\alpha x C_0 e^{-\alpha r^2}, -2\alpha y C_0 e^{-\alpha r^2}, -2\alpha z C_0 e^{-\alpha r^2}), \tag{19b}$$

where

$$r^2 = x^2 + y^2 + z^2 \tag{19c}$$

at $(1, -1, 1)$

$$r^2 = 3 \quad \text{and} \quad \nabla c = (-\gamma, \gamma, -\gamma),$$

where

$$\gamma = 2\alpha C_0 e^{-3\alpha}.$$

Furthermore, $\nabla c = 0$ only when $(x, y, z) = (0, 0, 0)$, so that the origin is a critical point. It is readily observed that this is a local *maximum,* since $c(x, y, z)$ is a function that decreases exponentially with distance from the origin. Thus, the maximal concentration of nutrient particles is $c(0, 0, 0) = C_0$. We further observe that equiconcentration loci are surfaces that satisfy

$$C_0 \exp -\alpha(x^2 + y^2 + z^2) = \text{constant}. \tag{20a}$$

After algebraic simplification this becomes

$$x^2 + y^2 + z^2 = K \qquad (K = \text{constant}), \tag{20b}$$

which represents spheres with centers at $(0, 0, 0)$ and radii $\sqrt{K}$.

Thus the organism will move in the direction of the gradient, and its path will eventually end at $(0, 0, 0)$.

In the next section our purpose is to understand the basic process through which a partial differential description of motion through space is obtained. We shall be concerned mainly with the dynamic processes that lead to changes in a spatial distribution over time. Some of the many examples cited pertain to the motion and continuous redistribution of animals, cells, and molecules through space. For this reason we shall deal with functions that depend on both space and time.

9.2 A QUICK DERIVATION OF THE CONSERVATION EQUATION

The *conservation equation* in its various forms is the most fundamental statement through which changes in spatial distributions are described. Most of the PDEs we will encounter are ultimately based on such *statements of balance*. To gain an easy familiarity with the basic concepts we will consider a rather special case and give an informal derivation first, later to be made more rigorous and more general.

Our initial assumptions are that

1. Motion takes place in a single space dimension (as, for example, in the thin tube of Figure 9.3a).
2. The cross-sectional area of the vessel or container is constant along its entire length.

We shall let x represent the distance along the tube from some arbitrary location. Fixing attention on the interval between x and $x + \Delta x$, let us describe changes

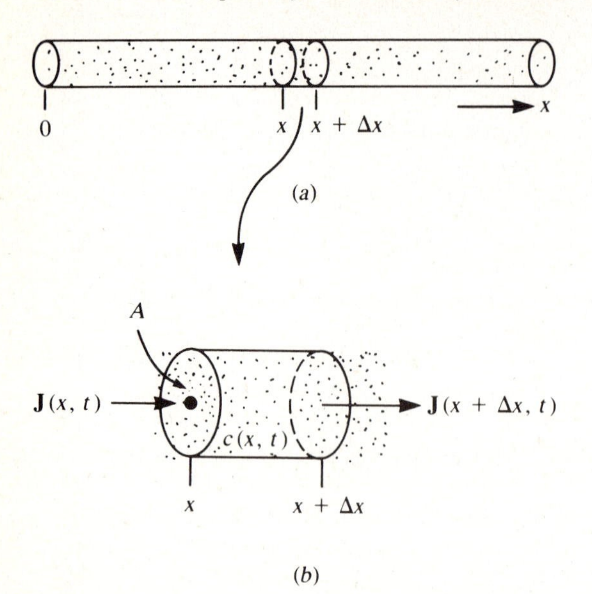

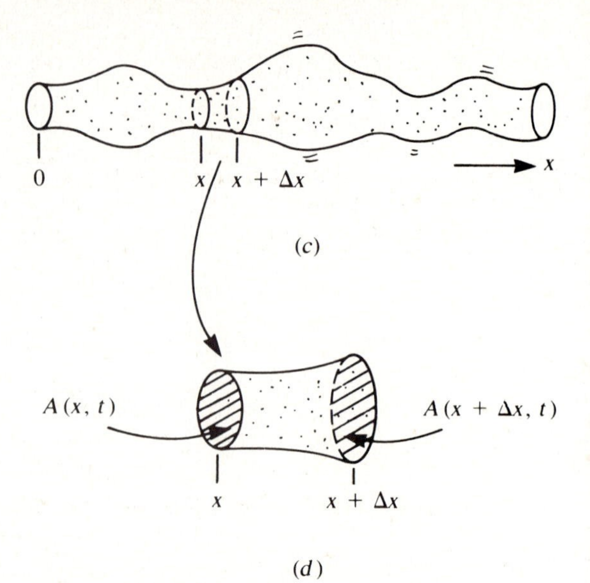

Figure 9.3 *Equations of balance are derived for flow of particles [concentration* c(x, t)*] along a tube. (a) If the tube has uniform cross-sectional area* A*, equation (24) results from a balance of flows into and out of a small section (b) of length* Δx*. (c) If the tube has spatially or temporally varying area* A(x, t)*, one obtains equation (29) by formulating the balance statement for the small region shown in (d).*

in the concentration by accounting for two possible effects: (1) flow of particles into and out of the interval $(x, x + \Delta x)$, and (2) processes that introduce new particles or degrade particles locally (such as through a chemical reaction). The balance equation can be written either in terms of *mass* or *number* of particles. We arbitrarily choose the latter description, and so our statement will be

$$\begin{pmatrix}\text{rate of change}\\ \text{of particle}\\ \text{population in}\\ (x, x+\Delta x)\\ \text{per unit time}\end{pmatrix} = \begin{pmatrix}\text{rate of entry}\\ \text{into } (x, x+\Delta x)\\ \text{per unit time}\end{pmatrix} - \begin{pmatrix}\text{rate of}\\ \text{departure}\\ \text{from}\\ (x, x+\Delta x)\\ \text{per unit time}\end{pmatrix} \pm \begin{pmatrix}\text{rate of}\\ \text{local degra-}\\ \text{dation or}\\ \text{creation per}\\ \text{unit time}\end{pmatrix} \tag{21}$$

To go further, define the following quantities:

$c(x, t)$ = concentration of particles (number per unit volume) at (x, t),

$\mathbf{J}(x, t)$ = flux of particles at (x, t) = number of particles crossing a unit area at x in the positive direction per unit time,

$\sigma(x, t)$ = sink/source density = number of particles created or eliminated per unit volume at (x, t).

We note that the only *flux* that changes the total population is that entering or leaving through the cross sections at x and $x + \Delta x$, namely, $\mathbf{J}(x, t)$ and $\mathbf{J}(x + \Delta x, t)$.

To now translate (21) into a dimensionally correct equation, it is necessary to take into account the following quantities:

A = cross-sectional area of tube,

ΔV = volume of length element $\Delta x = A\,\Delta x$.

Every term in the equation must have the same units as the terms on the LHS of (21): number per unit volume per unit time.

This leads to the following equation:

$$\frac{\partial}{\partial t}[c(x, t)A\,\Delta x] = \mathbf{J}(x, t)A - \mathbf{J}(x + \Delta x, t)A \pm \sigma(x, t)A\,\Delta x. \tag{22}$$

Note that since c depends on two variables, its derivative with respect to time is a partial derivative. Choosing to write equation (22) in terms of x, the coordinate of the *left* boundary of the interval, is entirely arbitrary since we are about to take the limit $\Delta x \to 0$.

We observe that a flux in the positive x direction tends to contribute to the net population positively at x but negatively at $x + \Delta x$; hence the signs of the terms in (22). See Figure 9.3(*b*).

Now dividing through by $A\,\Delta x$, which by assumption is constant, we obtain

$$\frac{\partial c(x, t)}{\partial t} = \frac{\mathbf{J}(x, t) - \mathbf{J}(x + \Delta x, t)}{\Delta x} \pm \sigma(x, t). \tag{23}$$

Taking a limit of this equation as $\Delta x \to 0$, that is, as the slice width gets vanishingly small, we arrive at a local statement, the *one-dimensional balance equation,*

$$\frac{\partial c(x, t)}{\partial t} = \underbrace{-\frac{\partial \mathbf{J}(x, t)}{\partial x}}_{\text{net motion}} \pm \underbrace{\sigma(x, t)}_{\text{source/sink}}. \tag{24}$$

The minus sign on $\partial \mathbf{J}/\partial x$ stems from the fact that the finite difference in (23) has a sign opposite to that in the definition of a derivative.

This is the basic form of the balance law that we shall soon apply to numerous specific problems. Before doing so, we will make a number of extensions and general statements. It is possible to skip this material and go on to Section 9.4 without loss of continuity.

9.3 OTHER VERSIONS OF THE CONSERVATION EQUATION

Tubular Flow

We shall drop assumption (2) and consider the possibility that the cross-sectional area of the tube may vary over space and time. To be somewhat more formal, we take the following definitions: By the *concentration* $c(x, t)$ we mean a quantity such that

$$\int_{x_1}^{x_2} c(x, t)A(x, t)\,dx = \text{total number of particles located within the tube in the interval } (x_1, x_2) \text{ at time } t. \tag{25a}$$

Similarly, the source density $\sigma(x, t)$ is defined by

$$\int_{x_1}^{x_2} \sigma(x, t)A(x, t)\, dx = \text{net rate of particle creation or degradation within the interval } (x_1, x_2) \text{ at time } t. \tag{25b}$$

The equation of balance can then be written in integral form (sometimes called the *weak form*), as follows:

$$\frac{\partial}{\partial t}\int_{x_0}^{x_0+\Delta x} c(x, t)A(x, t)\, dx = \mathbf{J}(x_0, t)A(x_0, t) - \mathbf{J}(x_0 + \Delta x, t)A(x_0 + \Delta x, t) \pm \int_{x_0}^{x_0+\Delta x} \sigma(x, t)A(x, t)\, dx. \tag{26}$$

(This is similar to a derivation in Segel (1980, 1984) for constant area.)

An *integral mean value theorem* allows one to conclude that at some locations (x_1, x_2) (where $x_0 \leq x_i \leq x_0 + \Delta x$ for $i = 1, 2$) the following is true:

$$\frac{\partial}{\partial t}[c(x_1, t)A(x_1, t)]\, \Delta x = \mathbf{J}(x_0, t)A(x_0, t) - \mathbf{J}(x_0 + \Delta x, t)A(x_0 + \Delta x, t) \pm [\sigma(x_2, t)A(x_2, t)]\, \Delta x. \tag{27}$$

Now dividing through by Δx and letting $\Delta x \to 0$, we get $x_1 \to x_0$ and $x_2 \to x_0$, so that in the limit equation (27) becomes

$$\frac{\partial}{\partial t}[c(x_0, t)A(x_0, t)] = -\frac{\partial}{\partial x}[\mathbf{J}(x_0, t)A(x_0, t)] \pm [\sigma(x_0, t)A(x_0, t)] \tag{28}$$

Special cases

1. If $A(x_0, t) = \tilde{A}$ is constant, dividing by $\tilde{A}$ reduces equation (28) to equation (24).
2. If $A(x, t) = \tilde{A}(x) \neq 0$ (that is, if the area does not change with time), then the equation can be written in the form

$$\frac{\partial c(x, t)}{\partial t} = -\frac{1}{\tilde{A}(x)}\frac{\partial}{\partial x}[\mathbf{J}(x, t)\tilde{A}(x)] \pm \sigma(x, t) \tag{29}$$

When the partial derivative is expanded, one obtains

$$\frac{\partial c(x, t)}{\partial t} = -\frac{\partial \mathbf{J}(x, t)}{\partial x} - \frac{\mathbf{J}(x, t)}{\tilde{A}(x)}\frac{\partial \tilde{A}(x)}{\partial x} \pm \sigma(x, t). \tag{30}$$

The equation is thus similar to (24) but contains an extra term which accounts for an effect similar to dilution; that is, a change in concentration that stems from local changes in the fluid volume "felt" by particles as they move along the length of the tube.

3. If $A(x, t) = \tilde{A}(t) \neq 0$ (if the area of the tube is uniform along its length but possibly time varying), then equation (28) leads to the following:

$$\tilde{A}(t)\frac{\partial c(x, t)}{\partial t} + c(x, t)\frac{\partial \tilde{A}(t)}{\partial t} = -\tilde{A}(t)\frac{\partial \mathbf{J}(x, t)}{\partial x} \pm \sigma(x, t)\tilde{A}(t). \tag{31}$$

After some rearranging, this becomes

$$\frac{\partial c(x, t)}{\partial t} = -\frac{c(x, t)}{\tilde{A}(t)} \frac{\partial \tilde{A}(t)}{\partial t} - \frac{\partial \mathbf{J}(x, t)}{\partial x} \pm \sigma(x, t). \tag{32}$$

Again the equation resembles (24), with an additional term that roughly speaking also describes a dilution effect as the tube expands or contracts.

4. In a situation where the cross-sectional area varies both spatially and temporally [$A = A(x, t)$] it follows that equation (28) can be written

$$\frac{\partial c(x, t)}{\partial t} = -\frac{\partial \mathbf{J}(x, t)}{\partial x} \pm \sigma(x, t) - \frac{1}{A(x, t)} \left[\mathbf{J}(x, t) \frac{\partial A(x, t)}{\partial x} + c(x, t) \frac{\partial A(x, t)}{\partial t} \right] \tag{33}$$

Flows in Two and Three Dimensions

To write a balance equation analogous to (24) in higher dimensions we consider a small rectangular element of volume $\Delta V = \Delta x \Delta y \Delta z$ situated in R^3 and account for motion of particles into and out of the region. First it proves necessary to extend somewhat our definition of flux, for now both direction *and* magnitude of flow have to be considered.

Let us focus attention on a point (x_0, y_0, z_0) in R^3. We shall define *flux* by counting the number of particles per unit time that traverse an imaginary unit area A suspended at (x_0, y_0, z_0) with some particular *orientation*. As the orientation of the "test area" is varied, the rate of crossings changes. Indeed, the highest rate of crossing is achieved when the predominant flow direction is orthogonal to the area that it must cross. This leads us to define *flux* as a vector in the direction **n** whose magnitude is given by:

$$|\mathbf{J}(x, y, z, t)| = \begin{array}{l}\text{net number of particles crossing}\\ \text{a unit area at } (x, y, z) \text{ per}\\ \text{unit time at time } t\end{array} \tag{34}$$

where **n** is the unit normal vector to that element of area that admits the greatest net crossings. [See Figure 9.4(*a*).]

We shall symbolize the components of **J** (in R^3) as follows:

$$\mathbf{J}(x, y, z) = (J_x, J_y, J_z). \tag{35}$$

Each of the components J_x, J_y, and J_z may in general depend on both space and time.

In three dimensions the magnitude of flux is given by the quantity

$$|\mathbf{J}| = (J_x^2 + J_y^2 + J_z^2)^{1/2} = (\mathbf{J} \cdot \mathbf{J})^{1/2} \tag{36}$$

(where $\cdot$ represents a vector dot product). Given some test area A, this definition of flux allows one to calculate the number of particle crossings N that take place in a given time t. If **m** is a unit vector perpendicular to the test area, one obtains

$$N = (\mathbf{J} \cdot \mathbf{m}) A \, \Delta t. \tag{37}$$

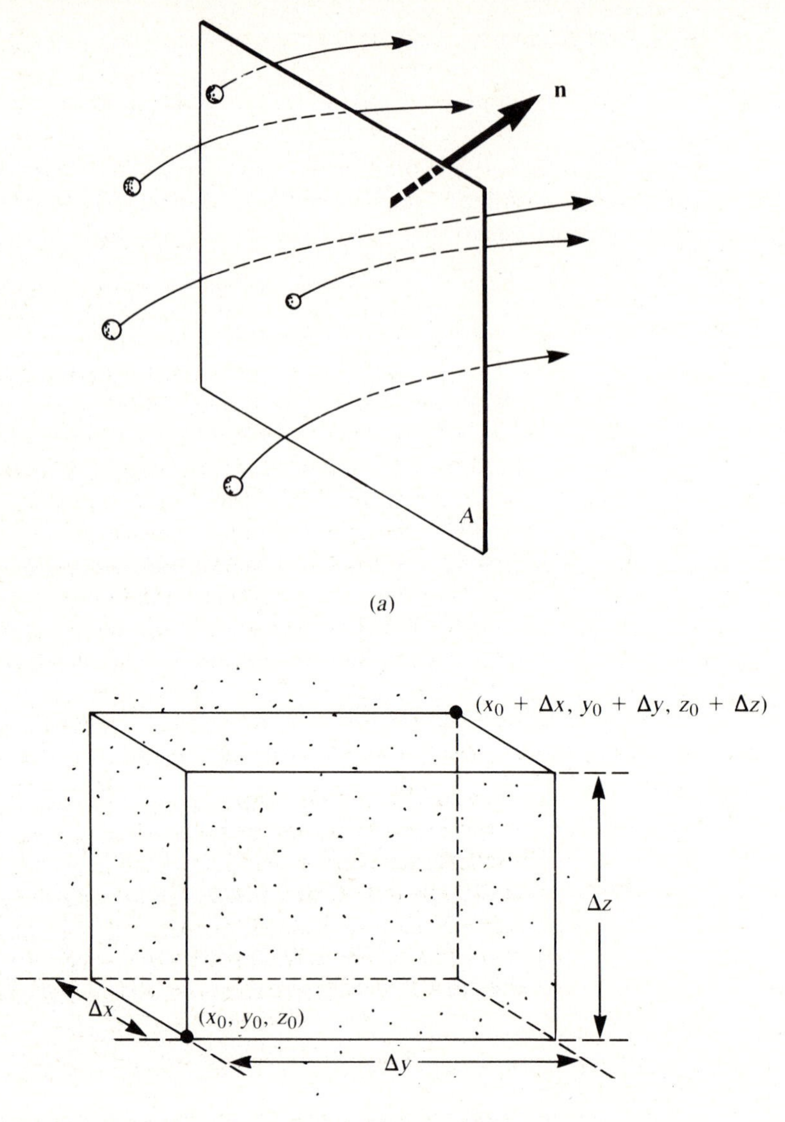

***Figure 9.4** (a) In R^3 flux **J** is a vector. Its magnitude represents the net number of particles crossing an imaginary unit area per unit time. Its direction is given by the normal vector **n** to the given area A. (b) The equation of conservation (39) is derived by considering net flow into a small rectangular volume of dimensions $\Delta x \times \Delta y \times \Delta z$.*

To illustrate this idea, consider a wall of unit area at the point $(-1, 0, 3)$ orthogonal to the direction $m = (1, 0, 0)$, and a flux $\mathbf{J} = (z - y, x - z, y - x)$. At the point in question,

$$\mathbf{J} = (3, -4, 1),$$

so that the number of crossings is

$$N = (\mathbf{J} \cdot \mathbf{m})\, 1\Delta t = [(3, -4, 1) \cdot (1, 0, 0)]\, \Delta t = 3\Delta t.$$

Thus three particles traverse the wall per unit time.

Given a small rectangular volume as shown in Figure 9.4(b), the statement of balance must accommodate, as before, local creation and entry or departure through *each of the six* planar surfaces. Since these planes are parallel to the coordinate planes, we can readily determine their normal vectors and calculate the net number of particles crossing (inwards) through each wall. (See Table 9.1.)

The net rate of change of concentration inside the volume that accrues from combining all these factors is the following:

$$\frac{\partial c}{\partial t} = \frac{J_x(x_0, y_0, z_0) - J_x(x_0 + \Delta x, y_0, z_0)}{\Delta x} + \frac{J_y(x_0, y_0, z_0) - J_y(x_0, y_0 + \Delta y, z_0)}{\Delta y} + \frac{J_z(x_0, y_0, z_0) - J_z(x_0, y_0, z_0 + \Delta z)}{\Delta z} \pm \sigma(x, y, z). \tag{38}$$

Taking the limit as $\Delta x \to 0$, $\Delta y \to 0$, and $\Delta z \to 0$, one obtains

$$\frac{\partial c}{\partial t} = -\left(\frac{\partial J_x}{\partial x} + \frac{\partial J_y}{\partial y} + \frac{\partial J_z}{\partial z}\right) \pm \sigma(x, y, z). \tag{39}$$

$$\frac{\partial c}{\partial t} = -\boldsymbol{\nabla} \cdot \mathbf{J} \pm \sigma \tag{40}$$

where $\boldsymbol{\nabla} \cdot \mathbf{J}$, called the *divergence of* $\mathbf{J}$, is the parenthetical term in equation (39). This scalar quantity can be described roughly as the net tendency of particles to leave an infinitesimal volume at the point (x, y, z). More details about the del operator $\boldsymbol{\nabla}$ are given in the box.

We have completed the derivation of the three-dimensional conservation equation. Note the similarity of equations (40) and (24). The two-dimensional case is left as an easy exercise for the reader. We must next turn to the question of what induces the motion of particles, molecules, or organisms so we can relate the idea of flux to the functions that describe spatial distributions.

Table 9.1 ***Particles Entering the Box (See Figure 9.4b.)***

Wall number	*Location on the plane*	*Inwards normal vector* $\mathbf{n}$	*Net inwards crossing* $\mathbf{J} \cdot \mathbf{n}$
1	$x = x_0$	$(1, 0, 0)$	$J_x(x_0, y_0, z_0)$
2	$x = x_0 + \Delta x$	$(-1, 0, 0)$	$-J_x(x_0 + \Delta x, y_0, z_0)$
3	$y = y_0$	$(0, 1, 0)$	$J_y(x_0, y_0, z_0)$
4	$y = y_0 + \Delta y$	$(0, -1, 0)$	$-J_y(x_0, y_0 + \Delta y, z_0)$
5	$z = z_0$	$(0, 0, 1)$	$J_z(x_0, y_0, z_0)$
6	$z = z_0 + \Delta z$	$(0, 0, -1)$	$-J_z(x_0, y_0, z_0 + \Delta z)$

***The Del Operator* ∇**

Loosely speaking, the quantity ∇ behaves like a vector whose components in R^3 are

$$\nabla = \left(\frac{\partial}{\partial x}, \frac{\partial}{\partial y}, \frac{\partial}{\partial z}\right). \tag{41}$$

We can think of the components as partial derivatives "hungry for a function to differentiate." The del operator can be combined with vector or scalar functions in several ways that parallel standard vector operations, as shown in Table 9.2.

Table 9.2 ***Analogies between Vector and Del Operations***

	Ordinary vector operations	*Analogous del operations*
Scalar multiplication	For $\mathbf{v} = (v_1, v_2, v_3)$ and a scalar α, $\alpha\mathbf{v} = (\alpha v_1, \alpha v_2, \alpha v_3)$. The result is a *vector*.	For ∇ as defined in (41) and a function $f(x, y, z)$, $\nabla f = \left(\frac{\partial}{\partial x}, \frac{\partial}{\partial y}, \frac{\partial}{\partial z}\right)f = \left(\frac{\partial f}{\partial x}, \frac{\partial f}{\partial y}, \frac{\partial f}{\partial z}\right)$. This is the *gradient* of f and is a *vector*.
Dot products	For $\mathbf{v} = (v_1, v_2, v_3)$ and $\mathbf{u} = (u_1, u_2, u_3)$, $\mathbf{v}\cdot\mathbf{u} = v_1u_1 + v_2u_2 + v_3u_3$. The result is a *scalar*.	For ∇ as defined in (41) and $\mathbf{v} = (v_1, v_2, v_3)$, $\nabla\cdot\mathbf{v} = \left(\frac{\partial}{\partial x}, \frac{\partial}{\partial y}, \frac{\partial}{\partial z}\right)\cdot(v_1, v_2, v_3) = \frac{\partial v_1}{\partial x} + \frac{\partial v_2}{\partial y} + \frac{\partial v_3}{\partial z}$. This is the *divergence* of the vector field $\mathbf{v}$ and is a *scalar* quantity.
Cross products	For two vectors $\mathbf{v}$ and $\mathbf{u}$ as defined above, $\mathbf{v}\times\mathbf{u} = \begin{pmatrix} \mathbf{i} & \mathbf{j} & \mathbf{k} \\ v_1 & v_2 & v_3 \\ u_1 & u_2 & u_3 \end{pmatrix} = (v_2u_3 - u_2v_3, v_3u_1 - v_1u_3, v_1u_2 - v_2u_1)$. The result is a *vector*.	For ∇ and $\mathbf{v}$ as defined above, $\nabla\times\mathbf{v} = \begin{pmatrix} \mathbf{i} & \mathbf{j} & \mathbf{k} \\ \frac{\partial}{\partial x} & \frac{\partial}{\partial y} & \frac{\partial}{\partial z} \\ v_1 & v_2 & v_3 \end{pmatrix} = \left(\frac{\partial v_3}{\partial y} - \frac{\partial v_2}{\partial z}, \frac{\partial v_1}{\partial z} - \frac{\partial v_3}{\partial x}, \frac{\partial v_2}{\partial x} - \frac{\partial v_1}{\partial y}\right)$. This is the *curl* of the vector field $\mathbf{v}$ and is a *vector*.

In descriptive terms, the vector ∇f measures local variations in a function and points to the direction of greatest steepness. The scalar quantity $\nabla \cdot \mathbf{v}$ measures the tendency of a vector field to represent the divergence (departure) of a fluid; the vector $\nabla \times \mathbf{v}$ (called the *curl* of $\mathbf{v}$) depicts a magnitude and axis of rotation (for example, of fluid in a vortex). Figure 9.5 demonstrates several vector fields that have net curl or divergence. We shall not concern ourselves too greatly with curl since in biological situations rotation is rarely encountered. (It does play a role in other physical sciences such as meteorology, where rotating atmospheric flows can be viewed as generating turbulent storms.) As equation (40) indicates, however, divergence is more relevant since we attempt to keep track of changes in densities of biological substances or populations. More details and basics about the properties of vector fields and the operations on them can be obtained in most standard calculus texts.

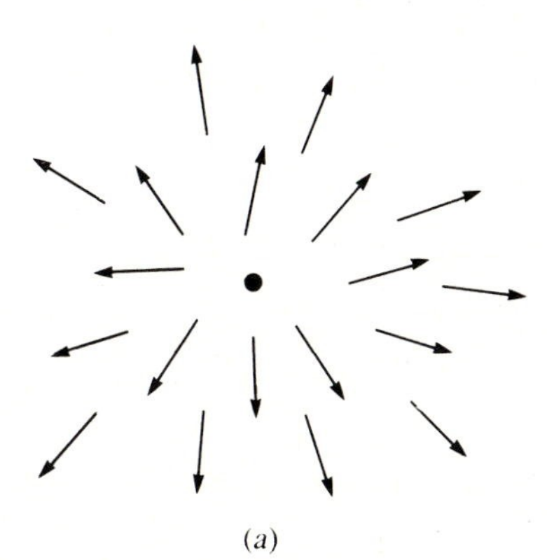

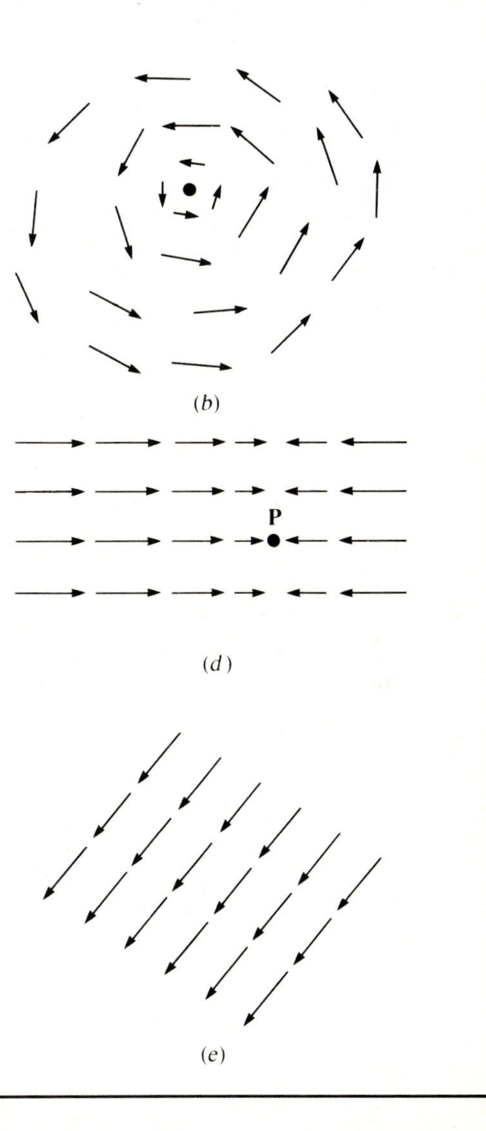

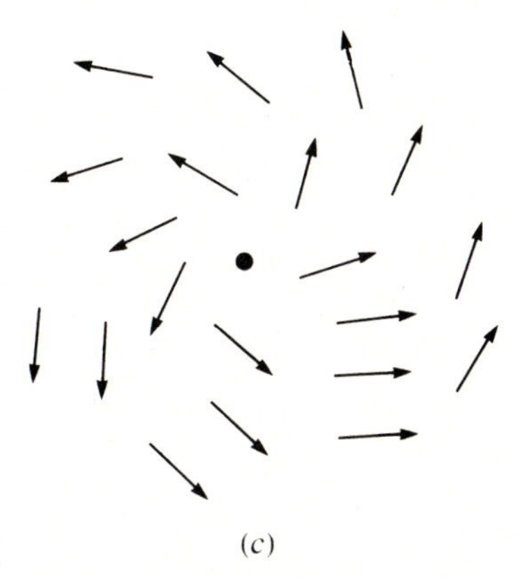

Figure 9.5 *Several examples of vector fields depicting such properties as divergence and rotation of particles. (a)* $\nabla \cdot \mathbf{v} \neq 0$ *and* $\nabla \times \mathbf{v} = \mathbf{0}$*. (b)* $\nabla \cdot \mathbf{v} = \mathbf{0}$ *and* $\nabla \times \mathbf{v} \neq \mathbf{0}$*. (c)* $\nabla \cdot \mathbf{v} \neq 0$ *and* $\nabla \times \mathbf{v} \neq \mathbf{0}$*. (d)* $\nabla \cdot \mathbf{v} \neq 0$ *and* $\nabla \times \mathbf{v} = \mathbf{0}$ *(at point* $\mathbf{P}$*). (e)* $\nabla \cdot \mathbf{v} = 0$ *and* $\nabla \times \mathbf{v} = \mathbf{0}$*.*

Example 4. Propagation of the Action Potential Along an Axon

In Section 8.1 equation (9) was derived for the action potential in the membrane of a voltage-clamped nerve axon. (Recall that voltage clamping means keeping the voltage the same all along the axon.) In real axons the action potential is a signal that propagates from the soma (cell body) along the axon to the *terminal synapses*. A space-dependent model must take this into account. Here we derive a balance equation for *charge* that incorporates the effect of transport in the axial direction. Define

x = distance along axon,

$q(x, t)$ = charge density per unit length inside axon at location x and time t,

$\mathbf{J}(x, t)$ = flux of charged particles (= current) at location x and time t.

$\sigma(x, t)$ = rate at which charge enters or leaves axon through its membrane at (x, t).

By referring to Figure 9.3 and to equation (24) one concludes that q is governed by the equation

$$\frac{\partial q}{\partial t} = -\frac{\partial \mathbf{J}}{\partial x} + \sigma. \tag{42}$$

(See problem 18.) In Section 8.1 we established that

$$q(x, t) = 2\pi a C v(x, t), \tag{43}$$

where

C = the capacitance of the axonal membrane,

a = the radius of the axon,

v = the voltage across the membrane.

It has further been shown that the rate at which charge enters the axon is

$$\sigma(x, t) = -2\pi a I_i, \tag{44}$$

where I_i is the net ionic current into the axon. Note that σ is analogous to a local source of charge. (It is the only term that leads to changes in q in the voltage-clamped equation (6) of Section 8.1.)

To find an expression for J we now use *Ohm's law,* which states that the current (in the axial direction) is proportional to a voltage gradient and inversely proportional to the resistance of the intracellular fluid. This implies that the net axial current in the axon would be

$$\mathbf{J} = -\left(\frac{\pi a^2}{R}\right)\frac{\partial v}{\partial x} \tag{45}$$

where $\partial v/\partial x$ is a local voltage gradient and R is the intracellular *resistivity* (ohm-cm).

Making the appropriate substitutions leads to the following equation for voltage:

$$\frac{\partial v}{\partial t} = \frac{a}{2RC}\frac{\partial^2 v}{\partial x^2} - \frac{I_i}{C}. \tag{46}$$

(See problem 18.) This equation with the appropriate assumptions about I_i is used to study propagated action potentials.

9.4 CONVECTION, DIFFUSION, AND ATTRACTION

Equations (24) and (40) are general statements that apply to numerous possible situations. To be more specific, it is necessary to select terms for **J** and σ capturing the particular forces and effects that lead to the motion, and to the creation or elimination of particles. The choices may be made on the basis of known underlying mechanisms, suitable approximations, or analogy with classical cases. We deal here with three classical forms of the flux term **J**.

Convection

Particles in a moving fluid take on the fluid's velocity and participate in a net collective motion. If $\mathbf{v}(x, y, z, t)$ is the velocity of the fluid, one can easily demonstrate that the flux of particles is given by

$$\mathbf{J} = c\mathbf{v}, \tag{47}$$

where all quantities may vary with space and time. (See problem 10 for the key idea in proving this.)

Substituting (47) into equation (24) leads to the following one-dimensional *transport equation:*

$$\frac{\partial c(x, t)}{\partial t} = -\frac{\partial}{\partial x}[c(x, t)\mathbf{v}(x, t)], \tag{48}$$

or, in arbitrary space dimensions,

$$\frac{\partial c}{\partial t} = -\nabla \cdot (c\mathbf{v}). \tag{49}$$

Attraction or Repulsion

Suppose ψ is a function that represents some source of attraction for particles. (For example, the particles could be charged, and ψ could be an electrostatic field.) An attractive force would pull particles towards the site of greatest attraction. The direction and the magnitude of motion would thus be determined by the gradient of ψ (for example, it might be $\alpha\nabla\psi$ for some scalar α); the net flux in that direction would be

$$\mathbf{J} = c\alpha\nabla\psi. \tag{50}$$

(In one dimension this is simply $\mathbf{J} = c\alpha(\partial\psi/\partial x)$.) Substituting into equation (24) results in the following one-dimensional equation for attraction to ψ:

$$\frac{\partial c(x, t)}{\partial t} = -\frac{\partial}{\partial x}\left[c(x, t)\alpha\frac{\partial\psi}{\partial x}\right], \tag{51}$$

or, in arbitrary space dimensions,

$$\frac{\partial c}{\partial t} = -\nabla \cdot (c\alpha \nabla \psi). \tag{52}$$

We will later encounter two realistic versions of this general form, one of which depicts the motion of organisms towards sites of high nutrient concentration, and another the avoidance of overcrowding.

Random Motion and the Diffusion Equation

One of the most important sources of collective motion on the molecular level is *diffusion,* which results from the perpetual random motion of individual molecules. Diffusion is an important "metabolically cheap" transport mechanism in biological systems, but as we shall see, its effectiveness decreases rapidly with distance. A familiar assumption made in the context of diffusion through cell membranes is that the rate of flow depends linearly on concentration differences. This is an approximation to a more complicated situation. An extension of this concept to more general situations is known as *Fick's law,* which states that the flux due to random motion is approximately proportional to the local gradient in the particle concentration:

$$\mathbf{J} = -\mathcal{D}\nabla c. \tag{53}$$

The constant of proportionality $\mathcal{D}$ is the *diffusion coefficient*. The net migration due to diffusion is "down the concentration gradient," in a direction away from the most concentrated locations. This makes sense since in most situations where there is a relatively large local concentration, more molecules *leave* on average than return (due to the random character of their motions).

In one dimension, diffusive flux is simply given by $J = -\mathcal{D}(\partial c/\partial x)$, so that upon substitution into equation (24) one gets

$$\frac{\partial c(x, t)}{\partial t} = \frac{\partial}{\partial x}\left[\mathcal{D}\frac{\partial}{\partial x}c(x, t)\right] \tag{54}$$

(If $\mathcal{D}$ is a constant and does not depend on c or x, it can be drawn outside the outer derivative, giving the most familiar version of the one-dimensional diffusion equation:

$$\frac{\partial c}{\partial t} = \mathcal{D}\frac{\partial^2 c}{\partial x^2}. \tag{55a}$$

In arbitrary dimensions this result would be written

$$\frac{\partial c}{\partial t} = \nabla \cdot (\mathcal{D}\nabla c), \tag{55b}$$

or, if $\mathcal{D}$ is constant, then

$$\frac{\partial c}{\partial t} = \mathcal{D}\,\Delta c = \mathcal{D}\nabla^2 c. \tag{55c}$$

Random Walk and the Diffusion Equation

A collection of particles shown in Figure 9.6 moves randomly with an average step length Δx every time unit τ. Assume that the probability of moving to the left, λ_l, and to the right, λ_r, are both equal; that is, $\lambda_l = \lambda_r = \frac{1}{2}$. The x axis is subdivided into segments of length Δx. We write a discrete equation that describes the change in the number of particles located at x.

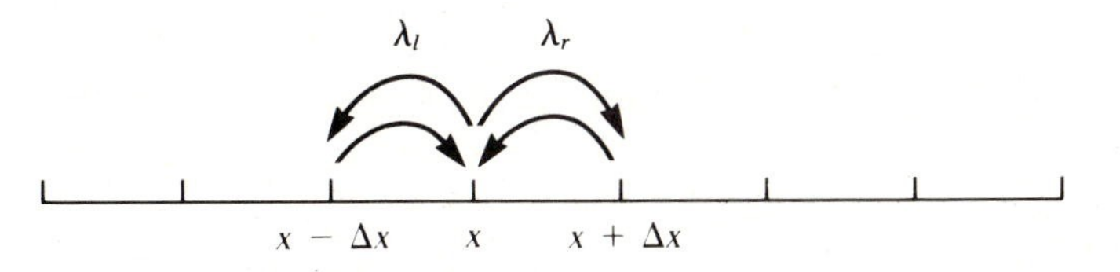

Figure 9.6 *Particles arrive at or depart from x randomly, with probabilities λ_l and λ_r of moving left or right.*

Let $C(x, t)\,\Delta x$ be the number of particles within the segment $[x, x + \Delta x]$ at time t. Then

$$C(x, t + \tau) = C(x, t) + \lambda_r C(x - \Delta x, t) - \lambda_r C(x, t) + \lambda_l C(x + \Delta x, t) - \lambda_l C(x, t). \tag{56}$$

Now we write Taylor-series expansions of these terms, as follows:

$$\begin{aligned} C(x, t + \tau) &= C(x, t) + \frac{\partial C}{\partial t}\tau + \frac{1}{2}\frac{\partial^2 C}{\partial t^2}\tau^2 + \dots, \\ C(x \pm \Delta x, \tau) &= C(x, t) \pm \frac{\partial C}{\partial x}\Delta x + \frac{1}{2}\frac{\partial^2 C}{\partial x^2}\Delta x^2 \pm \dots . \end{aligned} \tag{57}$$

Substituting into (56) and using the fact that $\lambda_r = \lambda_l = \frac{1}{2}$, we obtain

$$\frac{\partial c}{\partial t}\tau + \frac{1}{2}\frac{\partial^2 c}{\partial t^2}\tau^2 + \dots = \frac{1}{2}\frac{\partial^2 c}{\partial x^2}\Delta x^2 + \frac{1}{4}\frac{\partial^4 c}{\partial x^4}\Delta x^4 + \dots . \tag{58}$$

Dividing through by τ, we look at a limiting form of this equation for $\tau \to 0$, $\Delta x \to 0$ such that

$$\frac{(\Delta x)^2}{2\tau} = \text{constant} = \mathcal{D}. \tag{59}$$

Then the result is

$$\frac{\partial c}{\partial t} = \frac{(\Delta x)^2}{2\tau}\frac{\partial^2 c}{\partial x^2} = \mathcal{D}\frac{\partial^2 c}{\partial x^2}. \tag{60}$$

Note that (60) is equivalent to equation (55a).

Extensions of the random walk model for $\lambda_l \neq \lambda_r$ and for numerous other special cases are described by Okubo (1980).

The symbol Δ is the *Laplacian* of c; it stands for the combination $\nabla \cdot \nabla$ (read "div dot grad"), also written ∇^2. Equations (54) and (55a) are also known as *heat equa-*

tions since they describe equally well the diffusion of heat following *Newton's law of cooling*.

Fick's law is just one version of flux of diffusion and warrants several remarks. Clearly the term $-\mathcal{D}\nabla c$ gives a directionality to $\mathbf{J}$. The diffusion coefficient $\mathcal{D}$ represents the degree of random motion (how "motile" the particles are); $\mathcal{D}$ depends strongly on the size of the particles, the type of solvent, and the temperature.

While the assumption is common that diffusive flux takes the form of equation (53), this is not the only possibility. From a consideration of the Taylor series, diffusive flux can be appreciated as a reasonable first approximation. Since diffusion derives from concentration differences, consider the Taylor-series expansion

$$c(x + \Delta x, t) - c(x, t) = \Delta x \frac{\partial c}{\partial x} + \left(\frac{\Delta^2 x}{2}\right) \frac{\partial^2 c}{\partial x^2} + \cdots .$$

If flux depends linearly on differences in concentrations, for quite small Δx, it is approximately proportional to $\partial c/\partial x$, which is the one-dimensional version of (53).

In more complicated situations (chiefly for high concentrations when interactions between molecules become important), Fick's law is no longer accurate and other versions of diffusion are more applicable. It is a challenging physical problem to deal with such situations. A full treatment of the diffusion equation and of random walks is given in Okubo (1980) along with references and outlines of its extension to more complicated situations.

9.5 THE DIFFUSION EQUATION AND SOME OF ITS CONSEQUENCES

The one-dimensional diffusion equation derived in Section 9.4 is

$$\frac{\partial c}{\partial t} = \mathcal{D} \frac{\partial^2 c}{\partial x^2}. \tag{55a}$$

In radially and spherically symmetric cases in two and three dimensions the equation is slightly different: In two dimensions one obtains

$$\frac{\partial c}{\partial t} = \frac{\mathcal{D}}{r} \frac{\partial}{\partial r}\left(r \frac{\partial c}{\partial r}\right), \tag{61}$$

whereas in three dimensions the result is

$$\frac{\partial c}{\partial t} = \frac{\mathcal{D}}{R^2} \frac{\partial}{\partial R}\left(R^2 \frac{\partial c}{\partial R}\right), \tag{62}$$

where r and R are the distances away from the origin. (See problem 12 for an easy derivation.)

The methods one would apply to solving each of these equations would be somewhat different. However, even without solving them explicitly, certain interesting conclusions can be made. Based on dimensional considerations alone it follows from any one of equations (55a), (61), or (62) that $\mathcal{D}$ has the following units:

$$\mathcal{D} = \frac{(\text{distance})^2}{\text{time}}. \tag{63}$$

Table 9.3 Diffusion Coefficients of Biological Molecules

Temperature (°C)	*Substance*	*$\mathcal{D}$(cm² sec⁻¹)*	*Ref*
0	Oxygen in air	1.78×10^{-1}	1
20	Oxygen in air	2.01×10^{-1}	1
18	Oxygen in water	1.98×10^{-5}	1
25	Oxygen in water	2.41×10^{-5}	1
20	Sucrose in water	4.58×10^{-6}	2

Sources:
1. L. Leyton (1975), *Fluid Behavior in Biological Systems,* Clarendon Press, New York.
2. K. E. Van Holde (1971), *Physical Biochemistry,* Prentice-Hall, Englewood Cliffs, N. J.

From this simple observation follow a number of results whose consequences are important in numerous biological systems. First, as we shortly see, equation (63) implies that

1. The average distance through which diffusion works in a given time is proportional to $(\mathcal{D}t)^{1/2}$.
2. The average time taken to diffuse a distance d is proportional to $d^2/\mathcal{D}$.

The diffusion coefficients of several key biological substances are given in Table 9.3. As a typical magnitude for the diffusion coefficient of small molecules such as oxygen in a medium such as water, we shall take

$$\mathcal{D}_{\text{oxygen}} \simeq 10^{-5}\ \text{cm}^2\ \text{sec}^{-1}.$$

The dimensions of a single cell are roughly 1 to 10 microns ($1\ \mu = 10^{-4}$ cm $= 10^{-6}$ m). As shown in Table 9.4, the amount of time it takes to diffuse through a given distance increases rapidly with the length scale.

On the scale of intracellular structures, diffusion is an extremely rapid process and can thus act as a metabolically free transport mechanism, in the sense that no energy need be expended by the cell to maintain it. On somewhat larger scales, (such as 1 mm), diffusion is already inadequate for such critical functions as oxygen transport. The longest cells of the human body are neurons; some of these have axons that are at least 1 m in length. Transport of small molecules from one end to the other would take roughly 30 years if diffusion were the only available mechanism.

Table 9.4 Time Taken to Diffuse Through a Given Distance

Distance	*Diffusion time*
$1\ \mu\text{m} = 10^{-6}$ m	10^{-3} sec
$10\ \mu\text{m}$	0.1 sec
1 mm	10^3 sec $\simeq$ 15 min
1 cm	10^5 sec $\simeq$ 25 h $\simeq$ 1 day
1 m	10^8 sec $\simeq$ 27 years

The following simple arguments due to, for example, Haldane (1928) and LaBarbera and Vogel (1982) lead to a number of deductions about the limitations of diffusion.

Consider a spherical cell of radius r. The volume and surface area of such a cell are

$$V = \tfrac{4}{3}\pi r^3, \qquad S = 4\pi r^2. \tag{64}$$

Suppose that the cell metabolizes a given substance completely, so that its concentration at $r = 0$ is $c(r, t) = 0$, while its concentration at the surface is c_0. The gradient thus established is c_0/r (concentration difference per unit distance). Thus a diffusive flux of magnitude $\mathcal{D}\, c_0/r$ would admit molecules through the cell surface. The total number of molecules entering the cell per unit time would be

$$JS \simeq \mathcal{D}\frac{c_0}{r}4\pi r^2 = 4\pi\mathcal{D}\, c_0 r. \tag{65}$$

The rate of degradation of substance, however, is generally proportional to the volume of the cell:

$$\text{rate used} \simeq \frac{4}{3}\frac{\pi r^3}{\tau}, \tag{66}$$

where τ is the time constant for the degradation process. Thus

$$\frac{\text{rate supply}}{\text{rate used}} \simeq 3\mathcal{D}\, c_0\frac{\tau}{r^2}. \tag{67}$$

Since for a viable cell this ratio should not fall below 1, it is necessary that

$$1 \simeq 3\mathcal{D}\, c_0\frac{\tau}{r^2}, \tag{68}$$

or

$$c_0 = \frac{r^2}{3\tau\mathcal{D}}. \tag{69}$$

To match supply and demand the minimum external substance concentration must be proportional to the square of the cell radius. It is therefore unrealistic to expect spherical cells whose radii are large to rely solely on diffusion as a means of conveying crucial substances inside the cell.

LaBarbera and Vogel (1982) point out some of the most common ways adopted by organisms to reduce the limitations due to diffusion. These are highlighted below.

Size and shape

Geometry influences diffusion rates. Flat shapes (such as algal leaves) or long branched filaments (such as fungi, filamentous algae, roots, and capillaries) are ideally suited for organisms that rely heavily on direct absorption of substances from their environment; these shapes can increase in volume (for example, by getting longer) without changing the distance through which diffusion must act (that is, the radius of the cylinder). La Barbera and Vogel suggest a dimensionless *flatness index*

$$\gamma = \frac{S^3}{V^2} \tag{70}$$

as an appropriate description of shape; they point out that with increasing size, an organism relying on diffusion must increase its flatness.

Dimensionality

It can be shown that the diffusion time taken to reach some internal sink depends differently on length scales in one, two, and three dimensions; one obtains somewhat different results from the equations (55a), (61), or (62). With the geometries given in Figure 9.7 one can establish the results that the diffusion time is as follows:

in one dimension: $$\tau_1 \simeq \frac{L^2}{2\mathcal{D}_1}. \tag{71a}$$

in two dimensions: $$\tau_2 \simeq \frac{L^2}{2\mathcal{D}_2} \ln \frac{L}{a}. \tag{71b}$$

in three dimensions: $$\tau_3 \simeq \frac{L^2}{2\mathcal{D}_3} \frac{2}{3} \frac{L}{a}. \tag{71c}$$

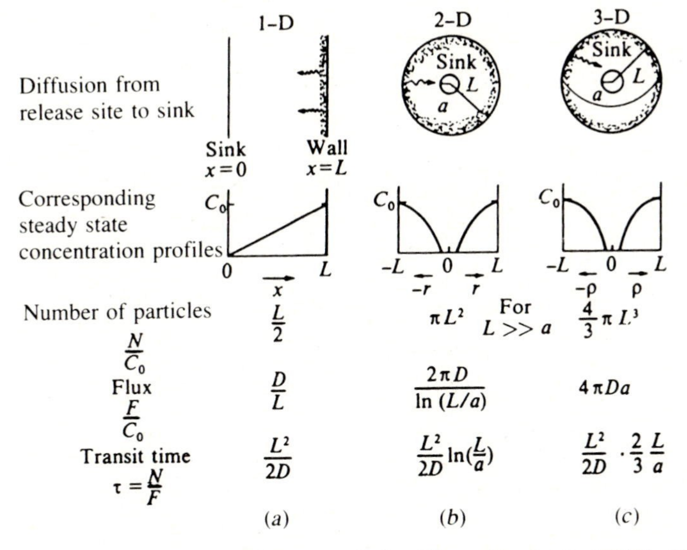

Figure 9.7 *The average time it takes a particle to diffuse from a source to a sink, called the transit time τ, depends on the dimensionality. (a) In one dimension, τ is proportional to* L^2 *where* L *is the distance. (b) In two dimensions, τ is greater by a factor of* ln (L/a) *where* a *is the radius of the sink. (c) In three dimensions the multiplicative factor is* L/a. *[From Hardt, S. (1980). Transit times. Fig. 6.2.1, p. 455. Copyright © 1980 by Cambridge University Press and reprinted with their permission.] In L.A. Segel,* Mathematical Models in Molecular and Cellular Biology. *Cambridge University Press, England.*

Here L is the cell radius and a is the radius of an internal sink (for example, an enzyme molecule that degrades substance). (See details in Figure 9.7 and Section 9.6.)

Murray (1977, chap. 3) gives an in-depth analysis and application of the effects of dimensionality to the antenna receptors of moths. In a set of papers, S. Hardt describes a convenient way of calculating transit times without explicitly solving the time-dependent diffusion equation.

We thus see that diffusion acts much more quickly in one- or two-dimensional settings than in three dimensions. This provides an advantage for intracellular organization of chemical reactions on membranes (which are two-dimensional) rather than on "loose" enzymes in the cytoplasm. Hardt (1978) compares the two- and three-dimensional cases where a represents the dimensions of an enzyme ($\sim$10 Å) and $\mathcal{D}_2 \simeq 100\mathcal{D}_3$. She concluded that for the cells of diameter larger than 1 μ, the organism benefits by arranging enzymes on internal membranes.

Circulatory systems

Where *geometric* solutions to diffusion limitations have failed, organisms have evolved ingenious mechanisms to convey substances to their desired destinations. From the intracellular *cytoplasmic streaming* and assorted mechanochemical methods to the circulatory system of macroscopic organisms, the ultimate purpose is to overcome the deficiency of long-distance diffusion and to transport substances efficiently. A fascinating account of the minimal design principle necessary to make a circulatory system work is given by LaBarbera and Vogel.

9.6 TRANSIT TIMES FOR DIFFUSION

Despite limitations on large distance scales, diffusion is of great importance in many processes on the cellular level. To give one example, communication between neighboring neurons is based on a chemical information system. Substances such as *acetylcholine* (called a *neurotransmitter*) are released by vesicles at the terminal branches of a given neuron, diffuse across the synapses, and relay messages to the neighboring neuron. An important consideration, particularly so in this example, is the average length of time taken to diffuse a given distance and how this time depends on particular features of the geometry.

Until a recent innovation suggested by Hardt, the problem of *diffusion transit times* was addressed by solving the time-dependent diffusion equation in the geometry of interest and using the resulting solution to derive a relationship. This process tends to be rather cumbersome for all but the simplest cases because solving diffusion equations in complicated geometries is a nontrivial task. Thus the approach was less than ideal.

A simpler method, proposed by Hardt (1978), is based on the observation that the *mean transit time* τ of a particle is independent of the presence or absence of other particles (given that no interactions occur) and can thus be computed in a steady-state situation. Hardt remarked that τ is then given by a simple ratio of two quantities that can be calculated in a straightforward way once the steady-state diffusion equation is solved. Solving the latter is always easier than solving the time-dependent problem, since it is an ordinary differential equation. To establish Hardt's result we define the following quantities:

N = total number of particles in the region,

F = total number entering the region per unit time,

λ = average removal rate of particles,

$\tau = 1/\lambda$ = average time of residency in the region.

Regardless of spatial variations, one can make an approximate general statement about the total number of particles in a given region. For instance, if particles enter at some constant rate F and are removed at a sink with rate λ, then

$$\frac{dN}{dt} = \frac{\text{entering}}{\text{rate}} - \frac{\text{removal}}{\text{rate}} = F - \lambda N. \tag{72}$$

In steady state ($dN/dt = 0$) we obtain the result that

$$F = \lambda N = \frac{N}{\tau} \Rightarrow \tau = \frac{N}{F}. \tag{73}$$

Example 5

Consider the one-dimensional geometry shown in Figure 9.7(*a*), with particles entering at $x = L$ and diffusing to $x = 0$. Then assuming particles cannot leave the region (the interval $[0, L]$), the mean residency time for a particle is the same as the mean time it takes to diffuse from the source (wall at $x = L$) to the sink (at $x = 0$). (It is assumed that particles are removed only at the sink.)

The time-dependent particle concentration is given by equation (55a). However, according to Hardt's observation, to compute the mean time for diffusion it suffices to find the steady-state quantities. To do so we consider the steady-state equation

$$\frac{\partial}{\partial x}\left(\mathcal{D}\frac{\partial c}{\partial x}\right) = 0, \tag{74}$$

and assume that $c(L) = C_0$, $c(0) = 0$. (These are *boundary conditions,* to be discussed in more detail in Section 9.8 and the Appendix. In the equation to be solved c depends only on x, so we have an ODE whose solution is easily found to be

$$c(x) = \alpha x + \beta. \tag{75}$$

By using the boundary conditions we find $\alpha = C_0/L$ and $\beta = 0$, so that

$$c(x) = C_0\frac{x}{L}. \tag{76}$$

(See problem 21.) The total number of particles is

$$N = \int_0^L C(x)\,dx = \frac{C_0}{L}\int_0^L x\,dx = \frac{C_0}{L}\frac{L^2}{2} = C_0\frac{L}{2}. \tag{77}$$

Particles enter through $x = L$ due to diffusive flux,

$$J = -\mathcal{D}\frac{\partial c}{\partial x}.$$

Assuming a wall of unit area at $x = L$, we obtain the result that

$$F = \text{flux} \times \text{area} = J \times 1 = \mathcal{D}\frac{C_0}{L}. \tag{78}$$

Note that in higher dimensions it will be necessary to take into account the *area* through which particles can enter, which depends in a less trivial way on the geometry of the region (see problems 19 and 20).

Thus the mean transit time from source to sink is

$$\tau = \frac{N}{F} = \frac{C_0 L}{2}\frac{L}{\mathcal{D}C_0} = \frac{L^2}{2\mathcal{D}}. \tag{79}$$

Derivations of similar results for two and three dimensions are outlined in the problems.

9.7 CAN MACROPHAGES FIND BACTERIA BY RANDOM MOTION ALONE?

Macrophages are cells that are implicated in a number of defense responses to infection in the body. One of their important roles is to clear the lung surface of the bacteria we inhale with every breath. Macrophages are motile, crawling about on the walls of *alveoli* (the small air sacs in the lung at which gas exchange with the blood takes place) until they locate and eliminate an invader. Although the whole process is a complicated one involving several types of cells and chemical intermediates, the basic goal can be summarized simply: the macrophage response must be sufficiently rapid and accurate to prevent the proliferation of invading microorganisms. A good summary of the macrophage response to the bacterial challenge is given by Lauffenburger (1986) and Fisher and Lauffenburger (1986).

These authors propose an interesting question regarding the motion of the defending macrophages: Is random motion by the macrophages adequate to find their bacterial targets before rapid population increase has occurred? To answer this question, Lauffenburger observes that a macrophage moves at a characteristic speed s. The direction of motion is typically fairly constant for a time duration τ, and then some reorientation may occur. If the motion is truly random, it is possible to define an "effective diffusion coefficient" for macrophages.

$$\mathcal{D} = \tfrac{1}{2}\tau s^2. \tag{80}$$

(This can be based on rigorous random-walk calculations; see problem 22.)

We now consider a simple two-dimensional geometry such as the one shown in Figure 9.7(b). The sink (or "target") will represent a bacterium, assumed to have an approximate radius of detection a, and the disk (with radius R) will depict the surface of an alveolus. We shall assume that a macrophage enters the region through its circular boundary and searches until it arrives at its target. The transit time based on a purely random motion is given by equation (71b). According to Lauffenburger and Fisher (1986), the values of constants that enter into consideration are as follows:

a = radius of bacterial cell = 20 μ,

s = speed of motion of macrophage = 3 μ min^{-1}

ϵ = time spent moving in given direction = 5 min,

A = area of alveolus = $2.5 \times 10^5\ \mu^2$,

N = number of bacteria = 1,

ν = reproductive rate of bacteria = 0.2 hr^{-1}.

Then the radius of the disk is

$$L = \left(\frac{A}{\pi}\right)^{1/2} = \left(\frac{2.5 \times 10^5}{\pi}\right)^{1/2} = 2.8 \times 10^2\ \mu. \tag{81}$$

The effective diffusion coefficient is

$$\mathcal{D} = \frac{s^2\epsilon}{2} = (3\ \mu)^2 \times \frac{5 \text{ min}}{2} = 22.5\ \mu^2 \text{ min}^{-1}. \tag{82}$$

Thus the average time to reach the bacterial cell is

$$\begin{aligned} \tau = \frac{L^2}{2\mathcal{D}} \ln \frac{L}{A} &= \frac{(2.8 \times 10^2)^2}{(2)(22.5)} \ln \frac{2.8 \times 10^2}{20}, \\ &= 1.74 \times 10^3\ (2.64) \simeq 4.6 \times 10^3 \text{ min} = 76 \text{ h}. \end{aligned} \tag{83}$$

However, the bacterial doubling time τ_d is given by

$$\tau_d = \frac{1}{\nu} = \frac{1}{0.2} = 5 \text{ h}. \tag{84}$$

Thus if random motion was the only means of locomotion, the macrophage would on average be unable to find its target before proliferation of bacteria takes place.

By contrast, if macrophages are perfectly sensitive to the relative location of their targets, the time taken to reach the bacteria would be

$$T = \frac{L}{s} = 2.8 \times \frac{10^2}{3} = 93 \text{ min} \simeq 1.5 \text{ h}. \tag{85}$$

In practice, neither one of these two extremes is totally accurate; the orientation of the macrophage is indeed governed by gradients in chemical factors produced as byproducts of infection, although some random motion is also present. We shall discuss chemotaxis more fully in the following chapters.

9.8 OTHER OBSERVATIONS ABOUT THE DIFFUSION EQUATION

In this section we make some general observations about the mathematical properties of the diffusion equation, leaving certain technical details to the Appendix at the end of this chapter.

Consider the one-dimensional diffusion equation

$$\frac{\partial c}{\partial t} = \mathcal{D} \frac{\partial^2}{\partial x^2} c. \tag{86}$$

By a solution to equation (86) we mean a real-valued function of (x, t) whose partial derivatives satisfy (86). We first remark that (86) is linear. Thus if $c = \phi_1(x, t)$ and $c = \phi_2(x, t)$ are two solutions of (86), then so is $c = A\phi_1(x, t) + B\phi_2(x, t)$. This follows from the superposition principle, as in linear difference or differential equations.

We can reinforce the connection between partial and ordinary differential equations by writing (86) in "operator notation":

$$\frac{\partial c}{\partial t} = \mathcal{L}c, \tag{87a}$$

where

$$\mathcal{L} = \mathcal{D}\frac{\partial^2}{\partial x^2}. \tag{87b}$$

$\mathcal{L}$ is a *linear operator,* also called the *diffusion operator;* it is an entity that takes a function c and produces another function ($\mathcal{D}$ × the second x partial derivative of c). It will soon be clear why such notation is helpful.

Equation (86) has two spatial derivatives and one time derivative. Thus, in order to select out a single (*unique*) solution from an infinite class of possibilities, it is necessary to specify, in addition to (86), two other spatial constraints (boundary conditions) and one time constraint (an initial condition). However, were this done in an arbitrary or haphazard way, the problem might be such that no sensible solutions to it would exist. ("Sensible" solutions are those that conform to real physical processes.) The problem is then said to be ill posed. What constitutes a *well-posed problem* depends on the character of the PDE and the combination of added conditions. (Mathematicians are particularly concerned with proving *well-posedness,* since this is essentially equivalent to guaranteeing that a unique and meaningful solution exists.) We shall avoid this issue entirely since it is beyond our scope.

Several examples of initial and boundary conditions typically applied to equation (86) are given in the Appendix. Physically such conditions specify the initial configuration [the concentration at time zero at every point in the region, $c(x, 0)$] and what happens at the boundary of the domain. It makes intuitive sense that both factors will influence the evolution of the concentration $c(x, t)$ with time. For example, a region for which particles are admitted through the boundaries will support different behavior than one that has impermeable boundaries.

In forming solutions to the diffusion equation, one finds especially useful functions $f(x)$ that satisfy the relation

$$\mathcal{L}f = \lambda f, \tag{88}$$

where $\mathcal{L}$ is given by (87b). Such functions are called *eigenfunctions,* and here again, in terminology previously encountered, λ is an *eigenvalue*. Eigenfunctions of the diffusion operator have the property that their second derivative is a multiple of the original function. Three familiar functions that fall into this category are the following:

$$f_1(x) = \exp(\pm\sqrt{\lambda}x), \tag{89a}$$

$$f_2(x) = \sin(\pm\sqrt{\lambda}x), \tag{89b}$$

$$f_3(x) = \cos(\pm\sqrt{\lambda}x). \tag{89c}$$

By straightforward partial differentiation, the reader may verify that

$$c(x, t) = e^{Kt} f_i(x) \qquad (i = 1, 2, 3) \tag{90}$$

are solutions to (86) if the constant K is chosen appropriately [see problem 15(a,b)]. We arrive at the same result in the Appendix using the technique of separation of variables.

To illustrate an important point, let us momentarily consider a finite one-dimensional domain of length L and assume that at the boundary of the region there is a sink that eliminates all particles. By this we mean that the concentration $c(x, t)$ is zero (and held fixed) at the ends of the interval so that for $x \in [0, L]$ the appropriate boundary conditions are

$$c(0, t) = 0, \tag{91a}$$

$$c(L, t) = 0. \tag{91b}$$

From the form of the solution in (90) it is readily verified that to satisfy (91a) one should select $f_1(x) = \sin(\pm\sqrt{\lambda}x)$, since neither of the other two eigenfunctions are zero at $x = 0$. Further restrictions are necessary to ensure that (91b) too is satisfied. This can be done by choosing

$$\sqrt{\lambda} = n\frac{\pi}{L} \qquad (n = 1, 2, \ldots), \tag{92}$$

since then $\sin(\sqrt{\lambda}L) = 0$.

Now consider a second situation. Suppose that this finite one-dimensional domain has impermeable boundaries, so that particles neither enter nor leave at the ends of the interval. This means that diffusive flux is zero at $x = 0$ and $x = L$. According to our definitions in Section 9.4,

$$\mathbf{J}(0) = \mathbf{J}(L) = \mathscr{D}\frac{\partial c}{\partial x}\bigg|_{\text{boundary}} = 0.$$

Thus *no-flux boundary conditions* are equivalent to the conditions

$$\frac{\partial c}{\partial x} = 0 \quad \text{at} \quad x = 0,$$

$$\frac{\partial c}{\partial x} = 0 \quad \text{at} \quad x = L.$$

To satisfy the first boundary condition we must choose in the solution (90) the eigenfunction $f_c(x) = \cos(\pm\sqrt{\lambda}\,x)$. (This has a "flat" graph at $x = 0$.) Similarly, to satisfy the second condition we need

$$\sqrt{\lambda} = n\frac{\pi}{L} \qquad (n = 0, 1, 2, \ldots). \tag{92}$$

Since then $\cos(\sqrt{\lambda}\,L) = \cos(n\pi) = \pm 1$. (In other words, the cosine has a "flat" graph also at $x = L$.)

This discussion illustrates the idea that imposing boundary conditions tends to weed out certain classes of solutions (for example, (89a,c) in the first example). Furthermore, in a given class of eigenfunctions only certain members are compatible. (For example, in the first case discussed, only those sine functions that go through zero at both ends of the interval are compatible.) This has important implications that will be touched on in later discussions.

The diffusion equation has many other types of solutions. Some of these will be described in the Appendix. In higher dimensions the geometry of the region may be much more complicated and difficult to treat analytically. At times certain features such as radial symmetry are exploited in solving the two- or three-dimensional diffusion equation. Crank (1979) and Carslaw and Jaeger (1959) describe methods of solution in such cases. An application to chemical bioassay is described in the next section.

9.9 AN APPLICATION OF DIFFUSION TO MUTAGEN BIOASSAYS

Chemical substances that are suspected of being carcinogens are frequently tested for *mutagenic properties* using a bioassay. Typically one seeks to determine whether a critical concentration of the substance causes *genetic mutations* (aberrations in the genetic material), for example in bacteria. The bacteria are grown on the surface of a solid agar nutrient medium to which a small amount of *mutagen* is applied. Generally, the chemical is applied on a presoaked filter paper at the center of a *petri dish* and spreads outwards gradually by diffusion. If the substance has an effect, one eventually observes concentric variations in the density and appearance of the bacterial culture that correlate with different levels of exposure to the substance.

While such qualitative tests have been commonly used for antibiotic, mutagenic, and other chemical tests, more recently quantitative aspects of the test were developed by Awerbuch et al (1979). These investigators noted that the radius of the observed zones of toxicity and mutagenesis (see Figure 9.8) could be used directly in obtaining good estimates of the threshold concentrations that produce these effects.

Working in radially symmetric situations, Awerbuch et al. (1979) used the radial form of the diffusion equation,

$$\frac{\partial c}{\partial t} = \mathcal{D}\left(\frac{\partial^2 c}{\partial r^2} + \frac{1}{r}\frac{\partial c}{\partial r}\right) - \frac{c}{\tau} \tag{93}$$

where

r = radial distance from the center of the dish,

$c(r, t)$ = the concentration at a radial distance r and time t,

$\mathcal{D}$ = diffusion coefficient of the mutagen,

$1/\tau$ = the rate of spontaneous decay of the mutagen. (See problem 17.)

Because the probability of a mutation taking place depends both on the exposure concentration and the exposure duration, a time-integrated concentration was defined as follows:

$$C(r) = \frac{1}{(T_2 - T_1)s}\int_{T_1}^{T_2} c(r, t)\, dt, \tag{94}$$

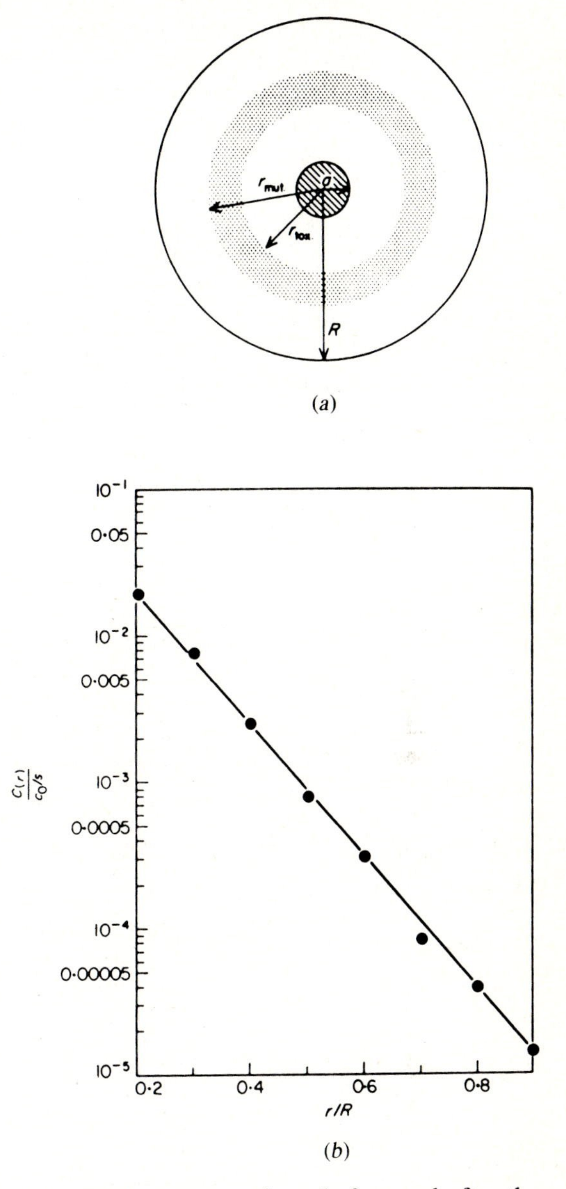

Figure 9.8 *(a) In a test for mutagenicity, Awerbuch et al. (1979) place a mutagen-soaked filter paper (radius = a) at the center of a petri plate (radius = R). The substance diffuses outwards. Beyond some threshold level the substance fails to be toxic but does cause changes in the appearance of the bacteria growing on the plate (due to increased mutation). (b) The time-integrated concentration of mutagen [equation (94)] can be computed as a function of radial distance by solving the diffusion equation; a plot of log* c(r) *versus* r *can then be used to determine the threshold concentrations for toxicity* $c(r_{tox})$ *and for mutagenesis* $C(r_{mut})$. *[From Awerbuch et al. (1979). A quantitative model of diffusion bioassays. J. Theor. Biol., 79, figs. 1 and 2; reproduced by permission of Academic Press Inc. (London)]*

where s is the width of the agar and T_1 and T_2 represent times before and after the diffusion wave arrives at the point r.

The initial situation, shown in Figure 9.8(a), corresponds to a constant mutagen level within the filter paper disk (radius a) at the center of the dish. Thus at $t = 0$ the concentration can be described by the equation

$$c(r, 0) = \begin{cases} c_0 & \text{for } r < a \\ 0 & \text{for } r \geq a \end{cases}$$

This statement is an *initial condition* (see Appendix).

Furthermore, because the walls of the dish (at radius $r = R$) are impermeable to chemical diffusion, there is no radially directed flux of particles at $r = R$. Thus an additional condition is that

$$\frac{\partial c}{\partial r} = 0 \qquad \text{for } r = R.$$

This is the radial equivalent of the one-dimensional no-flux condition discussed in the previous section. It is also trivially true that $\partial c/\partial r = 0$ at $r = 0$ in this example. More discussion of boundary and initial conditions is given in the Appendix.

We will not go into the details of how the radially symmetric diffusion equation (93) is solved (see Awerbuch et al., 1979, and Caslaw and Jaeger, 1959). The methods are well known but not of particular importance to our discussion. Once a solution is obtained, the quantity (94) can be computed and tabulated. Figure 9.8(*b*) demonstrates a typical relationship between the value of $C(r)/c_0 s$ and radial distance that can then be used directly in making a quantitative estimate of the time-averaged mutagen threshold. Observe that if bacteria were exposed to a uniform fixed chemical concentration C_0, the value of $C(r)$ would be the same for all r and would equal C_0. In this way a correspondence can be made between results of the *diffusion bioassay* and similar *homogeneous bioassay* concentrations.

To illustrate the method, Awerbuch et al. (1979) quote the following example for the bacteria *Salmonella typhimurium* and the mutagen N-methyl-N-nitro-N^1-nitrosoguanidine. Conditions of the bioassay were as follows:

a = radius of chemically treated filter paper = 0.318 cm,

R = radius of petri plate = 2.5 cm,

s = thickness of agar = 0.356 cm,

τ = decay time of mutagen = 2.25 h,

$\mathcal{D}$ = diffusion coefficient of chemical in agar = 7.2×10^{-6} cm^2 sec^{-1},

c_0/S = initial concentration of chemical (applied on filter paper) corrected for agar thickness = 221.04 μg cm^{-3}.

[*Note:* c_0, $c(r, t)$, and $c(r)$ have dimensions of grams per centimeter squared since only two-dimensional diffusion is being considered here. For this reason it is necessary to divide by agar thickness so as to obtain a concentration in grams per centimeter cubed.]

Under these conditions, a ring of mutated bacteria occurs at a radial distance of 2.17 cm. We thus have

$$\frac{r}{R} = \frac{2.17}{2.5} = 0.868.$$

From Figure 9.8(*b*) we observe that corresponding to this radius is a dimensionless time-averaged concentration,

$$\frac{C(r)}{(c_0/s)} = 1.95 \times 10^{-5}.$$

Thus the critical concentration for mutagenicity is

$$C_{\text{mut}} = 1.95 \times 10^{-5} \times 221.04\ \mu\text{g ml}^{-1},$$
$$= 4.31 \times 10^{-3}\ \mu\text{g ml}^{-1}.$$

The diffusion-based assay is of wide applicability. Considering the older methods of serial dilutions and tests of bacteria cultured at numerous mutagen concentrations, one appreciates the elegance of this simple and time-saving procedure.

PROBLEMS*

Problems 1 through 6 are suitable for reviewing the properties of functions of several variables.

1. For the following functions, sketch the surface corresponding to $z = f(x, y)$ and the level curves in the xy plane:

(a) $f(x, y) = x^2 + y^2$.

(b) $f(x, y) = -2x^2 - 2y^2$.

(c) $f(x, y) = \exp \dfrac{-(x^2 + y^2)}{2}$.

(d) $f(x, y) = 2x + y$.

(e) $f(x, y) = xy$.

(f) $f(x, y) = \sin x \cos y$.

2. For each function in problem 1 find the following:

(1) $\dfrac{\partial f}{\partial x}$,

(2) $\dfrac{\partial^2 f}{\partial x \partial y}, \dfrac{\partial^2 f}{\partial y \partial x}$, and

(3) ∇f.

3. For each function in problem 1, determine whether there are any critical points. Which if any are local maxima?

4. Sketch the level curves described by the following equations. Give an equation for a surface that has these level curves. Sketch the vector field corresponding to ∇f by using its property of orthogonality to level curves:

(a) $c = \dfrac{x^2}{a^2} + \dfrac{y^2}{b^2}$.

(b) $c = x^2 + y$.

(c) $c = x - y$.

(d) $c = \dfrac{x^2}{a^2} - \dfrac{y^2}{b^2}$.

(e) $c = (x^2 + y^2)$.

5. For the following vector fields, find $\nabla \times \mathbf{F}$, $\nabla \cdot \mathbf{F}$:

(a) (x, y, z).

(b) $(y - z, z - x, x - y)$.

(c) $(x^2 + 2y + z, y^2 - x, z^2 - y^2)$.

(d) $(\sin xyz, \cos xyz, e^{xyz})$.

(e) $\left(\dfrac{1}{x}, \dfrac{1}{y}, \dfrac{1}{z}\right)$.

(f) $(x + y, y + z, z + x)$.

*Problems preceded by an asterisk (*) are especially challenging.

*6. Determine whether the following vector fields are gradient fields. If so, find ϕ such that $\mathbf{F} = \nabla\phi$:
(a) (x, y).
(b) (y^2, x^2).
(c) $(\sin xy, \cos xy)$.
(d) $(x + y, x - y)$.
(e) (ye^{xy}, xe^{xy}).
(f) (x^2y, y^2x).

7. **(a)** Verify that terms in equation (22) carry the correct dimensions.
(b) Explain why the integral in equation (25a) represents the number of particles in the interval (x_1, x_2).
(c) Similarly, explain the integral in equation (25b).
(d) Give justification for equation (26).
(e) Verify that equation (27) leads to (28) when the appropriate limit is taken.

8. The cross-sectional area of the small intestine varies periodically in space and time due to peristaltic motion of the gut muscles. Suppose that at position x (where x = length along the small intestine) the area can be described by

$$A(x, t) = \frac{a}{2}[2 + \cos(x - vt)],$$

where v is a constant.
(a) Write an equation of balance for $c(x, t)$, the concentration of digested material at location x.
(b) Suppose there is a constant flux of material throughout the intestine from the stomach [that is, $\mathbf{J}(x, t) = 1$] and that material is absorbed from the gut into the bloodstream at a rate proportional to its concentration for every unit area of intestinal wall. Give the appropriate balance equation.
(c) Show that even if $\mathbf{J}(x, t) = 0$ and $\sigma(x, t) = 0$, the concentration $c(x, t)$ appears to change.

9. For a planar flow, consider a small rectangular region of dimensions $\Delta x \times \Delta y$. Carry out steps analogous to those of Section 9.3 (subsection "Flows in Two and Three Dimensions") to derive the two-dimensional form of the equation of conservation.

10. Consider the fluid shown in the accompanying diagram. Assume that every particle has the same velocity $\mathbf{v}$.
(a) What is the flux of particles through the unit area dA? (*Hint:* Consider all particles contained in an imaginary prism of length $v\Delta t$, where v is the magnitude of $\mathbf{v}$. During a time Δt they will have all crossed the wall dA. Now use the definition of flux to show that (47) holds.
(b) Extend your reasoning in part (a) to the case where $\mathbf{v}$ varies over space and time.

11. Suppose the diffusion coefficient of a substance is a function of its concentration; that is,

$$\mathcal{D} = f(c).$$

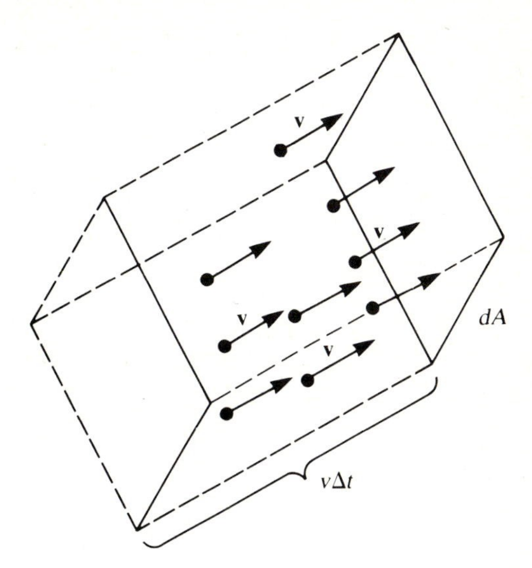

Figure for problem 10.

Show that c satisfies the equation

$$\frac{\partial c}{\partial t} = \mathcal{D}\frac{\partial^2 c}{\partial x^2} + g\left(\frac{\partial c}{\partial x}\right)^2,$$

where $g = f'(c)$.

12. **(a)** Consider a radially symmetric region $\mathcal{D}$ in the plane, and let $c(r, t)$ be the concentration of substance at distance r from the origin and $J(r, t)$ the radical flux. Use the derivation of the conservation laws in Section 9.3 (subsection "Tubular Flow") to show that $c(r, t)$ satisfies the following relation:

$$\frac{\partial c}{\partial t} = -\frac{1}{r}\frac{\partial}{\partial r}(Jr) \pm \sigma(r, t)$$

Hint: consider a pie-shaped "tube," that is, a small sliver removed from a circular disk—see Figure (a), and determine how its cross-sectional area changes with radial distance.

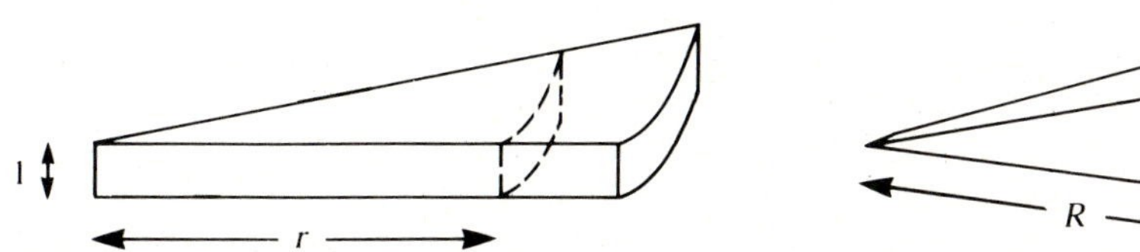

Figure (a) for problem 12. ***Figure (b) for problem 12.***

(b) Extend the idea in part (a) to a spherically symmetric region S in R^3, and show that $c(R, t)$, the concentration at distance R from the origin, satisfies the following:

$$\frac{\partial c}{\partial t} = -\frac{1}{R^2}\frac{\partial (JR^2)}{\partial R} \pm \sigma(R, t).$$

Hint: For the "tube," consider an element of volume of a sphere and determine how its cross-sectional area depends on radial distance. See Figure (*b*).

(c) Use parts (a) and (b) to obtain the radially and spherically symmetric diffusion equations given in Section 9.5.

13. The time to diffuse from source to sink depends on dimensionality as demonstrated in equations (71a–c). Give conditions on the ratio L/a for which $\tau_3 > \tau_2 > \tau_1$.

14. (a) Show that the conclusions regarding diffusion transport into a cell hold equally well if the cell is nonspherical (such as a cell that has length l, width w, and girth g), provided that as it grows all three dimensions are expanded.

(b) Find the flatness ratio γ for the following shapes:

(1) An ellipsoid of dimensions $a \times b \times c$.

(2) A sphere of radius R.

(3) A long cylinder of radius r and height h (neglect the top and bottom caps).

(4) A cone whose radius is r when its height is h (neglect the top).

(c) *Extended project*. Make a summary of the various ways in which organisms overcome diffusional limitations, and illustrate these with examples drawn from the biological literature.

15. (a) Verify that equation (90) is a solution to equation (86).

(b) Determine what the restrictions are on K.

(c) Show that equations (91a,b) can only be satisfied by choosing

$$f(x) = \sin(\pm\sqrt{\lambda}x), \text{ where } \sqrt{\lambda} = \frac{n\pi}{L}.$$

16. (a) Suppose that in a diffusion bioassay for the mutagen N-methyl-N-nitro-N^1-nitrosoguanidine, one finds that mutations occur at a radial distance $r = 0.4R$, where R is the radius of the petri dish. Using constants quoted in Section 9.9 determine C_{mut}, the threshold concentration for mutation.

(b) Repeat part (a) for $r = 0.6R$.

17. (a) Explain equation (93) by expanding equation (61).

(b) Suppose the bioassay devised by Awerbuch et al. (1979) is performed in a thin tube rather than a radially symmetric plate. What would the appropriate equations and conditions of the problem be?

(c) Referring to your answers to part (a), what solutions for $c(x, t)$ would then typically be encountered?

18. *Propagated action potentials.*

(a) Explain equation (42) based on the definitions given for charge density, current, and charge "creation" σ.

(b) Explain equations (43) and (44). What would I_i depend on? (See Section 8.6.)

(c) Ohm's law states that $V = IR$, where V = potential difference across a resistance R, and I = current. How is equation (45) related to this law?

(d) Show that equations (42–45) together imply equation (46).

Problems 19 and 20 suggest generalized versions of mean diffusion transit times. Solution requires familiarity with double and triple integration.

19. In two dimensions consider the radially symmetric region shown in Figure 9.7(*b*) with a sink of radius a and a source of radius L. Assume that

$$c(a) = 0, \qquad c(L) = C_0.$$

(a) Solve the steady-state two-dimensional equation of diffusion

$$0 = \frac{\mathcal{D}}{r}\frac{\partial}{\partial r}\left(r\frac{\partial c}{\partial r}\right).$$

(b) Define

$$N = \iint_{\text{disk}} c\, dA = \int_0^{2\pi}\int_a^L c(r) r\, dr\, d\theta.$$

Compute this integral and interpret its meaning.

(c) Define

$$F = \text{flux} \times \text{circumference of circle}$$
$$= \left(\mathcal{D}\frac{\partial c}{\partial r}\right)(2\pi L)$$

Calculate F.

(d) Find $\tau = N/F$, and compare this with the value given in Figure 9.7.

20. In three dimensions consider the spherically symmetric region of Figure 9.7(*c*), again taking the sink radius to be a and the source radius to be L, where $c(a) = 0$ and $c(L) = C_0$. Let ρ = radial distance from the origin.

(a) Solve the steady-state equation

$$0 = \frac{\mathcal{D}}{\rho^2}\frac{\partial}{\partial \rho}\left(\rho^2\frac{\partial c}{\partial \rho}\right).$$

(b) Define

$$N = \iiint_{\text{sphere}} c\, dV = \int_0^{2\pi}\int_0^{\pi}\int_a^L c(\rho)\rho^2 \sin\phi\, d\rho\, d\phi\, d\theta.$$

Compute this integral and interpret its meaning.

(c) Define

$$F = \text{flux} \times \text{surface area of sphere}$$
$$= \left(\mathcal{D}\frac{\partial c}{\partial \rho}\right)(4\pi L^2)$$

Find F.

(d) Find $\tau = N/F$.

21. *Transit times for diffusion*. Consider the steady-state equation

$$\frac{\partial}{\partial x}\left(\mathcal{D}\frac{\partial c}{\partial x}\right) = 0$$

with boundary conditions $c(L) = C_0$ and $c(0) = 0$.

(a) Show that the solution is

$$c(x) = C_0\frac{x}{L}.$$

(b) Explain why

$$N = \int_0^L C(x)\,dx$$

is the number of particles in $[0, L]$.

(c) If λ is the average removal rate at the sink, explain why $1/\lambda$ is the average diffusion transit time in example 2 in Section 9.6.

22. *Random versus chemotactic motion of macrophages.*

(a) Define

Δx = average distance traveled by a macrophage in a fixed direction,

ϵ = time taken to move this distance,

$s = \Delta x/\epsilon$ = speed of motion.

Justify the relationship $\mathcal{D} = \frac{1}{2}\epsilon s^2$ based on the results of the random-walk calculation in Section 9.4.

(b) How would the conclusions of Section 9.7 change under each of the following circumstances:

(1) The macrophage moves twice as fast.
(2) The target is twice as big.
(3) The area of the alveolus is half as big.
(4) The reproductive rate of the bacteria is twice as large.

(c) Define

τ = time to reach bacterium based on random motion,

T = time to reach bacterium based on direct motion towards the target,

$R_0 = \tau/T$.

Find an expression for R_0 based on parameters of the problem. Is R_0 ever equal to 1?

REFERENCES

The Mathematics of Diffusion

Boyce, W. E., and DiPrima, R. C. (1969). *Elementary Differential Equations and Boundary Value Problems*. 2d ed. Wiley, New York.

Cannon, J. R. (1984). *The One-Dimensional Heat Equation*. Addison-Wesley, Menlo Park, Calif.

Carslaw, H. S., and Jaeger, J. C. (1959). *Conduction of Heat in Solids*. 2d ed. Clarendon Press, Oxford.

Crank, J. (1979). *The Mathematics of Diffusion*. 2d ed. Oxford University Press, London.

Kendall, M. G. (1948). A form of wave propagation associated with the equation of heat conduction. *Proc. Camb. Phil. Soc., 44,* 591–593.

Shewmon, P. G. (1963). *Diffusion in Solids*. McGraw-Hill, New York.

Widder, D. V. (1975). *The Heat Equation*. Academic Press, New York.

Diffusion: Limitations and Geometric Considerations

Adam, G., and Delbruck, M. (1968). Reduction of dimensionality in biological diffusion processes. In A. Rich and N. Davidson, eds., *Structural Chemistry and Molecular Biology,* Freeman, San Francisco, Calif., pp. 198–215.

Berg, H. C. (1983). *Random Walks in Biology*. Princeton University Press, Princeton, N.J.

Haldane, J. B. S. (1928). On being the right size. In *Possible Worlds,* Harper & Brothers, New York. Reproduced in J. R. Newman (1967). *The World of Mathematics,* vol. 2. Simon & Schuster, New York.

Hardt, S. L. (1978). Aspects of diffusional transport in microorganisms. In S. R. Caplan and M. Ginzburg, eds. *Energetics and Structure of Halophilic Microorganisms,* Elsevier, Amsterdam.

Hardt, S. L. (1980). Transit times. In L. A. Segel, ed., *Mathematical Models in Molecular and Cellular Biology,* Cambridge University Press, Cambridge, England, pp. 451–457.

Hardt, S. (1981). The diffusion transit time: A simple derivation. *Bull. Math. Biol., 43,* 89–99.

Jones, D. S., and Sleeman, B. D. (1983). *Differential Equations and Mathematical Biology*. Allen & Unwin, Boston.

LaBarbera, M., and Vogel, S. (1982). The design of fluid transport systems in organisms. *Am. Sci., 70,* 54–60.

Murray, J. D. (1977). *Lectures on Nonlinear Differential Equation Models in Biology*. Clarendon Press, Oxford.

Okubo, A. (1980). *Diffusion and Ecological Problems: Mathematical Models*. Springer-Verlag, New York.

Segel, L. A., ed. (1980). *Mathematical Models in Molecular and Cellular Biology*. Cambridge University Press, Cambridge, U.K.

Diffusion Bioassays

Awerbuch, T. E., and Sinskey, A. J. (1980). Quantitative determination of half-lifetimes and mutagenic concentrations of chemical carcinogens using a diffusion bioassay. *Mut. Res., 74,* 125–143.

Awerbuch, T. E., Samson, R.; and Sinskey, A. J. (1979). A quantitative model of diffusion bioassays. *J. Theor. Biol., 79,* 333–340.

Macrophages and Bacteria on the Lung Surface

Fisher, E. S., and Lauffenburger, D. A. (1987). Mathematical analysis of cell-target encounter rates in two dimensions: the effect of chemotaxis. *Biophys. J., 51,* 705–716.

Lauffenburger, D. A. (1986). Mathematical analysis of the macrophage response to bacterial challenge in the lung. In R. van Furth, Z. Cohn, and S. Gordon, eds. *Mononuclear Phagocytes: Characteristics, Physiology, and Function*. Martinas Nijhoff, Holland.

APPENDIX TO CHAPTER 9
SOLUTIONS TO THE ONE-DIMENSIONAL DIFFUSION EQUATION

A.1 REMARKS ABOUT BOUNDARY CONDITIONS

Here we consider a number of boundary conditions that could be suitable for the diffusion equation (86).

1. *Infinite domains*. In some problems one is interested in observing the changes in a finite distribution of particles that are far away from walls or boundaries. See Figure 9.9(*a*). It is then customary to assume that the concentration is "zero at infinity":

$$c(x, t) \longrightarrow 0 \qquad \text{as} \qquad x \longrightarrow \pm\infty. \tag{95}$$

The approximation is valid provided that in the time scale of interest there is little or no reflection at the boundaries.

2. *Periodic boundary conditions*. If diffusion takes place in an annular tube of length L, the concentration at x has to equal that at $x + L$ [see Figure 9.9(*b*)]. Thus periodic boundary conditions lead to

$$c(x, t) = c(x + L, t),$$

or in particular,

$$c(0, t) = c(L, t). \tag{96}$$

3. *Constant concentrations at the boundary*. The hollow tube in Figure 9.9(*c*) is suspended between two large reservoirs whose concentrations are assumed to be fixed. This leads to the following boundary conditions:

$$c(0, t) = C_1, \tag{97a}$$
$$c(L, t) = C_2. \tag{97b}$$

If $C = C_2 = 0$, the boundary conditions are said to be homogeneous.

4. *No flux through the boundaries*. When the ends of the tube are sealed, no particles can cross the barriers at $x = 0$ and $x = L$. This means that the flux, defined by (53) must be zero; that is,

$$\frac{\partial c}{\partial x} = 0 \qquad \text{at} \begin{cases} x = 0, \\ x = L. \end{cases} \tag{98}$$

In general, it is true that the solution of the diffusion equation, or for that matter any PDE, depends greatly on the boundary conditions that are imposed. Carslaw and Jaeger (1959) show the derivations of solutions appropriate for many sets of boundary and initial conditions.

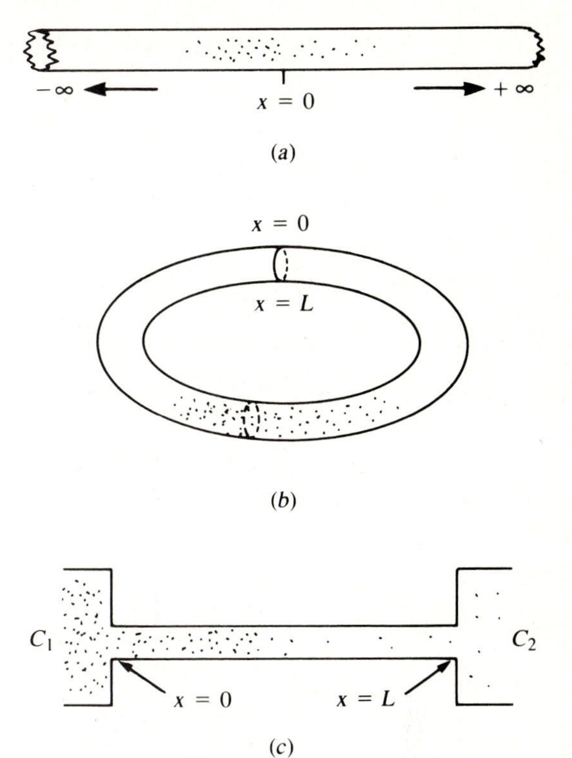

Figure 9.9 *Boundary conditions often used in solving the one-dimensional diffusion equation: (a) no particles at infinity [equation (95)]; (b) periodic boundary conditions [equation (96)]; (c) constant concentrations at one or both boundaries [equations (97a,b); (d) no flux at the boundaries, i.e., boundaries impermeable to particles [equation (98)].*

A.2 INITIAL CONDITIONS

Different initial configurations may be of interest in studying the process of diffusion. These may include the following:

1. *Particles initially absent:*

$$c(x, 0) = 0. \qquad (99)$$

This condition is suitable for problems in which particles are admitted through the boundaries.

2. *Single-point release*. If particles are initially "injected" at one location (considered theoretically of infinitesimal width), it is customary to write

$$c(x, 0) = C_0\delta(x). \qquad (100)$$

$\delta(x)$ is the *Dirac delta function,* actually a generalized function called a *distribution,* which has the property that

$$\delta(x) = \begin{cases} \infty & \text{if } x = 0, \\ 0 & \text{if } x \neq 0, \end{cases} \tag{101}$$

and

$$\int_{-\infty}^{\infty} \delta(x)\, dx = 1. \tag{102}$$

See Figure 9.10(*a*).

3. *Extended initial distribution*. The initial configuration shown in Figure 9.10(*b*) would be described by

$$c(x, 0) = \begin{cases} C_0 & x < 0, \\ 0 & x \geq 0. \end{cases} \tag{103}$$

4. *Release in a finite region*. If the concentration is initially constant within a small subregion of the domain [see Figure 9.10(*c*)], the appropriate initial condition is

$$c(x, 0) = \begin{cases} C_0 & -a < x < a, \\ 0 & \text{otherwise}. \end{cases} \tag{104}$$

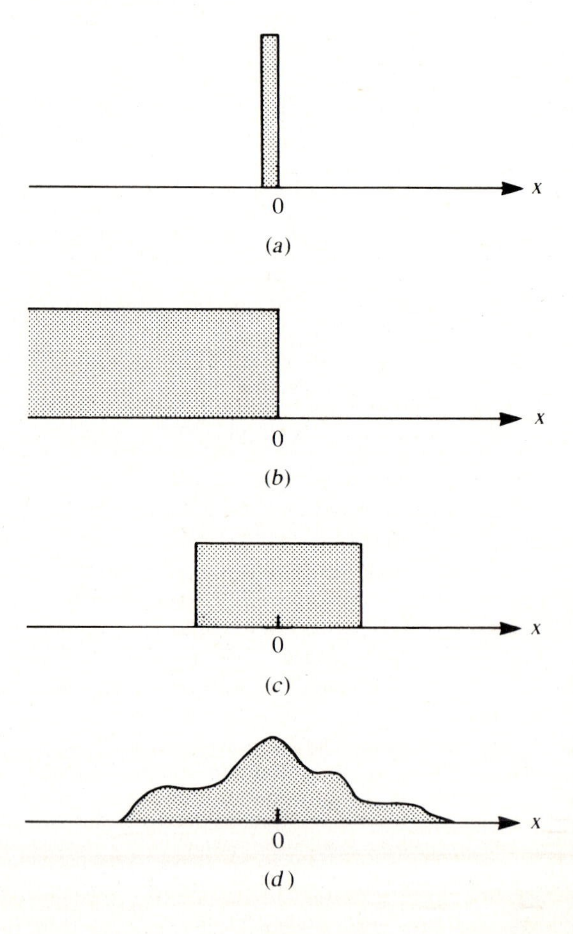

Figure 9.10 *Typical initial conditions for which the diffusion equation might be solved: (a) particles initially at* x = 0 *[equation (100)]; (b) extended initial distribution [equation (103)]; (c) release in a finite region [equation (104)]; (d) arbitrary initial distribution [equation (105)].*

5. *More general initial conditions*. It can more broadly be assumed that

$$c(x, 0) = f(x). \tag{105}$$

See Figure 9.10(*d*). Such initial conditions must generally be handled by Fourier-transform or Fourier-series methods unless $f(x)$ is of an especially elementary form. (Jones and Sleeman, 1983, discuss the details of this case.)

A.3 SOLVING THE EQUATION BY SEPARATION OF VARIABLES

In this section we briefly highlight a way of solving the one-dimensional diffusion equation given a set of boundary and initial conditions. Since this is not meant to be a self-contained guide but rather a quick introduction, the only method we discuss is separation of variables. Serious applied mathematics students should plan on taking a course on partial differential equations in which the more advanced and useful methods of Fourier transforms are taught.

We consider the equation

$$c_t = \mathcal{D} c_{xx}. \tag{106}$$

We will discuss problems in which the general solution takes the form

$$c(x, t) = c_T(x, t) + \bar{c}(x) \tag{107}$$

where $c_T(x, t)$ is a transient space- and time-dependent function that decays to zero and $\bar{c}(x)$ is the steady-state solution. We use separation of variables to find c_T.

Separation of Variables

Assume that $c_T(x, t)$ can be expressed as a product of two functions:

$$c_T(x, t) = S(x)T(t), \tag{108}$$

where S depends only on the spatial variable and T only on time. Substitute (108) into (106) to obtain

$$S(x)T'(t) = \mathcal{D} S''(x)T(t). \tag{109}$$

Rearranging (109) gives the following:

$$\frac{T'(t)}{T(t)} = \mathcal{D}\frac{S''(x)}{S(x)} = K. \tag{110}$$

This is called *separation of variables*. In equation (110) we have equated both sides to a constant K. This is the only possibility; otherwise by independently varying x and t, it would be possible to change one or the other side of the equation separately and a contradiction would be reached. Three distinct cases arise: $K = 0$, $K < 0$, and $K > 0$. In any of these the solutions to $T(t)$ and $S(x)$ can be obtained by solving

$$T'(t) = KT(t), \tag{111a}$$

$$S''(x) = \frac{K}{\mathcal{D}} S(x). \tag{111b}$$

These are both linear ODEs since each function depends on a single variable. The case $K = 0$ will not concern us since it leads to the somewhat uninteresting situation $T(t) =$ constant. If

$K > 0$, then by problem 24(c),

$$T(t) = \exp Kt, \tag{112a}$$

$$S(x) = \exp\left(\pm \sqrt{\frac{K}{\mathcal{D}}}x\right). \tag{112b}$$

Observe that transient solutions to equation (106) are thus of the form

$$\exp Kt \exp\left(\pm \sqrt{\frac{K}{\mathcal{D}}}x\right). \tag{113}$$

If $K < 0$ then a more convenient way of expressing these is to set

$$K = -\mathcal{D}\lambda, \qquad \frac{\sqrt{K}}{\sqrt{\mathcal{D}}} = -\lambda,$$

where λ is a constant (previously called the eigenvalue in section 9.8). Then complex exponentials lead to sinusoidal terms; one finds that by forming real-valued linear combinations, a (real-valued) transient solution can be written in the form

$$c_T(x, t) = \exp(-\lambda\mathcal{D}t)(A \cos \sqrt{\lambda}x + B \sin \sqrt{\lambda}x). \tag{114}$$

Note that $c_T(x, t) \to 0$ for $t \to \infty$. The values of λ, A, and B depend on the boundary and initial conditions of the problem.

Example 5

Consider equation (106) with the following boundary (BC) and initial (IC) conditions:

$$BC: \quad c(0, t) = c(L, t) = 0, \tag{115a}$$

$$IC: \quad c(x, 0) = f(x) \qquad \text{(to be specified below)}. \tag{115b}$$

It is easily verified that $\bar{c}(x) = 0$ is the steady-state solution (since the steady-state solution must satisfy the boundary condition). The boundary condition further implies that

$$T(t)S(0) = T(t)S(L) = 0. \tag{116}$$

Otherwise, if $T(t) = 0$, one would get $c_T \equiv 0$ for all t. Then observe that equation (111b) together with the first of these conditions leads to the conclusion that

$$S(x) = B \sin \sqrt{\lambda}x, \tag{117}$$

since $\sin(0) = 0$. Thus the separation constant K is necessarily negative in this case. The second condition can only be satisfied by choosing $\sqrt{\lambda}$ to be an integral multiple of π/L:

$$\lambda = \left(\frac{n\pi}{L}\right)^2 \qquad (n = 0, 1, 2, \ldots). \tag{118}$$

There is thus an infinite set of eigenfunctions that satisfy equation (106) and the homogeneous boundary condition. From the previous discussion we must conclude that the solution is of the form

$$c_T(x, t) = B \exp\left[-\left(\frac{n\pi}{L}\right)^2 \mathcal{D}t\right] \sin\left(\frac{n\pi x}{L}\right) \qquad (n = 0, 1, 2, \ldots). \tag{119}$$

In treating this problem, we have thus far neglected the initial condition from consideration. Notice that at $t = 0$ equation (119) reduces to the function

$$c_T(x, 0) = B \sin\left(\frac{n\pi x}{L}\right),$$

which is supposed to match the *a priori* specified initial condition (115b). Thus it would appear that the problem is consistent only with sinusoidal initial distributions. However, what makes its applicability much broader is the fact that all well-behaved functions $f(x)$ can be represented as a superposition of possibly infinitely many trigonometric functions such as sines, cosines or both. Such infinite superpositions are called *Fourier series.* We state this in the following important theorem.

The Fourier Theorem

If f and its derivative are continuous (or piecewise so) on some interval $0 \le x < L$, then on this interval f can be represented by an infinite series of sines:

$$f(x) = \sum_{n=1}^{\infty} (a_n \sin \beta_n x). \tag{120}$$

Equation (120) is called a *Fourier sine series,* and it converges to $f(x)$ at all points where f is continuous. The constants a_n are then related to f by the formula

$$a_n = \frac{2}{L}\int_0^L f(x) \sin \beta_n x \, dx. \tag{121}$$

See Boyce and DiPrima (1969) for more details and for similar theorems about cosine expansions.

Now recall that since equation (106) is linear, any linear superposition of solutions such as equation (119) will be a solution. With this in mind, we return to example 5.

Example 5 (continued)

From the observations in the previous box, we are led to consider solutions of the form

$$c_T(x, t) = \sum_{n=1}^{\infty} a_n \exp\left[-\left(\frac{n\pi}{L}\right)^2 \mathcal{D}t\right] \sin\left(\frac{n\pi x}{L}\right). \tag{122}$$

At $t = 0$ this must satisfy the initial condition. Then

$$\sum_{n=1}^{\infty} a_n e^0 \sin\frac{n\pi x}{L} = \sum_{n=1}^{\infty} a_n \sin\frac{n\pi x}{L} \equiv f(x).$$

According to Fourier's theorem this will hold provided that

$$a_n = \frac{2}{L}\int_0^L f(x) \sin\frac{n\pi x}{L}\, dx. \tag{123}$$

To find the full solution it then remains only to solve for the steady-state solution, $\bar{c}(x)$, which satisfies the equation

$$0 = \frac{\partial^2 c}{\partial x^2}. \tag{124}$$

Two integration steps lead to

$$\bar{c}(x) = \alpha x + \beta, \tag{125}$$

where α and β are integration constants. To satisfy the boundary conditions it is necessary to select $\bar{c}(0)$ and $\bar{c}(L)$ such that

$$\bar{c}(0) = \beta = 0, \tag{126a}$$

$$\bar{c}(L) = \alpha L = 0. \tag{126b}$$

Thus $\alpha = \beta = 0$, and the steady state of this problem is the trivial solution

$$\bar{c}(x) = 0.$$

The full solution is thus

$$c(x, t) = c_T(x, t), \tag{127}$$

where $c_T(x, t)$ is given by (122). For a particular $f(x)$ it is necessary to integrate the expression in equation (123) in order to obtain the values of the constants a_n. For examples and further details see Boyce and DiPrima (1969).

General Summary of Methods

1. Assume the transient solution $c_T(x, t) = S(x)T(t)$, and substitute this into the equation.
2. Separate variables to obtain ODEs for each part S and T separately.
3. Determine whether the separation constant K should be positive or negative by noting which eigenfunctions will satisfy the boundary conditions. ($K < 0 \Rightarrow$ sines or cosines; $K > 0 \Rightarrow$ exponentials.)
4. Further determine which eigenvalues λ will be consistent with the boundary conditions.
5. Write $c_T(x, t)$ as a (possibly infinite) superposition of solutions of the form $S_\lambda(x)T_\lambda(t)$ as in equation (122).
6. Find the constants a_n by using the initial condition of the problem along with equation (123).
7. Find the steady-state solution $\bar{c}(x)$ of equation (106).
8. The general solution is then

$$c(x, t) = c_T(x, t) + \bar{c}(x).$$

Note: Other boundary conditions may call for other eigenfunctions. (For example, boundary conditions of type 4 are only consistent with cosine eigenfunctions.) The general solution will then consist of Fourier cosine series or possibly of a full Fourier series. Such cases are described in greater detail in any text that treats boundary-value problems and the heat equation.

A.4 OTHER SOLUTIONS

New solutions to equation (106) can always be generated from preexisting ones by forming (1) linear combinations, (2) translations, or by (3) differentiation or integration with respect to a parameter. Also important are the following special classes of solutions that we describe without formal justification.

1. *Point release into an "infinite region."* With initial conditions (101) and boundary conditions (95), it can be shown that the solution to (106) is

$$c(x, t) = \frac{M}{2\sqrt{\pi \mathcal{D} t}} \exp \frac{-x^2}{4\mathcal{D} t}, \tag{128}$$

where M is the total number of particles:

$$M = \int_{-\infty}^{\infty} c \, dx.$$

See Figure 9.11 for the time behavior of this function, which is known as the *fundamental solution* of (106).

2. *Extended initial distributions.* An initial condition (103) with "far away" boundaries as in class 1 can be treated by considering the contributions of a whole array of point sources and summing these (integrating) over the appropriate region. This leads one to define a quantity known as the *error function:*

$$\text{erf } z = \frac{2}{\sqrt{\pi}} \int_0^z \exp(-x^2) \, dx. \tag{129}$$

with the properties that

$$\text{erf } (-z) = -\text{erf } z, \tag{130a}$$

$$\text{erf } 0 = 0, \tag{130b}$$

$$\text{erf } \infty = 1.$$

The solution of (106) can then be written in the form

$$c(x, t) = \frac{1}{2} c_0 \left(1 - \text{erf} \frac{x}{2\sqrt{\mathcal{D} t}}\right). \tag{131}$$

While the integral in equation (129) cannot be reduced to more elementary functions, it is a tabulated function of its argument. See most mathematics handbooks for such tables.

3. *Finite initial distributions.* If the initial distribution is concentrated in a finite interval $-h < x < h$, the integration of equation (129) over this domain leads to the solution

$$c(x, t) = \frac{1}{2} c_0 \left[\text{erf} \frac{h - x}{2\sqrt{\mathcal{D} t}} + \text{erf} \frac{h + x}{2\sqrt{\mathcal{D} t}}\right]. \tag{132}$$

The problem is more complicated if the effects of boundaries are to be considered. See Carslaw and Jaeger (1959) for the detailed treatment of such cases.

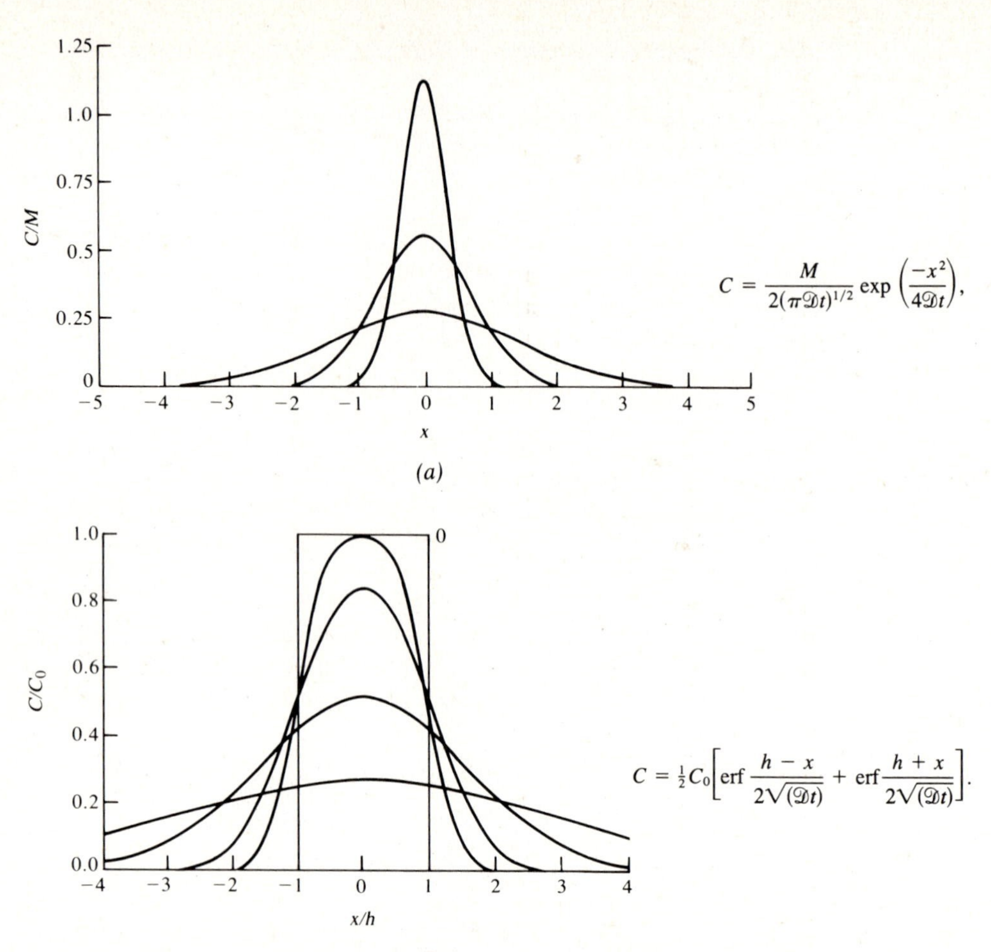

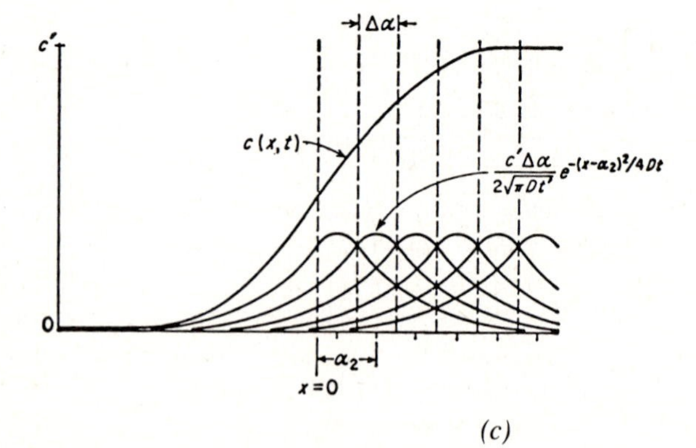

$$c(x, t) = \frac{c'}{2}\left\{1 + \operatorname{erf}\left[\frac{x}{2\sqrt{\mathscr{D}t}}\right]\right\}$$

$$\operatorname{erf} z = \frac{2}{\sqrt{\pi}} \int_0^z \exp(-\eta^2)\, d\eta$$

(c)

Figure 9.11 *Solutions of the diffusion equation for some discontinuous initial distributions: (a) point release [$c(x, 0) = \delta(x)$]; (b) finite extended distribution [$c(x, 0)] = 1$ for $-1 < x < 1$]; (c) infinite extended distribution [$c(x, 0) = 1$ for $x \geq 0$]. [Parts (a) and (b) from Crank, J. (1979).* The Mathematics of Diffusion. *2 ed. Oxford University Press, London, Figs. 2.1 and 2.4. Part (c) from Shewmon, P. G. (1963).* Diffusion in Solids, *McGraw-Hill, New York, Fig. 1.5.]*

PROBLEMS FOR THE APPENDIX

Problems 23 to 27 are based on the Appendix to Chapter 9.

23. Determine the boundary and initial conditions appropriate for the diffusion of salt in each of the following situations.

- **(a)** A hollow tube initially containing pure water connects two reservoirs whose salt concentrations are C_1 and 0 respectively.
- **(b)** The tube is sealed at one end. Its other end is placed in a salt solution of fixed concentration C_1.
- **(c)** The tube is sealed at both ends and initially has its greatest salt concentrations halfway along its length. Assume the initial distribution is a trigonometric function.

24. In this problem we investigate certain details that arise in solving the diffusion equation by separation of variables.

- **(a)** Show that equation (108) implies (109).
- **(b)** Determine the consequences of assuming $K = 0$ in equation (110).
- **(c)** Show that for $K > 0$ solutions to (110) are given by equations (112a,b).
- **(d)** Justify the assertion that for $K < 0$ solutions are of the form (114).

25.

- **(a)** What kind of boundary and initial conditions are used in (115)?
- **(b)** Show that the steady-state solution of equation (106) is then trivially $c(x) = 0$.
- **(c)** Show that equations (110) and (115) lead to (117).
- **(d)** Justify the assumption (118).

26. Solve the one-dimensional diffusion equation subject to the following conditions:

- **(a)** $c(0, t) = c(1, t) = 0,$
  $c(x, 0) = \sin \pi x.$
- **(b)** $c(0, t) = c(L, t) = 0,$
  $$c(x, 0) = \sin\left(\frac{2\pi}{L}x\right) + \sin\left(\frac{3\pi}{L}x\right).$$

27.

- **(a)** Verify that (128) is a solution to the diffusion equation by performing partial differentiation with respect to t and x.
- **(b)** Similarly, verify that (131) is also a solution.
- **(c)** Use your result in part (b) to argue that (132) is also a solution without calculating partial derivatives. (*Hint:* Use the properties of solutions described in the Appendix, Section A.4.)

10 Partial Differential Equation Models in Biology

Give me space and motion and I will give you a world
R. Descartes (1596–1650) quoted in E. T. Bell (1937) Men of Mathematics, Simon & Schuster

Because populations of molecules, cells, or organisms are rarely distributed evenly over a featureless environment, their motions, migrations, and redistributions are of some interest. At the level of the individual, movement might result from special mechanochemical processes, from macroscopic contractions of muscles, or from amoeboid streaming. On the population level, these different mechanisms may have less bearing on net migration than other aspects such as (1) variations in the environment, (2) population densities and degree of overcrowding, and (3) motion of the fluid or air in which the organisms live.

On the collective level it is often appropriate to make a *continuum* assumption, that is, to depict discrete cells or organisms by continuous density distributions. This leads to partial differential equation models that are quite often analogous to classical models for molecular diffusion, convection, or attraction.

Historically, biological models involving partial differential equations (PDEs) date back to the work of K. Pearson and J. Blakeman in the early 1900s. In the 1930s others, including R. A. Fisher, applied PDEs to the spatial spread of genes and of diseases. The 1950s witnessed several important developments including the work of A. M. Turing on pattern formation (see Chapter 11) and the analysis of Skellam (1951), who was among the first to formally apply the diffusion equation in modeling the random dispersal of a population in nature. Some of these models and several others drawn from molecular, cellular, and population biology are outlined in this chapter.

In Section 10.1 we begin with an account of Skellam's work and his analysis of spreading populations of animals and plants. Models for the collective motion of microorganisms are then developed in Section 10.2. There is an underlying parallel between the equations for moving populations and the conservation equations encountered in physical phenomena such as particle diffusion. However, there are some noteworthy differences; among these are models for density-dependent dispersal, which are mentioned in Section 10.3. Two applications of convection equations to growth in a branching network are described in 10.4.

Many models discussed in this chapter cannot be solved analytically in closed form. This is particularly true of the nonlinear equations. We must often work with relatively elementary solutions that do not address the full complexity inherent in the time-dependent evolution of the system. A standard first approach is to reduce the problem to one that can be solved, generally by converting the PDEs to ordinary differential equations (ODEs) that describe some simpler situation. Two types of solutions can be thus ascertained: steady-state distributions and *traveling waves*. Methods for finding such solutions are described in Section 10.5. In Section 10.6 we apply such techniques to Fisher's equation, which depicts the spread of an advantageous gene in a population. A second example of biological waves emerges from the study of microorganisms such as yeast growing on glucose. Remarkably, phase-plane analysis reemerges as a handy tool in this unlikely setting. The phenomenon of long-range transport of biological substances inside the neural axon is described in Section 10.7.

In Sections 10.8 and 10.9 our emphasis shifts slightly. Here again we aim to uncover the generality and power of abstract thinking by demonstrating that familiar equations can be applied in novel and surprising ways to seemingly unrelated settings. We begin with a calculation due to Takahashi that uncovers a connection between the aging of a cell and the processes of spatial redistribution previously studied. The analogy between age distributions and spatial distributions is then more fully explored. Section 10.9 is a do-it-yourself modeling venture that exploits such analogies, and Section 10.10 provides references and suggestions for further study.

10.1 POPULATION DISPERSAL MODELS BASED ON DIFFUSION

Among the first to draw an analogy between the random motion of molecules and that of organisms was Skellam (1951). He suggested that for a population reproducing continuously with rate α and spreading over space in a random way, a suitable continuous description would be

$$\frac{\partial P}{\partial t} = \mathcal{D}\nabla^2 P + \alpha P. \tag{1}$$

$\mathcal{D}$, called the *dispersion rate*, is analogous to a diffusion coefficient (also called the *mean square dispersion per unit time*). $P(\mathbf{x}, t)$ is the population density at a given time and location.

Skellam was particularly interested in the rate with which the area initially colonized by a population expands with time, and quoted two interesting examples based on biological data. Here we examine equation (1) in more detail but in a one-dimensional, rather than a full two-dimensional setting.

The growth term αP not only increases the density locally but also causes a faster spatial spread in the population than that anticipated by diffusion alone. Indeed, as Skellam pointed out, the outward propagation of the *equipopulation contours* (the level curves of the population density) eventually takes place at a constant radial rate.

To see why this is true, we consider a point release of the species at time $t = 0$ and location $x = 0$. Then it can be shown that a solution of equation (1) is

$$P(x, t) = \frac{P_0}{2(\pi \mathcal{D} t)^{1/2}} \exp\left(\alpha t - \frac{x^2}{4\mathcal{D} t}\right). \tag{2}$$

See problem 3.

To study the lateral population expansion one could observe the translation of those points x_i for which

$$P(x_i, t) = P_i = \text{constant}.$$

(These points are analogous to the equipopulation contours in higher dimensions.)

After initial spreading, the population will have achieved a detectable level at some distance $\hat{x}$ from its origin. The outer boundary of the population will continue to spread so that $\hat{x}$ will translate outwards. It is of interest to derive some estimate of this *rate of propagation*. (*Note:* We are investigating a one-dimensional setting for the sake of simplicity. In two dimensions we would be asking a similar question about the rate of propagation of level curves; see Figure 10.1 for an example.)

In problem 3 it is shown that by setting $P(x, t) = \tilde{P}$ (a constant) in equation (2) and looking at the large-time behavior, one obtains in the limit,

$$\hat{x} = 2(\alpha \mathcal{D})^{1/2} t, \tag{3}$$

so that the location at which the population density is $\tilde{P}$ travels asymptotically at a constant speed, $2(\alpha \mathcal{D})^{1/2}$. One reason this equation admits a more rapid advance rate than does a diffusion equation without the growth term stems from the fact that the birth rate *reinforces* gradients. Where the population is large, the population gets larger quickly so that the diffusion process is *driven* by internal growth.

A similar asymptotic expansion rate holds for radially outwards diffusion in a two-dimensional distribution; that is, one can verify that for large t, equipopulation contours expand at the rate

$$(\text{area})^{1/2} = 2(\alpha \mathcal{D})^{1/2} t. \tag{4}$$

Skellam used this result to demonstrate that the spread of certain populations can be explained on the basis of a diffusion approximation (in other words, an underlying random walk of individuals). Particularly noteworthy is his example of recorded spread of a muskrat population over central Europe after a Bohemian landowner mistakenly allowed several to escape in 1905. Because of uneven spatial terrain (presence of towns, for example) the population contours are not particularly regular. However, Skellam showed that the square root of the area increases linearly with time, as anticipated from equation (4).

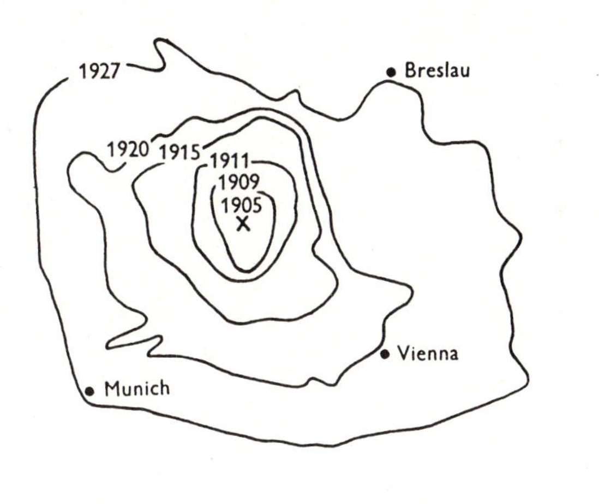

(*a*)

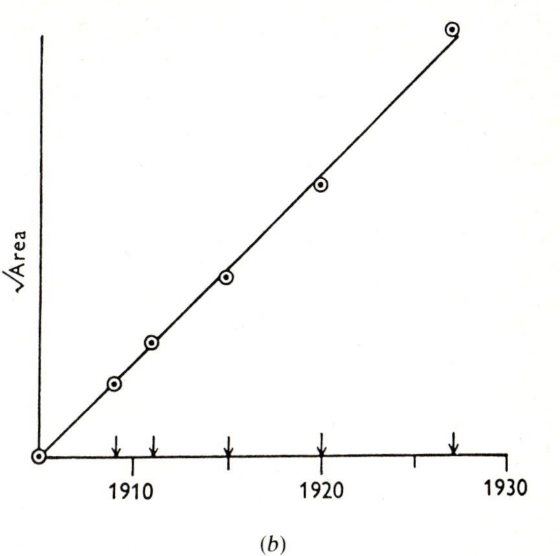

(*b*)

***Figure 10.1** Spread of muskrats over central Europe during a period of 27 years described by Skellam (1951) as a random dispersal. (a) Equipopulation contours (level curves of* $p(\mathbf{x}, t)$ *for the lowest detectable muskrat population. A graph of* $(area)^{1/2}$ *of the regions enclosed by these curves reveals linear dependence on time* t, *as predicted by the growth-dispersal model of equation (1). [From Skellam J. G. (1951). Random dispersal in theoretical populations.* Biometrika, 38, *figs. 1 and 2, p. 200. Reprinted with permission of the Biometrika Trustees.]*

Skellam applied similar conclusions to the spread of oak forests over Britain and by simple calculations argued that small animals must have played an important role in the dispersal of acorns (see problem 5).

In the recent literature, diffusion-like models for population dispersal have become quite common. The homing and migration of birds, fish, and other animals have been described by diffusion with a "sticky" target site (see Okubo, 1980, for survey and references). Smaller organisms such as insects have also been modeled by diffusion equations. Ludwig et al. (1979) describe the spread of the spruce budworm by the equation

$$\frac{\partial P}{\partial t} = \mathcal{D}\frac{\partial^2 P}{\partial x^2} + \alpha P\left(1 - \frac{P}{K}\right) - \beta\frac{P^2}{H^2 + P^2}, \tag{5}$$

where β is the rate of mortality due to predation, and K is a constant. See problem 2(d) for an exercise in interpreting the equation. Kareiva (1983) applied diffusion models to data for herbivorous insects under a number of conditions and derived rigorous tests for the validity of such approximations.

Aside from the motion of organisms, it has also been recently popular to describe by diffusion the spread of genes, disease, and other similar properties. A good general survey of many references is given by Okubo (1980), Fife (1979), and Murray (1977).

Table 10.1 Dispersal Rates

Equation	*Initial Conditions*	*Speed of Propagation*	*References*
$\frac{\partial P}{\partial t} = \mathcal{D}\frac{\partial^2 P}{\partial x^2} + \alpha P$ (dispersal with exponential growth).	$P(x, 0) = \delta(x)$ (initially all population at $x = 0$).	$\frac{x}{t} = \pm 2(\mathcal{D}\alpha)^{1/2}$ (for large t).	Kendall (1948) and problem 3.
Same as above.	$P(x, 0) = \frac{P_0}{(2\pi\sigma_0)^{1/2}} \exp\left(\alpha t - \frac{x^2}{2\sigma_0^2}\right)$ (initially a Gaussian distribution with variance).	Variance $\sigma^2 = \sigma_0^2 + 2\mathcal{D}t$, $\frac{x}{t} = 2(\alpha\mathcal{D})^{1/2}$.	Kendall (1948) and Okubo (1980).
	$P(x, 0) = P_0 \exp(-bx)$ $(b > 0)$.	$\frac{x}{t} > 2(\alpha\mathcal{D})^{1/2}$ when $b < \left(\frac{\alpha}{\mathcal{D}}\right)^{1/2}$.	Kendall (1948), Mollison (1977), and Okubo (1980).
$\frac{\partial P}{\partial t} = \mathcal{D}\frac{\partial^2 P}{\partial x^2} + \alpha P - \beta P^2$.	$P(x, t) = P(x - ct)$ (traveling wave with trailing end).	$\frac{x}{t} = c \geq 2(\alpha\mathcal{D})^{1/2}$.	Fisher (1937) and Kolmogorov et al. (1937).
$\frac{\partial P}{\partial t} = \mathcal{D}\frac{\partial^2 P}{\partial x^2} + F(P)$.	Same as above.	$\frac{x}{t} = c = 2\left(\frac{\partial F}{\partial P}\Big\vert_0 \mathcal{D}\right)^{1/2}$	Kolmogorov et al. (1937).

A preoccupation with propagation rates is still quite current. Many recent papers address these questions, with emphasis on the role of initial conditions and of growth rates other than simple exponential growth. Another interesting question is whether equations such as (1) or its various modifications admit traveling-wave solutions (solutions that move in space without changing their "profiles"). Table 10.1 lists some of the results regarding propagation speeds. In a later section we deal at greater length with traveling-wave solutions.

10.2 RANDOM AND CHEMOTACTIC MOTION OF MICROORGANISMS

Many unicellular organisms have elaborate patterns of locomotion that may include *ciliary beating* (synchronous motion of hair-like appendages on the cell surface,) helical swimming, crawling on surfaces, tumbling in three dimensions, and *pseudopodial extension* (protrusion of part of the cell and streaming of the cellular contents.) In the absence of overriding external cues, such motion may appear *saltatory* (jerky) or random, although of course, it is strictly determined by events on subcellular levels. At the population level, the pseudo-random motion could be approximately described as a process analogous to molecular diffusion. A one-dimensional equation that would represent changes in the spatial distribution of a large population of such microorganisms would then be

$$\frac{\partial b}{\partial t} = \mu \frac{\partial^2 b}{\partial x^2} + rb, \tag{6}$$

where

$b(x, t)$ = population density at location x and time t,

μ = coefficient that depicts the *motility* or *dispersal rate* of the organisms,

r = growth rate (if positive) or death rate (if negative).

Note that rb is a local source/sink term, previously denoted by σ, since it accounts for local addition or elimination of individuals; note also that equation (6) is a diffusion equation.

Segel et al. (1977) applied equation (6), where $r = 0$, to the dispersal of bacteria. Based on experimental observations, they calculated a value of μ of 0.2 $cm^2\ h^{-1}$ for *Pseudomonas fluorescens*. (See Segel, 1984, and problem 6 for a summary.)

Similar equations (with negative r) have been applied to *plankton* (microscopic marine organisms) by Kierstead and Slobodkin (1953). Bergman (1983) discusses a model for contact-inhibited cell division that reduces to a diffusion equation in the limit of unrestricted cell division.

When the microorganisms depend on some growth-limiting nutrient for their survival, the relative rate of nutrient diffusion to organism motility may be of some importance. An example of substrate-dependent growth is given by Gray and Kirwan (1974) for yeast growing on solid medium. A recent model for the effects of random motility on bacteria that consume a diffusible substrate is described by Lauffenburger et al. (1981), who suggest the following equations:

$$\frac{\partial b}{\partial t} = \mu \frac{\partial^2 b}{\partial x^2} + [f(s) - k_e]b, \tag{7a}$$

$$\frac{\partial s}{\partial t} = \mathcal{D} \frac{\partial^2 s}{\partial x^2} - \frac{1}{Y} f(s)b, \tag{7b}$$

where

$b(x, t)$ = bacterial density,
$s(x, t)$ = substrate concentration,
Y = the yield (mass of bacteria per unit mass of nutrient),
$f(s)$ = the substrate-dependent growth rate,
k_e = the bacterial death rate when s is depleted.

Some details of their model are explored in problem 8.

Keller and Segel (1970, 1971) were among the first to describe a continuum equation for the phenomenon of chemotaxis in microorganisms, discussed earlier in Chapter 9. *Taxis* refers to the purposeful motion of organisms in response to environmental cues. Some animals are known to be attracted to brighter light, warmer temperatures, and higher levels of certain chemical substances (for example, pheromones or nutrients) while being repelled from potentially damaging influences (such as toxins, extremes of temperature or extremes of pH).

Keller and Segel applied the idea of attraction and repulsion in deriving their equation for bacterial chemotaxis:

$$\frac{\partial B}{\partial t} = \underbrace{-\frac{\partial}{\partial x}\left(\chi B \frac{\partial c}{\partial x}\right)}_{\text{attraction}} + \underbrace{\frac{\partial}{\partial x}\left(\mu \frac{\partial B}{\partial x}\right)}_{\text{random motion}}, \tag{8}$$

where

$B(x, t)$ = bacterial density at location x and time t,
χ = chemotactic constant.

χ depicts the relationship between a gradient in the substance c and the velocity of migration of the population. In other words, the chemotactic flux is assumed to be proportional to a gradient $\partial c/\partial x$:

$$\mathbf{J}_{\text{chemotactic}} = \chi B \frac{\partial c}{\partial x}. \tag{9}$$

The second term of equation (8) contains, as before, the flux due to random motion

$$\mathbf{J}_{\text{random}} = -\mu \frac{\partial B}{\partial x}. \tag{10}$$

Underlying molecular mechanisms that might produce a chemotactic response in microorganisms have been studied by Segel (1977). The equations have also been applied in a model for aggregation of microorganisms that will be described more fully in Section 11.1.

Certain cells implicated in the immune response of higher organisms are also able to undergo chemotactic motion in response to substances associated with infection or inflammation. White blood cells known as *polymorphonuclear leukocytes* (PMNs) are responsible for engulfing small foreign bodies in a process called *phagocytosis*. To locate such bodies, PMNs first orient their motion chemotactically in response to chemical substances released by damaged tissue (Zigmond, 1977). Lauffenburger (1982) who developed several models for the chemotaxis of PMNs demonstrated that the accuracy of the response can only be explained by assuming that cells average local concentrations over a time scale of several minutes. A slight modification of the Keller-Segel equations has also been applied (Lauffenburger and Kennedy, 1983) to describe spatial properties of the immune response to bacterial infection (see problem 11).

Chemotactic equations can be rigorously derived from first principles once certain assumptions are made about the "choice" of step size and direction of motion of individuals in a population. Okubo (1980) discusses derivation of such equations from one-dimensional "biased" random-walk models. Alt (1980) and others have similarly made the connection in higher dimensions.

10.3 DENSITY-DEPENDENT DISPERSAL

In recent work, Gurtin and MacCamy (1977) have extended the more classical models to density-dependent population dispersal. They suggest that more realistic assumptions about dispersal might include a nonconstant rate of dispersal that increases when overcrowded conditions prevail. Typically this would lead to a modified diffusion flux

$$\mathbf{J} = -\mathcal{D}(p)\nabla p, \tag{11}$$

or, in one dimension,

$$\mathbf{J} = -\mathcal{D}(p)\frac{\partial p}{\partial x}. \tag{12}$$

A form the authors use for $\mathcal{D}(p)$ is

$$\mathcal{D}(p) = kp^m, \tag{13}$$

where k is positive, and $m \geq 1$. An increase in the population thus causes the dispersal rate to increase. In one dimension an equation describing the population movement would then be

$$\frac{\partial p}{\partial t} = k\frac{\partial}{\partial x}\left(p^m\frac{\partial p}{\partial x}\right) + F(p), \tag{14}$$

where F is the local growth rate. It is readily shown that an equivalent form is

$$\frac{\partial p}{\partial t} = K\frac{\partial^2(p^{m+1})}{\partial x^2} + F(p), \qquad (15)$$

where $K = k(m + 1)$. Several interesting and somewhat desirable properties of this equation include the following: (1) If the population initially occupies a finite region, it will always occupy a finite region. (2) The size of this region will increase if birth dominates over mortality; (3) however, if mortality is the stronger influence, the population will not expand spatially beyond certain bounds.

Similar equations have since been applied to epidemic models in which the disease spreads with a migrating population (see, for example, Busenberg and Travis, 1983).

A particularly striking extension of the idea of density-dependent dispersal appears in a recent publication by Bertsch et al. (1985). These authors consider a pair of interacting populations with densities $u(x, t)$ and $v(x, t)$, in which dispersal is a response to the total population at a location that is, to $[u(x, t) + v(x, t)]$. It is assumed that the velocities of motion are proportional to gradients in the total population:

$$\mathbf{q} = -k_1\nabla(u + v), \qquad (16a)$$

$$\mathbf{w} = -k_2\nabla(u + v), \qquad (16b)$$

where $\mathbf{q}$ and $\mathbf{w}$ are velocities, and k_1 and k_2 are constants. This means that individuals are moving *away* from sites of high total population at velocities proportional to the gradient of $u + v$. For the case of no net death or growth, the conservation statements are

$$\frac{\partial u}{\partial t} = -\nabla\cdot(u\mathbf{q}), \qquad (17a)$$

$$\frac{\partial v}{\partial t} = -\nabla\cdot(v\mathbf{w}), \qquad (17b)$$

so that in one-dimension the full equations are

$$\frac{\partial u}{\partial t} = k_1\frac{\partial}{\partial x}\left[u\frac{\partial(u + v)}{\partial x}\right], \qquad (18a)$$

$$\frac{\partial v}{\partial t} = k_2\frac{\partial}{\partial x}\left[v\frac{\partial(u + v)}{\partial x}\right]. \qquad (18b)$$

The following interesting result is proved by Bertsch et al (1985). If the initial populations colonize distinct regions without overlap (that is, if they are *segregated*), they will remain segregated for all future times by virtue of these interactions. This prediction is independent of k_1 and k_2 and of the details of the initial distributions, provided only that they are segregated.

Models such as this point to the rather nonintuitive and surprising features of fairly simple sets of PDEs. In the next chapter we briefly examine two models in which pairs of PDEs lead to rather interesting predictions.

10.4 APICAL GROWTH IN BRANCHING NETWORKS

We next consider a pair of related models in which growth of a branching organism or network is described as a *translation* of *apices,* (endpoints of branches, at which growth takes place). Collectively this kind of growth can sometimes be approximated as a convection, provided the appropriate definitions of variables are made.

If one is concerned with the spatial distribution of density in filamentous organisms, it often makes sense to define two densities, which are then used simultaneously in describing growth, branching, and other possible interactions (to be mentioned later). One model focuses on fungi, which often grow in densely branched *colonies* (see Figure 10.2). The model consists of the following variables:

$\rho(x, t)$ = length of filaments per unit area,

$n(x, t)$ = number of growing apices per unit area.

It is assumed that apical growth leads to a constant rate of elongation that makes apices move at a fixed velocity. Define

v = growth rate in length per unit time.

Using these assumptions, it can be shown that an appropriate system of equations in a one-dimensional setting is

$$\frac{\partial \rho}{\partial t} = nv - \gamma\rho, \tag{19a}$$

$$\frac{\partial n}{\partial t} = -\frac{\partial(nv)}{\partial x} + \sigma, \tag{19b}$$

where γ = the rate of filament mortality and σ = the rate of creation of new apices (which occurs whenever branching takes place). More details about this model can be found in Edelstein (1982), and a detailed derivation of these equations is given as a modeling exercise in problem 12.

Perhaps underscoring the generality of mathematics is a recent application of similar equations to the seemingly unrelated phenomenon of tumor-induced blood vessel growth. Balding and McElwain (1985) have modified (19) to describe the formation and growth of capillary sprouts into a tumor. It is known that a chemical intermediate secreted by tumors (called *tumor angiogenesis factor,* or TAF) promotes rapid growth of the endothelial cells that line the blood vessels. This leads to the sprouting of new capillaries, which apparently grow chemotactically toward high concentrations of TAF. In the model suggested by Balding and McElwain the diffusion of TAF [whose concentration is represented by $c(x, t)$] and its effect on capillary tip growth and sprouting are represented as follows:

$$\frac{\partial c}{\partial t} = \mathcal{D}\frac{\partial^2 c}{\partial x^2}, \qquad \text{(TAF)} \tag{20a}$$

$$\frac{\partial \rho}{\partial t} = nv - \gamma\rho, \qquad \text{(capillary density)} \tag{20b}$$

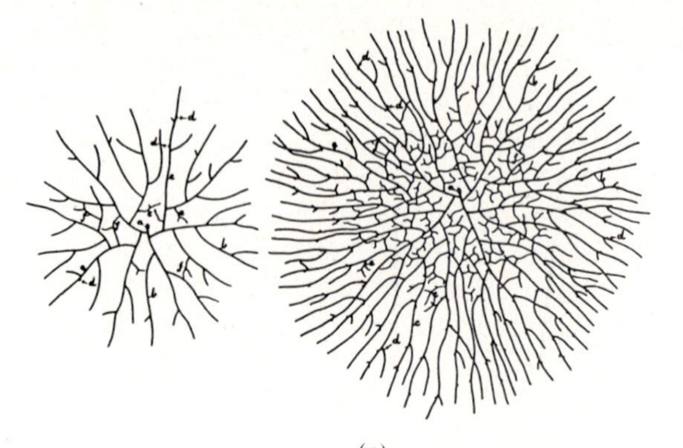

(*a*)

(*b*)

Figure 10.2 *Branching organisms such as fungi grow by extension and ramification of long slender filaments. Growth can take place in one, two, or three dimensions. A one-dimensional model also applicable to other networks such as blood vessels is given by equations (20a–c). (a) Two stages in the growth of* Coprinus. *(b) Stages in the development of* Pterula gracilis. *[(a) From A. R. M. Buller (1931),* Researches in Fungi, *vol. 4, Longmans, Green, London, figs. 87 and 88; (b) from J. T. Bonner (1974),* On Development; the biology of form, *fig. 17, reprinted by permission of Harvard University Press.]*

$$\frac{\partial n}{\partial t} = -\frac{\partial(nv)}{\partial x} + \sigma, \qquad \text{(capillary tips)} \tag{20c}$$

where

$$v = \chi\frac{\partial c}{\partial x}, \tag{20e}$$

$$\sigma = \alpha c\rho - \beta n\rho. \tag{20f}$$

In these equations, c may be determined independently and leads to a concentration field that acts as a chemotactic gradient. Equations (20b,c) include capillary-tip chemotaxis with rate χ, sprouting from vessels at a rate proportional to the concentration c, and loss of capillary tips due to *anastomosis* (reconnections that form closed networks). The equation for vessels (20b) includes growth by extension of tips and a rate γ of degradation of old vessels. This model illustrates the connection between the general concept of convective flux (as defined in Section 9.4) and the particular case of chemotaxis.

10.5 SIMPLE SOLUTIONS: STEADY STATES AND TRAVELING WAVES

Many models described in this chapter cannot be solved in full generality by analytic techniques, since they consist of coupled PDEs, some of which may be nonlinear. It is frequently challenging to make even broad generalizations about their time-dependent solutions, and abstract mathematical theory is called for in such endeavors.

We shall skirt these issues entirely and deal only with easier questions that can be settled by applying methods developed for ODEs to understand certain special cases. Two types of solutions can be obtained by such means: the first are steady states (time-independent distributions); a familiarity with the concept of steady states can thus be extended into the realm of spatially distributed systems. The second and distinctly new class of solutions are the *traveling waves,* distributions that move over space while maintaining a characteristic "shape" or profile. A special trick will be used to address the question of existence and properties of such solutions.

Nonuniform Steady States

By a steady state $\bar{c}(x)$ of a PDE model we mean a solution to the equations of the model that additionally satisfies the equation

$$\frac{\partial c}{\partial t} = 0.$$

If the problem is written for a single space dimension, setting the time derivative to zero turns the equations into a set of ODEs. Often, but not always, these can be solved to obtain an analytical formula for the steady-state spatial distribution $\bar{c}(x)$.

Homogeneous (Spatially Uniform) Steady States

A homogeneous steady state is a solution for which both time and space derivatives vanish. For example, in one space dimension

$$\frac{\partial c}{\partial t} = 0, \qquad \frac{\partial c}{\partial x} = 0.$$

These solutions satisfy algebraic relationships that are often relatively easy to solve explicitly. They describe spatially uniform unvarying levels of the population. Often such solutions are less interesting on their own merits but are rather significant for their special stability properties. The effects of spatially nonuniform perturbations of such homogeneous steady states forms a separate topic to be discussed in Chapter 11.

Example 1
Find a (nonuniform) steady-state solution of equation (8) for bacterial chemotaxis.

Solution
Setting $\partial B/\partial t = 0$ leads to

$$\frac{\partial}{\partial x}\left(-\chi B \frac{\partial c}{\partial x} + \mu \frac{\partial B}{\partial x}\right) = 0. \tag{21}$$

Integrating once then results in

$$\mathbf{J} = \text{constant} \tag{22}$$

where $\mathbf{J}$ is bacterial flux, the expression in parentheses in equation (21).

Suppose (21) is confined to a domain $[0, L]$ and that no bacteria enter or leave the boundaries. Then by equation (22), $\mathbf{J} = 0$ at $x = 0$ implies that $\mathbf{J} \equiv 0$ for all x.

Thus

$$-\chi B \frac{\partial c}{\partial x} + \mu \frac{\partial B}{\partial x} = 0, \qquad \text{or} \qquad \chi \frac{\partial c}{\partial x} = \frac{\mu}{B}\frac{\partial B}{\partial x} \qquad \text{for } B \neq 0.$$

Integrating once more leads to

$$\chi c(x) = \mu \ln B(x) + k, \tag{23}$$

where k is an integration constant. Thus

$$B(x) = \bar{k} \exp \frac{\chi c(x)}{\mu}, \tag{24}$$

where $\bar{k} = \exp(-k)$.

Observe from (24) that a steady-state distribution of bacteria $B(x)$ can be related to a given chemical concentration $c(x)$. In general, another equation might describe the distribution of this substance. For a simple example, consider plain diffusion, for which

$$\frac{\partial c}{\partial t} = \mathscr{D}\frac{\partial^2 c}{\partial x^2}, \tag{25}$$

$c(0) = C_0$, and $c(L) = 0$. This is an artificial example chosen purely for illustrative purposes: the chemical concentration is contrived to be fixed at the ends of the tube while bacteria are not permitted to enter or leave; furthermore, bacteria orient chemotactically according to the c gradient but do not consume the substance.

By results of Section 9.5, a steady-state solution of equation (25) is

$$c(x) = C_0\frac{x}{L}. \tag{26}$$

The corresponding bacterial profile would be

$$B(x) = \tilde{k}\exp\left(\frac{\chi C_0 x}{\mu L}\right). \tag{27}$$

From this solution it follows that at every location x the bacterial flux due to chemotactic motion is exactly equal and opposite to the flux due to random bacterial dispersion. For this reason a nonuniform distribution can be maintained.

Example 2

Find steady states of the density-dependent dispersal equations (18a,b).

Solution

Set $\partial u/\partial t = 0$ and $\partial v/\partial t = 0$ in equations (18a,b) to obtain

$$\frac{\partial}{\partial x}\left[u\frac{\partial(u+v)}{\partial x}\right] = 0, \tag{28a}$$

$$\frac{\partial}{\partial x}\left[v\frac{\partial(u+v)}{\partial x}\right] = 0. \tag{28b}$$

After integrating once, observe that

$$u\frac{\partial(u+v)}{\partial x} = C_1, \tag{29a}$$

$$v\frac{\partial(u+v)}{\partial x} = C_2. \tag{29b}$$

Note that C_1 and C_2 are constants and represent population flux terms for species u and v respectively. If $C_1 = C_2 = 0$, then two possibilities emerge:

1. $u = 0,\ v = K_2$ or $v = 0,\ u = K_1,$ (30)

where K_1 and K_2 are non-negative real numbers; or

2. $u \neq 0,\ v \neq 0;\ \dfrac{\partial(u + v)}{\partial x} = 0;\ u = K - v,$ (31)

where K is a positive constant.

If $C_1 \neq 0$ and $C_2 \neq 0$, then neither u or v can ever be zero, so that a third result is obtained:

3. $\dfrac{\partial(u + v)}{\partial x} = \dfrac{C_1}{u} = \dfrac{C_2}{v}.$ (32)

This means that $u = (C_1/C_2)v$, so that

$$\left(1 + \frac{C_1}{C_2}\right)\frac{\partial}{\partial x}v = \frac{C_2}{v},$$

$$\int v\,\partial v = \int \frac{C_2}{1 + C_1/C_2}\,dx,$$

$$v = \left(2x\frac{C_2}{1 + C_1/C_2} + K\right)^{1/2}, \tag{33}$$

where K is a constant.

These solutions are written for the case of infinite one-dimensional domains. If a finite domain is to be considered, the steady states just given may or may not exist depending on boundary conditions (see problem 15).

It is possible to patch together a mosaic of solutions (of type 1, for example) that would satisfy the given system of equations at all points save for a few singular locations where a transition between one species and the next occurs. (The spatial derivatives are undefined at such places.) Such mosaics depict a partitioning of the domain into separate habitats where either u or v (but not both) prevail. In this situation the populations are said to be *segregated*.

We next turn to traveling-wave solutions and indicate how a similar reduction to ODEs can be made.

Traveling-Wave Solutions

Let $f(x, t)$ be a function that represents a wave moving to the right at constant rate v while retaining a fixed shape; f is thus a traveling wave. An observer moving at the same speed in the direction of motion of the wave sees an unchanging picture, which he or she might describe alternately as $F(z)$. The connection between the stationary and moving observers is

$$F(z) = f(x, t) \qquad \text{provided} \qquad z = x - vt. \tag{34}$$

See Figure 10.3. We note that $F(z)$ is now a function of a single variable: distance along the wave from some point arbitrarily chosen to be $z = 0$. Using equation (34) and the chain rule of differentiation, we may conclude that

$$\frac{\partial f}{\partial x} = \frac{dF}{dz}\frac{\partial z}{\partial x} = \frac{dF}{dz}, \tag{35a}$$

$$\frac{\partial f}{\partial t} = \frac{dF}{dz}\frac{\partial z}{\partial t} = -v\frac{dF}{dz}. \tag{35b}$$

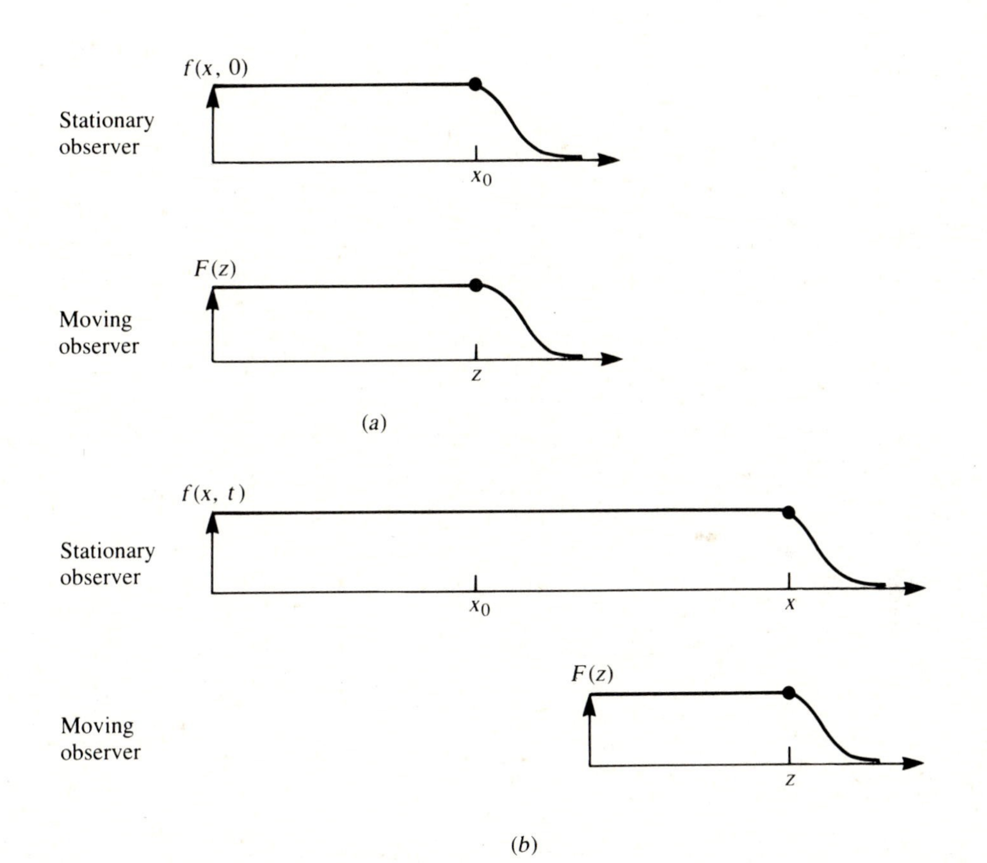

***Figure 10.3** Traveling waves in stationary and moving descriptions. A function* f(x, t) *is a traveling wave if there is a coordinate system moving with constant speed* v *such that* f(x, t) = F(z) *where* z = x − vt *(or* z = x + vt *for motion in the opposite direction). One exploits this fact in converting a PDE in* f *into an ODE or possibly a set of ODEs in* F. *(a) time* = 0, *(b) time* = t.

Example 3

Consider the equation

$$\frac{\partial p}{\partial t} = \mathscr{D} \frac{\partial^2 p}{\partial x^2} + \alpha p(1 - p). \tag{36}$$

The motivation for this equation is discussed in the next section. Letting

$$P(z) = p(x, t) \tag{37a}$$

for

$$z = x - vt, \tag{37b}$$

we obtain

$$-v \frac{dP}{dz} = \mathscr{D} \frac{d^2P}{dz^2} + \alpha P(1 - P). \tag{38}$$

This is a second-order ODE. We shall convert it to a system of first-order ODEs by making the substitution

$$-S = \frac{dP}{dz}. \tag{39}$$

The system we obtain is

$$\frac{dP}{dz} = -S \tag{40a}$$

$$\frac{dS}{dz} = +\frac{\alpha}{\mathscr{D}} P(1 - P) - \frac{v}{\mathscr{D}} S. \tag{40b}$$

If we can find a way of understanding this system of ODEs, then we can make a statement about the existence and properties of the traveling-wave solutions. This is our main topic in the next section.

10.6 TRAVELING WAVES IN MICROORGANISMS AND IN THE SPREAD OF GENES

In this section we describe two problems that can be approached by applying familiar techniques in a rather novel way. We first deal with a classic model due to Fisher that illustrates ideas in a simple, clear setting. A second modeling problem is then handled using similar methods.

Fisher's Equation: The Spread of Genes in a Population

Fisher (1937) considered a population of individuals carrying an advantageous allele (call it a) of some gene and migrating randomly into a region in which only the allele A is initially present. If p is the frequency of a in the population and $q = 1 - p$ the frequency of A, it can be shown that under Hardy-Weinberg genetics, the rate of change of the frequency p at a given location is governed by the equation

$$\frac{\partial p}{\partial t} = \mathcal{D}\frac{\partial^2 p}{\partial x^2} + \alpha p(1 - p), \tag{36}$$

where α is a constant coefficient that depicts the intensity of selection (see Hoppensteadt, 1975). Note that $0 < p < 1$. Historically this model elicited considerable interest and was investigated and generalized by a host of mathematicians. Good reviews of the historical perspective and of the mathematical methods can be found in Fife (1979), Murray (1977), and Hoppensteadt (1975). We remark that the equation can also describe a population $p(x, t)$ that reproduces logistically and disperses randomly.

Equation (36) has a variety of solutions depending on other constraints (such as boundary conditions). Here we shall deal exclusively with *propagating waves* on an infinite domain. The goal before us is to ascertain whether a process described by equation (36) can give rise to biologically realistic waves of gene spread in a population.

In Section 10.5 we observed that the strategy behind studying traveling-wave solutions is that the mathematical problem is thereby reduced to one of solving a set of ODEs. From example 3 it transpires that if equation (36) has traveling-wave solutions, these must satisfy equation (38), or equivalently the system of equations

$$\frac{dP}{dz} = -S, \tag{40a}$$

$$\frac{dS}{dz} = \frac{\alpha}{\mathcal{D}}P(1 - P) - \frac{v}{\mathcal{D}}S. \tag{40b}$$

This system of ODEs is nonlinear and therefore not necessarily analytically solvable. However, the system can be understood qualitatively by phase-plane methods, as follows:

Consider a PS phase plane corresponding to system (40). By our previous methods of attack we first deduce that nullclines are those curves for which

$$S = 0, \qquad (P \text{ nullcline}) \tag{41a}$$

$$S = \frac{\alpha}{v}P(1 - P), \qquad (S \text{ nullcline}) \tag{41b}$$

and that intersections ("steady states") occur at

$$(\bar{P}_1, \bar{S}_1) = (0, 0), \tag{42a}$$

$$(\bar{P}_2, \bar{S}_2) = (1, 0). \tag{42b}$$

The Jacobian of (40a,b) is

$$\mathbf{J}_i = \begin{pmatrix} 0 & -1 \\ \dfrac{\alpha}{\mathcal{D}}(1 - 2P) & -\dfrac{v}{\mathcal{D}} \end{pmatrix}_{(\bar{P}_i, \bar{S}_i)}, \tag{43a}$$

so that

$$\mathbf{J}_1 = \begin{pmatrix} 0 & -1 \\ \dfrac{\alpha}{\mathcal{D}} & -\dfrac{v}{\mathcal{D}} \end{pmatrix}, \tag{43b}$$

$$\mathbf{J}_2 = \begin{pmatrix} 0 & -1 \\ -\dfrac{\alpha}{\mathcal{D}} & -\dfrac{v}{\mathcal{D}} \end{pmatrix}. \tag{43c}$$

This makes $(\bar{P}_1, \bar{S}_1)$ a stable node and $(\bar{P}_2, \bar{S}_2)$ a saddle point provided $v > 2(\alpha\mathcal{D})^{1/2}$ (see problem 16). The *PS* phase plane is shown in Figure 10.4(*a*). On this figure arrows correspond to increasing *z* values, since *z* is the independent variable in equations (40a,b). A single trajectory emanates from the saddle point and approaches the node as *z* increases from $-\infty$ to $+\infty$. A trajectory that connects two steady states is said to be heteroclinic. Such trajectories have special significance to our analysis, as will presently be shown.

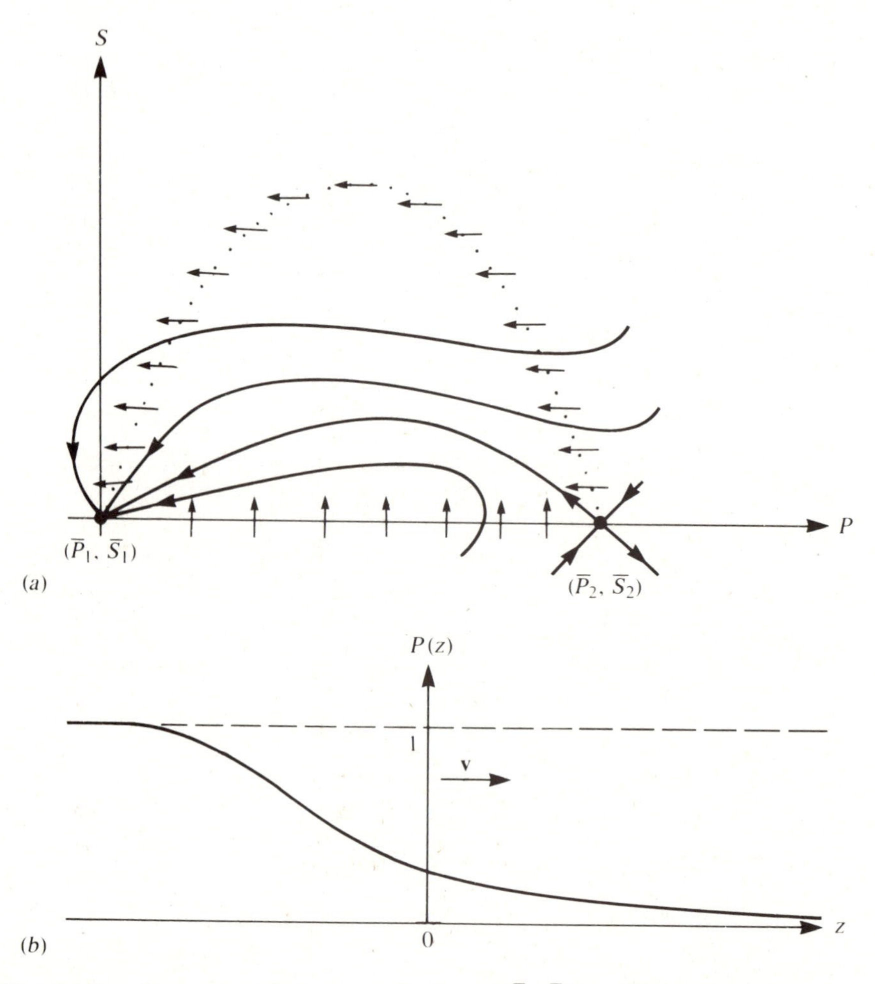

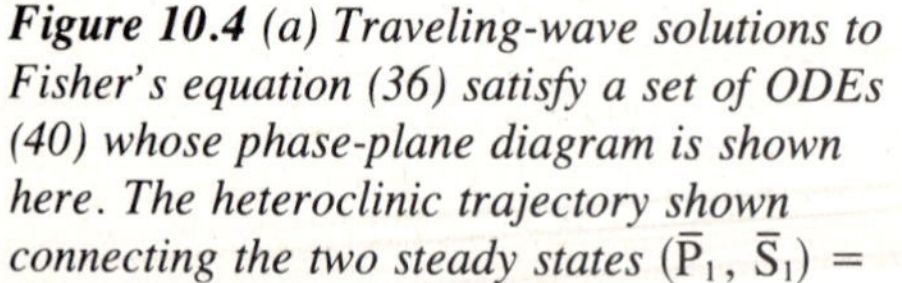
Figure 10.4 *(a) Traveling-wave solutions to Fisher's equation (36) satisfy a set of ODEs (40) whose phase-plane diagram is shown here. The heteroclinic trajectory shown connecting the two steady states* $(\bar{P}_1, \bar{S}_1) = (0, 0)$ *and* $(\bar{P}_2, \bar{S}_2) = (1, 0)$ *is the only bounded positive trajectory. (b) The qualitative shape of the wave.* $P(z) \to 1$ *(for* $z \to -\infty$*) and* $P(z) \to 0$ *(for* $z \to +\infty$*). The direction and speed of motion is indicated.*

To interpret Figure 10.4(*a*) in light of propagating waves we must first recall the interpretation given to the functions $P(z)$ and $S(z)$. Forfeiting a role previously played by time t, the variable z stands for distance along the length of a wave. Any one curve in the *PS* plane thus depicts the gene frequency (P) and its spatial variation (S) from one end of the wave ($z = -\infty$) to the other ($z = +\infty$). We are primarily interested in the former, $P(z)$. However, not all the phase-plane trajectories give reasonable depictions of a biological wave. We consider first the distinguished heteroclinic orbit mentioned earlier and observe the following properties of this curve:

$$P(z) \to \overline{P}_2 = 1 \qquad (z \to -\infty), \tag{44a}$$

$$P(z) \to \overline{P}_1 = 0 \qquad (z \to +\infty), \tag{44b}$$

$$\overline{P}_1 < P(z) < \overline{P}_2 \qquad (-\infty < z < +\infty). \tag{44c}$$

A sketch of P as a function of z derived exclusively from these observations is given in Figure (10.4(*b*). This wave has the shape of a moving *front*. At large positive z values $P(z)$ is very small (approaching zero for $z \to +\infty$), whereas at large negative z values $P(z)$ is very close to 1 (approaching 1 for $z \to -\infty$). This means that allele a has become dominant in the population at the left part of the domain, whereas allele A is still the only gene present towards the right.

Recall that $z = x - vt$ depicts a wave traveling from left to right. The arrow in Figure 10.4(*b*) indicates the direction that the given wave would move with respect to a stationary observer. We observe that the advantageous allele a becomes *dominant* in the population as the wave sweeps through the domain; that is, allele a spreads in the population towards fixation at any particular location.

Now examining other phase-plane trajectories shown in Figure 10.4(*a*) we encounter unrealistic features that lead us to reject these as possible candidates for biological traveling waves. Some of these trajectories tend to infinitely large P values for $z \to -\infty$. This would lead to unbounded levels of P that are inconsistent with the assumption that P is confined to the interval $0 \leq P \leq 1$. These waves are biologically meaningless. Other trajectories that lead to negative P values are equally unacceptable. It can be thus established that only the heteroclinic trajectory depicts a bounded positive wave consistent with a biological interpretation. We conclude the following:

> Biologically meaningful propagating solutions are only obtained if the phase plane corresponding to traveling waves admits a bounded trajectory that is contained entirely in the positive population quadrant.

In the Fisher equation (36) the only (nontrivial) bounded trajectory is the heteroclinic one. It remains in the positive P half-plane provided $v > 2(\alpha\mathcal{D})^{1/2}$ (see problem 16). This means that waves of the shape shown in Figure 10.4(*b*) must move at speeds that exceed the minimum velocity

$$v_{\min} = 2(\alpha\mathcal{D})^{1/2}. \tag{45}$$

It is generally true that propagating-wave solutions of PDEs, if they exist at all, must satisfy constraints on the speed of propagation. Less clear from this result is which of the infinitely many possible speeds is most *stable;* this turns out to be a rather formidable theoretical question. Fisher's equation is simple enough that an answer to this problem could be given. It was shown by Kolmogorov et al. (1937) that if suitable initial conditions are assumed, the solution of equation (36) would evolve into a traveling wave such as that of Figure 10.4(*b*) and would move at the minimal wavespeed $v_{\min}$. Readers interested in learning further details should consult Murray (1977, sec. 5.3).

Spreading Colonies of Microorganisms

A remarkable attribute of many living things is an ability to grow in size while maintaining a particular shape or geometry. This property is common in advanced multicellular organisms where strong intercellular communication links are present. It also occurs in much more primitive settings such as populations of microorganisms, although the underlying mechanisms might be rather different. Here we consider the nutrient-dependent growth of yeast cells and determine whether a colony can exhibit a coordinated spread over space.

Let us focus on the growth of yeast under normal laboratory conditions. A typical experiment begins with a petri dish containing a small volume of sterile nutrient-rich medium. Usually the medium is a solidified gel-like substance called *agar,* which permits free diffusion of small molecules and provides a convenient two-dimensional surface on which to grow microorganisms. A small number of yeast cells are placed on the agar surface. By absorbing nutrient from below, they grow and multiply to such an extent that the population gradually expands and spreads over the surface of the substrate. In many cases, the shape of the colony remains essentially unchanged as it grows in size.

Gray and Kirwan (1974) introduced a model for the spread of yeast colonies which, with some modifications, will serve as our example. A colony of yeast usually takes the form of a glossy disk, visible to the naked eye, that continually enlarges in diameter. We will find it more convenient to deal with a one-dimensional model of the colony, as depicted in Figure 10.5. Accordingly we define the following:

$n(x, t)$ = density of cells at location x at time t,

$g(x, t)$ = concentration of glucose in medium at location x at time t.

Assuming that yeast cells undergo slight random motion and that they produce progeny only when glucose is sufficiently abundant, a simple set of equations to describe the situation would be as follows:

$$\frac{\partial n}{\partial t} = \mathcal{D}_0 \frac{\partial^2 n}{\partial x^2} + kn(g - g_1), \tag{46a}$$

$$\frac{\partial g}{\partial t} = \mathcal{D} \frac{\partial^2 g}{\partial x^2} - ckn(g - g_1). \tag{46b}$$

In these equations g_1 is a constant, representing the minimal amount of glucose necessary for cell proliferation. The yeast reproduction rate is $k(g - g_1)$; that is, cells

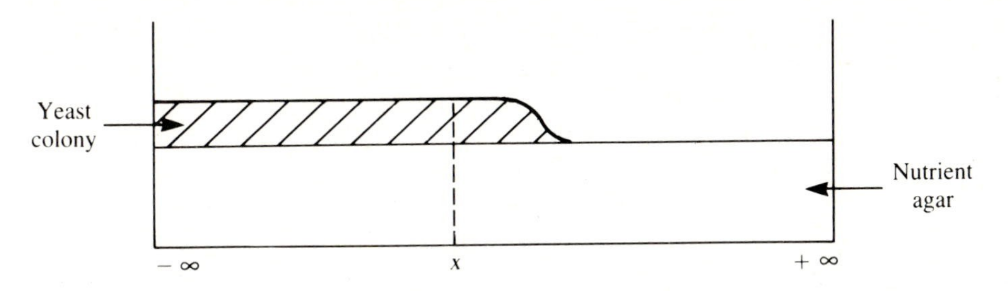

Figure 10.5 *Side view of petri dish with yeast colony.*

are assumed to increase proportionately with the amount of glucose in excess of g_1. Glucose undergoes diffusion with rate $\mathcal{D}$ and is depleted at a rate proportionate to the production rate of yeast cells (c units of glucose used per new cell made). Recall that, although both equations contain diffusion-like terms, these represent on the one hand the approximate nature of random yeast-cell motion, $\mathcal{D}_0(\partial^2 n/\partial x^2)$, and on the other hand true diffusion of glucose in a medium in which it is dissolved. (Agar is 90% or more water and permits essentially free diffusion.)

As a first step, let us define $\hat{g}$ for convenience as

$$\hat{g}(x, t) = g(x, t) - g_1.$$

Assuming that the motility of cells is very slow compared with diffusion of glucose and with the rate of budding of cells, we take the simplified version of the model proposed by Gray and Kirwan:

$$\frac{\partial n}{\partial t} = kn\hat{g}, \tag{47a}$$

$$\frac{\partial g}{\partial t} = \mathcal{D}\frac{\partial^2 \hat{g}}{\partial x^2} - ckn\hat{g}. \tag{47b}$$

This step, though recognized by the advanced reader as fraught with pitfalls, will considerably aid the analysis. In fact, Gray and Kirwan began with equations (47a,b) and to avoid these pitfalls assumed that yeast cells were nonmotile.

Let us now consider as possible traveling-wave solutions to equations (47a,b) functions N and G, where

$$N(z) = n(x, t), \qquad G(z) = \hat{g}(x, t), \qquad z = x - vt,$$

which satisfy the equations of the model.

Using the identities (35a,b), N and G would then have to satisfy

$$-v\frac{dN}{dz} = kNG, \tag{48a}$$

$$-v\frac{dG}{dz} = \mathcal{D}\frac{d^2G}{dz^2} - ckNG. \tag{48b}$$

Multiplying (48a) by the constant c and adding to (48b) yields

$$-vc\frac{dN}{dz} - v\frac{dG}{dz} = \mathcal{D}\frac{d^2G}{dz^2}. \tag{49}$$

Equation (49) can be integrated once, resulting in

$$\text{constant} - vcN - vG = \mathcal{D}\frac{dG}{dz}. \tag{50}$$

By considering values of N and G far behind the edge of the colony, it may be verified that the arbitrary constant introduced by integration has the value cvN_0, where N_0 is the maximal density of cells in the colony interior. This calculation is given in the boxed insert.

The system of equations to be considered is therefore

$$\frac{dG}{dz} = -\frac{v}{\mathcal{D}}G + \frac{Ncv}{\mathcal{D}} - \frac{vcN_0}{\mathcal{D}}, \tag{51a}$$

$$\frac{dN}{dz} = -\frac{kNG}{v} \tag{51b}$$

Conditions at $z = -\infty$

Far behind the leading edge of the colony where the colony has existed for a long time (large t, i.e., $z \to -\infty$) we might expect that cells have attained some limiting density N_0. (N_0 is in general limited by the amount of available glucose that could be used up). At $z = -\infty$ we would also expect $G = 0$, since glucose has been depleted to its lag concentration g_1. Furthermore, it is reasonable to assume that

$$\left.\frac{dG}{dz}\right|_{z=-\infty} = 0,$$

that is, the glucose concentration profile is relatively flat.

Integrating equation (49) from $-\infty$ to z we obtain

$$-v\int_{-\infty}^{z}\left(c\frac{dN}{dz} + \frac{dG}{dz}\right)dz = \int_{-\infty}^{z}\mathcal{D}\frac{d^2G}{dz^2}\,dz, \tag{52a}$$

$$(-vcN - vG)\Big|_{-\infty}^{z} = \mathcal{D}\left.\frac{dG}{dz}\right|_{-\infty}^{z}, \tag{52b}$$

$$-vcN(z) + vcN(-\infty) - vG(z) + vG(-\infty) = \mathcal{D}\left.\frac{dG}{dz}\right|_{z} - \mathcal{D}\left.\frac{dG}{dz}\right|_{-\infty}. \tag{52c}$$

Now using the conditions at $z = -\infty$, we obtain

$$vcN_0 - vcN(z) - vG(z) = \mathcal{D}\left.\frac{dG}{dz}\right|_{z}. \tag{53}$$

Thus equation (51a) is confirmed.

Thus, for the special solutions in which we are interested, it suffices to explore the behavior of these two ODEs. Although these are nonlinear by virtue of the NG term, again by recourse to phase-plane methods one may obtain qualitative solutions. It is left as an exercise for the reader to show that the result is the GN phase-plane diagram given in Figure 10.6(a).

It now remains to interpret what the information in Figure 10.6(a) reveals about $N(z)$ and $G(z)$. Again a single bounded trajectory extends between two points, $(0, N_0)$ and $(cN_0, 0)$ in the GN plane. This trajectory describes a transition from a situation in which glucose is absent and cell density is given by N_0 to one in which glucose is at its maximal level cN_0 and no cells are present. This transition is also depicted qualitatively in Figure 10.6(b). As the yeast colony propagates to the right, it depletes the nutrient that was initially available to it. Gray and Kirwan (1974) draw a parallel between this biological process of contagion and the propagation of a flame N in combustion of a fuel G. Further details of the analysis are suggested in problem 17.

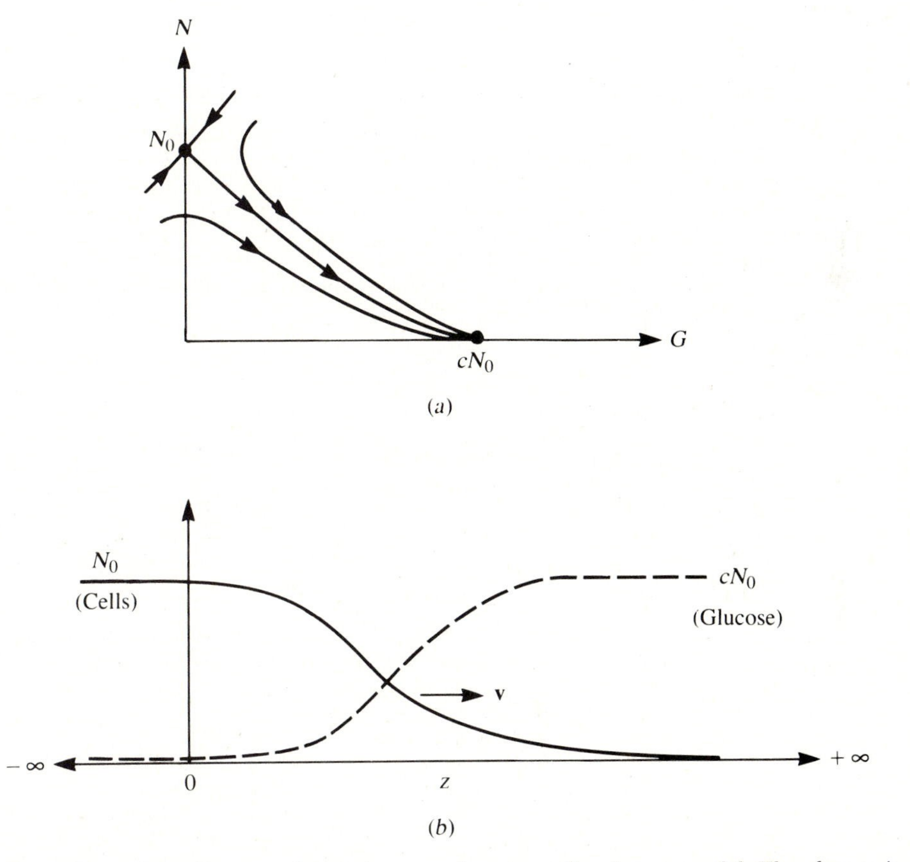

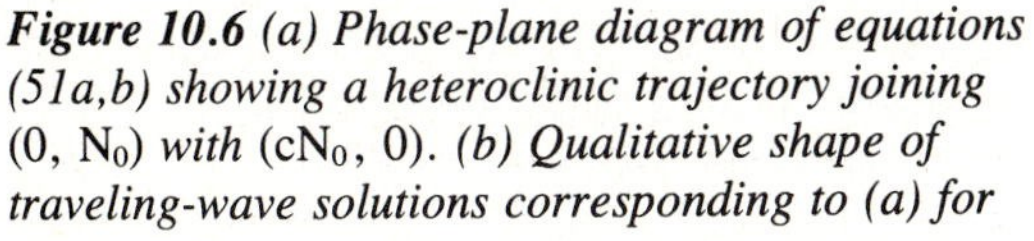
Figure 10.6 *(a) Phase-plane diagram of equations (51a,b) showing a heteroclinic trajectory joining* $(0, N_0)$ *with* $(cN_0, 0)$. *(b) Qualitative shape of traveling-wave solutions corresponding to (a) for the yeast cell–glucose model. The glucose is depleted in places over which the yeast colony has advanced.*

Some Perspectives and Comments

1. The two examples discussed in this section (Fisher's equation and spreading microbial colonies) are readily approached by phase-plane analysis. The versatility of such mathematical methods are thus gratifying, but can they always be expected to work? Inspection of both examples reveals that our analysis depended crucially on the fact that traveling waves satisfied a *pair* of ODEs. In the second example (spreading colonies of microorganisms) this was only true by virtue of a fortuitous integration step that reduced the order of equation (49). In many instances this good fortune does not occur. Higher-order equations or bigger systems of PDEs may have traveling-wave solutions that satisfy a larger system of ODEs. While the concepts are the same, the analysis is much harder. For bounded traveling waves, one would still seek bounded trajectories (but in a higher-dimensional phase space). As we already know, phase-plane analysis is a well-developed technique only in the plane, so the problem may be much harder when the ODE system is larger.

2. Both our examples also shared a phase-plane feature, the heteroclinic trajectory that depicts bounded waves. Two other types of waves can be encountered that correspond to other bounded trajectories. (a) A homoclinic orbit (one that emerges from and eventually returns to a single steady state) would result in a wave that asymptotically approaches the same value for $z \to \pm\infty$. (For example, the propagating action potential in the nerve axon is a peaked disturbance that tapers off to the resting voltage both far ahead and far behind its peak.) (b) A limit cycle would depict an infinite train of peaks or oscillations that propagate over space. (We have indicated that a train of action potentials can occur in nerve cells, given prolonged superthreshold stimulation.)

3. Traveling-wave solutions are sometimes abstractions of reality that give a good general description of the phenomenon of propagation but have unrealistic features as well. Both examples we have discussed share the inaccurate prediction that density of the propagating material (for example, genes or yeast cells) is *never* actually zero, even far ahead of the "front." (This stems from the fact that the solutions only *approach* zero for $z \to +\infty$.) In reality, of course, there are sharp transitions at the edge of an expanding population.

4. The question of wave speed was touched on but not carefully deliberated. In principle one would like to ascertain which of the possible family of waves (for different velocities) is the most stable. In practice the techniques for establishing this are rather advanced, and often such problems are too formidable to yield to analysis. (There are some instances when linear analysis predicts a *unique* wave speed. This occurs whenever phase-plane analysis reveals a heteroclinic trajectory connecting two saddle points. Such trajectories are easily disrupted by slight parameter changes.)

5. Even if analysis does not lead us to find bounded traveling-wave solutions in the exact sense described in this section, there may still be biologically interesting propagating solutions, such as those that undergo very slight changes in shape or velocity with time. In the next section we briefly discuss another biological setting in which long-range transport is important. A recent model for axonal transport due to Blum and Reed (1985) has been shown to lead to such *pseudowaves* (wavelike mov-

ing fronts of material that propagate down the length of the axon). The analysis of such examples is generally based heavily on computer simulations.

10.7 TRANSPORT OF BIOLOGICAL SUBSTANCES INSIDE THE AXON

The anatomy of a neural axon was described in Section 8.1, where our primary concern was the electrical property of its membrane. Other quite unrelated transport processes take place within the axonal interior. All substances essential for metabolism and for normal turnover of the components of the membrane are synthesized in the soma (cell body). Since these are to be used throughout the axon, which may be many centimeters or even meters in length, a transport process other than diffusion is called for. Indeed, with the aid of the light microscope one can distinguish motion of large particles called *vesicles* (macromolecular complexes in which smaller molecules such as acetylcholine are packaged). The motion is *saltatory* (discontinuous rather than smooth), with frequent stops along the way. There seem to be several operational processes, including the following:

1. Fast transport mechanism(s), which can convey substances at speeds on the order of 1 m day^{-1}.
2. Slow transport, with typical speeds of 1 mm day^{-1}.
3. *Retrograde transport* (movement in the reverse direction, from the terminal end to the soma), at speeds on the order of 1 m day^{-1}.

There have been numerous hypotheses for underlying mechanisms, mostly based in some way on the interaction of particles with microtubules. *Microtubules* are long cable-like macromolecules that are important structural components of a cell, and apparently have functional or organizational properties as well. It was held that microtubule sliding, paddling, or change of conformation might lead to fluid motion that would carry particles in *microstreams* within the axon (see Odell, 1977). Many of the original theories were thus based on bulk fluid flow inside the axon. There were problems with an understanding of the simultaneous forward and retrograde transport that led to rather elaborate explanations, none of which completely agreed with experimental observations.

A somewhat different concept has been suggested by Rubinow and Blum (1980) who propose that transport is not fluid-mediated but stems from reversible binding of particles such as vesicles to an intracellular "track" that moves at a constant velocity. They model this hypothetical mechanism by considering interactions of three intermediates, P, Q, and S, whose concentrations, p, q, and s, are defined as follows:

$$p(x, t) = \text{density of free particles at location } x \text{ and time } t,$$
$$q(x, t) = \text{density of particles bound to track at location } x \text{ and time } t,$$
$$s(x, t) = \text{density of unoccupied tracks at location } x \text{ and time } t.$$

It was assumed that particles bind reversibly to tracks as follows:

$$nP + S \underset{k_2}{\overset{k_1}{\rightleftharpoons}} Q.$$

The factor n is incorporated as a possible form of cooperativity in binding. Their assumptions lead to two equations:

$$\frac{\partial p}{\partial t} = -k_1 p^n s + k_2 q, \tag{54a}$$

$$\frac{\partial q}{\partial t} = -\frac{\partial q v_0}{\partial x} - k_2 q + k_1 p^n s. \tag{54b}$$

In these equations bound particles move with the tracks, whereas free particles are stationary. The authors include diffusion terms that are omitted from (54a,b) and that were in fact shown to be unimportant in later work. The Rubinow-Blum model is suggestive, but several conceptual errors made by the authors in analysis of the model have engendered some confusion in the literature. (For reasons outlined in their paper, they deduce that cooperativity is essential, so that $n \geq 2$, but the solutions they describe are not biologically accurate.)

In more recent work, Blum and Reed (1985) have corrected some of these errors and produced a more realistic description. The authors propose that particles can only move along the track if they are bound to intermediates that couple them to the microtubules. In a curious coincidence, the development of this model and new experimental observations simultaneously point to similar conclusions. Indeed, new technology such as enhanced videomicroscopy reveals that all moving particles have leglike appendages to which they are reversibly bound and which seem to propel them along the track. These intermediates have been called *kinesins*.

In their model, Blum and Reed define

$e(x, t)$ = concentration of kinesins ("legs" or "engines"),

and $p(x, t)$, $q(x, t)$, and $s(x, t)$ as previously defined. The assumed chemical interactions are as follows:

1. Binding of kinesins to free particles:

$$P + nE \rightleftharpoons C,$$

where C is the P—nE complex.

2. Binding of C complexes to track(s):

$$C + mS \rightleftharpoons Q.$$

3. Binding of kinesins directly to tracks:

$$E + S \rightleftharpoons ES.$$

4. Binding of free particles to the complex of kinesins and tracks, (ES):

$$P + m(ES) \rightleftharpoons Q.$$

By a combination of numerical simulation and analysis Blum and Reed demonstrated that the above mechanism adequately describes all accurate experimental observations for the fast transport system of the axon. (Some discussion of their model is given as a problem in this chapter.) Of particular interest is their finding that this system admits pseudowaves. These numerical solutions undergo slight changes in shape but are virtually indistinguishable in their behavior from traveling waves as defined earlier.

10.8 CONSERVATION LAWS IN OTHER SETTINGS: AGE DISTRIBUTIONS AND THE CELL CYCLE

We now turn to processes that have little apparent connection with spatial propagation or spatial distributions. Here we shall be concerned with age structure in a population and with changes that take place in these populations as death and birth occur. (Recall that such questions were briefly discussed in Chapter 1 within the context of difference equations and Leslie matrices.)

We begin with a description of cellular maturation and discuss a discrete model for a population of cells at different stages of maturity. Such problems have medical implications, particularly in the treatment of cancer by chemotherapy. Agents used to attack malignant cells are *cycle-specific* if their effect depends on the stage in the cell cycle (degree of maturation of the cell). After examining M. Takahashi's model for the cell cycle we turn to a continuous description of the phenomenon that uncovers a familiar underlying mathematical framework, the conservation equations.

The Cell Cycle

Cells undergo a process of maturation that begins the moment they are created from parent cells and continues until they themselves are ready to divide and give rise to daughter cells. This process, known as the *cell cycle,* is traditionally divided into five main stages. Mature cells that are not committed to division (such as nerve cells) are in the G_0 phase. G_1 is a growth phase characterized by rapid synthesis of RNA and proteins. Following this is the S phase, during which DNA is synthesized. The G_2 phase is marked by further RNA and protein synthesis, preparing for the M phase, in which mitosis occurs. See Figure 10.7.

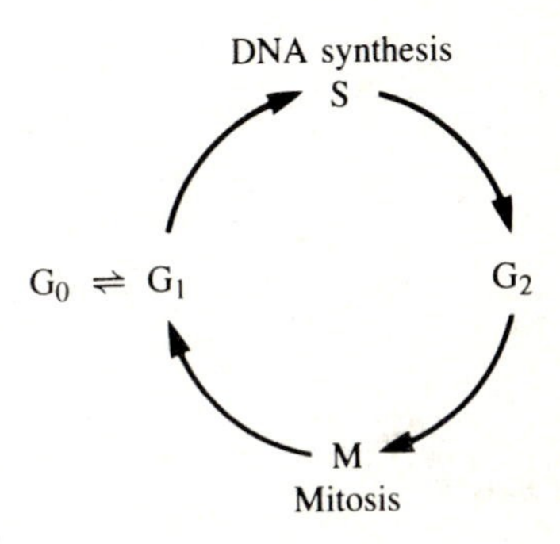

Figure 10.7 *Schematic diagram of the cell cycle.*

While these events are conveniently distinguishable to the biologist, they provide only broad subdivisions, since the journey for the cell from birth to maturity is a continuous one. However, at certain times during this process the cell's susceptibility to its environment may change. This, in fact, forms the basis for cycle-specific chemotherapy, a treatment using drugs that selectively kill cells at particular stages in their development.

It is far from obvious how a course of cycle-specific chemotherapy should be administered. Should it be continuous or intermittent? What frequency of treatments and doses works best, and how does one base one's appraisal on aspects of a given system? Here we shall not dwell on clinical problems associated with chemotherapy design. An excellent survey may be found in articles by Newton (1980) and Aroesty et al. (1973), who delineate mathematical approaches and their implementations in *oncology* (the study of tumors). The following simple discrete model for the cell cycle, which is due to Takahashi, will form our point of departure.

Let us suppose that the cell cycle can be subdivided into k discrete phases and that N_j represents the number of cells in phase j. The transition of a cell from the jth to the $(j + 1)$st phase will be identified with a probability per unit time, λ_j. Furthermore, the likelihood that a cell will die in the jth phase will be assigned the probability per unit time μ_i. See Figure 10.8.

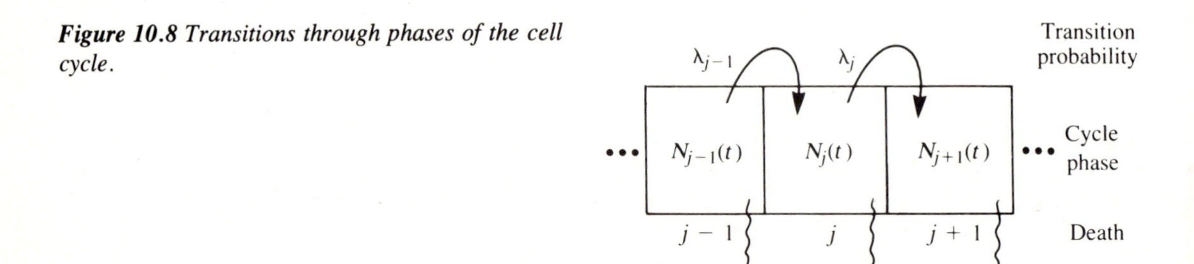

Figure 10.8 *Transitions through phases of the cell cycle.*

During a time interval Δt the number of cells entering phase j is $\lambda_{j-1}N_{j-1}(t)\Delta t$, and the number of cells leaving is $\lambda_j N_j(t)\Delta t$. With the death rate the process can be described by the equation

$$\frac{\Delta N_j}{\Delta t} = \lambda_{j-1}N_{j-1} - \lambda_j N_j - \mu_j N_j. \tag{55a}$$

Suppose that upon maturity each cell (in phase k) divides into β new cells of initial phase $j = 1$. This leads to a boundary condition for the problem:

$$\frac{\Delta N_1}{\Delta t} = \beta\lambda_k N_k(t) - \lambda_1 N_1(t) - \mu_1 N_1(t). \tag{55b}$$

In problem 21 it is shown that if transition probabilities are all equal, if $B = 2$, and if $\mu_j = 0$, the model can be written in the following way:

$$\frac{dN_j}{dt} = \lambda(N_{j-1} - N_j), \tag{56a}$$

$$\frac{dN_1}{dt} = \lambda(2N_k - N_1). \tag{56b}$$

The steady-state solution of (56) reveals that the fraction of cells in each stage follows a Γ distribution.

The discrete model leads to a set of k equations, one for each of the cell-cycle stages. Only the first of these contains a term for birth since we have assumed that, on cell division, daughter cells are in the initial stage of their cycle.

At this point two options are available: First, one could study these equations in their present form, as difference equations. A computer simulation program could then directly use this discrete recipe for generating the phase distributions. Cycle-specific death rates could optionally be included. However, a second approach proves rather illuminating in that it leads to insight based on familiarity with other physical processes. The approach is based on deriving a statement that represents the process of maturation as a continuous and gradual transition. Instead of subdividing the cycle strictly into discrete stages, suppose we represent the degree of maturity of a cell by a continuous variable α, which might typically range between 0 and 1. Instead of accounting for the number of cells in a given stage (that is, in one of the k compartments of Figure 10.8), let us consider a continuous description of cell density along the scale of maturity α.

We will define a *cell-age distribution frequency* in the following way:

$$N_j(t) = n(\alpha_j, t)\,\Delta\alpha \qquad (\alpha_j = j\,\Delta\alpha).$$

In other words, think of $n(\alpha, t)$ as a *cell density per unit age*. Then, provided that the compartment width $\Delta\alpha$ is small, a formal translation can be made from discrete to continuous language using Taylor-series expansions. We write

$$\begin{aligned} n(\alpha_{j-1}, t) &= n(\alpha_j - \Delta\alpha, t) \\ &= n(\alpha_j, t) - \left.\frac{\partial n}{\partial \alpha}\right|_j \Delta\alpha + \left.\frac{\partial^2 n}{\partial \alpha^2}\right|_j \frac{(\Delta\alpha)^2}{2!} - \left.\frac{\partial^3 n}{\partial \alpha^3}\right|_j \frac{(\Delta\alpha)^3}{3!} + \cdots, \end{aligned} \tag{57}$$

Neglecting cell death, omitting terms of order $(\Delta\alpha)^3$ or higher, and substituting into equation (56a) leads to

$$\frac{1}{\lambda}\frac{\partial n}{\partial t}(\alpha_j, t) = -\left.\frac{\partial n}{\partial \alpha}\right|_j \Delta\alpha + \left.\frac{\partial^2 n}{\partial \alpha^2}\right|_j \frac{(\Delta\alpha)^2}{2} + \cdots. \tag{58}$$

Note that for k equal subdivisions $\Delta\alpha = 1/k$, so that

$$\frac{\partial n}{\partial t}(\alpha_j, t) + \left.\frac{\partial n}{\partial \alpha}\right|_j \frac{\lambda}{k} = \left.\frac{\partial^2 n}{\partial \alpha^2}\right|_j \frac{1}{2}\frac{\lambda}{k^2} + \cdots. \tag{59}$$

Now let

$$\nu_0 = \frac{\lambda}{k}, \qquad d_0 = \frac{1}{2}\frac{\nu_0}{k}.$$

A somewhat familiar equation results from these substitutions:

$$\frac{\partial n}{\partial t} + \nu_0 \frac{\partial n}{\partial \alpha} = d_0 \frac{\partial^2 n}{\partial \alpha^2}. \tag{60}$$

This equation contains one term that resembles diffusion and a second that resembles convective transport. This suggests some kind of analogy between the process of cellular maturation and physical particle motion. We explore this more fully in the following subsection and discuss several details in the problems.

Analogies with Particle Motion

To approach the same problem in a more informal way, we abandon temporarily the detailed discrete derivation and view the process of cellular maturation as a continuous transition from birth to maturity of the cell. It is rather natural to picture cell maturation as "motion" of the cell along a scale α. We now make the analogy more precise.

Consider the physical motion of particles in a one-dimensional setting. For x = the distance and $c(x, t)$ = the density of particles (per unit length) we derived a conservation equation (24 in Chapter 9) to describe collective particle motion. If particles move at some velocity v, then the displacement of each individual particle might be described by

$$\frac{dx}{dt} = v(x, t). \tag{61}$$

Collectively, their flux would then be

$$J_c(x, t) = c(x, t)v(x, t). \tag{62}$$

Now replace (1) physical distance in space by "distance" along a scale of maturation ($x \to \alpha$) and (2) density of particles in space by density of cells along the maturation cycle $[c(x, t) \to n(\alpha, t)]$. Then the velocity of a particle would correspond to the rate of change of cellular maturity:

$$\frac{d\alpha}{dt} = \nu(\alpha, t). \tag{63}$$

In other words, we make the connection that $v(x, t) \to \nu(\alpha, t)$. This means that the number of cells that mature through a stage α_0 per unit time (the cell flux) would be

$$J_n = n(\alpha, t)\nu(\alpha, t), \tag{64}$$

that is, $J_c \to J_n$. Finally, a local loss of particles $\sigma_c(x, t)$ would be analogous to a local cellular loss rate $\sigma_n(\alpha, t)$. This could stem from a mortality μ, where μ might be a function of the cellular maturity:

$$\sigma_n(\alpha, t) = -\mu(\alpha)n(\alpha), \tag{65}$$

and $\sigma_c(x, t) \rightarrow \sigma_n(\alpha, t)$.

Based on these analogies the following connection emerges:

$$\underset{\text{particle conservation equation}}{\frac{\partial c}{\partial t} = -\frac{\partial J_c}{\partial x} + \sigma_c} \longrightarrow \underset{\text{cell conservation equation}}{\frac{\partial n}{\partial t} = -\frac{\partial J_n}{\partial \alpha} + \sigma_n}.$$

Thus, without further derivation we have arrived at a continuous description of the maturity distribution of the cell population, given by the equation

$$\frac{\partial n}{\partial t} = -\frac{\partial n\nu}{\partial \alpha} - \mu n. \tag{66}$$

See Figure 10.9. This equation merits several comments in light of the somewhat different result derived from Takahashi's model:

1. The maturation rate ν is not assumed to be a constant.
2. The age-dependent death rate μ is explicitly assumed.
3. Most notably, the diffusion term is missing.

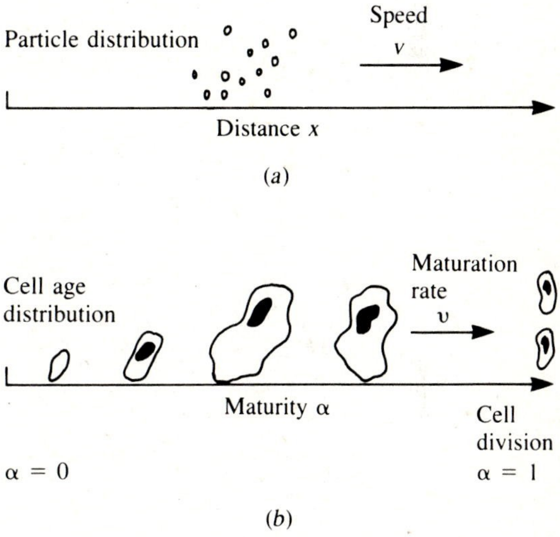

Figure 10.9 *Analogy between (a) particle motion through a distance* x *(at speed* v*) and (b) cellular maturation through maturity* α *(at speed* ν*) forms the basis for the derivation of equation (66).*

In many of the more recent models, diffusion terms such as that in the RHS of equation (60) are often omitted. Note that this term arises formally when maturation is depicted as a chain of random transitions rather than a deterministic "unidirectional flow." It is important to note that the constants ν_0 and d_0 in equation (60) are not independent! Letting $\nu_0 \rightarrow 0$ means that $d_0 \rightarrow 0$ simultaneously. Otherwise a curious contradiction is obtained: if $\nu_0 = 0$ and $d_0 > 0$, the cells appear to undergo pure diffusion, with some continually getting younger while others increase in age. Some modified versions of the diffusion term have recently been suggested to alleviate this conceptual problem, particularly when a terminal age is attained (see Thompson, 1982).

4. A term for birth of new individuals or increase in the population due to cell division again merits a separate equation. It cannot be incorporated directly into equation (66) because cells are only "born into" the lowest maturity class, $\alpha = 0$. Thus birth is specified as a boundary condition of the problem. For cells that divide only at maturity, when $\alpha = 1$, such birth terms could be given by

$$n(0, t) = \beta n(1, t), \tag{67}$$

where β is the number of divisions at mitosis. However, note that (66) could apply to more general age-structured populations, where $n(\alpha, t)$ is the "density" of individuals at different ages α (commonly called the *age distribution of a population*). In such cases females of different ages may give birth to newborns. Thus the number of newborns is a sum of all such contributions, given by

$$n(0, t) = \int_0^\infty n(\alpha, t)\beta(\alpha)\, d\alpha, \tag{68}$$

where β is the *age-specific fecundity* (average number of births from a female between the ages α and $\alpha + d\alpha$), and n is the number of females. If the initial age distribution $\phi(a)$ is known, equation (66) is supplemented with the initial condition

$$n(0, a) = \phi(a). \tag{69}$$

Further discussion of equation (66) and its associated conditions is given in problems 23 and 24.

To give some historical perspective to the model for age distributions, a long list of contributors deserve mention.

Apparently, the first formulation of a PDE for the age distribution of a population is due to McKendrick (1926). Much later, Von Foerster (1959) independently derived a similar equation and applied it to the dynamics of blood cell populations. (Equations such as (66) are often given his name despite the historical inaccuracy.) Analytic solutions of such equations were given by Trucco (1965) and Rubinow (1968), who somewhat generalized the original models.

In the last two decades, PDE models have been applied to problems stemming from demography. Hoppensteadt (1975) gives a good review of this area. Theoretical results have proliferated rapidly, many of them involving a considerable depth of mathematical analysis. Gurtin and MacCamy (1974) dealt with density-dependent birth and mortality. Reviews of this abstract topic may be found in Webb (1985), Heijmans (1985), and Metz and Diekmann (1986).

A further direction has been the generalization of the concept underlying the derivation of the age-distribution equation to rather different problems where variables other than age are of interest. We shall encounter a second example of this type in the do-it-yourself modeling problem outlined in Section 10.9.

A Topic for Further Study: Applications to Chemotherapy

Equations such as (55), (56) and (66) are now commonly applied to modeling the effect of chemotherapeutic agents on malignant cells. Several references provided in the "For Further Study" section of the References could be used as the basis for independent study or further discussion. One example, briefly indicated in the box, is explored in more detail in problem 25.

Example 4
Bischoff et al. (1971) suggested a simplified model for a particular course of chemotherapy of the leukemia strain L1210 using arabinose-cytosine. They assumed that the maturation rate of malignant cells is a constant v and that the drug results in cell death with rate constant μ that varies with drug concentration and cell age.

The equations of their model are:

$$\frac{\partial n}{\partial t} = -v\frac{\partial n}{\partial \alpha} - \mu n, \tag{70}$$

where

$$\mu(\alpha, t) = \frac{K_1(\alpha, t)c(t)}{K_2(\alpha, t) + c(t)}. \tag{71}$$

An equation that describes mitosis in their model is

$$n(0, t) = 2n(1, t). \tag{72}$$

The authors explored the limiting case when cycle specificity of the drug was low and found an asymptotic solution:

$$n(\alpha, t) = N_0(2 \ln 2)2^{-(\alpha - vt)} \exp \int_0^t -\mu(t')\, dt'. \tag{73}$$

In problem 25 this observation is used in showing that eventually the number of cells at lowest maturity level $\alpha = 0$ is a constant fraction of the total population. It is further shown that equation (70) implies that

$$\frac{dN}{dt} = aN - \mu N \tag{74}$$

where

$$N = \int_0^1 n(\alpha, t)\, d\alpha \quad \text{and} \quad a = v \ln 2.$$

Other references, notably those of Newton (1980) and Aroesty et al. (1973), discuss the role of mathematics in oncology in a much broader setting.

Summary

In this section we observed, by considering the aging of a cell, that conservation laws apply to a much broader class of problems than described in previous discussions. Here is a setting in which spatial position and motion in space play no role; another continuous variable, *age,* is more important in describing the system. Yet the ideas of *conservation* lead to equations that are essentially identical (except for renaming variables) to spatial balance equations. This stems from the fact that conservation equations are "bookkeeping" statements: together with their boundary and initial conditions, they serve to keep track of all the progeny of some initial population of parent cells.

In the next section another application of similar mathematical ideas is suggested. However, rather than spelling out all details, we approach a new problem in a sequence of reasoning steps in which reader participation is encouraged.

10.9 A DO-IT-YOURSELF MODEL OF TISSUE CULTURE

In this section you are invited to participate in developing a model as an aid for studying a rather simple biological question. The particular situation to be modeled is a new one, though some of the concepts presented in previous sections can be brought to bear on the problem. The derivation of the model is given in a step-by-step outline; however, you should attempt to use your own inventiveness before consulting the hints in the text.

A Statement of the Biological Problem

Figure 10.10 illustrates a common method for growing certain multicellular organisms. A flask containing nutrient medium is inoculated with numerous small pieces

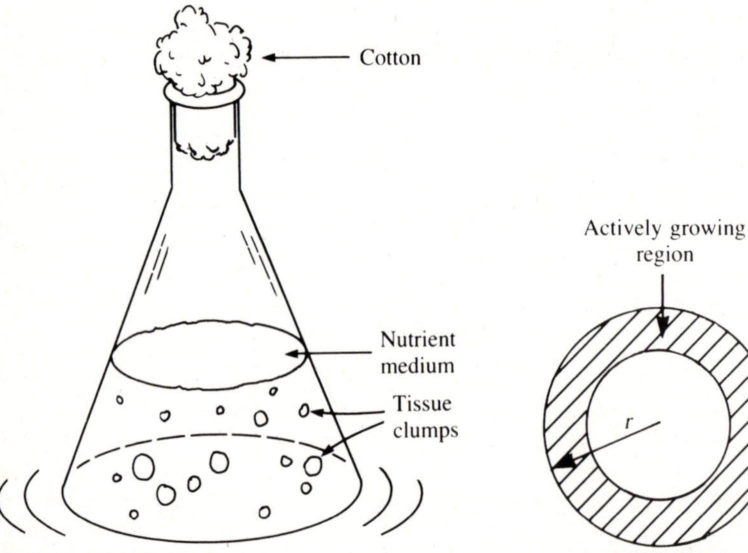

Figure 10.10 *Tissue culture.*

Figure 10.11 *Enlarged view of cross section of an idealized tissue clump.*

or clumps of tissue, which may each consist of many cells. Provided conditions are right (such as plentiful nutrient), each of the pieces will grow. Eventually they may be harvested and used in performing biological experiments.

Question

Suppose the initial biomass is known. What will be the total biomass at some later time t?

Step 1: A Simple Case

We start by making some drastically simplifying assumptions about the geometry of the tissue clumps:

Assumption 1a. All the clumps are spherical.
Assumption 1b. All pieces have the same radius r (see Figure 10.11).

At this point some information about how the tissue pieces grow is required. (This would generally come from empirical observations.) Often one finds that because the core of a tissue particle is not exposed to the nutrient medium, active growth can take place largely at its surface, so that the volume of a single spherical clump, $V(t)$, changes at a rate that is proportional to its surface area, $S(t)$.

Problem 1 (easy)
Relate this information to the radial growth of the particle.

Answer
Using the formulae for the volume and the surface area of a sphere,

$$V = \frac{4}{3}\pi r^3, \qquad S = 4\pi r^2,$$

and the relation

$$\frac{dV}{dt} = kS,$$

we get

$$\frac{4}{3}\pi\frac{dr^3}{dt} = 4\pi k r^2.$$

Differentiating r^3 and cancelling a factor of $4\pi r^2$ from both sides, we get

$$\frac{dr}{dt} = k.$$

We will take this last equation as our basic initial assumption about radial growth of the particles:

Assumption 1c. The radial growth rate of each tissue clump is given by

$$\frac{dr}{dt} = k. \tag{75}$$

Problem 2
Use the previous information to deduce the total volume of tissue after time t if initially its volume at $t = 0$ was V_0.

Answer
If there are N particles, their initial radial size is

$$r_0 = \left(\frac{3}{4\pi}\frac{V_0}{N}\right)^{1/3}.$$

The solution of equation (75) is simply

$$r(t) = kt + r_0,$$

implying that the volume at time t is

$$V(t) = N\frac{4\pi}{3}r^3(t) = N\frac{4\pi}{3}\left[kt + \left(\frac{3}{4\pi}\frac{V_0}{N}\right)^{1/3}\right]^3,$$

$$= \left[\left(\frac{N4\pi}{3}\right)^{1/3} kt + V_0^{1/3}\right]^3 \tag{76}$$

Conclusion

The total volume increases in a way that is cubic with time. This stems from the assumption that radial expansion is constant.

Step 2: A Slightly More Realistic Case

Rarely is it true that all tissue pieces will have an initially identical size. For example, in growing filamentous fungi for the purposes of experimental microbiology, the initial suspension of particles is prepared by grinding or blending the *mycelium* (the vegetative part of the fungus consisting of numerous interconnected branched filaments and resembling a furry disk). In that case many initial particle sizes are present in the suspension. As time passes, each small particle indeed grows to resemble a spherical clump (or *pellet*).

We now treat this more general situation of growth when a size distribution is present. Again, some simplifying assumptions are necessary:

Assumption 2a. All the clumps are spherical (same as assumption 1a).
Assumption 2b. Initially there is some distribution $\alpha(r)$ of pellet sizes.

Assumption 2c. Each pellet grows at a constant radial rate as given by equation (75).

Assumption 2d. There are no pieces smaller in size than some small radius ϵ.

> **Problem 3**
> Before continuing, you are encouraged to attempt to define some meaningful variables and write an equation or equations to describe the situation.

Step 3: Writing the Equations

Hint 1.

Think about the "number density" of particles with radius r at time t. In other words, define the distribution $p(r, t)$ such that

$$\int_r^{r+\Delta r} p(r, t)\, dr = \text{number of pellets whose radii range between } r \text{ and } r + \Delta r.$$

What kind of equation would p satisfy?

Hint 2.

See Figure 10.12.

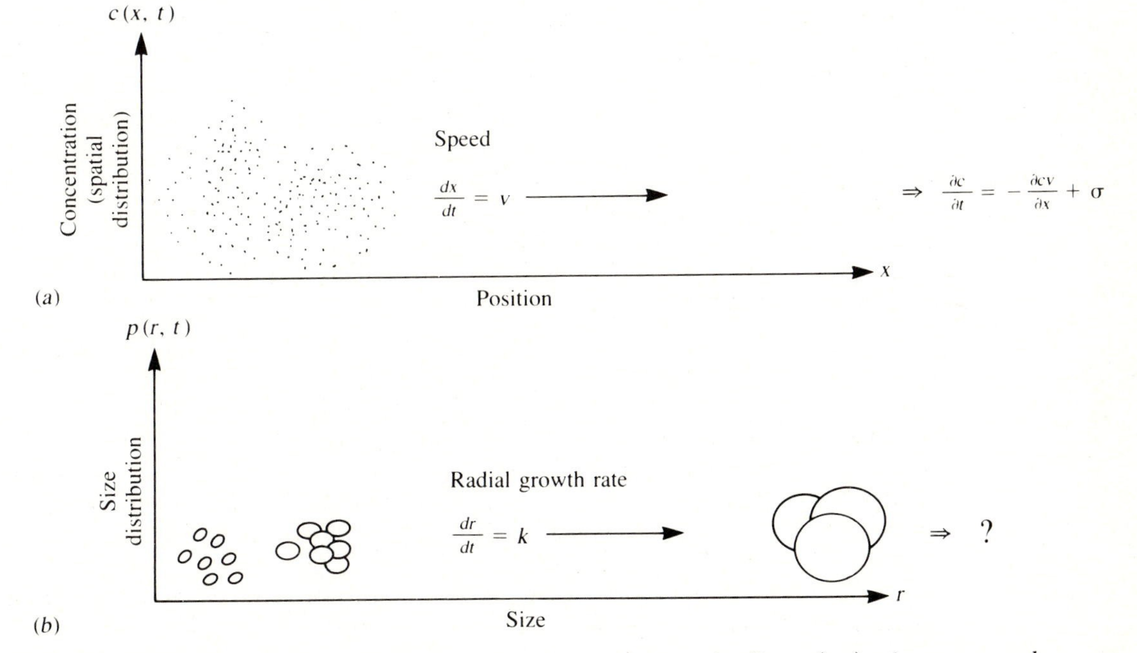

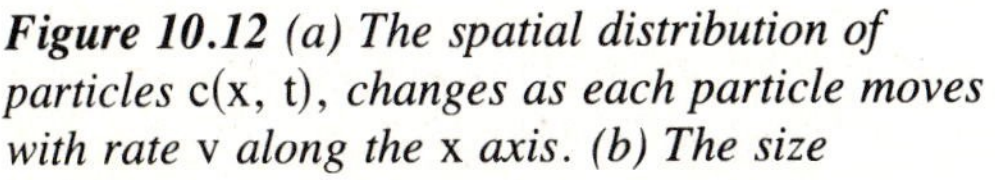

Figure 10.12 *(a) The spatial distribution of particles* c(x, t), *changes as each particle moves with rate* v *along the* x *axis. (b) The size distribution of pellets,* p(r, t), *changes as each pellet grows to a bigger size with rate* k.

Answer

Using a little imagination, we see that the rate of change of the size of a pellet is somewhat analogous to the speed or rate of change of location of a particle. Thus a direct translation from the spatial variables of particles to size variables for pellets would be as follows:

Spatial Variables		*Analogous Size Variables*	
Position,	x	Size,	r
Particle spatial distribution,	$c(x, t)$	Pellet size distribution	$p(r, t)$
Rate of change of position (speed),	$\frac{dx}{dt} = v$	Rate of change of size (growth rate),	$\frac{dr}{dt} = k$
Source/ sink term	σ	Source/ sink term	σ

With the above correspondence we deduce that an equation for $p(r, t)$ is:

$$\frac{\partial p}{\partial t} = -\frac{\partial pk}{\partial r} + \sigma. \tag{77}$$

σ would be present in the equation only if pellets are added ($\sigma > 0$) or eliminated ($\sigma < 0$) during growth. We will at present assume that $\sigma = 0$ and define the mathematical problem as follows:

main equation $$\frac{\partial p}{\partial t} = -\frac{\partial pk}{\partial r}, \tag{77a}$$

initial condition $$p(r, 0) = \alpha(r) \quad \text{(from assumption 2b)} \tag{77b}$$

boundary condition $$p(0, t) = 0, \quad \text{(from assumption 2d).} \tag{77c}$$

Problem 4

Find a solution to the above set of equations (77a–c).

Hint 1

If you are not familiar with equations such as (77), you might try looking for special solutions such as $p(z)$ for $z = r - ct$. This would be equivalent to a pellet size distribution that shifts to higher sizes without altering its basic shape; recall that in the context of motion in space such solutions were called traveling-wave solutions.

Answer

It may be verified by carrying out the appropriate differentiation that the equation

$$p(r, t) = \alpha(r - kt) \tag{78}$$

solves problem 4 provided the initial distribution $\alpha(r)$ satisfies $\alpha(0) = 0$ [that is, the boundary condition (77c) and the initial condition (77b) do not conflict]. This makes good sense when you remember that pellets are all growing in size at the same rate k. Figure 10.13 illustrates several successive size distribution profiles at times t_1, t_2, and so forth, based on the initial distribution at $t = 0$.

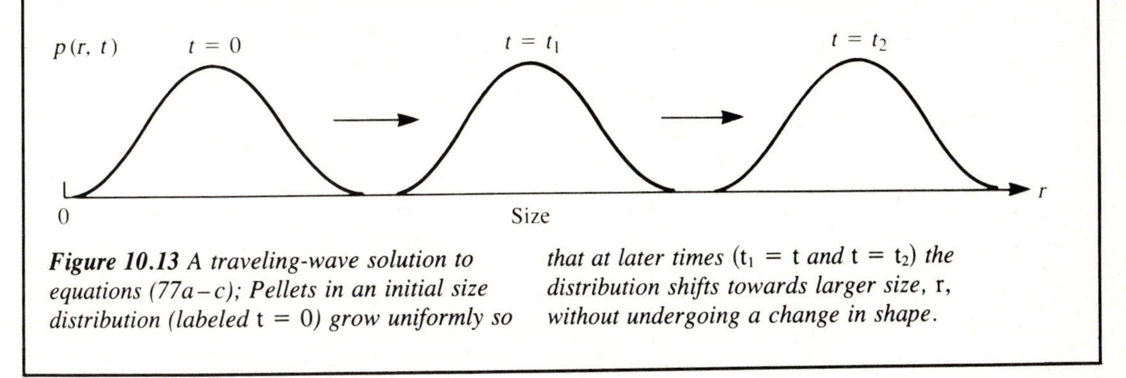

Figure 10.13 *A traveling-wave solution to equations (77a–c); Pellets in an initial size distribution (labeled* t = 0*) grow uniformly so that at later times* (t_1 = t *and* t = t_2) *the distribution shifts towards larger size,* r, *without undergoing a change in shape.*

The Final Step

Now close to our goal, it remains for us to perform several computations to obtain the mass accumulation. In preparation for this, consider the following definitions of quantities that are average properties of the tissue culture; in technical terminology these are *moments of the distribution* $p(r, t)$:

1. Total number of pellets:

$$N(t) = \int_0^\infty p(r, t)\, dr. \tag{79a}$$

2. Average radius of a pellet:

$$\hat{R}(t) = \frac{1}{N(t)} \int_0^\infty r p(r, t)\, dr. \tag{79b}$$

3. Average surface area of a pellet:

$$\hat{S}(t) = \frac{4\pi}{N(t)} \int_0^\infty r^2 p(r, t)\, dr. \tag{79c}$$

4. Average volume of a pellet:

$$\hat{V}(t) = \frac{4\pi}{3N(t)} \int_0^\infty r^3 p(r, t)\, dr. \tag{79d}$$

Problem 5

Use the fact that equation (78) is a solution to derive the dependence of the total volume $V(t)$ of the tissue on the moments $\hat{N}_0$, $\hat{R}_0$, $\hat{S}_0$, $\hat{V}_0$ of the initial distribution.

Answer

The computation to be carried out is integration of the following expression:

$$V(t) = \frac{4\pi}{3}\int_0^\infty r^3\alpha(r - kt)\,dr.$$

This can be done by making the substitution $u = r - kt$ [which implies that $du = dr$ and $r^3 = (u + kt)^3$] and expanding the cubic expression:

$$\begin{aligned} V(t) &= \frac{4\pi}{3}\int_0^\infty (u + kt)^3\,\alpha(u)\,du, \\ &= \frac{4\pi}{3}\int_0^\infty [u^3 + 3u^2(kt) + 3u(kt)^2 + (kt)^3]\alpha(u)\,du. \end{aligned}$$

After multiplying throughout by $\alpha(u)$, we find that

$$\begin{aligned} V(t) = {} & \frac{4\pi}{3}\int_0^\infty u^3\alpha(u)\,du + 4\pi(kt)\int_0^\infty u^2\alpha(u)\,du \\ & + 4\pi(kt)^2\int_0^\infty u\alpha(u)\,du + \frac{4\pi}{3}(kt)^3\int_0^\infty \alpha(u)\,du. \end{aligned}$$

The integrals in this equation are moments of the initial size distribution as given in the definitions (79a–d). Thus the answer to our problem may be stated as follows:

$$V(t) = \left[\hat{V}_0 + (kt)\hat{S}_0 + 4\pi(kt)^2\hat{R}_0 + \frac{4\pi}{3}(kt)^3\right]N_0, \tag{80}$$

for $\hat{V}_0$, $\hat{S}_0$, $\hat{R}_0$ (the average volume, surface area, and radius respectively of the initial pellets in the culture), and N_0 (the initial number of pellets). If the pellets are all of a single size, you should be able to demonstrate that equation (80) reduces to equation (76).

Discussion

The assorted computations which were stepping stones to the final answer contained in equation (80) should not cloud our vision. The single key step in the model is realizing that an equation such as (77a) can be applied to pellet growth as it previously was applied to cellular aging. This permits generalization of the simple situation of identical pellets to the more realistic case of many pieces in many sizes.

10.10 FOR FURTHER STUDY: OTHER EXAMPLES OF CONSERVATION LAWS IN BIOLOGICAL SYSTEMS

The ideas discussed in Sections 10.8 and 10.9 have appeared in a variety of models in the recent scientific literature. A small selection of references is given in the "For Further Study" section of the References. Some of these papers are rather sophisticated mathematically but many would be accessible to readers who have grasped the basic concepts of our discussion. These references have been subdivided into three topics, each suitable for further independent study and presentation to the class.

PROBLEMS*

1. Suggest a (set of) partial differential equations to describe the following processes:

(a) A predator-prey system in which both species move randomly in a one-dimensional setting.

(b) A predator-prey system in which the predator moves towards higher prey densities and the prey moves towards lower predator densities.

(c) A pair of reacting and diffusing chemicals such that species 1 activates the formation of both substances and species 2 inhibits the formation of both substances. Design your model so that it has a homogeneous steady state that is stable.

(d) A population of cells that secretes a chemical substance. Assume that the cells orient and move chemotactically in gradients of this chemical and also more randomly. Further assume that the chemical diffuses and is gradually broken down at some rate.

2. What kinds of processes might be described by the following equations?

(a) $\dfrac{\partial f(x, t)}{\partial t} = \nabla \cdot (f\mathbf{v}) - \mu f \qquad (f = \text{fish density}).$

(b) $\dfrac{\partial p_i}{\partial t} = \mathcal{D}_i \nabla^2 p_i + r p_i (1 - p_i - \beta_{ij} p_j) \qquad (i, j = 1, 2).$

(c) $\dfrac{\partial c}{\partial t} = \dfrac{\partial}{\partial x}\left[\mathcal{D}(c)\dfrac{\partial c}{\partial x} - cv\right] + rc(1 - c).$

(d) Equation (5) for the spruce budworm population.

3. Consider a population described by the equation

$$\frac{\partial P}{\partial t} = \mathcal{D}\frac{\partial^2 P}{\partial x^2} + \alpha P, \tag{1}$$

* Problems preceded by an asterisk (*) are especially challenging.

which at $t = 0$ is concentrated entirely at $x = 0$. Then the solution of this equation is

$$P(x, t) = \frac{P_0}{2(\pi \mathscr{D} t)^{1/2}} \exp\left(\alpha t - \frac{x^2}{4\mathscr{D} t}\right). \tag{2}$$

(a) Explain what is being assumed in equation (1).

(b) Verify that (2) is a solution by differentiating and showing that the PDE is satisfied.

(c) Now consider equipopulation contours: points (x, t) such that

$$P(x, t) = \tilde{P} = \text{constant}.$$

Show that on such a contour the ratio x/t is given by

$$\frac{x}{t} = \pm\left[4\alpha\mathscr{D} - \frac{2\mathscr{D}}{t}\ln t - \frac{4\mathscr{D}}{t}\ln\left(\sqrt{2\pi\mathscr{D}}\frac{\tilde{P}}{P_0}\right)\right]^{1/2}.$$

(d) Show that as $t \to \infty$ one can approximate the above by

$$\frac{x}{t} = \pm 2(\alpha\mathscr{D})^{1/2}.$$

4. Many of Skellam's arguments are based directly on random-walk calculations, not derived from the continuous PDEs such as equation (1). In his original article he defines the following:

a^2 = mean square dispersion per generation,

R = radial distance from point at which population was released,

λ = growth rate of the population,

n = number of elapsed generations,

p = proportion of the population lying outside a circle of radius R after n generations.

He proves that

$$p = \exp\frac{-R^2}{na^2}.$$

(a) If the population growth is described by

$$N_{n+1} = \lambda N_n,$$

and initially there is just a single individual, show that the population in the nth generation is $N_n = \lambda^n$.

(b) Now consider a region that contains all but a single individual. Show that the radius of this region at the nth generation is given by

$$R_n = (na^2 \ln N_n)^{1/2}.$$

[*Hint:* Why is it true that $1/N_n = \exp(-R^2/na^2)$ holds for this radius?]

(c) Use parts (a) and (b) to show that

$$R_n = na(\ln \lambda)^{1/2}.$$

(d) Show that, save for a proportionality factor, this result agrees with the rate of spread of a population given by equation (4).

5. This problem is based on the formula for R_n derived in problem 4. Skellam (1951) quotes the following sentence from Clement Reid (1899), *The Origin of the British Flora,* Dulow, London:

> . . . Few plants that merely scatter their seed could advance more than a yard in a year, for though the seed might be thrown further, it would be several seasons before an oak, for instance, would be sufficiently grown to form a fresh starting point.

He then illustrates, by a simple calculation, that the dispersal of oaks in Great Britain was assisted by small animals. The following estimates are used:

i. The generation time of an oak tree is roughly 60 years. This is the approximate age at which it produces acorns.
ii. The time available for dispersal (from the end of the ice age in 18,000 B.C. until records were first kept in Roman Britain) is roughly 18,000 years, or n generations.
iii. The approximate number of daughter oaks produced by a single parent during one generation is estimated as 9 million.

(a) Use these estimates to show that after n generations

$$\frac{R_n}{a} \simeq 1{,}200,$$

where R_n = the radius that encloses all but a single oak tree, and a^2 = the mean-square displacement defined in problem 4.
(b) From Reid's data, Skellam estimated that the actual radius of the oak forests in Roman Britain was 600 miles. What is the value of a (the root-mean-square distance of daughter oaks about their parents)?
(c) How did Skellam conclude that animals assisted in dispersing the acorns?

6. Segel et al. (1977) calculated the motility coefficients based on equation (6) (where $r = 0$) as follows. A capillary tube (cross-sectional area = A) is filled with fluid. At time $t = 0$ the open end is placed in a bacterial suspension of concentration C_0 and removed at time $t = T$. The number of bacteria in the tube is counted. Motility μ is then computed as follows:

$$\mu = \frac{\pi N^2}{4C_0^2A^2T}.$$

***(a)** Give justification for or derive their formula.
(b) Suppose that (1) the radius of the capillary is approximately 0.01 cm, (2) the bacterial suspension contains a density $C_0 = (1/7 \times 10^{-7})$ bacteria per milliliter, and (3) the following observations (from Segel et al., 1977) are made:

T (min)	N (bacteria)
2	1800
5	3700
10	4800
12.5	5500
15	6700

What do you conclude about μ?

7. Segel et al. (1977) consider the following more detailed description of bacterial motion. They point out that bacteria undergo a series of straight swimming (with average speed v) interrupted by random tumbles in which the orientation is changed. (See figure.) The *mean free path,* λ (average length of a straight swim) and the bacterial motility, μ satisfy the following relationship.

$$\mu = \frac{1}{3} v \lambda$$

[Lovely and Dahlquist (1975)].

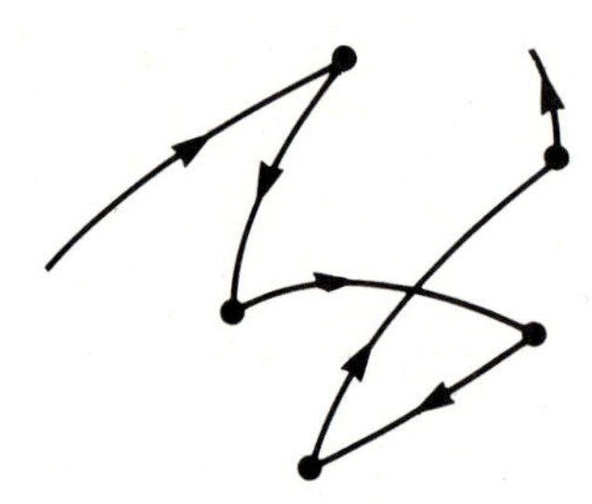

Figure for problem 7.

(a) Why is this a reasonable assumption based on dimensional considerations?

(b) Suppose that the speed v of a cell is proportional to the force F that drives it through a fluid of viscosity η. Show that

$$\mu = \frac{F\lambda}{3k} \frac{1}{\eta},$$

where k is a constant depicting frictional effects.

(c) If τ is the mean time between turns, show that μ can be written in terms of τ as follows:

$$\mu = \frac{\tau F^2}{2k^2} \frac{1}{\eta^2}.$$

8. In this problem we investigate several details of a model for bacterial motility and growth proposed by Lauffenburger et al. (1981) [see equations (7a,b)].

(a) A one-dimensional geometry is considered, with bacteria confined to a tube of length L. In the experiment the substrate level at $x = L$ is kept artificially constant at concentration s_0. At $x = 0$ the tube is sealed. What are the appropriate boundary conditions on $b(x, t)$ and $s(x, t)$?

(b) For the reproductive rate of the bacteria, G_b, the authors assume that

$$G_b = \frac{ks}{K + s} b - k_e b,$$

and for the rate of consumption of substrate, G_s is taken to be

$$G_s = \frac{1}{Y}\frac{ks}{K + s}b.$$

Explain these assumptions and give the meanings of Y, K, k, and k_e. Sketch G_b/b and G_s/b as functions of s.

(c) To simplify the model it is then assumed that a somewhat simpler relationship holds, namely that

$$G_b' = \begin{cases} (k - k_e)b & s > s_c \\ -k_e b & s \le s_c \end{cases},$$

$$G_s' = \begin{cases} \frac{1}{Y}kb & s > s_c \\ 0 & s \le s_c \end{cases}.$$

Explain these approximations and sketch G_b'/b and G_s'/b as functions of s.

(d) Use part (c) to explain equations (7a,b) and determine the functional form for the function $f(s)$ which appears in equations (7a,b).

(e) To reduce the number of parameters, the following dimensionless quantities are defined:

$$u = \frac{s}{s_0}, \quad v = \frac{b}{v_0}, \quad \xi = \frac{x}{L}, \quad \tau = \frac{\mathcal{D}t}{L^2},$$

$$\lambda = \frac{\mu}{\mathcal{D}}, \quad \theta = \frac{k_e L^2}{\mathcal{D}}, \quad K = \frac{kL^2}{\mathcal{D}}, \quad b_0 = \frac{Ys_0\mathcal{D}}{kL^2},$$

$$F(u) = \begin{cases} 1 & u \ge u_c, \\ 0 & u < u_c. \end{cases}$$

Write the equations and other conditions in terms of these quantities.

9. Consider the Keller-Segel equation for bacterial chemotaxis given by (8). Explain how the equation would be modified to incorporate the following further assumptions:

(a) The chemotactic sensitivity increases linearly with the chemical concentration.

(b) The random motion decreases as the cell density increases.

(c) The cell population increases logistically with carrying capacity proportional to the concentration of the chemical.

10. *Density-dependent dispersal*

(a) Interpret equation (13).

(b) Show that equation (14) is equivalent to equation (15).

(c) Show that if $m = 1$ and $F(p) = 0$, equations (18a,b) are analogous to equation (14) given that the total population density causes the dispersal of individuals.

(d) Find biological examples of density-dependent dispersal.

11. Polymorphonuclear (PMN) phagocytes (white blood cells) are generally the first defense mechanism employed in the body in response to bacterial invasion. PMN phagocytes are rapidly mobilized cells that emigrate across walls of *venules* (small veins that connect capillaries and systemic veins) to ingest and eliminate microbes and other foreign bodies in the tissue. Lauffenburger and Kennedy (1983) suggest a model to describe this process. They consider the density of bacteria (b) and of phagocytes (c) and assume the following:

i. Bacteria, microbes or other foreign bodies disperse randomly (motility coefficient μ_b = area/time).

ii. Phagocytes undergo both random motion (with motility coefficient μ_c) and chemotaxis towards relatively high bacterial densities (χ = chemotaxis coefficient).

iii. Bacteria grow at rate $f(b)$ and are eliminated at the rate $d(b, c)$, where b = bacterial density and c = phagocyte density.

iv. Phagocytes emigrate from venules at the rate $\mathcal{A}(c, b)$ and die with rate constant g.

(a) Write a set of equations to describe the motions and interactions of microbes b and phagocytes c.

(b) Additional assumptions made were that

$$f(b) = \frac{k_g b}{1 + b/K_i}, \qquad d(b, c) = \frac{-k_d bc}{K_b + b},$$

$$\mathcal{A}(c, b) = h_0 \frac{A}{V} C_b \left(1 + \frac{h_1}{h_0} b\right),$$

where k_g = bacterial growth rate constant,

k_d = phagocytic killing rate constant,

h_0 = rate of emigration from venules when inflammation is absent,

h_1 = inflammation-enhanced emigration rate,

A/V = ratio of venule wall-surface area to tissue volume,

C_b = phagocyte density in the venules.

(1) Explain the meaning of these assumptions.

(2) Define K_i and K_b and give dimensions of all parameters above.

(c) By dimensional analysis, it is possible to reduce the equations to the following form:

$$\frac{\partial v}{\partial \tau} = \rho \frac{\partial^2 v}{\partial \xi^2} + \frac{\gamma v}{1 + v} - \frac{uv}{\kappa + v},$$

$$\frac{\partial u}{\partial \tau} = \frac{\partial^2 u}{\partial \xi^2} - \delta \frac{\partial}{\partial \xi}\left(u \frac{\partial v}{\partial \xi}\right) + \alpha(1 + \sigma v - u).$$

where v, u, ξ, and τ are dimensionless varibles.

(1) Find the definitions of the parameters appearing in these equations in terms of the original parameters.

***(2)** Explain the meanings of these parameters.

(d) Show that the equations in part (c) have two types of uniform steady-state solutions:
(1) $v = 0, \quad u = 1.$
(2) $v > 0, \quad u = 1 + \sigma v.$
Identify the biological meaning of these steady states.

12. In this problem we investigate the model for branching discussed in Section 10.4. (See figure on page 484.)

(a) A given apex is assumed to move a distance $v\Delta t$ during Δt time units. This gives rise to elongation of a branch and thus contributes to an increment of the branch density ρ by an amount $v\Delta t$. (Note that a filament is thus deposited in the trail of the moving apex.) Elaborate on this reasoning to explain the term nv in equation (19a).

(b) Explain the term $-\gamma\rho$ in equation (19a).

(c) Now consider equation (19b). Explain the term $-\partial(nv)/\partial x$ by defining flux J_n of the moving particles.

(d) The term σ is actually a difference of two terms:

$$\sigma = \sigma_{\text{br}} - \sigma_{\text{mort}},$$

where σ_{br} is a branching rate and σ_{mort} is a loss rate. Explain why such terms appear in this equation.

(e) Branching can occur in several possible ways, including the following:
(1) Dichotomous branching: One apex produces a pair of daughter apices.
(2) Lateral branching: A filament produces a new branch apex somewhere along its length.
See Figure (*c*). Suggest appropriate forms for σ_{br} in each of these two cases.

(f) Apices disappear by anastomosis when the end of a filament forms a connection with a neighboring filament. This can happen in one of two possible ways:
(1) if two ends (apices) come into contact,
(2) if one apex forms a contact directly with a filament.
Suggest appropriate forms for σ_{mort} in each of these cases, and others shown in Figure (*c*).

13. Explain the Balding-McElwain equations for capillary growth. (See problem 12.)

14. Odell (1980) describes the following model for chemical wave fronts in a separation column (see figure on page 485). He defines:

$u_1(x, t)$ = concentration of free protein at distance x along the column,

$u_2(x, t)$ = concentration of protein bound (reversibly) to stationary beads at location x,

B = total number of binding sites on beads per unit distance (B is constant),

P_0 = protein concentration in the stock solution (at $x = 0$.)

Figure for problem 12. *Equation (19a,b) can be derived by considering changes in length of branches and in the numbers of apices at a given location, as shown in (a) and (b). The branching term σ would contain terms that account for the biological events shown in (c). [From Edelstein, (1982). The propagation of fungal colonies. A model for tissue growth.* J. Theor. Biol., 98, *fig. 1 and Table 1. Reprinted by permission of Academic Press.]*

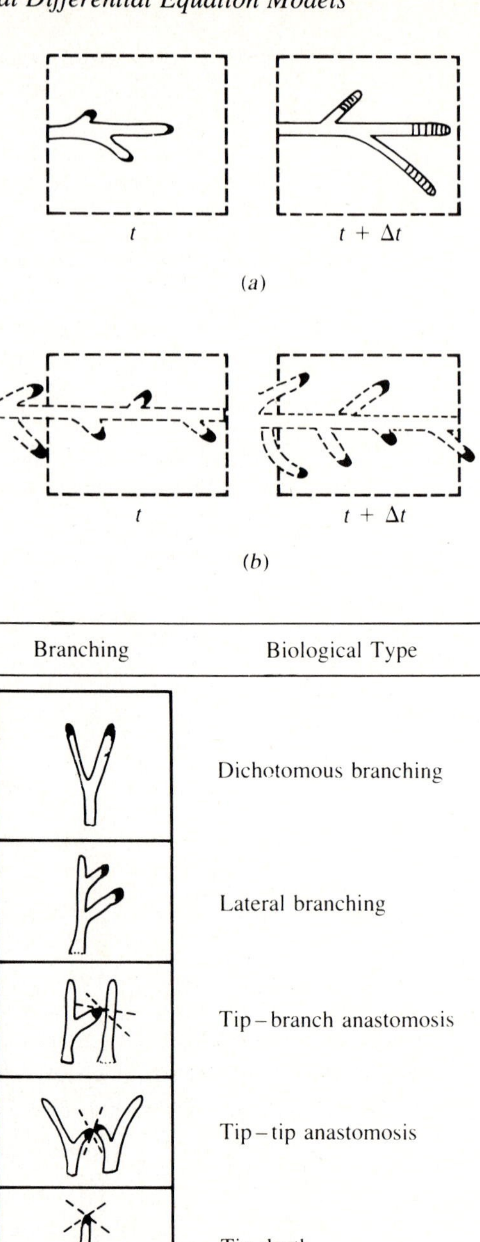

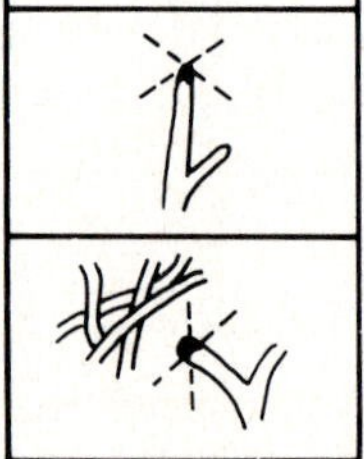

(c)

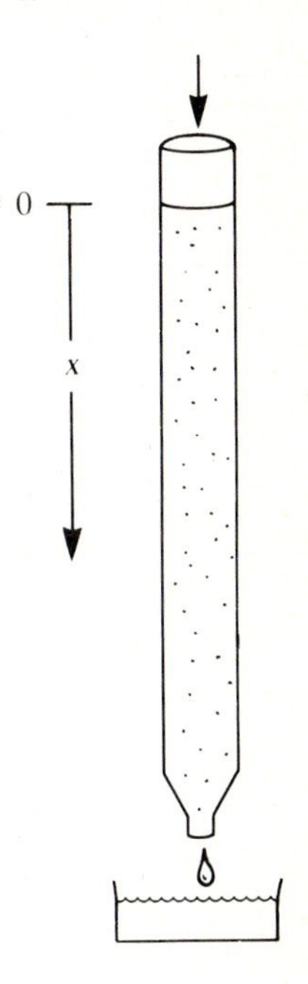

Figure for problem 14.

He derives the following equations for $u_i(x, t)$:

$$\frac{\partial u_1}{\partial t} = \frac{\partial}{\partial x}\left(D\frac{\partial u_1}{\partial x} - vu_1\right) - [k_1(B - u_2)u_1 - k_2u_2],$$

$$\frac{\partial u_2}{\partial t} = k_1(B - u_2)u_1 - k_2u_2.$$

(a) Interpret the terms in these equations and define k_1, k_2, D, and v. What boundary conditions apply at $x = 0$? (The problem is idealized as a half-infinite, one-dimensional domain.)

(b) Determine whether the system has nonuniform steady-state solutions.

(c) Determine whether the system has homogeneous steady-state solutions.

(d) Determine whether the system has bounded traveling-wave solutions. [*Note:* Readers consulting the original reference should be forewarned of a mistake in Odell's equation (49), which unfortunately significantly changes his conclusions.]

15. Discuss the effect of boundary conditions on the existence and nature of steady states that one might obtain for the density-dependent dispersal model given by equations (28a,b). Consider a finite domain $x \in [0, L]$ and let

(a) $J_1(0) = J_1(L) = \hat{J}_1, \quad J_2(0) = J_2(L) = \hat{J}_2,$
where $\hat{J}_1$ and $\hat{J}_2$ are constants representing fluxes of species 1 and 2.

(b) $J_1(0) = \hat{J}_{10}, \quad J_1(L) = \hat{J}_{1L} \quad \hat{J}_{10} \neq \hat{J}_{1L}$
$J_2(0) = \hat{J}_{20}, \quad J_2(L) = \hat{J}_{2L} \quad \hat{J}_{20} \neq \hat{J}_L.$

(c) $u(0) = 0, \quad u(L) = \hat{u}, \qquad v(0) = 0, \quad v(L) = \hat{v},$
where $\hat{u}$ and $\hat{v}$ are constants representing species densities.
(d) $J_1(0) = J_1(L) = \hat{J}_1, \qquad v(0) = v(L) = \hat{v}.$
(e) $u(0) = u(L) = \hat{u}, \qquad v(0) = v(L) = \hat{v}.$
(*Note:* Your conclusion for some of these might be that no steady state exists.)

16. *Fisher's equation*
(a) Show that traveling-wave solutions to Fisher's equation (36) must satisfy equations (40a,b).
(b) Verify the locations of steady states given in (42a,b) and show that the Jacobian matrices at these steady states are then given by (43a,b).
(c) Show that (P_1, S_1) is a stable node and (P_2, S_2) is a saddle point provided that

$$\left(-\frac{v}{\mathcal{D}}\right)^2 - 4\frac{\alpha}{\mathcal{D}} > 0.$$

(d) What happens if the condition in part (c) is not met? (Sketch the resulting phase-plane diagram and discuss why one cannot obtain biologically realistic traveling waves.)
(e) Conclude that the minimum wave speed is

$$v_{\min} = 2(\alpha\mathcal{D})^{1/2}.$$

17. **(a)** Show that traveling-wave solutions to equations (47a,b) for yeast cells on glucose medium would have to satisfy (48).
(b) Verify that these equations lead to the system of ODEs (51a,b).
(c) Find steady states of equations (51a,b), sketch nullclines, and compute the stability properties of the steady states.
(d) Determine whether there is any constraint on the speed v of the wave. (*Hint:* You must determine whether the heteroclinic trajectory always exists and remains in the positive *GN* quadrant.)

18. Equations for space-dependent voltage in the membrane of the neural axon were derived in Section (9.3). Determine what equations would be satisfied by traveling-wave solutions to these equations. (*Note:* The ionic current I_i is given in the Hodgkin-Huxley model in Section 8.1). Such solutions correspond to propagating action potential.

***19.** Rubinow and Blum (1980) studied propagating (traveling-wave) solutions $P(z)$ and $Q(z)$ to equations (54a,b), for $z = x - ct$.
(a) Find the equations satisfied by $P(z)$ and $Q(z)$.
(b) Sketch the phase-plane diagrams for the case $n = 1$.
(c) Sketch the phase-plane diagram for the case $n = 2$ (cooperativity).
(d) Why did they conclude that traveling waves exist only if binding is cooperative?
(e) Sketch the shape of waves they predicted based on part (c).
(f) Rubinow and Blum claim that the waves described in part (e) could represent the following observation. Substances *normally present* (and transported) by the axon can also be added artifically. (These could be radioactively labeled and injected into the axon.) One then observes a propagating front of radioactivity transported down the length of the

axon. Blum and Reed later found the mistake in this claim. Why are these traveling-wave solutions not consistent with the biological phenomenon?

20. **(a)** Write a system of equations for the Blum-Reed model for fast axonal transport. Assume that only the complex Q can move along the track and that velocity of motion v is constant.

(b) Show that the total number of units of E in various forms, E_0, remains fixed.

(c) Similarly show that the total number of tracks in various complexes, S_0, remains fixed.

(*Note:* The Blum-Reed model, like the Rubinow-Blum model admits pseudowaves—propagating solutions—without additional assumptions about cooperativity.)

21. *Takahashi's cell-cycle model.* Consider equations (55a,b). Assume that transition probabilities $\lambda_j = \lambda$ are the same for all phases, that cell death is negligible, and that the cell divides into two daughter cells.

(a) Show that the model can be written in the following way:

$$\frac{dN_j}{dt} = \lambda(N_{j-1} - N_j), \qquad \frac{dN_1}{dt} = \lambda(2N_k - N_1).$$

(b) Verify that the following Γ distribution is a solution:

$$N_j(t) = \left[\frac{\lambda^j}{(j-1)!}\right]^{j-1} e^{\lambda t}.$$

Note that this is a consequence of the way that cell cycle stages were discretized in the model, not in the biology.

22. **(a)** Interpret equation (60) derived on the basis of Takahashi's model. What are the constants v_0 and d_0 in relation to the process of maturation?

(b) Verify that

$$n(\alpha, t) = \frac{1}{\sqrt{4\pi d_0 t}} \exp -\left(\frac{\alpha - v_0 t}{4 d_0 t}\right)^2,$$

is a solution of equation (60).

(c) Give a boundary condition for equation (60) which would be analogous to equation (56b).

23. Equation (66) may be applied to age distributions in populations of cells, plants, humans, or other organisms. The assumptions about birth, death, and maturation rates would depend on the particular situation. How would you formulate boundary conditions (and/or change the maturation equation) for each of the following examples?

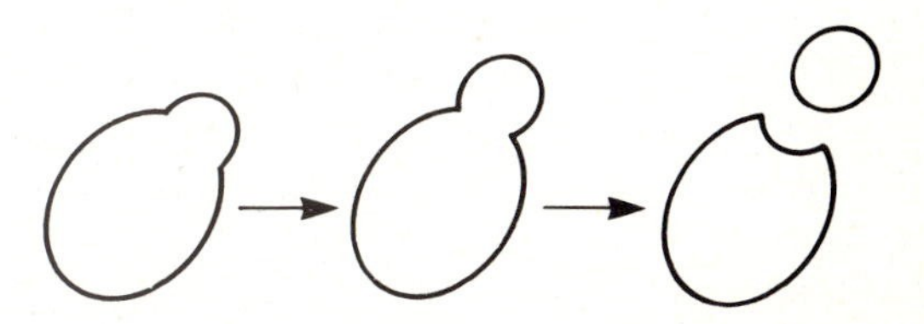

Figure for problem 23 (b).

(a) Bacterial cells divide into a pair of identical daughter cells when they mature.

(b) In yeast a mature cell undergoes "budding," eventually producing a small daughter cell and a larger parent cell. The parent cell is permanently marked by a bud scar for each daughter cell that it has produced. After a certain number of buds have been made, the parent cell dies.

24. The variable α in equation (66) may be identified with any one of a number of measurable cellular parameters. How would equation (66) be written if

(a) $\alpha =$ the chronologic age of a cell.

(b) $\alpha =$ the volume of a cell (assuming that the radius grows at a constant rate per unit time, $dr/dt = K$),

(c) α is the level of activity of an enzyme required in mitosis. (Assume rate of activity increases at a rate proportional to the current level of activity.)

(*Hint:* Consider $\nu(\alpha, t)$ in each case.)

25. *Model for chemotherapy of leukemia.* Consider the model due to Bischoff et al. (1971) given by equations (70) to (72).

(a) What has been assumed about the rate of maturation v?

(b) Assume that cycle specificity of the drug is low, in other words, that μ is nearly independent of maturity. Take

$$\mu(t) = \frac{K_1(t)c(t)}{K_2(t) + c(t)}.$$

Further assume a solution of equation (70) of the form

$$n(\alpha, t) = N_0 e^{\beta t} h(\alpha) e^{-A(t)}.$$

Show that

$$A(t) = \int_0^t \mu(t')\, dt'$$

and that boundary condition (72) implies that $n(\alpha, t)$ is given by equation (73).

(c) Show that eventually the number of cells of maturity $\alpha = 0$ is a constant fraction of the total population:

$$n(0, t) = 2 \ln 2\, N(t).$$

(*Hint:* Note that

$$N(t) = \int_0^1 n(\alpha, t)\, d\alpha.)$$

(d) Show that if N is the total population, then

$$\frac{dN}{dt} = aN - \mu N$$

where $a = v \ln 2$ and μ is defined as in part (b).

(e) Bischoff et al. (1971) estimate that the mean generation time of L1210 leukemia cells is $\tau = 14.4$ h. (What does this imply about v, the maturation speed?) The authors further take the parameters for mortality due to the chemotherapeutic drug arabinose-cytosine to be

$$K_1 = 0.25 \text{ h}^{-1}, \qquad K_2 = 0.3\ \mu\text{g ml}^{-1}.$$

How long would it take for the cells to fall off to 10^{-3} of their initial population if a constant drug concentration of $c(t) = 15$ mg kg^{-1} body weight is maintained in the patient?

26. *Other modeling problems related to tissue cultures*

(a) Suppose a tissue culture is grown in a chemostat, with constant outflow at some rate F. Give a set of equations to describe the problem.

(b) If particles of size r_{max} always break apart into n identical particles of size r_s, how would the problem be formulated?

***(c)** Suppose now that pellets can diminish in size due to shaving off of minute pieces (such as single cells) as a result of friction or turbulence. Assume this takes place at a rate proportional to the pellet radius. How would you model this effect in the following two situations?

(1) All minute pieces can then grow into bigger pieces (participate in the overall growth).

(2) All fragments die.

27. *Plant-herbivore systems and the quality of the vegetation.* In problem 17 of Chapter 3 and problem 20 of Chapter 5 we discussed models of plant-herbivore interactions that considered the *quality* of the vegetation. We now further develop a mathematical framework for dealing with the problem. We shall assume that the vegetation is spatially uniform but that there is a variety in the quality of the plants. By this we mean that some chemical or physical plant trait q governs the success of herbivores feeding on the vegetation. For example, q might reflect the succulence, nutritional content, or digestibility of the vegetation, or it may signify the degree of *induced chemical substances,* which some plants produce in response to herbivory. We shall be primarily interested in the mutual responses of the vegetation and the herbivores to one another.

(a) Define

$$q(t) = \text{quality of the plant at time } t,$$
$$h(t) = \text{average number of herbivores per plant at time } t.$$

Reason that equations describing herbivores interacting with a (single) plant might take the form

$$\frac{dq}{dt} = f(q, h), \qquad \frac{dh}{dt} = g(q, h) = hr(q, h).$$

What assumptions underly these equations?

(b) We wish to define a variable to describe the *distribution of plant quality* in the vegetation. Consider

$$p(q,t) = \text{biomass of the vegetation whose quality is } q \text{ at time } t.$$

Give a more accurate definition by interpreting the following integral:

$$\int_q^{q+\Delta q} p(q, t)\, dq = ?$$

(c) Show that the total amount of vegetation and total quality of the plants at time t is given by

$$P(t) = \int_0^\infty p(q, t)\, dq,$$

$$Q(t) = \int_0^\infty qp(q, t)\, dq.$$

What would be the *average quality* $\overline{Q}(t)$ of the vegetation at time t?

***(d)** Suppose that there is no removal (death) or addition of plant material. Write down an equation of conservation for $p(q, t)$ that describes how the distribution of quality changes as herbivory occurs (*Hint:* Use an analogy similar to that of Sections 10.8 and 10.9.)

(e) Suppose that the herbivores are only affected by the average plant quality, $\overline{Q}(t)$. What would this mean biologically? What would it imply about the equation for dh/dt?

(f) Further suppose that the function $f(q, h)$ is linear in q. (*Note:* This is probably an unrealistic assumption, but it will be used to illustrate a point.) Assume that

$$f(q, h) = f_1(h) + qf_2(h).$$

Interpret the meanings of f_1 and f_2.

***(g)** Show that the model thus far can be used to conclude that an equation for the average quality of the vegetation is

$$\frac{d\overline{Q}}{dt} = f_1(h) + f_2(h)\overline{Q}.$$

[Use the assumptions in parts (d) and (f), the equation you derived in (d), and integration by parts.]

(h) Explore what this model would imply about average quality and average number of herbivores per plant if f and r are given by

$$f(q, h) = K_1 - K_2\, qh(h - h_0),$$
$$r(q, h) = K_3(1 - K_4\, h/\overline{Q}).$$

[*Hint:* See problem 20 of Chapter 5 where $I(t) \to h(t)$.] Interpret your results.

(*Note:* This problem is based on Edelstein-Keshet, 1986. It can be extended into a longer project for more advanced students.)

PROJECTS

1. *Extended project.* Analyze the model given in problem 8, referring to methods outlined in the paper by Lauffenburger et al. (1981).

2. This project is suitable only for mathematically advanced students. Discuss the qualitative behavior of the traveling-wave solutions described in problem 18. For references, see Jones and Sleeman (1983), sec. 6.2, and other references in Rinzel (1981).

3. Write a short simulation program incorporating the discrete Takahashi model for the cell cycle, given the following assumptions:

(a) For $\lambda_i = \lambda =$ constant,

$$\mu = \frac{k_1 c}{k_2 + c}.$$

(b) For $\lambda_i = \lambda \alpha_i(1 - \alpha_i)$,

$$\mu = \frac{K_1(\alpha)c}{K_2 + c}, \qquad K_1(\alpha) = 1 - e^{-\alpha}.$$

(c) For any other set of assumptions that are of biological relevance.

REFERENCES

Dispersal in Biology

Bertsch, M.; Gurtin, M. E.; Hilhorst, D.; and Peletier, L. A. (1985). On interacting populations that disperse to avoid crowding: Preservation of segregation. *J. Math. Biol., 23,* 1–13.

Busenberg, S. N., and Travis, C. C. (1983). Epidemic models with spatial spread due to population migration. *J. Math. Biol., 16,* 181–198.

Fisher, R. A. (1937). The wave of advance of advantageous genes. *Ann. Eugen.* (London), *7,* 355–369.

Gurtin, M. E., and MacCamy, R. C. (1977). On the diffusion of biological populations. *Math. Biosci., 33,* 35–49.

Kareiva, P. M. (1983). Local movement in herbivorous insects: Applying a passive diffusion model to mark-recapture field experiments. *Oecologia, 57,* 322–327.

Kierstead, H., and Slobodkin, L. B. (1953). The size of water masses containing plankton blooms. *J. Mar. Res., 12,* 141–147.

Kolmogorov, A.; Petrovsky, I.; and Piscounov, N. (1937). Étude de l'équation de la diffusion avec croissance de la quantité de matière et son application à un problème biologique. *Moscow Univ. Bull. Ser. Internat. Sec. A, 1,* 1–25.

Levin, S. (1981). Models of population dispersal. In S. Busenberg and K. Cooke, eds., *Differential Equations and Applications in Ecology, Epidemics, and Population Problems.* Academic Press, New York.

Ludwig, D.; Aronson, D. G.; and Weinberger, H. F. (1979). Spatial patterning of the spruce budworm. *J. Math. Biol., 8,* 217–258.

Mollison, D. (1977). Spatial contact model for ecological and epidemic spread. *J. Roy. Statist. Soc. B, 39,* 283–326.

Okubo, A. (1980). *Diffusion and Ecological Problems: Mathematical Models.* Springer-Verlag, New York.

Skellam, J. G. (1951). Random dispersal in theoretical populations. *Biometrika, 38,* 196–218.

Chemotaxis

Alt, W. (1980). Biased random walk models for chemotaxis and related diffusion approximations. *J. Math. Biol. 9,* 147–177.

Bonner, J. T. (1974). *On Development: The Biology of Form.* Harvard University Press, Cambridge, Mass.

Keller, E. F., and Segel, L. A. (1970). Initiation of slime mold aggregation viewed as an instability. *J. Theor. Biol., 26,* 399–415.

Keller, E. F., and Segel, L. A. (1971). Model for chemotaxis. *J. Theor. Biol., 30,* 225–234.

Lauffenburger, D. A., (1982). Influence of external concentration fluctuations on leukocyte chemotactic orientation. *Cell Biophys., 4,* 177–209.

Lauffenburger, D. A., and Kennedy, C. R. (1983). Localized bacterial infection in a distributed model for tissue inflammation. *J. Math. Biol., 16,* 141–163.

Lovely, P. S., and Dahlquist, F. W. (1975). Statistical measures of bacterial motility and chemotaxis. *J. Theor. Biol., 50,* 477–496.

Okubo, A. (1986). Dynamical aspects of animal grouping: Swarms, schools, flocks, and herds. *Adv. in Biophys., 22,* 1–94.

Segel, L. A. (1977). A theoretical study of receptor mechanisms in bacterial chemotaxis. *SIAM J. Appl. Math., 32,* 653–665.

Segel, L. A.; Chet, I.; and Henis, Y. (1977). A simple quantitative assay for bacterial motility. *J. Gen. Microbiol., 98,* 329–337.

Segel, L. A., and Jackson, J. L. (1973). Theoretical analysis of chemotactic movement in bacteria. *J. Mechanochem. Cell. Motil., 2,* 25–34.

Zigmond, S. H. (1977). Ability of polymorphonuclear leukocytes to orient in gradients of chemotactic factors. *J. Cell Biol., 75,* 606–616.

Traveling Waves and Fisher's Equation

See Fisher (1937), Kolmogorov et al. (1937), and the following:

Fife, P. C. (1979). *Mathematical Aspects of Reacting and Diffusing Systems.* Springer-Verlag, New York.

Hoppensteadt, F. C. (1975). *Mathematical Theories of Populations: Demographics, Genetics, and Epidemics.* (Regional Conference Series in Applied Mathematics, no. 20). SIAM, Philadelphia.

Murray, J. D. (1977). *Lectures on Nonlinear Differential Equation Models in Biology.* Clarendon Press, Oxford, sec. 5.3.

Odell, G. M. (1980). Biological waves. In L. A. Segel, ed., *Mathematical Models in Molecular and Cellular Biology.* Cambridge University Press, Cambridge.

The Cell Cycle and Models for Cancer Chemotherapy

Aroesty, J.; Lincoln, T., Shapiro, N.; and Boccia, G. (1973). Tumor growth and chemotherapy: Mathematical methods, computer simulations, and experimental foundations. *Math. Biosci.*, 17, 243–300.

Bischoff, K. B.; Himmelstein, K. J.; Dedrick, R. L.; and Zaharko, D. S. (1973). Pharmacokinetics and cell population growth models in cancer chemotherapy. *Chem. Eng. Med. Biol., (Advances in Chemistry Series.), 118,* 47–64.

McKendrick, A. G. (1926). Application of mathematics to medical problems. *Proc. Edin. Math. Soc., 44,* 98–130.

Merkle, T. C.; Stuart, R. N.; and Gofman, J. W. (1965). *The Calculation of Treatment Schedules for Cancer Chemotherapy.* UCRL–14505. University of California Lawrence Livermore Radiation Laboratory, Livermore, Calif.

Newton, C. M. (1980). Biomathematics in oncology: Modeling of cellular systems. *Ann. Rev. Biophys. Bioeng., 9,* 541–579.
Rubinow, S. I. (1968). A maturity-time representation for cell populations. *Biophys. J., 8,* 1055–1073.
Stuart, R. M. and Merkle, T. C. (1965). *The Calculation of Treatment Schedules for Cancer Chemotherapy, II,* UCRL–14505, University of California Lawrence Livermore Radiation Laboratory, Livermore, Calif.
Takahashi, M. (1966). Theoretical basis for cell cycle analysis, I. labelled mitosis wave method. *J. Theor. Biol., 13,* 201–211.
Takahashi, M. (1966). Theoretical basis for cell cycle analysis, I. labelled mitosis wave method. *J. Theor. Biol., 13,* 202–211.
Trucco, E. (1965). Mathematical models for cellular systems. The von Foerster equation, I and III. *Bull. Math. Biophy., 27,* 285–304, and 449–471.
Von Foerster, J. (1959). Some remarks on changing populations. In F. Stohlman, ed. *The Kinetics of Cell Proliferation.* Grune & Stratton, New York, pp. 382–407.

Axonal Transport

Blum, J. J., and Reed, M. C. (1985). A model for fast axonal transport. *Cell Motil., 5,* 507–527.
Odell, G. M. (1977). *Theories of Axoplasmic Transport.* (*Lectures on Mathematics in the Life Sciences,* Vol. 9.) American Mathematical Society, Providence, R.I., pp. 141–186.
Rubinow, S. I., and Blum, J. J. (1980). A theoretical approach to the analysis of axonal transport. *Biophys. J., 30,* 137–148.

Branching

Balding, D., and McElwain, D. L. S. (1985). A mathematical model of tumour-induced capillary growth. *J. Theor. Biol., 114,* 53–73.
Edelstein, L. (1982). The propagation of fungal colonies: A model for tissue growth. *J. Theor. Biol., 98,* 679–701.
Edelstein, L., and Segel, L. A. (1983). Growth and metabolism in mycelial fungi. *J. Theor. Biol., 104,* 187–210.

Miscellaneous

See Fife (1979) and references below:

Bergman, M. (1983). Mathematical model for contact inhibited cell division. *J. Theor. Biol., 102,* 375–386.
Bird, R. B.; Stewart, W. E.; and Lightfoot, E. N. (1960). *Transport Phenomena.* Wiley, New York.
Gray, B. F., and Kirwan, N. A. (1974). Growth rates of yeast colonies on solid media. *Biophys. Chem., 1,* 204–213.
Jones, D. S., and Sleeman, B. D. (1983). *Differential Equations and Mathematical Biology.* Allen & Unwin, London.
Kendall, M. G. (1948). A form of wave propagation associated with the equation of heat conduction. *Proc. Cambridge Phil. Soc. 44,* 591–593.

Lauffenburger, D; Aris, R.; and Keller, K. H. (1981). Effects of random motility on growth of bacterial populations. *Microb. Ecol., 7,* 207–227.

Rinzel, J. (1981). Models in neurobiology. In F. C. Hoppensteadt, ed., *Mathematical Aspects of Physiology*. American Mathematical Society, Providence, R.I., pp 281–297.

Rubinow, S. I. (1975). *Introduction to Mathematical Biology*. Wiley, New York, chap. 5.

Segel, L. A. (1984). *Modelling Dynamic Phenomena in Molecular and Cellular Biology*. Cambridge University Press, Cambridge, chap. 7.

For Further Study

Age distributions

Gurtin, M. E., and MacCamy, R. C. (1974). Nonlinear age-dependent population dynamics. *Arch. Rat. Mech. Anal., 5,* 281–300.

Gyllenberg, M. (1982). Nonlinear age-dependent population dynamics in continuously propagated bacterial cultures. *Math. Biosci., 62,* 45–74.

Lee, K. Y.; Barr, R. O.; Gage, S. H.; and Kharkar, A. N. (1976). Formulation of a mathematical model for insect pest ecosystems: The cereal leaf beetle problem. *J. Theor. Biol., 59,* 33–76.

Oster, G., and Takahashi, Y. (1974). Models for age-specific interactions in a periodic environment. *Ecol. Mono., 44,* 483–501.

Saleem, M. (1983). Predator-prey relationships: Egg-eating predators. *Math. Biosci., 65,* 187–197.

Sinko, J. W., and Streifer, W. (1969). Applying models incorporating age-size structure of a population to *Daphnia*. *Ecology, 50,* 608–615.

Sinko, J. W., and Streifer, W. (1971). A model for populations reproducing by fission. *Ecology, 52,* 330–335.

Thompson, R. W., and Cauley, D. A. (1979). A population balance model for fish population dynamics. *J. Theor. Biol., 81,* 289–307.

Wang, Y.; Gutierrez, A. P.; Oster, G.; and Daxl, R. (1977). A population model for plant growth and development: Coupling cotton-herbivore interaction. *Can. Entomol., 109,* 1359–1374.

Webb, G. F. (1985). *Theory of Nonlinear Age-Dependent Population Dynamics,* Marcel Dekker, New York.

Size distributions

Edelstein, L., and Hadar, Y. (1983). A model for pellet size distributions in submerged Mycelial cultures. *J. Theor. Biol., 105,* 427–452.

Hara, T. (1984). A stochastic model and the moment dynamics of the growth and size distribution in plant populations. *J. Theor. Biol., 109,* 173–190.

Thompson, R. W. (1982). Comments on size dispersion in living systems. *J. Theor. Biol., 96,* 87–94.

Van Sickle, J. (1977). Mortality rates from size distributions. *Oecologia, 27,* 311–318.

Other Structured-Population Models

Edelstein-Keshet, L. (1986). Mathematical theory for plant-herbivore systems. *J. Math. Biol., 24,* 25–58.

Heijmans, H. J. A. M. (1985). Dynamics of structured populations. PhD thesis, University of Amsterdam.

Metz, J. A. J., and Diekmann, O., eds. (1986). *The dynamics of physiologically structured populations*. Lecture notes in Biomathematics 68, Springer-Verlag, Berlin.

11 Models for Development and Pattern Formation in Biological Systems

It is suggested that a system of chemical substances, called morphogens, reacting together and diffusing through a tissue, is adequate to account for the main phenomena of morphogenesis.
A. M. Turing, (1952).
The chemical basis for morphogenesis.
Phil. Trans. Roy. Soc. Lond., 237, 37–72.

. . . *Simple interactions can have consequences that are not predictable by intuition based on biological experience alone.*
L. A. Segel, (1980). ed.
Mathematical Models in Molecular and Cellular Biology,
Cambridge University Press, Cambridge, England.

This indicates a genuine developmental constraint, namely that it is not possible to have a striped animal with a spotted tail . . .
J. D. Murray (1981a).

The beauty of natural forms and the intricate shapes, structures, and patterns in living things have been a source of wonder for natural philosophers long before our time. Like the spiral arrangement of leaves or florets on a plant, the shapes of shells, horns, and tusks were thought to signify some underlying geometric concepts in Nature's designs. The teeming world of minute organisms was found to hold patterns no less amazing than those on the grander scales. Forms of living things were used from ancient times as a means of classifying relationships among organisms. The study of phenomena underlying such forms, although more recent, also dates back to previous centuries.

Initially, a primary fascination with static designs was characterized by attempts to fathom the secrets of natural forms with geometric concepts or simple physical analogies. *Phyllotaxis* (the study of the geometric arrangement of plant parts; e.g. leaves on a stem, scales on a cone, etc.) was then restricted to classification of spiral patterns on the basis of mathematical sequences (such as the Fibonacci numbers). The shapes of minute aquatic organisms and the structures created by successive cell divisions were compared to those formed by soap bubbles suspended on thin wire frames. (Forces holding these shapes together were described formally by F. Plateau and P. S. Laplace in the 1800s.) Spiral growth was explained in the early 1900s by D. W. Thompson as a continuous addition of self-similar increments.

In recent times the emphasis has shifted somewhat in our study of development, differentiation, and morphogenesis (*morpho* = shape or form; *genesis* = formation) of living things. We have come to recognize that the forms of organisms, as well as the patterns on their leaves, coats, or scales, arise by complex dynamic processes that span many levels of organization, from the subcellular through to the whole individual. While detailed understanding is incomplete, some broad concepts are now recognized as underlying principles.

First, at some level, forms of organisms are genetically determined. Second, the final shape, design, or structure is usually a result of multiple stages of development, each one involving a variety of influences, intermediates, and chemical or physical factors. Third, dramatic events during development are sometimes due to rather gradual changes that culminate in sudden transitions. (This can be likened to a stroll that takes one unexpectedly over a precipice.) Finally, environmental influences and interactions with other organisms, cells, or chemicals can play nontrivial roles in a course of development.

There are numerous unrelated theories and models for differentiation and morphogenesis, just as there are many aspects to the phenomena. Here it would be impossible to give all these theories their due consideration, although some brief indications of references for further study are suggested in a concluding section. Instead of dealing in generalities, the discussion will be based on two rather interesting models for development and morphogenesis that are of recent invention.

In the first model we focus on the phenomenon of aggregation, a specific aspect of one unusually curious developmental system, the *cellular slime molds*. A partial differential equation (PDE) model due to Keller and Segel is described and analyzed. An explanation of several observations then follows from the mathematical results.

A second topic is then presented, that of chemical *morphogens*, (the putative molecular prepatterns that form signals for subsequent cellular differentiation). The theory (due to A. M. Turing) is less than 40 years old but stands as one of the single most important contributions mathematics has made to the realm of developmental biology.

While these topics are somewhat more advanced than those in the earlier chapters of this book, a number of factors combine to motivate their inclusion. From the pedagogic point of view, these are good illustrations of some analysis of PDE models. (While the analysis falls short of actually solving the equations in full generality, it nevertheless reveals interesting results.) Furthermore, techniques of linear stability methods, and the important concepts of gradual parameter variations (which are familiar to the reader) are reapplied here. (Although certain subtleties may require some guidance, the underlying philosophy and basic steps are the same.) This then is a final "variation on a theme" that threads it way through the approach to the three distinctly different types of models (discrete, continuous, and spatially distributed).

Even more to the point, the models presented here are examples of genuine insight that mathematics can contribute to biology. These case studies point to fundamental issues that would be difficult, if not impossible, to resolve based on verbal arguments and biological intuition alone. These examples reinforce the belief that theory may have an important role to play in the biological sciences.

The chapter is organized as follows: Sections 11.1 to 11.3 are devoted to the problem of aggregation. In particular, Section 11.2 introduces the methods of analyzing (spatially nonuniform) deviations from a (uniform) steady state. Readers who have not covered Chapters 9 and 10 in full detail can nevertheless follow this analysis provided that the equations of the model are motivated and that the form of the perturbations given by equations (9a,b) is taken at face value.

Sections 11.4 and 11.5 then introduce reaction-diffusion systems and chemical morphogens. The general model requires a very cursory familiarity with the diffusion equation (or faith that the special choice of perturbations given by equation (27) are appropriate; see motivation in Section 11.2). Further analysis is essentially straightforward given familiarity with Taylor series, eigenvalues, and characteristic equations. (A somewhat novel feature encountered is that the growth rate σ of perturbations depends on their spatial "waviness" q.) It is possible to omit the details of the derivation leading up to the conditions for diffusive instability, (32a,b) and (38), in the interest of saving time or making the material more accessible. Section 11.6 is an important one in which we use simple logical deductions to make physically interesting statements based on the conditions derived in the preceding analysis. The concepts in Section 11.7 are more subtle but lead to an appreciation of the role of the domain size on the chemical patterns. Implications for morphogenesis and for other systems are then given in Sections 11.8 and 11.9.

11.1 CELLULAR SLIME MOLDS

Despite their mildly repelling name, slime molds are particularly fascinating creatures offering an extreme example of "split personality." In its native state a slime mold population might consist of hundreds or thousands of unicellular amoeboid cells. Each one moves independently and feeds on bacteria by *phagocytosis* (i.e. by engulfing its prey). There are many species of slime molds commonly encountered in the soil; one of the most frequently studied is *Dictyostelium discoideum*.

When food becomes scarce, the amoebae enter a phase of starvation and an interesting sequence of events ensues. First, an initially uniform cell distribution develops what appear to be centers of organization called *aggregation sites*. Cells are attracted to these loci and move towards them, often in a pulsating, wavelike manner. Contacts begin to form between neighbors, and streams of cells converge on a single site, eventually forming a shapeless multicellular mass. The aggregate undergoes curious contortions in which its shape changes several times. For a while it takes on the appearance of a miniature slug that moves about in a characteristic way. The cells making up the forward portion become somewhat different biochemically from their colleagues in the rear. Already a process of differentiation has occurred; if left undisturbed the two cell types (called *prespore* and *prestalk*) will have quite distinct fates: Anterior cells turn into stalk cells while posterior cells become spores.

At this stage the differentiation of the multicellular mass is as yet reversible. A fascinating series of experiments (see box in Section 11.3) has been carried out to demonstrate that the ratio of the two cell types is self-regulating; if a portion of the slug is excised, some of the cells change their apparent type so as to preserve the proper ratio.

The sluglike collection of cells executes a crawling motion; understanding of the underlying mechanism is just beginning to emerge (see Odell and Bonner, 1986). It then undertakes a sequence of shapes including that of a dome. As a culmination of this amazing sequence of events, cellular streaming resembling a "reverse fountain" brings all prestalk cells around the outside and down through the center of the mass. The result is a slender, beautifully sculptured stalk bearing a spore-filled capsule at its top. In order to provide a rigid structural basis, the stalk cells harden and eventually die. The spore cells are thereby provided with an opportunity to survive the harsh conditions, to be dispersed by air currents, and to thus propagate the species into more favorable environments. Since individual slime mold cells do not reproduce after the onset of aggregation, there is always some fraction of the population that is destined to die as part of the structural material. (There is a natural tendency to view this anthropomorphically as an example of self-sacrifice for the group interest.)

Slime molds have fascinated biologists for many decades, not simply for their amazing repertoire, but also because of their easily accessible and malleable developmental system. There are many intriguing questions to be addressed in understanding the complicated social behavior of this population of relatively primitive organisms. To outline just a few theoretical questions, consider:

1. What causes cells to aggregate, and how do cells "know" where an aggregation center should form?
2. What mechanisms underly motion in the slug?
3. What determines the prespore-prestalk commitment, and how is the ratio controlled? (This is a problem of size regulation in a pattern.)
4. How are cells sorted so that prestalk and prespore cells fall into appropriate places in the structure?
5. What forces lead to the formation of a variety of shapes including that of the final *sporangiophore* (the spore-bearing structure)?

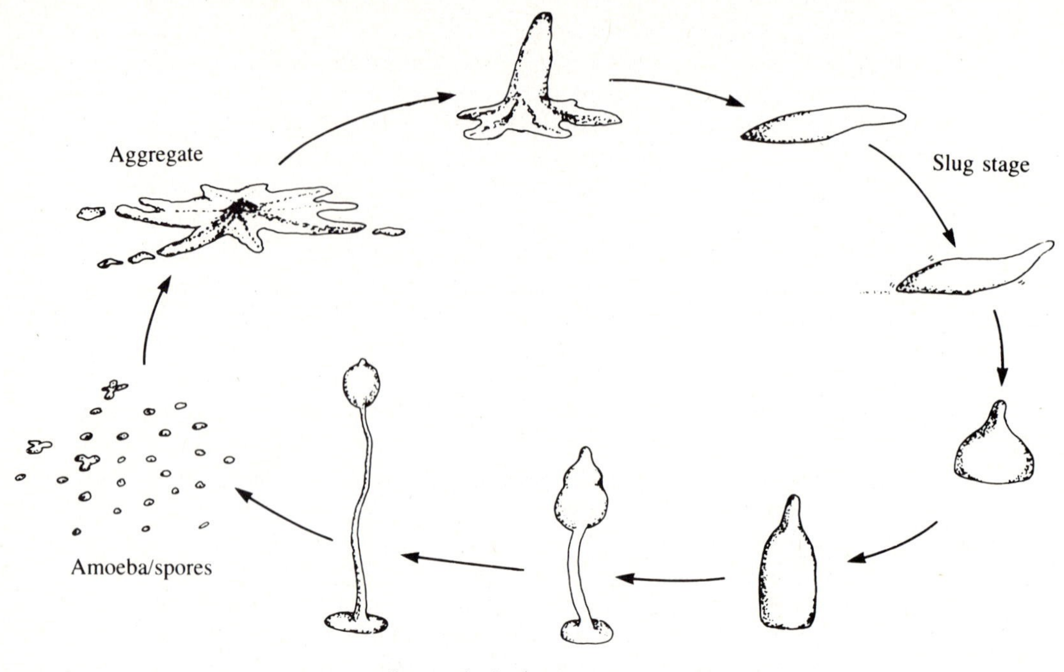

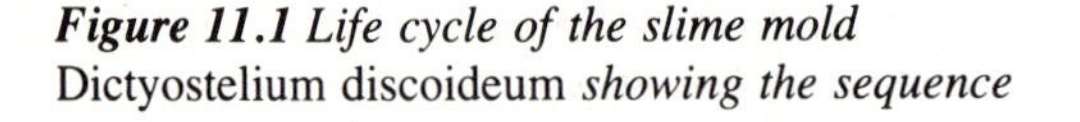

Figure 11.1 *Life cycle of the slime mold* Dictyostelium discoideum *showing the sequence of events leading to formation of the spore-bearing stalk, the sporangiophore.*

While the theoretical literature deals with such questions, we shall address only the first of these. (Other topics are excellent material for advanced independent study.) Perhaps best understood to date, the aggregation process has been rather successfully analyzed by Keller and Segel (1970) in a mathematical model that leads to a more fundamental appreciation of this key step in the longer chain of events.

For many years it has been known that starved slime mold amoebae secrete a chemical that attracts other cells. Initially named acrasin, it was subsequently identified as *cyclic AMP* (cAMP), a ubiquitous molecule whose role is that of an intracellular messenger. The cells also secrete *phosphodiesterase,* an enzyme that degrades cAMP. cAMP secretion is autocatalytic in the sense that free extracellular cAMP promotes further cAMP secretion by a chain of enzymatic reactions (in which a membrane-bound enzyme, *adenylate cyclase,* is implicated). The molecular system has been studied in some depth (see references).

To give a minimal model for the aggregation phase, Keller and Segel (1970) made a number of simplifying assumptions:

1. Individual cells undergo a combination of random motion and chemotaxis towards cAMP.
2. Cells neither die nor divide during aggregation.
3. The attractant cAMP is produced at a constant rate by each cell.

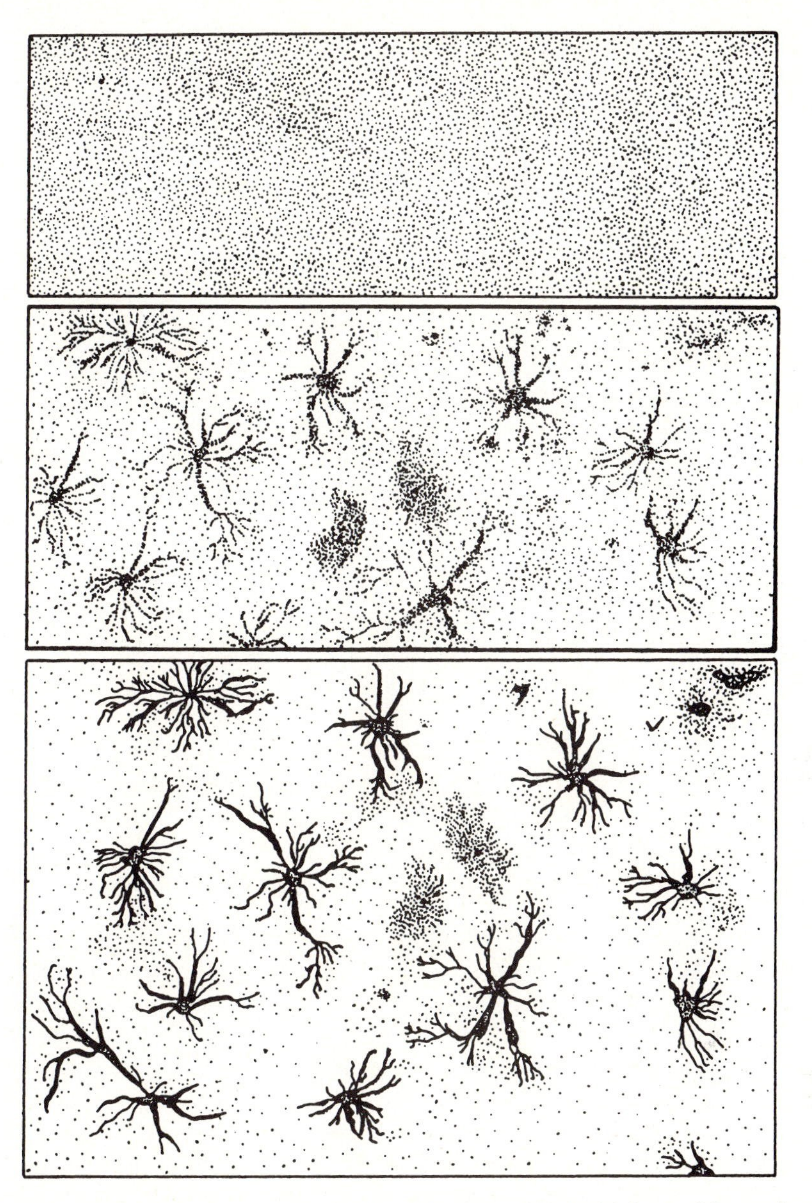

Figure 11.2 *States in the aggregation of an initially uniform distribution of slime molds (each dot represents a cell). [From Bonner, J. T. (1974).* On Development; the biology of form. *fig. 39. Reprinted by permission of Harvard University Press, Cambridge, Mass., fig. 39.]*

4. The rate of degradation of cAMP depends linearly on its concentration.
5. cAMP diffuses passively over the aggregation field.

Of this list, only assumptions 3 and 4 are drastic simplifications.

These assumptions lead to the following set of equations, given here for a one-dimensional domain:

$$\frac{\partial a}{\partial t} = -\frac{\partial}{\partial x}(J_{\text{random}} + J_{\text{chemotactic}}), \tag{1a}$$

$$\frac{\partial c}{\partial t} = -\frac{\partial}{\partial x}(J_{\text{diffusion}}) + \text{sources}, \tag{1b}$$

where

$a(x, t)$ = density of cellular slime mold amoebae per unit area at (x, t),
$c(x, t)$ = concentration of cAMP (per unit area) at (x, t),
J = flux.

By the results of Chapter 9, the appropriate assumptions for chemotactic and for random motion lead to flux terms that can be incorporated into these equations; we then obtain the following:

$$\frac{\partial a}{\partial t} = -\frac{\partial}{\partial x}\left(-\mu \frac{\partial a}{\partial x} + \chi a \frac{\partial c}{\partial x}\right), \tag{2a}$$

$$\frac{\partial c}{\partial t} = -\frac{\partial}{\partial x}\left(-D\frac{\partial c}{dx}\right) + fa - kc, \tag{2b}$$

where

μ = amoeboid motility,
χ = chemotactic coefficient,
D = diffusion rate of cAMP,
f = rate of cAMP secretion per unit density of amoebae,
k = rate of degradation of cAMP in environment.

The quantities μ, χ, f, and k may in principle depend on cellular densities and chemical concentrations, but for a first step they are assumed to be constant parameters of the system.

We shall study this model in two stages. First we find the simplest solution of the equations, that is, a solution constant in both space and time. This is called the homogeneous steady state. We will next examine its stability properties and thereby address the question of aggregation.

11.2 HOMOGENEOUS STEADY STATES AND INHOMOGENEOUS PERTURBATIONS

A *homogeneous steady state* of a PDE model is a solution that is constant in space and in time. For equations (2) such solutions must then satisfy the following:

$$c(x, t) = \bar{c}, \qquad a(x, t) = \bar{a}, \tag{3a}$$

where

$$\frac{\partial \overline{c}}{\partial t} = \frac{\partial \overline{a}}{\partial t} = 0, \tag{3b}$$

$$\frac{\partial \overline{c}}{\partial x} = \frac{\partial \overline{a}}{\partial x} = 0.$$

Now substitute these into equations (2a,b). Observe that, since all derivatives are zero, one obtains

$$0 = 0,$$
$$0 = 0 + f\overline{a} - k\overline{c},$$

In the homogeneous steady state it must follow that

$$f\overline{a} = k\overline{c}. \tag{4}$$

This means that in every location the amount of cAMP degraded per unit time matches the amount secreted by cells per unit time.

It is of interest to determine whether or not such steady states are stable. As before, we analyze stability properties by considering the effect of small perturbations. However, a somewhat novel feature of the problem to be exploited is its space dependence. In determining whether aggregation of slime molds is likely to begin, we must look at spatially nonuniform (also called *inhomogeneous*) perturbations, and explore whether these are amplified or attenuated.

If an amplification occurs, then a situation close to the spatially uniform steady state will *destabilize,* leading to some new state in which spatial variations predominate. Keller and Segel (1970) identified the *onset of aggregation* with a process of destabilization.

Starting close to $\overline{a}$ and $\overline{c}$ we take our distributions to be

$$a(x, t) = \overline{a} + a'(x, t), \tag{5a}$$

$$c(x, t) = \overline{c} + c'(x, t), \tag{5b}$$

where a' and c' are small. In preparation, we rewrite (2) in the expanded form

$$\frac{\partial a}{\partial t} = \mu \frac{\partial^2 a}{\partial x^2} - \chi\left(\frac{\partial a}{\partial x}\frac{\partial c}{\partial x} + a\frac{\partial^2 c}{\partial x^2}\right), \tag{6a}$$

$$\frac{\partial c}{\partial t} = D\frac{\partial^2 c}{\partial x^2} + fa - kc. \tag{6b}$$

In this form equations (6a,b) contain two nonlinear terms within parentheses. (All other terms contain no multiples or nonlinear functions of the dependent variables.) However, the fact that a' and c' *are small* permits us to neglect the nonlinear terms. To see this, substitute (5a,b) into (6a,b) and expand. Then using the fact that $\overline{a}$ and $\overline{c}$ are constant and uniform we arrive at the following equations:

$$\frac{\partial a'}{\partial t} = \mu\frac{\partial^2 a'}{\partial x^2} - \chi\left(\frac{\partial a'}{\partial x}\frac{\partial c'}{\partial x} + \overline{a}\frac{\partial^2 c'}{\partial x^2} + a'\frac{\partial^2 c'}{\partial x^2}\right), \tag{7a}$$

$$\frac{\partial c'}{\partial t} = D \frac{\partial^2 c'}{\partial x^2} + fa' - kc'. \tag{7b}$$

The terms $(\partial a'/\partial x)(\partial c'/\partial x)$ and $a'\partial^2 c'/\partial x^2$ are quadratic in the perturbations or their derivatives and consequently are of smaller magnitude than other terms, provided a' and c' and their derivatives are small (see problem 1).

We now rewrite the approximate equations:

$$\frac{\partial a'}{\partial t} = \mu \frac{\partial^2 a'}{\partial x^2} - \chi \bar{a} \frac{\partial^2 c'}{\partial x^2}, \tag{8a}$$

$$\frac{\partial c'}{\partial t} = D \frac{\partial^2 c'}{\partial x^2} + fa' - kc'. \tag{8b}$$

Note that they are now *linear* in the quantities a' and c'. Based on a similarity with the diffusion equation discussed in the previous chapter, we shall build solutions to equations (8a,b) from the basic functions $e^{\sigma t} \cos qx$ and $e^{\sigma t} \sin qx$. Without attempting to deal in full generality we restrict our attention to the following possibilities:

$$a'(x, t) = Ae^{\sigma t} \cos qx, \tag{9a}$$

$$c'(x, t) = Ce^{\sigma t} \cos qx, \tag{9b}$$

While this rather special assumption may seem at first sight surprising, it makes sense for several reasons.

1. The functions appearing as factors in equations (9a,b) are related to their own first and second partial derivatives (with respect to time and space respectively). These functions are thus good candidates for solutions to equations such as (8) in which $\partial^2/\partial x^2$ and $\partial/\partial t$ appear. (Indeed, such functions were shown to appear rather naturally in Chapter 9, where we encountered their spatial parts as eigenfunctions of the diffusion operator.)
2. The spatial dependence of $\cos qx$ is that of a function with maxima and minima—precisely descriptive of an aggregation field where there is depletion of cells in some places and accumulation in others.
3. Later we explicitly consider the effect of domain size and boundary conditions on the aggregation process. We saw in Section 9.8 that for impermeable boundaries, the eigenfunction $\cos qx$ (where the *wavenumber* q is suitably defined) is appropriate. This will soon be discussed more fully.
4. The time dependence of (9a,b), $e^{\sigma t}$ would be suitable for either increasing or decreasing perturbations. It is up to the analysis to determine whether $\sigma > 0$ (that is, instability of the uniform state) is compatible with the model.
5. In a given realistic example, $a'(x, t)$ and $c'(x, t)$ might have more complicated spatial forms. There is then a theorem (called the *Fourier theorem*) analogous to that in the Appendix to Chapter 9, which guarantees that the spatial functional form of the perturbations can be expressed as an infinite sum of cosines. (Such an expansion is called a *Fourier cosine series*.) For simplicity, we are merely isolating one component in (9). Since the approximate equations (8a,b) are linear, it is always possible to construct general solutions from linear superpositions of simpler ones.

Pursuing the consequences of assumptions (9) for the perturbations, we substitute (9) into (8) and differentiate. This leads to

$$A\sigma e^{\sigma t}\cos qx = -\mu q^2 A e^{\sigma t}\cos qx + \chi\bar{a}Cq^2 e^{\sigma t}\cos qx, \tag{10a}$$

$$C\sigma e^{\sigma t}\cos qx = -Dq^2 C e^{\sigma t}\cos qx + fAe^{\sigma t}\cos qx - kCe^{\sigma t}\cos qx. \tag{10b}$$

A factor of $e^{\sigma t}\cos qx$ can be cancelled to obtain a pair of equations in A and C:

$$A(\sigma + \mu q^2) - C(\chi\bar{a}q^2) = 0, \tag{11a}$$

$$A(-f) + C(\sigma + Dq^2 + k) = 0. \tag{11b}$$

Thus equations (9a,b) indeed constitute a solution to (8a,b) provided that algebraic equations (11a,b) are satisfied. One trivial way of solving these is to set $A = C = 0$. (This is in fact the unique solution unless the equations are redundant.) But this is not what we want since it means that the perturbations a' and c' are identically zero for all t. To study nonzero perturbations it is essential to have $A \neq 0$ and $C \neq 0$; the only way this can be achieved is by setting the determinant of equations (11a,b) equal to zero:

$$\det\begin{pmatrix} \sigma + \mu q^2 & -\chi\bar{a}q^2 \\ -f & \sigma + Dq^2 + k \end{pmatrix} = 0. \tag{12}$$

The resulting equation for σ is then

$$(\sigma + \mu q^2)(\sigma + Dq^2 + k) - \chi\bar{a}q^2 f = 0, \tag{13}$$

or, after rearranging terms, the quadratic equation in σ:

$$\sigma^2 + \beta\sigma + \gamma = 0, \tag{14a}$$

where

$$\beta = q^2(\mu + D) + k, \tag{14b}$$

$$\gamma = q^2[\mu(Dq^2 + k) - \chi\bar{a}f]. \tag{14c}$$

We can now address the question of whether the growth rate σ of the perturbations can ever have a positive real part (Re $\sigma > 0$). First note that the two possible roots of (14a) are

$$\sigma_{1,2} = \frac{-\beta \pm \sqrt{\beta^2 - 4\gamma}}{2}. \tag{15}$$

From (14b) we observe that β is always positive, so there is always one root whose real part is negative. In order for the second root to have a positive real part it is therefore necessary that the value of γ given by (14c) be *negative* [see problem (1e)]:

Condition for Aggregation of Cellular Slime Molds

$$\mu(Dq^2 + k) < \chi\bar{a}f. \tag{16}$$

(*Note:* This also implies that both σ_1 and σ_2 are real numbers.) Under this condition we may conclude that the homogeneous cell distribution is unstable to perturbations of the form (9) and that the uniform population will begin to aggregate.

11.3. INTERPRETING THE AGGREGATION CONDITION

We shall now pay closer attention to what is actually implied by inequality (16). To do so, first consider the effect of boundary conditions on the quantity q. Suppose the amoebae are confined to a region with length dimension L; that is, $0 < x < L$. This means that equations (2a,b) and therefore also equations (8a,b) are equipped with the boundary conditions

$$\frac{\partial c}{\partial x} = 0 \quad \text{at} \quad x = 0, x = L, \tag{17a}$$

$$\frac{\partial a}{\partial x} = 0 \quad \text{at} \quad x = 0, x = L. \tag{17b}$$

As we have seen in Section 9.8 such conditions can only be satisfied provided functions (9a,b) are chosen appropriately. In particular, it is essential that

$$\frac{\partial}{\partial x}(\cos qx) = 0 \quad \text{at} \quad x = L,$$

that is, $\sin qL = 0$. This is satisfied only for

$$q = \frac{n\pi}{L} \quad (n = 0, 1, 2, \ldots), \tag{18}$$

where the nonnegative integer n is called the *mode*. Thus inequality (16) implies that

$$\mu\left[\frac{D}{L^2}(n\pi)^2 + k\right] < \chi\bar{a}f. \tag{19}$$

To satisfy this it is necessary to have one or several of the following conditions met:

1. Values of μ, D, k, and/or n must be small.
2. Values of L must be large.
3. Values of χ, $\bar{a}$, and/or f must be large.

This means that the factors promoting the onset of aggregation (leading to instability) are as follows:

1. Low random motility of the cells and a low rate of degradation of cAMP.
2. Large chemotactic sensitivity, high secretion rate of cAMP, and a high density of amoebae, $\bar{a}$.

Indeed, it has been observed experimentally that the onset of *Dictyostelium* aggregation is accompanied by an increase in chemotactic sensitivity and cAMP production, so it appears that these changes indeed bring about the condition for instability, given by inequality (19), that leads to aggregation.

Several other factors apparently promote aggregation; of particular note is the prediction that large L and small n are favorable for instability. This gives the following somewhat surprising results:

1. Aggregation is favored more highly in larger domains than in smaller ones.
2. The perturbations most likely to be unstable are those with low wavelength. For example, $n = 1$ leads to the smallest possible LHS of inequality (19), all else being equal.

The perturbation whose wavelength is $q = \pi/L$ (that is, where $n = 1$) looks like the function shown in Figure 11.3(a). If this is the most unstable mode, we expect that the aggregation domain should first be roughly bisected, with amoebae moving away from one side of the dish and toward the other. Of course, if the parameter values gradually shift so that inequality (19) is also satisfied for $n = 2$, one can expect further subdivision of the domain into smaller *aggregation domains*. Biologists were at one time puzzled about the fact that the size of aggregation domains is not proportional to the density of amoebae $\overline{a}$. This result, too, is explained by inequality (19). (See problem 4.)

From this analysis one gains evidence for Segel's (1980) statement that even relatively simple interactions can have consequences that are not predictable based only on intuition or biological experience. With the hindsight that mathematical analysis gives, Keller and Segel (1970) were able to explain the counterintuitive aspects of the results as follows:

1. The forces of diffusion act most efficiently to smooth large gradients on small length scales. For this reason, in a large dish the long-range variation in cAMP concentrations and in cell densities has a greater chance of being self-reinforced before being obliterated by diffusion.
2. Similarly, the lower wavelengths present locally shallower gradients and are less prone to being effaced by diffusion and random motility.

Suggestions for Further Study or Independent Projects on Cellular Slime Molds

1. *The prespore-prestalk ratio*. See Bonner (1967, 1974) for surveys of older papers and a summary of the observed phenomena. For a more recent review of modeling efforts consult MacWillians and Bonner (1979) and Williams et al. (1981).
2. *cAMP activity and secretion in* D. discoideum. The substance cAMP can be secreted in an excitable or an oscillatory response as well as constant steady levels. In a series of models, Segel and Goldbetter outline the possible underlying molecular events. See Figure 8.16 and Segel (1984), Goldbetter and Segel (1980), and Devreotes and Steck (1979).
3. *Later developmental stages (shape of the aggregate)*. See Rubinow et al. (1981).
4. *Locomotion in the slug*. See Odell and Bonner (1981).

Figure 11.3 *Perturbations* a′ *or* c′ *of the homogeneous steady state,* $\bar{a}$ *(or* $\bar{c}$*). When a finite one-dimensional domain has imposed boundary conditions, there are restrictions on the wavenumber* q *of permissible perturbations. For example, no-flux boundaries imply that* $\partial(\cos qx)/\partial x = 0$ *at* $x = 0$ *and* $x = L$. *This can only be satisfied for* $q = n\pi/L$, *where* n *is an integer. Examples of the cases where* n = 1, 2, 3 *and* 4 *are shown here. The amplitudes have been exaggerated; we can only address the evolution of small-amplitude perturbations. Dotted lines represent a homogeneous steady state of equations (2a,b).*

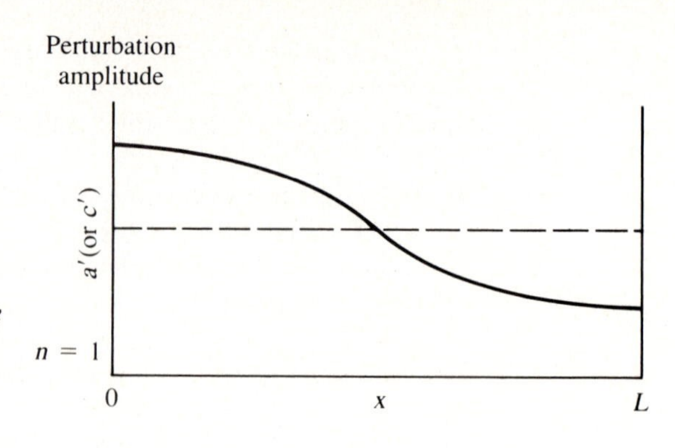

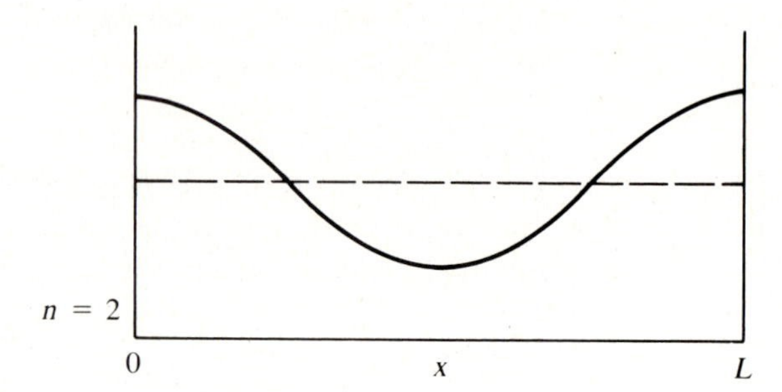

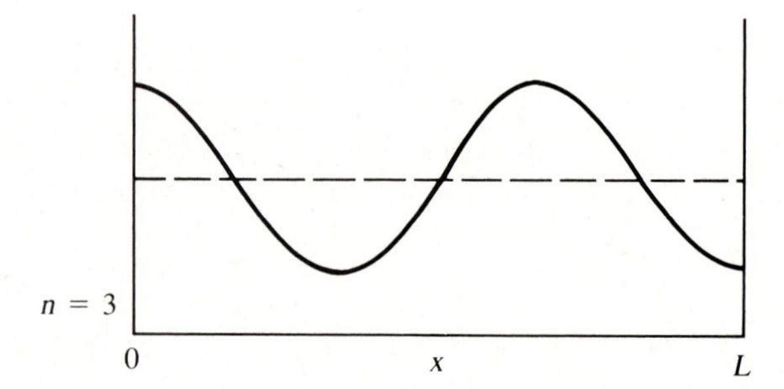

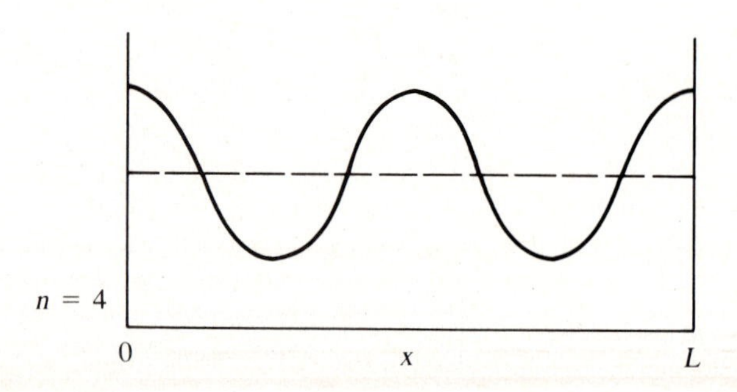

While it may be said that developmental biologists have to some extent resisted the intrusion of mathematics, the Keller-Segel model has become a classic in this field and is now frequently quoted in biological references. Many other aspects of the differentiation of cellular slime molds are equally, if not more, amazing than the aggregation phase. The topic continues to attract the most refined experimental and theoretical efforts.

11.4. A CHEMICAL BASIS FOR MORPHOGENESIS

Morphogenesis describes the development of shape, pattern, or form in an organism. The processes that underly morphogenesis are rather complex, spanning subcellular to multicellular levels within the individual. They have been studied empirically and theoretically for over a century. Among the first to devote considerable attention to the topic was D'Arcy Thompson, whose eclectic and imaginative book *On Growth and Form* could be considered as one of the first works on theoretical biology. Thompson emphasized the parallels between physical, inorganic, and geometric concepts and the shapes of a variety of plant and animal structures.

Some 35 years ago the theoretical study of morphogenesis received new impetus from a discovery made by a young British mathematician, Alan Turing. In a startling paper dated 1952, Turing exposed a previously little-known physical result that foretold the potential of diffusion to lead to "chemical morphogenesis."

From our daily experience most of us intuitively associate diffusion with a smoothing and homogenizing influence that eliminates chemical gradients and leads to uniform spatial distributions. It comes as some surprise, then, to be told that diffusion can have an opposite effect, engendering chemical gradients and fostering nonuniform chemical "patterns." Indeed, this is what the theory predicts. The following sections demonstrate the mathematical basis for this claim and outline applications of the idea to morphogenesis.

The key elements necessary for chemical pattern formation are:

1. Two or more chemical species.
2. Different rates of diffusion for the participants.
3. Chemical interactions (to be more fully specified shortly).

An appropriate combination of these factors can result in chemical patterns that arise as a destabilization of a uniform chemical distribution.

Turing recognized the implications of his result to biological morphogenesis. He suggested that during stages in the development of an organism, chemical constituents generate a *prepattern* that is later interpreted as a signal for cellular differentiation. Since his paper, chemical substances that play a role in cellular differentiation have been given the general label *morphogen*.

The steps given in this section are aimed at exposing the basic idea on which the Turing (1952) theory is built. To deal in specific terms we consider a set of *two* chemicals that diffuse and interact. For simplicity, a one-dimensional domain is considered, although it is later shown that the conclusions hold in higher dimensions as well. See Segel and Jackson (1972) for the source of this method.

Let

$$C_1 = C_1(x, t), \qquad C_2 = C_2(x, t),$$

be the concentrations of chemicals 1 and 2, and let

$$\begin{aligned} R_1(C_1, C_2) &= \text{rate of production of } C_1, \\ R_2(C_1, C_2) &= \text{rate of production of } C_2, \\ D_1 &= \text{diffusion coefficient of chemical 1}, \\ D_2 &= \text{diffusion coefficient of chemical 2}. \end{aligned}$$

The quantities R_1 and R_2 (*kinetic terms*) generally depend on concentrations of the participating molecules. By previous remarks we know that a set of equations for C_1 and C_2 consists of the following:

$$\frac{\partial C_1}{\partial t} = R_1(C_1, C_2) + D_1 \frac{\partial^2 C_1}{\partial x_2}, \tag{20a}$$

$$\frac{\partial C_2}{\partial t} = R_2(C_1, C_2) + D_2 \frac{\partial^2 C_2}{\partial x^2}. \tag{20b}$$

To underscore the role of diffusion in pattern formation we assume that in absence of its effects (when solutions are well mixed) the chemical system has some positive spatially uniform steady state, $(\overline{C}_1, \overline{C}_2)$. By definition this means that

$$\frac{\partial \overline{C}_i}{\partial t} = 0, \tag{21a}$$

$$(i = 1, 2).$$

$$\frac{\partial^2 \overline{C}_i}{\partial x^2} = 0. \tag{21b}$$

We therefore conclude that

$$R_1(\overline{C}_1, \overline{C}_2) = 0, \tag{22a}$$

$$R_2(\overline{C}_1, \overline{C}_2) = 0. \tag{22b}$$

As in Section 11.2 we now examine the effects of small *inhomogeneous perturbations of this steady state*. We are specifically interested in those perturbations that are amplified by the combined forces of reaction and diffusion in equations (1) and (2). When such perturbations are introduced, the uniform distributions $\overline{C}_1$ and $\overline{C}_2$ are upset.

Let

$$C_1'(x, t) = C_1(x, t) - \overline{C}_1, \tag{23a}$$

$$C_2'(x, t) = C_2(x, t) - \overline{C}_2, \tag{23b}$$

be small nonhomogeneous perturbations of the uniform steady state. Provided these are sufficiently small, it is again possible to linearize equations (20a,b) about the homogeneous steady state. Recall that this can be achieved by writing Taylor-series expansions for $R_1(\overline{C}_1, \overline{C}_2)$ and $R_2(\overline{C}_1, \overline{C}_2)$ about the values $\overline{C}_1$, and $\overline{C}_2$ and retaining the linear contributions (see problem 9). We then obtain the linearized version of equations (20a,b):

$$\frac{\partial C_1'}{\partial t} = a_{11}C_1' + a_{12}C_2' + D_1\frac{\partial^2 C_1'}{\partial x^2}, \tag{24a}$$

$$\frac{\partial C_2'}{\partial t} = a_{21}C_1' + a_{22}C_2' + D_2\frac{\partial^2 C_2'}{\partial x^2}, \tag{24b}$$

where

$$a_{11} = \left.\frac{\partial R_1}{\partial C_1}\right|_{\bar{C}_1,\bar{C}_2}, \qquad a_{12} = \left.\frac{\partial R_1}{\partial C_2}\right|_{\bar{C}_1,\bar{C}_2},$$
$$a_{21} = \left.\frac{\partial R_2}{\partial C_1}\right|_{\bar{C}_1,\bar{C}_2}, \qquad a_{22} = \left.\frac{\partial R_2}{\partial C_2}\right|_{\bar{C}_1,\bar{C}_2}, \tag{25}$$

and C_1' and C_2' are perturbations from $\bar{C}_1$ and $\bar{C}_2$. For convenience, these equations can be written in the shorthand matrix form:

$$\mathbf{C}_t' = \mathbf{MC}' + \mathbf{DC}_{xx}', \tag{26a}$$

where

$$\mathbf{C}' = \begin{pmatrix} C_1'(x,t) \\ C_2'(x,t) \end{pmatrix} = \begin{pmatrix} C_1(x,t) - \bar{C}_1 \\ C_2(x,t) - \bar{C}_2 \end{pmatrix} \tag{26b}$$

$$\mathbf{M} = \begin{pmatrix} a_{11} & a_{12} \\ a_{21} & a_{22} \end{pmatrix}, \tag{26c}$$

$$\mathbf{D} = \begin{pmatrix} D_1 & 0 \\ 0 & D_2 \end{pmatrix}. \tag{26d}$$

The resulting system of equations is linear. It can be solved by various methods, including separation of variables. One set of possible solutions is

$$\mathbf{C}' = \begin{pmatrix} C_1' \\ C_2' \end{pmatrix} = \begin{pmatrix} \alpha_1 \\ \alpha_2 \end{pmatrix} \cos qx\, e^{\sigma t}. \tag{27}$$

(See problem 9.) While this is a special form, considerations identical to those of Section 11.2 are again viewed as sufficient justification for restricting analysis to this set of solutions.

By substituting perturbations (27) into equations (24) or (26) we obtain

$$\alpha_1\sigma = a_{11}\alpha_1 + a_{12}\alpha_2 - D_1q^2\alpha_1,$$
$$\alpha_2\sigma = a_{21}\alpha_1 + a_{22}\alpha_2 - D_2q^2\alpha_2.$$

While the quantities α_1, α_2, q, and σ are *a priori* unknown to us, we are interested specifically in the situation in which small perturbations *grow with time*. Rewriting these as linear equations in α_1 and α_2, we obtain

$$\alpha_1(\sigma - a_{11} + D_1q^2) + \alpha_2(-a_{12}) = 0, \tag{28a}$$
$$\alpha_1(-a_{21}) + (\sigma - a_{22} + D_2q^2)\alpha_2 = 0, \tag{28b}$$

As in Section 11.2, we arrive at a set of algebraic equations in the perturbation amplitudes α_1 and α_2. Since the RHS is $(0,0)$, one solution is always $\alpha_1 = \alpha_2 = 0$.

This is dismissed as a trivial case; that is, perturbations are absent altogether for all t. But a nontrivial solution can only exist if the determinant of the coefficients appearing in equations (28a,b) is zero. It is thus essential that

$$\det\begin{pmatrix} \sigma - a_{11} + D_1q^2 & -a_{12} \\ -a_{21} & \sigma - a_{22} + D_2q^2 \end{pmatrix} = 0. \tag{29}$$

This leads to the following equation:

$$(\sigma - a_{11} + D_1q^2)(\sigma - a_{22} + D_2q^2) - a_{12}a_{21} = 0,$$

or

$$\sigma^2 + \sigma(-a_{22} + D_2q^2 - a_{11} + D_1q^2) \\ + [(a_{11} - D_1q^2)(a_{22} - D_2q^2) - a_{12}a_{21}] = 0. \tag{30}$$

Equation (30) is called the eigenvalue or the characteristic equation.

Our next goal is to determine whether for some q, the eigenvalue σ can have a positive real part, in other words, whether perturbations of particular "waviness" can cause instability to occur.

11.5. CONDITIONS FOR DIFFUSIVE INSTABILITY

As in previous analysis (of both ordinary and partial differential equations) the characteristic equation (30) will now be used to determine whether growing perturbations are possible. From equation (27) it is clear that the eigenvalue σ (the growth rate of the perturbations) should have a positive real part: Re $\sigma > 0$.

To concentrate solely on diffusion as a destabilizing influence we now incorporate the assumption that *in the absence of diffusion the reaction mixture is stable.* This implies that by setting $D_1 = D_2 = 0$ in equations (24) or in equation (30) one obtains only negative values of Re σ (consistent with stability). Eliminating D_1 and D_2 from equation (30) leads to

$$\sigma^2 - \sigma(a_{11} + a_{22}) + (a_{11}a_{22} - a_{12}a_{21}) = 0. \tag{31}$$

This quadratic equation is identical (save for the renaming of quantities) to the characteristic equation of the system of 2 ODEs obtained by omitting the diffusion terms in equations (20a,b). The conditions under which Re σ is negative are thus identical to stability conditions derived in Section 4.9:

Conditions for Stability of the Chemicals in the Absence of Diffusion

1. $a_{11} + a_{22} < 0,$ *(32a)*

2. $a_{11}a_{22} - a_{12}a_{21} > 0,$ *(32b)*

Strictly speaking, the second of these is required in the case where σ is real.

To appreciate how diffusion can act as a destabilizing influence, we consider analogous conditions obtained from equation (30) where $D_1, D_2 \neq 0$. By violating any *one* of these, a regime of instability would be created. The inequalities are as follows:

1. $(a_{11} + a_{22} - D_2q^2 - D_1q^2) < 0,$ *(33a)*

2. $[(a_{11} - D_1q^2)(a_{22} - D_2q^2) - a_{12}a_{21}] > 0.$ *(33b)*

Violation of either (1) or (2) leads to diffusive instability.

Because D_1 and D_2 and q^2 are positive quantities, it is clear that condition 33(a) always holds whenever 32(a) is true. The only other possibility then is that 33(b) may be reversed for certain parameter values. To study this more closely, we represent the LHS of 33(b) by H. We require that

$$H < 0. \tag{34}$$

By expanding the expression, H is found to be

$$H = D_1D_2(q^2)^2 - (D_1a_{22} + D_2a_{11})(q^2) + (a_{11}a_{22} - a_{12}a_{21}). \tag{35}$$

This expression seems complicated at first glance, but it can be understood in a straightforward way. Let us consider H as a function of q^2 and note that equation (35) is then simply a quadratic expression (with coefficients that depend on the a_{ij} and D_i).

Since the coefficient of $(q^2)^2$, D_1D_2, is positive, the graph of $H(q^2)$ is a parabola opening upwards (see Figure 11.4). The function thus has a minimum for some value of q^2. By simple calculus the value of this minimum can be determined. It is found to be

$$q^2_{\min} = \frac{1}{2}\left(\frac{a_{22}}{D_2} + \frac{a_{11}}{D_1}\right). \tag{36a}$$

(See problem 10.) Clearly, in order to satisfy (34) a minimal condition is that $H(q^2_{\min})$ should be negative:

$$H(q^2_{\min}) < 0, \tag{36b}$$

In problem 10 the reader is asked to evaluate $H(q^2_{\min})$ and thus demonstrate that (36) implies that

$$(a_{11}a_{22} - a_{12}a_{21}) - \frac{1}{4}\left(\frac{D_1a_{22} + D_2a_{11}}{D_1D_2}\right) < 0. \tag{37}$$

A more common way of writing this expression, after some rearranging and incorporating (32b), is as follows:

$$(a_{11}D_2 + a_{22}D_1) > 2\sqrt{D_1D_2}(a_{11}a_{22} - a_{12}a_{21})^{1/2} > 0. \tag{38}$$

(See problem 10.) When (38) is satisfied, $H(q^2_{\min})$ will be negative, so that for wavenumbers close to $q^2_{\min}$ the growth rate of perturbations σ can be positive. This in turn implies that diffusive instability to small perturbations of the form (27) will take place. Following is a summary of the key steps leading to this result.

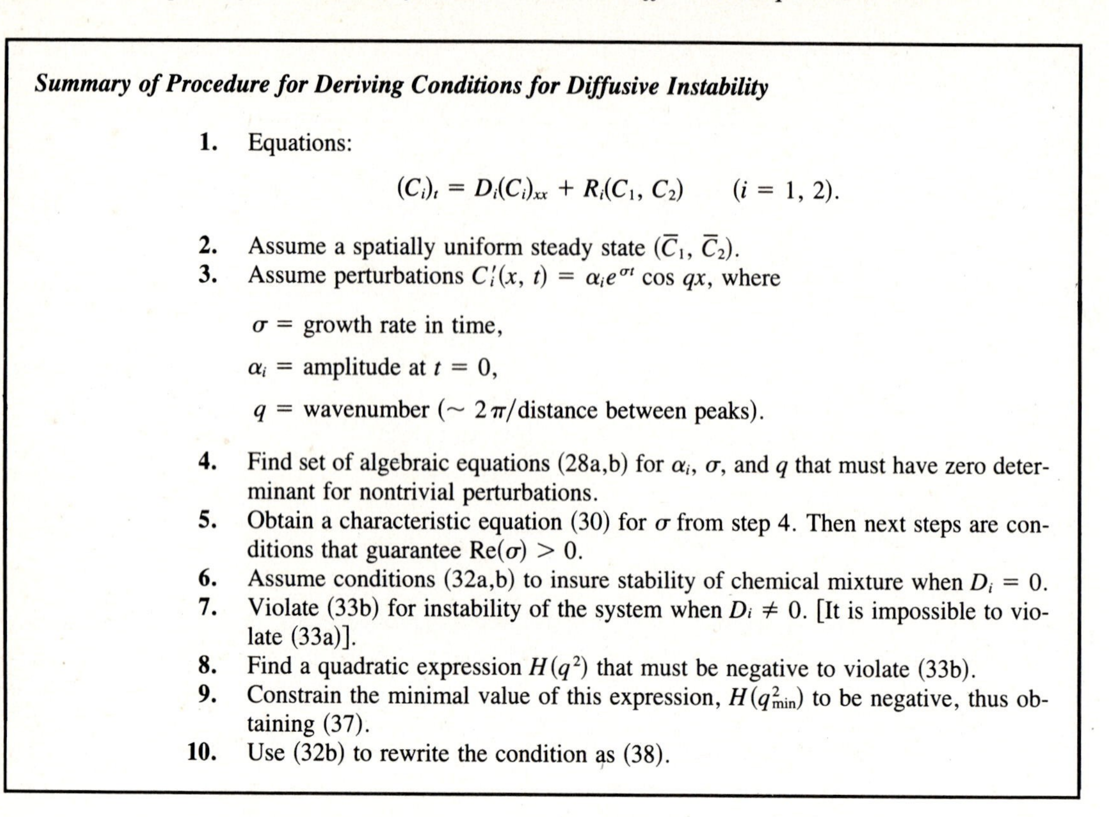

Summary of Procedure for Deriving Conditions for Diffusive Instability

1. Equations:

$$(C_i)_t = D_i(C_i)_{xx} + R_i(C_1, C_2) \qquad (i = 1, 2).$$

2. Assume a spatially uniform steady state $(\overline{C}_1, \overline{C}_2)$.
3. Assume perturbations $C_i'(x, t) = \alpha_i e^{\sigma t} \cos qx$, where

σ = growth rate in time,

α_i = amplitude at $t = 0$,

q = wavenumber ($\sim 2\pi$/distance between peaks).

4. Find set of algebraic equations (28a,b) for α_i, σ, and q that must have zero determinant for nontrivial perturbations.
5. Obtain a characteristic equation (30) for σ from step 4. Then next steps are conditions that guarantee $\text{Re}(\sigma) > 0$.
6. Assume conditions (32a,b) to insure stability of chemical mixture when $D_i = 0$.
7. Violate (33b) for instability of the system when $D_i \neq 0$. [It is impossible to violate (33a)].
8. Find a quadratic expression $H(q^2)$ that must be negative to violate (33b).
9. Constrain the minimal value of this expression, $H(q^2_{\text{min}})$ to be negative, thus obtaining (37).
10. Use (32b) to rewrite the condition as (38).

The three conditions for diffusive instability are then as follows:

Necessary and Sufficient Conditions for Diffusive Instabilty

1. $a_{11} + a_{22} < 0,$ *(32a)*
2. $a_{11}a_{22} - a_{12}a_{21} > 0,$ *(32b)*
3. $a_{11}D_2 + a_{22}D_1 > 2\sqrt{D_1D_2}\,(a_{11}a_{22} - a_{12}a_{21})^{1/2} > 0.$ *(38)*

It can be shown by simple rearrangement that the last of these may also be written in the following dimensionless form:

3. $\beta^{1/2} + \alpha\beta^{-1/2} > 2(\alpha - \epsilon)^{1/2} > 0$, where *(39)*

$$\beta = \frac{D_2}{D_1}, \qquad \alpha = \frac{a_{22}}{a_{11}} \quad (a_{11} > 0), \qquad \epsilon = \frac{a_{12}a_{21}}{a_{11}^2}.$$

(See problem 10.) We thus see that the condition for diffusive instability depends on dimensionless ratios of diffusion constants (and of kinetic terms) and not on any absolute magnitude.

In the next section we follow a physical interpretation of conditions (38) proposed by Segel and Jackson (1972).

Figure 11.4 *This sequence illustrates how the graph of* $H(q^2)$ *given by equation (35) might change as a parameter variation causes the onset of diffusive instability. (a)* H *is positive for all* q^2. *(b)* $H = 0$ *at* q^2_{min}, *which would then be the wavenumber of destabilizing perturbations. (c) A whole range of* q^2 *values makes* H *negative. Any one of these wavenumbers would lead to diffusive instability.*

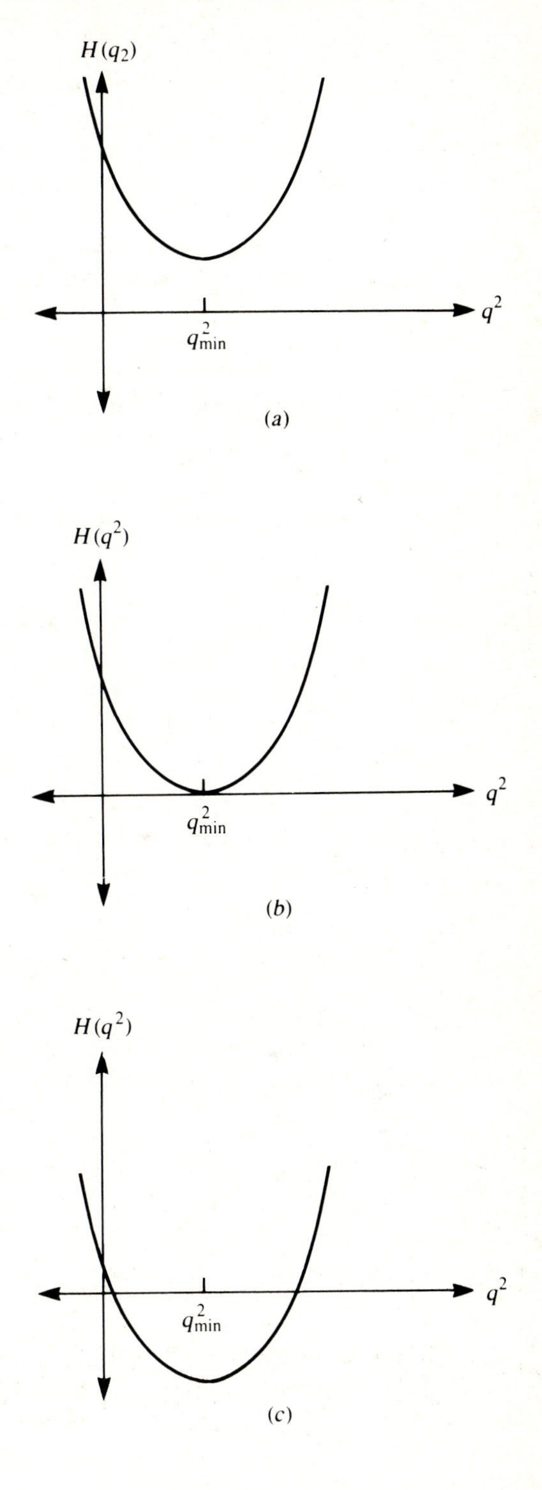

11.6 A PHYSICAL EXPLANATION

Turing's (1952) paper was followed a decade later by generalizations and extensions of the theory. The Brussels school, including G. Nicolis, I. Prigogine, and coworkers, based most of their ideas on mathematical and thermodynamic arguments, many of which would be inaccessible to most readers [see, for example, Nicolis and Prigogine, (1977)]. Segel and Jackson (1972) were apparently the first team to derive necessary and sufficient conditions for diffusive instability [equations (32a,b) and (38)] and then explain the meaning in an elegantly simple way. The logical progression of steps leading to their conclusions is reproduced here.

1. By condition (32a) at least one of the two coefficients, a_{11} or a_{22}, is negative. Suppose a_{22} is negative.

Interpretation: $\partial R_2/\partial c_2 < 0$; chemical 2 inhibits its own rate of formation. We shall call this substance an *inhibitor*.

2. From condition (38), $a_{11}D_2 + a_{22}D_1 > 0$, so clearly a_{11} and a_{22} cannot both be zero. Therefore a_{11} must be positive.

Interpretation: $\partial R_1/\partial c_1 > 0$; chemical 1 promotes or activates its own formation. This chemical species is called an *activator*.

3. Steps 1 and 2 together imply that

$$a_{11}a_{22} < 0. \tag{40}$$

4. From step 3 it follows that the inequality (32b) ($a_{11}a_{22} - a_{12}a_{21} > 0$) can only be met if

$$a_{12}a_{21} < 0. \tag{41}$$

This means that one of the two quantities, a_{12} or a_{21}, but not both, is negative. There are then two possibilities, each giving a distinct sign pattern to the Jacobian matrix **M** [equation (26c)]:

1. *Activator-inhibitor:*

$$a_{12} < 0, \qquad a_{21} > 0 \qquad \mathbf{M} = \begin{pmatrix} + & - \\ + & - \end{pmatrix}$$

2. *Positive feedback:*

$$a_{12} > 0, \qquad a_{21} < 0 \qquad \mathbf{M} = \begin{pmatrix} + & + \\ - & - \end{pmatrix}.$$

These are precisely the two cases that were explored in connection with pairs of chemically interacting substances in Section 7.8. Recall that these were called, partly for historic reasons, an activator-inhibitor and a positive-feedback system respectively.

By Segel and Jackson's reasoning, two chemical species that have the attribute of diffusive instability can only fall under one of these two classes. Moreover, in order for the chemical system to be stable when diffusion is absent (for example, in a well-stirred solution), a steady state of the (spatially) homogeneous system

$$\frac{dC_i}{dt} = R_i(C_1, C_2), \tag{42}$$

must be stable.

Necessary and sufficient geometric conditions for stability of such steady states were derived in Section 7.8. There we discovered that the nullclines of the present system must satisfy certain intersection properties (see Figure 7.10). If a given set of reactants has such qualitative geometry in the C_1C_2 phase plane, it makes a strong candidate for diffusive instability *provided that the condition given by (38) is also satisfied*. (Please note that axes in Figure 7.10(b) should be renamed in case 2 since the roles of the two chemicals have been permuted in the positive-feedback case.)

We continue to draw further conclusions about the two chemicals:

5. Dividing (38) through by D_2 leads to the following:

$$a_{11} + a_{22}\frac{D_1}{D_2} > 0. \tag{43}$$

This condition is met only if $D_1 \neq D_2$ because otherwise the inequality $a_{11} + a_{22}1 = 0$ contradicts (32a).

Interpretation: The diffusion coefficients of chemicals 1 and 2 must be dissimilar for diffusive instability to occur.

6. In both cases 1 and 2, signs of a_{11} and a_{22} are opposite.

Now define

$$\tau_1 = |a_{11}|, \tag{44a}$$

$$\tau_2 = |a_{22}|, \tag{44b}$$

where τ_1 and τ_2 are time constants associated with activation and inhibition. One can deduce from inequality (38) that

$$\frac{D_1}{\tau_1} > \frac{D_2}{\tau_2} \tag{45}$$

(see problem 11). Based on dimensional considerations and the discussion in Chapter 9, the ratios in (45) have units of area or, more precisely, mean square displacement during the doubling time of the activator or the half-life of the inhibitor (see problem 11). Such expressions are referred to as the *ranges* of activation or inhibition. They represent a physical area over which a peak concentration of chemical tends to exert its effect. Thus inequality (45) can be restated in words as follows:

The range of inhibition is larger than the range of activation.

Based on these conclusions, it can also be shown that $D_2 > D_1$ (problem 11). Together these observations give us the following picture of how diffusive instability is caused. Consider in particular the activator-inhibitor case. As a result of random perturbations, a small peak concentration of the activator is created at some location. This causes an enhanced local production of the inhibitor that, were it not for diffusion, would halt and reverse the process. However, the inhibitor diffuses away more rapidly than the activator ($D_2 > D_1$), so it cannot control the local activator production; thus the peak will grow). Now the region surrounding the initial peak will contain sufficient levels of inhibition to prevent further peaks of activation. This leads to the typical peak areas characterized by the expression D_2/τ_2 in inequality (45). (Case 2 is left as an exercise for the reader.)

The idea described in the previous paragraph is actually a special case of a much more general and ubiquitous pattern-forming mechanism called *lateral inhibition*. Briefly, positive reinforcement on a local scale together with a longer range of inhibition is universally applied in a variety of pattern-generating mechanisms, many of which are not driven by diffusion per se. Several examples for further study are given in the concluding sections of the chapter.

Can any other statements be made regarding the actual chemical patterns that are produced by these reaction-diffusion schemes? Here some care must be taken; our analysis will work *only* as long as the perturbations are sufficiently small to render the linear approximation of equations (24a,b) a valid representation of the truly nonlinear system of equations (20a,b). When the perturbations have been amplified beyond a small size, the analysis is no longer adequate. As in previous examples, *linear stability theory* applies only to states *close* to a steady state.

To extract even more information from our analysis we now consider how parameter variations bring about the onset of diffusive instability and what might be anticipated as a natural progression of events. It is well known that, as in the example of cellular slime molds, the process of development is often accompanied by gradual variations in characteristics of cells or tissues that could be depicted by key parameters. Such changes may stem from enhanced enzyme activity, increased affinity of reactions, or changes in cell–cell contacts (*gap junctions*) that permit intercellular communication.

It is often the case that dependence on cellular parameters enter into expressions such as (38) in nontrivial ways since the coefficients a_{ij} may depend in a complicated way on any given coefficient in the original nonlinear equations. Nevertheless, to take a simplified view, consider the following plausible (but not necessarily exclusive) progression brought about as a single parameter Γ varies through a *critical bifurcation value* Γ^*:

1. For $\Gamma < \Gamma^*$, diffusive instability is not possible. The parabola $H(q^2)$ is positive for all wavelengths q. No perturbations of any wavelength can lead to pattern formation.
2. For $\Gamma = \Gamma^*$, $H(q^2_{\text{min}}) = 0$. This is the threshold situation. Just beyond it lies the realm of diffusive instability; that is, for slightly larger values of Γ, H is negative but only for values of q^2 very close to q^2_{min}. Only perturbations whose wavenumber is q_{min} will be amplified and expressed in the final patterns.
3. For $\Gamma > \Gamma^*$, $H(q^2) < 0$ for a whole range of q values:

$$q^2_{\text{min}} - \Delta \leq q^2 \leq q^2_{\text{min}} + \Delta.$$

Any perturbation whose wavenumber falls within this range will be amplified.

Situation 2 is often called the *onset of diffusive instability*. At this critical point one would expect patterns with a typical spacing between chemical concentration peaks $\hat{d}$, where

$$\hat{d} = \frac{2\pi}{q_{\text{min}}}. \tag{46}$$

Summary of Two-Species Chemical Interactions Leading to Pattern Formation

Necessary conditions

1. The system must have a nontrivial spatially uniform steady state $\bar{S}$.
2. The pair must interact as an activator-inhibitor *or* a positive feedback system (in the sense that the sign pattern of the Jacobian of the system at $\bar{S}$ is the same as one of the cases given in this section).
3. (a) The steady state $\bar{S}$ should be *stable* in the well-mixed system. This is *equivalent* to the following:
   (b) The configuration of nullclines $R_1(C_1, C_2) = 0$ and $R_2(C_1, C_2) = 0$ at their nontrivial intersection should be of the type shown in Figures 7.10(*a* or *b*).
4. The rate of diffusion of the inhibitor (in case 1) or of the substance receiving negative feedback (in case 2) must be larger than that of the other activating substance.

Necessary and sufficient conditions

5. (a) Conditions (32a,b) and (38) should be satisfied, or
   (b) All necessary conditions 1 through 4 hold and, in addition,

$$\frac{a_{22}}{D_2} + \frac{a_{11}}{D_1} > 0.$$

This is equivalent to the statement that the *range of activation is smaller than the range of inhibition*.

Beyond the onset of diffusive instability the situation becomes somewhat more complicated. A whole range of possible wavenumbers can have a destabilizing effect. In general, perturbations might consist of many modes superimposed on one another, for example,

$$\begin{pmatrix} C_1' \\ C_2' \end{pmatrix} = \sum_{k=1}^{\infty} \begin{pmatrix} \alpha_{1k} \\ \alpha_{2k} \end{pmatrix} e^{\sigma_k t} \cos q_k x. \tag{47}$$

Once nonlinear effects become important, different modes may compete for dominance. Those that initially grow fastest may gain an advantage over others. It is possible to solve equation (30) for the growth rate σ_k associated with a given wavenumber q_k and thus determine the maximum possible growth rate. See Segel (1984) and problem 12. Beyond this, little more can be said analytically about the final patterns that diffusive instability creates. Further understanding requires considerably more elaborate *nonlinear* analysis and is beyond our scope.

An alternative to further analytical techniques is numerical simulation. In Section 11.8 we describe several references that present results based solely on computer-aided numerical solutions.

11.7 EXTENSION TO HIGHER DIMENSIONS AND FINITE DOMAINS

The analysis of reaction-diffusion systems has thus far been restricted to infinite one-dimensional domains. We now extend it to higher dimensions and indicate what new features arise when the system is confined to a bounded domain, for example, a rectangle in the plane.

Consider the following general reaction-diffusion system:

$$\frac{\partial C_1}{\partial t} = R_1(C_1, C_2) + D_1 \nabla^2 C_1, \tag{48a}$$

$$\frac{\partial C_2}{\partial t} = R_2(C_1, C_2) + D_2 \nabla^2 C_2, \tag{48b}$$

with homogeneous steady state $(\overline{C}_1, \overline{C}_2)$.

It is a straightforward matter to carry out the linearization of (48). Results are similar to (24) where $\nabla^2 C_i$ replaces $(\partial^2 C_i / \partial x^2)$. The form of perturbations to be used in testing stability is slightly different since variation in both spatial dimensions may occur. In problem 14 the reader is asked to verify the following set of possible solutions to a *two*-dimensional linear system:

$$C_i'(x, y; t) = \alpha_i e^{\sigma t} \cos q_1 x \cos q_2 y \tag{49}$$

where q_1 and q_2 are the wavenumbers for variations in the x and y directions respectively.

Now consider a finite rectangular domain of size $L_x \times L_y$. By way of example, suppose that the boundaries are impermeable (that is, no flux boundary conditions apply). Then

$$\frac{\partial C_i'}{\partial x} = 0 \qquad \text{at} \qquad x = 0, L_x, \tag{50a}$$

$$\frac{\partial C_i'}{\partial y} = 0 \qquad \text{at} \qquad y = 0, L_y,. \tag{50b}$$

To satisfy these additional constraints, q_i can only take on one of a discrete set of values:

$$q_1 = \frac{m\pi}{L_x} \qquad (m = 0, 1, \ldots), \tag{51a}$$

$$q_2 = \frac{n\pi}{L_y} \qquad (n = 0, 1, \ldots). \tag{51b}$$

(See problem 19.)
Define

$$Q^2 = q_1^2 + q_2^2. \tag{52}$$

Then it may be shown that one obtains a set of algebraic equations for the perturbation amplitudes α_i identical to (28) where Q^2 replaces q^2. This leads to an identical condition for diffusive instability after a procedure analogous to that of Section 11.5 (The dimensionality and geometry of the region have no influence on the stability condition.) However, one finds that the most destabilizing perturbations are those for which

$$Q^2 = Q_{\min}^2 = \frac{1}{2}\left(\frac{a_{22}}{D_2} + \frac{a_{11}}{D_1}\right). \tag{53}$$

Thus, at the onset of diffusive instability the waves that are amplified are those that satisfy

$$Q^2 = \pi^2\left(\frac{m^2}{L_x^2} + \frac{n^2}{L_y^2}\right) = \frac{1}{2}\left(\frac{a_{11}}{D_1} + \frac{a_{22}}{D_2}\right). \tag{54}$$

As we shall soon see, this implies that the wavenumbers q_1 and q_2 are interdependent. For a given *value* of the RHS of equation (53), increasing q_1 must mean decreasing q_2. To examine the effects of geometry and chemistry on the most excitable waves, it proves convenient to use the following dimensionless quantities:

$$\gamma = \frac{L_y}{L_x}, \qquad \beta = \frac{D_2}{D_1}, \qquad \alpha = \frac{a_{22}}{a_{11}}. \tag{55}$$

Then equation (54) can be rewritten in the following dimensionless form:

$$m^2 + \frac{n^2}{\gamma^2} = \frac{L^2 a}{2\pi^2 D}\left(1 + \frac{\alpha}{\beta}\right) \equiv E^2, \tag{56}$$

where we have written $L = L_x$, $a = a_{11}$, and $D = D_1$ for a more convenient notation. Ratios appearing in (56) bear important physical interpretations:

$$L^2 \simeq \text{area characteristic of the domain } (\equiv \text{area of domain if } \gamma = 1),$$

$$\frac{D}{a} \simeq \text{area of range of inhibitor,}$$

$$\frac{L^2 a}{D} \simeq \text{ratio of area characterizing domain to range of inhibition,}$$

$$\frac{\alpha}{\beta} = \frac{a_{22}/a_{11}}{D_2/D_1} = \frac{D_1/a_{11}}{D_2/a_{22}}$$

$$= \text{ratio of range of inhibition to range of activation.}$$

Thus the quantity E^2, defined by the RHS of equation (56), stands for

$$E^2 = \frac{\text{area of domain}}{\text{range of inhibition}}\left(1 + \frac{\text{range of inhibition}}{\text{range of activation}}\right).$$

Note that, by (54), $E^2 = Q^2 L_x^2/\pi^2$.

In the following example of parameter variations it will be assumed that α and β [and ϵ appearing in (39)] are held fixed so that condition (39) is just satisfied. Thus we consider a system at the *onset* of diffusive instability when *only* the modes satisfying (54) or alternatively (56) lead to the formation of patterns. From equation (56) it can be seen that the value of E^2 can be changed even though α and β might be held constant. For example, this can be done by

1. Increasing the diffusion rates D_1 and D_2 of both chemicals but keeping $\beta = D_2/D_1$ fixed.
2. Altering all the chemical reaction rates a_{ij} while keeping $\alpha = a_{22}/a_{11}$ and $\epsilon = a_{12}a_{21}/a_{11}^2$ fixed.
3. Increasing or decreasing the domain size L.

The effect of any such parameter variation is that the relative ratios of the domain size and the chemical range size will change. As such changes occur, the value of E^2 changes. One then expects abrupt transitions in the values of m and n that satisfy (56).

In the following example we take $\gamma = 2$ for illustrative purposes. We compute the expression

$$E^2 = m^2 + \frac{n^2}{\gamma^2} \qquad \begin{cases} m = 0, 1, 2, \ldots . \\ n = 0, 1, 2, \ldots , \end{cases}$$

and then rank the pairs of integers (m, n) in order of increasing values of E^2 (see Figure 11.5). The succession of modes thus generated might arise from any one of the above parameter variations (as long as E^2 is thereby made to vary gradually and monotonically). A different value of γ would lead to a somewhat different sequence of modes. Some of the chemical patterns corresponding to such modes are then displayed schematically in Figure 11.6.

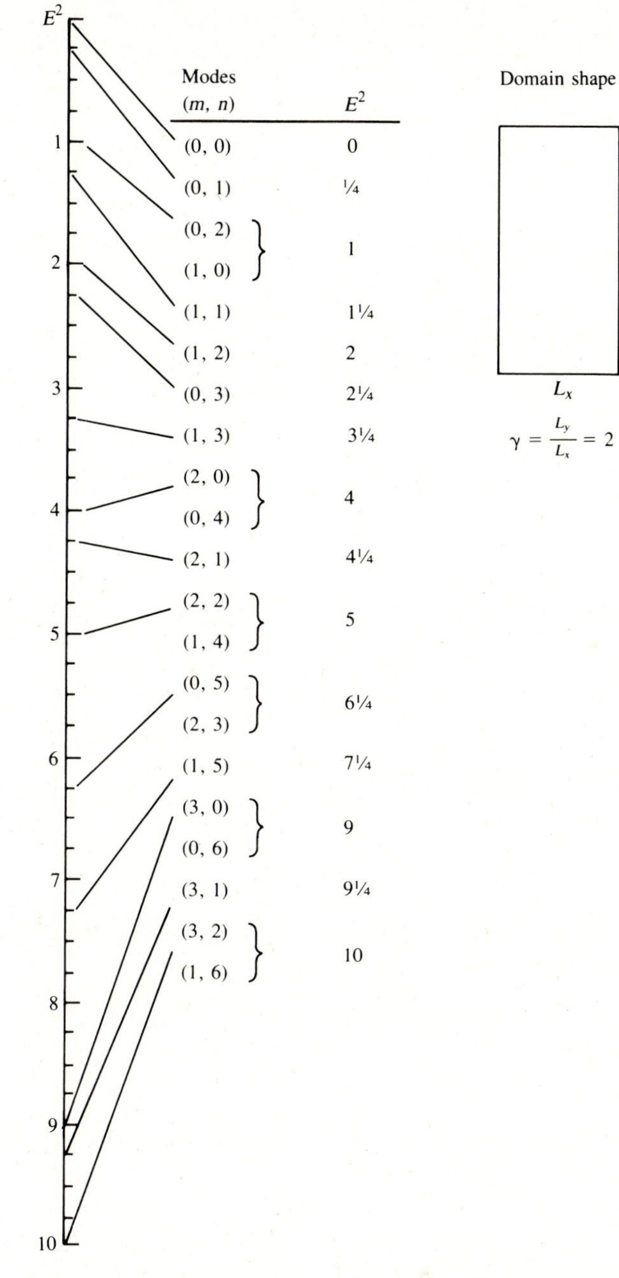

Figure 11.5 *In a two-dimensional region the most excitable modes* (m, n) *of a given reaction-diffusion system would depend on the interplay between geometric and chemical aspects. For a given ratio of sides* ($\gamma = L_y/L_x$) *in a rectangular domain, the number of wave peaks in the* x *and* y *directions* (m, n) *changes as the ratio of domain size to range of inhibitor changes, that is, as* E^2 *is altered. Here we have taken* $\gamma = 2$ *by way of example.*

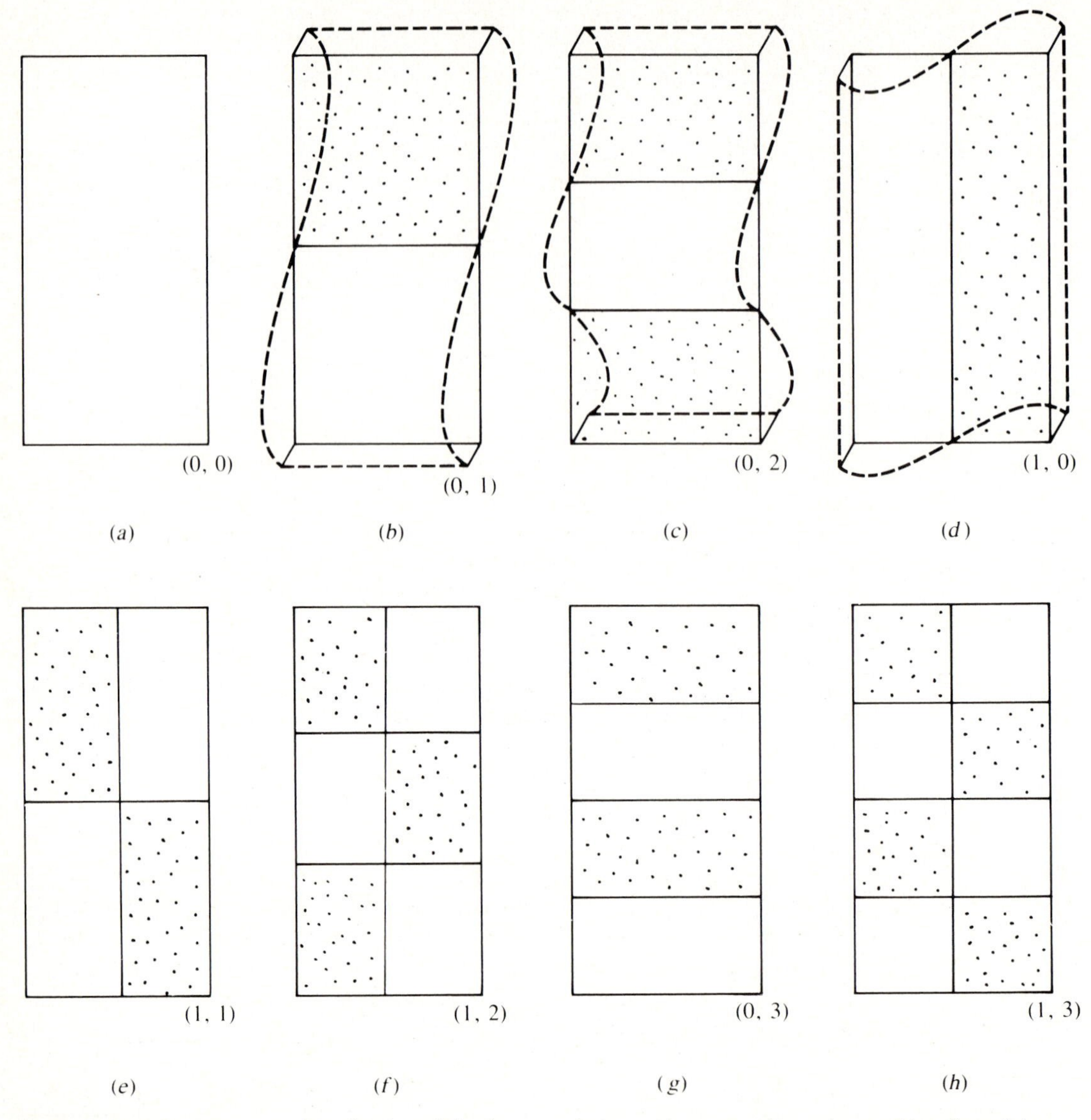

Figure 11.6 *The first several modes* (m, n) *in the sequence predicted in Figure 11.5.* (b ≃ d) *The functional shapes of the excitable perturbations are shown by dotted outlines. Stippled areas are those in which chemical concentrations are higher than their steady-state values. A transition from (a) to (h) (and beyond) would occur if the size of the domain was increased or the size of the chemical range was decreased.*

From such results several observations can be made:

1. The transitions from one mode to the next do not necessarily occur at equal increments of the parameter E^2; some transitions are closer than others.
2. There are certain values of the parameter E^2 that correspond to more than one possible mode. Which mode is actually expressed would then depend on initial

conditions and nonlinear interactions that might tend to stabilize one pattern at the expense of a second.

3. When the domain is wider in one direction (such as in the example in Figure 11.5 $L_y = 2L_x$), there is a tendency for successive subdivisions to occur in the longer direction *before* they occur in the shorter one.

Such results bear several interesting implications in developmental systems. The first of these is that the interactions of diffusing morphogens in a *growing domain* will lead to a discrete succession of patterns. Kauffman et. al. (1978) have proposed an intriguing application of such ideas to the sequence of compartmental subdivision in the imaginal disks of *Drosophila* (see box "Chemical Patterns and Compartments in *Drosophila*"). Murray (1981) has similarly applied these concepts to the spatial succession of patterns that occur on animal coats as the geometry of the domain changes (for example, from an extremity to a broader region such as the body). His model is discussed in greater detail in section 11.8.

Similar ideas with constant domain size and other spatially varying parameters have been used in a number of different contexts. Berding et al. (1983) have shown, for example, that spiral patterns typical of those found on sunflower heads can be generated by a reaction-diffusion system in a circular domain with diffusivities that vary as r^2, where r is the radial distance. Other implications for pattern formation systems are suggested in the problems as topics for independent exploration.

To keep the limitations of this theory in perspective, one must remember that we have been operating under the admittedly artificial assumption that the patterns develop at or close to the critical bifurcation (that is, at the onset of diffusive instability). When a whole *range* of Q^2 values are excitable, one must again address the possible nonlinear effects that lead to considerably more complicated problems.

Although linear analysis of two- or three-dimensional cases is an immediate generalization of the one-dimensional case, there are genuinely novel effects that arise beyond the predictive power of these theories. For large-amplitude perturbations, nonlinear interactions of the waves in the x and y directions play an increasingly important role. The final patterns may be different from those anticipated solely on the basis of the linear approximations.

We summarize some of our findings in the following statements:

1. Pattern formation can only occur if the *range of inhibition* is greater than the *range of activation:*

$$\left|\frac{D_1}{a_{11}}\right| > \left|\frac{D_2}{a_{22}}\right|.$$

2. The type of patterns that then occur depends on two factors:

a. $\gamma = \dfrac{L_y}{L_x}$ = the geometry of the domain,

b. $\dfrac{L^2}{D/a}$ = the ratio of the size of the domain to the range of the inhibitor.

3. Parameter variations can lead to changes in the factors described in 2 without affecting the inequality given in 1 or vice versa.

***Chemical Patterns and Compartments in* Drosophila**

Drosophila melanogaster, the common fruitfly, has a number of distinct life stages including egg, larva, pupa, and adult. In the transition from egg to larva, certain groups of cells are reserved in the structures known as *imaginal discs*. Specific parts of these objects will eventually undergo growth and differentiation to produce adult structures. In the larval tissues there are pairs of imaginal discs for eyes, legs, wings, and other body parts. Moreover, each of these may be subdivided into groups of cells destined for specific parts of the final structure. For example, the wing disc has compartments corresponding to subdivisions of a wing including (1) anterior-posterior wing parts, (2) dorsal-ventral wing parts, (3) wing-thorax wing parts, and others.

Based on experimental evidence, it is held that commitment to a given fate is attained as a result of a sequence of stages, each of which increases the restrictions on a given group of cells. It is possible to map cells (according to their positions on the wing disc) with their eventual destinations. (See compartments in Figure 11.7.) Moreover, the sequence of subdivision of the wing discs can also be ascertained (successive boundaries labeled 1 to 5 in the figure). Such compartmentalization appears spontaneously as the imaginal disc grows.

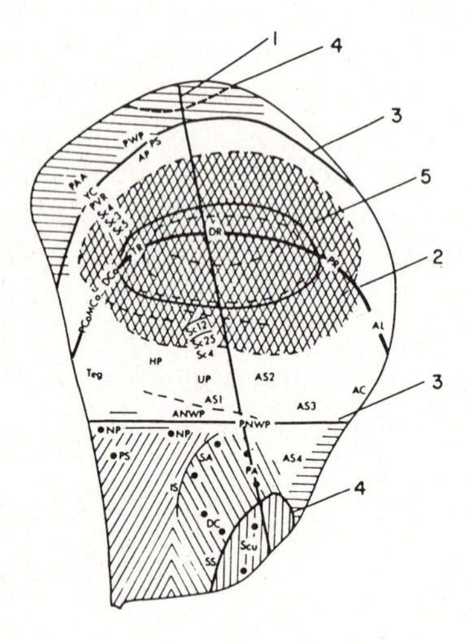

Figure 11.7. *Actual shape and compartmental boundaries of imaginal disc in* Drosophilia. *[From Figure 5 in Kauffman (1977), American Zoologist 17:631–648.]*

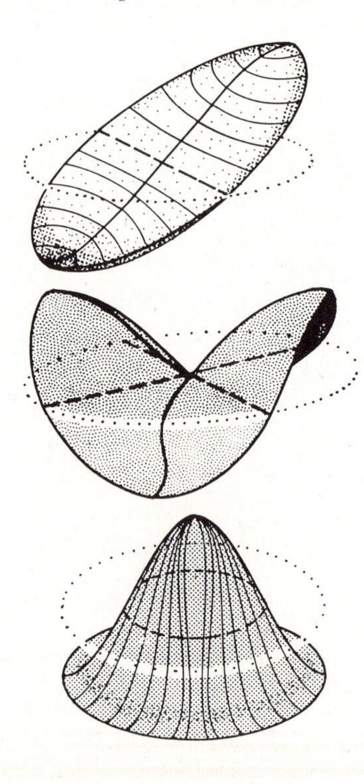

Figure 11.8. *Wave patterns on a disk-shaped domain. The peaks may correspond to places where certain chemical concentrations are high. [From Figure 2 in Kauffman (1977), American Zoologist 17:631–648.]*

On the basis of such information, Kauffman et. al. (1978) has proposed the following model of successive compartmentalization. Identify the imaginal disc of a wing with an approximate shape of an elliptical region. Consider a hypothetical reaction-diffusion system on this elliptical domain. As the domain grows in size, a discrete sequence of chemical wave patterns is created. Figure 11.8 demonstrates typical wave patterns on circular or elliptical domains. Note a basic similarity to Figure 11.6. The nodal lines of successive patterns that fit on an ellipse as it enlarges are shown in the sequence in Figure 11.9. Projecting all five such predicted boundaries onto a single ellipse results in the compartmentalization shown in Figure 11.10.

This type of sequential process of compartmentalization could, then, arise *spontaneously* by virtue of the fact that an increasing domain size [analogous to increasing L^2 in the expression on the RHS of (56)] causes a succession of modes (m, n) to appear in the chemical patterns that are expressed.

In this example the geometry of the domain is elliptical; however, the basic idea of a succession of patterns (or rather compartmental subdivisions) resulting from a gradual shift in a key parameter (in this case the size) is very much the same as that in a rectangular domain. The actual nodal lines are computed by solving the diffusion equation on a circular disk.

While there is as yet no direct evidence that a reaction-diffusion system underlies the differentiation of *Drosophila* imaginal discs, the time sequence and geometry of compartmentalization proves quite suggestive of some underlying wave phenomenon.

Figure 11.9. *Successive nodal lines on a growing ellipse. [From Figure 3 in Kauffman (1977), American Zoologist 17:631–648.]*

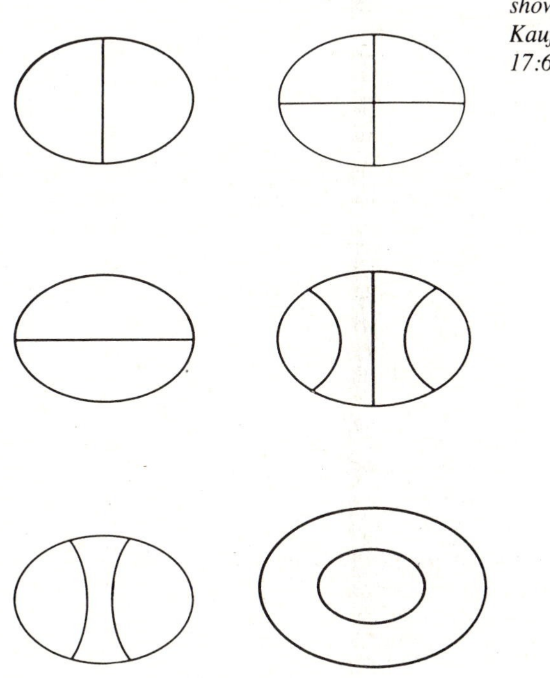

Figure 11.10. *Successive boundaries formed in the idealized elliptical domain. Compare with the actual compartmental boundaries shown in Figure 11.7. [From Figure 6 in Kauffman (1977), American Zoologist 17:631–648.]*

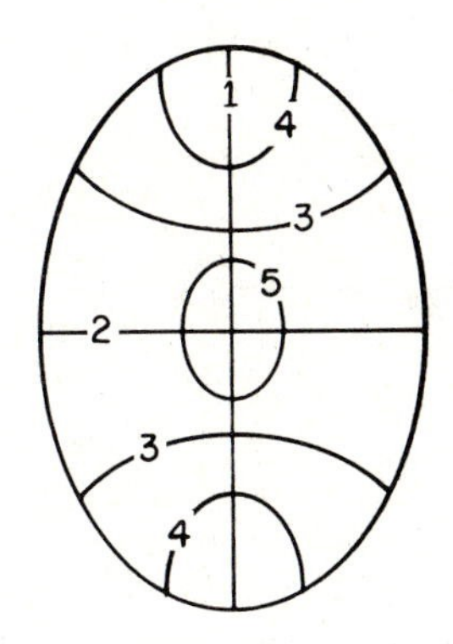

11.8. APPLICATIONS TO MORPHOGENESIS[1]

Since Turing's (1952) paper, two parallel directions have arisen in the research on reaction-diffusion systems. A highly abstract mathematical development characterizes one branch (see, for example, Smoller, 1983 and Rothe, 1984); a more biologically oriented approach characterizes the other. The abstract theories have addressed numerous questions, including the existence of nonstationary (travelling-wave) patterns, spirals, solitary peaks, and fronts. See Fife (1979) for a summary. The more biological directions have formed the connection between specific models and developmental patterns found in biological systems.

The interest of biologists was largely aroused due to contributions by Meinhardt and Gierer in the 1970s. Their work consists predominantly of numerical simulations of reaction-diffusion systems in various geometries. The results often bear a realistic likeness to patterns commonly found in nature. (See the survey in Meinhardt, 1982, and Figures 11.11 to 11.13.)

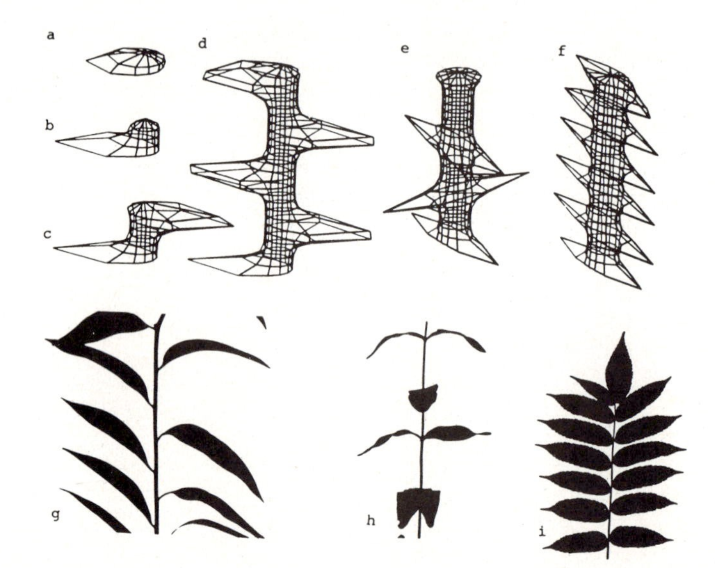

Figure 11.11. *Phyllotaxis, the regular arrangement of leaves on the stem of a plant, was modeled by Meinhardt using an activator-inhibitor mechanism with reaction-diffusion occurring on the surface of the cylindrical stem. (a–f) Locations of activator peaks (obtained by computer simulation); (g–i) Comparable leaf arrangements. [From Meinhardt, H. (1978). Models for the ontogenetic development of higher organisms. Rev. Physiol. Biochem. Pharmacol., 80, 47–104, fig. 5. Reprinted by permission of Springer Verlag.]*

Biology students will find the Meinhardt-Gierer papers readily accessible since very little mathematical analysis is given. Indeed, in the initial stages of their research these authors were unaware of the Turing theory and based their ideas entirely on a familiarity with lateral inhibition in other contexts. They have since de-

1. Part of the material in this section is based on a review by Marjorie Buff.

veloped considerable insight into the combinations of geometry, chemical kinetics, and other effects that lead to interesting patterns. (Not always is their insight imparted to readers.) We now discuss two examples of models proposed by Gierer and Meinhardt.

Consider their first set of equations:

$$\frac{\partial A}{\partial t} = \frac{\rho A^2}{(1 + \kappa A^2)H} - \mu A + D_A \nabla^2 A + \rho_0 \rho, \tag{57a}$$

$$\frac{\partial H}{\partial t} = \rho' A^2 - \nu H + D_H \nabla^2 H + \rho_1. \tag{57b}$$

In these equations A and H are the concentrations of activator and inhibitor respectively. The parameters ρ_0, ρ, μ, ρ', ρ_1, ν, κ, D_A, and D_H reflect source terms, decay rates, reaction rates, and diffusional properties.

One interpretation of terms in equations (57a,b) is as follows. In (57a) the activator A decays spontaneously with rate μ, diffuses at the rate $D_A \nabla^2 A$, and is created in two ways: by a small source term $\rho_0 \rho$, which could vary over space, and by an

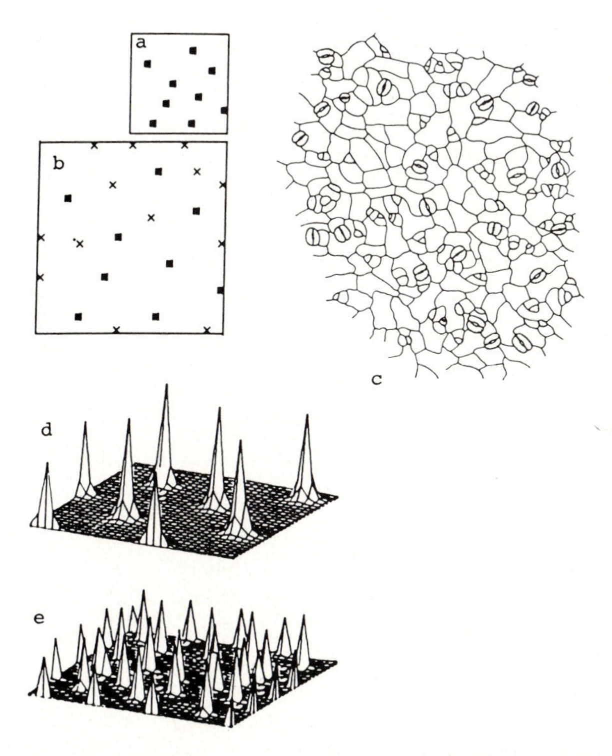

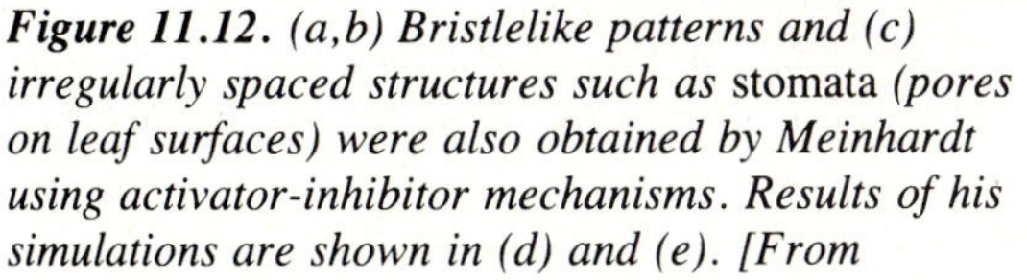

Figure 11.12. *(a,b) Bristlelike patterns and (c) irregularly spaced structures such as* stomata *(pores on leaf surfaces) were also obtained by Meinhardt using activator-inhibitor mechanisms. Results of his simulations are shown in (d) and (e). [From Meinhardt, H. (1978). Models for the ontogenic development of higher organisms. Rev. Physiol. Biochem. Pharmacol., 80, 47–104, fig. 6. Reprinted by permission of Springer Verlag.]*

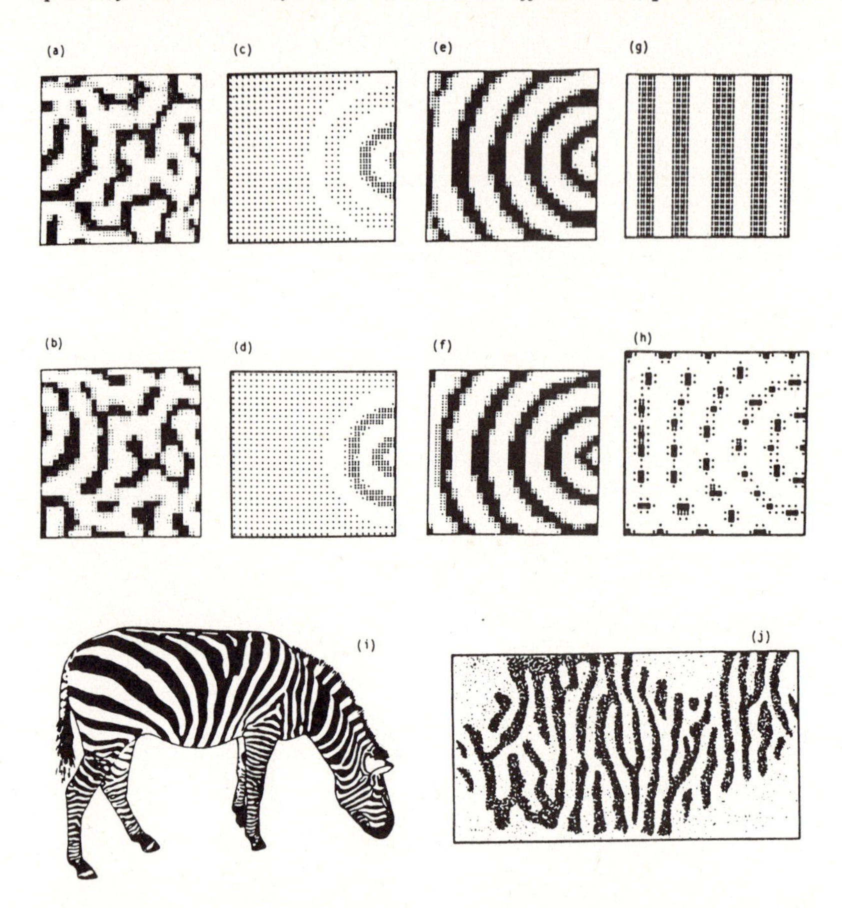

Figure 11.13. *(a–h) stripes and other patterns that can be produced by reaction-diffusion mechanisms in a planar domain (under a variety of initial conditions and chemical interactions). Examples of naturally occurring stripes (i) in zebras and (j) in the visual cortex. [From Meinhardt, H. (1982).* Models of Biological Pattern Formation. *Academic Press, New York, fig. 12.2.]*

autocatalytic reaction represented by the first term of (57a). The sigmoidal kinetic term $A^2/(1 + \kappa A^2)$ depicts a cooperativity effect (an enhanced reaction rate when two molecules of A are present). (See Chapter 7 for the derivation of such terms.) The presence of H in the denominator indicates that the inhibitor would tend to decrease the rate of production of the activator. The effect of these two terms is to limit the maximal activator production rate so that an activator peak will not grow indefinitely.

In equation (57b) the inhibitor H is produced as a result of the cross-catalytic influence of the activator A^2. Hence A indirectly results in its own inhibition. H also decays spontaneously at the rate ν, diffuses ($D_H \nabla^2 H$), and has a small activator-independent source term ($\rho^1 A^2$) that prevents pattern formation at low concentrations of the activator. To produce patterns with particular polarity, Gierer and Meinhardt have often biased the source distribution by assuming a convenient spatial dependence. There has been some controversy regarding the validity of this approach.

A second simpler set of equations they have used is given in the box and analyzed as an example illustrating the techniques developed in this chapter. (An interpretation is left as a problem for the reader.)

Example

Gierer and Meinhardt applied the following set of equations to a pattern-forming process.[2] Define

$a(x, y; t)$ = concentration of activator at location (x, y) and time t,
$h(x, y; t)$ = concentration of inhibitor at location (x, y) and time t.

Then

$$\frac{\partial a}{\partial t} = \frac{c_1 a^2}{h} - a\mu + D_a \nabla^2 a, \tag{58a}$$

$$\frac{\partial h}{\partial t} = c_2 a^2 - \nu h + D_h \nabla^2 h. \tag{58b}$$

A homogeneous steady state $(\bar{a}, \bar{h})$ *satisfies*

$$\left.\begin{aligned} \frac{c_1 \bar{a}^2}{\bar{h}} - \bar{a}\mu = 0 \\ c_2 \bar{a}^2 - \nu \bar{h} = 0 \end{aligned}\right\} \Rightarrow \begin{aligned} \bar{a} &= \frac{c_1 \nu}{c_2 \mu}, & (59a) \\ \bar{h} &= \frac{c_1{}^2 \nu}{c_2 \mu^2}. & (59b) \end{aligned}$$

The Jacobian for this system is

$$\mathbf{J} = \begin{pmatrix} \dfrac{2c_1 a}{h} - \mu & \dfrac{-c_1 a^2}{h^2} \\ 2c_2 a & -\nu \end{pmatrix}_{ss} = \begin{pmatrix} \mu & \dfrac{-\mu^2}{c_1} \\ \dfrac{2c_1 \nu}{\mu} & -\nu \end{pmatrix}, \tag{60}$$

so

$$\det \mathbf{J} = \mu\nu(-1 + 2) = \mu\nu. \tag{61}$$

The condition for diffusive instability is

$$\mu D_a - \nu D_h > 2(\mu\nu)^{1/2}(D_a D_h)^{1/2}, \tag{62a}$$

2. In the original paper $c_1 = c_2 = c_j$; this leads to inconsistency in the dimensions and so has been modified here.

or

$$\epsilon\delta - \frac{1}{\epsilon\delta} > 2, \tag{62b}$$

where

$$\epsilon = \left(\frac{\mu}{\nu}\right)^{1/2}, \qquad \delta = \left(\frac{D_h}{D_a}\right)^{1/2}. \tag{62c}$$

Problem 17 shows that (62b) implies that $\epsilon\delta$, which must be a positive quantity, satisfies

$$\epsilon\delta > 1 + \sqrt{2}. \tag{63}$$

At the onset of instability, the most excitable modes are characterized by

$$Q^2 = q_x^2 + q_y^2 = \frac{1}{2}\left(\frac{\mu}{D_a} - \frac{\nu}{D_h}\right). \tag{64}$$

Note that the instability condition depends only on dimensionless ratios such as ϵ and δ, whereas Q^2 carries dimensions of $(1/\text{distance})^2$ and thus depends on absolute magnitudes of the diffusion coefficients.

Figures 11.11 through 11.13 are a sample of some of the elegant patterns produced over the years by Meinhardt and Gierer. Unfortunately, rarely do they specify the exact conditions and parameter values used in their simulations. It has been shown (see Murray, 1982) that the sets of parameter values leading to pattern formation in these models are unrealistically restrictive.

Bridging the gap between the totally abstract and the predominantly biological literature are several partly theoretical papers whose main concern is *explaining* properties of patterns on the basis of chemical interactions and geometric considerations. Among these are several classic contributions by Murray, including his model for animal coat patterns briefly highlighted here.

Murray (1981*a,b*) describes a hypothetical mechanism for *melanogenesis* (synthesis and deposition of *melanin* granules, which are responsible for the dark pigmentation of mammalian skin and fur). It is assumed that reactions and diffusion occur on the plane of an active membrane and that the kinetics stem from a substrate-inhibited enzymatic reaction between two substances, S and A whose concentrations we represent by s and a (see problem 21).

In dimensionless form the model equations are as follows:

$$\frac{\partial s}{\partial t} = \gamma g(s, a) + \nabla^2 s, \tag{65a}$$

$$\frac{\partial a}{\partial t} = \gamma f(s, a) + \beta\nabla^2 a, \tag{65b}$$

where

$$\beta = D_a/D_s = \text{ratio of diffusion coefficients,}$$

$$\gamma = \text{dimensionless parameter proportional to the area of the domain,}$$

$$g, f = \text{the functions given in Figure 8.14.}$$

These phase-plane drawings reveal a cubic nullcline configuration, with an intersection that depends on the size of a parameter K, which represents the degree of substrate inhibition. Decreasing K results in the transition of (*a*) to (*b*) to (*c*) in Figure 8.14. While these equations do not permit easy analysis since their steady state cannot be explicitly obtained, it transpires that only in one configuration, namely the one shown in Figure 8.14(*b*), is diffusive instability possible. (See problem 21.) This leads to a suggestive but entirely hypothetical mechanism for the development of the patterns.

Suppose that there is a substance that inhibits melanogenesis whose concentration early in embryonic development is high. Suppose that, at later stages, a decrease in the level of this inhibitor is brought about. At that point, diffusive instability leading to spatial patterns will occur, resulting in the prepattern for coat markings. If, however, there is further decrease in inhibition, diffusive instability will no longer be possible, and no further development of pattern can take place.

A second prediction concerns the interchange between size or geometry and the nature of the patterns. Variations in the area of the domain can be conveniently depicted by variation in the parameter γ (see Figure 11.14). Murray demonstrates that as γ increases, there is a succesion of excitable modes, as we discovered in Section 11.7. If one of the dimensions (for example, L_x, length in x direction) is sufficiently small compared to the other (L_y), there is a tendency for a succession of patterns with modes characterized by $m = 0$ and $n = 1, 2, \ldots, k$ *before* the first pattern with $m \geq 1$ is obtained. This means that stripes predominate on this narrow domain and spots are more characteristic of wider regions. (See Figure 11.6(*g*) for "stripes" and Figure 11.6(*f*) for "spots".) A common feature in many spotted animals is that their slender extremities retain a striped pattern, as predicted by this theory.

As an endpoint of his model Murray terminates with a whimsical but ingenious prediction about pigmentation patterns on tails. Tails are often broader at their base than at their end, so that their shape is roughly conical. If the fur is removed and the cone opened by a longitudinal bisection, one obtains a triangular domain with a narrow end and a broader base of radius r_0. This geometry can be represented by letting the size parameter γ vary gradually across the length of the domain (see Figure 11.15). According to our previous discussion, such variation is consistent with a transition from stripes at the thin one-dimensional end to spots at the broader base. The opposite transition would be inconsistent and thus contradictory of the theory. To date, animals with the proper tail patterns have been found in great abundance. As yet no single possessor of an inconsistent tail has reportedly been sighted. (See Figure 11.16.)

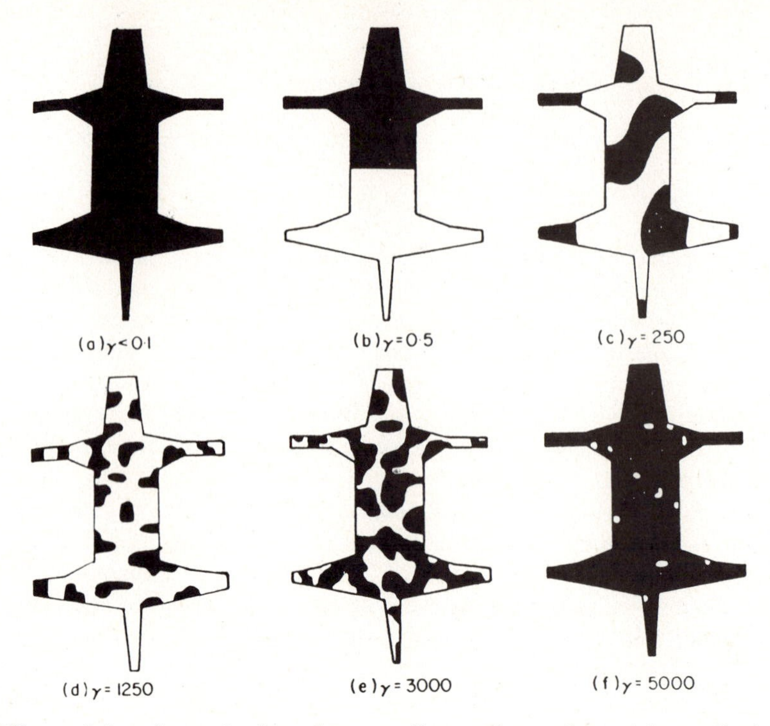

Figure 11.14. *Effects of size of an animal on the patterns formed on its coat by a reaction-diffusion prepattern mechanism proposed by Murray (1981). As the size parameter* γ *changes from (a) to (f), a succession of patterns (typical of different animals) occurs. For very large domains the nonlinear effects make a substantial contribution, so that patterns are no longer predictable based on linear theory. [From Murray, J. D. (1981). A prepattern formation mechanism for animal coat markings.* J. Theor. Biol. 88, *161–199, fig. 8. Reprinted by permission of Academic Press.]*

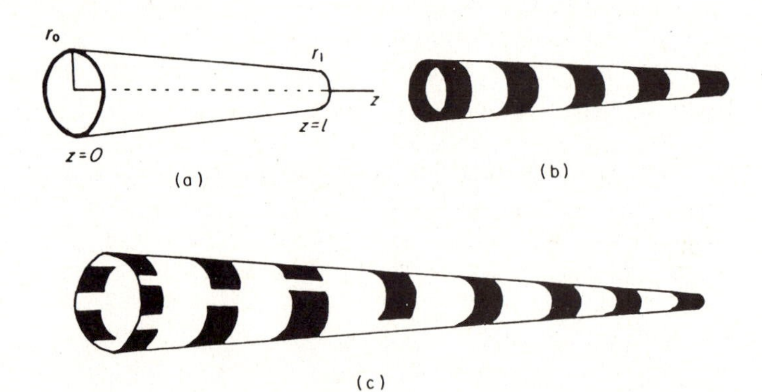

Figure 11.15. *Effects of geometry on patterns. Reaction diffusion on a conical surface (such as a tail) may result in a progression from stripes to spots in only one way. The spots tend to appear if (a)* r_0*, the radius of the base, is large enough to admit several chemical "peaks" around its circumference, as in (c) [From Murray, J. D. (1981a). A prepattern formation mechanism for animal coat markings.* J. Theor. Biol., 88, *161–199, fig. 5. Reprinted by permission of Academic Press.]*

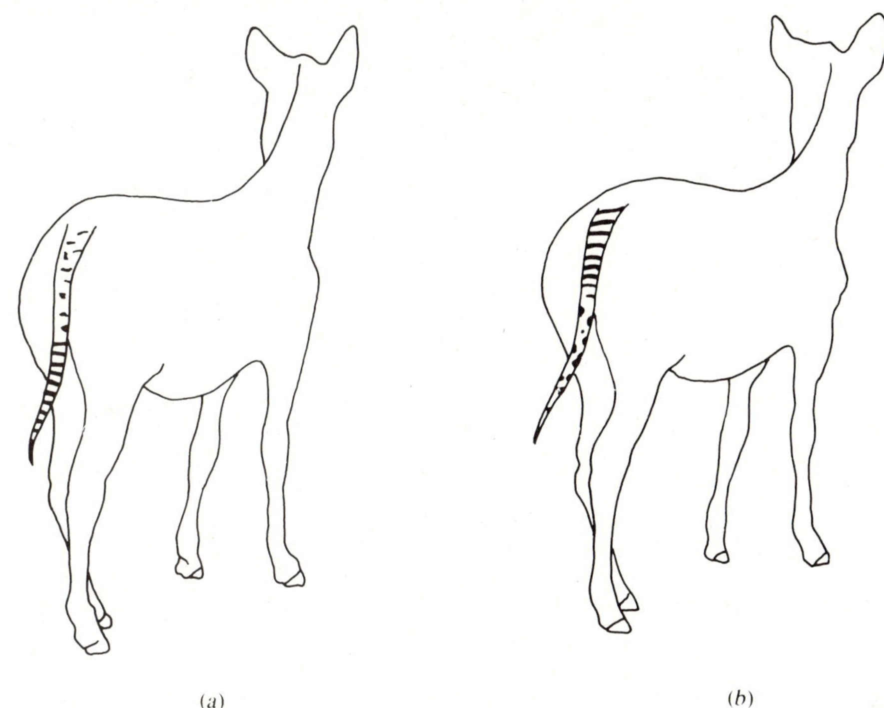

Figure 11.16. (a) Tail pattern predicted by linear theory; (b) tail pattern impossible by linear theory. *[Drawn by Marjorie Buff.]*

11.9. FOR FURTHER STUDY

Patterns in Ecology

A recurring theme in this book is that mathematical models lead us to draw parallels between situations that may seem totally unrelated on first inspection. Analogies between the microscopic molecular realm and the macroscopic population level have appeared repeatedly in previous discussions. For this reason it is to be anticipated that spatial patterns emerging from unstable uniform distributions may occur in ecological settings as well, particularly in species that interact and disperse at different rates.

The first prediction that this may indeed occur appears in Segel and Jackson (1972). The authors realized that a similarity between activator-inhibitor chemicals and prey-predator species exists. However, merely "tinkering" with the Lotka-Volterra models, augmented by dispersal terms, was unsuccessful in producing diffusive instability. One of the attractive features of the Segel-Jackson paper is the detailed discussion of the motivation that led to their specific model. (Rarely are authors as candid about the development of a theory.) A good topic for advanced students would be to independently address the following question posed by the authors:

> [What are] the characteristics of a situation where uneven geographic distribution of predator and prey would be mutually advantageous? (1972, p. 553.)

As a second stage, students might attempt to write their own model and test the Turing criteria (32a,b) and (38) before turning to the papers provided in the references by way of a comparison. For reasons outlined in their paper, Segel and Jackson (1972) considered the following set of equations:

$$\text{Prey:} \quad \frac{\partial V}{\partial t} = VR(V) - AVE + \mu_1 \nabla^2 V, \tag{66a}$$

$$\text{Predators:} \quad \frac{\partial E}{\partial t} = BVE - ME - CE^2 + \mu_2 \nabla^2 E, \tag{66b}$$

where $V(x, t)$ = the prey ("victims"), $E(x, t)$ = the predators ("exploiters"), and

$$R(V) = K_0 + K_1 V. \tag{67}$$

Interpretation of the model and its parameters is left as an exercise.

Patchy distributions of populations have been observed in nature under numerous conditions. One well-documented example is *plankton,* the microscopic aquatic organisms often found in uneven distributions at or close to the surface of the water. (See Okubo, 1980, for review and references.) Plankton actually consists of a multitude of uni- and multicellular organisms, which are frequently characterized simply as *zooplankton* or *phytoplankton*. The latter are capable of photosynthesis; like higher plants they are in a sense self-sufficient, relying mainly on sunlight for their energy. The former are predatory, feeding on phytoplankton and on each other.

Patchy distributions of plankton may arise from different mechanisms, and a conclusive explanation has not been given. However, the idea that the natural dispersal rate of these microscopic organisms might lead to instability of the type described in this chapter is rather intriguing.

Mimura and Murray (1978) discuss a slightly different theoretical model given by the equations

$$\frac{\partial P}{\partial t} = [f(P) - Q]P + \mathcal{D}_P \frac{\partial^2 P}{\partial x^2}, \tag{68a}$$

$$\frac{\partial Q}{\partial t} = -[g(Q) - P]Q + \mathcal{D}_Q \frac{\partial^2 Q}{\partial x^2}, \tag{68b}$$

where P is phytoplankton and Q is zooplankton. A particular assumption made is that the graph of $f(P)$ has a "hump" and that $g(Q)$ has a positive slope (see Figure 11.17). A typical set of functions proposed in the paper is the following:

$$f(P) = K(K_0 + K_1 P - P^2), \tag{69a}$$

$$g(Q) = K_2 + K_3 Q. \tag{69b}$$

Noting that $f(P)$ represents the prey growth rate, one might attribute its hump

Figure 11.17. The predator-prey phase plane for equations (68a,b) with f(P) and g(Q) given by equations (69a,b) in a model by Mimura and Murray (1978). Note the hump effect in f(P).

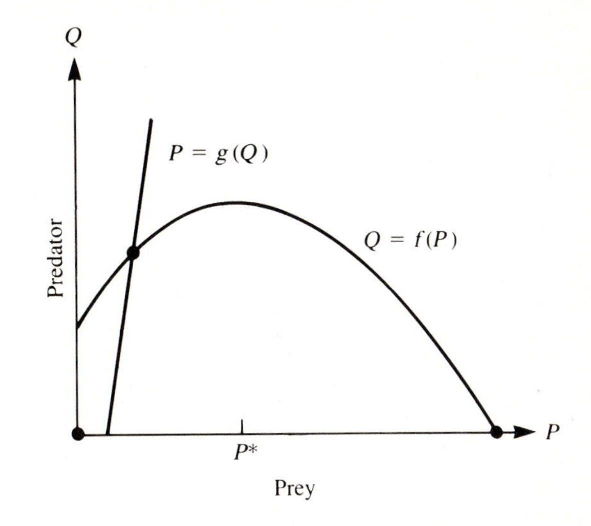

to an Allee effect (see Section 6.1). The authors prove that the hump in *f* leads to a property crucial for realistic solutions; namely, that spatial patterns resulting from diffusive instability will in certain limiting cases tend to have alternately low and high *plateaus* of population density. Models in which the hump effect is absent tend to have sharply peaked spikelike patterns of prey density. These are considered to be less realistic. (Proof rests on singular perturbation analysis and may be too advanced for some readers.)

Evidence for Chemical Morphogens in Developmental Systems

How has the chemical theory for morphogenesis fared in light of empirical research? The true test of such theories would be to isolate and characterize chemical substances that have demonstrable morphogenetic effects. This is a challenging task given that such substances, if they are present, may be found in minute quantities.

Work in this direction is being carried out. It is said that in an attempt to isolate such substances in one accessible system, the *hydra*, it was necessary to process and analyze material from several tons of sea anemone, which are relatives of this organism. (H. C. Schaller 1973, 1976).

The hydra is a small freshwater organism (3 to 4 mm long) with the ability to regenerate body parts [see Figure 11.18(*a*)]. Any small piece excised from its body column (a roughly cylindrical shape) can form a whole new hydra by making a tentacled "head" and a "foot" in the appropriate ends. Chemical studies by Schaller and others have revealed several molecular morphogenetic substances, among them the *head activator* and the *head inhibitor*. Characterization of the nature and interactions of such substances is proceeding, but many details are still missing.

Based on this partial characterization, Kemmner (1984) has formulated a model for a possible reaction-diffusion system in the hydra [see Figure 11.18(*b*) and Table 11.1]. The evidence indicates a somewhat more subtle underlying mechanism,

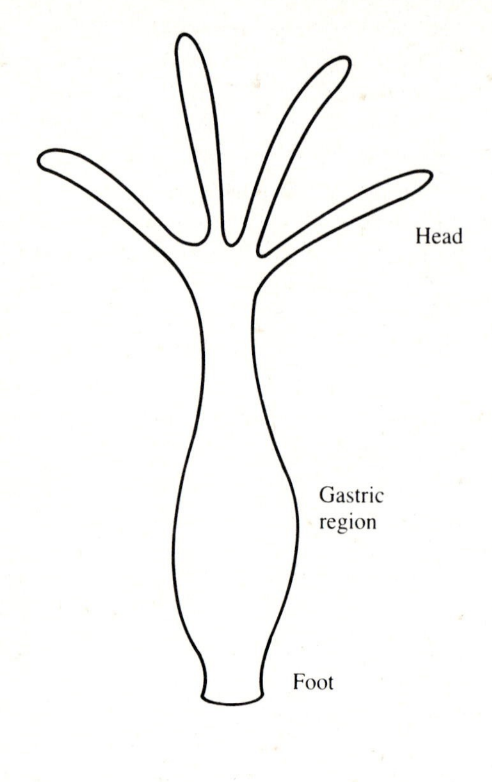

(a)

Release control:

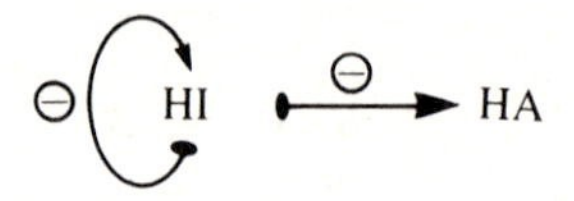

Production of sources:

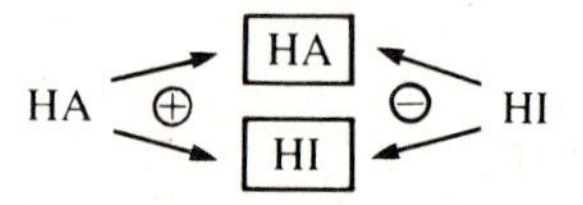

Interaction of head inhibitor and head activator

(b)

Figure 11.18. *(a) Simple diagram of the hydra, a freshwater polyp. (b) Chemicals identified as morphogenetic substances in hydra. The head inhibitor (HI) inhibits both its own release and the release of head activator (HA) from sources. The formation of sources depends on the presence of HA. In this sense HA is autocatalytic. [(b) From Kemmner, W. (1984). Head regeneration in* Hydra: *Biological studies and a model. In W. Jager and J. D. Murray, eds.,* Modelling of Patterns in Space and Time. *Springer-Verlag, New York, fig. 3.]*

Table 11.1 **Properties of morphogenetic substances from hydra controlling head and foot formation.**

Morphogen	*Molecular Weight*	*Nature*	*Purification (x-fold)*	*Active Concentration*	*Gradient*
Head activator	1142	Peptide	10^9	10^{-13} M	
Head inhibitor	<500	Nonpeptide	10^5	$<10^{-9}$ M	
Foot activator	~1000	Peptide	10^5	$<10^{-9}$ M	
Foot inhibitor	<500	Nonpeptide	10^4	$<10^{-3}$ M	

Source: Kemmner, W. (1984). Head regeneration in *Hydra:* Biological studies and a model. In W. Jager and J. D. Murray, eds., *Modelling of Patterns in Space and Time*. Springer-Verlag, New York, table 1.

namely that one of the two substances inhibits the release of both from preexisting sources. Moreover, self-enhancement arises not by direct autocatalysis but rather by virtue of the fact that sources of the substances are restored (if removed) by the presence of the activator. The paper by Kemmner and other references provided therein are recommended as good sources for further independent investigation.

Few other systems have so readily revealed their secrets to us. It would at present appear that many developmental systems are not governed (even partially) by simple pairs of chemical species. The Turing theory as yet remains a vivid paradigm rather than an accurate description of any one real morphogenetic event.

A Broader View of Pattern Formation in Biology

A recent review paper by Levin and Segel (1985) provides a general survey and further recent references on pattern-generating processes. While the Turing theory still ranks among the top contenders for pattern-forming mechanisms, a variety of different theories have been formulated for special systems. It has been shown in recent papers that neural networks (with excitatory and inhibitory elements) can generate patterns of various sorts. A number of self-organizing systems such as cellular automata and clonal organisms (both described in boxes to come), which are governed by simple recursive rules, have been studied. Mechanochemical theories have addressed the morphogenesis of tissues formed by migration and movement or deformation of cells. Some of these theories are entirely unrelated to those described in this chapter. Others do share certain common conceptual features; for instance, many are based on the property of lateral inhibition. One example, to be described here, is drawn from neural interactions.

As previously mentioned, nerves communicate with one another over great distances. One can define the *range of activation* and the *range of inhibition* in a neural network as the average distance over which one neuron transmits stimulatory or inhibitory signals to its neighbors via synapses. The effect modulated by a synapse can be either positive (excitatory) or negative (inhibitory) to the neuron on which it impinges. Thus, interactions over a distance in a neural network are *analogous* to interactions over the diffusional range of activator-inhibitor chemicals in the Turing system we described. The details of the mechanism and its mathematical description, though, are different.

The boxes in this section give a brief survey of recent work in other theories of pattern formation. For details and further sources, consult the appropriate References.

Patterns in neural networks: **Visual Hallucinations**

The patterns shown in Figure 11.19 are experienced during drug-induced visual hallucinations. In the figure, designs on the left-hand side are visual images that are perceived to be centered at the *fovea* (the center of the visual field in the *retina* of the eye). The

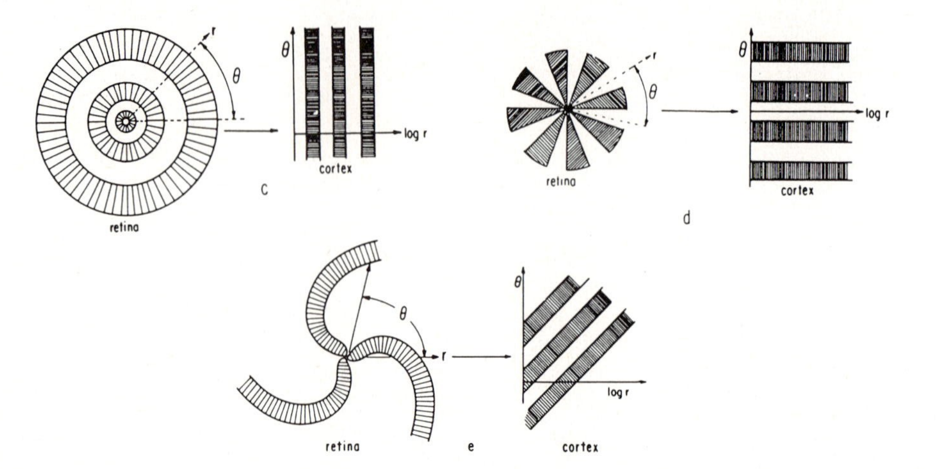

Figure 11.19. *Visual hallucination patterns in retinal and cortical coordinates: On the RHS are patterns that arise in some possibly small region of the visual cortex as a result of long-range inhibition and short-ranged activation within the cortical neural network. On the LHS are the actual hallucination patterns perceived as though they are on the retina. (A log-polar to rectangular coordinate transformation governs the correspondence between the patterns on the LHS and those on the RHS.) [From Ermentrout, G. B., and Cowan, J. (1979). Mathematical theory of visual hallucination patterns.* Biol. Cybernet., 34, *137–150, fig. 2.]*

corresponding designs on the right-hand side are patterns of neural excitation on the *visual cortex* (an area of the brain that processes visual signals). The correspondence, called the *retinocortical map*, has been established empirically according to Ermentrout and Cowan (1979). It is essentially a coordinate transformation from polar coordinates in the retina to cortical rectangular coordinates in the cortex.

It is believed that hallucination patterns form in the cortex, which is a complex neural network, even though no visual stimuli are present on the retina. Ermentrout and Cowan (1979) proposed the theory that patterns of excitation in the cortex arise spontaneously as a result of instability of a uniform resting state. They further suggest that instability occurs when enhanced excitatory and decreased inhibitory effects of neural synapses occurs, for example, as a result of chemical changes in the brain. In their 1979 paper they demonstrate that spatial patterns such as those on the right side of Figure 11.19 are possible in neural networks with the property of lateral inhibition. An informal presentation of these results is to be found in Ermentrout (1984).

For other neural network patterns, see Swindale (1980) and Ermentrout et al. (1986).

A Spatially Discrete Lateral-Inhibition Model

In a discrete model for vertebrate skin patterns proposed by Young (1984), a differentiated cell located at $R = 0$ exerts an effect on its neighbors according to the activation-inhibition field shown in Figure 11.20. w_1 and w_2 are respectively the net positive and negative effects; R is radial distance; and R_1 and R_2 are analogous to the ranges of activation and inhibition respectively. If the net effect on some undifferentiated cell (summed up over contributions of all its neighbors) is positive, the cell will differentiate into a pigmented cell. Patterns thus produced by computer simulations for $R_1 = 2.3$, $R_2 = 6.01$, $w_1 = 1$, and four values of w_2 are shown. Note that, although diffusion is not explicitly depicted, the mechanism for pattern formation is that of lateral inhibition: local activation and long-range inhibition.

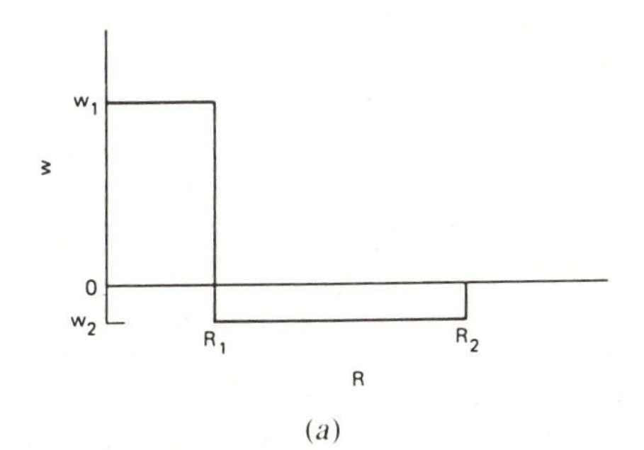

(*a*)

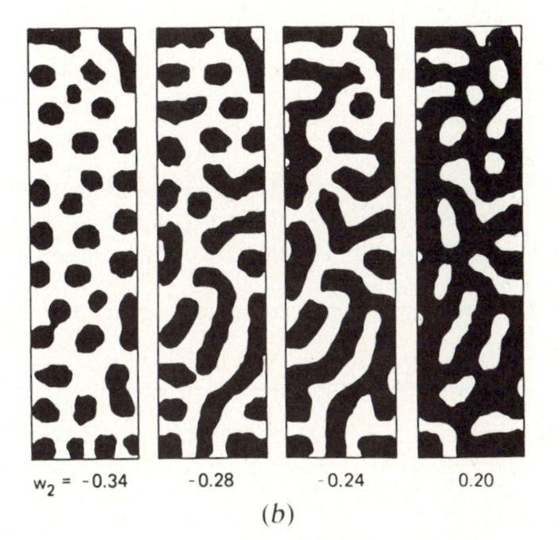

(*b*)

Figure 11.20. *(a) A discrete activation-inhibition field. (b) Patterns produced from an initially random distribution of differentiated "cells" (not shown) by successive iterations using the spatial effects given in (a). [Reprinted by permission of the publisher from "A local activator-inhibitor model of vertebrate skin patterns" by D. A. Young,* Math. Biosci., 72, *figs. 1b and 2, pp. 51–58. Copyright 1984 by Elsevier Science Publishing Co., Inc.]*

Mechanochemical Patterns

In many developing systems, groups of cells are seen to migrate or move collectively in forming the final shape of some part of an organism. The elasticity, adhesiveness, and relative affinities of different cells for one another then governs the succession of shapes or patterns that result in these systems. Models based on underlying physical, mechanical, and chemical processes have been explored recently by Murray and Oster (1984) and Odell et al. (1981).

Odell et al. (1981) explored the process of *gastrulation*, part of the early development of an embryo. They modeled each cell as an element that could stretch and distend in shape, subject to properties of its internal structural components. In the computer simulations (shown in Figure 11.21) the group of cells is seen to undergo a sequence of shapes eventually forming the *gastrula* in which an invagination is present. (These simulations depict a two-dimensional analog of a three-dimensional shape.)

For other references, see Oster et al. (1983), Murray and Oster (1984), also Sulsky et al. (1984).

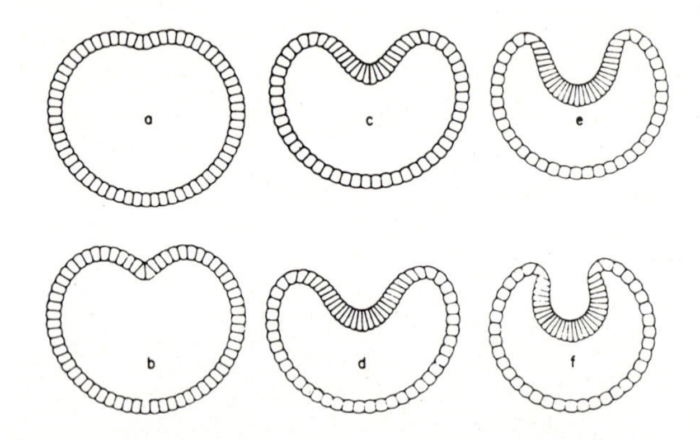

Figure 11.21 *(a–f) Computer simulation of gastrulation in the sea urchin. [From Odell, G. M., Oster, G., Alberch, P. and Burnside, B. (1981). The mechanical basis of morphogenesis. Develop. Biol., 85, 446–462, fig. 7.] Academic Press, New York.*

Patterns in Cellular Automata

One class of models for self-organizing systems is the *cellular automata*. These theoretical systems consist of a discrete number of "cells" (arranged, for example, in a row). Each cell has some initial *state* (represented by an integer). The automaton is governed by a *transition rule*, which assigns a new state to every cell based on the current state of the cell and of its neighbors. The mathematical properties of cellular automata have been described in numerous papers by S. Wolfram (1984a,b). Such systems are more popularly known for their striking patterns, which are readily studied by simple computer simulations (see Figure 11.22). See Wolfram (1984*a*) for a general review and biological applications.

Figure 11.22. *In these one-dimensional cellular automata, the initial state of a row of cells (at the top of each frame) changes successively as the transition rule is applied. The resulting space-time patterns are created.* *[From Figures selected from Wolfram, S. (1984). Universality and complexity in cellular automata,* Physica, 10D. *Reprinted by permission of North-Holland Physics Publishing.]*

Patterns in clonal organisms

Many simple organisms grow by adding new segments to preexisting structures (see Sections 1.9 and 10.4). Branching of filaments is one instance of clonal growth. The patterns shown in Figure 11.23 are produced by a simple set of recursive rules applied successively in a numerical simulation of the growth of such networks. Here the final patterns result from local interactions of parts of the structure and the continual random formation of new segments by branching. For other references on a similar subject see Cohen (1967) and Jackson et al. (1986).

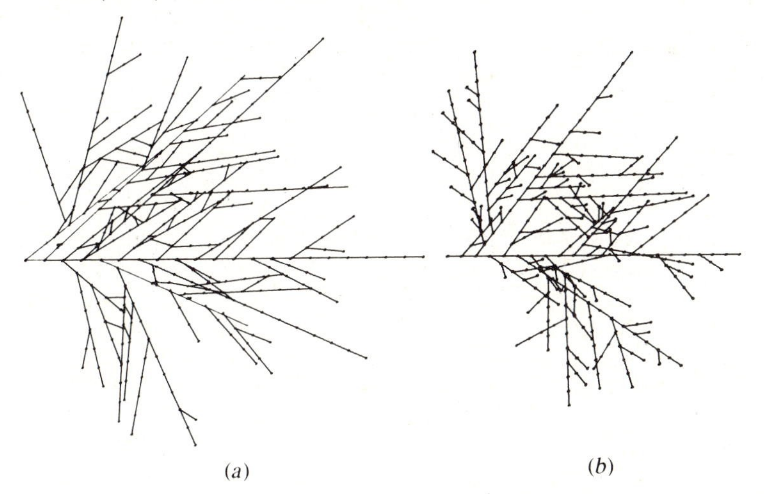

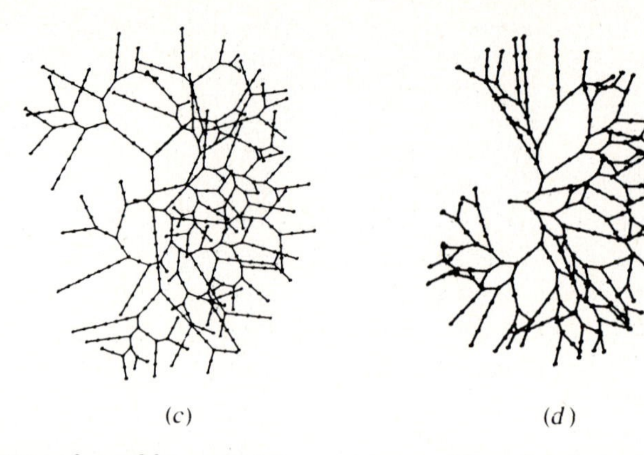

Figure 11.23. Patterns formed by computer simulations of simple branching rules. (a,b) Branching occurs only along a preexisting branch. (c,d) Branching is apical. Anastomoses *(reconnections) occur between apices and branches in (a,d) and between apices only in (b,c). [Simulation program written and figures produced by Richard Fogel.]*

PROBLEMS*

1. *Cellular aggregation*

(a) Justify each of the terms appearing in equations (2a,b).

(b) Verify that equations (6a,b) and (7a,b) are obtained by expanding (2a,b) and then linearizing about the steady state $(\overline{a}, \overline{c})$.

***(c)** In order for linearization to be a valid approximation it is necessary to assume that the perturbations a' and c' are small. What else must be assumed about the perturbations so that nonlinear terms in (7a,b) can be neglected? How does this assumption influence the validity of the special forms (9a,b)?

(d) Verify that equations (14a–c) are obtained from (13).

(e) Explain the reasoning used to deduce condition (16) from equations (14a–c).

(f) Show that if equations (2a,b) for $a(x, t)$ and $c(x, t)$ have the no-flux boundary conditions of equations (17a,b) imposed on them, then the perturbations $a'(x, t)$ and $c'(x, t)$ must also satisfy the same no-flux boundary conditions.

*Problems preceded by an asterisk (*) are especially challenging.

(g) The functions $\sin x$ and $\cos x$ have periods of 2π. Why then are permissible values of the wavenumber q given by equation (18) rather than by $q = n(2\pi)/L \, (n = 0, 1, 2, \ldots)$?

2. Which of the following procedures would tend to promote the onset of aggregation?
(a) Placing barriers so that the domain is subdivided into several subregions.
(b) Increasing the ambient cAMP concentration in the initial homogeneous state.
(c) Causing the amoeba to reproduce vigorously before entering the starved state.
(d) Raising the temperature.
(e) Adding *phosphodiesterase* to the medium.

3. For $n = 1$, inequality (19) can also be written in the form

$$\frac{1}{f}\frac{(L/\pi)^2}{\chi\bar{a}} < \frac{1}{k + (\pi/L)^2 D}\frac{(L/\pi)^2}{\mu}.$$

(See Segel, 1980.)
(a) Verify this fact.
(b) Interpret the meaning of each of the four terms in the equation.
(c) Give a verbal interpretation of the above aggregation condition.

4. What quantity in the model discussed in Section 11.3 depicts the size of the aggregation domain? (Observe that the amoeba density does not enter into this quantity.)

5. *Possible modification of the cellular aggregation model*
(a) Suggest how such quantities as χ, μ, f, and k (assumed constant in the simplest model) might depend on the variables a and c.
(b) How would equations (2a,b) and (6a,b) change under the assumption that the parameters in (a) are nonconstant? [You may wish to take general forms such as $\chi = \chi(c)$, $k = k(a)$, $f = f(c)$, and so forth.
(c) How would equations (8a,b) change?
***(d)** By assuming small perturbations of the form shown in (9a,b) and proceeding with a similar analysis, derive a stability condition analogous to (16).
***(e)** Use your result to argue what properties of the functions χ, μ, f, and k might promote aggregation.

6. Generalize the Keller-Segel model to rectangular two-dimensional aggregation domain of dimensions $L_x \times L_y$.

7. Lauffenburger and Kennedy (1983) suggest a model for the chemotaxis of phagocytes (white blood cells) towards high bacterial densities (part of the tissue inflammatory response to bacterial infection). A set of dimensionless equations that they studied are:

$$\frac{\partial v}{\partial t} = \rho \frac{\partial^2 v}{\partial x^2} + \frac{\gamma v}{1 + v} - \frac{uv}{\kappa + v}$$

$$\frac{\partial u}{\partial t} = \frac{\partial^2 u}{\partial x^2} - \delta \frac{\partial}{\partial x}\left(u \frac{\partial v}{\partial x}\right) + \alpha(1 + \sigma v - u)$$

where

v = dimensionless bacterial density,
u = dimensionless phagocyte density.
γ = ratio of maximum bacterial growth rate to maximum phagocyte killing rate;
σ = ratio of enhanced phagocyte emigration rate to normal "background" emigration rate;
κ = ratio of inhibition effect of increasing bacteria density on bacterial growth to inhibition effect of killing;
α = ratio of phagocyte death rate to maximum phagocytic killing rate.

(See problem 11 of Chapter 10.)

(a) Show that the stability of a uniform steady state to uniform (i.e. space-independent) perturbations is governed by the Jacobian matrix

$$\mathbf{J}_l = \begin{bmatrix} \dfrac{\gamma}{(1+v)^2} - \dfrac{\kappa u}{(\kappa + v)^2} & -\dfrac{v}{\kappa + v} \\ \alpha\sigma & -\alpha \end{bmatrix}.$$

(b) What would be the corresponding matrix **J** governing stability of nonuniform perturbations?

(c) For the steady state $(\bar{u}, \bar{v}) = (1, 0)$ show that the eigenvalues of the matrix in part (b) are

$$\lambda_1 = -\rho q^2 + \gamma - \frac{1}{\kappa}, \qquad \lambda_2 = -q^2 - \alpha.$$

(d) Which mode (i.e. which value of q) is the most likely to cause instability? What is the implication?

(e) For the second steady state of the equations, $\bar{v} > 0$, $\bar{u} = 1 + \sigma\bar{v}$, show that the stability matrix is

$$\mathbf{J} = \begin{bmatrix} -\rho q^2 + F(\bar{v}) & -H(\bar{v}) \\ \delta \bar{u} q^2 + \alpha\sigma & -q^2 - \alpha \end{bmatrix},$$

where

$$F(v) = \frac{v(1 + \sigma v)(1 - \kappa)}{(1 + v)(\kappa + v)^2}$$

and

$$H(v) = \frac{v}{\kappa + v}.$$

and that eigenvalues have negative real parts provided the following inequalities are satisfied:

$$\text{tr } \mathbf{J}_l - (1 + \rho)q^2 < 0,$$

$$\det \mathbf{J}_l + (\rho q^2 + \rho\alpha + \delta\overline{u}H(\overline{v}) - F(\overline{u}))q^2 > 0.$$

***(f)** Further discuss how diffusive *instability* might arise in this model. You may wish to refer to the analysis in Lauffenburger and Kennedy (1983).

8. Cells in the sluglike phase of *Dictyostelium discoideum* are either prespore or prestalk, and the two cell classes maintain a definite ratio despite experimental manipulations such as excision of part of the mass. Use a simple model outlined in Chapter 7 to suggest one type of mechanism that might lead to this ratio regulation. Comment on ways of testing such a model.

9. *Reaction-diffusion systems*

(a) Use Taylor-series expansions for R_1 and R_2 to show that equations (20a,b) lead to equations (24a,b) for *small* perturbations.
(b) Show that equations (24a,b) are equivalent to system (26a–d).
(c) Show that by assuming perturbations of the form (27) one arrives at equations (28a,b).
(d) Verify that equation (30) results.
(e) Use the inequality (33b) directly to reason that diffusion has the capacity to act as a destabilizing influence [*Hint:* note the effect of $D_i q^2$ on the self-influencing terms a_{ii}.]

10. **(a)** Show that the expression on the LHS of inequality (33b) corresponds to the quantity defined in equation (35).
(b) Demonstrate that $H(q^2)$ has a minimum, and prove that $q^2_{\min}$ is given by (36a) (*Hint:* Set $dH/dq^2 = 0$ and solve for $q^2_{\min}$.)
(c) Evaluate $H(q^2_{\min})$.
(d) Show that (34) implies (37).
(e) Derive inequality (38) from (37) and (32b).
(f) Verify that (38) can be written in the dimensionless form given in (39).

11. **(a)** Justify inequalities (40) and (41).
(b) Verify inequality (45). What are the dimensions of τ_1 and τ_2? of D_2/τ_2 and D_1/τ_1?
(c) Explain what is meant physically by the statement that the range of inhibition should be larger than the range of activation.
(d) Show that for (45) to hold it must also be true that $D_2 > D_1$.
(e) Explain the deduction that spacing between patterns will be approximately given by (46) at the onset of diffusive instability.
(f) Suggest what sort of biological influences might bring about the parameter variation that leads to diffusive instability.

***12.** The growth rate of a given wavenumber q_i is σ_i, and is given by (30). Describe how one could use this equation to solve for σ_i, and find the wavenumber of the *fastest*-growing perturbation.

13. Generalize the physical arguments given in Section 11.6 for the case of activator-inhibitor systems to the case of positive-feedback systems.

14. *Reaction-diffusion in two dimensions*

(a) Write out the linearized system corresponding to equations (48a,b).

(b) By explicitly differentiating equation (49) show that these are solutions to the linearized equations. Obtain a set of algebraic equations in α_1 and α_2 (involving σ, q_i, and other parameters.)

(c) An alternate assumption about perturbations is that they take the following form:

$$C_i'(x, t) = \alpha_i e^{\sigma t} \cos (q_1 x + q_2 y).$$

Use a trigonometric identity and boundary conditions (50a,b) to show that this is equivalent to equation (49).

(d) Verify that equation (53) is obtained for the quantity Q^2 characterizing the most destabilizing perturbations.

15. In the following problems you are given a set of reaction terms $R_1(c_1, c_2)$ and $R_2(c_1, c_2)$. Determine whether or not a homogeneous steady state can be obtained and whether the system is capable of giving rise to diffusive instability. If so, give explicit conditions for instability to arise, and determine which modes would be most destabilizing.

(a) Lotka-Volterra:
$R_1 = ac_1 - bc_1c_2,$
$R_2 = -ec_2 + dc_1c_2.$

(b) Species competition:
$R_1 = \mu_1 c_1 - \alpha_1 c_1^2 - \gamma_{12} c_1 c_2,$
$R_2 = \mu_2 c_2 - \alpha_2 c_2^2 - \gamma_{21} c_1 c_2,$
$\left(\dfrac{\mu_2}{\mu_1} < \dfrac{\gamma_{21}}{\alpha_1}, \quad \dfrac{\mu_2}{\mu_1} < \dfrac{\alpha_2}{\gamma_{12}}\right).$

(c) Glycolytic oscillator:
$R_1 = \delta - kc_1 - c_1c_2^2,$
$R_2 = kc_1 + c_1c_2^2 - c_2.$

(d) Schnakenberg chemical system:
$R_1 = c_1^2c_2 - c_1 + b,$
$R_2 = -c_1^2c_2 + a.$

(e) Van der Pol oscillator:
$R_1 = c_2 - \dfrac{c_1^3}{3} + c_1,$
$R_2 = -c_1.$

(f) Phytoplankton-herbivore system (Levin and Segel, 1976):
$R_1 = ac_1 + ec_1^2 - b_1c_1c_2,$
$R_2 = -dc_2^2 + b_2c_1c_2.$

(g) Tyson-Fife model for Belousov-Zhabotinsky reaction:
$R_1 = \dfrac{1}{\epsilon}\dfrac{c_1(1 - c_1) - bc_1(c_2 - a)}{c_2 + a},$
$R_2 = c_1 - c_2.$

(h) Meinhardt (1983) model:
$R_1 = ec_1^2c_2 - \mu c_1,$
$R_2 = e_0 - ec_1^2c_2.$

16. In this problem you are asked to explore the mechanism Meinhardt and Gierer suggested [equations (57a,b)] by using linear stability theory. Set $\rho_0 = \rho_1 = 0$ and assume that ρ' and ρ are constant.

(a) Suppose diffusion is absent. Draw a phase-plane portrait of the reaction scheme in equations (57a,b) and show the location of the nullclines and steady state $(\bar{A}, \bar{H})$.

(b) Using the conditions derived in this chapter, show that diffusive instability of a homogeneous steady state $(\bar{A}, \bar{H})$ is possible, and give the necessary conditions on the parameters.

17. (a) Interpret equations (58a,b) of the Gierer-Meinhardt model in terms of a hypothetical set of interacting chemicals a and h. Show that $c_1 = c_2 = c$ leads to dimensional inconsistency.
(b) Explain conditions (62a,b) in physical terms.
(c) Verify that (62b) implies (63). (*Hint:* Consider the equation $x - 1/x = 2$ and solve for x.)

18. Interpret the Segel-Jackson equations (66a,b) for a spatially distributed predator-prey system. Why would the term $-CE^2$ be described as a "combat" term?
(a) The authors claim that for $C = 0$ the model will not yield a diffusive instability. Justify and interpret this result.
(b) The assumption $M = 0$ does not prevent instability from occurring. Why?
(c) Assume that $M = 0$, in other words, that predator mortality from "combat" predominates over that due to other causes. Show that the equations can be written in the following dimensionless form:

$$\frac{\partial v}{\partial t} = (1 + kv)v - aev + \delta^2 \nabla^2 v,$$

$$\frac{\partial e}{\partial t} = ev - e^2 + \nabla^2 e,$$

where $e = EC/K_0$, $v = VB/K_0$, distances are scaled in units of $(\mu_2/K_0)^{-1/2}$ and time is scaled in units of K_0.
(d) Show that the dimensionless parameters appearing in part (c) are $k = K_1/B$, $a = A/C$, and $\delta^2 = \mu_1/\mu_2$. Interpret the meanings of these quantities.
(e) Show that the nontrivial steady state of this system is

$$e = v = \frac{1}{a - k}.$$

(f) Show that the instability condition is

$$k - \delta^2 > 2(a - k)^{1/2},$$

and that the wavenumber of the excitable modes is given by

$$Q^2 = \frac{1}{\delta(a - k)^{1/2}}.$$

19. *Effects of geometry on patterns formed*. Consider a rectangle of sides L_x and L_y where

$$L_x = L, \qquad L_y = \gamma L.$$

(a) Suppose that the boundaries are impermeable (that is, no flux boundary conditions apply). Show that the wavenumbers q_1 and q_2 of a perturbation in the x and y directions satisfy (51a,b).

(b) Suppose that the quantities L, D_{ij}, and a_{ij} are kept fixed but that the size proportion γ is gradually decreased. Interpret what effect this would have on the shape of the domain.

(c) Now assume that the *only* excitable modes are those satisfying (56). If initially $m = n = 4$ and $\gamma = 1$, determine the succession of modes (m_j, n_j) corresponding to a sequence of increasing values of γ.

(d) Draw several of these patterns.

20. *Effects of boundary conditions on patterns formed.* Now suppose that in problem 19 the boundaries are not impermeable but rather are artificially kept at the steady-state concentrations $\overline{C}_1$ and $\overline{C}_2$:

$$C_1(x, y, t) = \overline{C}_1 \qquad (x = 0, x = L, y = 0, y = \gamma L),$$

$$C_2(x, y, t) = \overline{C}_2 \qquad (x = 0, x = L, y = 0, y = \gamma L).$$

(a) How would this change the assumed form of the perturbations given by equation (49)?

(b) How would this affect the form of the wavenumbers given by (51a,b)?

(c) What effect would this have on expressions in (54) and (56)?

(d) Now consider mixed boundary conditions: constant concentrations along the x boundaries and no flux along the y boundaries. Repeat steps (a) through (c).

(e) In each of the above two cases, assume that $\gamma = 2$ and that the size of the region, L^2, is gradually increased (all else being held constant). Describe the succession of modes you might expect to encounter by drawing up a table analogous to the one in Figure 11.5.

21. *A model for animal coat patterns (Murray, 1981a)* In melanogenesis the precursor tyrosine is oxidized to dihydrophenylalanine (DOPA) in the presence of the enzyme tyrosinase. DOPA oxidizes further before the final polymerization to melanin. Murray (1981a) proposes that prepattern formation occurs before the arrival of *melanoblasts* (precursors of the melanin-producing *melanocytes*) at the epidermal surface. He considers two species, a high-molecular-weight substrate S and a cosubstrate A. For example, the substrate could be associated with tyrosine or the enzyme tyrosinase, and the cosubstrate might be an inhibitor governing melanogenesis initiation. The kinetics are taken to be that of substrate inhibition of an immobile enzyme. It is assumed that the two reactive species S and A are maintained at constant concentrations S_0 and A_0 in a reservoir. From this reservoir they diffuse through an inactive membrane of thickness L_1 onto an active membrane of thickness L_2. (See accompanying diagram.) The active membrane contains the immobile enzyme tyrosinase, which allows S and A to react at the following rate:

$$R = \frac{V_m AS}{K_m + S + S^2/K_S}, \qquad (70)$$

where V_m, K_m, and K_S are assumed to be constant. Murray then suggests the following reaction-diffusion equations:

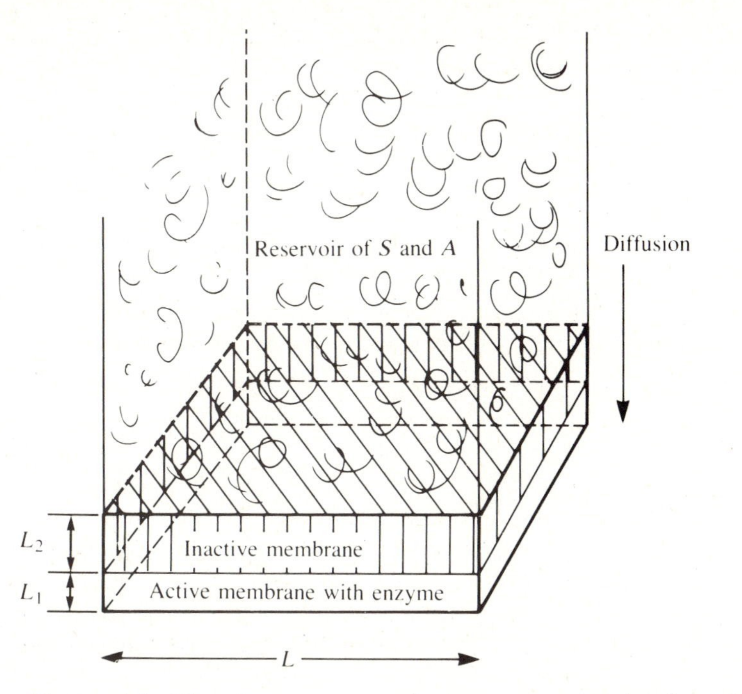

Figure for problem 21. S and A *diffuse from a reservoir to the active membrane, where they undergo an enzyme-catalyzed reaction. [Based on a drawing by Marjorie Buff.]*

$$\frac{\partial S}{\partial T} = \frac{D_S'}{L_1 L_2}(S_0 - S) - \frac{V_m AS}{K_m + S + S^2/K_S} + D_s \nabla^2 S, \tag{71a}$$

$$\frac{\partial A}{\partial T} = \frac{D_A'}{L_1 L_2}(A_0 - A) - \frac{V_m AS}{K_m + S + S^2/K_S} + D_A \nabla^2 A, \tag{71b}$$

where D_S' and D_A' are the diffusion coefficients for S and A in the inactive membrane, and D_S and D_A are the diffusion coefficients for S and A in the active layer.

(a) Explain reaction mechanism (70). (*Note:* You may find it useful to refer to problem 22 in Chapter 7.)

(b) Explain the various terms in equations (71a,b). [*Note:* Diffusive flux into the active membrane from the reservoir is proportional to $D_S'(S_0 - S)$ and $D_A'(A_0 - A)$.] Why does the product $L_1 L_2$ appear in this expression?

(c) Murray assumes that $D_S < D_A$. On what property of the substance S does he base this assumption? Why is such an assumption necessary?

(d) Define dimensionless variables as follows:

$$s = \frac{S}{K_m}, \qquad a = \frac{A}{K_m}, \qquad t^* = \frac{tD_S}{L_2}, \qquad \nabla^{*2} = L^2 \nabla^2.$$

Show that equations (71a,b) can be written in the dimensionless form given by equations (65a,b) of Section 11.8, where

$$f(s, a) = \alpha(a_0 - a) - \rho F(s, a), \tag{72a}$$

$$g(s, a) = s_0 - s - \rho F(s, a), \tag{72b}$$

$$F(s, a) = \frac{sa}{1 + s + Ks^2}. \tag{72c}$$

(e) What are the dimensionless parameters α, β, γ, K, and ρ that appear in equations (65) and (72)? (For example, find that $K = K_m/K_s$, and so forth.)

(f) Interpret the biological significance of these parameters.

(g) Why is it not possible to obtain a simple analytical solution for the homogeneous steady state $(\bar{s}, \bar{a})$ of this model?

(h) Find the Jacobian matrix

$$\mathbf{J} = \begin{pmatrix} g_s & g_a \\ f_s & f_a \end{pmatrix}$$

for f and g given in part (d).

(i) Show that $\partial f/\partial a < 0$ and $\partial g/\partial a < 0$. What must be true about $\partial f/\partial s$ and $\partial g/\partial s$ at the steady state $(\bar{s}, \bar{a})$ if this chemical reaction is to exhibit diffusive instability? Which of the two types of chemical interactions given in Section 11.6 is consistent with the signs of the partial derivatives in **J**?

(j) Show that the requirements found in part (i) are consistent with $\bar{s} > 1\sqrt{K}$. Can you interpret this result biologically?

(k) It is not an easy task to graph the curves corresponding to the nullclines of this system. However, given that for three different K values the graphs have the qualitative features shown in Figures 8.6(a), it is a straightforward process to deduce which of these corresponds to a situation that implies diffusive instability. Use results of Section 7.8 to make this deduction.

22. *Diffusive predator-prey model (Mimura and Murray, 1978).* Consider the general model suggested by Mimura and Murray [equations (68a,b)]. Suppose $(\bar{P}, \bar{Q})$ stands for the homogeneous steady state of these equations.

(a) What are the equations of the nullclines in the spatially homogeneous equations? (What equations do $\bar{P}$ and $\bar{Q}$ satisfy?)

(b) Give conditions for stability of this steady state in the absence of diffusion. Determine the possible sign patterns of elements in the Jacobian.

(c) The two nullclines shown in Figure 11.17 intersect to the left of the hump in the P-nullcline. Is this configuration consistent with diffusive instability?

(d) Give the full condition for diffusive instability in this system.

REFERENCES

Reaction Diffusion and Models for Developmental Biology

Bard, J. B. L. (1977). A unity underlying the different zebra striping patterns. *J. Zool.* (London), *183,* 527–539.

Berding, C.; Harbich, T.; and Haken, H. (1983). A prepattern formation mechanism for the spiral-type patterns of the sunflower head. *J. Theor. Biol., 104,* 53–70.

Gierer, A., and Meinhardt, H. (1972). A theory of biological pattern formation. *Kybernetik, 12,* 30–39.

Kauffman, S. A. (1977). Chemical patterns, compartments, and a binary epigenetic code in *Drosophila. Am. Zool., 17,* 631–648.

Kauffman, S. A., Shymko, R. M., Trabert, K. (1978). Control of sequential compartment formation in Drosphila. *Science, 199,* 259.

Lacalli, T. C. (1981). Dissipative structures and morphogenetic pattern in unicellular algae. *Phil. Trans. Roy. Soc. Lond. B (Biol. Sci.), 294,* 547–588.

Lacalli, T. C., and Harrison, L. G. (1979). Turing's conditions and the analysis of morphogenetic models. *J. Theor. Biol., 76,* 419–436.

Meinhardt, H. (1978). Models for the ontogenetic development of higher organisms. *Rev. Physiol. Biochem. Pharmacol., 80,* 47–104.

Meinhardt, H. (1982). *Models of Biological Pattern Formation.* Academic Press, New York.

Meinhardt, H., and Gierer, A. (1974). Applications of a theory of biological pattern formation based on lateral inhibition. *J. Cell. Sci., 15,* 321–346.

Meinhardt, H., and Gierer, A. (1980). Generation and regeneration of sequence of structures during morphogenesis. *J. Theor. Biol., 85,* 429–450.

Murray, J. D. (1981*a*). A prepattern formation mechanism for animal coat markings. *J. Theor. Biol., 88,* 161–199.

Murray, J. D. (1981*b*). On pattern formation mechanisms for Lepidopteran wing patterns and mammalian coat markings. *Phil. Trans. R. Soc. Lond. B, 295,* 473–496.

Segel, L. A., ed. (1980). *Mathematical Models in Molecular and Cellular Biology.* Cambridge University Press, Cambridge.

Segel, L. A. (1984). *Modeling Dynamic Phenomena in Molecular and Cellular Biology.* Cambridge University Press, Cambridge, chap. 8.

Reaction Diffusion and Ecological Patterns

Conway, E. D. (1983). Diffusion and the predator-prey interaction: Pattern in closed systems. Tulane University Report in Computing and Applied Mathematics, New Orleans.

Mimura, M., and Murray, J. D. (1978). On a diffusive prey-predator model which exhibits patchiness. *J. Theor. Biol., 75,* 249–262.

Okubo, A. (1980). *Diffusion and Ecological Problems: Mathematical Models.* Springer-Verlag, New York.

Segel, L. A., and Jackson, J. L. (1972). Dissipative structure: An explanation and an ecological example. *J. Theor. Biol., 37,* 545–59.

Cellular Slime Molds and Mathematical Models

Bonner, J. T. (1944). A descriptive study of the development of the slime mold *Dictyostelium discoideum, Amer. J. Bot., 31,* 175–182.

Bonner, J. T. (1959). *The Cellular Slime Molds.* Princeton University Press, Princeton, N.J.

Bonner, J. T. (1974). *On Development; the biology of form.* Harvard University Press, Cambridge, Mass.

Devreotes, P. N., and Steck, T. L. (1979). Cyclic 3′5′ AMP relay in *Dictyostelium discoideum. J. Cell Biol., 80,* 300–309.

Goldbeter, A., and Segel, L. A. (1980). Control of developmental transitions in the cyclic AMP signalling system of *Dictyostelium discoideum*. *Differentiation, 17,* 127–135.

Keller, E. F., and Segel, L. A. (1970). The initiation of slime mold aggregation viewed as an instability. *J. Theor. Biol., 26,* 399–415.

Odell, G. M., and Bonner, J. T. (1986). How the Dictyostelium discoideum grex crawls. *Phil. Trans. Roy. Soc. Lond. B,* 312, 487–525.

Pate, E. F., and Odell, G. M. (1981). A computer simulation of chemical signaling during the aggregation phase of *Dictyostelium discoideum*. *J. Theor. Biol., 88,* 201–239.

Pate, E. F., and Othmer, H. G. (1986). Differentiation, cell sorting, and proportion regulation in the slug stage of *Dictyostelium discoideum*. *J. Theor. Biol., 118,* 301–319.

Rubinow, S. I., Segel, L. A.; Ebel, W. (1981). A mathematical framework for the study of morphogenetic development in the slime mold. *J. Theor. Biol., 91,* 99–113.

Segel, L. A. ed. (1980). *Mathematical Models in Molecular and Cellular Biology,* Cambridge University Press, Cambridge.

Segel, L. A. (1981). Analysis of population chemotaxis. Sect. 6.5 in Segel, L. A. ed. *Mathematical Models in Molecular and Cellular Biology,* Cambridge University Press, Cambridge, pp. 486–501.

Segel, L. A. (1984). *Modeling dynamic phenomena in molecular and cellular Biology,* (Chapter 6), Cambridge University Press, Cambridge.

Segel, L. A., and Stoeckly, B. (1972). Instability of a layer of chemotactic cells, attractant, and degrading enzyme. *J. Theor. Biol., 37,* 561–585.

Mathematical and General Sources

Fife, P. C. (1979). *Mathematical Aspects of Reacting and Diffusing Systems*. Springer-Verlag, New York.

Murray, J. D. (1982). Parameter space for Turing instability in reaction diffusion mechanisms: A comparison of models. *J. Theor. Biol., 98,* 143–163.

Rothe, F. (1984). *Global Solutions of Reaction-Diffusion Systems*. Springer-Verlag, Berlin.

Smoller, J. (1983). *Shock Waves and Reaction-Diffusion Equations*. Springer-Verlag, New York.

Tyson, J. J., and Fife, P. C. (1980). Target patterns in a realistic model of the Belousov-Zhabotinski reaction. *J. Chem. Phys., 73,* 2224–2237.

Reviews

Bard, J., and Lauder, I. (1974). How well does Turing's theory of morphogenesis work? *J. Theor. Biol., 45,* 501–531.

Levin, S., ed. (1983). *Modelling of Patterns in Space and Time*. Springer-Verlag, New York.

Levin, S. A., and Segel, L. A. (1985). Pattern generation in space and aspect. *SIAM Rev., 27,* 45–67.

Chemical Morphogens in Hydra

Kemmner, W. (1984). Head regeneration in *Hydra:* Biological studies and a model. In W. Jager and J. D. Murray, eds., *Modelling of Patterns in Space and Time*. Springer-Verlag, New York. (See also references therein.)

Other Pattern Formation Models

Ermentrout, G. B.; Campbell, J.; and Oster, G. (1986). A model for shell patterns based on neural activity. *Veliger, 28,* 369–388.

Ermentrout, G. B., and Cowan, J. D. (1979). A mathematical theory of visual hallucination patterns. *Biol. Cybernet., 34,* 137–150.

Ermentrout, G. B. (1984). Mathematical models for the patterns of premigrainous auras, in A. V. Holden and W. Winlow, eds., *The Neurobiology of Pain,* Manchester University Press, Manchester, England.

Cohen, D. (1967). Computer simulation of biological pattern generation processes. *Nature, 216,* 246–248.

Jackson, J. B. C.; Buss, L. W.; and Cook, R. E., ed. (1985). *Population Biology and Evolution of Clonal Organisms.* Yale University Press, New Haven, Conn.

Murray, J. D., and Oster, G. F. (1984). Cell traction models for generating pattern and form in morphogenesis. *J. Math. Biol., 19,* 265–279.

Nijhout, H. F. (1981). The color patterns of butterflies and moths. *Sci. Am.,* November 1981, pp. 140–151.

Odell, G. M.; Oster, G.; Alberch, P.; and Burnside, B. (1981). The mechanical basis of morphogenesis. *Develop. Biol., 85,* 446–462.

Oster, G. F.; Murray, J. D.; and Harris, A. K. (1983). Mechanical aspects of mesenchymal morphogenesis. *J. Embryol. Exp. Morph., 78,* 83–125.

Sulsky, D.; Childress, S.; and Percus, J. K. (1984). A model of cell sorting. *J. Theor. Biol., 106,* 275–301.

Swindale, N. V. (1980). A model for the formation of ocular dominance stripes. *Proc. R. Soc. Lond. B, 208,* 243–264.

Thompson, D'Arcy, W. (1942). *On Growth and Form.* Cambridge University Press, Cambridge.

Young, D. A. (1984). A local activator-inhibitor model of vertebrate skin patterns. *Math. Biosci., 72,* 51–58.

Wolfram, S. (1984*a*). Cellular automata as models of complexity. *Nature, 311,* 419–424.

Wolfram, S. (1984*b*). Universality and complexity in cellular automata. *Physica, 10D,* 1–35.

Miscellaneous

Lauffenburger, D. A., and Kennedy, C. R. (1983). Localized bacterial infection in a distributed model for tissue inflammation. *J. Math. Biol., 16,* 141–163.

Schaller, H. C. (1973). Isolation and characterization of a low molecular weight substance activating head and bud formation in hydra. *J. Embryol. Exp. Morphol., 29,* 27–38.

Schaller, H. C. (1976). Action of the head activator on the determination of interstitial cells in hydra. *Cell. Differ. 5,* 13–20.

Nicolis, G., and Prigogine, I. (1977). *Self-Organization in Nonequilibrium Systems.* Wiley, New York.

Selected Answers

CHAPTER 1

1. (b) $x_n = 10 \cdot 2^n$.

2. (b) $x_n = 1 + 2 \cdot (4)^n$; (c) $x_n = (-1)^n + 4$; (e) $x_n = 5 + (-2)^n$.

3. (b) (i) $A_1 + nA_2$; (ii) $(-1)^n[A_1 + nA_2]$; (iii) $3^n[A_1 + nA_2]$.

6. (a) $A(5)^n + B(2)^n$; (b) $A(\frac{1}{2})^n + B(-\frac{1}{2})^n$; (e) $A(\frac{1}{3})^n + B(-1)^n$.

7. (b) $c_1\begin{bmatrix}4\\1\end{bmatrix}(\frac{1}{2})^n + c_2\begin{bmatrix}4\\-3\end{bmatrix}(-\frac{1}{2})^n$;

(d) $c_1\begin{bmatrix}1\\-2\end{bmatrix}(-1)^n + c_2\begin{bmatrix}1\\1\end{bmatrix}(2)^n$;

(f) $c_1\begin{bmatrix}4\\1\end{bmatrix}(\frac{1}{2})^n + c_2\begin{bmatrix}6\\1\end{bmatrix}(\frac{1}{4})^n$.

8. (a) $(\sqrt{2})^n e^{i\pi n/4}$; (c) $(10)^n e^{i\pi n/2}$; (e) $\left(\frac{1}{\sqrt{2}}\right)^n e^{i5\pi n/4}$.

9. (a) $c_1 \cos\left(\frac{n\pi}{2}\right) + c_2 \sin\left(\frac{n\pi}{2}\right)$; (c) $\sqrt{2}^n\left[c_1 \cos\left(\frac{3\pi n}{4}\right) + c_2 \sin\left(\frac{3\pi n}{4}\right)\right]$.

10. (b) $f > 1/r(1 - m)$.

11. (b) $K = d/(a + b + c)$.

13. (a) 2,1,3,4,7,11,18,29,47,76,123.

14. (b) $R_n^0 = \frac{1}{\sqrt{5}}\left[\left(\frac{1+\sqrt{5}}{2}\right)^{n+1} - \left(\frac{1-\sqrt{5}}{2}\right)^{n+1}\right]$.

18. (a) $C_{n+1} = C_n - \beta V_n + m$,
$V_{n+1} = \alpha C_n$.

(c) $4\alpha\beta < 1 \Rightarrow$ amount of CO_2 lost per breath is less than $\frac{1}{4}$ (amount of CO_2 that induces a unit volume of breathing).
$4\alpha\beta > 1$: $\lambda = \frac{1}{2}\{1 \pm \gamma i\}$, $\gamma = (4\alpha\beta - 1)^{1/2}$. $|\lambda| \geq 1$ when
$\gamma \geq \sqrt{3} \Rightarrow \alpha\beta > 1$. Frequency $\phi = \frac{\pi}{3}$.

20. (c)

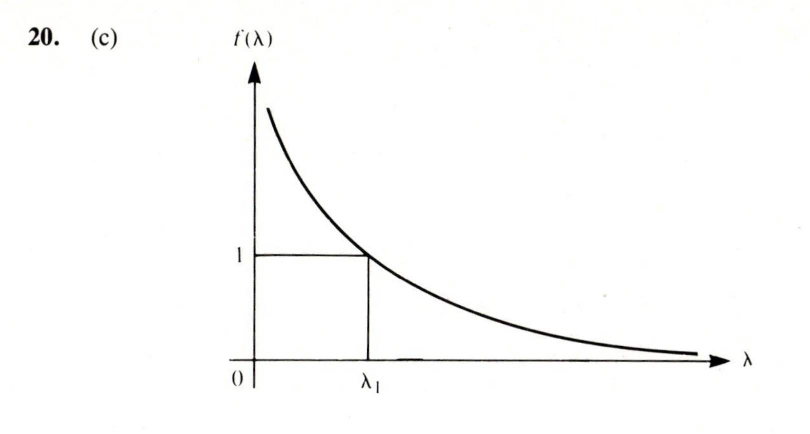

CHAPTER 2

1. (a) Linear, $x_n = \left(\frac{1-\alpha}{1-\beta}\right)^n, \quad \beta \neq 1.$

(c) Nonlinear, $\bar{x} = 0.$

(e) Nonlinear, $\bar{x} = (K - k_2)/k_1 \quad k_1 \neq 0.$

2. (a) Stable for $|r| < 1$. (b) Unstable. (c) Stable. (d) Unstable.

4. (a)

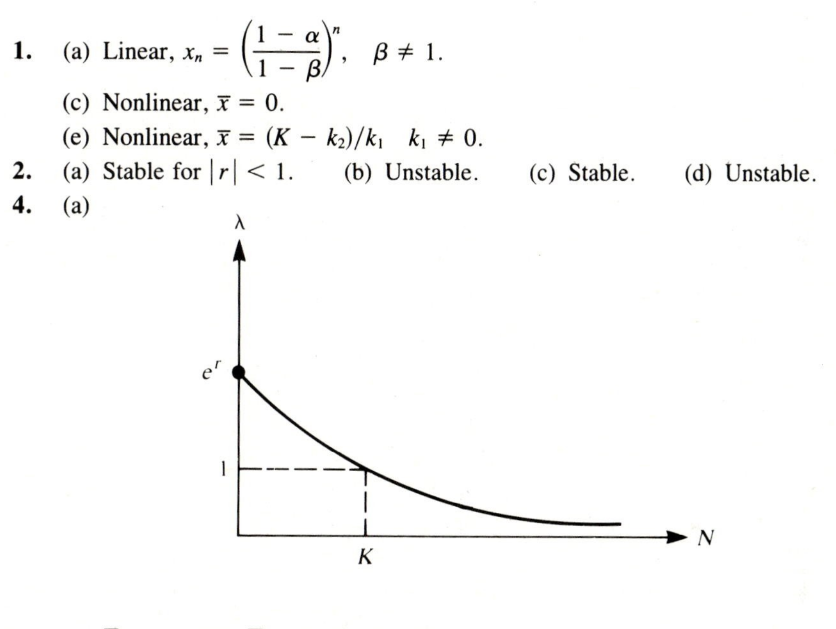

8. (b) $\bar{N}_1 = 0$ or $\bar{N}_2 = (\lambda^{1/b} - 1)/a.$

$\bar{N}_1$ stable iff $\lambda < 1$; $\bar{N}_2$ stable iff $0 < b(1 - \lambda^{-1/b}) < 2.$

11. Steady states: (0, 0) and $\left(\frac{Fk(F^{1/k} - 1)}{a(F-1)}, \frac{k}{a}(F^{1/k} - 1)\right).$

14. $f'(x) = kb/(b+x)^2, \quad f'(0) = k/b > 1.$

16. (b) $C_{t+1} = fC_tS_t; \quad S_{t+1} = S_t - fC_tS_t + B.$

17. (b) $C_{n+1} = C_n - \beta C_n V_n + m, \qquad V_{n+1} = \alpha C_n.$

(c) Need $(m\alpha\beta) < 1$ for stability.

(d) yes, for $|x - 3| < 2\sqrt{2}$ where $x = (m\alpha\beta)^{1/2}.$

(h) $\bar{C} = \frac{1}{2}\{\gamma \pm \sqrt{\gamma^2 + 4k\gamma}\}$ where $\gamma = m/\beta\alpha,$
$\bar{V} = \alpha\bar{C}/(K + \bar{C}).$

CHAPTER 3

4. (c) $\overline{N}$ stable for $|1 - b(\lambda^{-1/b} - 1)| < 1$.

5. (b) 1-stable, 2-stable, 3-stable, 4-unstable, 5-stable.

7. (a) $n_{t+1} = \lambda n_t e^{-p_t}$, $p_{t+1} = n_t(1 - e^{-p_t})$ for $\overline{N} = \dfrac{1}{(ac)}$, $\overline{P} = \dfrac{1}{a}$.

9. (b) $a_{11} = (1 - r\overline{N}/K)$, $a_{12} = -\overline{N}a$, $a_{21} = \overline{P}/\overline{N}$, $a_{22} = a(\overline{N} - \overline{P})$.

10. (b) $N_{t+1} = \lambda N_t \exp[-(aP_t)^{1-m}]$, $P_{t+1} = N_t(1 - \exp[-(aP_t)^{1-m}])$.

(c) $\overline{P} = (\ln \lambda/a)^s$, $\overline{N} = \lambda\overline{P}/(\lambda - 1)$.

15. (a) Steady states $\overline{h} = (\ln f)/a$, $\overline{q} = \overline{h}\Big/\left(\delta - \dfrac{1}{r}\right)$.

(b) $k = \ln f$, $b = r\left[\delta - \dfrac{1}{r}\right]$.

(d) For $F(Q, H) = Qe^{k(1-H)}$, $G(Q, H) = bH\left(1 + \dfrac{1}{b} - \dfrac{H}{Q}\right)$.

$F_Q(\overline{Q}, \overline{H}) = 1$, $F_H(\overline{Q}, \overline{H}) = -k$, $R_x(\overline{Q}, \overline{H}) = -b$

(where $R = G/H$, $x = H/Q$).

19. $u_{n+1} = u_n^2 + \frac{1}{4}v_n^2$, $\quad v_{n+1} = \frac{1}{2}v_n^2$, $\quad w_{n+1} = \frac{1}{4}v_n^2 + w_n^2$.

CHAPTER 4

5. (a) r = intrinsic growth rate. B = carrying capacity.

(d) For $t \to \infty$, $e^{-rt} \to 0$ so $N(t) \to N_0B/N_0 = B$.

For N_0 small, $N(t) \approx N_0B/Be^{-rt} = N_0e^{rt}$.

6. (b) (mass nutrient)/(number of bacteria).

8. $\alpha_1 = \dfrac{v}{F}K_{\max}$ = ratio of [emptying time of chamber] to [(1/ln 2) times bacterial doubling time].

$\alpha_2 = c_0/K_n$ = ratio of [stock nutrient concentration] to [concentration which produces a half-maximal bacterial growth rate].

10. (a) $\dfrac{dN}{dt} = \left(\dfrac{C}{a + C}\right)N - bN$, $\dfrac{dC}{dt} = -\left(\dfrac{C}{a + C}\right) - bC + 1$.

for $a = K_nVK_{\max}/FC_0$, $\quad b = F/K_{\max}V$.

11. Increase C_0, V, decrease F. (No local maximum).

15. (a) $y'' = -\cot(x)$: linear, order 2, not homogeneous, constant coefficients.

(d) $(2y + 2)y' - y = 0$; nonlinear, order 1, homogeneous, nonconstant coefficients.

(i) $\dfrac{dy}{dt} + y = \dfrac{1}{t}$ $(t \neq 0)$; linear, order 1, nonhomogeneous, constant coefficients.

16. (d) Steady states (0, 0), (1, 1); $\quad J(0, 0) = \begin{bmatrix} 1 & 0 \\ 0 & -1 \end{bmatrix}$, $\quad J(1, 1) = \begin{bmatrix} 0 & -1 \\ 1 & 0 \end{bmatrix}$.

18. (b) $y(t) = .001e^{10t}$, $\quad$ (d) $y(t) = \frac{5}{2}(e^{3t} + e^{-3t})$, $\quad$ (e) $y(t) = \frac{1}{5}(3 + 2e^{5t})$.

22. (a) $\begin{bmatrix} x(t) \\ y(t) \end{bmatrix} = c_1\begin{bmatrix} 1 \\ 0 \end{bmatrix}e^{-t} + c_2\begin{bmatrix} 0 \\ 1 \end{bmatrix}e^{t}$,

(c) $\begin{bmatrix} x(t) \\ y(t) \end{bmatrix} = c_1\begin{bmatrix} 7 \\ -2 \end{bmatrix}e^{-4t} + c_2\begin{bmatrix} 1 \\ 1 \end{bmatrix}e^{5t}$,

(f) $\begin{bmatrix} x(t) \\ y(t) \end{bmatrix} = c_1\begin{bmatrix} 1 \\ 2 - \sqrt{7} \end{bmatrix}e^{-(2+\sqrt{7})t} + c_2\begin{bmatrix} 1 \\ 2 + \sqrt{7} \end{bmatrix}e^{(-2+\sqrt{7})t}$.

29. (b) $\overline{x}_1 = \dfrac{B}{F}\Big/\left[\dfrac{C}{F - E} - \dfrac{A}{F}\right]$, $\overline{x}_3 = \dfrac{C\overline{x}_1}{F - E}$, $\overline{x}_2 = A\overline{x}_1 - B$.
where $A = (u + k_{12})/k_{21}$, $B = D/k_{21}$, $C = k_{12}/(k_{21} + s + k_{23})$,
$E = k_{32}/(k_{21} + s + k_{23})$, $F = k_{32}/k_{23}$.

CHAPTER 5

1. (b)

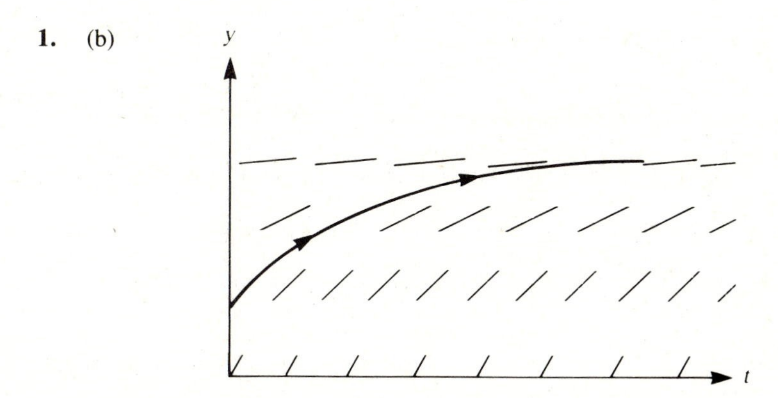

(e)

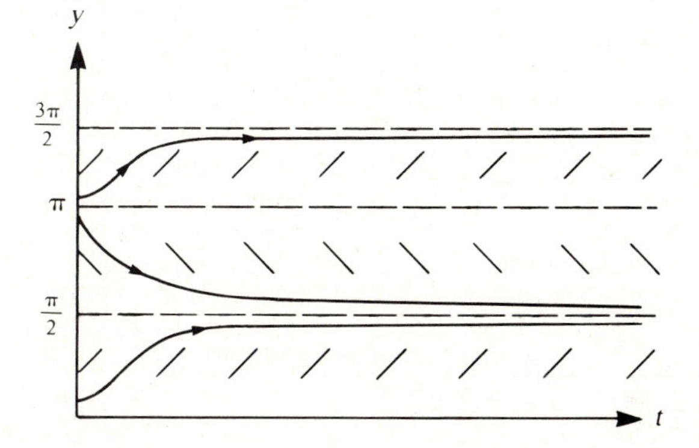

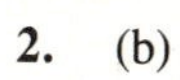
2. (b)

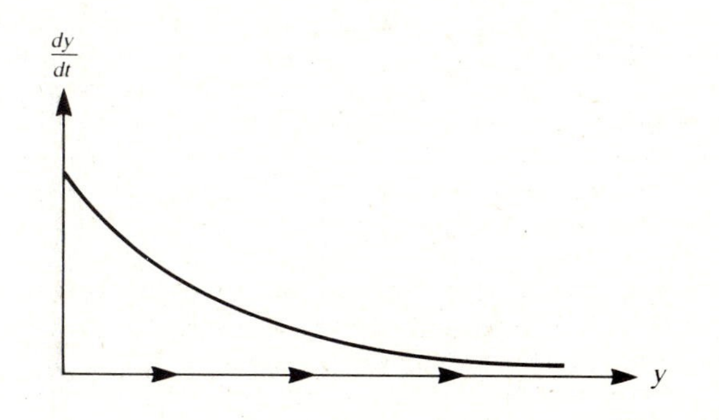

(e)

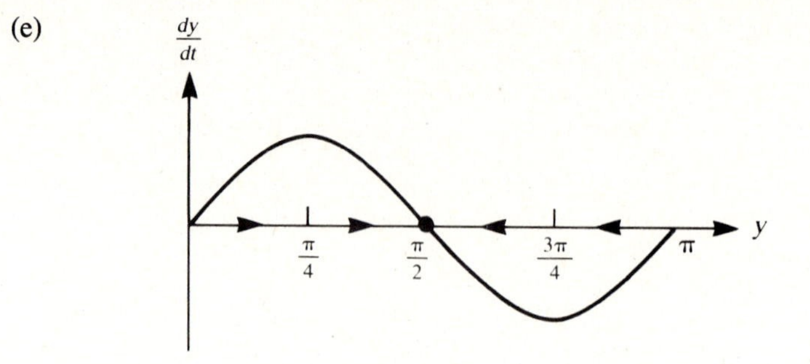

4. (b) (1) $(x, y) = (t, t(t-1)), \left(\frac{dx}{dt}, \frac{dy}{dt}\right) = (1, 2t-1).$

(4) $(x, y) = (\cos t, \sin t), \left(\frac{dx}{dt}, \frac{dy}{dt}\right) = (-\sin t, \cos t).$

(6) $(x, y) = (t, 4t^2), \left(\frac{dx}{dt}, \frac{dy}{dt}\right) = (1, 8t).$

5. (a)

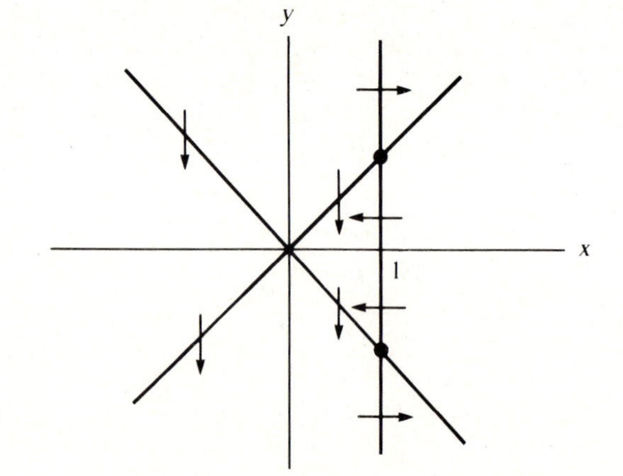

(e)

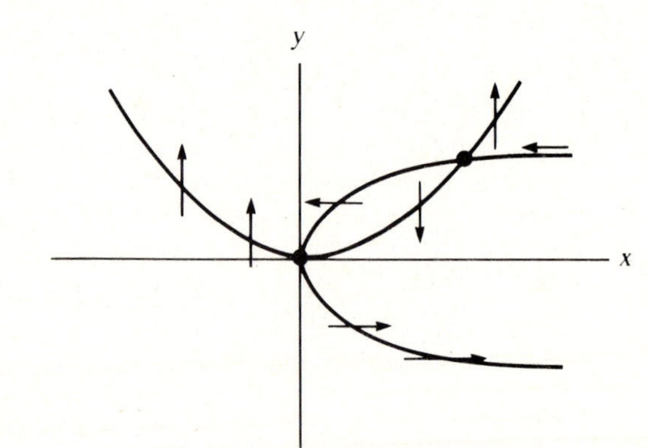

(g)

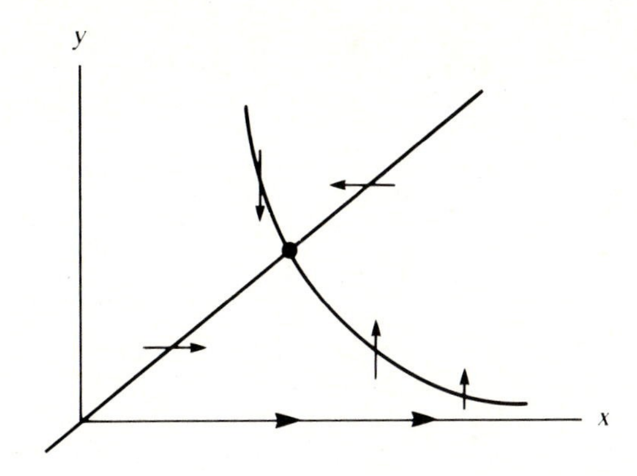

6. (a) $J(1, 1) = \begin{bmatrix} -2 & 2 \\ 1 & 0 \end{bmatrix}$, saddle; $J(1, -1) = \begin{bmatrix} -2 & -2 \\ 1 & 0 \end{bmatrix}$ stable spiral.

(d) $J(0, 1) = \begin{bmatrix} -1 & 0 \\ 0 & -1 \end{bmatrix}$ stable node; $J(-1, 0) = \begin{bmatrix} 0 & 1 \\ 1 & 0 \end{bmatrix}$ saddle point.

7. (a) neutral center, (b) saddle, (c) unstable node,
(d) saddle, (e) stable spiral, (f) unstable spiral.

12. (a) $A = \dfrac{\alpha_1}{\alpha_2}(\alpha_1 - 1)^2 - \dfrac{(\alpha_1 - 1)}{\alpha_1}$.

(b) $\beta = -(A + 1),\ \gamma = A,\ \lambda_{12} = \frac{1}{2}\{-(A + 1) \pm \sqrt{((A + 1)^2 - 4A)}\} = -A, -1.$

$$\frac{N - \alpha_1(\alpha_2 - \bar{C}_1)}{C - \bar{C}_1} = -\alpha_1 \Rightarrow N - \alpha_1\alpha_2 = -\alpha_1 C.$$

15. (c)

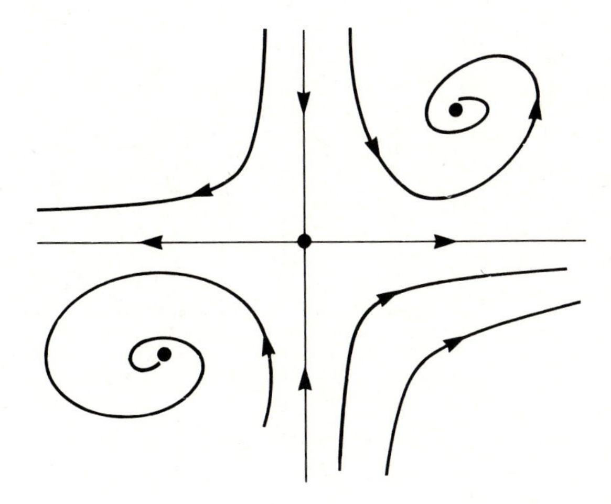

(e)

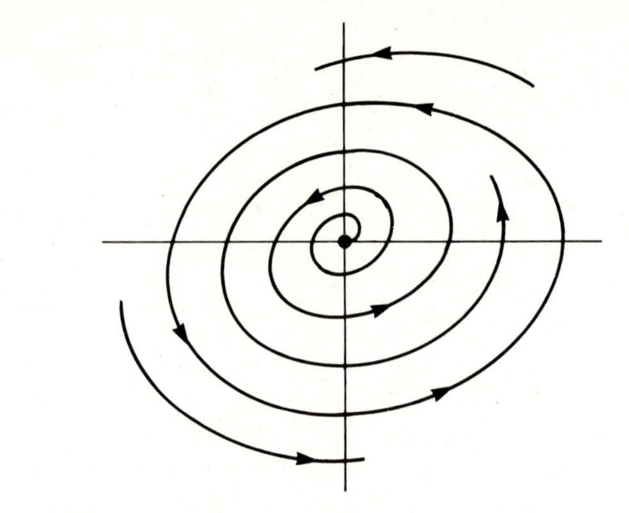

20. (b) $K = K_2K_4I_0^3$, $\alpha = K_3K_4I_0$.

(c) steady state: $\bar{q} = \bar{I}$, $\bar{I}^2(\bar{I} - 1) = \frac{1}{K}$, $(\Rightarrow \bar{I} > 1)$.

$$J = \begin{bmatrix} -KI(I-1) & -Kq(2I-1) \\ \alpha & -\alpha \end{bmatrix}.$$

$Tr(J) < 0 \quad Det(J) > 0$ for $\bar{I} > 1$ $\therefore$ stable.

21. (a) $\epsilon = 1/\Delta t$.

(b) $\frac{dC}{dt} = -\beta V + m$, $\frac{dV}{dt} = \alpha C - \epsilon V$.

steady state: $\bar{V} = m/\beta$, $\bar{C} = \frac{\epsilon m}{\alpha\beta}$ (stable).

(c) steady state: $\bar{V} = \frac{m}{\beta C}$, $\bar{C} = \left(\frac{\epsilon m}{\alpha\beta}\right)^{1/2}$ (stable).

Decaying oscillations if $\left[\epsilon + \frac{\delta}{\epsilon}\right]^2 < 8\delta$ for $\delta^2 = (\epsilon m\beta\alpha)$.

22. (e) $m = 2$, $2\alpha\beta < 1$.

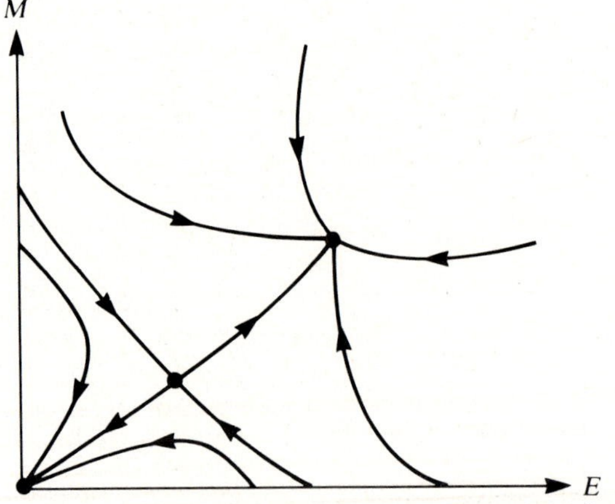

CHAPTER 6

2. (a) $a_1 = -rKM$, $a_2 = r(K + M)$, $a_3 = -r$.

(c) $\dfrac{1}{N^{1/KM}} \dfrac{|N - M|^C}{|K - N|^B} = Pe^{rt}$, P = constant.

3. (b) Beverton-Holt solution: $N^{\alpha}e^{N} = P\, e^{rt}$, P = constant.
(steady state $N = O$ is unstable).

4. (a) not stabilizing. (d) not stabilizing.

6. (d) $N = K$ is a stable steady state.

9. (b) $\dfrac{dx}{dt} = (a - \phi)x - bxy$, $\dfrac{dy}{dt} = -(c + \phi)y + dxy$;

steady states: (0, 0) and $\left(\dfrac{c + \phi}{d}, \dfrac{a - \phi}{d}\right)$.

14. Oscillations when $ac < (2dK)^2\left(1 - \dfrac{c}{dK}\right)$.

17. (a) $(K_1 + \alpha N_2)$ is the carrying capacity of species 1. Thus the presence of species 2 contributes positively to the carrying capacity of species 1.

(b) steady states: (0, 0) (unstable node), $(K_1, 0)$, $(0, K_2)$ (saddle points).
$\left(\dfrac{K_1 + \alpha K_2}{1 - \alpha\beta}, \dfrac{K_2 + \beta K_1}{1 - \alpha\beta}\right)$ (stable node).

(c) (Last steady state exists only if $\alpha\beta < 1$).

18. $H_1 = (a_1)$, $H_2 = \begin{bmatrix} a_1 & 1 \\ a_3 & a_2 \end{bmatrix}$, $H_3 = \begin{bmatrix} a_1 & 1 & 0 \\ a_3 & a_2 & a_1 \\ 0 & 0 & a_3 \end{bmatrix}$

$\det H_1 = a_1$, $\det H_2 = a_1a_2 - a_3$, $\det H_3 = a_3(\det H_2)$.

23. (1) (a) $\begin{bmatrix} 0 & - & 0 & 0 \\ + & 0 & 0 & 0 \\ 0 & + & 0 & - \\ 0 & 0 & - & - \end{bmatrix}$. (e) $\begin{bmatrix} 0 & + & 0 & 0 & 0 \\ - & 0 & 0 & 0 & 0 \\ 0 & + & 0 & 0 & 0 \\ 0 & 0 & + & 0 & + \\ 0 & 0 & 0 & - & 0 \end{bmatrix}$;

(2) (a) {1, 2}. (e) {1, 2}, {4, 5}.

(3) (a) not qualitatively stable; (e) not qualitatively stable.

24. (b)

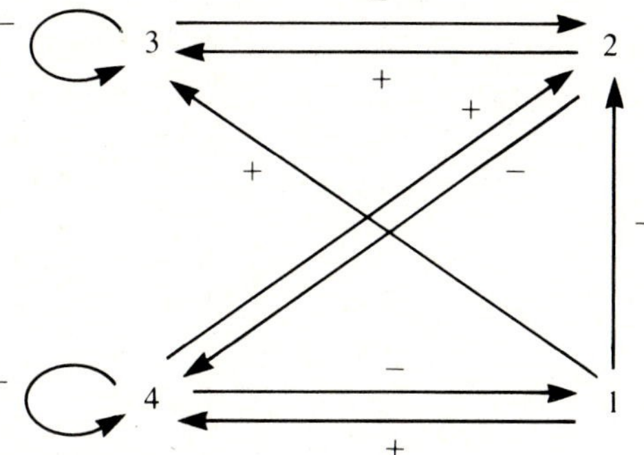

predation community:
{1, 4, 2, 3}.
Not qualitatively stable.

29. For $\lambda > 0$, the only equilibrium of S, I equations is $(S, I) = (0, 0)$.
But $J(0, 0) = \begin{bmatrix} -\lambda & 0 \\ 0 & -\gamma \end{bmatrix} \Rightarrow (0, 0)$ stable.

32. Steady states: (K, 0) (saddle point), $(\bar{x}_2, b\bar{x}_2)$ where $\bar{x}_2 = r \Big/ \left(ab + \frac{r}{K}\right)$ (stable).

CHAPTER 7

2. (a) $x_1 = \dfrac{k_1 rc}{(k_1 + k_2) + k_1 c}$.

4. $K \ln c + c = -\lambda t + \text{constant}$.

6. $x_1(t) = (1 - e^{-(K+1)t})/(K + 1)$, $x_1(t) \uparrow \dfrac{1}{(K + 1)}$ for $t \to \infty$.

7. (a) $K = r$, $a = (k_{-1}/k_1)$, $b = (k_{-1} + k_2)/k_1$.
(b)

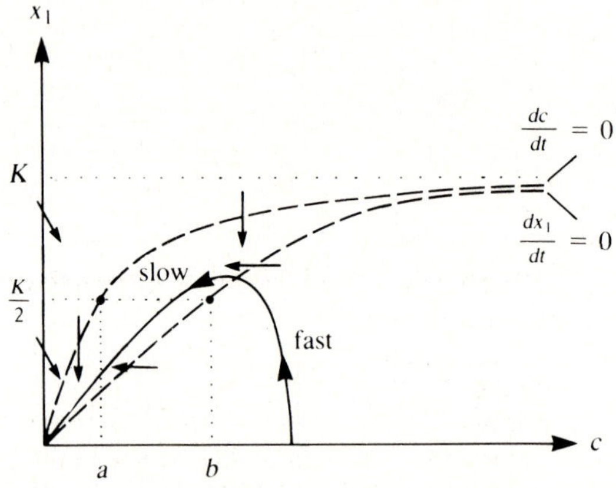

12. $\dfrac{da}{dt} = -K_1 ab + K_{-1}x$, $\dfrac{db}{dt} = -K_1 ab + K_{-1}x$.
$\dfrac{dx}{dt} = -K_{-1}x + K_1 ab$. At equilibrium $\dfrac{da}{dt} = \dfrac{db}{dt} = \dfrac{dx}{dt} = 0 \Rightarrow x = \dfrac{K_1}{K_{-1}}(ab)$.

16. (b) $J = \begin{bmatrix} ds_1 & -b \\ -ds_2 & d \end{bmatrix}$.
(c) $Tr\, J = bs_1 + d < 0$, $det\, J = bd(s_1 - s_2) > 0$.

19. (b) $J = \begin{bmatrix} -(k + \delta^2) & \dfrac{-2\delta^2}{k + \delta^2} \\ (k + \delta^2) & \dfrac{\delta^2 - k}{\delta^2 + k} \end{bmatrix}$; sign pattern $\begin{bmatrix} - & - \\ + & + \end{bmatrix}$ if $k < \delta^2$.

20. (a) Let $k_1 = k_2 = k_3 = k_4 = 1$.
(b) $\bar{x} = A$, $\bar{y} = B/A$.
(c) $J = \begin{bmatrix} B - 1 & A^2 \\ -B & -A^2 \end{bmatrix}$.

21. (a) Steady state: $(\rho + \gamma, (\rho + \gamma)^2/\gamma)$.

23. (b) $k_{11} = -F_x$, $k_{21} = \alpha F_x$, $k_{12} = -F_y$, $k_{22} = \alpha(F_y - G_y)$.

CHAPTER 8

2. (b) $\left[\dfrac{\partial f}{\partial x} + \dfrac{\partial g}{\partial y}\right] = b > 0.$

4. (a–c) No limit cycle possible.
(d) Cannot rule out limit cycle.

6. (b)

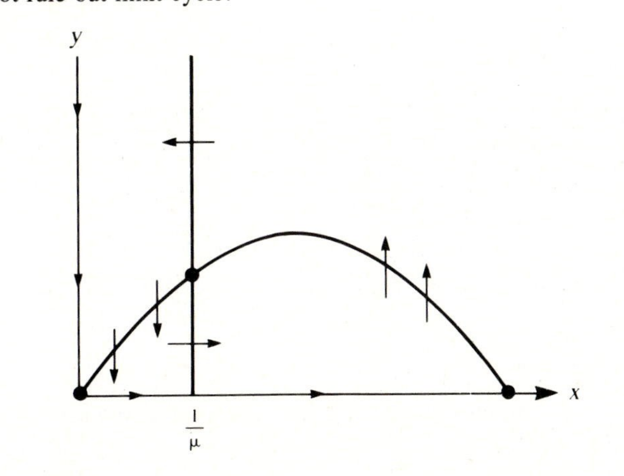

Flow can escape to (or emanate from) arbitrarily large y values. Poincaré–Bendixson theory inconclusive.

7. (b) No limit cycle.
(c) limit cycle exists.

10. (a) Functions have no maxima or minima, only an inflection point.

13. (b) Figure b. (c) Cannot solve cubic equation.

17. (c) Condition guarantees $f = 0$ nullcline intersects x-axis to the right of the intersection of $g = 0$ with x-axis.

18. (a) $J = \begin{bmatrix} xf_x + f & xf_y \\ yg_x & yg_y + g \end{bmatrix}$. But $f = g = 0$ at the nontrivial steady state.

21. (a)

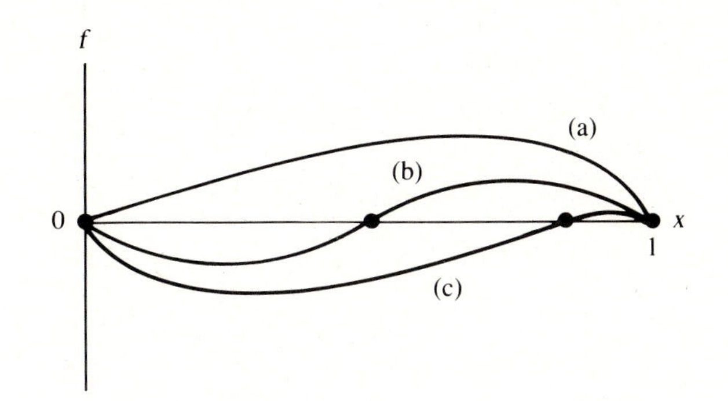

(a) $\alpha(y - 1) > 1$.
(b), (c) $0 < \alpha(y - 1) < 1$.

(b) $g(x, y) = \beta(Kx - y)$
(c, d)

(h) periodic solution about an unstable steady state $(\gamma, K\gamma)$ will exist.

26. (a) Let $r^2 = x^2 + y^2$. $\dfrac{dr^2}{dt} = 2x\dfrac{2x}{dt} + 2y\dfrac{dy}{dt} = 2xy - 2xy = 0$

so r^2 = Constant is a solution (neutrally stable).

30. (b) $\dfrac{dr^2}{dt} = 2r^2[1 - r^2]$. $\dfrac{d\theta}{dt} = 1$.

CHAPTER 9

1. (a) Level curves are circles. Surface is a paraboloid

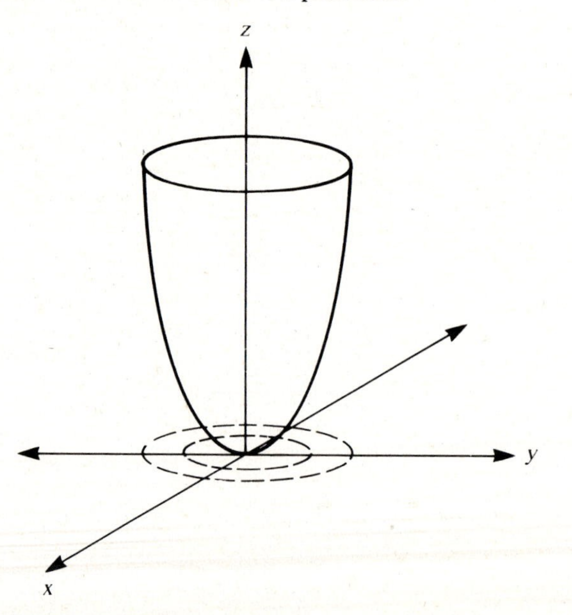

(b) Level curves are circles. Surface is a Gaussian, with maximum at (0, 0).

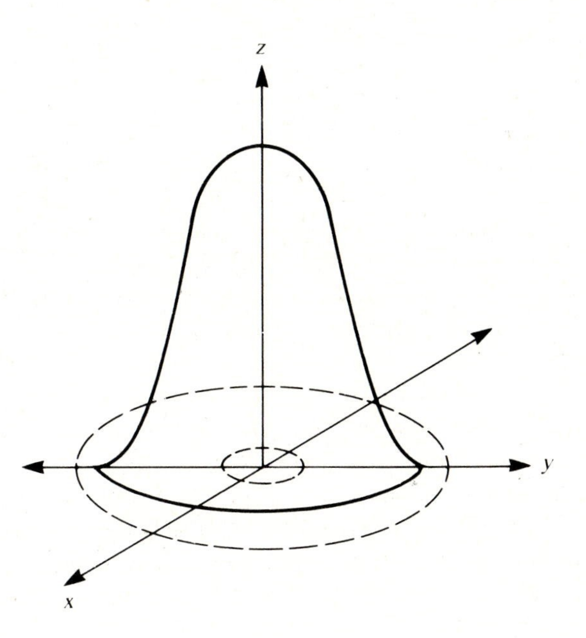

2.

	$\frac{\partial f}{\partial x}$	$\frac{\partial^2 f}{\partial x\,\partial y}$	∇f	**3.** critical points
(a)	$2x$	0	$(2x, 2y)$	minimum at (0, 0).
(c)	$-xe^{-R^2/2}$	$yxe^{-R^2/2}$	$(-x, -y)e^{-R^2/2}$	maximum at (0, 0), $R^2 = x^2 + y^2$.
(e)	y	1	(y, x)	(0, 0) is a saddle point.

4. (c)

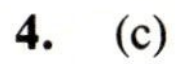

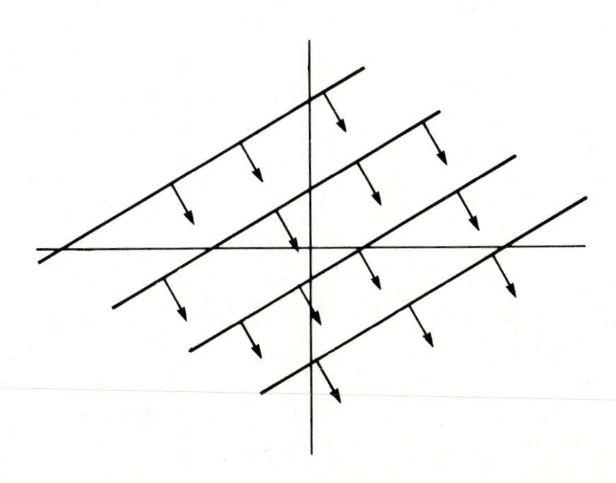

(d)

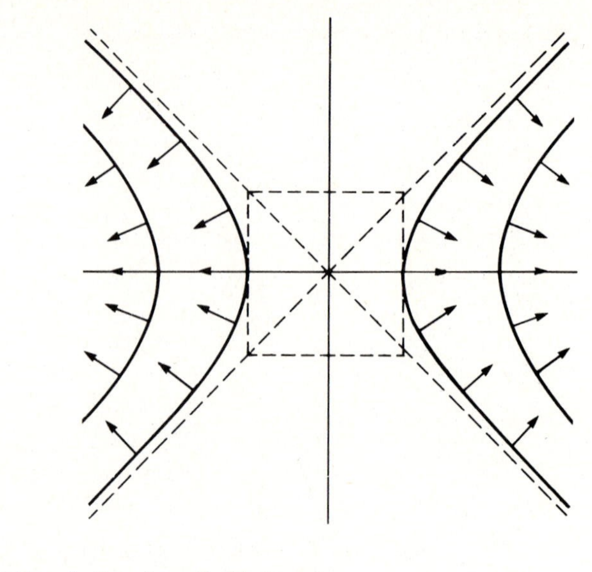

5. (b) $\nabla \times F = (-2, -2, -2)$, $\nabla \cdot F = 0$.

(c) $\nabla \times F = (-2y, 1, -3)$, $\nabla \cdot F = 2x + 2y + 2z$.

6. (a) $\phi = \frac{1}{2}(x^2 + y^2) + C$.

(d) Not a gradient field.

(e) $\phi = e^{xy} + C$.

8. (b) $\dfrac{\partial c}{\partial t} = -\alpha c - \dfrac{a}{2A(x, t)}([2 - \sin(x - vt)] + c[2 + v \sin(x - vt)])$.

12. (a) Hint: Let $A = r\Delta\theta\Delta h$ (where $\Delta\theta$, Δh are constant).

(b) Let $A = R^2(\Delta\theta\Delta\phi)$.

14. (b) sphere: $\gamma = 36\pi$,

cylinder: $\gamma = 8\pi\dfrac{h}{r}$.

16. (a) $C(0.4R) \approx 3 \times 10^{-3} \times (C_0/s) = .663\mu$g/ml (see Figure 9.7b).

17. (b) $\dfrac{\partial c}{\partial t} = \mathcal{D}\dfrac{\partial^2 c}{\partial x^2} - \dfrac{c}{\tau}$, $c(x, 0) = \begin{cases} c_0 & x < a \\ 0 & x \geq a \end{cases}$,

$\dfrac{\partial c}{\partial x} = 0$ for $x = L$ where L is length of tube.

$c(0, t) = c_0$.

19. (a) $c(r)\ln\left(\dfrac{L}{a}\right) = c_0 \ln\left(\dfrac{r}{a}\right)$.

22. (b) (i) $s_0 = 2s \Rightarrow \tau_0 = \dfrac{1}{4}\tau$.

(ii) $a_0 = 2a \Rightarrow \tau_0 = \tau \ln\left(\dfrac{L}{2a}\right) \Big/ \ln\left(\dfrac{L}{a}\right) = \tau\left(1 - \dfrac{\ln 2}{\ln(L/a)}\right)$

CHAPTER 10

1. (a) For v = prey (victim), e = predator (exploiter)

$\dfrac{\partial v}{\partial t} = \mathcal{D}_v\dfrac{\partial^2 v}{\partial x^2} + F(v, e)$, $\dfrac{\partial e}{\partial t} = \mathcal{D}_e\dfrac{\partial^2 e}{\partial x^2} + G(v, e)$,

with F, G any predator-prey kinetic terms.

2. (b) Two species competition with random motion of each of the species.

6. (b) $\mu \approx 0.2\ \text{cm}^2 h^{-1}$.

7. (b) Use $F = kv\eta$.

(c) The mean time between turns is $\tau = \lambda/v$.

8. (a) $\dfrac{\partial b}{\partial x} = 0,\ s = s_0$ at $x = L$,

$\dfrac{\partial b}{\partial x} = 0,\ \dfrac{\partial s}{\partial x} = 0$ at $x = 0$.

(e) $\dfrac{\partial v}{\partial \tau} = \lambda \dfrac{\partial^2 v}{\partial \xi^2} + [KF(u) - \theta]v,$

$\dfrac{\partial u}{\partial \tau} = \dfrac{\partial^2 u}{\partial \xi^2} - F(u)v.$

[with boundary conditions $\dfrac{\partial v}{\partial \xi} = 0,\ u = 1$ at $\xi = 1$, and $\dfrac{\partial v}{\partial \xi} = 0,\ \dfrac{\partial u}{\partial \xi} = 0$ at $\xi = 0$].

11. (c) $\rho = \dfrac{\mu_b}{\mu_c},\ \delta = \dfrac{\chi K_i}{\mu_c},\ \alpha = \dfrac{gK_i}{k_d c_0},\ \sigma = \dfrac{h_1 K_i}{k_d c_0},\ \kappa = K_b/K_i.$

12. (b) $-\gamma\rho$ = rate of branch mortality.

(e) dichotomous: $\sigma_{br} = \alpha n$;

lateral: $\sigma_{br} = \alpha\rho$.

14. (a) k_1 = rate of molecular binding to free sites,

k_2 = rate of unbinding.

D = effective diffusivity of free molecules.

v = speed of buffer through column.

(c) $\overline{u}_2 = k_1 B\overline{u}/(k_1\overline{u}_1 + k_2),\ \overline{u}_1 > 0.$

(d) Traveling Waves:

Require $U_1, U_2 \to 0,\ \dfrac{dU_1}{dz} \to 0$ for $z \to \infty$

waves satisfy $\dfrac{dU_1}{dz} = [(v - c)/D]U_1 - (c/D)U_2,$

$\dfrac{dU_2}{dz} = -(k_1/c)(B - U_2)(U_1) + (k_2/c)U_2.$

24. (a) $\dfrac{\partial n}{\partial t} = -\dfrac{\partial n}{\partial \alpha} - \mu n.$

(c) $\dfrac{\partial n}{\partial t} = -\partial \dfrac{(kn\alpha)}{\partial \alpha} - \mu n$, k a constant.

25. (e) $v = 1/\tau = .0694\ \text{hr}^{-1},\ a = v \ln 2 = .048,$

$\mu \approx 0.25\ \text{hr}^{-1},\ dN/dt \approx -.2N,$

$t_{10^{-3}} \approx 34.5$ hr.

27. (b) Biomass of vegetation whose quality is within the range $(q, q+\Delta q)$ at time t.

(c) $\overline{Q} = Q/P$.

(d) $\dfrac{\partial p}{\partial t} = -\dfrac{\partial pf}{\partial q}.$

(e) $\dfrac{dh}{dt} = h\, r(\overline{Q}, h).$

(g) Take $\dfrac{dQ}{dt} = \displaystyle\int q \frac{\partial p}{\partial t}\, dq = -\int q \frac{\partial fp}{\partial q}\, dq.$

Integrate *RHS* by parts.

CHAPTER 11

2. (a) No. (b) Yes. (c) Yes. (d) inconclusive. (e) No.

3. (b) $\left[\frac{1}{f}\right]$ = time to build up a significant local cAMP concentration,

$\left[\frac{1}{k + D(\pi/L)^2}\right]$ = time to efface a cAMP perturbation by chemical decay and diffusion.

7. (b) $J(1, 0) = \begin{bmatrix} pq^2 + \gamma - (1/K) & 0 \\ \alpha\sigma & -q^2 - \alpha \end{bmatrix}$

15. (a) No diffusive instability possible.
(b) No diffusive instability possible.
(f) Diffusive instability if $b_1 b_2 > de$, $b_2 > e$
$D_2/D_1 > [(b_1/d)^{1/2} - (b_1/d - e/b_2)^{1/2}]^{-2}$.

19. (b) Increasing γ tends to stretch domain in the y direction.
(c) $E^2 = 32$, assumed fixed.

m	n	γ
4	4	1
5	3	1.28
4	5	1.56
4	6	2.25
5	4	2.28

20. (a) $C_i' = \alpha_i e^{\sigma t} \sin(q_1 x) \sin(q_2 y)$.
(b) Same form of wavenumbers.
(c) $C_i' = \alpha_i e^{\sigma t} \cos(q_1 x) \sin(q_2 y)$,
$q_1 = m\pi/L$, $q_2 = n\pi/\gamma L$.

21. (e) $\alpha = D_A'/D_S'$, $\beta = D_A/D_S$, $\gamma = L^2 D_S'/L_1 L_2 D_S$,
$\rho = L_1 L_2 V_m K_m/D_S'$.

22. (a) Nullclines: $Q = f(P)$, $P = g(Q)$.
(b) $J = \begin{bmatrix} Pf'(P) & P \\ Q & -Qg'(Q) \end{bmatrix} = \begin{bmatrix} + & - \\ + & - \end{bmatrix}$ for P to the left of hump.
(d) $Pf' - Qg' < 0, f'g' < 1$, and
$(Pf' D_2 - Qg' D_1) > 2\sqrt{D_1 D_2}\,(PQ(1 - f'g'))^{1/2} > 0$.

Author Index

Subject Index